What Is Quantitative Reasoning?
or Why Am I in This Course?

The ability to interpret and reason with quantitative information—information that involves numbers or mathematical ideas—is a crucial aspect of literacy. This so-called **quantitative reasoning** is essential to making important decisions and understanding issues that appear in the news every day.

Topics: The topics covered in this text will help you work with quantitative information and make critical decisions. For example,

- You should possess strong **critical and logical thinking skills** so you can navigate the media and be an informed citizen;

- You should have a strong **number sense** and be proficient at **estimation** so you can put numbers from the news into a context that makes them understandable;

- You should be able to read news reports of **statistical studies** in a way that will allow you to evaluate them critically and decide whether they should affect your personal beliefs;

- You should be familiar with the basic ideas of **probability** and **risk**, and be aware of how they impact your life;

- You should possess the mathematical tools needed to make **basic financial decisions**; and

- You should understand how mathematics helps us study **important social issues**, such as the growth of populations, the depletion of resources, and methods of voting.

In sum, this course will focus on **understanding** and **interpreting** mathematical topics to help you develop the quantitative reasoning skills you will need for **college**, **career**, and **life**.

USING AND UNDERSTANDING

MATHEMATICS

A QUANTITATIVE REASONING APPROACH

THIRD EDITION

JEFFREY BENNETT
University of Colorado at Boulder

WILLIAM BRIGGS
University of Colorado at Denver

PEARSON

Addison
Wesley

Boston San Francisco New York
London Toronto Sydney Tokyo Singapore Madrid
Mexico City Munich Paris Cape Town Hong Kong Montreal

Publisher: Greg Tobin
Senior Acquisitions Editor: Anne Kelly
Editorial Assistant: Cecilia Fleming
Associate Project Editor: Joanne Ha
Senior Production Supervisor: Karen Wernholm
Senior Designer: Barbara T. Atkinson
Cover Design: Studio Montage
Production Services: Lifland et al., Bookmakers
Senior Marketing Manager: Becky Anderson
Marketing Coordinator: Carolyn Buddeke
Associate Media Producer: Sara Anderson
Software Development: David Malone and Marty Wright
Senior Manufacturing Buyer: Evelyn Beaton
Rights and Permissions Advisor: Dana Weightman
Composition: WestWords, Inc.
Illustrations: Techsetters, Inc.
Cover Photo: © Pascal Rondeau/GettyImages

For permission to use copyrighted material, grateful acknowledgment is made to the copyright holders on pages 741–742, which are hereby made part of this copyright page.

Library of Congress Cataloging-in-Publication Data

Bennett, Jeffrey O.
 Using and understanding mathematics : a quantitative reasoning approach / Jeffrey O. Bennett, William L. Briggs.—3rd ed.
 p. cm.
 Includes index.
 ISBN 0-321-22773-5
 1. Mathematics. I. Briggs, William L. II. Title.

QA39.3.B46 2005
510—dc22

2003061179

4 5 6 7 8 9 10 — VH — 0706

This book is dedicated to everyone who wants a better understanding of our world, and especially to those who have struggled with mathematics in the past. We hope this book will help you achieve your goals.

And it is dedicated to those who make our own lives brighter, especially Lisa, Julie, Katie, Grant, and Brooke.

CONTENTS

Part 4 *Modeling*

8 EXPONENTIAL ASTONISHMENT 489

9 MODELING OUR WORLD 531

10 MODELING WITH GEOMETRY 575

Part 5 *Further Applications*

11 MATHEMATICS AND THE ARTS 623

12 MATHEMATICS AND POLITICS 655

PREFACE

TO THE STUDENT

*T*here is no escaping the importance of mathematics in the modern world. However, for most people, the importance of mathematics lies not in its abstract ideas, but in its application to personal and social issues. This book is designed with such practical considerations in mind. In particular, we've designed this book with three specific purposes:

- To prepare you for the mathematics you will encounter in other **college** courses, particularly core courses in social and natural sciences.
- To develop your ability to reason with quantitative information in a way that will help you achieve success in your **career.**
- To provide you with the critical thinking and quantitative reasoning skills needed to understand major issues in **life.**

We hope this book will be useful to everyone, but it is designed primarily for those who are not planning to major in a field that requires advanced mathematical skills. In particular, if you've ever felt any fear or anxiety about mathematics, we've written this book with you in mind. Perhaps you will discover that mathematics is much more important and relevant to your life than you guessed.

Wherever your interests lie—social sciences, environmental issues, politics, business and economics, art and music, or any of many other topics—you will find relevant examples in this book. But the most important idea to take away from this book is that mathematics can help you understand a variety of topics and issues. Once you have completed your study of this book, you should be prepared to understand most quantitative issues that you will encounter.

TO THE INSTRUCTOR

This book represents a new approach to mathematics for general education students—that is, students who are planning to major in fields such as liberal arts, fine arts, business, and education. Most students have never been taught how to combine critical thinking and mathematical skills in a way that applies to their lives. In the jargon adopted by MAA, AMATYC, and other mathematical organizations,

students need to learn **quantitative reasoning.** This book is designed to help you succeed in teaching quantitative reasoning to your students.

THE CHALLENGE: WINNING OVER YOUR STUDENTS

Perhaps the greatest challenge in teaching mathematics to general education students lies in winning them over—that is, convincing them that you have something useful to teach them. This challenge arises because, by the time they reach college, many students dislike or fear mathematics. Indeed, the vast majority of students in general education mathematics courses are there not by choice, but because such courses are required for graduation. Reaching your students therefore requires that you teach with enthusiasm and convince them that mathematics is useful and enjoyable.

We've built this book around two important strategies that are designed to help you win students over:

1. Confront negative attitudes about mathematics head on, showing students that their fear or loathing is ungrounded and that mathematics actually is relevant to their lives. This strategy is embodied in the Prologue of this book (pages 1–12), which we urge you to emphasize in class. It continues implicitly throughout the rest of the text.

2. Focus on goals that are meaningful to students—namely, on the goals of learning mathematics for *college, career,* and *life.* Your students will then learn mathematics because they will see how it affects their lives.

THE KEY TO SUCCESS: A CONTEXT-DRIVEN APPROACH

Broadly speaking, approaches to teaching mathematics can be divided into two categories:

- A *content-driven* approach is organized by mathematical ideas. After each mathematical topic is presented, examples of its applications are shown.
- A *context-driven* approach is organized by practical contexts. Applications drive the course, and mathematical ideas are presented as needed to support the applications.

The same content can be covered through either approach, but the context-driven approach has an enormous advantage for general education students: It shows them directly how mathematics is important to college, career, and life. In contrast, the content-driven approach tends to come across as "learn this content because it's good for you," causing many students to tune out before you ever reach the practical applications. For more details, see our article "General Education Mathematics: New Approaches for a New Millennium" (*AMATYC Review,* Fall 1999).

MODULAR STRUCTURE OF THE BOOK

We'd all love to have a year or more to teach mathematics to general education students. Unfortunately, most schools have only a one-quarter or one-semester mathematics requirement, so we can cover only a fraction of the material we'd cover in an ideal course. This book is therefore organized with a modular structure that allows

you to create a course to meet your particular concerns and constraints. The 13 chapters are organized broadly by contextual areas. Each chapter, in turn, is divided into a set of self-contained *units* that focus on particular concepts or applications. In most cases, you can cover chapters in any order; and while the units within each chapter build sequentially in terms of sophistication, in many cases you can skip certain units, particularly those toward the end of the chapter.

PREREQUISITE MATHEMATICAL BACKGROUND

Because of its modular structure, this book can be used by students with a wide range of mathematical backgrounds. Many of the units require nothing more than arithmetic and a willingness to think about quantitative issues in new ways. Other units use techniques of algebra or geometry, which are reviewed as they arise. Thus, this book should be accessible to any student who has completed two or more years of high school mathematics. Note, however, that this book is *not* remedial: While much of the book relies on mathematical techniques from secondary school, the techniques always arise in applications that require students to demonstrate a high level of critical thinking.

PEDAGOGICAL FEATURES

Besides the main narrative of the text, this book includes the following features, each designed with a specific pedagogical purpose in mind.

Chapter Overview. Appearing at the beginning of each chapter, this feature offers a unit-by-unit overview of the chapter. It is designed both to show students how the chapter is organized and to help instructors decide which units to cover in class.

Time Out to Think. Appearing throughout the book, the "Time Out to Think" features pose short conceptual questions designed to help students reflect on important new ideas. They also serve as excellent starting points for classroom discussions.

Summary Boxes. Flowing right along with the narrative, the book contains boxes that summarize key ideas, definitions, or formulas.

Examples and Case Studies. Numbered examples are designed to build understanding and to offer practice with the types of questions that appear in the exercises. Each example is accompanied by a "Now try" tag that relates the example to specific similar exercises. Occasional case studies go into more depth than the numbered examples.

Margin Features. This book includes a wide margin so that students have room to make notes while studying. The margin also contains material that should spur student interest but is peripheral to the main discussion. This material comes in three basic forms:

- **By the Way** features contain interesting notes and asides relevant to the topic at hand.
- **Historical Note** remarks give historical context to the ideas presented in the chapter.
- **Technical Notes** contain details that are important mathematically, but generally do not affect students' understanding of the material.

A Brief Review. These boxed features review key mathematical skills that students should have learned previously, but in which many students still need review and practice. They appear in the book wherever a particular skill is first needed, and exercises based on the review boxes can be found at the end of the unit.

Practical Matters. These boxed features focus on topics of particularly high practical importance, such as how to watch out for bad deals on telephone service, how to avoid credit card trouble, and how to save money and energy at the same time.

Thinking About. These boxed features build upon ideas in the main narrative but go somewhat beyond the level of other material in the book. Examples include boxes on the proof of the Pythagorean theorem and on Zeno's paradox.

Chapter Summary. Appearing at the end of each chapter, this feature offers a detailed outline of the chapter that students can use as a study guide.

ASSESSMENT TOOLS

Each unit ends with questions designed for student assessment, organized into the following sections:

Review Questions. These questions ask students to summarize the important ideas covered in the unit. They are designed primarily for self-study and generally can be answered simply by reviewing the text.

NEW! **Does It Make Sense?** These qualitative questions test conceptual understanding by asking students to decide whether the given statements are sensible and to explain why or why not.

Basic Skills & Concepts. These questions offer practice with the concepts covered in the unit. The exercises can be used for homework assignments or for self-study (answers to most odd-numbered exercises appear in the back of the book).

NEW! **Further Applications.** Through additional applications, these exercises extend the ideas and techniques covered in the text.

Web Projects. The Web Projects require searching for data or other information on the Web. They can be used for extended projects, discussion, group activities, or essays.

In the News. These questions ask students to find examples from the news or from daily life of the ideas presented in the unit. These questions may be assigned as homework or used for class discussions.

CHANGES IN THE THIRD EDITION

We've been pleased by the positive response of so many users to the second edition of this text. Nevertheless, a book that relies heavily on facts and data always requires a major updating effort to keep it current. In addition, we've received numerous valuable suggestions for improvement from instructors and reviewers. We have sought to incorporate all these updates and improvements into the new edition while keeping

the major features of the prior edition intact. The major changes that second edition users will notice are the following:

- All our data have been updated to keep them relevant and current, and some units have undergone wholesale changes to reflect changes in society—especially the units on income taxes and the federal budget.
- We have heavily revised the exercise sets, adding the "Does It Make Sense?" exercises, revising and replacing many other exercises, and organizing the exercise sets in a way that should make it easier for you to select questions for assignment. The new "Further Applications" exercises extend the ideas and techniques covered in the text. All told, we've added more than 300 new exercises to the new edition and modified hundreds of others.
- At the end of examples in the text, we've added "Now try" tags, which point to specific exercises related to the example.
- We have expanded our coverage of geometry in the new Chapter 10, called Modeling with Geometry. This chapter contains two full units on basic geometry and its applications, as well as a unit on fractal geometry.
- As a result of adding the new chapter, Mathematics and the Arts, Mathematics and Politics, and Mathematics and Business are now Chapters 11 through 13, rather than 10 through 12 as they were in the second edition.

SUPPLEMENTS

SUPPLEMENTS FOR THE INSTRUCTOR

Instructor's Edition
ISBN 0-321-22774-3
This special edition of the text includes answers to all of the problems.

Instructor's Guide and Solutions Manual
Lori Rosenthal, Austin Community College
ISBN 0-321-22776-X
This manual contains model curricula, unit-by-unit teaching hints, answers or discussion points for all the "Time Out to Think" questions, and detailed, worked-out solutions to all of the exercises in the text.

Instructor's Testing Manual
Linda Rogers, Florida State University
ISBN 0-321-22777-8
The Testing Manual includes four alternative tests per chapter. These items may be used as actual tests or as references for creating tests. Answer keys are included.

TestGen-EQ with QuizMaster-EQ
ISBN 0-321-22780-8
TestGen enables instructors to build, edit, print, and administer tests using a computerized bank of questions developed to cover all the objectives of the text. TestGen is algorithmically based so that multiple yet equal versions of the same question or test can be generated at the click of a button. Instructors can also modify

test bank questions or add new questions by using the built-in question editor, which allows users to create graphs, import graphics, insert math notation, and insert variable numbers or text. Tests can be printed or administered online via the Web or other network. Many questions in TestGen can be expressed in a short-answer or multiple-choice form, giving instructors greater flexibility in their test preparation. TestGen comes packaged with QuizMaster, which allows students to take tests on a local area network. The software is available on a dual-platform Windows/Macintosh CD-ROM.

Videotapes
ISBN 0-321-22778-6
The videotapes feature an engaging team of mathematics instructors who present comprehensive coverage of each section of the text. The presentations support an approach that emphasizes visualization and problem-solving. The videos are also included in MyMathLab.

PowerPoint Slides
This classroom presentation software is geared specifically to the sequence and philosophy of this textbook. Lecture content and key graphics from the book are included. Available within MyMathLab or on Supplements Central Web site at http://suppscentral.aw.com.

NEW! MyMathLab

MyMathLab is a complete online course available with this text and is perfect for a lecture-based or self-paced course. This site offers instructors and students a wide variety of resources from dynamic multimedia—video clips, downloads, and more—to course management tools. With MyMathLab, instructors can customize their course and help students increase their comprehension and success.

- MyMathLab provides a powerful system for creating and assigning tests, as well as a gradebook for tracking all student performance.
- The entire textbook is available online and is supplemented by multimedia content, such as videos, which are used to further explain concepts.

For more information, visit our Web site at www.mymathlab.com.

SUPPLEMENTS FOR THE STUDENT

Student's Study Guide and Solutions Manual
Lori Rosenthal, Austin Community College
ISBN 0-321-22775-1
This manual contains hints for learning and studying, as well as detailed, worked-out solutions to all the odd-numbered unit exercises.

Digital Video Tutor
ISBN 0-321-23716-1
The video lectures for this text are also available on CD-ROM, making it easy and convenient for students to watch videos from a computer at home or on campus. The complete video set, affordable and portable, is ideal for distance learning or supplemental instruction. The videos are also included in MyMathLab.

Addison-Wesley Math Tutor Center

www.aw-bc.com/tutorcenter

The Addison-Wesley Math Tutor Center is staffed by qualified college mathematics instructors who tutor students on examples and exercises from the textbook. Tutoring is provided via toll-free telephone, toll-free fax, e-mail, or the Internet, and White Board technology allows tutors and students to actually see the problems worked out while they "talk" in real time over the Internet during tutoring sessions. For more information, please visit our Web site at www.aw-bc.com/tutorcenter.

NEW! **Research Navigator**

www.researchnavigator.com

Research Navigator is designed to help students understand the research process and complete a research paper or research assignment. Research Navigator includes an overview of the research process and four credible and reliable sources of content: EBSCO's ContentSelect Academic Journal and Abstract Database, The New York Times Search by Subject Archive, Pearson's "Best of the Web" Link Library, and new content from the Financial Times.

NEW! **MyMathLab** *MyMathLab*

MyMathLab is a complete online course available with this text and is perfect for a lecture-based or self-paced course. This site offers instructors and students a wide variety of resources from dynamic multimedia—video clips, downloads, and more—to course management tools. With MyMathLab, instructors can customize their course and help students increase their comprehension and success.

- MyMathLab provides a powerful system for creating and assigning tests and homework, as well as a gradebook for tracking all student performance.
- The entire textbook is available online and is supplemented by multimedia content, such as videos, which are used to further explain concepts.

For more information, visit our Web site at www.mymathlab.com.

ACKNOWLEDGMENTS

A textbook may carry author names, but it is the result of hard work by hundreds of committed individuals. This book has been under development for nearly 20 years, and even its beginnings were a group effort, as one of the authors was a member of a committee at the University of Colorado that worked to establish one of the nation's first courses in quantitative reasoning. Since that beginning, the book has benefited from input and feedback from many faculty members and students.

First and foremost we extend our thanks to Bill Poole, whose faith in this project allowed it to grow from typewritten class notes into a true textbook. We'd also like to thank the other members of our outstanding publishing team, including Greg Tobin, Anne Kelly, Karen Guardino, Karen Wernholm, Becky Anderson, Barbara Atkinson, Joanne Ha, Sara Anderson, Cecilia Fleming, and Carolyn Buddeke. Special thanks go to Sally Lifland, for her outstanding efforts in producing this book. We also thank Shane Goodwin, L. Taylor Ollmann, and Sybil Smith-Darlington for an excellent job on accuracy checking.

We'd like to thank the following people for their help with one or more editions of this book (reviewers of the third edition are marked with an "*"):

Lou Barnes, *Boulder West Financial Services*
Carol Bellisio, *Monmouth University*
Bob Bernhardt, *East Carolina University*
* Terence R. Blows, *Northern Arizona University*
W. Wayne Bosché, Jr., *Dalton College*
Michael Bradshaw, *Caldwell Community College and Technical Institute*
W. E. Briggs, *University of Colorado, Boulder*
Margaret Cibes, *Trinity College*
Walter Czarnec, *Framingham State College*
Adrian Daigle, *University of Colorado, Boulder*
Andrew J. Dane, *Angelo State University*
Greg Dietrich, *Florida Community College at Jacksonville*
Marsha J. Driskill, *Aims Community College*
John Emert, *Ball State University*
Kathy Eppler, *Salt Lake Community College*
Kellie Evans, *York College of Pennsylvania*
Fred Feldon, *Coastline Community College*
David E. Flesner, *Gettysburg College*
Pat Foard, *South Plains College*
* Shane Goodwin, *Brigham Young University—Idaho*
Barbara Grover, *Salt Lake Community College*
Louise Hainline, *Brooklyn College*
Ward Heilman, *Bridgewater State University*
Peg Hovde, *Grossmont College*
Andrew Hugine, *South Carolina State University*
Lynn R. Hun, *Dixie College*
Hal Huntsman, *University of Colorado, Boulder*
Joel Irish, *University of Southern Maine*
David Jabon, *DePaul University*
Melvin F. Janowitz, *University of Massachusetts, Amherst*
Craig Johnson, *Brigham Young University—Idaho*
Vijay S. Joshi, *Virginia Intermont College*
Anton Kaul, *University of South Florida*
Bonnie Kelly, *University of South Carolina*
William Kiley, *George Mason University*
Jim Koehler, *University of Colorado, Denver*
Robert Kuenzi, *University of Wisconsin, Oshkosh*
Deann Leoni, *Edmonds Community College*
Linda Lester, *Wright State University*
Paul Lorczak, *MathSoft, Inc.*
Jay Malmstrom, *Oklahoma City Community College*

Erich McAlister, *University of Colorado, Boulder*
Patricia McNicholas, *Robert Morris College*
Elaine Spendlove Merrill, *Brigham Young University— Hawaii*
* Carrie Muir, *University of Colorado*
Colm Mulcahy, *Spelman College*
Paul O'Heron, *Broome Community College*
* L. Taylor Ollmann, *Austin Community College*
* A. Dean Palmer, *Pima Community College*
Mary K. Patton, *University of Illinois at Springfield*
Frank Pecchioni, *Jefferson Community College*
* Jonathan Prewett, *University of Wyoming*
Evelyn Pupplo-Cody, *Marshall University*
Scott Reed, *College of Lake County*
Anne Roberts, *University of Utah*
* Sylvester Roebuck Jr., *Olive Harvey College*
Lori Rosenthal, *Austin Community College*
Hugo Rossi, *University of Utah*
Doris Schraeder, *McLennan Community College*
Dee Dee Shaulis, *University of Colorado, Boulder*
Judith Silver, *Marshall University*
* Sybil Smith-Darlington, *Middlesex County College*
Alu Srinivasan, *Temple University*
John Supra, *University of Colorado, Boulder*
Scott Surgent, *Arizona State University*
Timothy C. Swyter, *Frederick Community College*
David Theobald, *University of Colorado, Boulder*
Terry Tolle, *Southwestern Community College*
Christina Vertullo, *Marist College*
Pam Wahl, *Middlesex Community College*
Thomas Wangler, *Benedictine University*
Richard Watkins, *Tidewater Community College*
Emily Whaley, *DeKalb College*
David Wilson, *University of Colorado, Boulder*
Robert Woods, *Broome Community College*
Fred Worth, *Henderson State University*
Margaret Yoder, *Eastern Kentucky University*
Fredric Zerla, *University of Southern Florida*
Donald J. Zielke, *Concordia Lutheran College*

Jeffrey Bennett
William Briggs

Prologue

Literacy for the Modern World

Mathematics is the key to opportunity. No longer just the language of science, mathematics now contributes in direct and fundamental ways to business, finance, health, and defense. For students, it opens doors to careers. For citizens, it enables informed decisions. For nations, it provides knowledge to compete in a technological economy. To participate fully in the world of the future, America must tap the power of mathematics.

—FROM *EVERYBODY COUNTS*

WHAT IS QUANTITATIVE LITERACY?

Literacy is the ability to read and write, and it comes in varying degrees. Some people can recognize only a few words and write only their names; others read and write in many languages. A primary goal of our educational system is to provide citizens with a level of literacy sufficient to read and write about the important issues of our time.

Today, the abilities to interpret and reason with **quantitative information**—information that involves mathematical ideas or numbers—are crucial aspects of literacy. This so-called **quantitative literacy** is essential to understanding issues that appear in the news every day. The process of interpreting and reasoning with quantitative information is called **quantitative reasoning.** The purpose of this book is to help you gain skills in quantitative reasoning as it applies to issues you will encounter in

- your subsequent course work
- your career, and
- your daily life.

QUANTITATIVE REASONING AND CULTURE

Quantitative reasoning enriches the appreciation of both ancient and modern culture. The historical record shows that all great cultures devoted substantial energy to mathematics and to science (or to observational studies that predated modern science). Without a sense of how quantitative concepts are used in art, architecture, and science, you cannot fully appreciate the incredible achievements of the Mayans in Central America, the builders of the great city of Zimbabwe in Africa, the ancient Egyptians and Greeks, the early Polynesian sailors, and countless others.

A ROCK, A RIVER, A TREE
Hosts to species long since departed
Marked the mastodon,
The dinosaur, who left dry tokens
Of their sojourn here
On our planet floor,
Any broad alarm of their hastening doom
Is lost in the gloom of dust and ages.

—MAYA ANGELOU, *ON THE PULSE OF THE MORNING*

Similarly, quantitative concepts can help you understand and appreciate the great works of the modern artists. You can more fully appreciate the excerpt here from a Maya Angelou poem if you know that dinosaurs have been extinct for 65 million years and are able to put this amount of time into perspective. The ties between mathematics and music can be found in modern and classical music, as well as in the digital production of music. Indeed, it is hard to find popular works of art, film, or literature that do not make at least some indirect reference to mathematics.

QUANTITATIVE REASONING IN THE WORK FORCE

Quantitative reasoning is important in the work force. A lack of quantitative skills puts out of reach many of the most challenging and highest paying jobs. Table 1 defines skill levels in language and mathematics on a scale of 1 to 6, and Table 2 shows the typical levels needed in many jobs.

TABLE 1 Skill Levels

LEVEL	LANGUAGE SKILLS	MATH SKILLS
1	Recognizes 2500 two- or three-syllable words. Reads at a rate of 95–120 words per minute. Writes and speaks simple sentences.	Adds and subtracts two-digit numbers. Does simple calculations with money, volume, length, and weight.
2	Recognizes 5000–6000 words. Reads 190–215 words per minute. Reads adventure stories and comic books, as well as instructions for assembling model cars. Writes compound and complex sentences with proper grammar and punctuation.	Adds, subtracts, multiplies, and divides all units of measure. Computes ratio, rate, and percentage. Draws and interprets bar graphs.
3	Reads novels and magazines, as well as safety rules and equipment instructions. Writes reports with proper format and punctuation. Speaks well before an audience.	Understands basic geometry and algebra. Calculates discount, interest, profit and loss, markup, and commissions.
4	Reads novels, poems, newspapers, and manuals. Prepares business letters, summaries, and reports. Participates in panel discussions and debates. Speaks extemporaneously on a variety of subjects.	Has true quantitative reasoning abilities. Understands logic, problem solving, ideas of statistics and probability, and modeling.
5	Reads literature, book and play reviews, scientific and technical journals, financial reports, and legal documents. Can write editorials, speeches, and critiques.	Knows calculus and statistics. Is able to deal with econometrics.
6	Same types of skills as level 5, but more advanced.	Works with advanced calculus, modern algebra, and statistics.

TABLE 2 Skill-Level Requirements

OCCUPATION	LANGUAGE LEVEL	MATH LEVEL	OCCUPATION	LANGUAGE LEVEL	MATH LEVEL
Biochemist	6	6	Web page designer	5	4
Computer engineer	6	6	Corporate executive	5	5
Mathematician	6	6	Computer sales agent	4	4
Cardiologist	6	5	Athlete's agent	4	4
Social psychologist	6	5	Management trainee	4	4
Lawyer	6	4	Insurance sales agent	4	4
Tax attorney	6	4	Retail store manager	4	4
Newspaper editor	6	4	Cement mason	3	3
Accountant	5	5	Poultry farmer	3	3
Personnel manager	5	4	Tile setter	3	3
Corporate president	5	5	Travel agent	3	3
Weather forecaster	5	5	Janitor	3	2
Secondary teacher	5	5	Short-order cook	3	2
Elementary teacher	5	4	Assembly-line worker	2	2
Financial analyst	5	5	Toll collector	2	2
Journalist	5	4	Laundry worker	1	1

SOURCE: *Both tables adapted from* The Wall Street Journal.

Calvin and Hobbes by Bill Watterson

Note that the occupations requiring high skill levels are generally the most prestigious and highest paying. Note also that most of these occupations call for high skill levels in *both* language and math, refuting the myth that if you're good at language you don't have to be good at mathematics, and vice versa.

MISCONCEPTIONS ABOUT MATHEMATICS

Do you consider yourself to have "math phobia" (fear of mathematics) or "math loathing" (dislike of mathematics)? We hope not; but if you do, you aren't alone. Many adults harbor fear or loathing of mathematics, and, unfortunately, these attitudes are often reinforced by classes that present mathematics as an obscure and sterile subject.

In reality, mathematics is not nearly so dry as it sometimes seems in school. Indeed, attitudes toward mathematics often are directed not at what mathematics really is but at some common misconceptions about mathematics. Let's investigate a few of these misconceptions and the reality behind them.

MISCONCEPTION ONE: MATH REQUIRES A SPECIAL BRAIN

One of the most pervasive misconceptions is that some people just aren't good at mathematics because learning mathematics requires special and rare abilities. The reality is that nearly everyone can do mathematics. All it takes is self-confidence and hard work— the same qualities needed to learn to read, to master a musical instrument, or to become skilled at a sport. Indeed, the belief that mathematics requires special talent found in a few elite people is peculiar to the United States. In other countries, particularly in Europe and Asia, *all* students are expected to become proficient in mathematics.

Of course, different people learn mathematics at different rates and in different ways. For example, some people learn by concentrating on concrete problems, others by thinking visually, and still others by thinking abstractly. No matter what type of thinking style you prefer, you can succeed in mathematics.

In a sense, we are all mathematicians— and superb ones. It makes no difference what you do. Your real forte lies in navigating the complexities of social networks, weighing passions against histories, calculating reactions, and generally managing a system of information that, when all laid out, would boggle a computer.

—A. K. DEWDNEY, *200% OF NOTHING*

*T*hinking about . . .

PEOPLE WHO STUDIED MATHEMATICS

The critical thinking skills developed through the study of mathematics are valuable in many careers. The following is only a small sample of people who studied mathematics but became famous for work in other fields. (Most of the names on this list are from an article by Steven G. Buyske, "Famous Nonmathematicians," *American Mathematical Monthly,* November 1993.)

Corazon Aquino, former president of the Philippines, a mathematics minor

Harry Blackmun, former Supreme Court justice, summa cum laude in mathematics, Harvard University

Lewis Carroll (Charles Dodgson), author of *Alice in Wonderland,* a mathematician

David Dinkins, former mayor of New York City, BA in mathematics, Howard University

Alberto Fujimori, former president of Peru, MS in mathematics, University of Wisconsin

Art Garfunkel, musician, MA in mathematics, Columbia University

Grace Hopper, computer pioneer and first woman Rear Admiral in the U.S. Navy, Ph.D. in mathematics, Yale

Mae Jemison, first African-American woman in space, studied mathematics as part of her degree in chemical engineering from Stanford University

John Maynard Keynes, economist, MA in mathematics, Cambridge University

Lee Hsien Loong, politician in Singapore, BA in mathematics, Cambridge University

Edwin Moses, three-time Olympic champion in the 400-meter hurdles, studied mathematics as part of his degree in physics from Morehouse College

Florence Nightingale, pioneer in nursing, studied mathematics and applied it to her work

William Perry, former secretary of defense, Ph.D. in mathematics, Pennsylvania State University

Sally Ride, first American woman in space, studied mathematics as part of her Ph.D. in physics from Stanford University

David Robinson, basketball star, bachelor's degree in mathematics, U.S. Naval Academy

Alexander Solzhenitsyn, Nobel prize winning Russian author, degrees in mathematics and physics from the University of Rostov

Bram Stoker, author of *Dracula,* studied mathematics at Trinity University, Dublin

Laurence Tribe, Harvard law professor, summa cum laude in mathematics, Harvard University

Virginia Wade, Wimbledon champion, bachelor's degree in mathematics, Sussex University

MISCONCEPTION TWO: THE MATH IN MODERN ISSUES IS TOO COMPLEX

Some people claim that the advanced mathematical concepts underlying many modern issues are too complex for the average person to understand. It is true that only a few people receive the training needed to work with or discover advanced mathematical concepts. However, most people are capable of understanding enough about the mathematical basis of important issues to develop informed and reasoned opinions.

The situation is similar in other fields. For example, years of study and practice are required to become a proficient professional writer, but most people can read a

Equity for all requires excellence for all; both thrive when expectations are high.

—FROM *EVERYBODY COUNTS*

Math phobic's nightmare

book. It takes hard work and a law degree to become a lawyer, but most people can understand how the law affects them. And though few have the musical talent of Mozart, anyone can learn to appreciate his music. Mathematics is no different. If you've made it this far in school, you can understand enough mathematics to succeed as an individual and a concerned citizen. Don't let anyone tell you otherwise!

MISCONCEPTION THREE: MATH MAKES YOU LESS SENSITIVE

Some people believe that learning mathematics will somehow make them less sensitive to the romantic and aesthetic aspects of life. In fact, understanding the mathematics that explains the colors of a sunset or the geometric beauty in a work of art can only enhance aesthetic appreciation. Furthermore, many people find beauty and elegance in mathematics itself. It's no accident that people trained in mathematics have made important contributions to art, music, and many other fields.

MISCONCEPTION FOUR: MATH MAKES NO ALLOWANCE FOR CREATIVITY

The "turn the crank" nature of the problems in many textbooks may give the impression that mathematics stifles creativity. Some of the facts, formalisms, and skills required for mathematical proficiency are fairly cut and dried, but *using* these mathe-

matical tools takes creativity. Consider designing and building a home. The task demands specific skills to lay the foundation, frame in the structure, install plumbing and wiring, and paint walls. But building the home involves much more: Creativity is needed to develop the architectural design, respond to on-the-spot problems during construction, and factor in constraints based on budgets and building codes. The mathematical skills you've learned in school are like the skills of carpentry or plumbing. Applying mathematics is like the creative process of building a home.

MISCONCEPTION FIVE: MATH PROVIDES EXACT ANSWERS

A mathematical formula will yield a specific result, and in school that result may be marked right or wrong. But when you use mathematics in real-life situations, answers are never so clear-cut. For example:

> *A bank offers simple interest of 5%, paid at the end of one year (that is, after one year the bank pays you 5% of your account balance). If you deposit $1000 today and make no further deposits or withdrawals, how much will you have in your account after one year?*

A straight mathematical calculation seems simple enough: 5% of $1000 is $50; so you should have $1050 at the end of a year. But will you? How will your balance be affected by service charges or taxes on interest earned? What if the bank fails? What if the bank is located in a country in which the currency collapses during the year? Choosing a bank in which to invest your money is a *real* mathematics problem that doesn't necessarily have a simple or definitive solution.

MISCONCEPTION SIX: MATH IS IRRELEVANT TO MY LIFE

No matter what your path in college, career, and life, you will find mathematics involved in many ways. A major goal of this text is to show you hundreds of examples in which mathematics applies to everyone's life. We hope you will find that mathematics is not only relevant but also interesting and enjoyable.

WHAT *IS* MATHEMATICS?

In discussing misconceptions, we identified what mathematics is *not*. Now let's look at what mathematics *is*. The word *mathematics* is derived from the Greek word *mathematikos,* which means "inclined to learn." Thus, literally speaking, to be mathematical is to be curious, open-minded, and interested in always learning more! Today, we tend to look at mathematics in three different ways: as the sum of its branches, as a way to model the world, and as a language.

MATHEMATICS AS THE SUM OF ITS BRANCHES

As you progressed through school, you undoubtedly learned to associate mathematics with some of its branches. Among the better known branches of mathematics are these:

Probably the most harmful misconception is that mathematics is essentially a matter of computation. Believing this is roughly equivalent to believing that writing essays is the same as typing them.

—JOHN ALLEN PAULOS

Skills are to mathematics what scales are to music or spelling is to writing. The objective of learning is to write, to play music, or to solve problems—not just to master skills.

—FROM *EVERYBODY COUNTS*

- **logic**—the study of principles of reasoning;
- **arithmetic**—methods for operating on numbers;
- **algebra**—methods for working with unknown quantities;
- **geometry**—the study of size and shape;
- **trigonometry**—the study of triangles and their uses;
- **probability**—the study of chance;
- **statistics**—methods for analyzing data; and
- **calculus**—the study of quantities that change.

One can view mathematics as the sum of its branches. Indeed, most "traditional" mathematics books focus on one branch of mathematics at a time. In this book, we emphasize the connections between the branches rather than their differences. Nevertheless, familiarity with the various branches of mathematics is helpful, even if only to know when you are using them.

MATHEMATICS AS A WAY TO MODEL THE WORLD

Mathematics also may be viewed as a tool for creating models, or representations of real phenomena. Modeling is not unique to mathematics. For example, a road map is a model that represents the roads in some region.

Mathematical models can be as simple as a single equation that predicts how the money in your bank account will grow or as complex as a set of thousands of interrelated equations and parameters used to represent the global climate. By studying models, we gain insight into otherwise unmanageable problems. A global climate model, for example, can help us understand weather systems and ask "what if" questions about how human activity may affect the climate. When a model is used to make a prediction that does *not* come true, it points out areas where further research is needed. Today, mathematical modeling is used in nearly every field of study. Figure 1 indicates some of the many disciplines that use mathematical modeling to solve problems.

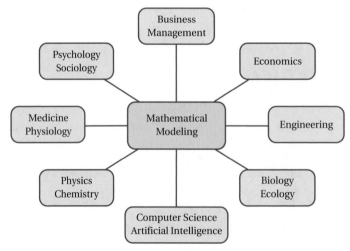

FIGURE 1

MATHEMATICS AS A LANGUAGE

A third way to look at mathematics is as a language with its own vocabulary and grammar. Indeed, mathematics often is called "the language of nature" because it is so useful for modeling the natural world. As in any language, different degrees of fluency are possible. From this point of view, quantitative literacy is the level of fluency required for success in today's world.

The idea of mathematics as a language also is useful in thinking about how to *learn* mathematics. Table 3 compares learning mathematics to learning a language and learning art.

Mathematics is a way of thinking that can help make muddy relationships clear. It is a language that allows us to translate the complexity of the world into manageable patterns.

—K. C. Cole, *The Universe and the Teacup*

TABLE 3	Learning Mathematics: An Analogy to Language and Art	
LEARNING A LANGUAGE	LEARNING THE LANGUAGE OF ART	LEARNING THE LANGUAGE OF MATHEMATICS
Become familiar with many styles of speaking and writing, such as essays, poetry, and drama.	Become familiar with many styles of art, such as classical, renaissance, impressionist, and modern.	Become familiar with techniques from many branches of mathematics, such as arithmetic, algebra, and geometry.
Place literature in context by studying its history and the social conditions under which it was created.	Place art in context by studying its history and the social conditions under which it was created.	Place mathematics in context by studying its history, purposes, and applications.
Learn the elements of language—such as words, parts of speech (nouns, verbs, etc.), and rules of grammar—and practice their proper use.	Learn the elements of visual form—such as lines, shapes, colors, and textures—and practice using them in your own art work.	Learn the elements of mathematics—such as numbers, variables, and operations—and practice using them to solve simple problems.
Develop the ability to analyze the language in complex forms critically—novels, short stories, essays, poems, speeches, debates, and similar works.	Develop the ability to analyze works of art critically—painting, sculpture, architecture, photography, and similar works.	Develop the ability to analyze quantitative information critically—mathematical models, statistical studies, economic forecasts, investment strategies, and similar works.
Use language creatively for your own purposes, such as writing a term paper or story or engaging in debate.	Use your sense of art creatively, such as in designing your house, taking a photograph, or making a sculpture.	Use mathematics creatively to solve problems you encounter and to help you understand important issues in the world around you.

HOW TO SUCCEED IN MATHEMATICS

If you are reading this book, you probably are enrolled in a mathematics course. The keys to success in your course include approaching the material with an open and optimistic frame of mind, paying close attention to how useful and enjoyable mathematics can be in *your* life, and studying effectively and efficiently. The following sections offer a few specific hints that may be of use as you study.

USING THIS BOOK

Before we get into more general strategies for studying, here are a few guidelines that will help you use *this* book most effectively.

- Before doing any assigned exercises, read assigned material *twice:*
 —On the first pass, read quickly to gain a "feel" for the material and concepts presented.
 —On the second pass, read the material in more depth and work through the examples carefully.
- During the second reading, take notes that will help you when you go back to study later. In particular:
 —*Use the margins!* The wide margins in this textbook are designed to give you plenty of room for making notes as you study.
 —Don't highlight—underline! Using a pen or pencil to underline material requires greater care than highlighting and therefore helps to keep you alert as you study.
- After you complete the reading, and again when studying for exams, make sure you can answer the review questions at the end of each unit.
- You'll learn best by *doing,* so do plenty of the end-of-unit exercises. In particular, try some of the exercises that have answers in the back of the book, in addition to those assigned by your instructor.

BUDGETING YOUR TIME

A general rule of thumb for college classes is that you should expect to study about 2 to 3 hours per week *outside* class for each unit of credit. For example, a student taking 15 credit hours should spend 30 to 45 hours each week studying outside of class. Combined with time in class, this works out to a total of 45 to 60 hours per week—not much more than the time required of a typical job, and you get to choose your own hours. Of course, if you are working or taking care of a family while you attend school, you will need to budget your time carefully.

The following table gives some rough guidelines for how you might divide your studying time. If you are spending fewer hours than these guidelines suggest, you could probably improve your grade by studying more. If you are spending more hours than these guidelines suggest, you may be studying inefficiently; in that case, you should talk to your instructor about how to study more effectively for a mathematics class.

If Your Course Is	Time for Reading the Assigned Text (per Week)	Time for Homework Assignments (per Week)	Time for Review and Test Preparation (Average per Week)	Total Study Time (per Week)
3 credits	1 to 2 hours	3 to 5 hours	2 hours	6 to 9 hours
4 credits	2 to 3 hours	3 to 6 hours	3 hours	8 to 12 hours
5 credits	2 to 4 hours	4 to 7 hours	4 hours	10 to 15 hours

GENERAL STRATEGIES FOR STUDYING

- Don't miss class. Listening to lectures and participating in discussions is much more effective than reading someone else's notes. Active participation will help you retain what you are learning.

- Budget your time effectively. An hour or two each day is more effective, and far less painful, than studying all night before homework is due or before exams. It's also good to study shortly after class, while the lessons are still fresh in your mind.

- If a concept gives you trouble, do additional reading or problem solving beyond what has been assigned. If you still have trouble, ask for help. You surely can find friends, colleagues, or teachers who will be glad to help you learn.

- Working together with friends can be valuable, as you will solidify and enhance your own understanding when discussing concepts with others. However, be sure that you learn *with* your friends and do not become dependent on them.

PREPARING FOR EXAMS

- Rework exercises and other assignments. Try additional exercises to be sure you understand the concepts. Study your performance on assignments, quizzes, or exams from earlier in the semester.

- Study your notes from lectures and discussions. Pay attention to what your instructor expects you to know for an exam.

- Reread the relevant sections in the textbook, paying special attention to notes you have made in the margins.

- Study individually before joining a study group with friends. Study groups are effective only if *every* individual comes prepared to contribute.

- Don't stay up too late before an exam. Don't eat a big meal within an hour of the exam (thinking is more difficult when blood is being diverted to the digestive system).

- Try to relax before and during the exam. If you have studied effectively, you are capable of doing well. Staying relaxed will help you think clearly.

PROLOGUE

Discussion Questions

1. **Mathematics in Modern Issues.** Describe at least one way that mathematics is involved in each issue below.

 Example: The spread of AIDS: Mathematics is used to study the probability of contracting AIDS.

 a. The long-term viability of the Social Security system
 b. The appropriate level for the federal gasoline tax
 c. The fairness of medical insurance rates
 d. Job discrimination against women or ethnic minorities
 e. Effects of population growth (or decline) on your community
 f. Possible bias in standardized tests (e.g., the SAT)
 g. The degree of risk posed by carbon dioxide emissions
 h. Immigration policy of the United States
 i. Violence in public schools
 j. An issue of your choice from today's newspaper

2. **Quantitative Concepts in the News.** Identify the major unresolved issue discussed on the front page of today's newspaper. List at least three areas in which quantitative concepts play a role in the policy considerations of this issue.

3. **Mathematics and the Arts.** Choose a well-known historical figure in a field of art in which you have a personal interest (e.g., a painter, sculptor, musician, or architect). Briefly describe how mathematics played a role in or influenced that person's work.

4. **Quantitative Literature.** Choose a favorite work of literature (poem, play, short story, or novel). Describe one or more instances in which quantitative reasoning is helpful in understanding the subtleties intended by the author.

5. **Your Quantitative Major.** Identify ways in which quantitative reasoning is important within your major field of study. (If you haven't yet chosen a major, pick a field that you are considering for your major.)

6. **Career Preparation.** Realizing that most Americans change careers several times during their lives, identify at least three occupations in Table 2 that interest you. Do you have the necessary skills for them at this time? If not, how can you acquire these skills?

7. **Attitudes Toward Mathematics.** Most children have a natural affinity for mathematics; they take pride in their counting skills and enjoy puzzles, building blocks, and computers. Unfortunately, this natural interest seems to be snuffed out in most people by the time they reach adulthood. What is your attitude toward mathematics? If you have a negative attitude, can you identify when in your childhood that attitude developed? If you have a positive attitude, can you explain why? How might you encourage someone with a negative attitude to become more positive?

Civilized life depends upon the success of reason in social intercourse, the prevalence of logic over violence in interpersonal conflict.

—JULIANA GERAN PILON

CHAPTER

1

Thinking Critically

Critical thinking is an approach to problem solving and decision making that involves careful reading (or listening), sharp thinking, logical analysis, good visualization, and healthy skepticism. You probably already apply critical thinking to many problems that affect you personally, but you may never have studied it in a course. In this first chapter, you will study a variety of skills—many of them skills of logic—that will be useful to you not only in the rest of this book, but also in the thinking you do in daily life.

UNIT **1A** **Recognizing Fallacies:** We will begin our study of critical thinking by looking at deceptive arguments, or fallacies.

UNIT **1B** **Propositions and Truth Values:** We will define and study basic components of logic, including propositions, truth values, and truth tables. We will also consider the logical connectors *and, or,* and *if . . . then.*

UNIT **1C** **Sets and Venn Diagrams:** We will explore how many propositions express relationships between sets, or categories, and how Venn diagrams can help us visualize these relationships.

UNIT **1D** **Analyzing Arguments:** We will study the basic differences between inductive and deductive arguments and how to evaluate each type.

UNIT **1E** **Critical Thinking in Everyday Life:** The skills built in the first four units are useful in and of themselves, but our primary goal is to become proficient at analyzing information carefully. We will put the experience gained in earlier units to work in studying critical thinking situations encountered in everyday life.

UNIT 1A

RECOGNIZING FALLACIES

We often try to persuade other people of our views. These discussions take many different forms, some of which are more useful than others. Consider the following "argument" between two classmates.

BOB: The death penalty is immoral.
SUSAN: No it isn't.
BOB: Yes it is! Judges who give the death penalty should be impeached.
SUSAN: You don't even know how the death penalty is decided.
BOB: I know a lot more than you know!
SUSAN: I can't talk to you; you're an idiot!

People generally quarrel because they cannot argue.

—G. K. CHESTERTON (1874–1936), ENGLISH AUTHOR

This type of heated conversation is fairly common, but it accomplishes little. Rather than helping Bob or Susan sway the other, it is likely to leave both of them upset and angry. Fortunately, there is a better way to argue. We can use skills of **logic**—the study of the methods and principles of reasoning. Arguing logically may not change Bob's or Susan's position, but it can help them understand each other.

In this book, the term **argument** refers to reasoned or thoughtful processes. Specifically, an argument uses a set of facts or assumptions, called **premises,** to support a **conclusion.** Some arguments provide strong support for their conclusions, but others do not. An argument that fails to make a compelling case for its conclusion may contain some error in reasoning, or **fallacy** (from the Latin for "deceit" or "trick"). In other words, a fallacious argument tries to persuade in a way that doesn't really make sense when analyzed carefully.

by the way

U.S. businesses spend more than $250 billion per year on advertising. The average American sees over 1 million advertisements by age 18. The best advertising firms do extensive research before embarking on new campaigns.

DEFINITIONS

Logic is the study of the methods and principles of reasoning.

An **argument** uses a set of facts or assumptions, called **premises,** to support a **conclusion.**

A **fallacy** is a deceptive argument—an argument in which the conclusion is not well supported by the premises.

You might expect fallacies to be rare, but they are not. If you study advertisements, political campaigns, or editorial commentaries, you'll find that many of them are filled with fallacies. Moreover, fallacies are not always easy to recognize, in part because public relations specialists have spent billions of dollars researching how subtle fallacies can be used to persuade us to buy products, vote for candidates, or support particular policies.

Because fallacies are so common, it is important to be able to recognize them. Thus, we begin our study of critical thinking with examples of a few of the most com-

mon fallacies. The fallacy in each example has a fancy name, but learning the names is far less important than learning to recognize the faulty reasoning. The experience you gain by analyzing fallacies will provide a foundation upon which to build additional critical thinking skills.

►EXAMPLE 1 Appeal to Popularity

"Ford makes the best automobile in America. After all, more people drive Fords than any other American car."

Analysis The first step in dealing with any argument is recognizing which statements are premises and which are conclusions. This argument tries to make the case that *Ford makes the best automobiles in America,* so this statement is its conclusion. The only evidence it offers to support this conclusion is the statement *more people drive Fords than any other American car.* This is the argument's one and only premise. Overall, this argument has the form

Premise: More people drive Fords than any other American car.
Conclusion: Ford makes the best automobiles in America.

Note that the original written argument states the conclusion before the premise. Such "backwards" structures are common in everyday speech and are perfectly legitimate as long as the argument is well reasoned. In this case, however, the reasoning is faulty. The fact that more people drive Fords does not necessarily mean that Fords are the best cars.

This argument suffers from the fallacy of *appeal to popularity* (or *appeal to majority*), in which the fact that large numbers of people believe or act some way is used inappropriately as evidence that the belief or action is correct. We can represent the general form of this fallacy with a diagram in which the letter *p* stands for a particular statement (Figure 1.1). In this case, *p* stands for the statement *Ford makes the best automobiles in America.* <u>Now try Exercise 11.</u> ◄

Many people believe
p is true;
therefore ...

↓

p is true.

FIGURE 1.1
The fallacy of appeal to popularity. The letters *p* and *q* (used in later diagrams) represent statements.

►EXAMPLE 2 False Cause

"I placed the quartz crystal on my forehead, and in five minutes my headache was gone. The crystal made my headache go away."

Analysis We identify the premises and conclusion of this argument as follows:

Premise: I placed the quartz crystal on my forehead.
Premise: Five minutes later my headache was gone.
Conclusion: The crystal made my headache go away.

The premises tell us that one thing (crystal on forehead) happened before another (headache went away), but they don't prove any connection between them. That is, we cannot conclude that the crystal *caused* the headache to go away.

This argument suffers from the fallacy of *false cause,* in which the fact that one event came before another is incorrectly taken as evidence that the first event *caused* the second event. We can represent this fallacy with a diagram in which *A* and *B* represent two different events (Figure 1.2). In this case, *A* is the event of putting the crystal

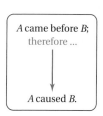

A came before *B*;
therefore ...

↓

A caused *B*.

FIGURE 1.2
The fallacy of false cause. The letters *A* and *B* represent events.

on the forehead and *B* is the event of the headache going away. (We'll discuss how cause *can* be established in Chapter 5.) *Now try Exercise 12.* ◄

►EXAMPLE 3 Appeal to Ignorance

"Scientists have not found any concrete evidence of aliens visiting Earth. Therefore, anyone who claims to have seen a UFO must be hallucinating."

Analysis If we strip the argument to its core, it says this:

Premise: There's no proof that aliens have visited Earth.

Conclusion: Aliens have not visited Earth.

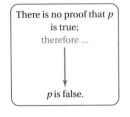

FIGURE 1.3
The fallacy of appeal to ignorance.

The fallacy should be clear: A *lack* of proof of alien visits does not mean that visits have not occurred. This fallacy is called *appeal to ignorance* because it uses ignorance (lack of knowledge) about the truth of a proposition to conclude the opposite (Figure 1.3). We sometimes sum up this fallacy with the statement "An absence of evidence is not evidence of absence." *Now try Exercise 13.* ◄

Time out to think _____

Suppose a person is tried for a crime and found *not* guilty. Can you conclude that the person is innocent? Why or why not? Why do you think our legal system demands that prosecutors prove guilt, rather than demanding that defendants (suspects) prove innocence? How is this idea related to the fallacy of appeal to ignorance?

►EXAMPLE 4 Hasty Generalization

"Two cases of childhood leukemia have occurred along the street where the high-voltage power lines run. The power lines must be the cause of these illnesses."

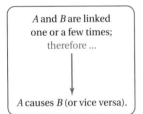

FIGURE 1.4
The fallacy of hasty generalization.

Analysis The premise of this argument cites two cases in which leukemia occurred on the street with the power lines. But two cases are not enough to establish a pattern, let alone to conclude that the power lines are the cause of the illnesses.

This argument illustrates the fallacy of *hasty generalization,* in which a conclusion is drawn from an inadequate number of cases or cases that have not been sufficiently analyzed. If any connection between power lines and leukemia exists, it will have to be established with far more evidence than is provided in this argument (see Unit 5E). We can represent this fallacy with a diagram in which *A* and *B* represent two linked events (Figure 1.4). *Now try Exercise 14.* ◄

►EXAMPLE 5 Limited Choice

"You don't support the President, so you are not a patriotic American."

Analysis This argument has the form

Premise: You don't support the President.

Conclusion: You are not a patriotic American.

The argument suggests that there are only two types of Americans: patriotic ones who support the President and unpatriotic ones who don't. But there are many other possibilities, such as being patriotic while disliking a particular president.

This fallacy is called *limited choice* (or *false dilemma*) because it artificially precludes choices that ought to be considered. Figure 1.5 shows one common form of this fallacy. Limited choice also arises with questions such as "Have you stopped smoking?" Both *yes* and *no* answers imply that you smoked in the past. Thus, the question precludes the possibility that you never smoked. (In legal proceedings, questions of this type are disallowed because they attempt to "lead the witness.") Another simple and common form of this fallacy is "You're wrong, so I must be right." *Now try Exercise 15.* ◀

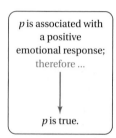

FIGURE 1.5
The fallacy of limited choice.

▶**EXAMPLE 6** Appeal to Emotion

In a commercial for Michelin tires, a picture of a baby is shown with the words "because so much is riding on your tires."

Analysis If we can consider this an argument at all, it has the form

Premise: You love your baby.
Conclusion: You should buy Michelin tires.

The argument makes no appeal to logic. Instead, the advertisers hope that the love you feel for a baby will make you want to buy their tires. This attempt to evoke an emotional response as a tool of persuasion represents the fallacy of *appeal to emotion*. Figure 1.6 shows its form when the emotional response is positive. Sometimes the appeal is to negative emotions. For example, the statement *if my opponent is elected, your tax burden will rise* tries to convince you that electing the other candidate will lead to consequences you won't like. (In this negative form, the fallacy is sometimes called *appeal to force*.) *Now try Exercise 16.* ◀

FIGURE 1.6
The fallacy of appeal to emotion.

▶**EXAMPLE 7** Personal Attack

GWEN: You should stop drinking because it's hurting your grades, endangering people when you drink and drive, and destroying your relationship with your family.
MERLE: I've seen you drink a few too many on occasion yourself!

Analysis Gwen's argument is well reasoned, with premises offering strong support for her conclusion that Merle should stop drinking. Merle rejects this argument by noting that Gwen sometimes drinks too much herself. Even if Merle's claim is true, it is irrelevant to Gwen's point. Merle has resorted to attacking Gwen personally rather than arguing logically, so we call this fallacy *personal attack* (Figure 1.7). (It is also called *ad hominem*, Latin for "to the person.")

The fallacy of personal attack can also apply to groups. For example, you might hear someone say, "This new bill will be an environmental disaster because its sponsors received large campaign contributions from heavy industrial polluters." This argument is fallacious because it doesn't challenge the provisions of the bill. Instead, it questions the motives of the sponsors.
 Now try Exercise 17. ◀

FIGURE 1.7
The fallacy of personal attack.

Time out to think _____

A person's (or group's) character, circumstances, and motives occasionally *are* logically relevant to an argument. That is why, for example, witnesses in criminal cases often are asked questions about their personal lives. If you were a judge, how would you decide when to allow such questions?

➤ **EXAMPLE 8** Circular Reasoning

"Society has an obligation to shelter the homeless because the needy have a right to the resources of the community."

Analysis This argument states the conclusion (*society has an obligation to shelter the homeless*) before the premise (*the needy have a right to the resources of the community*). But if you read them carefully, you'll see that the premise and the conclusion both say essentially the same thing. After all, society has an obligation to provide shelter only if the needy have a right to societal resources. Thus, this argument suffers from *circular reasoning* (Figure 1.8). *Now try Exercise 18.* ◄

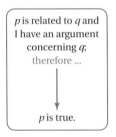

FIGURE 1.8
The fallacy of circular reasoning.

➤ **EXAMPLE 9** Diversion (Red Herring)

"We should not continue to fund genetic research because there are so many ethical issues involved. Ethics is at the heart of our society, and we cannot afford to have too many ethical loose ends."

Analysis The argument begins with its conclusion—*we should not continue to fund genetic research.* However, the discussion is all about ethics. This argument represents the fallacy of *diversion* (Figure 1.9) because it attempts to divert attention from the real issue (funding for genetics) by focusing on another issue (ethics). The issue to which attention is diverted is sometimes called a *red herring*. (A herring is a fish that turns red when rotten. Use of the term *red herring* to mean a diversion can be traced back to the 19th century, when British fugitives discovered that they could divert bloodhounds from their pursuit by rubbing a red herring across their trail.) Note that personal attacks (see Example 7) are often used as diversions. *Now try Exercise 19.* ◄

FIGURE 1.9
The fallacy of diversion.

➤ **EXAMPLE 10** Straw Man

Columnist William F. Buckley has written many articles arguing in favor of legalizing drugs. A speaker who opposes legalization begins by stating, "Mr. Buckley doesn't think there's anything wrong with drug use, but I do."

Analysis William F. Buckley has never said that drug use is acceptable. His arguments for legalization are based on considerations such as effects on the crime rate, cost to society, and difficulties of overcoming addiction. Thus, the speaker has distorted Buckley's views. Any argument based on a distortion of someone's ideas or beliefs is called a *straw man* (Figure 1.10). The term comes from the idea that the speaker has used a poor representation of a person's beliefs in the same way that a straw man is a poor representation of a real man. A straw man is similar to a diversion. The primary difference is that a diversion argues against an unrelated issue, while the straw man argues against a distorted version of the real issue. *Now try Exercise 20.* ◄

FIGURE 1.10
The straw man fallacy.

◼ EXERCISES 1A

Review Questions

1. What is logic? Briefly explain how logic can be useful.

2. How do we define *argument*? What is the basic structure of an argument?

3. What is a fallacy? Why is it important to be able to recognize fallacies?

4. Choose three of the examples from this unit and, in your own words, describe how the given argument is deceptive.

Does It Make Sense?

Decide whether each of the following statements makes sense (or is clearly true) or does not make sense (or is clearly false). Explain your reasoning.

5. Bob and Susan couldn't have been making an argument, because they weren't shouting at each other.

6. I persuaded my father that I was right with a carefully constructed argument that contained no premises at all.

7. I couldn't even figure out the premises on which he based his argument, so he clearly didn't convince me of his conclusion.

8. She convinced me she was right, even though she stated her conclusion before supporting it with any premises.

9. I disagree with your conclusion, so your argument must contain a fallacy.

10. Your argument suffers from a fallacy, but I think your conclusion is true.

Basic Skills & Concepts

Understanding Fallacies. Exercises 11–20 each give an example of one of the fallacies described in the text. For each exercise, do the following:

 a. Identify the premise(s) and conclusion of the stated argument.

 b. The argument contains an example of the fallacy named in parentheses. In one or two sentences, explain how the fallacy occurs in the argument.

 c. Make up your own example of another argument that suffers from the same fallacy.

11. (Appeal to Popularity) Polls show that 80% of the population supports the Governor's proposal, so it must be a good idea.

12. (False Cause) President Clinton raised taxes in 1993, so this tax increase must have been responsible for the great increase in government revenue that followed just a few years later.

13. (Appeal to Ignorance) There's no proof that global warming will have any bad consequences for our society, so we have no reason to be concerned about it.

14. (Hasty Generalization) I've talked to five friends who started to get better grades after they quit eating fast food, so I'm sure my grades will improve if I can just avoid fast foods.

15. (Limited Choice) He refused to testify by pleading the Fifth Amendment, so he must be guilty.

16. (Appeal to Emotion) I know that Dennis is a sweet and loving father, so there's no way he could have hurt that child.

17. (Personal Attack) Lawyer to jury: The prosecution wants you to believe this witness, but you cannot trust him because he is a convicted drug dealer.

18. (Circular Reasoning) Schools must implement a zero tolerance policy toward drug use, because any tolerance of drugs is unacceptable.

19. (Diversion) We should not build more highways, because there's too much pollution already.

20. (Straw Man) A politician states: The opponents of my health care plan think that it's OK for poor people to lack medical care.

Further Applications

Recognizing Fallacies. Each of the arguments given or described in Exercises 21–36 involves some kind of fallacy. Identify the premises and the conclusion of the argument, and then explain why the argument is deceptive. If the fallacy happens to fit one or more of the types described in the examples of this unit, name the types and explain how they apply.

21. Following President Reagan's defense buildup, the Soviet Union began the process of democratization that ultimately led to its breakup. Therefore, Reagan is responsible for the changes that led to the demise of the Soviet Union.

22. The Golden Rule is a sound ethical principle because it is basic to every system of ethics in every culture.

23. At a hearing to decide a child custody case, a lawyer asks the mother: "Have you stopped using drugs?"

24. Everyone I know is voting for the Governor, so you should vote for her too.

25. My mom says that I should never smoke, but I'm not going to pay any attention because she smoked when she was my age.

26. Most of the great mathematicians in history have been men, so men must be better at mathematics than women.

27. I believe in telepathy because no one has ever proven that it doesn't exist.

28. My two best teachers were both women, proving that women make better teachers than men.

29. There's no way that Senator Smith's bill can help the cause of gun control because he is one of the biggest recipients of campaign contributions from the National Rifle Association.

30. Violent crime by youth has risen in virtual lockstep with increased violence on television, thus proving that television violence leads to real violence.

31. Since 1960, the percentage of the population over 18 who smoke has decreased from 40% to about 20%. During the same period, the percentage of overweight people has increased from 25% to 35%. Clearly, quitting smoking leads to overeating.

32. On average, boys score higher than girls on standardized mathematics tests (such as the SAT), which proves that boys have a greater aptitude for mathematics than girls. Similarly, girls score higher on verbal tests, proving that girls are better at verbal skills.

33. My little boy loves dolls and my little girl loves trucks, so there's no truth to the claim that boys are more interested in mechanical toys while girls prefer maternal toys.

34. In the 1964 presidential campaign, President Johnson ran a television ad against Barry Goldwater (the Republican candidate, who ultimately lost the election) that showed a little girl picking daisy petals and counting them, "One, two, three. . . ." In the background, a gruff male voice began a countdown, "Ten, nine, eight . . . ," slowly drowning out the little girl's voice. As the countdown ended, the screen lit up with an atomic explosion. President Johnson's voice then said, "These are the stakes: to make a world in which all of God's children can live or go into the dark."

35. During the 2000 presidential campaign, George W. Bush favored an end to the estate tax. Because estate taxes fall most heavily on the wealthy, the wealthy would realize the greatest benefits from this change. The Gore campaign often argued against Bush's position with statements that began, "George Bush thinks that rich people aren't rich enough. . . ."

36. During the 2000 presidential campaign, Al Gore proposed regulatory changes that he claimed would reduce the danger of global warming. The Bush campaign countered with statements that began, "Al Gore believes that government is the solution to all our problems. . . ."

Web Projects

Find useful links for Web Projects on the text Web site: www.aw.com/bennett-briggs

37. **Campaign Site.** Find the Web site for a candidate or initiative (for or against) in an upcoming or recent election. Study the arguments given on the site, looking for examples of fallacies. Write a brief summary of the fallacies. Overall, does the site make a strong or weak case for its candidate or position? Explain.

38. **Gun Control Debate.** One of the most contentious debates in American politics concerns the question of whether the United States should enact stricter gun control laws. Visit one or more of the many Web sites for each side in the gun control debate. Identify at least two fallacies given by each side. Without stating which side you support, give your opinion of the logical quality of the arguments on the Web sites for the two sides—that is, do the arguments on one involve more fallacies than those on the other side? Explain.

39. **Death Penalty Debate.** The death penalty is another contentious issue in American politics. Visit one or more of the many Web sites for each side in the death penalty

debate. Identify at least two fallacies given by each side. Without stating which side you support, give your opinion of the logical quality of the arguments on the Web sites for the two sides—that is, do the arguments on one side involve more fallacies than those on the other side? Explain.

40. **Health Care Debate.** Choose some aspect of the health care debate in the United States, such as whether there should be coverage for prescription drugs or national insurance for all citizens. Find a Web site for each side of the debate. Without stating which side you support, give your opinion of the logical quality of the arguments on the Web sites for the two sides—that is, do the arguments on one side involve more fallacies than those on the other side? Explain.

41. **Fallacy Web Sites.** There are many Web sites devoted to the study of fallacies. Visit one and choose a fallacy of a type not covered in this unit. Explain the fallacy and give an example of it.

In the News

42. **Editorial Fallacies.** Examine editorials and letters to the editor in your local newspaper. Find at least three examples of fallacies. In each case, describe how the argument is deceptive. If the fallacy represents one or more of the common types described in this unit, describe the types.

43. **Fallacies in Advertising.** Pick a single night and a single commercial television channel and analyze the advertisements shown over a 1-hour period. Describe how each advertisement tries to persuade the viewer, and discuss whether the argument is fallacious. What fraction of the advertisements involve fallacies? Are any fallacy types more common than others?

44. **Fallacies in Politics.** Discuss the tactics used by both sides in a current or recent political campaign. How much of the campaign is/was based on fallacies? Describe some of the fallacies used in the campaign. Overall, do you believe that fallacies influenced (or will influence) the outcome of the vote?

45. **Personal Fallacies.** Describe an instance in which you were persuaded of something that you later decided was untrue. Explain how you were persuaded and why you later changed your mind. Did you fall victim to any fallacies? If so, how might you prevent the same thing from happening in the future?

UNIT 1B

PROPOSITIONS AND TRUTH VALUES

Having discussed fallacies in Unit 1A, we now study proper arguments. The building blocks of arguments are called **propositions**—statements that make (propose) a claim that may be either true or false. A proposition must have the structure of a complete sentence and must make a distinct assertion or denial. For example:

- *Joan is sitting in the chair* is a proposition because it is a complete sentence that makes an assertion.
- *I did not take the pen* is a proposition because it is a complete sentence that makes a denial.
- *Are you going to the store?* is not a proposition because it is a question. It does not assert or deny anything.
- *three miles south of here* is not a proposition because it does not make any claim and it is not a complete sentence.
- 7 + 9 = 2 is a proposition, even though it is false. It can be read as a complete sentence and it makes a distinct claim. *Now try Exercises 13–18.*

"Contrariwise,"
continued Tweedledee,
"if it was so, it might
be; and if it were so, it
would be; but as it isn't,
it ain't. That's logic."

—Lewis Carroll,
*Through the Looking
Glass*

DEFINITION

A **proposition** makes a claim (either an assertion or a denial) that may be either true or false. It must have the structure of a complete sentence.

NEGATION (OPPOSITES)

The opposite of a proposition is called its **negation.** For example, the negation of *Joan is sitting in the chair* is *Joan is not sitting in the chair,* and the negation of $7 + 9 = 2$ is $7 + 9 \neq 2$. If we represent a proposition with a letter such as p, then its negation is *not p* (sometimes written ~p). Note that negations are also propositions, since they have the structure of a complete sentence and may be either true or false.

A proposition has a **truth value** of either true (T) or false (F). If a proposition is true, its negation must be false, and vice versa. Thus, a proposition and its negation cannot both be true or both be false. We can represent these facts with a simple **truth table**—a table that has a row for each possible set of truth values. The following truth table shows the possible truth values for a proposition p and its negation *not p*. It has two rows because there are only two possibilities.

p	*not p*	
T	F	← This row shows that if p is true (T), *not p* is false (F).
F	T	← This row shows that if p is false (F), *not p* is true (T).

DEFINITIONS

Any proposition has two possible **truth values:** T = true or F = false.

The **negation** of a proposition p is another proposition that makes the opposite claim of p. It is written *not p* (or ~p) and has the opposite truth value of p.

A **truth table** is a table with a row for each possible set of truth values for the propositions being considered.

➤ **EXAMPLE 1** Negation

Consider the proposition *Betsy is the fastest runner on the team.* Write its negation. If the negation is false, is Betsy really the fastest runner on the team?

Solution The negation of the given proposition is *Betsy is not the fastest runner on the team.* If the negation is false, the original statement must be true, meaning that Betsy *is* the fastest runner on the team. <u>*Now try Exercises 19–22.*</u> ◄

I cannot say that I do not disagree with you.

—GROUCHO MARX

Double Negation

The Groucho Marx quotation in the margin may be an extreme example, but many everyday statements contain double (or multiple) negations. We've already seen that the negation *not p* has the opposite truth value of the original proposition p. The

double negation *not not p* must therefore have the same truth value as the original proposition *p*. We can show this fact with a truth table. The first column contains the two possible truth values for *p*. Two additional columns show the corresponding truth values for *not p* and *not not p*.

p	*not p*	*not not p*
T	F	T
F	T	F

In ordinary language, double negations rarely involve phrases like "not not." As the following two examples show, we must analyze wording carefully to recognize double negations.

If people don't want to come out to the baseball park, then nobody's going to stop them.

—YOGI BERRA

> ►**EXAMPLE 2** Radiation and Health

After reviewing data showing an association between low-level radiation and cancer among older workers at the Oak Ridge National Laboratory, Dr. Steve Wing (a health scientist from the University of North Carolina, Chapel Hill) was asked about the possibility of a similar association among younger workers at another national laboratory. He was quoted as saying (*Boulder Daily Camera*):

> *My opinion is that it's unlikely that there is no association.*

Does Dr. Wing think there is an association between low-level radiation and cancer?

Solution Because of the words "unlikely" and "no association," Dr. Wing's statement contains a double negation. To see the effects of these words clearly, let's start with a simpler proposition:

> *p = it's likely that there is an association (between low-level radiation and cancer).*

The word "unlikely" gives us the statement *it's unlikely that there is an association,* which we identify as *not p.* The words "no association" transform this last statement into Dr. Wing's statement, *it's unlikely that there is no association,* which we recognize as *not not p.* Because the double negation has the same truth value as the original proposition, we conclude that Dr. Wing believes it likely that there *is* an association between low-level radiation and cancer. *Now try Exercises 23–24.* ◄

> ►**EXAMPLE 3** The 2000 Miranda Ruling

In a June 2000 decision (*Dickerson v. United States*), the U.S. Supreme Court voted 7–2 to uphold the basic requirements of the 1966 Miranda decision (*Miranda v. State of Arizona*). That decision required that suspects taken into custody be informed of their constitutional rights, such as the right to remain silent and the right to legal counsel. In his majority opinion, Chief Justice William Rehnquist wrote:

> . . . *[legal] principles weigh heavily against overruling [Miranda] now.*

According to this statement, did the Chief Justice feel that legal principles support or oppose the original Miranda decision?

by the way

Ernesto Miranda confessed to and was convicted of a 1963 rape and kidnapping. His lawyer argued that the confession should not have been admitted as evidence because Miranda had not been told of his right to remain silent. The Supreme Court agreed and overturned his conviction. He was then retried, and again convicted (on the basis of evidence besides the confession). Miranda was stabbed to death during a bar fight after his release from prison. A suspect in his killing chose to remain silent upon arrest, and police never filed charges.

Solution Analyzed carefully, the Chief Justice's statement contains a double negation. The first negation comes from the term "overruling," which alone would imply opposition to the original decision. But the statement argued *against* overruling. Thus, the statement concludes that legal principles support the original Miranda decision. *Now try Exercises 25–28.* ◄

Logical Connectors

Propositions are often joined together with *logical connectors*—words such as *and, or,* and *if . . . then.* For example, consider the following two propositions:

$$p = \textit{the test was hard}$$
$$q = \textit{I got an A}$$

If we join the two propositions with *and,* we get a new proposition (sometimes called a *compound proposition*) that reads *the test was hard and I got an A.* If we join them with *or,* we get the statement *the test was hard or I got an A.* Although such statements are familiar from everyday speech, we must analyze them carefully.

And Statements (Conjunctions)

In formal logic, an *and* statement is called a **conjunction.** According to the rules of logic, the conjunction *p and q* is true only if *p* and *q* are *both* true. For example, the statement *the test was hard and I got an A* is true only if it *was* a hard test and you *did* get an A.

by the way

In symbolic logic, the connector *and* is denoted by \wedge. Thus, the conjunction *p and q* is written $p \wedge q$.

The Logic of *And*

Given two propositions *p* and *q,* the statement *p and q* is called their **conjunction.** It is true only if *p* and *q* are *both* true.

To make a truth table for the conjunction *p and q,* we analyze all possible combinations of the truth values of the individual propositions *p* and *q.* Because *p* and *q* each have two possible truth values (true or false), there are $2 \times 2 = 4$ cases to consider. The four cases become the four rows of the truth table.

Truth Table for Conjunction *p and q*

p	*q*	*p and q*	
T	T	T	← case 1: *p, q* both true
T	F	F	← case 2: *p* true, *q* false
F	T	F	← case 3: *p* false, *q* true
F	F	F	← case 4: *p, q* both false

Now try Exercises 29–30.

Note that the *and* statement is true only in the first case shown in the table, where both individual propositions are true.

➤ **EXAMPLE 4** *And* Statements

Evaluate the truth value of the following two statements.

 a. The capital of France is Paris and Antarctica is cold.
 b. The capital of France is Paris and the capital of America is Madrid.

Solution

 a. The statement contains two distinct propositions: *The capital of France is Paris* and *Antarctica is cold.* Because both propositions are true, their conjunction is also true.
 b. The statement contains two distinct propositions: *The capital of France is Paris* and *the capital of America is Madrid.* Although the first proposition is true, the second is false. Thus, their conjunction is false.

 Now try Exercises 31–36. ◄

➤ **EXAMPLE 5** Triple Conjunction

Suppose you are given three individual propositions p, q, and r. Make a truth table for the conjunction p *and* q *and* r. Under what circumstances is the conjunction true?

Solution When we considered the statement p *and* q, we saw that there were four possible cases for truth values. For each of these four cases, proposition r can be either true or false. Thus, the statement p *and* q *and* r has $4 \times 2 = 8$ possible cases for truth values. The following truth table contains a row for each of the eight cases. Note that the four cases for p *and* q each appear twice, once with r true and once with r false.

p	q	r	p and q and r
T	T	T	T
T	T	F	F
T	F	T	F
T	F	F	F
F	T	T	F
F	T	F	F
F	F	T	F
F	F	F	F

The conjuction p *and* q *and* r is true only if all three statements are true, as shown in the first row. You can see why if you remember that the conjunction of two statements is true only if both statements are true. Because we can think of p *and* q as a single statement (which is true only when p and q are both true), the conjunction of this statement with r can be true only if r is also true.

 Now try Exercises 37–38. ◄

HISTORICAL NOTE
The Greek philosopher Aristotle (384–322 B.C.) made the first known attempt to put logic on a rigorous foundation. He believed that truth could be established from three basic laws: (1) A thing is itself. (2) A statement is either true or false. (3) No statement is both true and false. Aristotle's laws were used by Euclid (c. 325–270 B.C.) to establish the foundations of geometry, and logic remains an important part of mathematics.

Time out to think _____

Given four propositions *p*, *q*, *r*, and *s*, how many rows are required for a truth table of the conjunction *p and q and r and s*? Under what circumstances is the conjunction true?

Or Statements (Disjunctions)

The connector *or* can have two different meanings. If a health insurance policy says that it covers hospitalization in cases of illness *or* injury, it probably means that it covers either illness or injury *or both*. This is an example of the **inclusive *or*** that means "either or both." In contrast, when a restaurant offers a choice of soup *or* salad, you probably are not supposed to choose both. This is an example of the **exclusive *or*** that means "one or the other."

Two Types of *Or*

The word *or* can be interpreted in two distinct ways:

- An **inclusive *or*** means "either or both."
- An **exclusive *or*** means "one or the other, but not both."

➤ **EXAMPLE 6** Inclusive or Exclusive?

Kevin's insurance policy states that his house is insured for earthquakes, fire, *or* robbery. Imagine that a major earthquake levels much of his house, the rest burns in a fire, and his remaining valuables are looted in the aftermath. Would Kevin prefer that the *or* in his insurance policy be inclusive or exclusive? Why?

Solution He would prefer an inclusive *or* so that his losses from all three events (earthquake, fire, looting) would be covered. If the *or* were exclusive, the insurance would cover only one set of losses. *Now try Exercises 39–44.* ◄

In general, the only way to determine whether an *or* statement is inclusive or exclusive is by its context. However, in logic we always assume that *or* is *inclusive* unless told otherwise.

A compound statement made with *or* is called a **disjunction.** Because we assume that the *or* is inclusive, the disjunction *p or q* is true if either or both propositions are true. Thus, a disjunction *p or q* is false only if both individual propositions are false.

by the way

In symbolic logic, the connector *or* is denoted by \vee. Thus, the disjunction *p or q* is written $p \vee q$.

The Logic of *Or*

Given two propositions *p* and *q*, the statement *p or q* is called their **disjunction.** In logic, we always assume that *or* is *inclusive* unless told otherwise. Thus, the disjunction is true if either or both propositions are true, which means that it is false only if *both* propositions are false.

These rules lead to the following truth table.

Truth Table for Disjunction *p or q*

p	q	p or q
T	T	T
T	F	T
F	T	T
F	F	F

Now try Exercises 45–50.

by the way

Logical rules lie at the heart of modern computer science. Computers generally use electric current to represent only two numbers: 0 and 1. A 1 means a circuit carries current and a 0 means a circuit carries no current. Computer scientists think of 1 as true and 0 as false and use logical connectors to design circuits. For example, an *and* circuit allows current to pass (its value is 1 = true) only if both incoming circuits carry current.

▶**EXAMPLE 7** Smart Cows?

Consider the statement *airplanes can fly or cows can read.* Is it true?

Solution The statement is a disjunction of two propositions: *airplanes can fly* and *cows can read.* The first proposition is clearly true, while the second is clearly false. As shown in the second row of the truth table for the disjunction, *p or q* is true in this case. Thus, the statement *airplanes can fly or cows can read* is true. *Now try Exercises 51–56.* ◀

▶**EXAMPLE 8** Key Word Search

You are doing a research report for which you need to find articles about the federal debt that appeared in either *Time* or *Newsweek* between 2002 and 2004. You have access to a database that allows you to search on date, magazine, and title. How should you structure the search?

Solution In general, using OR widens the search whereas using AND narrows the search. For the date, the key word search

2002 OR *2003* OR *2004*

will generate a list with articles published in any of the three years (because *or* is assumed to be inclusive). Similarly, for the magazine, you would search on

Time OR *Newsweek*

to ensure that both magazines are included on the list.

The title requires a bit more thought. Searching on *federal* OR *debt* would lead to a list with many articles unrelated to the federal debt, such as articles on federal crimes or corporate debt. Searching on *federal* AND *debt* would limit the search to articles containing both words in the title, but might miss relevant articles with different key words in their titles. For example, it would miss an article with the title "the United States budget debt." One way to solve this problem is to use a more complex set of search criteria, such as

(*federal* OR *United States* OR *government*) AND (*debt* OR *budget*)

This search would find articles with any of the following word combinations in their titles: *federal debt, United States debt, government debt, federal budget, United States budget,* or *government budget.* *Now try Exercises 57–60.* ◀

If . . . Then Statements (Conditionals)

Another common way to connect propositions is with the words "if . . . then," as in the statement "If all politicians are liars, then Representative Smith is a liar." Statements of this type are called **conditional propositions** (or *implications*) because they propose something to be true (the *then* part of the statement) on the *condition* that something else is true (the *if* part of the statement).

We can represent a conditional proposition as *if p, then q*. Proposition p is called the **hypothesis** (or *antecedent*) and proposition q is called the **conclusion** (or *consequent*).

Let's use an example to examine the truth table for a conditional proposition. Suppose that while running for Congress, candidate Jones claimed:

If I am elected, then the minimum wage will increase.

This proposition has the standard form *if p, then q*, where p = *I am elected* and q = *the minimum wage will increase*. Because each individual proposition can be either true or false, we must consider four possible cases for the truth value of *if p, then q*:

1. **p and q both true.** In this case, Jones *was* elected (p true) and the minimum wage increased (q true). Jones kept her campaign promise, so her claim, *if I am elected, then the minimum wage will increase,* was true.
2. **p true and q false.** In this case, Jones was elected, but the minimum wage did not increase. Because things did not turn out as she promised, her claim was false.
3. **p false and q true.** This is the case in which Jones was *not* elected, yet the minimum wage still increased. The conditional statement makes a claim about what should happen only in the event that Jones is elected. Because she was not elected, she surely did not break any campaign promise, regardless of whether or not the minimum wage increased. According to the rules of logic, Jones's original statement was true.
4. **p and q both false.** Now we have the case in which Jones was not elected and the minimum wage did not increase. Again, because she was not elected, she surely did not violate her campaign promise, even though the minimum wage did not increase. The rules of logic dictate that the original statement was true.

In summary, the statement *if p, then q* is true in all cases except when the hypothesis p is true and the conclusion q is false. Here is the truth table.

Truth Table for Conditional *if p, then q*

p	q	*if p, then q*
T	T	T
T	F	F
F	T	T
F	F	T

Now try Exercises 61–62.

The Logic of *If . . . Then*

A statement of the form *if p, then q* is called a **conditional proposition** (or *implication*). Proposition *p* is called the **hypothesis** and proposition *q* is called the **conclusion.** The conditional *if p, then q* is true in all cases except the case in which *p* is true and *q* is false.

Time out to think

Suppose candidate Jones had made the following campaign promise: *If I am elected, I will personally eliminate all poverty on Earth.* According to the rules of logic for cases (3) and (4) above, we consider this statement to be true in the event that Jones is *not* elected. Is this logical definition of truth the same one that you would use if you heard her make this promise? Explain.

▶**EXAMPLE 9** Conditional Truths

Evaluate the truth of the statement *if* $2 + 2 = 5$, *then* $3 + 3 = 4$.

Solution The statement has the form *if p, then q*, where *p* is $2 + 2 = 5$ and *q* is $3 + 3 = 4$. Both *p* and *q* are clearly false. However, according to the rules of logic, the conditional *if p, then q* is true any time *p* is false, regardless of what *q* says. Thus, the statement *if* $2 + 2 = 5$, *then* $3 + 3 = 4$ is true. *Now try Exercises 63–70.* ◀

Alternative Phrasings of Conditionals

In ordinary language, conditional statements don't always appear in the standard form *if p, then q*. In such cases, it can be useful to rephrase the statements in the standard form. For example, the claim *I'm not coming back if I leave* has the same meaning as *if I leave, then I'm not coming back.* Similarly, the statement *more rain will lead to a flood* can be recast as *if there is more rain, then there will be a flood.*

Two particularly important ways of phrasing conditionals use the words *necessary* and *sufficient*, respectively. The statement *if p, then q* means that *p* cannot be true unless *q* is true. Therefore, the statement *if p, then q* is equivalent to the statement *q is necessary for p*. For example, the statement *breathing is necessary for living* can be rephrased as *if you are living, then you must breathe.* This does not mean that breathing is the only necessity for living. It is one of many necessities (such as eating, breathing, and having a heartbeat).

The statement *if p, then q* also means that *q* will be true if *p* is true. Therefore, the statement *if p, then q* is equivalent to the statement *p is sufficient for q*. For example, the statement *being in Denver is sufficient for being in Colorado* can be rephrased as *if you are in Denver, then you are in Colorado.* This does not mean that being in Denver is necessary for being in Colorado, because many other places are also in Colorado.

HISTORICAL NOTE
The system of logic presented in this chapter was first developed by the ancient Greeks and is called *binary logic* because a proposition is either true or false, but not both and certainly nothing in between. Recently, mathematicians have developed systems of logic in which other truth values are possible. One form, called *fuzzy logic*, allows for a continuous range of values between absolutely true and absolutely false. Fuzzy logic is used in many new technologies.

Alternative Phrasings of Conditionals

The following are common alternative ways of stating *if p, then q:*

p is sufficient for q	*p will lead to q*	*p implies q*
q is necessary for p	*q if p*	*q whenever p*

➤ **EXAMPLE 10** Rephrasing Conditional Propositions

Recast each of the following statements in the form *if p, then q.* Discuss whether the statement is true or false.

 a. A rise in sea level will devastate Florida.
 b. A red tag on an item is sufficient to mean it's on sale.
 c. Being male is necessary for being president of the United States.

Solution

 a. This statement is equivalent to *if sea level rises, then Florida will be devastated.* The truth of the statement depends on how much sea level rises. A few inches might have only limited effects, while a few feet could flood most of the state.

 b. This statement can be written as *if an item is marked with a red tag, then it is on sale.* This statement is true only if all red tag items really are on sale.

 c. This statement can be expressed as *if a person is president, then the person is male.* While it is true that all American presidents to date have been male, many other nations have had female presidents and the United States may have a female president in the future. Thus, the statement is false.

Now try Exercises 71–76. ◄

Converse, Inverse, and Contrapositive

The order of the propositions does not matter in a conjunction or disjunction. For example, *p and q* has the same meaning as *q and p,* and *p or q* has the same meaning as *q or p.* However, when we switch the order of the propositions in a conditional, we create a completely different proposition called the **converse.** The following box summarizes definitions for the converse and two other variations on a conditional proposition *if p, then q.*

VARIATIONS ON THE CONDITIONAL

Name	Form	Example
Conditional	*if p, then q*	If you are sleeping, then you are breathing.
Converse	*if q, then p*	If you are breathing, then you are sleeping.
Inverse	*if not p, then not q*	If you are not sleeping, then you are not breathing.
Contrapositive	*if not q, then not p*	If you are not breathing, then you are not sleeping.

We can determine the truth values for the converse, inverse, and contrapositive with a truth table. Because all the statements use the same two propositions (*p, q*), the table has four rows. The first two columns show the truth values for *p* and *q*, respectively. The next two columns show the truth values for the negations *not p* and *not q*, which are needed for the inverse and contrapositive. The fifth column shows the truth values found previously for the conditional proposition *if p, then q*. We then find the truth values for the converse, inverse, and contrapositive by applying the rule that a conditional is false only when the hypothesis is true and the conclusion is false.

p	*q*	*not p*	*not q*	*if p, then q*	*if q, then p* (converse)	*if not p, then not q* (inverse)	*if not q, then not p* (contrapositive)
T	T	F	F	T	T	T	T
T	F	F	T	F	T	T	F
F	T	T	F	T	F	F	T
F	F	T	T	T	T	T	T

Note that the truth values for the conditional *if p, then q* are the same as those for its contrapositive. We therefore say that a conditional and its contrapositive are **logically equivalent:** If one is true, so is the other, and vice versa. Similarly, the table shows that the converse and inverse are logically equivalent.

DEFINITION

Two statements are **logically equivalent** if they share the same truth values: If one is true, so is the other, and if one is false, so is the other.

▶ **EXAMPLE 11** Logical Equivalence

Consider the statement *if a creature is a whale, then it is a mammal*. Write its converse, inverse, and contrapositive. Evaluate the truth of each statement. Which statements are logically equivalent?

Solution The statement has the form *if p, then q*, where *p = a creature is a whale* and *q = a creature is a mammal*. Thus, we find

converse (*if q, then p*): *if a creature is a mammal, then it is a whale.* This statement is false, because most mammals are not whales.
inverse (*if not p, then not q*): *if a creature is not a whale, then it is not a mammal.* This statement is also false; for example, dogs are not whales, but they are mammals.
contrapositive (*if not q, then not p*): *if a creature is not a mammal, then it is not a whale.* This statement is true, because all whales are mammals.

Note that the original proposition and its contrapositive have the same meaning and are logically equivalent. Similarly, the converse and inverse have the same meaning and are logically equivalent. *Now try Exercises 77–82.* ◀

TECHNICAL NOTE
The contrapositive is sometimes said to be the "inverse of the converse" You can see why by starting with the converse of *if p then q*, which is *if q then p*. Now, making the inverse of this converse gives us the statement *if not q, then not p*, which is the contrapositive of the original conditional *if p, then q*.

■ EXERCISES 1B

Review Questions

1. What is a proposition? Give a few examples, and explain why each is a proposition.

2. What do we mean by the negation of a proposition? Make up your own example of a proposition and its negation.

3. Define *conjunction, disjunction,* and *conditional,* and give an example of each in words.

4. What is the difference between an inclusive *or* and an exclusive *or*? Give an example of each.

5. Make a truth table for each of the following: *p and q; p or q; if p, then q.* Explain all the truth values in the tables.

6. Describe how to make the converse, inverse, and contra-positive of a conditional proposition. Make a truth table for each. Which statement is logically equivalent to the original conditional?

Does It Make Sense?

Decide whether each of the following statements makes sense (or is clearly true) or does not make sense (or is clearly false). Explain your reasoning.

7. My logical proposition to you is a question that you must answer.

8. If the "undead" means anyone who is still alive, then I'm certainly not a member of the undead.

9. We intend to catch him, dead or alive.

10. I can have my cake and eat it too. (Note: The expression "I can have my cake" implies that you will continue to have all of it to save for later.)

11. Now that I've studied logic, I can always determine the truth of any statement by making a truth table for it.

12. The fact is that higher taxes lead to lower revenues and, conversely, lower taxes lead to higher revenues. (Hint: Focus on whether the word "conversely" is used properly.)

Basic Skills & Concepts

Is It a Proposition? For Exercises 13–18, determine whether each statement is a proposition and give your reasons.

13. High taxes penalize the wealthy.

14. Over the hill and far away.

15. All cabbages are vegetables.

16. Some of her whiskers are gray.

17. Have you seen her latest movie?

18. My dog is a Rottweiler.

Negation. For Exercises 19–22, write the negation of each proposition. Then state the truth value of the original proposition and of its negation.

19. London is the capital of England.

20. The Gettysburg Address was given by George Washington.

21. Caesar was a Roman.

22. Mark Twain wrote *Tom Sawyer.*

Multiple Negations. Each of the statements in Exercises 23–28 contains a double or multiple negation. Analyze each statement, explaining what the statement means.

23. The city council did not approve the no-confidence vote for the police chief.

24. The Senator opposes the ban on anti-war demonstrations.

25. Congress voted against overriding the veto of the tax in-crease bill.

26. The Senate voted to override the President's veto on the disaster relief bill.

27. The overturning of the anti-discrimination policy violates the Constitution.

28. The ban on anti-nuclear demonstrations was lifted.

Truth Tables. In Exercises 29–30, make a truth table for the given statement. The letters p, q, r, and s represent propositions.

29. q and r

30. p and s

And Statements. Each of the statements in Exercises 31–36 connects two individual propositions with the word *and*. State the two individual propositions and, if possible, whether each is true or false. Then state whether the entire statement is true or false and explain why.

31. Quebec is the capital of Canada and Moscow is the capital of Russia.

32. Emily Dickinson was an American poet and Boris Yeltsin was a Russian president.

33. Ben is married and Ben is a bachelor.

34. $3 \times 7 = 21$ and $6 + 3 = 9$.

35. Grass is green and skies are blue.

36. Some people are nice and some people are mean.

Truth Tables. In Exercises 37–38, make a truth table for the given statement. The letters p, q, r, and s represent propositions.

37. q and r and s

38. p and q and r and s

Interpreting Or. For each of the statements in Exercises 39–44, state whether *or* is used in the exclusive or inclusive sense. Explain.

39. The menu offers a choice of appetizer or dessert.

40. The car warranty covers parts for 3 years or 36,000 miles.

41. If I win the lottery, I will go to Brazil or Nepal.

42. I will wear either boots or shoes.

43. The road will be made of asphalt or concrete.

44. The insurance policy covers fire or theft.

Truth Tables. In Exercises 45–50, make a truth table for the given statement. The letters p, q, r, and s represent propositions.

45. r or s

46. p or r

47. p and (*not p*)

48. q or (*not q*)

49. p or q or r

50. p or (*not p*) or q

Or Statements. Each of the statements in Exercises 51–56 connects two individual propositions with the word *or*. State the two individual propositions and, if possible, whether each is true or false. Then state whether the entire statement is true or false and explain why.

51. Seattle is the capital of the United States or Paris is the capital of France.

52. Julia Roberts was an 18th-century monk or Mark Twain was a writer.

53. Bill Clinton or Martin Luther King was a U.S. president.

54. Ben is married or Ben is a bachelor.

55. $8 \times 4 = 32$ or $7 \times 3 = 22$.

56. $3 \times 7 = 21$ or $6 + 3 = 9$.

Key Word Searches. Describe how you would structure a key word search to find the information described in each of Exercises 57–60.

57. Articles about earthquakes in California

58. Articles that mention either earthquakes or volcanic eruptions (or both) that occurred in Japan

59. Articles about Martin Luther King's assassination

60. Articles about how the Cold War affected relations between the United States and Argentina

Truth Tables. In Exercises 61–62, make a truth table for the given statement. The letters p, q, r, and s represent propositions.

61. if p, then r

62. if r, then s

If . . . Then Statements. For Exercises 63–70, identify the hypothesis and conclusion in each statement, and state their truth values. Then determine whether the entire statement is true or false and explain why.

63. If San Francisco is in California, then San Francisco is in the Pacific time zone.

64. If London is the capital of England, then Paris is the capital of China.

65. If Paris is the capital of China, then London is the capital of England.

66. If the Pope is unmarried, then the Pope is a bachelor.

67. If sparrows can fly, then sparrows are birds.

68. If butterflies can fly, then butterflies are birds.

69. If $2 + 2 = 5$, then primates are not mammals.

70. If $5 \times 7 = 0$, then fish can fly.

Rephrasing Conditionals. Rephrase each statement in Exercises 71–76 as a conditional proposition with the form *if p, then q*. Identify *p* and *q*. Then state whether the conditional proposition is true or false and explain why.

71. A resident of Miami is a resident of the United States.

72. She's a lawyer if she's a member of Congress.

73. Playing the saxophone is necessary for being considered a musician.

74. Passing a biology course is sufficient to meet a one-course science requirement.

75. Being a male is necessary for having prostate cancer.

76. Living in Santa Fe is sufficient to be living in New Mexico.

Converse, Inverse, and Contrapositive. For Exercises 77–82, write the converse, inverse, and contrapositive of the given proposition and state whether each is true or false. Which statements are logically equivalent?

77. If Marco lives in Chicago, then he lives in the United States.

78. If the patient is alive, then the patient is sleeping.

79. If I go swimming, then I will get wet.

80. If it's a penguin, then it can fly.

81. If the car's engine is running, there is gas in the tank.

82. If you live near the ocean, then you live in Hawaii.

Further Applications

Famous Quotes. Rephrase each of the quotations in Exercises 83–86 using one or more conditional forms (*if p, then q*). Then state whether you think each proposition is true or false and defend your opinion.

83. The young man who has not wept is a savage, and the old man who will not laugh is a fool. —GEORGE SANTAYANA

84. Laugh and the world laughs with you. —ANTHONY BURGESS

85. It is my experience that folks who have no vices have very few virtues. —ABRAHAM LINCOLN

86. The man who reads nothing at all is better educated than the man who reads nothing but newspapers. —THOMAS JEFFERSON

Writing Conditional Propositions. In Exercises 87–90, write your own example of a proposition that has the indicated characteristics. If it is not possible to write such a proposition, explain why.

87. A true conditional proposition whose converse is false

88. A true conditional proposition whose converse is true

89. A false conditional proposition whose converse is true

90. A false conditional proposition whose converse is false

91. Library Search. Suppose that you wanted to find the call number of the book *Finnegan's Wake* by James Joyce. For each of the following key word search combinations, state whether the book would be on the search list. Explain.

 a. Finnegan

 b. Finnegan and Joyce

 c. Joyce or Yeats

 d. Joyce and James

 e. Joyce and Yeats

 f. Finnegan and (Joyce or Yeats)

92. Alimony Tax Laws. The federal tax policy on alimony payments is as follows:

 1. Alimony you receive after you remarry is taxable if the payer did not know you had remarried.

 2. Alimony you receive after you remarry is not taxable if the payer did know you had remarried.

 3. Alimony you pay is never deductible by the payer.

 Rephrase these three statements as conditional propositions.

93. Artificial Intelligence. Consider the following quote from *TIME Magazine* (Ray Kurzweil, May 22, 2000):

 Let's consider the requirements for a computer to exhibit human-level intelligence. . . . A necessary (but not sufficient) condition is the requisite processing power, which I estimate at about 20 million billion calculations per second.

 Rephrase this statement as an *if . . . then* proposition and explain what it means.

Truth Tables and Logical Equivalence. Exercises 94–99 each list two statements in which *p, q,* and *r* represent propositions. Make a truth table for each of the two given state-

ments. Are the two statements logically equivalent? Explain. (Hint: Evaluate expressions within the parentheses first.)

94. *not* (*p* and *q*); (*not p*) or (*not q*)

95. *not* (*p* or *q*); (*not p*) and (*not q*)

96. *not* (*p* and *q*); (*not p*) and (*not q*)

97. *not* (*p* or *q*); (*not p*) or (*not q*)

98. (*p* and *q*) or *r*; (*p* or *r*) and (*p* or *q*)

99. (*p* or *q*) and *r*; (*p* and *r*) or (*q* and *r*)

Web Project

Find useful links for Web Projects on the text Web site: www.aw.com/bennett-briggs

100. **Internet Searches.** Investigate rules for key word searches on several Web sites (such as Yahoo or Britannica) or search engines (such as Google or Lycos). Are the rules consistent with the laws of logic described in this chapter? Of the sites you checked, which offers the most "user-friendly" search? Explain why you like it.

In the News

101. **Logical *Or*.** Find a news article or advertisement in which the connector *or* is used. Is it used in an inclusive or exclusive sense? Explain its meaning in the given context.

102. **Multiple Negation.** Find a news article or advertisement in which a double (or multiple) negation is used. Explain the meaning of the sentence in which it occurs.

103. **Conditional News.** Find a news article or editorial in which a conditional statement is used. Does it appear in *if . . . then* form or can it be rephrased in this form? From the context of the article, can you comment on the truth of the propositions involved? Is the entire conditional true or false? Explain.

UNIT 1C

SETS AND VENN DIAGRAMS

We've seen that propositions come in many forms. The only general requirement is that a proposition must make a clear claim (assertion or denial). In this unit, we focus on propositions that claim a relationship between two categories of things. For example, the proposition *all whales are mammals* makes the claim that the category *whales* is entirely contained within the category *mammals.*

Propositions of this type are most easily studied with the aid of two key ideas. The first is the idea of a *set,* which is really just another word for a collection. The second is the idea of a *Venn diagram,* which is a simple visual way of illustrating relationships between sets. Both ideas are very useful for organizing information and hence are important tools of critical thinking.

RELATIONSHIPS AMONG SETS

A **set** is a collection of objects, living or nonliving. The **members** of a set are the specific objects within it. For example:

- The members of the set *days of the week* are the individual days Sunday, Monday, Tuesday, Wednesday, Thursday, Friday, and Saturday.
- The members of the set *American military services* are the Army, Navy, Air Force, and Marine Corps.

by the way

The four military services are organized under the United States Department of Defense. The United States Coast Guard is often grouped with these services, but it is administered by the Department of Homeland Security.

- The members of the set *Academy Award winning actresses* are the individual actresses who have won Academy Awards.

Set Notation

Sets are often described by listing all their individual members within a pair of braces, { }. For example, the set *American military services* can be written as

<p style="text-align:center">{Army, Navy, Air Force, Marine Corps}</p>

Every member of the set is listed within the braces, with each member separated from the next by a comma.

Some sets have so many members that it would be difficult or impossible to list all of them. In that case, we can use three dots, . . . , to indicate that the list continues in the same basic manner. (The three dots are called an *ellipsis,* but most people just say "dot-dot-dot" when reading them.) If the dots come at the end of a list, they indicate that the list continues indefinitely. For example, we can write the set of *dog breeds* as

<p style="text-align:center">{Rottweiler, German Shepherd, Poodle, . . . }</p>

The three dots indicate that the list continues as you would expect—in this case, with the names of all other dog breeds. It does not matter how many members you list, as long as the pattern is clear. Listing three members, as in the dog breed list, is usually enough to make the point.

In other cases, the dots may come in the middle of the list, representing members that were not explicitly written. For example, the set

<p style="text-align:center">{A, B, C, . . . , Z}</p>

represents the entire alphabet in capital letters. The dots are simply a convenience, saving us from having to write every letter.

Although it is less common, the three dots can also come at the beginning of a list. For example, the set

$$\{\ldots, -3, -2, -1\}$$

represents all negative integers. Here, the dots indicate that the list continues to the left, to ever smaller (more negative) numbers.

The whole is more than the sum of its parts.

— ARISTOTLE, *METAPHYSICA*

DEFINITIONS

A **set** is a collection of objects; the individual objects are the **members** of the set. We often describe sets by listing their members within a pair of braces, { }. If there are too many members to list, we can use three dots, . . . , to indicate a continuing pattern.

Time out to think

How would you use braces notation to describe the set of students in your mathematics class? How about the set of countries you have visited? Choose one more example of a set that affects you personally, and state how you would describe it with braces notation.

SETS OF NUMBERS

In mathematics, we commonly work with the following important sets of numbers.

The set of **natural numbers** (or *counting numbers*) is

$$\{1, 2, 3, \dots\}$$

We can represent the natural numbers on a *number line* with equally spaced dots beginning at 1 and continuing to the right forever.

The set of **whole numbers** is the same as the set of natural numbers except it includes zero. Thus, its members are

$$\{0, 1, 2, 3, \dots\}$$

We can represent the whole numbers on a number line with equally spaced dots beginning at 0 and continuing to the right forever.

The set of **integers** includes the whole numbers and their negatives. Thus, its members are

$$\{\dots, -3, -2, -1, 0, 1, 2, 3, \dots\}$$

On the number line, the integers extend forever both to the left and to the right.

The set of **rational numbers** includes the integers and the fractions that can be made by dividing one integer by another, as long as we don't divide by zero. (The word *rational* refers to a *ratio* of integers.) In other words, rational numbers can be expressed in the form

$$\frac{x}{y},$$ where x and y are integers and $y \neq 0$

(Recall that the symbol \neq means "is not equal to.")

When expressed in decimal form, rational numbers are either terminating decimals with a finite number of digits (such as 0.25, which is $\frac{1}{4}$) or repeating decimals in which a pattern repeats over and over (such as 0.333 . . . , which is $\frac{1}{3}$).

Irrational numbers are numbers that *cannot* be expressed in the form x/y. When written as decimals, irrational numbers neither terminate nor have a repeating pattern. For example, the number $\sqrt{2}$ is irrational because it cannot be expressed exactly in a form x/y; as a decimal, we can write it as 1.414213562 . . . , where the dots mean that the digits continue forever with no pattern. The number π is also an irrational number, which as a decimal is written 3.14159265. . . .

The set of **real numbers** consists of both rational and irrational numbers; hence it is represented by the entire number line. Each point on the number line has a corresponding real number, and each real number has a corresponding point on the number line. In other words, the real numbers are the integers and "everything in between." A few selected real numbers are shown on the number line below.

Examples:
- The number 25 is a natural number, which means it is also a whole number, an integer, a rational number, and a real number.
- The number -6 is an integer, which means it is also a rational number and a real number.
- The number $\frac{2}{3}$ is a rational number, which means it is also a real number.
- The number 7.98418 . . . is an irrational number; the dots indicate that the digits continue forever with no particular pattern. It is also a real number.

Now try Exercises 13–28.

➤ **EXAMPLE 1** Set Notation

Use braces to write the contents of each of the following sets:

 a. the set of *planets in our solar system*
 b. the set of *years of the Cold War,* generally taken to have started in 1945 and ended in 1991
 c. the set of *natural numbers greater than 5*

Solution

 a. The set of *planets in our solar system* is {Mercury, Venus, Earth, Mars, Jupiter, Saturn, Uranus, Neptune, Pluto}.
 b. The set of *years of the Cold War* is {1945, 1946, 1947, . . . , 1991}; the dots indicate that the list includes all the years between 1947 and 1991, even though they are not listed explicitly.
 c. The set of *natural numbers greater than 5* is {6, 7, 8, . . . }; the dots indicate that the list continues to ever-higher numbers.

Now try Exercises 29–36. ◀

Illustrating Relationships with Venn Diagrams

The English logician John Venn (1834–1923) invented a simple visual way of describing relationships between sets. His diagrams, now called **Venn diagrams,** use circles to represent sets. Venn diagrams are fairly intuitive and best learned through examples.

 Consider the sets *whales* and *mammals.* Because every member of the set *whales* is also a member of the set *mammals,* we say that the set *whales* is a **subset** of the set *mammals.* As shown in Figure 1.11, we represent this relationship in a Venn diagram by drawing the circle for whales *inside* the circle for mammals. Notice that the diagram illustrates only the relationship between the sets. The sizes of the circles do not matter.

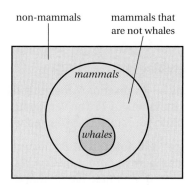

FIGURE 1.11 The set *whales* is a subset of the set *mammals*.

TECHNICAL NOTE
Some logicians distinguish between Venn diagrams and Euler (pronounced "oiler," after Leonhard Euler) diagrams, depending on the precise way the circles are used. In this book, we use the term *Venn diagrams* for all circle diagrams.

The circles are enclosed by a rectangle, so this diagram has three regions:

- The inside of the *whales* circle represents all whales.
- The region outside the *whales* circle but inside the *mammals* circle represents mammals that are not whales (such as cows, bears, and people).
- The region outside the *mammals* circle represents non-mammals; from the context, we interpret this region to represent animals (or living things) that are not mammals, such as birds, fish, and insects.

Next, consider the sets *dogs* and *cats*. A pet can be either a dog or a cat, but not both. We draw the Venn diagram with separated circles that do not touch, and we say that *dogs* and *cats* are **disjoint sets** (Figure 1.12). Again, we enclose the circles in a rectangle. This time, the context suggests that the region outside both circles represents pets that are neither dogs nor cats, such as birds and hamsters.

For our last general case, consider the sets *nurses* and *women*. As shown in Figure 1.13, these are **overlapping sets** because it is possible for a person to be both a woman and a nurse. Because of the overlapping region, this diagram has four regions:

- The overlapping region represents people who are both women and nurses—that is, female nurses.
- The non-overlapping region of the *nurses* circle represents nurses who are not women—that is, male nurses.
- The non-overlapping region of the *women* circle represents women who are not nurses.
- From the context, we interpret the region outside both circles to represent people who are neither nurses nor women—that is, men who are not nurses.

Note that the sizes of the regions are not important. For example, the small size of the overlapping region does not imply that female nurses are less common than male nurses. In fact, it is possible that some of the regions might have *no* members. For example, imagine that Figure 1.13 represents a single clinic where all four nurses are male. In that case, the overlapping region, which represents female nurses, would have no members. Speaking more generally, we use overlapping circles whenever two sets *might* have members in common.

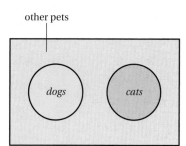

FIGURE 1.12 The set *dogs* is disjoint from the set *cats.*

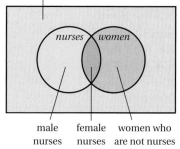

FIGURE 1.13 The sets *nurses* and *women* are overlapping.

Set Relationships and Venn Diagrams

Two sets A and B may be related in three basic ways:

TECHNICAL NOTE
Mathematically, there
is a fourth way that
two sets can be
related: They can be
equal, meaning that
they have precisely the
same members.

- A may be a **subset** of B, meaning that all members of A are also members of B. The Venn diagram for this case shows the circle for A inside the circle for B.

- A may be **disjoint** from B, meaning that the two sets have no members in common. The Venn diagram for this case consists of separated circles that do not touch.

- A and B may be **overlapping** sets, meaning that the two sets share some of the same members. The Venn diagram for this case consists of two overlapping circles. We also use overlapping circles for cases in which the two sets might share common members.

▶ **EXAMPLE 2** Venn Diagrams

Describe the relationship between the given pairs of sets and draw a Venn diagram showing this relationship. Interpret all the regions of the Venn diagram.

 a. *Democrats* and *Republicans* (party affiliations)
 b. *Nobel Prize winners* and *Pulitzer Prize winners*

Solution

 a. A person can be registered for only one political party, so the sets *Democrats* and *Republicans* are disjoint. Figure 1.14a shows the Venn diagram. The region outside both circles represents people who are neither Democrats nor Republicans—that is, people who are registered for other political parties, who are independent, or who are not registered.

by the way

Among the people who have won both a Nobel Prize (in literature) and a Pulitzer Prize are Toni Morrison, Ernest Hemingway, William Faulkner, John Steinbeck, and Saul Bellow.

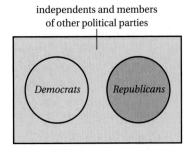

independents and members
of other political parties

Democrats Republicans

(a)

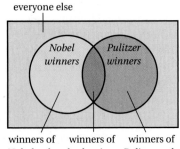

everyone else

Nobel winners Pulitzer winners

winners of Nobel only winners of both prizes winners of Pulitzer only

(b)

FIGURE 1.14

b. Several people have won both a Nobel Prize and a Pulitzer Prize, so these are overlapping sets. Figure 1.14b shows the Venn diagram. The region outside both circles represents people who have not won either prize, which is most of humanity.

Now try Exercises 37–42. ◄

➤ **EXAMPLE 3 Sets of Numbers**

Draw a Venn diagram showing the relationships among the sets of *natural numbers, whole numbers, rational numbers,* and *real numbers.* Where are irrational numbers found in this diagram? (If you've forgotten the meanings of these number sets, see *A Brief Review* on p. 37.)

Solution Natural numbers are also whole numbers, which means that the natural numbers are a subset of the whole numbers. Similarly, all whole numbers are integers, so the whole numbers are a subset of the integers. The integers, in turn, are a subset of the rational numbers, and the rational numbers are a subset of the real numbers. Figure 1.15 shows these relationships with a set of nested circles. Because irrational numbers are real numbers that are not rational, they are represented by the region outside the *rational numbers* circle but inside the *real numbers* circle.

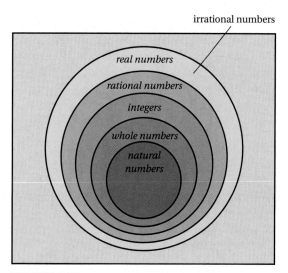

FIGURE 1.15

Now try Exercises 43–44. ◄

CATEGORICAL PROPOSITIONS

Now that we have discussed general relationships between sets, we are ready to study propositions that make claims about sets. For example, the proposition *all whales are mammals* makes a very specific claim—namely, that the set *whales* is a subset of the set *mammals.* Propositions of this type are called **categorical propositions** because they claim a particular relationship between two categories or sets.

Like all propositions, categorical propositions must have the structure of a complete sentence. Categorical propositions also have another important general

feature. Of the two sets in a categorical proposition, one set appears in the subject of the sentence and the other appears in the predicate. For example, in the proposition *all whales are mammals,* the set *whales* is the **subject set** and the set *mammals* is the **predicate set.** We usually use the letter *S* to represent the subject set and *P* for the predicate set, so we can rewrite *all whales are mammals* as

all S are P, where *S = whales* and *P = mammals*

Categorical propositions come in four standard forms, all shown in the following box.

THE FOUR STANDARD CATEGORICAL PROPOSITIONS

Form	Example	Subject Set (*S*)	Predicate Set (*P*)
All *S* are *P*	All whales are mammals	whales	mammals
No *S* are *P*	No fish are mammals	fish	mammals
Some *S* are *P*	Some doctors are women	doctors	women
Some *S* are not *P*	Some teachers are not men	teachers	men

Because categorical propositions express a relationship between two sets, we can use what we've already learned about Venn diagrams to make visual representations of categorical propositions. However, in the case of overlapping sets, we must be specific in stating which parts of the overlap are addressed by the proposition. The following example should make this idea clear.

▶ **EXAMPLE 4** Venn Diagrams for Categorical Propositions

Draw a Venn diagram to represent each of the four sample propositions in the box above.

Solution The proposition *all whales are mammals* claims that the set *S = whales* is a subset of *P = mammals.* Figure 1.16 shows the Venn diagram. Note that all propositions of the form *all S are P* would have the same basic Venn diagram.

The proposition *no fish are mammals* tells us that the set *S = fish* is disjoint from the set *P = mammals.* That is, they have no common members. The Venn diagram therefore shows two separated circles (Figure 1.17).

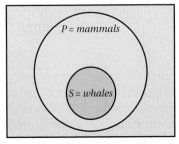

FIGURE 1.16 The Venn diagram for *all S are P.*

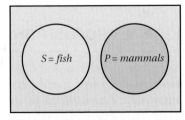

FIGURE 1.17 The Venn diagram for *no S are P.*

The Venn diagram for the proposition *some doctors are women* requires overlapping circles for *S = doctors* and *P = women*. However, we must be specific about the claim made in the proposition. In this case, the proposition asserts that there *are* some people (at least one) who are both doctors and women. We indicate this fact by putting an X in the overlapping region of the diagram (Figure 1.18). Note that the proposition does not tell us whether there are also doctors who are not women, nor does it tell us whether there are women who are not doctors. Without further information, we do not know whether the non-overlapping regions contain any members.

The proposition *some teachers are not men* also requires overlapping circles (Figure 1.19). This proposition asserts that there are some people (at least one) who are in the *S = teachers* circle but not in the *P = men* circle. Therefore, we put an X in the non-overlapping region of the *teachers* circle. No claim is made about whether there are teachers who are also men or men who are not teachers.

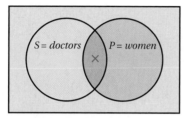

FIGURE 1.18 The Venn diagram for *some S are P.* The X indicates that the overlapping region has at least one member. No claim is made about whether other regions also have members.

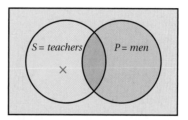

FIGURE 1.19 The Venn diagram for *some S are not P.* The X indicates the region claimed to have at least one member. No claim is made about other regions.

Now try Exercises 45–46. ◄

Time out to think _____

The principal of an elementary school states that at her school some teachers are not men. Can you conclude that some of the teachers *are* men? Why or why not? Can you conclude that *none* of the teachers are men? Why or why not?

Putting Categorical Propositions in Standard Form

Many statements in everyday speech make claims about relationships between two categories, but don't look precisely like one of the four standard forms for categorical propositions. It's often useful to rephrase such statements in one of the standard forms. For example, the statement *all diamonds are valuable* can be rephrased to read *all diamonds are things of value.* Now the statement has the form *all S are P,* where *S = diamonds* and *P = things of value.*

➤**EXAMPLE 5** Rephrasing in Standard Form

Rephrase each of the following statements in one of the four standard forms for categorical propositions. Then draw the Venn diagram.

 a. Some birds can fly.
 b. Elephants never forget.

Solution

 a. *Some birds can fly* can be rephrased as *some birds are animals that can fly.* It now has the form *some S are P,* where *S = birds* and *P = animals that can fly.* Figure 1.20a shows the Venn diagram, with an X in the overlapping region to indicate the claim that there *are* some birds that can fly.

 b. *Elephants never forget* can be rephrased as *no elephants are creatures that forget.* This proposition has the form *no S are P,* shown in Figure 1.20b, where *S = elephants* and *P = creatures that forget.*

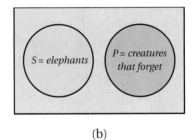

(a)

(b)

FIGURE 1.20

Now try Exercises 47–52. ◄

VENN DIAGRAMS WITH THREE SETS

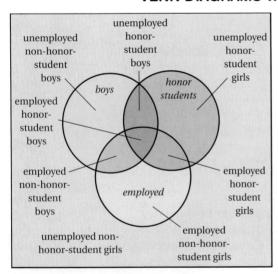

FIGURE 1.21 A Venn diagram for three overlapping sets has $2^3 = 8$ regions.

Venn diagrams are particularly useful for dealing with three sets that may overlap one another. For example, suppose you are conducting a study to learn how teenage employment rates differ between boys and girls and between honor students and others. For each teenager in your study, you need to record the answers to these three questions:

- Is the teenager a boy or a girl?
- Is the teenager an honor student or not?
- Is the teenager employed or not?

Figure 1.21 shows a Venn diagram that can help you organize this information. Notice that it has three circles, representing the sets *boys, honor students,* and *employed.* Because the circles overlap one another, they form a total of eight regions (including the region outside all three circles). Each region is labeled in the figure. Be sure that you understand how the labels correspond to the overlaps shown.

Time out to think _____

Could the same information be recorded in a diagram in which the three circles represented the sets *girls* (rather than boys), *honor students*, and *employed?* Could it be recorded in a diagram with the three circles representing the sets *girls, boys,* and *unemployed?* Explain.

►EXAMPLE 6 Makeup of the Study

You hire an assistant to help you with the study of teenage employment described above. He focuses on a small group of teenagers who are all enrolled in the same school. He reports the following facts about this group:

- Some of the honor-student boys are unemployed.
- Some of the non-honor-student girls are employed.

Put X's in the appropriate places in Figure 1.21 to indicate the regions that you can be sure have members. Based on this report, do you know whether any of the school's honor-student girls are unemployed? Why or why not?

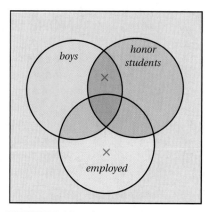

Solution Figure 1.22 shows the X's in the correct regions of the diagram (compare with the labels in Figure 1.21). The region corresponding to unemployed honor-student girls lies in the pink region of the *honor students* circle. There is no X in this region because the given information does not tell us whether this region has members. Thus, among the teenagers in the school, we do not know whether any honor-student girls are unemployed. *Now try Exercises 53–55.* ◄

FIGURE 1.22 The X's mark regions known to have members.

►EXAMPLE 7 Color Monitors

Color television and computer monitors make all the colors you see by combining *pixels* (short for *picture elements*) that display just three colors: red, green, and blue. In pairs, these combinations of colors give the following results (with colors at equal strength):

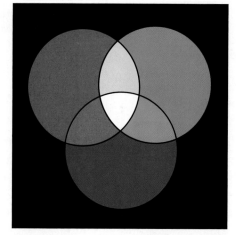

Combination	Result
Red-green	Yellow
Red-blue	Purple (or magenta)
Blue-green	Light blue (or cyan)

White is made by combining all three colors, and black is made by using none of the colors. Draw a Venn diagram to represent all this color information.

Solution The Venn diagram in Figure 1.23 represents the color combinations with three overlapping circles. The basic colors (red, green, and blue) appear in regions where there is no overlap. The red-green, red-blue, and blue-green combinations appear in

FIGURE 1.23 The color combinations possible from red, green, and blue (in equal strength).

regions where two circles overlap. White appears in the central region, where all three colors are present, while black appears outside all three circles, where no colors are present. A real monitor can make a much wider range of colors by varying the relative strengths of the colors. *Now try Exercises 56–58.* ◄

Time out to think _____

If you have a magnifying glass, hold it close to your television to see the individual red, blue, and green dots. How do you think a high-resolution monitor (such as a monitor for HDTV) differs from a lower-resolution one?

MORE USES OF VENN DIAGRAMS

So far we have used Venn diagrams only to describe relationships, such as whether two sets overlap or whether overlapping sets share common members. Venn diagrams can be even more useful when we add specific information, such as the number of members in each set or overlapping region.

Consider the study summarized in Table 1.1. This study was designed to learn whether a pregnant mother's status as a smoker or nonsmoker affects whether she delivers a low or normal birth weight baby. Notice that the table shows four numbers, which correspond to the four possible combinations of the baby's birth weight status and the mother's smoking status. To be sure that you are reading the table correctly, make sure you see how it shows the following facts:

- 18 babies were born with low birth weight to smoking mothers.
- 132 babies were born with normal birth weight to smoking mothers.
- 14 babies were born with low birth weight to nonsmoking mothers.
- 186 babies were born with normal birth weight to nonsmoking mothers.

TABLE 1.1 **Distribution of 350 Births by Birth Weight Status and Mother's Smoking Status**

		BABY'S BIRTH WEIGHT STATUS	
		Low Birth Weight	Normal Birth Weight
MOTHER'S SMOKING STATUS	Smoker	18	132
	Nonsmoker	14	186

U.S. National Center for Health Statistics.

Although a table is a good way to summarize data, a pictorial representation may be easier to understand. Because there are four combinations of birth weight and smoking status, we can use a Venn diagram with two overlapping circles (creating four regions on the diagram). Figure 1.24 shows one way of making the diagram. The circles represent the sets *smoking mothers* and *low birth weight babies*. The labels

SETS AND VENN DIAGRAMS

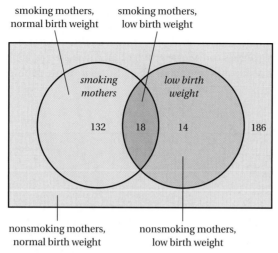

FIGURE 1.24 Venn diagram for the data in Table 1.1.

show how each region corresponds to one of the combinations in Table 1.1. This Venn diagram differs from those we have studied earlier in that it also shows the number of members in each region, taken from Table 1.1.

The Venn diagram makes it easy to see how smoking affected babies in the study. Notice that normal birth weight babies were much more common than low birth weight babies among both smokers and nonsmokers. However, the smoking mothers had a lower proportion of normal birth weight babies and a higher proportion of low birth weight babies. This suggests that smoking increases the risk of having a low birth weight baby, a fact that has been borne out by careful statistical analysis of this and other studies.

Time out to think _____

Explain why the Venn diagram also could be drawn with circles for the sets *nonsmoking mothers* and *normal birth weight babies.* Draw the diagram for this case, and put the four numbers in the right places.

▶ **EXAMPLE 8** Blood Types

Human blood is often classified according to whether three antigens, A, B, and Rh, are present or absent. Blood type is stated first in terms of the antigens A and B: Blood containing only A is called type A, blood containing only B is called type B, blood containing both A and B is called type AB, and blood containing neither A nor B is called type O. The presence or absence of Rh is indicated by adding the word *positive* (present) or *negative* (absent) or its symbol. Table 1.2 shows the eight blood types that result and the percentage of people with each type in the U.S. population. Draw a Venn diagram to illustrate these data.

TABLE 1.2	Blood Types in U.S. Population
BLOOD TYPE	PERCENTAGE OF POPULATION
A positive	34%
B positive	8%
AB positive	3%
O positive	35%
A negative	8%
B negative	2%
AB negative	1%
O negative	9%

Solution We can think of the three antigens as three sets *A, B,* and *Rh* (positive) and draw a Venn diagram with three overlapping circles. Figure 1.25 shows the eight regions, each labeled with its type and percentage of the population. For example, the central region corresponds to the presence of all three antigens (AB positive), so it is labeled with 3%. You should check that all eight regions are labeled according to the data from Table 1.2. *Now try Exercises 59–60.* ◄

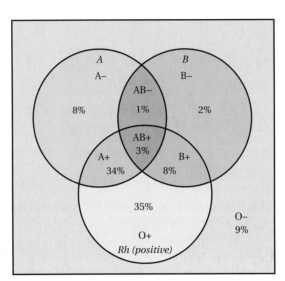

FIGURE 1.25 Venn diagram for blood types.

EXERCISES 1C

Review Questions

1. What is a set? Describe the use of braces notation for listing the members of a set.

2. What is a Venn diagram? How do we show that one set is a subset of another in a Venn diagram? How do we show disjoint sets? How do we show overlapping sets?

3. List the four standard categorical propositions. Give an example of each type, and draw a Venn diagram for each of your examples.

4. Briefly discuss how you can put a categorical proposition into one of the standard forms if it is not in such a form already.

5. Explain how to draw a Venn diagram for three overlapping sets. Discuss the types of information that can be shown in such diagrams.

6. Explain how to read a table such as Table 1.1 and how to show the information in a Venn diagram.

Does It Make Sense?

Decide whether each of the following statements makes sense (or is clearly true) or does not make sense (or is clearly false). Explain your reasoning.

7. The payments we make to the electric company are a subset of the payments we make to the phone company.

8. I don't know what you mean by the sets "jabbers" and "wocks," but I can still draw a Venn diagram for them if you just tell me if and how they share members.

9. I counted an irrational number of students in my class.

10. The number of students in my class is a real number.

11. My professor asked me to draw a Venn diagram for a categorical proposition, but I couldn't do it because the proposition was clearly false.

12. I used a Venn diagram to prove that your opinion is false.

Basic Skills & Concepts

Classifying Numbers. In Exercises 13–28, consider the sets *natural numbers, whole number, integers, rational numbers,* and *real numbers.* Identify from the list the *simplest* set that describes each number given. Explain.

13. 2.3

14. 23.8

15. $-\frac{3}{2}$

16. $\sqrt{3}$

17. 3

18. 15

19. -5

20. -25

21. 100.1

22. $\frac{4}{5}$

23. -6.1

24. -14.1

25. $\sqrt{5}$

26. $-\pi/2$

27. 2005

28. $-\pi$

Set Notation. In Exercises 29–36, use braces to write the members of each of the following sets, or state that the set has no members. (Dots may also be helpful.)

29. The months of the year

30. The natural numbers from 1 to 100 (inclusive)

31. The first three U.S. presidents

32. The states whose names start with N

33. The states whose names start with C

34. The letters in the second half of the alphabet

35. The natural numbers from 100 to 200 (inclusive)

36. The last three U.S. presidents

Venn Diagrams for Two Sets. For Exercises 37–44, draw a Venn diagram for the given sets. In words, explain why you drew one set as a subset of the other, disjoint sets, or overlapping sets.

37. doctors and women

38. reptiles and mammals

39. words and verbs

40. pianists and musicians

41. painters and artists

42. genetics courses and science courses

43. negative integers and natural numbers

44. whole numbers and positive integers

Categorical Propositions. Each of Exercises 45–52 gives a categorical proposition. If it is not already in standard form, rephrase in standard form. State the subject and predicate sets, and draw a Venn diagram for the proposition. Label all regions of the diagram clearly.

45. All bachelors are men.

46. Some pilots are women.

47. All U.S. presidents have been men.

48. Some mammals swim.

49. Fish don't fly.

50. Dogs can't swim.

51. Every nurse knows CPR.

52. Lawyers are highly educated.

Venn Diagrams for Three Sets. For Exercises 53–58, draw a Venn diagram with three overlapping circles for the three given sets. On your diagram, label the contents of every region. If a region has no members, state that fact clearly.

53. women, dentists, and airline pilots

54. basketball players, professional athletes, and men

55. songs, novels, and published works

56. oceans, bodies of salt water, and bodies of fresh water

57. clothes, hats, and shirts

58. states of the United States, places in North America, and places with coastlines on the Pacific Ocean

A Venn Diagram with Numbers. In Exercises 59–60, consider the following Venn diagram that represents people in a room. The number in each region is the number of people matching the characteristics of that region.

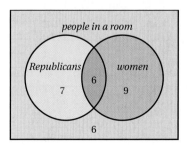

59. **a.** How many people are in the room?

 b. How many women are in the room?

 c. How many men are in the room?

 d. How many Republicans are in the room?

60. **a.** How many men are Republicans?

 b. How many men are not Republicans?

 c. How many women are Republicans?

 d. How many women are not Republicans?

Venn Diagrams for Sets. In Exercises 61–64, draw a Venn diagram to represent the given information.

61. Of the children on a school bus, 12 play soccer only, 6 play softball only, 2 play soccer and softball, and 4 play neither softball nor soccer.

62. A movie critic rated 15 feature movies and 10 documentaries. He gave 7 feature movies a "thumbs up" and the other feature movies a "thumbs down." He gave 9 of the documentaries a "thumbs up."

63. A doctor testing a new antibiotic puts 15 patients on a daily dose of the drug and 15 patients on a daily dose of a placebo (a substance that looks like the actual drug, but has no active ingredients). Twelve of the patients on the actual drug show improvement after a week, while 6 of the patients on the placebo show improvement.

64. A survey of 100 Toyota and Nissan owners revealed that all owners drove one vehicle and it was either an SUV or a pickup. Half of the 60 Toyota owners drove SUVs, while one-fourth of the Nissan owners drove SUVs.

Further Applications

Venn Diagram Answers. In Exercises 65–66, draw a two-circle Venn diagram that contains all the listed information. Label each region clearly. Then use the diagram to answer the questions.

65. Of the 43 students in the Cosmos Club, 16 are taking at least mathematics, 23 are taking at least biology, and 9 are taking mathematics only. How many students are taking biology only? How many are taking neither biology nor mathematics?

66. Of the 17 students in the Cycling Club, 12 own at least a road bike, 8 own at least a mountain bike, and 0 own no bike. How many students own both types of bike?

67. **Majors and Gender.** A scan of all biology and business majors at a college shows the following breakdown by gender. Draw a Venn diagram for the data.

	Biology	Business
Women	32	110
Men	21	87

68. **Tomatoes and Cancer.** A 1999 study by the Harvard Medical School (*Journal of the National Cancer Institute*, February 17, 1999) reviewed 72 previous studies of the effect of tomatoes on cancer. The data showed convincingly that "high consumers of tomatoes and tomato products are at substantially decreased risk of numerous cancers, although probably not all cancers." Consider the following table that shows the incidence of oral cancer for a group of people who ate an average of one tomato a day and another group of people who ate fewer than three tomatoes per week. Draw a Venn diagram for the data.

	No oral cancer	Oral cancer
One tomato per day	191	9
Fewer than three tomatoes per week	164	16

69. **Coffee and Gall Stones.** A study on the effect of coffee on gall stones (*Journal of the American Medical Association*, June 9, 1999) resulted (in small part) in the data shown below. The category *Coffee* means more than four cups of caffeinated coffee per day. The category *No coffee* means no caffeinated coffee. Draw a Venn diagram for the data.

	Gall stone disease	No disease
No coffee	385	14,068
Coffee	91	4,806

70. **Interpreting a Survey.** A survey asked newspaper readers which of the following three newspapers they read daily: *New York Times (NYT)*, *Washington Post (WP)*, and *Wall Street Journal (WSJ)*. The results are shown below. Answer the following questions, interpreting the connecting words carefully.

Paper(s)	Readers	Paper(s)	Readers
NYT only	24	NYT and WSJ only	14
WSJ only	27	NYT and WP only	16
WP only	26	WP and WSJ only	13
None	15	All three	8

a. Draw a Venn diagram for the table.

b. How many people read *WP* and *WSJ*?

c. How many people read *NYT* and *WSJ* and *WP*?

d. How many people read *NYT* or *WSJ* or *WP*?

e. How many people read *NYT* but not *WP*?

71. **Interpreting Sales.** A review of sales at a large automobile dealership for the past month shows that the following numbers of buyers chose air conditioning (AC), four-wheel-drive (4WD), and CD players (CD) in their cars.

Option(s)	Buyers	Option(s)	Buyers
AC only	45	AC and 4WD only	21
4WD only	8	AC and CD only	16
CD only	12	CD and 4WD only	8
None	3	All three	10

a. Draw a Venn diagram for the table.

b. How many people chose AC and 4WD?

c. How many people chose AC but not 4WD?

d. How many people chose AC or 4WD or CD?

e. How many people chose exactly two options?

72. **Analyzing Diets.** The following table shows ten common foods and whether they are significant sources of protein, fat, and carbohydrate. Organize these data in a Venn diagram. (Hint: You'll need circles for protein, fat, and carbohydrate; then put the name of each food in the appropriate region of the diagram.)

Food	Protein	Fat	Carbohydrate
Meat	X	X	X
Rice	X		X
Nuts	X	X	X
Potatoes			X
Apples			X
Cheese	X	X	
Eggs	X	X	
Lettuce			
Coffee			
Water			

More Than Three Sets. In Exercises 73–76, use your knowledge of the listed sets to draw a Venn diagram illustrating the relationships among them.

73. animals, house pets, dogs, cats, canaries

74. things that fly, birds, eagles, airplanes, jets, hang gliders

75. food, ice cream, dessert, fruit, citrus fruit, strawberries

76. people, men, women, surgeons, NFL football players, NFL quarterbacks, lawyers

Organizing Propositions. Each of Exercises 77–79 lists a set of propositions. Draw a Venn diagram that represents all the information in the propositions and use it (and no other assumptions) to answer the questions that follow. Explain your reasoning.

77. All hairy animals are mammals. No mammals are fish. Some mammals can swim. No fish can walk on land.

Questions: Could there be hairy fish? Could there be hairy animals that swim? Could there be walking mammals? Could there be hairy animals that walk on land?

78. All meat has protein. All dairy products have protein. Some beans have protein. All beans, but no meat or dairy products, are plants.

Questions: Could there be beans that are dairy products? Could there be meat that is a dairy product? Could there be dairy products that are plants? Could there be plants with protein?

79. No Republicans are Democrats. No Republicans are Green Party members. All Republicans are conservative. Some liberals are Democrats. No liberals are conservatives.

Questions: Could there be conservative Democrats? Could there be liberal Green Party members? Could there be liberal Republicans?

80. **Organizing Politicos.** You are at a conference attended by men and women of various political parties. The conference organizer tells you that none of the women are Republicans and some (but not all) of the Democrats are women.

a. Draw a Venn diagram to organize the given information.

b. Based on the given information, is it possible to meet a woman who is neither a Republican nor a Democrat?

c. Based on the given information, is it possible that there are any male Republicans?

81. **Organizing Literature.** In reviewing for an exam in your literature survey course, you notice the following facts about the writers that you studied:

- Some of the novelists are also poets.

- None of the novelists are playwrights.

- All of the novelists were born in the 20th century.
- All of the writers born in the 19th century are play-wrights.

 a. Organize these facts in a Venn diagram.

 b. Could you have studied a novelist born in the 19th century?

 c. Could you have studied a poet born in the 19th century?

 d. Could you have studied a writer born in the 20th century who was both a playwright and a poet?

82. **Is It Possible?** Shoppers are given a choice of one of three brands of vanilla ice cream and then asked if they approve or disapprove of the flavor. Is it possible to draw a Venn diagram with two circles that shows how many shoppers approved or disapproved of each brand? Explain.

83. *N Sets.* A computer store is offering a basic computer with four options A, B, C, and D, each of which buyers can select.

 a. How many different sets of options can buyers choose? (For example, one choice is no options, another choice is only options A and C, and another choice is all options.)

 b. Is it possible to make a Venn diagram using circles to show the number of buyers selecting each of the choices?

 c. Suppose the store offered five options A, B, C, D, and E. How many different sets of options would be available?

 d. Can you generalize parts a–c? If the store offers n options, each of which can be either selected or not selected, how many different sets of options are available?

Web Projects

Find useful links for Web Projects on the text Web site: www.aw.com/bennett-briggs

84. **State Politics.** Determine from the Web how many states have a Republican majority in the State House and how many states have a Republican majority in the State Senate. Draw a Venn diagram to illustrate the situation.

85. **U.S. Presidents.** Use the Web to learn the following about each past American president:

 - Bachelor or married (classify as married if married for part of the term)
 - Inaugurated before or after age 50
 - Served one term (or less) or more than one term

 Make a three-circle Venn diagram to represent your results.

86. **Uses of Venn Diagrams.** Visit the Web site for VennDiagram.com and read about the top 10 uses of Venn diagrams. Choose at least one of these uses and apply it to a specific problem in your own life.

In the News

87. **Categorical Propositions.** Find at least three examples of categorical propositions in news articles or advertisements. State the sets involved in each proposition and draw a Venn diagram for each proposition.

88. **Venn Diagrams in Your Life.** Describe a situation in your own life that could be described or organized using a Venn diagram.

89. **Quantitative Diagram.** Find a news article or research report with a table similar to Table 1.1 or 1.2. Draw a Venn diagram to represent the data in the table.

UNIT **1D**

ANALYZING ARGUMENTS

With the skills acquired in the previous units, we are now ready to analyze arguments. Recall that an argument begins with a set of premises that are intended to support one or more conclusions (see Unit 1A). If an argument is well constructed, its conclusions follow from its premises in a compelling way. But how do we determine whether an argument is well constructed and compelling? And, ultimately, how do we decide if we should accept an argument? These are the issues that we'll study in this unit.

TWO TYPES OF ARGUMENT: INDUCTIVE AND DEDUCTIVE

Arguments come in two basic types, known as *inductive* and *deductive*. The following two arguments illustrate the two types. As you read each one carefully, ask yourself whether its premises lead to its conclusion in a compelling way.

Argument 1 (Inductive)

Premise:	Birds fly up into the air but eventually come back down.
Premise:	People who jump into the air fall back down.
Premise:	Rocks thrown into the air come back down.
Premise:	Balls thrown into the air come back down.
Conclusion:	What goes up must come down.

Argument 2 (Deductive)

Premise:	All politicians are married.
Premise:	Senator Harris is a politician.
Conclusion:	Senator Harris is married.

Note that Argument 1 begins with a set of fairly specific premises, each of which makes a claim about a specific type of object. The conclusion of Argument 1 is a more general statement about how *any* object might behave. This type of argument, in which the conclusion is formed by *generalizing* from more specific premises, is called an **inductive argument.** (The term *inductive* has the root *induce*, which means "to lead by persuasion.")

In contrast, Argument 2 begins with a general statement about politicians and then draws a specific conclusion about a particular politician. Argument 2 is called a **deductive argument** because it allows us to *deduce* a specific conclusion from more general premises.

DEFINITIONS

An **inductive argument** makes a case for a general conclusion from more specific premises.

A **deductive argument** makes a case for a specific conclusion from more general premises.

Evaluating Inductive Arguments

Let's examine Argument 1 in greater detail. Its premises are clearly true, and each premise lends support to the conclusion. The variety of specific examples cited by its premises makes the argument appear quite compelling—and, indeed, people long believed the conclusion *what goes up must come down* to be true. Nevertheless, we now know that the conclusion is false because a rocket launched with sufficient speed can leave the Earth permanently. Thus, we see a key fact about inductive arguments: No matter how strong an inductive argument may seem, it does not *prove* that its conclusion is true.

Speaking more formally, we evaluate an inductive argument in terms of its **strength.** A strong argument makes the case for its conclusion seem quite convincing, even though it does not prove the conclusion true. A weak argument is one in which the premises do not seem to lend much support to the conclusion. Note that evaluating strength involves personal judgment. An argument that one person finds strong might appear weak to someone else.

Note also that the strength of an inductive argument is not necessarily related to the truth of its conclusion. Argument 1 was strong enough to convince people throughout most of history, but its conclusion turned out to be false. Conversely, a weak argument may have a true conclusion. For example, the simple argument *the sky is blue because I said so* is extremely weak, even though its conclusion is true.

Evaluating an Inductive Argument

An inductive argument cannot prove its conclusion true, so it can be evaluated only in terms of its **strength.** An argument is strong if it makes a compelling case for its conclusion. It is weak if its conclusion is not well supported by its premises.

> **EXAMPLE 1** Hit Movie

A movie director tells her producer (who pays for the movie) not to worry—her film will be a hit. As evidence, she cites the following facts: She's hired big stars for the lead roles, she has a great advertising campaign planned, and it's a sequel to her last hit movie. Explain why this argument is inductive, and evaluate its strength.

Solution Each of the three pieces of evidence is a specific characteristic of her movie. She uses them to support the more general conclusion that her movie will be a hit. Because the conclusion is more general than the premises, the argument is inductive. In this case, her argument is relatively weak. As all producers know, even the best-planned movies can flop. *Now try Exercises 15–22.* ◄

I think, therefore I am.

—RENÉ DESCARTES (1596–1650), FRENCH PHILOSOPHER AND MATHEMATICIAN

Time out to think _____

Suppose you are the director, and the producer says she needs more evidence that your movie will be a hit. How might you gather additional evidence to strengthen the case for your movie?

▶**EXAMPLE 2** Earthquake

Evaluate the following argument and discuss the truth of its conclusion.

> *Geological evidence shows that, for thousands of years past, the San Andreas Fault has suffered a major earthquake at least once every hundred years. Therefore, we should expect another earthquake on the fault during the next one hundred years.*

Solution This argument is inductive because it cites many specific past events as evidence that another earthquake will occur. The fact that the pattern has held for thousands of years suggests a strong likelihood that it will continue to hold. Thus, while the argument does not prove that another earthquake will occur, it makes another earthquake seem quite likely. The argument is strong.
Now try Exercises 23–28. ◀

Data from U.S Geological Survey.

Evaluating Deductive Arguments

Now let's turn to Argument 2 (p. 53). You may have the sense that it is a "solid" argument and that the conclusion follows in a convincing way from the premises. That is, if you accept its two premises—*all politicians are married* and *Senator Harris is a politician*—then it follows necessarily that Senator Harris is married.

Of course, you are not required to accept the premises, in which case you might not accept the conclusion either. In Argument 2, the first premise is clearly false: It is not true that all politicians are married. As a result, we cannot be sure that the conclusion (*Senator Harris is married*) is true. Thus, we see that evaluating a deductive argument requires answering *two* key questions:

• Does the conclusion follow necessarily from the premises?
• Are the premises true?

We can be sure that the conclusion is true only if the answer to both questions is yes.

Let's put this discussion in more formal terms. The first question above deals with the **validity** of the argument. A deductive argument is **valid** if its conclusion follows necessarily from its premises. Note that validity is concerned only with the logical structure of the argument. Validity involves no personal judgment and has nothing to do with the truth of the premises or conclusions.

If a deductive argument is valid *and* its premises are true, then we say that the argument is **sound.** Soundness represents the highest test of reliability of a deductive argument because, at least in principle, a sound argument proves its conclusion true beyond all doubt. However, soundness often involves personal judgment, because people may disagree about the truth of an argument's premises.

Argument 2 is valid because the conclusion that Senator Harris is married follows necessarily from its premises. But it is not sound, because the first premise (*all politicians are married*) is false.

KEY DISTINCTIONS: INDUCTIVE AND DEDUCTIVE ARGUMENTS

Inductive Arguments	Deductive Arguments
A conclusion is formed by *generalizing* from a set of more specific premises.	A *specific* conclusion is deduced from a set of more general (or equally general) premises.
An inductive argument can be analyzed only in terms of its *strength*. Evaluating strength involves personal judgment about how well the premises support the conclusion.	A deductive argument can be analyzed in terms of its *validity* and *soundness*. It is valid if its conclusion follows necessarily from its premises. It is sound if it is valid *and* its premises are true.
An inductive argument cannot prove its conclusion true. At best, a strong inductive argument shows that its conclusion *probably* is true.	Validity concerns only logical structure. Thus, a deductive argument can be valid even when its conclusion is blatantly false. A sound deductive argument provides definitive proof that its conclusion is true. (However, evaluating soundness often involves personal judgment.)

Evaluating a Deductive Argument

We apply two criteria when evaluating a deductive argument. The argument is **valid** if its conclusion follows necessarily from its premises, regardless of the truth of the premises or conclusions. The argument is **sound** if it is valid *and* its premises are all true.

Time out to think _____

Consider several decisions you made recently. For each case, decide whether the reasoning you used to reach the decision was inductive or deductive, and explain your reasoning process.

TESTS OF VALIDITY

We used intuition to determine validity in the case of Senator Harris. But deductive arguments can be subtle, and it's not always wise to rely on intuition alone. Instead, we can test validity by examining the structure of a deductive argument in a systematic way. One of the most common ways of testing validity uses Venn diagrams (see Unit 1C).

Let's begin with a Venn diagram test of Argument 2 on page 53. The first premise, *all politicians are married,* tells us that all members of the set *politicians*

are also members of the set *married people;* that is, it claims that *politicians* is a subset of *married people.* We represent this proposition by drawing the circle for *politicians* inside the circle for *married people* (Figure 1.26). The second premise tells us that Senator Harris is a member of the set *politicians.* We indicate this fact by putting an X, representing the Senator, inside the *politicians* circle. We test validity by checking to see whether the conclusion is contained within the Venn diagram for the premises. In this case, the X is also inside the *married people* circle, meaning that Senator Harris is a married person—just as the conclusion claims. Thus, the Venn diagram shows that the premises lead necessarily to the conclusion, demonstrating that the argument is valid.

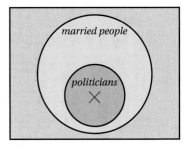

FIGURE 1.26 The X indicates that Senator Harris is a member of the *politicians* set, which means he is also a member of the *married people* set.

A Venn Diagram Test of Validity

To test the validity of a deductive argument with a Venn diagram:

1. Draw a Venn diagram that represents all the information contained in the premises.
2. Check to see whether the Venn diagram also contains the conclusion. If it does, then the argument is valid. Otherwise, the argument is not valid.

▶ **EXAMPLE 3** Invalid Argument

Evaluate the validity and soundness of the following argument.

Premise: All fish live in the water.
Premise: Whales are not fish.

Conclusion: Whales do not live in the water.

Solution Both premises are true, but the conclusion is false. The argument must therefore have a flaw in its logical structure, making it invalid. We can see the flaw by drawing a Venn diagram. The first premise tells us that the set *fish* is a subset of the set *things that live in the water,* so we draw the *fish* circle inside the circle for *things that live in the water.* The second premise tells us that whales are not fish. We can indicate this fact by putting an X, representing whales, outside the *fish* circle. However, because the second premise does not tell us whether whales live in the water, we do not know whether the X should be inside or outside the *things that live in the water* circle. Therefore, to be as general as possible, we place the X on the border of this circle, indicating that it may actually be either inside or outside (Figure 1.27).

We have now captured all the information in the two premises, so we next check whether the diagram contains the information in the conclusion. The conclusion states that whales do not live in the water, which means the X representing whales should be outside the circle for *things*

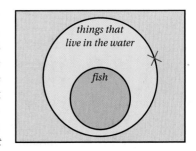

FIGURE 1.27 The X represents whales. The premises tell us only to put the X outside the *fish* circle; they do not say whether it should be outside the *things that live in the water* circle, so we place it on the border of this circle.

that live in the water. But it isn't; it is on the border, which means we don't have enough information to know whether it is inside or outside the circle. Therefore, the conclusion does not follow necessarily from the premises, and the argument is invalid. *Now try Exercises 29–32.* ◄

▶**EXAMPLE 4** Invalid but True Conclusion

Evaluate the validity and soundness of the following argument.

Premise: All 20th-century U.S. presidents were men.
Premise: Bill Clinton was a man.

Conclusion: Bill Clinton was a 20th-century U.S. president.

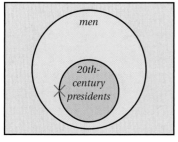

FIGURE 1.28 The X represents Bill Clinton. The premises tell us that the X belongs inside the *men* circle, but do not tell us whether it also belongs inside the *20th-century presidents* circle. We therefore place the X on the border of the inner circle.

Solution We start by drawing a Venn diagram for the first premise, which requires placing the circle for *20th-century presidents* inside the circle for *men.* The second premise tells us to put an X, representing Bill Clinton, inside the *men* circle. However, it does *not* tell us whether the X also belongs inside the *20th-century presidents* circle, so we place it on the border of this circle (Figure 1.28). The conclusion states that the X should be inside the *20th-century presidents* circle, but its border location means we lack the information needed to reach this conclusion. The argument is invalid and therefore cannot be sound.

Note that this argument is invalid even though its premises and conclusion are all true. If you are not convinced, try substituting another man, such as Albert Einstein, for Bill Clinton. The argument then reads

Premise: All 20th-century U.S. presidents were men.
Premise: Albert Einstein was a man.

Conclusion: Albert Einstein was a 20th-century U.S. president.

The premises are still true, but the conclusion is now false, showing that the argument's structure is invalid. *Now try Exercises 33–36.* ◄

Conditional Deductive Arguments

Consider the following argument:

Premise: If a person lives in Chicago, then the person likes windy days.
Premise: Carlos lives in Chicago.

Conclusion: Carlos likes windy days.

This type of deductive argument, in which the first premise is a conditional statement *if p, then q,* is among the most common and important types of argument. In this case, $p = a$ *person lives in Chicago* and $q =$ *the person likes windy days.* The second premise asserts that, for the person named Carlos, p is true. The conclusion asserts that q is also true for Carlos. You can probably see that this argument is valid. If it is really true that people who live in Chicago like windy days and that Carlos lives in Chicago, then it must be true that Carlos likes windy days.

Conditional arguments come in four basic forms. Each has a special name, which will make sense if you remember that p is the hypothesis and q is the conclu-

sion in *if p, then q*. For example, the second premise of the above argument about Carlos asserts the truth of the hypothesis, so the argument is called *affirming the hypothesis*. Arguments with this form are always valid. The following box summarizes the four forms of conditional arguments and states their validity.

TECHNICAL NOTE
The two valid forms are also known by Latin terms. Affirming the hypothesis is called *modus ponens*, and denying the conclusion is called *modus tollens*.

FOUR BASIC CONDITIONAL ARGUMENTS

	Affirming the Hypothesis (antecedent)	Affirming the Conclusion (consequent)	Denying the Hypothesis (antecedent)	Denying the Conclusion (consequent)
Structure	If p, then q. p is true. q is true.	If p, then q. q is true. p is true.	If p, then q. p is not true. q is not true.	If p, then q. q is not true. p is not true.
Validity	Valid	Invalid	Invalid	Valid

We can use a variation on standard Venn diagrams to show why each type of conditional argument is valid or invalid. Instead of using the circles to represent sets, we use them to represent the propositions p and q. We then draw a Venn diagram for *if p, then q* by putting the p circle inside the q circle (Figure 1.29). This tells us that whenever p is true, q is also true.

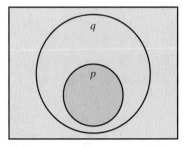

FIGURE 1.29 A Venn diagram for *if p, then q*. The p circle is inside the q circle, telling us that whenever p is true, q must also be true.

►EXAMPLE 5 Affirming the Hypothesis

Use a Venn diagram test to show that the argument above, concerning Carlos and Chicago, is valid.

Solution We begin with the Venn diagram shown in Figure 1.29 for *if p, then q*. In this case, p = *a person lives in Chicago* and q = *the person likes windy days*. The second premise asserts that p is true for the person named Carlos. We show this by putting an X in the p circle (Figure 1.30). We next check whether this diagram of the premises contains the conclusion. Notice that the X is also inside the q circle, meaning that q is also true for Carlos—that is, he *does* like windy days, just as the conclusion claims. The Venn diagram shows that the premises lead necessarily to the conclusion, so the argument is valid. *Now try Exercises 37–38.* ◄

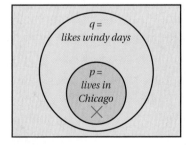

FIGURE 1.30 We put an X in the p circle to show that p is true for Carlos. Because the X is also in the q circle, q must also be true for him.

►EXAMPLE 6 Affirming the Conclusion

Use a Venn diagram to test the validity of the following argument.

Premise: If an employee is regularly late, then the employee will be fired.

Premise: Sharon was fired.

Conclusion: Sharon was regularly late.

Solution We begin with the Venn diagram for *if p, then q*. In this case, p = *an employee is regularly late* and q = *the employee will be fired*. The second premise

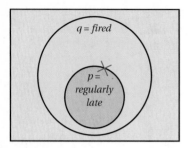

FIGURE 1.31 We put an X in the q circle to show that q is true for Sharon. The premises do not tell us whether p is also true, so we place the X on the border of the p circle. Because the conclusion demands an X *inside* the p circle, it does not follow from the premises.

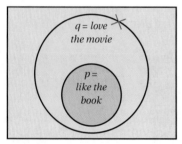

FIGURE 1.32 We put an X *outside* the p circle to show that p is false for you. We are not given enough information to determine whether the X should also be outside the q circle, so we place it on the border. Thus, we *cannot* conclude that q is also false for you.

asserts that Sharon was fired, meaning that *q* is true for the person named Sharon. We can show this by putting an X in the *q* circle. However, the premise says nothing about whether Sharon was regularly late. We therefore put the X on the border of the *p* circle to indicate that we don't know whether the X belongs inside or outside this circle (Figure 1.31). Now we check whether the diagram contains the conclusion. The conclusion states that Sharon was regularly late, meaning that *p* is true for Sharon. Therefore, if the conclusion is supported by the premises, our diagram should have an X in the *p* circle. But it doesn't—the X is on the border. Therefore, the argument is invalid, and we cannot conclude that Sharon was fired because she was regularly late. She might have been fired for some other reason. *Now try Exercises 39–40.* ◄

► **EXAMPLE 7 Denying the Hypothesis**

Use a Venn diagram to test the validity of the following argument.

Premise: If you liked the book, then you'll love the movie.
Premise: You did not like the book.

Conclusion: You will not love the movie.

Solution We have *p = you liked the book* and *q = you'll love the movie*. The second premise asserts that *p* is false for you, which we can show by putting an X outside the *p* circle. However, we do not know whether the X should also be outside the *q* circle, because the premise says nothing about *q* (that is, it says nothing about *you'll love the movie*). Thus, we put the X on the border of the *q* circle (Figure 1.32). Now we check whether the diagram contains the conclusion. The conclusion states that you will not love the movie, meaning that *q* is false for you. For the conclusion to be supported by the premises, the X must be outside the *q* circle, not on the border. Therefore, the argument is invalid. It's possible that, despite not liking the book, you will still love the movie. *Now try Exercises 41–42.* ◄

► **EXAMPLE 8 Denying the Conclusion**

Analyze the following argument:

Premise 1: A narcotic is habit-forming.
Premise 2: Heroin is not habit-forming.

Conclusion: Heroin is not a narcotic.

Solution The first premise can be rephrased to read *if a substance is a narcotic, then it is habit-forming*. We identify *p = a substance is a narcotic* and *q = the substance is habit-forming* in the Venn diagram (Figure 1.33). We illustrate the second premise by placing an X, representing heroin, outside the *q* circle. This placement indicates that heroin is *not* habit-forming. We now ask whether this diagram is consistent with the conclusion. The diagram shows the X outside the *narcotic* circle, meaning that heroin is not a narcotic—just as the conclusion claims. Thus, the argument is valid.

To test for soundness, we must assess the truth of the premises. The first premise is generally true, because narcotics are habit-forming for most people.

But the second premise is false. Heroin *is* habit-forming. Because it has a false premise, the argument is not sound. *Now try Exercises 43–44.* ◀

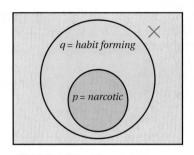

Time out to think _____

Suppose you replace heroin with aspirin in Example 8. Is the argument valid? Is it sound?

Deductive Arguments with a Chain of Conditionals

Another common type of deductive argument involves a chain of three or more conditionals. Such arguments have the following form:

Premise: If p, then q.
Premise: If q, then r.

Conclusion: If p, then r.

FIGURE 1.33 We put an X, representing heroin, *outside* the q circle to show that q is false. Because the X is also outside the p circle, p must also be false.

This particular chain of conditionals is valid: If p implies q and q implies r, it must be true that p implies r.

▶EXAMPLE 9 A Chain of Conditionals

Analyze this argument: "If elected to the school board, Maria Lopez will force the school district to raise academic standards, which will benefit my children's education. Therefore, my children will benefit if Maria Lopez is elected."

Solution This argument can be rephrased as a chain of conditionals:

Premise: If Maria Lopez is elected to the school board, then the school district will raise academic standards.
Premise: If the school district raises academic standards, then my children will benefit.
Conclusion: If Maria Lopez is elected to the school board, then my children will benefit.

Cast in this form, the conditional propositions form a clear chain from $p = Maria$ *Lopez is elected* to $q = the\ district\ will\ raise\ academic\ standards$ to $r = my\ children$ *will benefit.* Therefore, the argument is valid. *Now try Exercises 45–46.* ◀

▶EXAMPLE 10 Invalid Chain of Conditionals

Analyze the following argument: "We agreed that if you shop, I make dinner. We also agreed that if you take out the trash, I make dinner. Therefore, if you shop, you should take out the trash."

Solution Let's assign $p = you\ shop$, $q = I\ make\ dinner$, $r = you\ take\ out\ the\ trash$. Then this argument has the following form:

Premise: If p, then q.
Premise: If r, then q.

Conclusion: If p, then r.

The conclusion is invalid as there is no chain from p to r. *Now try Exercises 47–48.* ◀

INDUCTION AND DEDUCTION IN MATHEMATICS

Perhaps more than any other subject, mathematics relies on the idea of *proof.* A mathematical proof is a deductive argument that demonstrates the truth of a certain claim, or *theorem.* A theorem is considered proven if it is supported by a valid and sound proof. Although mathematical proofs use deduction, theorems are often discovered by induction.

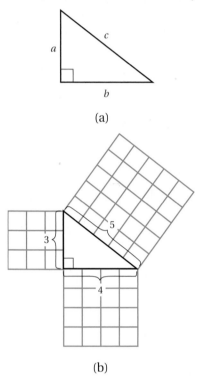

(a)

(b)

FIGURE 1.34

Consider the Pythagorean theorem, which applies to right triangles (those with one 90° angle). It says $a^2 + b^2 = c^2$, where c is the length of the longest side, or **hypotenuse,** and a and b are the lengths of the other two sides (Figure 1.34a). A geometric construction in the case $a = 3$, $b = 4$, and $c = 5$ shows how the squares of the sides are related (Figure 1.34b). The same relationship can be found in *any* right triangle, so there is good inductive evidence to suggest that the theorem is true for *all* right triangles.

The Pythagorean theorem is named for the Greek philosopher Pythagoras (c. 580–500 B.C.) because he was the first person known to have proved it deductively. However, the theorem was known to many ancient cultures and already had been used for at least a thousand years before the time of Pythagoras. These ancient cultures probably never bothered to prove it. Instead, they simply noticed that it was true every time they tested it and that it was very useful in art, architecture, and construction. The many test cases essentially formed an inductive argument, each one lending more support to the conclusion that the theorem is true.

The principle of seeking inductive evidence can be very useful when you are having difficulty remembering whether a particular theorem or mathematical rule applies. It often helps to try a few test cases and see if the rule works. Although test cases can never constitute a proof, they often are enough to satisfy yourself of a rule's truth. However, the rule cannot be true if even one test case fails.

➤ **EXAMPLE 11** Inductively Testing a Mathematical Rule

Test the following rule: For all numbers a and b, $a \times b = b \times a$.

Solution We begin with some test cases, using a calculator as needed.

$$\text{Does } 7 \times 6 = 6 \times 7? \qquad\qquad \Rightarrow \text{Yes!}$$

$$\text{Does } (-23.8) \times 9.2 = 9.2 \times (-23.8)? \qquad \Rightarrow \text{Yes!}$$

$$\text{Does } 4.33 \times \left(-\frac{1}{3}\right) = \left(-\frac{1}{3}\right) \times 4.33? \qquad \Rightarrow \text{Yes!}$$

The three test cases are each somewhat different (mixing fractions, decimals, and negative numbers), yet the rule works for all three. This outcome offers a strong inductive argument in favor of the rule. Although we have not proved the rule $a \times b = b \times a$, we have good reason to believe that it is true. Our belief would be strengthened by additional test cases that confirm the rule. *Now try Exercises 49–50.* ◄

Thinking about . . .

DEDUCTIVE PROOF OF THE PYTHAGOREAN THEOREM

Over the centuries, mathematicians found many different ways to prove the Pythagorean theorem deductively. One of the simplest proofs is attributed to a 12th-century Hindu mathematician named Bhaskara. His proof begins with a diagram of a large square, inside of which is a smaller square surrounded by four right triangles (Figure 1.35). Note that the diagram divides the large square into five separate regions (the small square and the four triangles) so that

$$\begin{pmatrix} \text{area of} \\ \textit{large} \text{ square} \end{pmatrix} = \begin{pmatrix} \text{area of} \\ \textit{small} \text{ square} \end{pmatrix} + \begin{pmatrix} \text{area of} \\ \text{the four} \\ \text{right triangles} \end{pmatrix}$$

FIGURE 1.35

Now we can write each of these areas as follows:

- The side length of the *large* square is c, so its area is c^2.
- The side length of the *small* square is $a - b$, so its area is $(a - b)^2$.
- The area of any triangle is given by the formula $\frac{1}{2} \times$ base \times height (see Unit 10A). Each of the four triangles has base length a and height b, so each has area $ab/2$.

Substituting these areas into the preceding equation gives

$$\underbrace{c^2}_{\substack{\text{area of} \\ \text{large square}}} = \underbrace{(a - b)^2}_{\substack{\text{area of} \\ \text{small square}}} + 4 \times \underbrace{\frac{ab}{2}}_{\substack{\text{area of} \\ \text{triangle}}} = (a - b)^2 + 2ab$$

We continue by using a bit of algebra to show that $(a - b)^2 = a^2 - 2ab + b^2$. Substituting this expression into the previous equation, we get

$$\begin{aligned} c^2 &= (a - b)^2 + 2ab \\ &= a^2 - 2ab + b^2 + 2ab \\ &= a^2 + b^2 \end{aligned}$$

We have arrived at the Pythagorean theorem by following a deductive chain of logic from the starting diagram. Legend has it that, when Bhaskara showed his proof to others, he accompanied it with just a single word: "Behold!"

▶ **EXAMPLE 12** Invalidating a Proposed Rule

Suppose you cannot recall whether adding the same amount to both the numerator and the denominator (top and bottom) of a fraction such as $\frac{2}{3}$ is legitimate. That is, you are wondering whether it is true that, for any number a,

$$\frac{2}{3} \overset{?}{=} \frac{2 + a}{3 + a}$$

Solution Again, we can check the rule with test cases.

Suppose that $a = 0$. Is it true that $\dfrac{2}{3} = \dfrac{2+0}{3+0}$? \Rightarrow Yes!

Suppose that $a = 1$. Is it true that $\dfrac{2}{3} = \dfrac{2+1}{3+1}$? \Rightarrow No!

Although the rule worked in the first test case, it failed in the second. Generalizing (inductively), we conclude that it is not legitimate to add the same value to the top and bottom of *any* fraction. *Now try Exercises 51–52.* ◄

■ EXERCISES 1D

Review Questions

1. Summarize the differences between deductive and inductive arguments. Give an example of each type.

2. Briefly explain the idea of strength and how it applies to inductive arguments. Can an inductive argument prove its conclusion true? Can an inductive argument be valid? Can it be sound?

3. Briefly explain the ideas of validity and soundness and how they apply to deductive arguments. Can a valid deductive argument be unsound? Can a sound deductive argument be invalid? Explain.

4. Describe the procedure used to test the validity of a deductive argument with a Venn diagram.

5. Create your own example of each of the four basic conditional arguments. Then explain why your argument is valid or invalid.

6. What is a chain of conditionals? Give an example of a valid argument made from such a chain.

7. Can inductive logic be used to prove a mathematical theorem? Explain.

8. How can inductive testing of a mathematical rule be useful? Give an example.

Does It Make Sense?

Decide whether each of the following statements makes sense (or is clearly true) or does not make sense (or is clearly false). Explain your reasoning.

9. Based on the testimonials of dozens of people who have lost weight following my diet, I will prove to you that my diet works for everyone.

10. The many examples of people whose cancer went away following chemotherapy make a strong case for the idea that chemotherapy can cure cancer.

11. Through the logic of deduction, I will show you that if you accept the truth of my premises, you must also accept the truth of my conclusion.

12. You can see that my argument is valid, and you must therefore accept the truth of my conclusion.

13. We should never accept an argument unless its conclusion has been proven true beyond all doubt.

14. Even before Fermat's Last Theorem was proved deductively, mathematicians were sure it was true.

Basic Skills & Concepts

Everyday Logic. For Exercises 15–22, explain whether the argument is inductive or deductive.

15. I see the same tall man leaving that office building every day at 5:00. He must work in the building.

16. If I carry a heavy load in the trunk of my car, the rear wheels squeak. If the rear wheels squeak, then I get a headache. Therefore, if I carry a heavy load in the trunk of my car, I get a headache.

17. Scott Turow's first five novels were great. I'm sure his next novel will be great too.

18. I've never had a bad meal at Dot's Diner, so I recommend it to everyone.

19. Any restaurant that gives free refills of coffee is OK by me, and since Fred's Food gives free refills, I'll go there any day.

20. When the wind blows, the garden dries out, and when the garden dries out, I need to water it. So whenever the wind blows, I can expect to water the garden.

21. The last four times I went skiing, the traffic was light on Tuesdays and heavy on Saturdays. Weekdays must have lighter traffic than weekends.

22. I like Chinese cuisine and I like Thai cuisine, so I'd also like any other Asian cuisine.

Analyzing Inductive Arguments. For Exercises 23–28, determine the truth of the premises, discuss the strength of the argument, and assess the truth of the conclusion.

23. Premise: Cows have four limbs, and they are mammals.
 Premise: Monkeys have four limbs, and they are mammals.
 Premise: Lions have four limbs, and they are mammals.
 Conclusion: All animals with four limbs are mammals.

24. Premise: $(-6) \times (-4) = 24$
 Premise: $(-2) \times (-1) = 2$
 Premise: $(-27) \times (-3) = 81$
 Conclusion: Whenever we multiply two negative numbers, the result is a positive number.

25. Premise: Grant Hill wears expensive shoes, and he is a great basketball player.
 Premise: Tim Duncan wears expensive shoes, and he is a great basketball player.
 Premise: Shaquille O'Neal wears expensive shoes, and he is a great basketball player.
 Conclusion: All people who wear expensive shoes are great basketball players.

26. Premise: Bach, Buxtehude, Beethoven, Brahms, Berlioz, and Britten are great composers.
 Conclusion: All great composers have names that begin with B.

27. Premise: Students from Ms. Callahan's Algebra I class do better in Algebra II than students from Mr. McIntyre's Algebra I class.
 Premise: Students from Ms. Callahan's Algebra I class do better in Algebra II than students from Mr. Murphy's Algebra I class.
 Conclusion: Ms. Callahan is the best of the three Algebra I teachers.

28. Premise: Sparrows are birds and they fly.
 Premise: Eagles are birds and they fly.
 Premise: Hawks are birds and they fly.
 Premise: Larks are birds and they fly.
 Conclusion: All birds fly.

Deductive Arguments. For Exercises 29–36, start by rephrasing the argument, if necessary, so that the first premise has the form *all S are P* or *no S is P.* Then draw a Venn diagram to determine whether the argument is valid. If possible, discuss the truth of the premises and whether the argument is sound.

29. Premise: All islands are tropical.
 Premise: Iceland is an island.
 Conclusion: Iceland is tropical.

30. Premise: All dairy products contain protein.
 Premise: Soybeans contain protein.
 Conclusion: Soybeans are dairy products.

31. Premise: All salty foods cause high blood pressure.
 Premise: Apples do not cause high blood pressure.
 Conclusion: Apples are not salty foods.

32. Premise: All U.S. presidents have been men.
 Premise: George Washington was a man.
 Conclusion: George Washington was a U.S. president.

33. Premise: All states in the Eastern Standard time zone are east of the Mississippi River.
 Premise: Maine is in the Eastern Standard time zone.
 Conclusion: Maine is east of the Mississippi River.

34. Premise: Doctors know anatomy.
 Premise: Cardiologists know anatomy.
 Conclusion: Cardiologists are doctors.

35. Premise: All opera singers can whistle a Mozart tune.
 Premise: Pavarotti is an opera singer.
 Conclusion: Pavarotti can whistle a Mozart tune.

36. Premise: No movie stars do their own laundry.
 Premise: Jack Applebee does his own laundry.
 Conclusion: Jack Applebee is not a movie star.

Conditional Deductive Arguments. For Exercises 37–44, start by rephrasing the argument, if necessary, so that the first premise is in the conditional form *if p, then q.* Identify the type of argument and determine its validity with a Venn diagram. If possible, discuss the truth of the premises and whether the argument is sound.

37. Premise: If an animal is a horse, then it is a mammal.
Premise: Clydesdales are horses.
Conclusion: Clydesdales are mammals.

38. Premise: In the United States, we have the right to say anything at any time.
Premise: Yelling "fire!" in a theater is saying something.
Conclusion: In the United States, we have the right to yell "fire!" in a theater.

39. Premise: If I eat breakfast, then I don't eat lunch.
Premise: I ate breakfast.
Conclusion: I didn't eat lunch.

40. Premise: If I eat breakfast, then I don't eat lunch.
Premise: I didn't eat lunch.
Conclusion: I ate breakfast.

41. Premise: If I don't eat breakfast, then I eat lunch.
Premise: I ate breakfast.
Conclusion: I didn't eat lunch.

42. Premise: It's necessary for nurses to know CPR.
Premise: Tom is a nurse.
Conclusion: Tom knows CPR.

43. Premise: When interest rates decline, the bond market improves.
Premise: Last week the bond market improved.
Conclusion: Interest rates must have declined.

44. Premise: If we can put a man on the moon, we can build a computer operating system that works.

Premise: We can build a computer operating system that works.
Conclusion: We can put a man on the moon.

Chains of Conditionals. For Exercises 45–48, start by rephrasing the given propositions, if necessary, in *if p, then q* form. Then write the argument as a chain of conditionals and evaluate its validity.

45. Premise: If you shop, I make dinner.
Premise: If I make dinner, you take out the trash.
Conclusion: If you shop, you take out the trash.

46. Premise: If fish have fins, then fish can fly.
Premise: If fish can fly, then dogs can bark.
Conclusion: If fish have fins, then dogs can bark.

47. Premise: If taxes are cut, then taxpayers will have more disposable income.
Premise: With more disposable income, taxpayer spending will fuel the economy.
Conclusion: A tax cut will fuel the economy.

48. Premise: If taxes are cut, the U.S. government will have less revenue.
Premise: If there is less revenue, then the deficit will be larger.
Conclusion: Tax cuts will lead to a larger deficit.

Testing Mathematical Rules. For Exercises 49–52, test the statement with several different numbers and state whether you think the statement is true.

49. Is it true for all numbers a and b that $a + b = b + a$?

50. Is it true for all numbers a and b that $a \div b = b \div a$?

51. Is it true for all numbers a and b that $a^2 + b^2 = (a + b)^2$?

52. Is it true for all positive integers n that

$$1 + 2 + 3 + \cdots + n = \frac{n \times (n + 1)}{2}?$$

Further Applications

Validity and Soundness. For Exercises 53–57, state whether it is possible for a deductive argument to have the given properties. If so, make up an example that has these properties.

53. Valid and sound

54. Not valid and sound

55. Valid and not sound

56. Valid with false premises and a true conclusion

57. Not valid with true premises and a true conclusion

Your Own Conditional Arguments. For Exercises 58–61, create a simple three-line argument of the given form. Choose your example so that it illustrates clearly whether or not the argument is valid.

58. Affirming the hypothesis

59. Affirming the conclusion

60. Denying the hypothesis

61. Denying the conclusion

62. **The Goldbach Conjecture.** Recall that a prime number is any number whose only factors are itself and 1 (for example, 2, 3, 5, 7, 11, . . .). The Goldbach conjecture, posed in 1742, claims that every even number greater than 2 can be expressed as the sum of two primes. For example, $4 = 2 + 2$, $6 = 3 + 3$, and $8 = 5 + 3$. A *deductive* proof of this conjecture has never been found. Test the conjecture for at least 10 even numbers and present an *inductive* argument for its truth. Do you think the conjecture is really true? Why or why not?

Web Projects

Find useful links for Web Projects on the text Web site: www.aw.com/bennett-briggs

63. **Fermat's Last Theorem.** One of the most famous mathematical theorems of all time is called Fermat's Last Theorem. For over 350 years, the proof eluded mathematicians; it was finally cracked by Andrew Wiles in the 1990s. Through Web research, learn what the theorem claims and why it became so famous. Briefly summarize how inductive evidence suggested its truth for centuries before Wiles finally found a deductive proof.

64. **The Pythagorean Theorem.** Learn more about the history of the Pythagorean theorem, and write a short report on one aspect of its history. For example, you might write about its use in cultures that predated Pythagoras or describe another proof of the theorem with its historical context.

In the News

65. **Deductive Reasoning in Your Life.** Give an example of a situation in which you used deductive reasoning in everyday life. Explain the situation, describe the steps in your thinking, and explain why it was deductive reasoning.

66. **Inductive Reasoning in Your Life.** Give an example of a situation in which you used inductive reasoning in everyday life. Explain the situation, describe the steps in your thinking, and explain why it was inductive reasoning.

67. **Editorial Arguments.** Find three simple arguments in editorials. State whether each is deductive or inductive and evaluate it accordingly.

68. **Arguing Your Side.** Choose an issue that you feel strongly about and create an argument in support of your position. Is your argument inductive or deductive? Evaluate your argument.

69. **Arguing the Other Side.** Choose an issue that you feel strongly about and create an argument that tends to *contradict* your position. That is, try to create an argument for the other side. Can you make the argument convincing? Does the argument help you understand the other side of the issue?

UNIT 1E

CRITICAL THINKING IN EVERYDAY LIFE

*T*he skills discussed in the preceding units are all useful in their own right. But critical thinking involves much more than isolated skills. It also involves knowing when to use particular skills and when to invent entirely new strategies to analyze an argument or make a decision. Indeed, as we stated at the very beginning of this chapter, critical thinking involves careful reading (or listening), sharp thinking, logical analysis, good visualization, and healthy skepticism.

Because it is so wide-ranging, critical thinking cannot be described by any simple step-by-step procedure. Instead, it is developed through experience and by questioning

and analyzing every argument or decision you face. Nevertheless, a few general guidelines or hints can be useful. In this unit, we offer a few such hints.

HINT 1: READ (OR LISTEN) CAREFULLY

Language can be used in complex ways that require careful effort to understand. Always read (or listen) carefully to make sure you've grasped precisely what was said. Also be sure that you've distinguished what was actually said, what was assumed, and what must be determined.

➤ **EXAMPLE 1** Colorado's Amendment 2

In 1992, Colorado voters were asked to vote *yes* or *no* on the following ballot question.

> *Shall there be an amendment to the Colorado constitution to prohibit the state of Colorado and any of its political subdivisions from adopting or enforcing any law or policy which provides that homosexual, lesbian, or bisexual orientation, conduct, or relationships constitutes or entitles a person to claim any minority or protected status, quota preferences, or discrimination?*

Explain the meaning of *yes* and *no* votes.

Solution The question contains a lot of obscure wording, making it difficult to see the key point. But, if we read carefully, we find that the question essentially asks: *Should the state prohibit laws that grant gay people minority status or protection from discrimination?* Even in this simplified form, the question is still somewhat confusing because it is asked in negative terms. That is, a *yes* vote is a vote *against* gay rights, while a *no* vote is a vote *for* gay rights. Some evidence suggests that the wording confused many voters. Pre-election polls indicated that the proposed amendment would be defeated. However, the amendment passed (53% to 47%), which means that the polls were inaccurate or that many voters were confused and voted opposite to their beliefs.

Now try Exercises 11–12. ◄

by the way

Despite passing, the Colorado amendment never took effect and was declared unconstitutional by the U.S. Supreme Court. The justices said the amendment denied gays a political right enjoyed by everyone else—the chance to seek protection from discrimination in employment, housing, and public accommodations.

Time out to think _____

If voters really did misunderstand the wording of the Colorado ballot initiative, would the confusion have affected both sides equally, or would it have favored one side more than the other? Explain.

HINT 2: LOOK FOR HIDDEN ASSUMPTIONS

The arguments we studied in Unit 1D consisted of clear premises leading to the conclusion. But many real arguments lack such clarity, relying instead on ambiguous terms or hidden assumptions. Often, the speaker (or writer) may think these premises are "obvious," but listeners (or readers) may not agree. Indeed, an argument that seems very convincing to the speaker may actually be quite weak to a listener who is not aware of the hidden assumptions.

➤ **EXAMPLE 2** Building More Prisons

Analyze the following argument:

> *We should build more prisons because incarcerating more criminals will reduce the crime rate.*

Solution This argument looks deceptively simple because it is so short. It concludes that we should build more prisons, based on a single premise—that incarcerating more criminals will reduce the crime rate. If we recognize the premise as a conditional statement, the argument has the form

Premise: If we incarcerate more criminals, then the crime rate will be reduced.

Conclusion: We should build more prisons.

Viewed this way, the argument hardly makes any sense at all, because the premise says nothing about building prisons. Clearly, the speaker must have intended some hidden assumptions to be "obvious" to the listeners.

We can begin to make sense of the argument by identifying possible hidden assumptions. For example, a plausible argument might begin like this:

Hidden Assumption 1: If we build more prisons, then more criminals can be incarcerated.

Stated Premise: If we incarcerate more criminals, then the crime rate will be reduced.

Hidden Assumption 2: If the crime rate is reduced, then we will have a more desirable society.

Conclusion: We should build more prisons.

With the two hidden assumptions, this part of the argument becomes a chain of conditionals with the following conclusion: *If we build more prisons, then we will have a more desirable society.* However, this still does not automatically lead to the argument's overall conclusion that we should build more prisons. For that, we must add one more hidden assumption—namely, *if a policy will lead to a more desirable society, then it should be enacted.*

We have now added three hidden assumptions to the one stated premise. With all of these assumptions, the argument is a long chain of conditionals that is deductively valid. But it is sound only if you accept the truth of the stated premise and hidden assumptions.

In fact, the premises are debatable. For example, many people would argue with the first hidden assumption, because incarcerating more criminals requires not only more prison space, but also a more efficient court system. The stated premise is also open to debate: Studies do not agree on whether a higher incarceration rate reduces the crime rate. And one could counter the last hidden assumption by arguing that crime reduction alone might not make a more desirable society, particularly if it requires policies that infringe on personal freedom or other basic rights or involve excessive cost.

In summary, the original argument makes sense only if we recognize several hidden assumptions, and some of these assumptions are open to debate. As a result, the truth of the conclusion is also debatable. *Now try Exercises 13–16.* ◀

HINT 3: IDENTIFY THE REAL ISSUE

People often argue about one question when there is a different issue involved. Identifying the real issue in a debate is often difficult because people can be experts at hiding their true intentions. Fortunately, by analyzing arguments carefully, we can often determine whether the real issue is hidden, even if we don't know exactly what it is.

HISTORICAL NOTE

Can logic settle all arguments? The German mathematician Gottfried Wilhelm von Leibniz (1646–1716) thought so and took the first steps in a two-century search for a "calculus of reasoning." An outgrowth of this search called symbolic logic is now used extensively in mathematics and computing. But Leibniz's dream of resolving arguments through logic was not to be. In 1931, the Austrian mathematician Kurt Gödel discovered that no system of logic can solve all mathematical problems, let alone problems of ethics or morality.

➤ **EXAMPLE 3** Banning Rock Concerts

Analyze the following segment of an editorial from a local newspaper.

> *With last Saturday's sellout crowd at the Moonlight Amphitheater, it is clear that the parking problem has become hopeless. Concert goers parked along residential streets up to a mile away from the amphitheater, badly overcrowding sidewalks, blocking driveways, and disrupting traffic. In light of this parking problem, future rock concerts should be canceled.*

Solution The argument makes several claims about parking problems, but stripped of the details the argument boils down to this:

Premise: There is a parking problem for rock concerts at the theater.

Conclusion: Future rock concerts at the theater must be canceled.

The rest of the argument simply lists reasons why the parking problem is serious. But is a parking problem really a good enough reason to cancel future rock concerts? After all, a parking problem could have many solutions. For example, it might be possible to create new parking lots, to use shuttle buses, to step up enforcement of parking violations, or to encourage carpooling. The weakness of the argument should make us wonder whether the editorial writers are *really* concerned about parking or whether they are using this issue to oppose the rock concerts themselves. Or perhaps they are responding to a few vocal residents who complained about the parking problem. It's hard to know for sure, but it seems unlikely that parking is the real issue.

Now try Exercises 17–18. ◄

HINT 4: USE VISUAL AIDS

When arguments, instructions, or options become complex, it can be useful to organize information visually. Useful tools include tables, Venn diagrams, flow charts, and other types of pictures.

➤ **EXAMPLE 4** Reading IRS Guidelines

According to Internal Revenue Service (IRS) guidelines on who must file a federal income tax return, a single person who can be claimed as a dependent on another person's tax return and is *either* 65 or older *or* blind must file a return if

- **(i)** earned income was more than $5350 ($6400 if 65 or older and blind); or
- **(ii)** unearned income was more than $1750 ($2800 if 65 or older and blind); or
- **(iii)** gross income was more than the total of earned income (up to $4300) or $700, whichever is greater, plus $1050 ($2100 if 65 or over and blind).

Decide whether these guidelines apply to each of the following individuals; if the guidelines do apply, use them to decide whether the individual must file a return.

- **a.** Chase is a 15-year-old boy who is claimed as a dependent on his parent's tax return. He has an earned income of $6500. Does this policy apply to Chase? If so, must he file an income tax return?
- **b.** Carmen is a single 24-year-old blind woman who is claimed as a dependent on her parent's tax return. Her earned income is $6500, and her unearned

income is $750. Does this policy apply to Carmen? If so, must she file an income tax return?

c. Serena is a single 68-year-old blind woman who is claimed as a dependent on her son's tax return. Her gross income is $2500, and her earned income is $550. Does this policy apply to Serena? If so, must she file an income tax return?

Solution As this example demonstrates, IRS policies can be notoriously difficult to interpret. A diagram or a flow chart, as shown in Figure 1.36, can simplify matters. Notice that conditions (i)–(iii) are separated by *or*, so if any one condition is satisfied, then a tax return must be filed. We can use the flow chart to see how the guidelines affect the three individuals.

a. Chase is neither over 65 nor blind, so the guidelines do not apply to him.

b. Because Carmen is blind, single, and claimed as a dependent on her parent's tax return, the guidelines apply to her. Because her earned income exceeds $5350, she must file a tax return by condition (i).

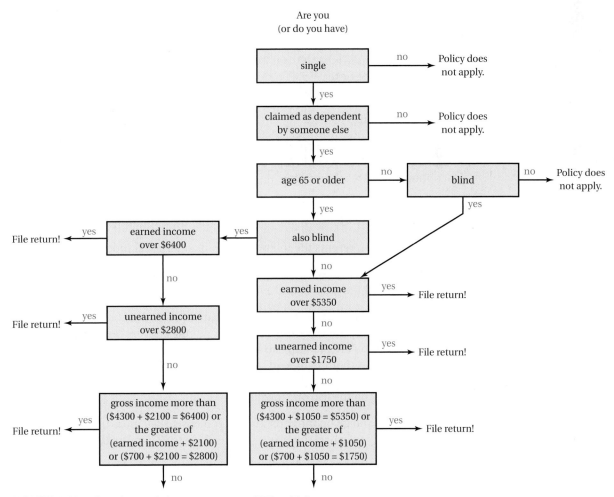

FIGURE 1.36 Flow chart to help interpret a set of IRS guidelines.

c. The guidelines apply to Serena because she is single, blind, over 65, and claimed as a dependent on her son's tax return. Condition (iii) applies to her case. The greater of Serena's earned income ($550) and $700 is $700. Thus, because Serena is blind and over 65, she must file a tax return if her gross income exceeds $700 + $2100 = $2800. But her gross income is only $2500, so she does not need to file a return. *Now try Exercises 19–22.* ◄

HINT 5: UNDERSTAND ALL THE OPTIONS

We regularly make decisions in situations where we have several options. For example, we face decisions about which insurance policy to choose, which auto loan to take, or which model of new computer to buy. The key to such decisions is making sure that you understand how each option would affect you.

►**EXAMPLE 5** Which Airline Ticket to Buy?

Airlines typically offer many different prices for the same trip. Suppose you are planning a trip six months in advance and discover that you have two choices in purchasing an airline ticket:

(A) The lowest fare is $1100, but 25% of the fare is nonrefundable if you change or cancel the ticket.

(B) A fully refundable ticket is available for $1900.

Analyze the situation.

Solution We can think of each of the two options as a pair of conditional propositions. Under option (A), you will lose 25% of $1100, or $275, if you cancel your trip. Thus, option A represents the following pair of conditional propositions:

(1A) If you purchase ticket A and *go* on the trip, then you will pay $1100.
(2A) If you purchase ticket A and *cancel* the trip, then you will pay $275.

Similarly, option B represents the following pair of conditional propositions:

(1B) If you purchase ticket B and *go* on the trip, then you will pay $1900.
(2B) If you purchase ticket B and *cancel* the trip, then you will pay $0.

Figure 1.37 represents the four possibilities. Clearly, option A is the better buy if you go on the trip, and option B is the better buy if you end up canceling your trip. However, because you are planning six months in advance, it's impossible to foresee all the circumstances that might lead you to cancel your trip. Therefore, you might want to analyze the *difference* between the two tickets under the two possibilities (going on the trip or canceling).

FIGURE 1.37

If you go on the trip: ticket B costs $800 more than ticket A.
If you cancel the trip: ticket A costs $275 more than ticket B.

In effect, you must decide which ticket to purchase by balancing the risk of spending an extra $800 if you go on the trip against spending an extra $275 if you cancel. How would *you* decide? *Now try Exercises 23–24.* ◄

Time out to think _____

Consider the two ticket options in Example 5 from the point of view of the airline. How does offering the two options help the airline maximize its revenue?

HINT 6: WATCH FOR FINE PRINT AND MISSING INFORMATION

There's an old saying that "the devil is in the details," and it is probably nowhere more applicable than in the fine print of offers, deals, and contracts. What looks like a great deal without the fine print may be a poor one with it. Worse yet, sometimes even the fine print neglects to state important information. Use your powers of critical thinking to decide what information you need. If anything is missing, be sure to ask before you act.

➤ **E X A M P L E 6** Shipping and Handling

Katie received a mail offer for a CD club. In large print, it offered "12 CDs for 1¢." The reply card asked Katie to sign her agreement to the following terms:

> *I've enclosed $1.50 (that's 1¢ for 12 CDs, plus $1.49 for enrollment charge). I agree to buy 6 CDs at regular Club prices within the next two years.*
>
> A shipping/handling charge is added to each shipment. Applicable sales tax added to each order.

Katie signed up and sent in $1.50, but when the CDs arrived they came with an invoice for $36.50. What happened?

Solution The fine print warned of a shipping/handling charge, but did not state how much it would be. It turned out to be quite significant. To fulfill her agreement, Katie now needs to purchase 6 more CDs at "regular Club prices." Based on this first experience, Katie should be concerned that these prices may turn out to be much higher than she would pay elsewhere. Furthermore, each of these CDs will have a significant shipping/handling charge, and perhaps sales tax as well. It's possible that the end result will be a decent deal, but certainly not 12 CDs for 1¢.

Now try Exercises 25–26. ◄

HINT 7: ARE OTHER CONCLUSIONS POSSIBLE?

You should never accept the first argument or choice you hear. Even when an argument proves its conclusion, it's still possible that there are other conclusions that have been left unstated. And remember that many real arguments are inductive, which means the conclusion is never proved no matter how strong the argument may seem.

➤ **E X A M P L E 7** Nuclear Deterrence

Evaluate the following argument and discuss its conclusion.

> *The development of nuclear weapons changed the way world leaders think about potential conflicts. A single nuclear weapon can kill millions of people, and the arsenals of the United States and the Soviet Union contained enough power to kill everyone on Earth many times over. This potential for catastrophic damage led to the idea of nuclear deterrence, which held that the United States and the Soviet Union would*

be deterred from direct warfare by the fear of nuclear war. For the more than 45 years of the Cold War, the United States and the Soviet Union never did fight directly. This was one of the longest periods in human history during which two major enemies avoided direct war. We can only conclude that nuclear deterrence prevented war between the United States and the Soviet Union.

Solution The conclusion of this argument is the claim that nuclear deterrence prevented war between the United States and the Soviet Union. The facts cited in support of this conclusion are all true and all appear relevant to the conclusion. Thus, the argument appears quite strong. Nevertheless, it does not *prove* that nuclear deterrence worked, and we can imagine other possible explanations for why the United States and the Soviet Union avoided war. For example, it might have been because of economic factors, or the emergence of a united Western Europe after World War II, or changes in non-nuclear weaponry. Indeed, because the argument concerns why history unfolded in a particular way, it is probably impossible for us to ever know for certain whether the conclusion is true or false. ◄

Time out to think _____

Can you think of any facts that would tend to undercut the argument that nuclear deterrence prevented war between the United States and the Soviet Union? Overall, in your judgment, is the argument in Example 7 convincing? Why or why not?

PUTTING IT ALL TOGETHER

The examples we've studied show only a few of the many ways in which critical thinking affects our daily lives. We live in a complex world, and unless you are careful in all your thinking and decision making, you are liable to be led astray. Learn to pick apart every argument or choice you are given. With time, you will gain the experience needed to make wise decisions in school, in your career, and in your life.

Practical matters

BEWARE OF "UP TO" DEALS

You've probably heard offers for cheap long distance calling, such as 99¢ for calls of "up to 20 minutes." At first, it might sound like a great deal, because most people focus on 20 minutes and 99¢. Indeed, if you make a 20-minute phone call, the rate works out to about 5¢ per minute, which is quite low. But the words "up to" hold hidden danger. Suppose you make a call and reach only your friend's answering machine. Even though it may take you only 30 seconds (or less) to leave your message and hang up, you'll still pay 99¢ for the call under this plan—which works out to about $2 per minute in this case. If your calls are longer than 20 minutes, there's usually a higher rate that kicks in after 20 minutes.

More generally, you should beware of any deal that contains the words "up to." For example, you may receive offers promising that you can earn "up to $5000 a month" while working from home or that you can get "up to five free oil changes" if you pay an upfront charge that is much more expensive than a single oil change. Such deals are usually structured to be great if you just touch the "up to" limit, but are often quite poor in all other circumstances. Unless you've carefully thought out how you will take advantage of the good parts of the deal, it's best to assume that "up to" deals are up to no good.

EXERCISES 1E

Review Questions

1. Describe critical thinking and why it is important to everyone.

2. Summarize the hints given in this unit, and explain how each is important to critical thinking.

3. Give a few examples of situations in which you used critical thinking in your own life.

4. Give at least one case you know of in which a poor decision was made (by you or someone else) because of a lack of good critical thinking.

Does It Make Sense?

Decide whether each of the following statements makes sense (or is clearly true) or does not make sense (or is clearly false). Explain your reasoning.

5. Reed was relieved because his insurance company chose not to deny his claim.

6. Although the plane crashed in Nevada, the survivors were buried in California.

7. Sue prefers the Red Shuttle because it gets her to the airport in an hour and a half, while the Blue Shuttle takes 80 minutes.

8. Alan decided to buy his ticket from Ticketmaster for $33 plus surcharge rather than from the box office, where it costs $35 with no additional charges.

9. There was no price difference, so Michael chose the tires with a 5-year, 40,000-mile warranty over the tires with a 50-month, 35,000-mile warranty.

10. Auto policy A has $3000 worth of collision insurance with an annual premium of $400. Auto policy B has $3000 worth of collision insurance with an annual premium of $300. Clearly policy B is the better policy.

Basic Skills & Concepts

11. **Reading a Ballot Initiative.** Consider the following ballot initiative that appeared in a statewide election:

 Shall there be an amendment to the state constitution stating that the state and any city, town, or county may control the promotion of obscenity to the full extent permitted by the First Amendment to the U.S. Constitution, and thereby preventing the state courts from interpreting the right of free expression more broadly under the state constitution than under the First Amendment to the U.S. Constitution in the area of obscenity?

 a. According to the text of this initiative, is the state constitution more or less strict about obscenity than the First Amendment to the U.S. Constitution? Explain.

 b. Explain the effect of a *yes* vote on this initiative.

 c. True or false: If the initiative passed, then the state would have a stricter law on obscenity than specified in the U.S. Constitution. Explain.

12. **Reading a Ballot Initiative.** Consider the following ballot initiative that appeared in a statewide election:

 Shall there be an amendment to the state constitution concerning term limits declarations that may be voluntarily submitted by candidates for the U.S. Congress, and in connection therewith, specifying when such declarations must be submitted to the secretary of state; providing that a candidate shall not be refused placement on the ballot if the candidate does not submit a declaration; providing that candidates may voluntarily declare that the candidate will not serve more than three terms as a U.S. Representative or more than two terms as a U.S. Senator or may voluntarily declare that the candidate has chosen not to accept term limits?

 a. Does the state have mandatory term limits for representatives and senators?

 b. If the initiative is passed, must a candidate accept term limits if elected? Explain.

 c. If the initiative is passed, will there be a penalty if a candidate refuses to accept term limits? Explain.

Exercises 13–16 each state a seemingly simple argument. In each case, identify at least two hidden assumptions.

13. We should stop renting and buy a house now because interest rates are so low.

14. I saw Jenny riding in a limousine, so she must be rich.

15. I support the United Nations because all countries need to learn to get along.

16. I support increased military spending because we need a strong America.

Exercises 17–18 each state several reason for a particular political position. While the reasons may be perfectly legitimate, there may be other "real reasons" that are left unstated. In each case, identify at least one unstated issue that, for at least some people, could be the real issue of concern. Explain your reasoning clearly.

17. I oppose the President's proposed tax cuts. These tax cuts will reduce the tax bill by thousands of dollars for the wealthiest among us, but provide no relief to minimum-wage earners. They may cause the federal deficit to grow, and there's no need for them anyway, since the overall rate of taxation today is already lower than it has been for more than a decade.

18. I support school vouchers (which are designed to help people send children to private rather than public schools) because our public schools have too many problems. They have not served our neediest children well, they are often in poor physical condition, and school administrators are too beholden to teachers' unions and other special interest groups.

19. **IRS Guidelines.** IRS guidelines state that a single person under 65 years of age who can be claimed as a dependent on another person's tax return must file a return if

 (i) unearned income was over $700; or

 (ii) earned income was over $4300; or

 (iii) total of earned and unearned income was more than the greater of $700 or earned income (up to $4050) plus $250.

 For each situation below, state whether the person must file a return.

 a. Trip had earned income of $600 and unearned income of $200. Must he file a federal tax return?

 b. Sally had earned income of $680 and no unearned income. Must she file a federal tax return?

 c. Monica had earned income of $3900 and unearned income of $300. Must she file a federal tax return?

 d. Horace had earned income of $4000 and unearned income of $200. Must he file a federal tax return?

 e. Delila had earned income of $4200 and unearned income of $200. Must she file a federal tax return?

20. **IRS Guidelines.** Consider the following IRS instructions:

 You must make estimated tax payments if you expect to owe at least $500 in tax and you expect your withholding and credits to be less than the smaller of:

1. *90% of the tax shown on next year's tax return, or*

2. *The tax shown on last year's tax return (110% of that amount if you are not a farmer or fisherman and the adjusted gross income shown on that return is more than $150,000 or, if filing separately next year, more than $75,000).*

For each situation below, state whether you must make estimated tax payments during the course of a year.

a. You expect to pay $450 in taxes next year.

b. You are a teacher and expect to pay more than $500 in taxes. Your adjusted gross income is $35,000. You expect your withholding and credits to be 85% of your tax for next year, and you expect next year's taxes to equal last year's taxes.

c. You are a teacher and expect to pay more than $500 in taxes. Your adjusted gross income is $35,000. You expect your withholding and credits to be 95% of your tax for next year and 120% of your tax for last year.

d. You are a lawyer and expect to pay more than $500 in taxes. Your adjusted gross income is $185,000. You expect your withholding and credits to be 80% of your tax for next year and 115% of your tax for last year.

e. You are a farmer and expect to pay more than $500 in taxes. Your adjusted gross income is $40,000. You expect your withholding and credits to be 95% of your tax for next year and 95% of your tax for last year.

21. **What Is a Farmer?** The Internal Revenue Service (IRS) provides the following test for who is considered a farmer:

 A landowner is materially participating in farming when any one of the following four tests are met:

 (1) the owner does any three of the following four activities: (a) advances, pays, or assumes liability for at least half of the direct costs of crop production, (b) furnishes at least half of the tools, equipment, and livestock used in crop production, (c) periodically advises and consults with the tenant, and (d) periodically inspects the production activities; (2) regularly and frequently makes or takes an important part in making management decisions, substantially contributing to or affecting the success of the enterprise; (3) works 100 hours or more in activities connected with crop production over a period of five weeks or more; or (4) otherwise engages in activities that demonstrate that the landowner is materially participating in farming.

a. Make a flow chart that illustrates the various branches of this test.

b. Jack works 150 hours every 4 weeks in activities connected with crop production. Is the test clear on Jack's situation? Does Jack qualify as participating in farming?

c. Jill works 150 hours every 8 weeks in activities connected with crop production. Is the test clear on Jill's situation? Does Jill qualify as participating in farming?

d. Ivan assumes liability for 75% of the direct costs of crop production and furnishes 60% of the tools used in crop production. Does Ivan qualify as participating in farming?

e. Joan assumes liability for 65% of the direct costs of crop production, meets weekly with the tenant, and inspects production activities monthly. Does Joan qualify as participating in farming?

22. **Reading a Lease.** Consider the following excerpt from the contract for the lease of an apartment:

Landlord shall return the security deposit to resident within one month after termination of this lease or surrender and acceptance of the premises, whichever occurs last.

a. Suppose your lease terminates on June 30, you move out of the apartment on June 5, and you receive your security deposit on June 28. Has the landlord complied with the terms of the lease? Explain.

b. Suppose your lease terminates on June 30, you move out of the apartment on June 5, and you receive your security deposit on July 2. Has the landlord complied with the terms of the lease? Explain.

c. Suppose your lease terminates on June 30, you move out of the apartment on June 5, and you receive your security deposit on July 7. Has the landlord complied with the terms of the lease? Explain.

23. **Choosing Airline Tickets.** You are planning a trip to Japan six months in advance and discover that you have two choices in purchasing an airline ticket:

(A) The lowest fare is $1600, but 25% of the fare is non-refundable if you change or cancel the ticket.

(B) A fully refundable ticket is available for $2900.

Describe all of your options, and discuss how you would decide which ticket to buy.

24. **Choosing Long Distance Carriers.** Suppose you currently use a long distance carrier that offers a uniform rate of $0.10 per minute on all calls (any time of day or night). You receive an offer to switch to a long distance carrier that offers the following rates:

- $0.25 per minute for daytime calls on a weekday

- $0.10 per minute for nighttime calls on a weekday

- $0.05 per minute for all weekend calls

What additional information would you like to have in order to make a decision about switching phone companies? For your own personal situation, which option is best? Describe the factors that led to your decision.

25. **More Fine Print.** Consider Example 6 from the text. What else is unstated in the CD club deal that should be of concern? Explain.

26. **You've Won!** You receive the following e-mail notification: "Through a random selection from more than 20 million e-mail addresses, you've been selected as the winner of our grand prize—a two-week vacation in the Bahamas. To claim your prize, please call our toll free number. Have your credit card ready for identification and a small processing fee." Does this sound like a deal worth taking? Explain.

Further Applications

Ambiguity in the News. Exercises 27–30 are direct quotes from news sources. In each case, describe how the quote is ambiguous. What further information would you need to remove the ambiguity?

27. LOS ANGELES—Two jumbo jets with more than 350 people aboard nearly collided over the city during a landing attempt . . .

28. There has been an alarming 25% rise in [child abuse] cases between 1988 and 1993, a year when 2.9 million incidents were reported to child welfare agencies.

29. France must decrease its public deficit from 6% of its gross domestic product, or $83 billion, to 3%.

30. The United States was found to regularly engage in human rights abuses.

31. **Comparing Candidates.** Two assistant district attorneys, Alice and Zack, are running for the position of district attorney. The candidates make the following statements:

 ZACK: In the last five years, Alice has prosecuted 126 defendants charged with sexual assault on children. Of those, only nine cases even went to trial, and of those, only four cases resulted in conviction and a prison sentence.

 ALICE: I have prosecuted 253 cases involving sexual assault on children. More than 40% of the 253 cases resulted in prison or jail sentences as part of the original plea agreement.

 a. Comment on whether the two candidates' numbers for cases prosecuted by Alice are consistent.

 b. Comment on whether the two candidates' numbers for convictions obtained by Alice are consistent.

 c. Would you use these statements as grounds for making a decision on whom to vote for in the district attorney race? If not, what would you do to obtain more complete information?

32. **A Financial Decision.** Consider the following statement:

 I need a special computer for a project I will be working on for the next three months. After that, I will no longer need this computer. I can lease the computer for $350 per month, or I can buy it for $2100. If I resell it after the three months, I can expect to get $1200.

 Identify all the essential facts for both options. Distinguish between "hard facts" and estimates that could affect the outcome of the argument. What is the most economical choice for acquiring a computer? What would you recommend in this situation? Why?

Decision Making. In Exercises 33–37, analyze each situation and explain how you would make a decision.

33. You need three sticks of butter for baking. Your local store sells individual sticks for 40¢ each and packages of four sticks for $1.25.

34. You and your spouse are expecting a baby. Your current health insurance costs $115 per month, but doesn't cover prenatal care or delivery. You can upgrade to a policy that will cover your prenatal care and delivery, but your new premium will be $275 per month. The cost of prenatal care and delivery is approximately $4000.

35. You are planning a trip to visit your family two months from now. The airline offers you two choices: (1) You may purchase a ticket for $350, and the ticket price can be refunded for a $75 penalty; or (2) you may purchase a fully refundable ticket for $600.

36. You own a small business and need to be in another city on Monday for a meeting. You could fly to the meeting on Monday morning and return Monday evening on a ticket priced at $750. However, if you included a Saturday night stay in your trip, the ticket price would be only $335. A hotel will cost $105 per night, and you estimate that meals away from home will cost you $55 per day.

37. You fly frequently between two cities 1500 miles apart. Average round-trip cost on Airline A is $350. Airline B offers the same trip for only $325. However, Airline A has a frequent flyer program in which you earn a free round-trip ticket after you fly 25,000 miles. Airline B does not have a frequent flyer program.

Critical Thinking. Exercises 38–46 give short arguments in support of a specific conclusion. Use all the critical thinking tools at your disposal to analyze the argument and determine if the conclusion is reached in a convincing way. Write out your analysis carefully and cite all aspects of the analysis.

38. **Athletic Priority.** The athletic program is given more money than any academic department, so this university must value athletics over academics. Also, the football coach is the highest paid university employee.

39. **Drugs and Athletes.** Recent track and field performances have been so unbelievable that athletes must be using performance-enhancing drugs.

40. **Oil Crisis.** Dependency on foreign oil has put the U.S. economy at great risk, yet the nation has at least a decade's worth of oil untouched in wilderness areas and offshore. The federal government should immediately open all potential oil fields to drilling.

41. **Health Care.** That we have a health care crisis is clear: Large numbers of Americans are uninsured, and the cost of health care to our nation is skyrocketing. I believe that only the government has the power to step in and solve this crisis, so I urge you to support candidates who will make health care reform their top priority.

42. **Global Warming.** The populations of China and India dwarf that of the United States, yet they use far less

energy. In the future, the people of China and India no doubt will demand a standard of living comparable to that in the United States. To achieve it, the governments of China and India are likely to exploit their enormous reserves of coal as an energy source. Their burning of so much fossil fuel undoubtedly will cause a worldwide environmental catastrophe.

43. **Human Population.** Overpopulation is not a real problem. Modern technology, especially in bioengineering, will enable scientists to develop far more efficient agriculture. In addition, advances in irrigation technology, along with the development of crops that can grow in salt water, will enable the conversion of much of the world's desert wastelands into productive farms. As a result, agribusiness will be able to produce enough food for at least 50 billion people, or some eight times the current world population. Some people claim that such large-scale agriculture would cause environmental damage, but this result is unlikely: The human ingenuity of a much larger population undoubtedly will prevent such damage.

44. **Classroom Calculators.** Fourth-grade students should use pencil and paper to learn to add fractions and strengthen their mental skills. Drill problems also induce good discipline. Therefore, fourth-grade students should not be allowed to use calculators. Anyway, their teachers and parents never used calculators when they were learning.

45. **Classroom Computers.** Word processors with spell checkers allow students to grow up without any spelling skills. Now, word processors have grammar checkers, so students will not learn grammar. What's next? Word processors should not be allowed in the schools.

46. **College Enrollment.** American public universities are approaching a crisis. State funding is declining as priorities shift to public school systems and prisons. At the same time, population projections indicate a 30% increase in high school graduates in the next 15 years. Many universities are at capacity and could not expand even if funds were available. Even if the number of university graduates remains level, the United States will have one of the highest university graduation rates of any country in the world. The solution is to raise both admission standards and tuition at four-year colleges and universities, thereby limiting the number of students admitted to these institutions.

47. **Poetry and Mathematics.** Consider the following poem by the English poet A. E. Housman (1859–1936).

Loveliest of Trees

> *Loveliest of trees, the cherry now*
> > *Is hung with bloom along the bough*
> *And stands about the woodland ride*
> > *Wearing white for Eastertide.*
> *Now, of my threescore years and ten,*
> > *Twenty will not come again,*
> *And take from seventy springs a score,*
> > *It only leaves me fifty more.*
> *And since to look at things in bloom*
> > *Fifty springs are little room,*
> *About the woodlands I will go*
> > *To see the cherry hung with snow.*

How old was the poet at the time he wrote this poem? (Hint: A score is 20.) Based on your reading of this poem, how much longer does the poet expect to live? Explain.

Web Projects

Find useful links for Web Projects on the text Web site: www.aw.com/bennett-briggs

48. **Interpreting the Second Amendment.** Much of the debate over gun control revolves around interpretations of the Second Amendment of the U.S. Constitution. It reads:

> *A well-regulated militia being necessary to the security of a free State, the right of the people to keep and bear arms shall not be infringed.*

Gun rights advocates tend to focus on the "right of the people to keep and bear arms." Gun control advocates tend to focus on "a well-regulated militia." Visit a few of the many Web sites on each side of this issue to find interpretations of the Second Amendment that both support and oppose gun control. Based on what you learn, do you believe that the Second Amendment allows for gun control laws? Defend your opinion.

49. **Ballot Initiatives.** Investigate a particular ballot initiative in your state, county, or city (or choose one in another area from a Web site). Explain the important arguments on each side of the issue. If possible, find a statement of the initiative as it appeared on the ballot. Is the statement of the initiative clear? Explain.

50. **Fuzzy Logic.** Use the Web to learn about *fuzzy logic*, which uses a continuous range of values between absolutely true and absolutely false (as opposed to the traditional logic discussed in this chapter, in which a proposition can be only true or false, and nothing in between). Discuss situations in which fuzzy logic would be useful and more realistic than traditional logic. Then investigate and report on at least one use of fuzzy logic in modern technology.

In the News

51. **Argument Analysis.** Find a news article or editorial in which a clear position is taken on a controversial issue. Analyze the argument using the hints and strategies given in this unit. Whether or not you agree with the final conclusion, state whether the argument is effective.

52. **Personal Decisions.** Discuss a situation in your own life in which you needed to think carefully before making a critical decision. Did you use critical thinking strategies? In retrospect, would you have made a different decision using critical thinking strategies?

53. **Ambiguous Terms.** Many arguments and promotions are difficult to analyze because of vague, ambiguous, or overused terms (for example, "natural food" or "human rights"). In news articles or advertising, find three terms that you believe are vague or ambiguous. Explain how the terms could be more clearly defined.

CHAPTER 1 SUMMARY

UNIT	KEY TERMS	KEY IDEAS AND SKILLS
1A	logic argument premises, conclusions fallacy	**Identify** an argument's premises and conclusions. **Recognize** fallacious arguments, which contain logical flaws so that their conclusions are not well supported by their premises.
1B	proposition truth values, truth tables negation: *not p* conjunction: *p and q* disjunction: *p or q* conditional: *if p, then q* logical equivalence	**Understand** truth tables for negation, conjunction, disjunction, conditional. **Inclusive versus exclusive *or*** Inclusive *or* means *either or both.* Exclusive *or* means *one or the other, but not both.* **Variations on the conditional** *if p, then q* **converse:** *if q, then p.* **inverse:** *if not p, then not q.* **contrapositive:** *if not q, then not p.*
1C	set categorical propositions Venn diagrams	**Set relationships** subsets, disjoint sets, overlapping sets **Four categorical propositions** All *S* are *P*; No *S* are *P*; Some *S* are *P*; Some *S* are not *P* **Uses of Venn diagrams** illustrating set relationships organizing information
1D	inductive argument strength deductive argument validity soundness	**Evaluating arguments** inductive arguments in terms of strength deductive arguments in terms of validity and soundness Venn diagram tests of validity chains of conditionals **Induction and deduction in mathematics** Propose a theorem inductively. Prove a theorem deductively.
1E	critical thinking	1. Read (or listen) carefully. 2. Look for hidden assumptions. 3. Identify the real issue. 4. Use visual aids. 5. Understand the options. 6. Watch for fine print and missing information. 7. Are other conclusions possible?

Nothing in life is to be feared. It is only to be understood.

—MARIE CURIE

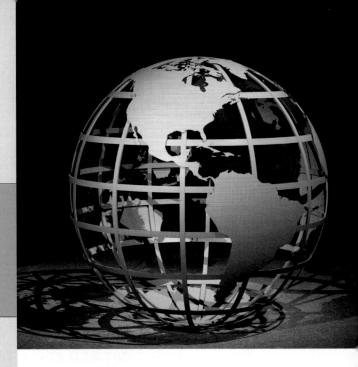

CHAPTER 2

Approaches to Problem Solving

In your past mathematics classes, it might have seemed that mathematics problems involved only numbers and symbols. But the mathematical problems that you encounter in other classes, jobs, and daily life are almost always posed in words. That's why Chapter 1 focused on logic and critical thinking: They help you find the key ideas buried in problems that are posed in words. In this chapter, we will begin to consider *quantitative* problem solving, in which the problems involve words *and* numbers or equations.

THE PROBLEM-SOLVING POWER OF UNITS

J ust as critical thinking can be approached in many different ways, there's no single best approach to solving quantitative problems. However, people have spent lifetimes coming up with strategies and guidelines that can help. The overall goal of this chapter is to help you learn some of these strategies and guidelines.

We will begin by looking at how the units of a problem can be useful in figuring out what the problem is about, as well as for solving the problem and checking its answer. This work will also provide valuable experience with problem solving—experience that we will build upon when we consider general guidelines for problem solving in Unit 2C.

YOU CAN'T ADD APPLES AND ORANGES

TECHNICAL NOTE

When abbreviating units, we do not distinguish between singular and plural. For example, we use *mi* for both mile and miles.

It may be a cliché, but mathematically it is true that you can't add apples and oranges. The reason is that we can add or subtract only quantities of the same type. For example, we can add 5 apples and 3 apples to get 8 apples. But apples and oranges are different things, so if we try to add 5 apples and 3 oranges, all we can say is that we have 5 apples and 3 oranges.

More generally, numbers in real problems usually represent a quantity of *something,* which we can describe by its **units.** If we count the number of apples in a basket, the result has units of *apples.* If we measure the length of a room, the result may have units of *feet.* If we ask how fast a car is traveling, the answer may have units of *miles per hour.*

DEFINITION

The **units** of a quantity describe what is being measured or counted.

FIGURE 2.1 Each small square is one foot on a side and therefore has an area of 1 square foot. The floor of the room contains a total of 120 of these squares, so its total area is 120 square feet.

We can add or subtract numbers only when they have the same units. However, we can multiply and divide quantities with the same or different units, much as we multiply and divide numbers. A few examples should clarify the idea:

- If you drove 100 miles in two hours, your average speed was the distance traveled (100 mi) divided by the time it took (2 hr):

$$\frac{100 \text{ mi}}{2 \text{ hr}} = \frac{50 \text{ mi}}{1 \text{ hr}} = 50 \frac{\text{mi}}{\text{hr}}$$

We read this answer as "50 miles per hour." Note that the word *per* means "divided by" (or "for every").

- To find the *area* of a room, we multiply its length by its width (Figure 2.1). If the room is 12 feet long and 10 feet wide, its area is

$$12 \text{ ft} \times 10 \text{ ft} = 120 \text{ (ft} \times \text{ft)} = 120 \text{ ft}^2$$

We read this area as "120 square feet," remembering that the word *square* implies something raised to the second power. Note that we multiply the numbers ($12 \times 10 = 120$) separately from the units ($\text{ft} \times \text{ft} = \text{ft}^2$) but keep track of both.

- To find the *volume* of a box, we multiply its width, depth, and height (Figure 2.2). If the box is 6 inches wide, 4 inches deep, and 10 inches long, its volume is

$$6 \text{ in.} \times 4 \text{ in.} \times 10 \text{ in.} = 240 \,(\text{in.} \times \text{in.} \times \text{in.}) = 240 \text{ in.}^3$$

We read this volume as "240 cubic inches," remembering that the word *cubic* implies something raised to the third power. (The abbreviation "in." includes a period.)

- To find the energy used by a light bulb, we multiply its power rating (in watts or kilowatts) by the length of time it is turned on. If a movie set uses a 0.5-kilowatt light bulb for 6 hours, its energy usage is

$$0.5 \text{ kilowatt} \times 6 \text{ hr} = 3 \text{ kilowatt} \times \text{hr}$$

We read this answer as "3 kilowatt-hours," where the hyphen between the units implies multiplication. (See Unit 2B for further discussion of energy units.)

FIGURE 2.2 Each small cube is one inch on a side and therefore has a volume of 1 cubic inch. The box contains a total of 240 of these cubes, so its total volume is 240 cubic inches.

READING UNITS

Operation	Key word or symbol	Example
Division	*per*	Read miles ÷ hours as "miles per hour."
Raising to second power	*square*	Read ft × ft, or ft², as "square feet" or "feet squared."
Raising to third power	*cube* or *cubic*	Read ft × ft × ft, or ft³, as "cubic feet" or "feet cubed."
Multiplication	hyphen	Read kilowatts × hours as "kilowatt-hours."

➤**EXAMPLE 1** Identifying Units

Identify the units of the answer for each of the following cases.

 a. The price you paid for gasoline, found by dividing its total cost in dollars by the number of gallons of gas that you bought.

 b. The area of a circle, found with the formula πr^2, where r is the radius of the circle measured in centimeters. (Note that π is a pure number and has no units.)

 c. A volume found by multiplying an area measured in acres by a depth measured in feet.

Solution

a. The price of the gasoline has units of dollars divided by gallons, which we write as $/gal and read as "dollars per gallon."

b. The area of the circle has units of centimeters to the second power, which we write as cm² and read as "square centimeters."

c. In this case, the volume has units of acres × feet, which we read as "acre-feet." This unit of volume is commonly used by hydrologists (water engineers) in the United States. *Now try Exercises 19–28.* ◄

Time out to think

Look for numbers in a newspaper or magazine. For at least five cases, identify the units of the numbers.

WORKING WITH FRACTIONS

We can express a fraction in three basic ways: as a *common fraction* such as $\frac{1}{2}$; as a fraction in *decimal form* such as 0.5; and as a *percentage* such as 50%. Here, we review the first two forms (see p. 135 for a review of percentages).

Common Fractions

Common fractions have the form a/b, where a and b can be any numbers as long as b is not zero. The number on top is called the **numerator** and the number on the bottom is called the **denominator.** A fraction represents division:

$$\text{numerator} \rightarrow \quad \frac{a}{b} \quad \text{means} \quad a \div b$$
$$\text{denominator} \rightarrow$$

We can write an integer as a fraction with a denominator of 1. For example:

$$3 = \frac{3}{1} \quad \text{or} \quad -4 = \frac{-4}{1}$$

Adding and Subtracting Fractions

If two fractions have a common (same) denominator, we can add or subtract them by adding or subtracting their numerators. For example:

$$\frac{1}{5} + \frac{2}{5} = \frac{1+2}{5} = \frac{3}{5} \quad \text{or} \quad \frac{7}{9} - \frac{2}{9} = \frac{7-2}{9} = \frac{5}{9}$$

Otherwise, we must write the fractions with a common denominator before adding or subtracting. For example, we can add $\frac{1}{2} + \frac{1}{3}$ by writing them with a common denominator of 6 as $\frac{3}{6}$ and $\frac{2}{6}$, respectively:

$$\frac{1}{2} + \frac{1}{3} = \frac{3}{6} + \frac{2}{6} = \frac{3+2}{6} = \frac{5}{6}$$

Multiplying Fractions

To multiply fractions, we multiply the numerators and denominators separately. For example:

$$\frac{1}{3} \times \frac{2}{5} = \frac{1 \times 2}{3 \times 5} = \frac{2}{15}$$

Sometimes we can simplify fractions at the same time we multiply them by *canceling* terms that occur in both the numerator and the denominator. For example:

$$\frac{3}{4} \times \frac{5}{3} = \frac{3 \times 5}{4 \times 3} = \frac{5}{4}$$

Reciprocals and Division

Two numbers are **reciprocals** if their product is 1. For example:

2 and $\frac{1}{2}$ are reciprocals because $2 \times \frac{1}{2} = 1$

$\frac{4}{3}$ and $\frac{3}{4}$ are reciprocals because $\frac{4}{3} \times \frac{3}{4} = 1$

We find the reciprocal of any fraction by inverting it (interchanging the numerator and the denominator). For an integer like 2, we think of it as $\frac{2}{1}$ to find that its reciprocal is $\frac{1}{2}$. In general:

UNIT CONVERSIONS

Many everyday problems involve nothing more than converting one set of units to another. For example, you might want to convert a distance from kilometers to miles or a recipe measure from quarts to cups.

The basic "trick" of unit conversions is to devise an appropriate way of multiplying by 1. For example, the following are all different ways of writing 1.

$$1 = \frac{1}{1} = \frac{8}{8} = \frac{\frac{1}{4}}{\frac{1}{4}} = \frac{1 \text{ kilogram}}{1 \text{ kilogram}} = \frac{1 \text{ week}}{7 \text{ days}} = \frac{12 \text{ inches}}{1 \text{ foot}}$$

The last expression shows the necessity of stating units: 12 ÷ 1 is *not* 1, but 12 in. ÷ 1 ft *is* 1 because 12 inches and 1 foot are equal.

- The reciprocal of a is $\frac{1}{a}$ ($a \neq 0$).

- The reciprocal of $\frac{a}{b}$ is $\frac{b}{a}$ ($a \neq 0$, $b \neq 0$).

To divide fractions, we replace the division with multiplication by the reciprocal. This process is sometimes called *invert* and *multiply,* as shown in the following examples:

$$10 \div \frac{1}{2} = 10 \times \underbrace{\frac{2}{1}}_{\substack{\text{invert} \\ \text{and multiply}}} = 20$$

$$\frac{3}{4} \div \frac{2}{5} = \frac{3}{4} \times \underbrace{\frac{5}{2}}_{\substack{\text{invert} \\ \text{and multiply}}} = \frac{15}{8}$$

Rules Summary

Addition/subtraction (common denominator required):

$$\frac{a}{c} + \frac{b}{c} = \frac{a+b}{c} \quad \text{or} \quad \frac{a}{c} - \frac{b}{c} = \frac{a-b}{c}$$

Multiplication:

$$\frac{a}{b} \times \frac{c}{d} = \frac{a \times c}{b \times d}$$

Division (invert and multiply):

$$\frac{a}{b} \div \frac{c}{d} = \frac{a}{b} \times \frac{d}{c}$$

Decimal Form

For a fraction in decimal form, each digit corresponds to a certain place value, which is always a power of 10 (such as 10, 100, 1000, . . .). The following example shows values for the decimal places in the number 3.141.

3	.	1	4	1
ones		tenths	hundredths	thousandths
(1)		($0.1 = \frac{1}{10}$)	($0.01 = \frac{1}{100}$)	($0.001 = \frac{1}{1000}$)

Converting to Common Form

Converting a fraction from decimal to common form requires only recognizing the value of the last digit in the decimal. For example:

$$0.4 = \frac{4}{10} \qquad 3.15 = \frac{315}{100} \qquad 0.097 = \frac{97}{1000}$$

Converting to Decimal Form

To convert a common fraction into decimal form, we carry out the division implied by the fraction. For example:

$$\frac{1}{4} = 1 \div 4 = 0.25$$

Many common fractions cannot be written *exactly* in decimal form, but instead have an endless repeating pattern of digits. For example, the decimal form of $\frac{1}{3}$ contains an endless string of 3s:

$$\frac{1}{3} = 0.3333333 \ldots$$

Mathematically, we represent the repeating digits by putting a bar over them. For example, the bar in $0.\overline{3}$ means that the 3 repeats indefinitely. In practice, we usually round repeating decimals to one or a few digits. For example, we might round the decimal form of $\frac{1}{3}$ to 0.33.

Now try Exercises 13–18.

Conversion Factors

The statement 12 in. = 1 ft is an example of a **conversion factor.** We can write this conversion factor in three equivalent ways (the latter two forms come from dividing on both sides of the first equation):

$$12 \text{ in.} = 1 \text{ ft} \qquad \text{or} \qquad \frac{12 \text{ in.}}{1 \text{ ft}} = 1 \qquad \text{or} \qquad \frac{1 \text{ ft}}{12 \text{ in.}} = 1$$

The key to unit conversions is to use the correct form of the conversion factor for the problem at hand.

➤ **EXAMPLE 2** Feet to Inches

Convert a distance of 7 feet into inches.

Solution We begin with *feet* and want to end up with *inches.* We therefore need the conversion factor between *inches* and *feet* in a form that has *feet* in the denominator, so that *feet* will cancel:

$$7 \text{ ft} = 7 \,\cancel{\text{ft}} \times \frac{12 \text{ in.}}{\underbrace{1 \,\cancel{\text{ft}}}_{1}} = 84 \text{ in.}$$

Note that we really just multiplied 7 feet by 1 in a form that allowed us to change the units. Seven feet is the same as 84 inches. *Now try Exercises 29–30.* ◄

➤ **EXAMPLE 3** Inches to Feet

Convert a length of 102 inches to feet.

Solution This time we use the conversion factor in the form with *inches* in the denominator, so that *inches* will cancel:

$$102 \text{ in.} = 102 \,\cancel{\text{in.}} \times \frac{1 \text{ ft}}{\underbrace{12 \,\cancel{\text{in.}}}_{1}} = 8.5 \text{ ft}$$

Thus, 102 inches is equivalent to 8.5 feet. *Now try Exercises 31–32.* ◄

➤ **EXAMPLE 4** Using a Chain of Conversions

How many seconds are there in one day?

Solution Most of us don't immediately know the answer to this question, but we do know that 1 day = 24 hr, 1 hr = 60 min, and 1 min = 60 s. We can answer the question by setting up a *chain* of unit conversions in which we start with *day* and end up with *seconds:*

$$1 \,\cancel{\text{day}} \times \frac{24 \,\cancel{\text{hr}}}{1 \,\cancel{\text{day}}} \times \frac{60 \,\cancel{\text{min}}}{1 \,\cancel{\text{hr}}} \times \frac{60 \text{ s}}{1 \,\cancel{\text{min}}} = 86{,}400 \text{ s}$$

By using the conversion factors needed to cancel the appropriate units, we are left with the answer in *seconds.* There are 86,400 seconds in one day.

Now try Exercises 33–40. ◄

FOXTROT © 1995 by Bill Amend. Reprinted with permission of UNIVERSAL PRESS SYNDICATE. All rights reserved.

Conversions with Units Raised to Powers

We must take special care when converting units raised to powers. For example, suppose we want to know the number of square feet in a square yard. We may not know the conversion factor between square yards (yd^2) and square feet (ft^2), but we know that 1 yd = 3 ft. Therefore, we can replace 1 yard by 3 feet when we write out 1 square yard:

$$1\ yd^2 = 1\ yd \times 1\ yd = 3\ ft \times 3\ ft = 9\ ft^2$$

That is, 1 square yard is the same as 9 square feet. We can also find this conversion factor by squaring both sides of the yards-to-feet conversion:

$$1\ yd = 3\ ft \xrightarrow[\text{both sides}]{\text{square}} (1\ yd)^2 = (3\ ft)^2 \xrightarrow[\text{each term}]{\text{square}} 1\ yd^2 = 9\ ft^2$$

Figure 2.3 confirms that, indeed, 9 square feet fit exactly into 1 square yard. As usual, we can write the conversion factor in three equivalent forms:

$$1\ yd^2 = 9\ ft^2 \quad \text{or} \quad \frac{1\ yd^2}{9\ ft^2} = 1 \quad \text{or} \quad \frac{9\ ft^2}{1\ yd^2} = 1$$

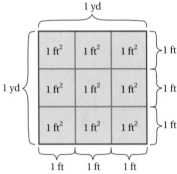

FIGURE 2.3 This diagram shows that 1 square yard is the same as 9 square feet.

▶ EXAMPLE 5 Carpeting a Room

You want to carpet a room that measures 10 feet by 12 feet, making an area of 120 square feet. But carpet is usually sold by the square yard rather than by the square foot. How many square yards of carpet do you need?

Solution We need to convert the room's area from units of *square feet* to *square yards*, so we use the conversion factor in the form that has square feet (ft^2) in the denominator:

$$120\ \cancel{ft^2} \times \frac{1\ yd^2}{9\ \cancel{ft^2}} = \frac{120}{9}\ yd^2 \approx 13.3\ yd^2$$

(The symbol ≈ means "approximately equal to.") Note that we rounded the answer to the nearest 0.1 square yard. Most stores will not sell fractions of a square yard, so you will need to buy 14 square yards of carpet. *Now try Exercises 41–44.* ◀

➤**EXAMPLE 6** Cubic Units: Purchasing Garden Soil

You are preparing a vegetable garden that is 40 feet long and 16 feet wide, and you need enough soil to fill it to a depth of 1 foot. The landscape supply store sells soil by the cubic yard. How much soil should you order?

Solution To find the volume of soil that you need, we multiply the garden's length (40 feet), width (16 feet), and depth (1 foot):

$$40 \text{ ft} \times 16 \text{ ft} \times 1 \text{ ft} = 640 \text{ ft}^3$$

Because soil is sold by the cubic yard, we need to convert this volume from units of *cubic feet* to *cubic yards*. We know that 1 yd = 3 ft, so we find the cubic conversion factor by cubing both sides of this equation:

$$1 \text{ yd} = 3 \text{ ft} \xrightarrow{\text{cube both sides}} (1 \text{ yd})^3 = (3 \text{ ft})^3 \xrightarrow{\text{cube each term}} 1 \text{ yd}^3 = 27 \text{ ft}^3$$

In the last step, we recognized that $3^3 = 3 \times 3 \times 3 = 27$. As usual, we can write this conversion factor in three equivalent forms:

$$1 \text{ yd}^3 = 27 \text{ ft}^3 \quad \text{or} \quad \frac{1 \text{ yd}^3}{27 \text{ ft}^3} = 1 \quad \text{or} \quad \frac{27 \text{ ft}^3}{1 \text{ yd}^3} = 1$$

To convert the soil volume from cubic feet to cubic yards, we use the conversion factor in the form that has cubic feet (ft^3) in the denominator:

$$640 \text{ ft}^3 \times \frac{1 \text{ yd}^3}{27 \text{ ft}^3} = \frac{640}{27} \text{ yd}^3 \approx 23.7 \text{ yd}^3$$

You will need to order about 24 cubic yards of soil for your garden.

Now try Exercises 45–48. ◀

$1 = 10.6045 pesos

1 peso = $0.0943

Currency Conversions

Different countries use different money, or **currency.** If you travel internationally, you'll need to change money from one currency to another. Converting between currencies is a unit conversion problem in which the conversion factors are the exchange rates.

Table 2.1 shows a typical table of currency exchange rates. The *Dollars per Foreign* column gives the conversion factor in the form

TABLE 2.1	Sample Currency Exchange Rates (June 2003)	
CURRENCY	DOLLARS PER FOREIGN	FOREIGN PER DOLLAR
British pound	1.678	0.5958
Canadian dollar	0.7483	1.336
European euro	1.169	0.8554
Japanese yen	0.008482	117.9
Mexican peso	0.0943	10.6045

useful for converting from foreign currency into U.S. dollars. For example, the row for Mexico shows 0.0943 in this column, which means

$$1 \text{ peso} = \$0.0943$$

The *Foreign per Dollar* column gives the conversion factor in the form useful for converting U.S. dollars into foreign currency. In this case,

$$\$1 = 10.6045 \text{ pesos}$$

Time out to think

Find the reciprocals of the numbers in the *Dollars per Foreign* column of Table 2.1. Are your results the numbers in the *Foreign per Dollar* column? Explain.

►EXAMPLE 7 Price Conversion

At a French department store, the price for a pair of Levi jeans is 45 euros. What is the price in U.S. dollars? Use the exchange rates in Table 2.1.

Solution From the *Dollars per Foreign* column in Table 2.1, we see that

$$1 \text{ euro} = \$1.169$$

As usual, we can write this conversion factor in two other equivalent forms:

$$\frac{1 \text{ euro}}{\$1.169} = 1 \qquad \text{or} \qquad \frac{\$1.169}{1 \text{ euro}} = 1$$

We use the latter form to convert the price from euros to dollars:

$$45 \text{ euro} \times \frac{\$1.169}{1 \text{ euro}} \approx \$52.60$$

The price of 45 euros is equivalent to $52.60. *Now try Exercises 49–52.* ◄

►EXAMPLE 8 Buying Currency

You are on holiday in Mexico and need cash. How many pesos can you buy with $100? Use the exchange rates in Table 2.1 and assume the transaction does not involve any fees.

Solution From the *Foreign per Dollar* column in Table 2.1, we see that

$$\$1 = 10.6045 \text{ pesos}$$

Two other equivalent forms of this conversion are

$$\frac{10.6045 \text{ peso}}{\$1} = 1 \qquad \text{or} \qquad \frac{\$1}{10.6045 \text{ peso}} = 1$$

The first form allows us to convert dollars to pesos:

$$\$100 \times \frac{10.6045 \text{ peso}}{\$1} = 1060.45 \text{ peso}$$

Your $100 will buy 1060.45 pesos. *Now try Exercises 53–56.* ◄

by the way

Currency exchange rates change nearly every day. Sometimes, especially when markets are volatile, currency rates change hour by hour or even minute by minute. You can find exchange rates listed daily in many newspapers and up-to-the-minute rates on the Web.

HISTORICAL NOTE

The euro came into being as a new currency for 11 countries of the European Union in January 1999. These countries immediately tied their currency values to the euro but, to ease the transition, decided to keep their national currencies in circulation for three additional years. Euro bills and coins were put in circulation in January 2002.

Practical matters

CHANGING MONEY IN FOREIGN COUNTRIES

It can cost money to change money from one currency to another, so you should always look for the best deal. Two factors affect the cost of changing currency: (1) the exchange rate and (2) fees for the exchange.

Published exchange rates are usually "wholesale" rates available only to banks. Most money changers—including airport change stations, hotels, and exchange booths on streets—make money by giving you a rate that is not as good as the wholesale rate. For example, if the wholesale rate is 10.6 pesos to the dollar, a money changer might give you only 10.3 pesos per dollar.

Once you find a good exchange rate, you must still watch out for additional fees. For example, many money changers (especially in hotels, stores, and street booths) charge a fee every time you make an exchange. Here are a few general hints for changing currency:

- Before you travel, get a small amount of your destination's currency from your local bank. That way, if you need cash upon arrival, you won't be forced to accept a poor exchange rate or high fees.

- Once you get to your destination, banks usually offer better exchange rates than other money changers. The best deal may be your ATM card, which generally gets bank exchange rates with the added convenience of ATM machines (at least in countries where ATM machines are common). But be sure you're aware of any fees: Before your trip, find out whether your bank charges a fee for foreign ATM transactions. Any fee charged by the ATM owner will usually (but not always) be stated during your transaction.

- Consider using a credit card for purchases and hotel and restaurant bills. Credit cards generally offer good exchange rates and, unlike cash, can be replaced if lost or stolen. But be sure to check whether your credit card company adds any fees for foreign purchases. And avoid using your credit card to get cash unless it is an emergency, because most credit cards charge high fees and interest for cash advances.

- Travelers' checks also offer safety in case of loss or theft, but watch out for fees. Some banks charge a fee when you purchase travelers' checks at home, and you're likely to pay another fee when you change the travelers' checks to local currency at your destination.

Time out to think _____

Find the current value of the Mexican peso in dollars. How does it compare to the value listed in Table 2.1? If it is significantly different, suggest some possible economic reasons for the change in the peso's value.

PROBLEM SOLVING WITH UNITS

Now that we have covered basic ideas of units and conversions, we are ready to apply them to problem solving. In some cases, units can help us decide how to solve a problem. In all cases, units offer a useful way of checking answers. The following box summarizes how we work with units. As you study, notice how these ideas are applied in the examples throughout this chapter and the rest of this book.

Working with Units

Step 1: Identify the units involved in the problem. Use the units to help you decide how to approach the problem and what units to expect in the answer.

Step 2: Perform any operations (such as multiplication or division) on both the numbers and their associated units. Remember:

- You cannot add or subtract numbers with different units (such as apples and oranges), but you can combine different units through multiplication, division, or raising to powers.
- To make your work with units easier, replace division with multiplication by the reciprocal. For example,

$$\text{instead of dividing by } 60\,\frac{\text{s}}{\text{min}}, \text{ multiply by } \frac{1\,\text{min}}{60\,\text{s}}.$$

Step 3: When you complete your calculations, make sure that your answer has the units you expected. If it doesn't, then you've done something wrong.

> **by the way**
> You can use your familiarity with *miles per hour* to find the important general relationship among distance, time, and speed. Simply remember that miles are a distance and hours are a time, so a speed like miles per hour must be a distance (such as miles) divided by a time (such as hours):
> $$\text{speed} = \frac{\text{distance}}{\text{time}}$$

▶**EXAMPLE 9** Distance, Time, and Speed

A car is traveling 25 miles every half-hour. How fast is it going?

Solution The "how fast" suggests that the final answer should be a speed. In this case, since the problem involves units of miles and hours, we should expect a speed in *miles per hour*. Knowing that these units mean miles *divided by* hours helps us realize that we should divide the distance traveled by the time it takes:

$$25\,\text{mi} \div \frac{1}{2}\,\text{hr} = 25\,\text{mi} \times \frac{2}{1\,\text{hr}} = 50\,\frac{\text{mi}}{\text{hr}}$$

Note that we replaced the division with multiplication by the reciprocal: Be sure you understand why the reciprocal of $\frac{1}{2}$ hr is $\frac{2}{1\,\text{hr}}$. The car is traveling at a speed of 50 miles per hour. *Now try Exercises 57–60.* ◀

▶**EXAMPLE 10** Buying Farm Land

You are buying 30 acres of farm land at a cost of $12,000 per acre. What is the total cost?

Solution The question asks about total cost, so the answer should be in *dollars*. We multiply the acreage by the cost per acre:

$$30\,\text{acres} \times \frac{\$12,000}{1\,\text{acre}} = \$360,000$$

Note how the unit *acre* "cancels" to leave the answer in units of dollars, as we expected. The land will cost $360,000. *Now try Exercises 61–70.* ◀

> **by the way**
> An *acre* is a unit of area originally defined as the amount of land a pair of oxen could plow in a day. Today it is defined as 43,560 square feet. One square mile is 640 acres.

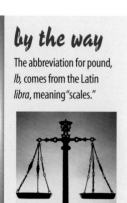
Using Units to Check Answers

Checking answers is one of the most important reasons for clearly showing all units. If the answer does not come out with the expected units, then something has gone wrong. Of course, you should always check your method and your calculations as well as your units. An answer with the correct units could still contain other errors.

▶**EXAMPLE 11** Exam Check

You are a grader for a math course. An exam question reads: "Eli purchased 5 pounds of apples at a price of 50 cents per pound. How much did he pay for the apples?" On the paper you are grading, a student has written: "$50 \div 5 = 10$. He paid 10 cents." Write a note to the student explaining what went wrong.

Solution Dear student—First, notice that your answer does not make sense. If one pound of apples costs 50¢, how could five pounds cost only 10¢? You could have prevented your error by keeping track of the units. In the exam question, the number 50 has units of *cents per pound* and the number 5 has units of *pounds*. Thus, your calculation of "$50 \div 5 = 10$" means

$$50\frac{¢}{\text{lb}} \div 5\,\text{lb} = 50\frac{¢}{\text{lb}} \times \frac{1}{5\,\text{lb}} = 10\frac{¢}{\text{lb}^2}$$

(As usual, we replaced the division with multiplication by the reciprocal.) Your calculation gives units of "cents per square pound," so it cannot be correct for a question that asks for a simple price. The correct calculation multiplies the price per pound by the weight in pounds:

$$50\frac{¢}{\cancel{\text{lb}}} \times 5\,\cancel{\text{lb}} = 250¢ = \$2.50$$

The units now work out as they should: The 5 pounds of apples cost $2.50.

Now try Exercises 71–74. ◀

Using Units to Help You Solve Problems

As we've seen, knowing the units to expect in the answer can help determine how to solve a problem. The following two examples illustrate how powerful this idea can be. The first shows how units can help us get "unstuck" on an everyday sort of problem, while the second shows how units can help us come up with a strategy for a more complex problem—in this case, a problem relevant to the issue of global warming.

▶**EXAMPLE 12** Gas Mileage

After a long day of driving, your destination is only 90 miles away. You know that your car gets 25 miles per gallon. How much gas do you need in your tank if you are to reach your destination without stopping? If you have a 12-gallon tank and the fuel gauge shows it is one-quarter full, will you make it? Show how units can help answer this question.

Solution The question asks "how much gas," so we are looking for an answer with units of *gallons*. We have two other pieces of information available: the distance of 90 *miles* and the car's gas mileage of 25 *miles per gallon*. Thus, we are looking for a way to combine these two pieces of information to get an answer in gallons. We can do it by dividing the distance by the mileage:

$$90 \text{ mi} \div \frac{25 \text{ mi}}{1 \text{ gal}} = 90 \text{ mi} \times \frac{1 \text{ gal}}{25 \text{ mi}} = 3.6 \text{ gal}$$

You'll need 3.6 gallons of gas to drive the 90 miles. One-quarter of a 12-gallon tank is only 3 gallons, so you'd better stop for gas. *Now try Exercises 75–78.* ◄

► **EXAMPLE 13** Melting of the Ice Caps

We usually think of "ice ages" as times when glaciers covered much of North America and Europe. But in some sense the Earth is still in an ice age today, because geological evidence suggests that the Earth had no polar ice at all during much of its history. Some time in the future, it is likely that the Earth will again have no polar ice. Measurements of polar ice show that, if all the ice melts, about 25 million cubic kilometers of water will be added to the oceans—most of it coming from Antarctica (Figure 2.4). How much will sea level rise as a result? Useful data: The total surface area of the Earth's oceans is about 340 million square kilometers.

Solution Because the question asks how much sea level will rise, we are looking for an answer with units of height, such as kilometers or meters. We are given two pieces of information that seem relevant: a volume of water in *cubic kilometers* and a surface area for the oceans in *square kilometers*. If we divide these two numbers, we get a result with units of kilometers:

FIGURE 2.4

$$\frac{25 \text{ million km}^3}{340 \text{ million km}^2} \approx 0.074 \text{ km} = 74 \text{ m}$$

Of course, the fact that this result has the expected units doesn't make it right. We must also ask whether it makes sense. In this case, it does. Dividing a volume by an area does indeed give a depth (because units of km³ divided by units of km² give units of km), and the depth we've found represents the rise in sea level as long as the melting does not appreciably change the surface area of the oceans. (Because most continents have coastal hills or mountains that are higher than the expected sea level rise, ocean water would not spread very far inland. Thus, the ocean surface area would increase only slightly.) Perhaps the greatest surprise is the size of the answer: Complete polar melting would cause sea level to rise by some 74 meters, which is more than 240 feet. Thus, if this ever occurs, future archaeologists will have to be extreme deep sea divers to visit the ruins of modern coastal cities.

Now try Exercise 79. ◄

Time out to think _____

Although it would probably take thousands of years for the ice caps to melt fully, some melting may already be occurring due to global warming. Suppose that 1% of the ice melts during the next 100 years, raising sea level by about 2.4 feet. How do you think such a sea level rise would affect human civilization? Explain.

by the way

The Antarctic ice sheet contains more than 90% of the Earth's ice. This ice sheet covers an area of 13 million square kilometers, with an average depth of 2 kilometers (1.2 mi). In some places, the ice is more than 4.5 kilometers (2.8 mi) thick.

EXERCISES 2A

Review Questions

1. What are units? Describe how to read units that involve division, multiplication, squares, and cubes.

2. Explain why a unit conversion really involves just multiplying by 1.

3. Describe the three forms in which we can write any conversion factor. If given the conversion in one form, such as 1 lb = 16 oz, how do you find the other two forms?

4. Explain in words or with a picture why there are 9 square feet in 1 square yard and 27 cubic feet in 1 cubic yard. Then describe generally how to find conversion factors involving squares or cubes.

5. Describe how to read and use the currency data in Table 2.1.

6. Briefly describe how units can help you check your answers and solve problems. Give examples.

Does It Make Sense?

Decide whether each of the following statements makes sense (or is clearly true) or does not make sense (or is clearly false). Explain your reasoning.

7. I drove really fast—my speed was 50 miles.

8. Our house has a floor area of 1500 square feet.

9. I have a box with a volume of two square feet.

10. I needed five acres of water to fill the swimming pool.

11. I figured out how long the airplane will take to reach Beijing by dividing the airplane's speed by the distance to Beijing.

12. I figured out how long the airplane will take to reach Beijing by dividing the distance to Beijing by the airplane's speed.

Basic Skills & Concepts

Working with Fractions. Exercises 13–18 use skills covered in the Brief Review on pp. 86–87.

13. Evaluate each of the following.
 a. $\dfrac{4}{3} \times \dfrac{1}{2}$
 b. $\dfrac{4}{3} + \dfrac{1}{2}$
 c. $\dfrac{4}{3} \div \dfrac{1}{2}$
 d. $\dfrac{4}{3} - \dfrac{1}{2}$
 e. $\dfrac{7}{20} + \dfrac{3}{5}$
 f. $\dfrac{12}{13} - \dfrac{1}{4}$
 g. $\dfrac{7}{17} \times \dfrac{2}{7}$
 h. $\dfrac{1}{3} + \dfrac{1}{5}$

14. Evaluate each of the following.
 a. $\dfrac{14}{5} - \dfrac{3}{10}$
 b. $\dfrac{5}{9} \div \dfrac{4}{9}$
 c. $\dfrac{2}{3} + \dfrac{1}{4}$
 d. $\dfrac{1}{2} + \dfrac{1}{3} + \dfrac{1}{4}$
 e. $\dfrac{1}{6} \times \dfrac{6}{11}$
 f. $\dfrac{1}{8} - \dfrac{1}{16}$
 g. $\dfrac{3}{5} \div \dfrac{5}{3}$
 h. $\dfrac{1}{2} \times \dfrac{1}{3} \times \dfrac{1}{4}$

15. Write each of the following as a common fraction.
 a. 0.3
 b. 0.124
 c. 0.78
 d. 0.005
 e. 1.84
 f. 3.009
 g. 0.0001
 h. 0.1001

16. Write each of the following as a common fraction.
 a. 0.7
 b. 1.4
 c. 0.45
 d. 0.988
 e. 0.401
 f. 1.001
 g. 3.07
 h. 0.00002

17. Convert each of the following common fractions into decimal form. If necessary, round to the nearest thousandth.
 a. $\dfrac{3}{2}$
 b. $\dfrac{4}{5}$
 c. $\dfrac{1}{6}$
 d. $\dfrac{1}{7}$
 e. $\dfrac{2}{9}$
 f. $\dfrac{7}{11}$
 g. $\dfrac{88}{91}$
 h. $\dfrac{122}{30}$

18. Convert each of the following common fractions into decimal form. If necessary, round to the nearest thousandth.
 a. $\dfrac{7}{12}$
 b. $\dfrac{27}{70}$
 c. $\dfrac{24}{45}$
 d. $\dfrac{18}{25}$
 e. $\dfrac{537}{912}$
 f. $\dfrac{243}{81}$
 g. $\dfrac{5}{6}$
 h. $\dfrac{16}{3}$

Identifying Units. In Exercises 19–28, identify the units you would expect for the given quantity. State the units both in words (e.g., dollars per gallon) and mathematically (e.g., $/gal).

19. The price of apples, found by dividing their total cost in dollars by their total weight in pounds.

20. A speed, found by dividing a distance measured in kilometers by a time measured in seconds.

21. The installation cost of floor tile, found by dividing the total cost of the installation in dollars by the area of the room in square feet.

22. The flow rate of a river in which 5000 cubic feet of water flow past a particular location every second.

23. The atmospheric pressure, measured by finding the weight of atmosphere (in pounds) above each square inch of surface on the ground.

24. The price of fabric, found by dividing its cost in dollars by its area in square feet.

25. The gas mileage of a car, found by dividing the distance in miles that it travels by the amount of gas in gallons that it uses.

26. The cost for grass seed when you buy enough to cover 80 square yards at a total price of $160.

27. The density of a rock, found by dividing its weight in grams by its volume in cubic centimeters.

28. A car engine torque, calculated by multiplying a force in pounds by a distance in feet.

Unit Conversions. In Exercises 29–38, carry out the indicated unit conversion.

29. Convert a distance of 18 yards into feet.

30. Convert a distance of 18 yards into inches.

31. Convert a distance of 7 miles into yards (1760 yd = 1 mi).

32. Convert a weight of 142 ounces into pounds (16 oz = 1 lb).

33. Convert a lot size of $\frac{1}{5}$ acre to square feet (1 acre = 43,560 ft^2).

34. Convert a park size of 3.5 square miles to acres (1 acre = $\frac{1}{640}$ mi^2).

35. Using the fact that there are 1760 yards in a mile and 3 feet in a yard, convert a distance of 3 miles into feet.

36. There are 8 ounces in a cup, 4 cups in a quart, and 4 quarts in a gallon. Using a chain with these conversions, convert 6 gallons into ounces.

37. Use a chain of conversions with familiar measures of time to convert 4 weeks into minutes.

38. A car is driving at 100 kilometers per hour. What is its speed in kilometers per second?

39. **Area and Volume Calculations.**

 a. A swimming pool 3 meters deep, 10 meters long, and 5 meters wide is filled with water. What is the area of the water's surface? What volume of water does the pool contain?

 b. A box measures 22 inches by 15 inches by 12 inches. What is the area of its largest sides? What is the volume of the box?

 c. A skyscraper is 1000 feet high with a 25,000-square-foot base. What is its total volume?

40. **Area and Volume Calculations.**

 a. A warehouse is 40 yards long and 25 yards wide and piled with cartons to a height of 3 yards. What is the area

of the warehouse floor? What is the total volume of the cartons? (Assume there is no space between cartons.)

 b. The bed of a pickup truck is 3.5 feet deep, 12 feet long, and 5 feet wide. What is the area of the bed's floor? What is the volume of the bed?

 c. A can has a circular base with an area of 6 square inches and is 4 inches tall. What is its total volume?

Conversions with Units Raised to Powers. Use the appropriate rules to answer Exercises 41–48.

41. Find a conversion factor between square inches and square feet. Write it in three forms.

42. There are 1000 meters in 1 kilometer. Find a conversion factor between square meters and square kilometers. Write it in three forms.

43. How many square inches are in 5 square feet?

44. A football field is 100 yards long and 60 yards wide. Find its area in square yards and square feet.

45. Find a conversion factor between cubic inches and cubic feet. Write it in three forms.

46. There are 100 centimeters in 1 meter. Find a conversion factor between cubic centimeters and cubic meters. Write it in three forms.

47. How many cubic inches are in 3 cubic yards?

48. A cargo container is 50 feet long, 10 feet wide, and 8 feet tall. Find its volume in cubic feet and in cubic yards.

Currency Conversions. Answer the questions in Exercises 49–56 using the exchange rates in Table 2.1.

49. Which is worth more, 1 British pound or 1 dollar? Explain.

50. Which is worth more, 1 Mexican peso or 1 Japanese yen? Explain.

51. You return from a trip with 2500 Mexican pesos. How much are your pesos worth in dollars?

52. You return from a trip with 75 British pounds. How much are your pounds worth in dollars?

53. How many Japanese yen can you buy for $100?

54. How many Canadian dollars can you buy for $100?

55. Gasoline in Germany sells for about 5.50 euros per gallon. What is this price in dollars per gallon?

56. Apples in Japan sell for about 75 yen each. If you buy 4 apples, how much have you spent in dollars?

Working with Units. **Use units to help you answer the questions in Exercises 57–70.**

57. A car travels 14 miles in 15 minutes. How fast is it going in miles per hour?

58. An airplane travels 95 miles in 10 minutes. How fast is it going in miles per hour?

59. You purchased 40 acres of farm land for $200,000. How much did you pay per acre?

60. A 40-acre orchard produces 12,000 apples. What is its yield in apples per acre?

61. You are buying 4.7 pounds of apples priced at $1.29 per pound. How much will you pay?

62. You are buying 2.8 kilograms of cherries priced at $3.50 per kilogram. How much will you pay?

63. You need 10 square yards of cloth priced at $2 per square yard. How much will the cloth cost?

64. You are buying floor tile to cover a room that measures 20 feet by 25 feet. The tile is priced at $7.50 per square foot. How much will the tile cost?

65. You are buying carpet to cover a floor area of 1520 square feet. The carpet costs $18 per square yard. How much will the carpet cost?

66. You are buying artificial turf to cover a game field that is 150 feet long and 100 feet wide. The turf costs $7.50 per square yard. How much will the turf cost?

67. You take a trip in which you drive 1200 miles in 20 hours. What is your average speed for the entire trip?

68. You work 40 hours per week and are paid $13.50 per hour. If you work all 52 weeks in a year, how much will you earn?

69. If you sleep an average of 7.5 hours each night, how many hours do you sleep in a year?

70. A human heart beats about 60 times per minute. If an average human being lives to the age of 75, how many times does the average heart beat in a lifetime?

What Went Wrong? **Exercises 71–74 include an exam question and a solution given by a student. State whether each solution is right or wrong. If it is wrong, write a note to the student explaining why the answer is wrong and how to solve the problem correctly.**

71. *Exam Question:* A candy store sells chocolate for $7.70 per pound. The piece you want to buy weighs 0.11 pound. How much will it cost, to the nearest cent? (Neglect sales tax.)
Student Solution: $0.11 \div 7.70 = 0.014$. It will cost 1.4¢.

72. *Exam Question:* You ride your bike up a steep mountain road at 5 miles per hour. How far do you go in 3 hours?
Student Solution: $5 \div 3 = 1.7$. I rode 1.7 miles.

73. *Exam Question:* You can buy a 50-pound bag of flour for $11 or you can buy a 1-pound bag for $0.39. Compare the per pound cost for the large and small bags.
Student Solution: The large bag price is $50 \div \$11 = \4.50 per pound, which is much more than the 39¢ per pound price of the small bag.

74. *Exam Question:* The average person needs 1500 Calories a day. A can of Coke contains 140 Calories. How many Cokes would you need to drink to fill your daily caloric needs? (Note: This diet may not meet other nutritional needs!)
Student Solution: $1500 \times 140 = 210,000$. You would need to drink 210,000 Cokes to meet your daily caloric needs.

Gas Mileage. **Exercises 75–78 involve practical gas mileage calculations.**

75. Suppose you drive a car with an average gas mileage of 28 miles per gallon. If you plan to take a 2500-mile cross-

country trip, how many gallons of gasoline should you expect to use?

76. Suppose you drive a car with an average gas mileage of 35 miles per gallon. If you plan to take a 2500-mile cross-country trip, how much should you budget (in dollars) for gasoline if the price of gasoline averages $1.65 per gallon?

77. Gas mileage actually varies slightly with the driving speed of a car (as well as with highway vs. city driving). Suppose your car averages 38 miles per gallon on the highway if your average speed is 55 miles per hour and averages 32 miles per gallon on the highway if your average speed is 70 miles per hour.

 a. What is the driving time for a 2000-mile trip if you drive at an average speed of 55 miles per hour? What is the driving time at 70 miles per hour?

 b. Assume a gasoline price of $1.65 per gallon. What is the gasoline cost for a 2000-mile trip if you drive at an average speed of 55 miles per hour? What is the gasoline cost at 70 miles per hour?

78. Suppose your car averages 32 miles per gallon on the highway if your average speed is 60 miles per hour and averages 25 miles per gallon on the highway if your average speed is 75 miles per hour.

 a. What is the driving time for a 1500-mile trip if you drive at an average speed of 60 miles per hour? What is the driving time at 75 miles per hour?

 b. Assume a gasoline price of $1.80 per gallon. What is the gasoline cost for a 1500-mile trip if you drive at an average speed of 60 miles per hour? What is the gasoline cost at 75 miles per hour?

79. **Filling a Pool.** A swimming pool is 75 feet long and 54 feet wide. It is only partially filled with water, so the water surface is 6 inches below where it is supposed to be. How much water would it take, in cubic feet, to raise the water level by 6 inches?

Further Applications

80. **House Footprint.** A local zoning ordinance says that a house's "footprint" (area of its ground floor) cannot occupy more than $\frac{1}{4}$ of the lot it is built on. Suppose you own a $\frac{1}{4}$-acre lot. What is the maximum allowed footprint for your house, in square feet (1 acre = 43,560 ft^2)?

81. **Full of Hot Air.** The average person breathes 6 times per minute (at rest), inhaling and exhaling half a liter of air

each time. How much "hot air" (the air is warmed by the body), in liters, does the average person exhale each day?

82. **E-Books.** Computer memory is measured in units of bytes, where one byte is enough memory to store one character (a letter in the alphabet or a number). How many typical pages of text can be stored on a 500-megabyte hard drive (a megabyte is one million bytes)? Assume 2000 bytes per page.

83. **Dog Years.** Sometimes the age of dogs is described in a unit called "dog years." A commonly used conversion is that 1 real year equals 7 dog years.

 a. If your dog is 15 real years old, what is her age in dog years?

 b. People often refer to the third year in the life of a human child as the "terrible twos" stage. If dogs have a terrible twos stage in their third dog year, how old are they, in real time, during this stage?

 c. Based on what you know about dogs, do you think the common conversion of 1 real year to 7 dog years seems reasonable? Explain.

84. **Glen Canyon Flood.** The Department of the Interior released a "spike flood" from the Glen Canyon Dam on the Colorado River. Its purpose was to restore the river and the habitats along its banks, particularly in the Grand Canyon. The reservoir behind the dam contains about 1.2 trillion (1,200,000,000,000) cubic feet of water. The release from the dam lasted a week at a rate of 25,800 cubic feet of water per second. About how much water was released during the 1-week flood? What fraction of the total water in the reservoir was released during the flood?

85. **Home Project.** Assume you are building a simple shed and doing some landscaping. The shed will be the shape

of a box that is 10 feet long, 10 feet wide, and 8 feet tall. The area you are landscaping is 75 feet long and 40 feet wide. For each of the following questions, find the price of the needed item at a local store.

a. How much would it cost to plant the entire landscaping area with grass seed? (Hint: Grass seed is usually rated by how many square feet can be covered by each pound of seed.)

b. Suppose you decide to cover the landscaping area with sod (i.e., rolls of pre-grown grass). How much will the sod cost?

c. You decide that before any planting, you need to add 8 inches of high-quality top soil to the entire landscape area. How much will the soil cost?

d. How much will it cost to paint the exterior walls of the shed? (Hint: Paint cans generally tell you how much area can be covered with the paint they contain.)

Web Projects

Find useful links for Web Projects on the text Web site: www.aw.com/bennett-briggs

86. South American Adventure. Suppose you are planning an extended trip through many countries in South America. Use one of the many currency exchange sites on the Web to get all the exchange rates you'll need. Make a brief table showing each currency you'll need and the value of each currency in dollars.

87. Polar Ice Melting. Starting with a search on "glaciers" or "glaciology" (the study of glaciers), use the Web to learn more about polar ice melting. Focus on one aspect of the issue, such as whether global warming is causing melting, or the environmental impacts of melting, or the geological history of ice ages. Write a one-page summary of what you learn.

In the News

88. Are the Units Clear? Find a news story that involves numerical data. Are all the numbers in the story given with meaningful units, or is the meaning of some of the units unclear? Briefly summarize how well (or not well) the article uses units.

89. Units on the Highway. Next time you are on the highway, look for three signs that use numbers (such as speed limits or distances to nearest exits). Are the units of the numbers given? If not, how are you expected to know the units? In cases where the units are not given, do you think the units would be obvious to everyone? Why or why not?

90. False Advertising? A Goodyear tire commercial began by stating that the Goodyear Aquatread tire can "channel away" 1 gallon of water per second. The announcer then said: "One gallon per second—that's 396 gallons per mile." What's wrong with this statement? What point do you think the advertisement was trying to make? Can you find other examples of advertisements that misstate units?

UNIT 2B

Thumb-width
(inch)

foot pace
(foot) (1000 paces = 1 mile)

STANDARDIZED UNITS:
MORE PROBLEM-SOLVING POWER

We've seen that working with units can be a powerful problem-solving technique—but only if the units are clear and meaningful. When units (such as feet, miles, or kilograms) are *standardized*, everyone agrees on their meaning. We will now review the two systems of standardized units in common use—the U.S. system and the international metric system—and how to convert between them. We'll also explore standardized units for temperature, energy, density, and concentration, which will allow us to study a much broader range of problems.

STANDARDIZED UNIT SYSTEMS: U.S. AND METRIC

Today, only one system of standardized units enjoys wide international use: the international metric system, also known as SI (from the French *Système International d'Unités*). However, the United States still uses an ancient measurement system sometimes called the English system, but more formally known as the *U.S. customary system* (USCS). Let's briefly review each system.

The U.S. Customary System

The roots of U.S. customary units reach back to ancient Middle Eastern civilizations, including the Egyptians, Sumerians, Babylonians, and Hebrews. Units of length come from ancient units based on body parts. For example, the *foot* was once the length of the foot of whoever was doing the measuring. The Romans discovered that most adult feet are about as long as 12 thumb-widths. Our word *inch* comes from the Latin *uncia*, meaning "thumb-width." The Romans paced out longer distances. Our word *mile* comes from the Latin *milia passum*, meaning "one thousand paces."

Time out to think

How many of *your* thumb-widths fit along your bare foot? Were the Romans correct in believing there are 12 thumb-widths in a foot?

As you might imagine, lengths that vary from person to person can lead to difficulties. For example, if you are buying 10 feet of rope, is it the length of your foot or the seller's foot that should be used to measure the rope? Many cultures therefore standardized units according to body measures of one particular person, such as a king. But such measures still changed every time one king was replaced by another. The first permanent standardization began when English King Henry I (1100–1135) set the *yard* as the measurement from the tip of *his* nose to the tip of *his* thumb on *his* outstretched arm. A few centuries later, the English brought their units to the American colonies, and they have since undergone further modification and standardization. Today, all lengths in the U.S. customary system are based on the inch, which is defined by law to be exactly 2.54 centimeters. Table 2.2 summarizes units of length in the U.S. customary system.

TABLE 2.2	USCS Lengths (common abbreviations in parentheses)		
1 inch (in.) = 2.54 centimeters		1 furlong	= 40 rods = $\frac{1}{8}$ mile
1 foot (ft) = 12 inches		1 mile (mi)	= 1760 yards = 5280 feet
1 yard (yd) = 3 feet		1 nautical mile	= 1.852 km ≈ 6076.1 feet
1 rod = 5.5 yards		1 land league	= 3 miles
1 fathom = 6 feet		1 marine league	= 3 nautical miles

For measuring weight, the U.S. customary system actually has three distinct sets of units. Traditionally, jewelers used *troy* measures of weight, pharmacists used *apothecary* measures of weight, and most other commerce was conducted with *avoirdupois* measures of weight. Table 2.3 summarizes all three systems, but only

HISTORICAL NOTE

The *avoirdupois* measures were standardized by London merchants around the year 1300. The term *avoirdupois* comes from the French for "goods of weight." The *troy* system was probably named for the town of Troyes, France, and was established in the 15th century. The apothecary system was developed by medieval pharmacists (*apothecary* is a synonym for *pharmacy*).

TABLE 2.3	USCS Weights (common abbreviations in parentheses)				
AVOIRDUPOIS MEASURES		**TROY MEASURES**		**APOTHECARY MEASURES**	
1 grain	= 0.0648 gram	1 grain	= 0.0648 gram	1 grain	= 0.0648 gram
1 ounce (oz)	= 437.5 grains	1 carat = 0.2 gram = 3.086 grains		1 scruple	= 20 grains
1 pound (lb)	= 16 ounces	1 pennyweight	= 24 grains	1 dram	= 3 scruples
1 ton	= 2000 pounds	1 troy ounce	= 480 grains	1 apoth. ounce	= 8 drams
1 long ton	= 2240 pounds	1 troy pound	= 12 troy ounces	1 apoth. pound	= 12 ounces

the avoirdupois measures are commonly used today. The basic unit of weight in all three sets is the *grain,* an ancient unit originally based on the weight of a typical grain of wheat.

The basic U.S. customary unit of volume is the cubic inch (in.3), but from here the system becomes complicated, because units with the same name represent different volumes depending on whether we're working with dry or liquid materials. For example, a *dry* pint is 33.60 cubic inches, while a *liquid* pint is only 28.88 cubic inches (Figure 2.5). Thus, a container that holds one pint of water is too small for one pint of flour. (If you travel, it can be even more complicated, because British customary measures differ from U.S. measures. For example, a British pint of beer is about 20% larger than a U.S. pint of beer.) Table 2.4 summarizes USCS measures of volume.

Flour Water

Volume of dry pint Volume of liquid pint
= 33.60 in.3 = 28.88 in.3

FIGURE 2.5 In the U.S. customary system, a dry pint (such as a pint of flour) contains a greater volume than a liquid pint (such as a pint of water).

TABLE 2.4	USCS Volumes (common abbreviations in parentheses)		
LIQUID MEASURES		**DRY MEASURES**	
1 tablespoon (tbsp or T)	= 3 teaspoons (tsp or t)	1 in.3	≈ 16.387 cm^3
1 fluid ounce (fl oz)	= 2 tablespoons = 1.805 in.3	1 ft^3	= 1728 in.3 = 7.48 gallons
1 cup (c)	= 8 fluid ounces	1 yd^3	= 27 ft^3
1 pint (pt)	= 16 fluid ounces = 28.88 in.3	1 dry pint (pt)	= 33.60 in.3
1 quart (qt)	= 2 pints = 57.75 in.3	1 dry quart (qt)	= 2 dry pints = 67.2 in.3
1 gallon (gal)	= 4 quarts	1 peck	= 8 dry quarts
1 barrel of petroleum	= 42 gallons	1 bushel	= 4 pecks
1 barrel of liquid	= 31 gallons	1 cord	= 128 ft^3

Practical matters

GEMS AND GOLD JEWELRY

If you've ever bought jewelry, you've probably seen labels stating *karats* or *carats*. But you may be surprised to know that these are not just alternative spellings of the same word. A *carat* is a unit of weight defined to be exactly 0.2 gram (see Table 2.3). A *karat* is a measure of the purity of gold: 24-karat gold is 100% pure, 18-karat gold is 75% pure, 14-karat gold is 58% pure (because $\frac{14}{24} \approx 0.58$), and so on.

If you are buying gold jewelry, look for a label stating both its purity in karats and its weight (usually given in grams); the amount of gold you are actually buying depends on both. For example:

- Ten grams of 18-karat gold actually contain 7.5 grams of pure gold, since 18-karat gold is 75% pure.
- Ten grams of 14-karat gold actually contain 5.8 grams of pure gold, since 14-karat gold is 58% pure.

Although a higher karat value is purer, it comes with an important practical tradeoff. Gold is a soft metal, and the metals (usually silver and copper) mixed into lower-karat gold give it added strength. Thus, for example, a 14-karat gold ring is stronger and more durable than an 18-karat gold ring.

You'll deal with carats rather than karats if you are buying gems such as diamonds or emeralds. Because 1 carat is 0.2 gram, the number of carats tells you the precise weight of the gem. However, several factors besides weight are important to the price, such as the gem's shape and color.

If you are gem shopping, you'll need to make trade-offs among weight, color, and clarity (and sometimes other factors as well) to stay within your budget. For example, if you are looking for a round diamond, $10,000 might buy either a 1-carat diamond with good color and clarity or a 2-carat diamond with poorer color and clarity. Given the cost of most gems, you should spend some time doing price comparisons before buying.

▶ **EXAMPLE 1** The Kentucky Derby

In the Kentucky Derby, horses race a distance of 10 furlongs. How many miles is the race?

Solution From Table 2.2, 1 furlong $= \frac{1}{8}$ mi, which is the same as 0.125 mile. We can write this conversion factor in two other equivalent forms:

$$\frac{1 \text{ furlong}}{0.125 \text{ mi}} = 1 \qquad \text{or} \qquad \frac{0.125 \text{ mi}}{1 \text{ furlong}} = 1$$

The second form allows us to convert furlongs to miles:

$$10 \text{ furlong} \times \frac{0.125 \text{ mi}}{1 \text{ furlong}} = 1.25 \text{ mi}$$

The Kentucky Derby is a race of 1.25 miles. *Now try Exercises 35–38.* ◀

▶ **EXAMPLE 2** 20,000 Leagues Under the Sea

In Jules Verne's novel *20,000 Leagues Under the Sea*, does the title refer to an ocean depth? How do you know?

Solution From Table 2.2, 1 marine league = 3 nautical miles. Thus, 20,000 leagues is

$$20{,}000 \text{ leagues} \times \frac{3 \text{ naut. mi}}{1 \text{ league}} = 60{,}000 \text{ naut. mi}$$

Because the oceans are at most a few miles deep, 20,000 leagues cannot possibly refer to an ocean depth. The book's title refers to the distance traveled by submarine under the sea. *Now try Exercises 39–42.* ◀

The International Metric System

The international metric system was invented in France late in the 18th century and is now used throughout the world. The metric system was designed for two primary reasons: (1) to replace many customary units with just a few basic units and (2) to simplify conversions through use of a decimal (base 10) system. Similar ideas had already been applied to currency. For example, in 1785, the United States replaced the customary currencies of the individual states with a national currency, the dollar (devised by Thomas Jefferson). Like the metric system, the dollar is based on a decimal relationship (because 100¢ = $1).

The basic units of length, mass, time, and volume in the metric system are

- the **meter** for length, abbreviated m
- the **kilogram** for mass, abbreviated kg
- the **second** for time, abbreviated s
- the **liter** for volume, abbreviated L

These basic units can be combined with a prefix that indicates multiplication by a power of 10. For example, *kilo* means 1000 so a kilometer is 1000 meters, and *micro* means *one-millionth* so a microgram is a millionth of a gram. Table 2.5 lists common metric prefixes.

TABLE 2.5	Metric Prefixes					
	SMALL VALUES			LARGE VALUES		
Prefix	Abbrev.	Value		Prefix	Abbrev.	Value
deci	d	10^{-1} (one-tenth)		deca	da	10^{1} (ten)
centi	c	10^{-2} (one-hundredth)		hecto	h	10^{2} (hundred)
milli	m	10^{-3} (one-thousandth)		kilo	k	10^{3} (thousand)
micro	μ	10^{-6} (one-millionth)		mega	M	10^{6} (million)
nano	n	10^{-9} (one-billionth)		giga	G	10^{9} (billion)
pico	p	10^{-12} (one-trillionth)		tera	T	10^{12} (trillion)

NOTE: The abbreviation for micro, μ, is the Greek letter mu.

Time out to think _____

People often say that expensive things cost "megabucks." What does this statement mean literally? Can you think of other cases where metric prefixes have entered popular language?

 A brief review . . .

POWERS OF 10

Powers of 10 indicate how many times to multiply 10 by itself. For example:

$$10^2 = 10 \times 10 = 100$$
$$10^6 = 10 \times 10 \times 10 \times 10 \times 10 \times 10 = 1,000,000$$

Negative powers indicate reciprocals of corresponding positive powers. For example:

$$10^{-2} = \frac{1}{10^2} = \frac{1}{100} = 0.01$$
$$10^{-6} = \frac{1}{10^6} = \frac{1}{1,000,000} = 0.000001$$

Thus, powers of 10 follow two basic rules:

1. A positive exponent tells how many 0s follow the 1. For example, 10^0 is a 1 followed by no 0s; 10^8 is a 1 followed by eight 0s.
2. A negative exponent tells how many places are to the right of the decimal point, including the 1. For example, $10^{-1} = 0.1$ has one place to the right of the decimal point; $10^{-6} = 0.000001$ has six places to the right of the decimal point.

Multiplying and Dividing Powers of 10

Multiplying powers of 10 simply requires adding exponents. For example:

$$10^4 \times 10^7 = \underbrace{10,000}_{10^4} \times \underbrace{10,000,000}_{10^7} = \underbrace{100,000,000,000}_{= 10^{4+7} = 10^{11}}$$

$$10^5 \times 10^{-3} = \underbrace{100,000}_{10^5} \times \underbrace{0.001}_{10^{-3}} = \underbrace{100}_{= 10^{5+(-3)} = 10^2}$$

$$10^{-8} \times 10^{-5} = \underbrace{0.00000001}_{10^{-8}} \times \underbrace{0.00001}_{10^{-5}} = \underbrace{0.0000000000001}_{= 10^{-8+(-5)} = 10^{-13}}$$

Dividing powers of 10 requires subtracting exponents. For example:

$$\frac{10^5}{10^3} = \underbrace{100,000}_{10^5} \div \underbrace{1000}_{10^3} = \underbrace{100}_{= 10^{5-3} = 10^2}$$

$$\frac{10^3}{10^7} = \underbrace{1000}_{10^3} \div \underbrace{10,000,000}_{10^7} = \underbrace{0.0001}_{= 10^{3-7} = 10^{-4}}$$

$$\frac{10^{-4}}{10^{-6}} = \underbrace{0.0001}_{10^{-4}} \div \underbrace{0.000001}_{10^{-6}} = \underbrace{100}_{= 10^{-4-(-6)} = 10^2}$$

Powers of Powers of 10

We can use the multiplication and division rules to raise powers of 10 to other powers. For example:

$$(10^4)^3 = 10^4 \times 10^4 \times 10^4 = 10^{4+4+4} = 10^{12}$$

Note that we get the same result by simply multiplying the two powers:

$$(10^4)^3 = 10^{4 \times 3} = 10^{12}$$

Adding and Subtracting Powers of 10

There is no shortcut for adding or subtracting powers of 10, as there is for multiplication or division. The values must be written in longhand notation. For example:

$$10^6 + 10^2 = 1,000,000 + 100 = 1,000,100$$
$$10^8 + 10^{-3} = 100,000,000 + 0.001 = 100,000,000.001$$
$$10^7 - 10^3 = 10,000,000 - 1000 = 9,999,000$$

Summary

We summarize our findings as follows, using n and m to represent any numbers:

- To *multiply* powers of 10, *add* exponents:
$$10^n \times 10^m = 10^{n+m}$$

- To *divide* powers of 10, *subtract* exponents:
$$\frac{10^n}{10^m} = 10^{n-m}$$

- To *raise* powers of 10 to other powers, multiply exponents:
$$(10^n)^m = 10^{n \times m}$$

Now try Exercises 19–34.

▶ EXAMPLE 3 Using Metric Prefixes

a. Convert 2759 centimeters to meters.

b. How many nanoseconds are in a microsecond?

Solution

a. Table 2.5 shows that *centi* means 10^{-2}, so 1 cm $= 10^{-2}$ m or, equivalently, 1 m $= 100$ cm. Thus, 2759 centimeters is the same as

$$2759 \ \cancel{\text{cm}} \times \frac{1 \text{ m}}{100 \ \cancel{\text{cm}}} = 27.59 \text{ m}$$

b. We compare the quantities by dividing the longer time (microsecond) by the shorter time (nanosecond):

$$\frac{1 \ \mu\text{s}}{1 \text{ ns}} = \frac{10^{-6} \ \cancel{s}}{10^{-9} \ \cancel{s}} = 10^{-6-(-9)} = 10^{-6+9} = 10^{3}$$

A microsecond is 10^{3}, or 1000, times as long as a nanosecond, so there are 1000 nanoseconds in a microsecond. *Now try Exercises 43–48.* ◀

Metric–USCS Conversions

We carry out conversions between metric and USCS units like any other unit conversions. Table 2.6 lists a few handy conversion factors. It's useful to memorize approximate conversions, particularly if you plan to travel internationally or if you work with metric units in sports or business. For example, if you remember that a kilometer is about 0.6 mile, you will know that a 10-kilometer road race is about 6 miles. Similarly, if you remember that a meter is about 10% longer than a yard, you'll know that a 100-meter race is about the same as a 110-yard race.

TABLE 2.6	USCS–Metric Conversions
USCS TO METRIC	**METRIC TO USCS**
1 in. = 2.540 cm	1 cm = 0.3937 in.
1 ft = 0.3048 m	1 m = 3.28 ft
1 yd = 0.9144 m	1 m = 1.094 yd
1 mi = 1.6093 km	1 km = 0.6214 mi
1 lb = 0.4536 kg	1 kg = 2.205 lb
1 fl oz = 29.574 mL	1 mL = 0.03381 fl oz
1 qt = 0.9464 L	1 L = 1.057 qt
1 gal = 3.785 L	1 L = 0.2642 gal

▶ EXAMPLE 4 Gas Price per Liter

At a gas station in Mexico, the price of gasoline is 8 pesos per liter. What is the price in dollars per gallon? Use the currency exchange rate in Table 2.1.

Solution We use a chain of conversions to convert from pesos to dollars and then from liters to gallons. From Table 2.1, the currency conversion is $0.0943 per peso.

From Table 2.6, there are 3.785 liters per gallon.

$$\frac{8 \text{ peso}}{1 \text{ L}} \times \frac{\$0.0943}{1 \text{ peso}} \times \frac{3.785 \text{ L}}{1 \text{ gal}} \approx \frac{\$2.86}{1 \text{ gal}}$$

The price of the gasoline is about $2.86 per gallon. *Now try Exercises 49–54.* ◄

►**EXAMPLE 5** Square Kilometers to Square Miles

How many square kilometers are in one square mile?

Solution We square both sides of the conversion factor, 1 mi = 1.6093 km.

$$(1 \text{ mi})^2 = (1.6093 \text{ km})^2 \quad \Rightarrow \quad 1 \text{ mi}^2 = 2.5898 \text{ km}^2$$

One square mile is 2.5898 square kilometers. *Now try Exercises 55–58.* ◄

Temperature Units: Fahrenheit, Celsius, and Kelvin

Three temperature scales are commonly used today (Figure 2.6). The **Fahrenheit** scale is used in the United States. It is defined so that water freezes at 32°F and boils at 212°F. Internationally, temperature is usually measured on the **Celsius** scale, which places the freezing point of water at 0°C and the boiling point at 100°C. In science, temperature is usually measured on the **Kelvin** scale, which is the same as the Celsius scale except for its zero point. A temperature of 0 K is the coldest possible temperature, known as **absolute zero.** It is approximately −273.15°C or −459.67°F. The following box summarizes conversions among the three scales.

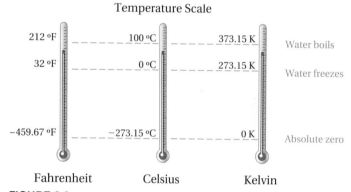

Temperature Scale

FIGURE 2.6

TEMPERATURE CONVERSIONS

The conversions are given both in words and with formulas in which C, F, and K are Celsius, Fahrenheit, and Kelvin temperatures, respectively.

To Convert from	Conversion in Words	Conversion Formula
Celsius to Fahrenheit	Multiply by 1.8 (or $\frac{9}{5}$). Then add 32.	$F = 1.8C + 32$
Fahrenheit to Celsius	Subtract 32. Then divide by 1.8 (or multiply by $\frac{5}{9}$).	$C = \dfrac{F - 32}{1.8}$
Celsius to Kelvin	Add 273.15.	$K = C + 273.15$
Kelvin to Celsius	Subtract 273.15.	$C = K - 273.15$

Where do these conversion formulas come from? It's easy to see their origin by studying Figure 2.6. For conversions between Kelvin and Celsius, note that the scales are the same except for the 273.15° difference in their zero points. Thus, we simply add 273.15 to a Celsius temperature to get a Kelvin temperature. For conversions between Celsius and Fahrenheit, look at the differences in temperature between the freezing and boiling points of water. The Celsius scale has 100°C between these points, while the Fahrenheit scale has 212°F − 32°F = 180°F between them. Thus, each Celsius degree represents $\frac{180}{100} = 1.8$ Fahrenheit degrees, which explains the factor of 1.8 (or $\frac{9}{5}$) in the conversions. The 32 appears in the Celsius–Fahrenheit conversions to account for the difference in their zero points.

▶ **EXAMPLE 6** Human Body Temperature

Average human body temperature is 98.6°F. What is it in Celsius and Kelvin?

Solution We convert from Fahrenheit to Celsius by subtracting 32 and then dividing by 1.8:

$$C = \frac{F - 32}{1.8} = \frac{98.6 - 32}{1.8} = \frac{66.6}{1.8} = 37.0$$

We find the Kelvin equivalent by adding 273.15 to the Celsius temperature:

$$K = C + 273.15 = 37 + 273.15 = 310.15 \text{ K}$$

Human body temperature is 37°C or 310.15 K. *Now try Exercises 59–62.* ◀

Time out to think
The local weather report says that tomorrow's temperature will be 59°, but does not specify whether it is in Celsius or Fahrenheit. Can you tell which it is? How?

UNITS OF ENERGY AND POWER

We pay energy bills to power companies, we use energy from gasoline to run our cars, and we argue about whether nuclear energy is a sensible alternative to fossil fuels. But what *is* energy?

Broadly speaking, energy is what makes matter move or heat up. We need energy from food to keep our hearts beating, to maintain our body temperatures, and to walk or run. A car needs energy to move the pistons in its engine, which turn the wheels. A light bulb needs energy to generate the heat that makes its filament glow.

For Americans, the most familiar energy unit is the food *Calorie* (uppercase C) used to measure the energy our bodies can draw from food. A typical adult uses about 2500 Calories of energy each day. The international metric unit of energy is the **joule.** Outside the United States, most foods and beverages are labeled with their energy content in joules (or kilojoules) rather than Calories. One Calorie is equivalent to 4184 joules.

The words *energy* and *power* are often used together, but they are not quite the same. Power is the *rate* at which energy is used, which means it has units of energy divided by time. The most common unit of power is the **watt,** defined as 1 joule per second.

Energy and Power

Energy is what makes matter move or heat up. The international metric unit of energy is the **joule.**

Power is the *rate* at which energy is used. The international metric unit of energy is the **watt,** defined as

$$1 \text{ watt} = 1\frac{\text{joule}}{\text{s}}$$

➤**EXAMPLE 7** Pedal Power

You are riding an exercise bicycle at a fitness center. The readout states that you are using 500 Calories per hour. Are you generating enough power to light a 100-watt bulb (1 Calorie = 4184 joules)?

Solution We use a chain of conversions to go from Calories per hour to joules per second:

$$\frac{500 \text{ Cal}}{1 \text{ hr}} \times \frac{4184 \text{ joule}}{1 \text{ Cal}} \times \frac{1 \text{ hr}}{60 \text{ min}} \times \frac{1 \text{ min}}{60 \text{ s}} \approx 581\frac{\text{joule}}{\text{s}}$$

Your pedaling generates energy at a rate of 581 joules per second, which is a power of 581 watts—enough to light five (almost six) 100-watt bulbs.

Now try Exercises 63–64. ◄

Electric Utility Bills

If you check an electric utility bill, you'll probably find that it refers to electric usage in units of **kilowatt-hours.** The hyphen in this unit tells us that it means 1 kilowatt × 1 hour, which is equivalent to 1000 watts × 1 hour. Replacing 1000 watts with 1000 joule/s, we find

$$1 \text{ kilowatt-hour} = \frac{1000 \text{ joule}}{1 \text{ s}} \times 1 \text{ hr} \times \frac{60 \text{ min}}{1 \text{ hr}} \times \frac{60 \text{ s}}{1 \text{ min}} = 3{,}600{,}000 \text{ joule}$$

DEFINITION

A **kilowatt-hour** is a unit of energy:

1 kilowatt-hour = 3.6 million joules

➤**EXAMPLE 8** Operating Cost of a Light Bulb

Your utility company charges 8¢ per kilowatt-hour of electricity. How much does it cost to keep a 100-watt light bulb on for a week?

by the way

If you purchase a gas appliance, such as a gas stove or a kerosene heater, its energy requirements may be labeled in *British thermal units,* or BTUs. One BTU is equivalent to 1055 joules.

Solution One watt is one joule per second, so a 100-watt light bulb consumes energy at a rate of 100 joules per second. The energy used by the bulb in a week is 100 joule/s × 1 week. Using a chain of conversions to convert a week to seconds, we have

$$\frac{100 \text{ joule}}{1 \text{ s}} \times 1 \text{ week} \times \frac{7 \text{ day}}{1 \text{ week}} \times \frac{24 \text{ hr}}{1 \text{ day}} \times \frac{60 \text{ min}}{1 \text{ hr}} \times \frac{60 \text{ s}}{1 \text{ min}} = 60{,}480{,}000 \text{ joule}$$

Next, we convert this result to kilowatt-hours, remembering that 1 kilowatt-hour is 3,600,000 joules:

$$60{,}480{,}000 \text{ joule} \times \frac{1 \text{ kilowatt-hour}}{3{,}600{,}000 \text{ joule}} = 16.8 \text{ kilowatt-hour}$$

The light bulb uses 16.8 kilowatt-hours of energy in a week. At a price of 8¢ per kilowatt-hour, the total cost is

$$16.8 \text{ kilowatt-hr} \times 8 \frac{\text{¢}}{\text{kilowatt-hr}} = 134.4\text{¢} \approx \$1.34$$

The electricity to keep the 100-watt light bulb on all week costs about $1.34 (assuming a price of 8¢ per kilowatt-hour). *Now try Exercises 65–66.* ◄

Time out to think

Check a utility bill (yours or a friend's). Is the electricity usage metered in units of kilowatt-hours (often abbreviated kWh)? If not, what units are used? If so, what is the price per kilowatt-hour?

Practical matters

SAVE MONEY AND SAVE THE EARTH

You can save both money and energy by replacing standard (incandescent) light bulbs with more efficient bulbs, such as compact fluorescent bulbs. The compact fluorescent bulbs save energy because virtually all the energy they use goes into making light, whereas most of the energy used by standard bulbs goes into heat (that's why standard bulbs get so hot when they are turned on). As a result, a compact fluorescent bulb needs much less energy to generate the same light as a standard bulb. For example, a typical 25-watt compact fluorescent produces the same light as a 100-watt standard bulb.

To see how much you can save, consider a 100-watt light bulb that is left on all the time. At an energy cost of 8¢ per kilowatt-hour, this light bulb costs $1.34 to operate for a week (see Example 8), which trans-

lates into about $70 per year. Replacing the 100-watt bulb with a 25-watt compact fluorescent means your annual energy cost will be only one-fourth (25%) as much, or $17.50. Thus, the light bulb replacement saves some $52.50 per year in energy costs while providing just as much light. This savings more than makes up for the fact that compact fluorescent bulbs are more expensive than standard bulbs. (Compact fluorescent bulbs also last typically ten times longer than standard bulbs.)

Of course, you probably don't (or shouldn't) leave 100-watt light bulbs on all the time, so your savings for a single light bulb will be less than calculated above. Nevertheless, your savings may still be quite substantial. And, while you save money by switching to more efficient bulbs, you also reduce total energy usage, which helps protect the environment.

UNITS OF DENSITY AND CONCENTRATION

You'll encounter many other standardized units in everyday life. In most cases, these units will be variations on units with which you are already familiar. Thus, you should be able to make sense of units from their context. Units that describe various types of density or concentration are particularly common.

Density describes compactness or crowding. Here are a few of the many ways that the idea of density is used:

- *Material density* is given in units of mass per unit volume, such as grams per cubic centimeter (g/cm^3). A useful reference is the density of water—about $1\ g/cm^3$. Objects with densities less than $1\ g/cm^3$ float in water, while higher-density objects sink.
- *Population density* is given by the number of people per unit area. For example, if 750 people live in a square region that is 1 mile on a side, the population density of the area is 750 people/mi^2.
- *Information density* is often used to describe how much memory can be stored by digital media. For example, each square inch on the surface of a DVD holds about 100 megabytes of information, so we say that a DVD has an information density of 100 mB/in^2.

Time out to think

Use the concept of density to explain why you float better in a swimming pool when your lungs are filled with air than when you fully exhale.

Concentration describes the amount of one substance mixed with another. Here are two of the many ways that the idea of concentration is used:

- The *concentration of an air pollutant* is often measured by the number of molecules of the pollutant per million molecules of air. For example, if there are 12 molecules of carbon monoxide each 1 million molecules of air, we state the carbon monoxide concentration as 12 parts per million (ppm). (The U.S. Environmental Protection Agency says that air is unhealthy if the carbon monoxide concentration is above 9 ppm.)
- **Blood alcohol content** (BAC) describes the concentration of alcohol in a person's body. It is usually measured in units of grams of alcohol per 100 milliliters of blood. For example, in most states a person is considered legally intoxicated if his or her blood alcohol content is at or above 0.08 gram of alcohol per 100 milliliters of blood (written as 0.08 g/100 mL).

▶**EXAMPLE 9** *New York City*

Manhattan Island has a population of about 1.5 million people living in an area of about 57 square kilometers. What is its population density? If there were no high-rise apartments, how much space would be available per person?

Solution We divide the population by the area to find that

$$\text{population density} = \frac{1{,}500{,}000\ \text{people}}{57\ km^2} \approx 26{,}000\ \frac{\text{people}}{km^2}$$

Manhattan's population density is about 26,000 people per square kilometer. If there were no high rises, each resident would have 1/26,000 square kilometer of land. This number is easier to interpret if we convert from square kilometers to square meters:

$$\frac{1 \text{ km}^2}{26{,}000 \text{ people}} \times \left(\frac{1000 \text{ m}}{1 \text{ km}}\right)^2 = \frac{1 \text{ km}^2}{26{,}000 \text{ people}} \times \frac{1{,}000{,}000 \text{ m}^2}{1 \text{ km}^2} \approx 38 \frac{\text{m}^2}{\text{person}}$$

Without high rises, each person would have only 38 square meters, equivalent to a room about 6 meters, or 20 feet, on a side—and this does not include any space for roads, schools, or other common properties. Clearly, Manhattan Island could not fit so many residents without high rises. *Now try Exercises 67–72.* ◄

▶**EXAMPLE 10** Blood Alcohol Content

An average-sized man has about 5 liters (5000 milliliters) of blood, and an average 12-ounce can of beer contains about 15 grams of alcohol (assuming the beer is about 6% alcohol by volume). If all the alcohol were immediately absorbed into the bloodstream, what blood alcohol content would we find in an average-sized man who quickly drank a single can of beer?

Solution If the alcohol were absorbed immediately, the man's 5000 milliliters of blood would contain the entire 15 grams of alcohol from the beer. We express this concentration as

$$\frac{15 \text{ g alcohol}}{5000 \text{ mL blood}}$$

We can convert this concentration to the standard units for blood alcohol (g/100 mL) in two steps. First, we first carry out the division to find the concentration in grams per milliliter:

$$\frac{15 \text{ g}}{5000 \text{ mL}} = 0.003 \frac{\text{g}}{\text{mL}}$$

Now, we multiply both the top and bottom of the fraction by 100 to find the concentration in grams per 100 milliliters:

$$0.003 \frac{\text{g}}{\text{mL}} \times \frac{100}{100} = 0.3 \frac{\text{g}}{100 \text{ mL}}$$

The man's blood alcohol concentration would be 0.3 gram per 100 milliliters of blood—almost four times the legal limit of 0.08 g/mL! In reality, the man's blood alcohol content won't get this high from a single beer because it takes some time for the alcohol to be absorbed into the bloodstream and because metabolic processes gradually eliminate the absorbed alcohol. (For a typical adult, alcohol is absorbed into the bloodstream over a period of between 30 minutes and a couple of hours, depending on how much food the person has eaten. It is eliminated from the bloodstream at a rate of about 10 to 15 grams per hour.) Nevertheless, a single beer is certainly enough to cause impaired brain function—making it unsafe to drive—and this example points out how quickly and easily a person can become dangerously intoxicated. *Now try Exercises 73–74.* ◄

by the way

According to the American Medical Association, brain function is impaired when the blood alcohol content reaches 0.04 g/100 mL of blood—which is half the legal limit of 0.08 g/100 mL. A blood alcohol content at or above 0.4 g/100 mL usually leads to coma and sometimes leads to death.

Time out to think _____

Many college students have lost their lives by rapidly consuming several "shots" of strong alcoholic drinks. Explain why such rapid consumption of alcohol can lead to death, even when the total amount of alcohol consumed may not sound like a lot.

■ EXERCISES 2B

Review Questions

1. Briefly describe the origin and use of common units in the U.S. customary system.

2. Briefly describe the origin and use of metric units. What are the basic metric units of length, mass, time, and volume? How are the metric prefixes used?

3. Using examples, show how to convert among the Fahrenheit, Celsius, and Kelvin temperature scales.

4. What is *energy?* List at least three common units of energy. Under what circumstances do the different units tend to be used?

5. What is the difference between *energy* and *power?* What are the standard units for power?

6. What do we mean by *density?* What do we mean by *concentration?* Describe common units of density and concentration, including blood alcohol content, with examples.

Does It Make Sense?

Decide whether each of the following statements makes sense (or is clearly true) or does not make sense (or is clearly false). Explain your reasoning. Hint: Be sure to consider whether the units are appropriate to the statement, as well as whether the stated amount makes any sense. For example, a statement that someone is 15 feet tall uses the units (feet) appropriately, but does not make sense because no one is that tall.

7. I drank 2 liters of water today.

8. I know a professional bicyclist who weighs 300 kilograms.

9. Today I drove along the interstate at 100 kilometers per hour.

10. I know someone who can run 35 liters per second.

11. A guy on our track team can high-jump 7 meters.

12. My friend ran 10,000 meters in less than an hour.

13. The book I sent you weighs 3 milligrams.

14. My car's gas tank holds 12 meters of gasoline.

15. My daily food intake gives me about 10 million joules of energy.

16. Our utility company charges 10¢ per watt for the electricity we use.

17. The beach ball we played with had a density of 10 grams per cubic centimeter.

18. I know someone who went into a coma after drinking only three shots of hard liquor.

Basic Skills & Concepts

Powers of 10. Do Exercises 19–34 using skills covered in the Brief Review on p. 105.

19. $10^6 \times 10^5$

20. $10^4 \times 10^{-3}$

21. $\dfrac{10^6}{10^5}$

22. $\dfrac{10^4}{10^{-3}}$

23. $10^{-2} \times 10^{-4}$

24. $\dfrac{10^{-7}}{10^{-9}}$

25. $10^6 + 10^5$

26. $10^7 - 10^2$

27. $10^{12} \times 10^9$

28. $10^6 \times 10^{-8}$

29. $\dfrac{10^{15}}{10^{12}}$

30. $\dfrac{10^{-7}}{10^4}$

31. $10^{-4} \times 10^{-6}$

32. $\dfrac{10^{-10}}{10^{-14}}$

33. $10^8 + 10^5$

34. $10^4 - 10^{-1}$

USCS Units. Exercises 35–42 involve conversions within the U.S. customary system.

35. What is your height in inches?

36. What is your weight in ounces (avoirdupois)? in tons?

37. A gallon of water weighs about 128 ounces. How many pounds is that?

38. A boat is moving at 30 knots (nautical miles per hour). What is its speed in miles per hour?

39. You have a $\frac{1}{2}$-gallon milk jug. How many liquid pints can it hold? How many dry pints can it hold?

40. Most soda cans contain 12 fluid ounces. How many cubic inches do they contain?

41. If 150 million bushels of wheat are traded on one day, how many cubic inches are traded? how many cubic feet?

42. A small city produces 500,000 cubic feet of garbage per week. If all of this garbage were stacked neatly (in a nice vertical pile) on a 100-yard by 60-yard football field, how high would the pile be (in feet)?

Metric Prefixes. For each pair of units in Exercises 43–48, state how much larger or smaller the first unit is than the second.

43. millimeter, meter

44. milliliter, deciliter

45. gram, kilogram

46. kilometer, micrometer

47. square millimeter, square kilometer

48. cubic meter, cubic centimeter

USCS–Metric Conversions. Convert each measurement in Exercises 49–58 to the units specified.

49. 10 meters to feet

50. 105 centimeters to yards

51. 880 yards to kilometers

52. 150 pounds to kilograms

53. 20 gallons to liters

54. 100 kilometers per hour to miles per hour

55. 5 milliliters to cubic inches

56. 5.5 grams per cubic centimeter to pounds per cubic foot

57. 1200 square feet to square meters

58. 25 miles per hour to kilometers per hour

Celsius-Fahrenheit Conversions. In Exercises 59–60, convert Celsius temperatures into Fahrenheit or Fahrenheit temperatures into Celsius.

59. **a.** 45°F **b.** 20°C **c.** −15°C **d.** −30°C
 e. 70°F

60. **a.** −8°C **b.** 15°F **c.** 15°C **d.** 75°F
 e. 20°F

Celsius-Kelvin Conversions. In Exercises 61–62, convert Celsius temperatures into Kelvin or Kelvin temperatures into Celsius.

61. **a.** 50 K **b.** 240 K **c.** 10°C

62. **a.** −40°C **b.** 400 K **c.** 125°C

63. **Basketball Power.** You burn 800 Calories while playing in a basketball game for an hour. What is your average power during the game, in watts? Is it enough to keep a 100-watt light bulb shining? Explain.

64. **Aerobics Power.** You burn 500 Calories while doing an aerobics class for 45 minutes. What is your average power during the class, in watts? Is it enough to keep a 100-watt light bulb shining? Explain.

Electric Bills. For the electric bills described in Exercises 65–66, do the following:

 a. Determine your total electrical energy use, in joules.

 b. Determine your average power use, in watts.

 c. Assuming the power company generated the energy by burning oil, calculate the amount of oil needed to provide the energy shown on your bill. Give your answer in both liters and gallons. (Hint: Burning 1 liter of oil releases 12 million joules of energy.)

65. Your electric bill states that you used 1250 kilowatt-hours of energy in June.

66. Your electric bill states that you used 970 kilowatt-hours of energy in September.

Densities. In Exercises 67–72, give the densities in appropriate units (population data from 2000 census).

67. A 40-gram pebble has a volume of 10 cubic centimeters. What is its density? Will it sink or float in water?

68. A jug has a total volume of 8 liters (which is 8000 cubic centimeters) and a mass of 6 kilograms. What is its density? Will it sink or float in water?

69. The land area of the United States is about 3.5 million square miles, and the population is about 280 million people. What is the average population density?

70. Find the population and the area of your home town. Calculate its population density.

71. New Jersey and Wyoming have areas of 7419 and 970,000 square miles, respectively, and populations of 8.4 million and 490,000, respectively. Calculate and compare their population densities.[7]

72. A new computer hard drive holds 250 gigabytes of information on a surface area of 40 square centimeters. Calculate the information density on the disk.

73. **Blood Alcohol Content: Wine.** A typical glass of wine contains about 20 grams of alcohol. Consider a 110-pound woman, with approximately 4 liters (4000 milliliters) of blood, who drinks two glasses of wine.

 a. If all the alcohol were immediately absorbed into her bloodstream, what would her blood alcohol content be? Explain why it is fortunate that, in reality, the alcohol is not absorbed immediately.

 b. Again assume that all the alcohol is absorbed immediately, but now assume that her body eliminates the alcohol (through metabolism) at a rate of 10 grams per hour. What is her blood alcohol content 3 hours after drinking the wine? Is it safe for her to drive at this time? Explain.

74. **Blood Alcohol Content: Hard Liquor.** Eight ounces of a hard liquor (such as whiskey) typically contain about 70 grams of alcohol. Consider a 200-pound man, with approximately 6 liters (6000 milliliters) of blood, who quickly drinks 8 ounces of hard liquor.

 a. If all the alcohol were immediately absorbed into his bloodstream, what would his blood alcohol content be? Explain why it is fortunate that, in reality, the alcohol is not absorbed immediately.

 b. Again assume that all the alcohol is absorbed immediately, but now assume that his body eliminates the alcohol (through metabolism) at a rate of 15 grams per hour. What is his blood alcohol content 4 hours after drinking the liquor? Is it safe for him to drive at this time? Explain.

Further Applications

75. **The Metric Mile.** In track and field, the 1500-meter race is sometimes called the *metric mile*.

 a. Compare the metric mile to a USCS mile. How much longer or shorter is it in terms of percentage?

 b. Look up the current men's and women's world records for the mile. If you assume that the runners maintain the same pace for the metric mile, what should their times be for the metric mile? Compare your predicted records at this pace to the actual current world records for the metric mile.

76. **Metric Tools.** Many tools come in both USCS and metric standards. In a standard socket set, the smallest USCS subdivision is $\frac{1}{16}$ inch; the smallest metric subdivision is 0.5 millimeter. Are the tools interchangeable? Explain.

77. **Tallest Mountain?** Mauna Kea, the highest mountain on the island of Hawaii, rises 13,796 feet above sea level. It extends an additional 18,200 feet from sea level to its base on the ocean floor. How tall is Mauna Kea from its base to its peak, in feet, miles, and kilometers? Compare its total extent to the height of Mt. Everest above sea level (29,023 ft). Would it be fair to call Mauna Kea the highest mountain in the world? Why or why not?

78. **The Cullinan Diamond and the Star of Africa.** The largest single rough diamond ever found, the Cullinan diamond, weighed 3106 carats. It was used to cut the world's largest diamond gem, the Star of Africa (530.2 carats), which is part of the British crown jewels collection. How much did the Cullinan diamond weigh in milligrams? in (avoirdupois) pounds? How much does the Star of Africa weigh in milligrams? in (avoirdupois) pounds?

Gems and Gold. Exercises 79–82 use karats and carats (see the Practical Matters box).

79. You find a nugget that is 25% gold. What is its purity in karats?

80. You purchase a 14-karat gold chain that weighs 15 grams. How much gold have you purchased (in grams)? Bonus: At the current price of gold, how much is the gold in the chain worth?

81. How much does a 2.5-carat diamond weigh in grams?

82. Is it possible to have jewelry made of 30-karat gold? Why or why not?

83. **Refrigerator Cost.** Your refrigerator uses an average power of 350 watts, and your utility company charges 8¢ per kilowatt-hour of energy. How much does it cost to run your refrigerator for a year? Explain.

84. **Hair Dryer Cost.** You have an 1800-watt hair dryer, which you use for an average of 10 minutes per day. Your utility company charges 9¢ per kilowatt-hour of energy. How much does it cost to run the hair dryer each day? each year?

85. **Compact Fluorescent Light Bulbs.** You replace a 100-watt standard light bulb with a 25-watt compact fluorescent bulb that supplies the same light. Over a 10,000-hour life (typical for a compact fluorescent bulb), how much energy, in kilowatt-hours, do you save? If electricity costs 8¢ per kilowatt-hour, how much money do you save?

86. **Human Wattage.** Suppose you require 2500 food Calories per day (which is about average).

 a. What is your average power, in watts? Compare your answer to the wattage of some familiar appliance.

 b. How much energy, in joules, do you require from food in a year? Counting all forms of energy (such as gasoline, electricity, and energy for heating), the average U.S. citizen consumes about 400 billion joules of energy each year. Compare this value to the energy needed from food alone.

87. **Coal Power Plant.** A new coal-burning power plant can generate 1 gigawatt (billion watts) of power. Burning 1 kilogram of coal yields about 450 kilowatt-hours of energy. How much energy, in kilowatt-hours, can the plant generate each month? How much coal, in kilograms, is needed by this power plant each month? If a typical home uses 1000 kilowatt-hours per month, how many homes can this power plant supply with energy?

88. **Nuclear Power Plant.** Operating at full capacity, the Fort St. Vrain Nuclear Power Station in Colorado can generate 330 megawatts of power. Nuclear fission of 1 kilogram of uranium (in the form of uranium–235) releases 16 million kilowatt-hours of energy. How much energy, in kilowatt-hours, can the plant generate each month? How much uranium, in kilograms, is needed by this power plant each month? If a typical home uses 1000 kilowatt-hours per month, how many homes can this power plant supply with energy?

Solar Energy. Use the following facts in Exercises 89–90: Solar (photovoltaic) cells convert sunlight directly into electricity. If solar cells were 100% efficient, they would generate about 1000 watts of power per square meter of surface area when exposed to direct sunlight. With lower efficiency, they generate proportionally less power. For example, 10% efficient cells generate 100 watts of power in direct sunlight.

89. Suppose a 1-square-meter panel of solar cells has an efficiency of 20% and receives the equivalent of 6 hours of direct sunlight per day. How much energy, in joules, can it produce each day? What average power, in watts, does the panel produce?

90. Suppose you want to supply 1 kilowatt of power to a house (the average household power requirement) by putting solar panels on its roof. For the solar cells described in Exercise 89, how many square meters of solar panels would you need? Assume that you can make use of the average power from the solar cells (by, for example, storing the energy in batteries until it is needed).

91. **Wind Power: One Turbine.** Modern wind energy "farms" use large wind turbines to generate electricity from the wind. At a typical installation, a single modern turbine can produce an average power of about 200 kilowatts. (This average takes wind variations into account.) How much energy, in kilowatt-hours, can such a turbine generate in a year? Given that the average household uses about 10,000 kilowatt-hours of energy each year, how many households can be powered by a single wind turbine?

92. **California Wind Power.** California currently has wind farms capable of generating a total of about 2000 megawatts (2 gigawatts) of power.

 a. How much energy, in kilowatt-hours, can these wind farms generate each year? Given that the average household uses about 10,000 kilowatt-hours of energy each year, how many households can be powered by these wind farms?

 b. One of the great advantages of wind power is that it does not produce the carbon dioxide emissions that contribute to global warming. On average, energy produced from fossil fuels generates about 1.5 pounds of carbon dioxide for every kilowatt-hour of energy. Suppose California did not have its wind farms and the energy were instead produced from fossil fuels. How much more carbon dioxide would be entering the atmosphere each year?

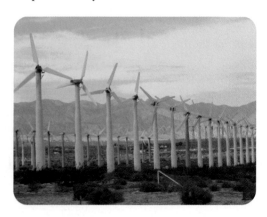

Currency Conversions. Find today's currency conversion rates, and use them to answer Exercises 93–98. Be sure to state clearly the currency conversion you use.

93. You see oranges in an Italian market priced at 0.60 euro per kilogram. What is the price in dollars per pound?

94. Gasoline at a German gas station costs 9.50 euros per liter. What is the price in dollars per gallon?

95. A French car manufacturer claims that its newest economy model has gas "mileage" of 15 kilometers per liter. What is the gas mileage in miles per gallon?

96. A supermarket in Mexico sells milk for 6.5 pesos per liter. What is the price in dollars per quart?

97. A fine ale in an English pub sells for 3.5 pounds (currency) per liter. What is the price in dollars per ounce?

98. A piece of land in the Belgian countryside with an area of 0.1 square kilometer is priced at 8000 euros. What is the price in dollars per acre?

99. **Project: Personal Energy Audit.** Do a thorough electrical energy audit of your home, apartment, or dormitory. That is, determine the energy used in a typical month by each of your electrical appliances, and your total energy usage per month for electricity. Based on the cost of electricity in your area, calculate the average monthly cost for your electricity. After studying your findings, propose at least three energy conservation strategies that you would consider implementing to save money and energy. Discuss the pros and cons of each strategy.

Web Projects

Find useful links for Web Projects on the text Web site: www.aw.com/bennett-briggs

100. **Metric History.** Research some aspect of the history of the metric system, such as how it came to be adopted around the world or how various units have been scientifically defined. Write a short report on your findings.

101. **Energy Issues.** The question of how we will continue to meet our energy needs is one of the most important issues of our time. Research one aspect of this issue using information available at the Web site for the U.S. Energy Information Administration. Write a short report on your findings.

102. **Pollution Progress.** Investigate the average concentrations of various pollutants in a major city of your choice.

Find the EPA standards for each pollutant, and find some of the hazards associated with exposure to each pollutant. Track how the levels of pollution in this city have changed over the past 20 years. Based on your findings, do you think it is likely that pollution in this city will get better or worse over the next decade? Summarize your findings and your conclusions in a one- or two-page report.

103. **Alcohol Poisoning.** Research some aspect of the dangers of alcohol, such as drunk driving or alcohol poisoning (in which a person dies from drinking too much too fast). Find statistics related to this issue, especially data that relate the blood alcohol content to dangers. Summarize your findings in a short report about how society might combat the danger.

104. **Wind Power.** Learn more about both the currently installed wind power and the future potential for wind power in your state. Overall, do you expect wind power to be important in the future in your state? Why or why not?

In the News

105. **Everyday Metric.** Describe three ways that you use metric units in your everyday life.

106. **Should the United States Go Metric?** Discuss the pros and cons of having the United States switch fully to the metric system. Do you think it will ever happen?

107. **Energy.** Look for a news article concerning energy or power. What units are used to describe energy or power? Summarize the article and explain the meaning of the units.

108. **Density and Concentration.** Look for a news article that uses units of density or concentration in any context. Summarize the article and explain the meaning of the units.

109. **Utility Bill.** Analyze a utility bill. Explain all the units shown, and determine the relative costs of different energy uses. What changes would you recommend if the recipient of the bill wanted to lower energy costs?

110. **Pollution Problems.** What (if any) pollutants are of concern in the area where you live? What units are used to describe their concentrations? At what concentrations are the pollutants considered dangerous? Suggest ways that these pollution problems might be addressed.

PROBLEM-SOLVING GUIDELINES AND HINTS

So far in this chapter, we've focused on how units help us solve problems. But if you look back through the examples we've studied, you'll see that we've actually done much more. We studied each problem in order to find a good strategy for solving it, and we carefully interpreted our results to make sure they had the correct units and made sense. In a few cases, we used pictures or diagrams to help us understand a problem. You may have noticed that while every problem is different, the *process* of problem solving always has a few common features.

The four-step process on the next page provides a useful guide to problem solving. Study it carefully and look back to see how these steps were applied in earlier examples. Keep in mind that the four steps offer *general* advice so they will not automatically lead to a solution. In this sense, problem solving is more of an art than a science, and the four steps are more like instructions for painting than a recipe for the Mona Lisa. Note also that while these steps apply to virtually all problem solving, it is not necessary (or particularly useful) to write them out step by step in solutions.

Time out to think

Choose two examples from Units 2A and 2B, and identify how each of the four steps was carried out in those examples. More generally, explain how these four steps are implicit in the solutions to the examples we've studied, even though we have not written them out in a step-by-step list.

Using the four-step process will help you become better organized. But the only sure way to become more creative and improve your problem solving is through practice. We now offer several hints about problem solving, each illustrated with an example. The hints themselves relate more to *attitudes* about problem solving than to actual techniques. Indeed, their primary purpose is to help you develop a "mind set" that is conducive to enjoyable and successful problem solving.

"I THINK YOU SHOULD BE MORE EXPLICIT HERE IN STEP TWO."

Copyright © 1998, 2004 by Sidney Harris.

HINT 1: THERE MAY BE MORE THAN ONE ANSWER

How can society best reduce the total amount of greenhouse gases emitted into the atmosphere? We won't even attempt to answer this question, but it should make the point that no *single* best answer may be available. Indeed, many different political and economic strategies could yield similar reductions in greenhouse gas emissions.

Most people recognize that policy questions do not have unique answers, but the same is true of many mathematical problems. For example, both $x = 4$ and $x = -4$ are solutions to the equation $x^2 = 16$. Without further information and context, we have no way to determine whether both solutions are valid for a particular problem. Nonunique solutions often occur because not enough information is available to distinguish among a variety of possibilities.

A Four-Step Problem-Solving Process

Step 1: Understand the problem. Be sure that you understand the nature of the problem. For example:

- Think about the context of the problem (that is, how it relates to other problems in the real world) to gain insight into its purpose.
- Make a list or table of the specific information given in the problem, including units for numerical data.
- Draw a picture or diagram to help you make sense of the problem.
- Restate the problem in different ways to clarify its question.
- Make a mental or written model of the solution, into which you can insert details as you work through the problem.

Step 2: Devise a strategy for solving the problem. Finding an appropriate strategy requires creativity, organization, and experience. In seeking a strategy, try any or all of the following:

- Obtain needed information that is not provided in the problem statement, using recall, estimation, or research.
- Make a list of possible strategies and hints that will help you select your overall strategy.
- Map out your strategy with a flow chart or diagram.

Step 3: Carry out your strategy, and revise it if necessary. In this step you are likely to use analytical and computational tools. As you work through the mathematical details of the problem, be sure to do the following:

- Keep an organized, neat, and written record of your work, which will be helpful if you later need to review your solution.
- Double-check each step so that you do not risk carrying errors through to the end of your solution.
- Constantly reevaluate your strategy as you work. If you find a flaw in your strategy, return to step 2 and create a revised strategy.

Step 4: Look back to check, interpret, and explain your result. Although you may be tempted to think you have finished after you find a result in step 3, this final step is the most important. After all, a result is useless if it is wrong or misinterpreted or cannot be explained to others. Always do all of the following:

- Be sure that your result makes sense. For example, be sure that it has the expected units, that its numerical value is sensible, and that it is a reasonable answer to the original problem.
- Once you are sure that your result is reasonable, recheck your calculations or find an independent way of checking the result.
- Identify and understand potential sources of uncertainty in your result.
- Write your solution clearly and concisely, including discussion of any relevant uncertainties or assumptions.
- Consider and discuss any pertinent implications of your result.

HISTORICAL NOTE
These four steps are modified from a process in *How to Solve It,* by George Polya (Princeton University Press, Princeton, N.J., 1957). First published in 1945, this book has sold more than 1 million copies and is available in at least 17 different languages.

How to Solve It

A New Aspect of Mathematical Method

G. Polya

SECOND EDITION

➤ **EXAMPLE 1** Box Office Receipts

Tickets for a fundraising event were priced at $10 for children and $20 for adults. Sally worked the first shift at the box office, selling a total of $130 worth of tickets. However, she did not keep careful count of how many tickets she sold for children and adults. How many tickets of each type (child and adult) did she sell?

Solution Let's proceed by trial and error. Suppose Sally sold just one $10 child ticket. In that case, she would have sold $130 − $10 = $120 worth of adult tickets. Because the adult tickets cost $20 apiece, this means she would have sold $120/ $120/($20 per adult ticket) = 6 adult tickets. Thus, we have found an answer to the question: Sally could have collected the $130 by selling 1 child and 6 adult tickets. Unfortunately, it is not the only answer, as we can see by testing other values. For example, suppose she sold three of the $10 child tickets, for a total of $30. Then she would have sold $130 − $30 = $100 worth of adult tickets, which means 5 of the $20 adult tickets. Thus, we have a second possible answer—3 child and 5 adult tickets—and no way to know which answer tells us the actual number of tickets sold.

In fact, there are seven possible answers to our question. In addition to the two answers we've already found, other possible answers are 5 child tickets and 4 adult tickets; 7 child tickets and 3 adult tickets; 9 child tickets and 2 adult tickets; 11 child tickets and 1 adult ticket, and 13 child tickets with 0 adult tickets. Without further information, we do not know which combination represents the actual ticket sales.

Now try Exercises 7–8. ◄

Time out to think _____

Verify that each of the child/adult ticket combinations listed as a possible solution in Example 1 does indeed total to $130. Then explain why there aren't any solutions that have an even number of child tickets.

HINT 2: THERE MAY BE MORE THAN ONE STRATEGY

Just as we must often admit more than one right answer, we should also expect more than one strategy for finding an answer. Given how rich and diverse mathematics and the human mind are, it would be unreasonable to expect everyone to follow the same path to a solution. As illustrated in the following example, an efficient strategy can save a lot of time and work.

➤ **EXAMPLE 2** Jill and Jack's Race

Jill and Jack ran a 100-meter race. Jill won by 5 meters; that is, Jack had run only 95 meters when Jill crossed the finish line. They decide to race again, but this time Jill starts 5 meters behind the starting line. Assuming that both runners run at the same pace as before, who will win?

Solution Strategy 1 One approach to this problem is analytical—we analyze each race quantitatively. We were not told how fast either Jill or Jack ran, so we can choose some reasonable numbers. For example, we might assume that Jill ran the 100 meters in the first race in 20 seconds. In that case, her pace was 100 m ÷ 20 s = 5 m/s, or 5 meters per second. Because Jack ran only 95 meters in the same 20 seconds, his pace was 95 m ÷ 20 s = 4.75 m/s, or 4.75 meters per second.

For the second race, Jill must run 105 meters (because she starts 5 meters behind the starting line) to Jack's 100 meters. We predict their times by dividing their respective race distances by their speeds from the first race:

$$\textit{Jill:} \quad 105 \text{ m} \div 5\frac{\text{m}}{\text{s}} = 105 \text{ m} \times \frac{1 \text{ s}}{5 \text{ m}} = 21 \text{ s}$$

$$\textit{Jack:} \quad 100 \text{ m} \div 4.75\frac{\text{m}}{\text{s}} = 100 \text{ m} \times \frac{1 \text{ s}}{4.75 \text{ m}} \approx 21.05 \text{ s}$$

Thus, Jill will win the second race by a slim margin.

Strategy 2 Although the analytical method works, we can use a much more intuitive and direct solution. In the first race, Jill runs 100 meters in the same time that Jack runs 95 meters. Therefore, in the second race, Jill will pull even with Jack 95 meters from the starting line. In the remaining 5 meters, Jill's faster speed will allow her to pull away and win. Note how this insight avoids the calculations needed in Strategy 1. *Now try Exercises 9–10.* ◄

HINT 3: USE APPROPRIATE TOOLS

You don't need a computer to check your tab in a restaurant, and you don't need calculus to find the area of a rectangular room. For any given task there is an appropriate level of power that is needed, and it is a matter of style and efficiency neither to underestimate nor to overestimate that level. You usually will have a choice of tools to use in any problem. Choosing the tools most suited to the job will make your task much easier.

►**EXAMPLE 3** The Cars and the Canary

Two cars, 120 miles apart, begin driving toward each other on a long straight highway. One car travels 20 miles per hour and the other 40 miles per hour (Figure 2.7). At the same time, a canary, starting on one car, flies back and forth between the two cars as they approach each other. If the canary always flies 150 miles per hour and turns around instantly at each car, how far has it flown when the cars collide?

FIGURE 2.7 Set-up for the cars and canary problem.

Solution Because the problem asks "how far," we might be tempted to calculate the *distance* traveled by the canary on each back and forth trip between the cars. However, these trips get shorter as the cars approach each other and we would have to add up all the individual distances. In principle, we would need to add up an *infinite* number of ever-smaller distances—a task that involves the mathematics of calculus.

\mathcal{T}hinking about . . .

ZENO'S PARADOX

The Greek philosopher Zeno of Elea (c. 460 B.C.) posed several paradoxes that defied solution for thousands of years. (A *paradox* is a situation or statement that seems to violate common sense or to contradict itself.) One paradox begins with an imaginary race between the warrior Achilles and a slow-moving tortoise. The tortoise is given a head start, but our common sense says that the swift Achilles will soon overtake the tortoise and win.

Zeno suggested a different way to look at the race. Suppose that, as shown in the figure, Achilles starts from point P0 and the tortoise starts from P1. During the time it takes Achilles to reach P1, the slow-moving tortoise will move ahead a little bit to P2. While Achilles continues on to P2, the tortoise will move ahead to P3. And so on. Thus, Achilles must cover an infinite set of ever-smaller distances to catch the tortoise (that is, from P0 to P1, from P1 to P2, etc.). From this point of view, it seems that Achilles will never catch the tortoise.

This paradox puzzled philosophers and mathematicians for more than 2000 years. Its resolution depends on a key insight that became clear only with the invention of calculus in the 17th century: It does not necessarily require an infinite amount of time to cover an infinite set of distances. For example, imagine that the infinite set of distances covered by Achilles begins with 1 mile, then $\frac{1}{2}$ mile, then $\frac{1}{4}$

mile, and so on. Then the total distance he covers, in miles, is

$$1 + \frac{1}{2} + \frac{1}{4} + \frac{1}{8} + \frac{1}{16} + \frac{1}{32} + \frac{1}{64} + \frac{1}{128} + \frac{1}{256}$$
$$+ \frac{1}{512} + \frac{1}{1024} + \frac{1}{2048} + \cdots$$

This sum is called an **infinite series** because it is the sum of an infinite number of terms. You might guess that an infinite series would sum to infinity. But note what happens when we add just the first four terms, then the first eight terms, and then the first twelve terms. You can confirm the following results with a calculator.

$$1 + \frac{1}{2} + \frac{1}{4} + \frac{1}{8} = 1.875$$

$$1 + \frac{1}{2} + \frac{1}{4} + \frac{1}{8} + \frac{1}{16} + \frac{1}{32} + \frac{1}{64} + \frac{1}{128} = 1.9921875$$

$$1 + \frac{1}{2} + \frac{1}{4} + \frac{1}{8} + \frac{1}{16} + \frac{1}{32} + \frac{1}{64} + \frac{1}{128} + \frac{1}{256}$$
$$+ \frac{1}{512} + \frac{1}{1024} + \frac{1}{2048} = 1.99951171875$$

If you continue to add more terms in this infinite series, you will find that the sum gets closer and closer to 2, but never exceeds it. In fact, it can be proven deductively that the sum of this infinite series is 2. Thus, even though the paradox has Achilles covering an infinite number of ever-shorter distances, the *total* distance he runs is finite—in this example, it is 2 miles. Clearly, it won't take him long to run a finite distance of 2 miles, so he will pass the slower tortoise and win the race.

(a) Starting point for Achilles — Starting point for tortoise — Finish. P0 P1 P2. While Achilles runs from P0 to P1... ...tortoise runs to P2

(b) Finish. P0 P1 P2 P3. Achilles runs from P1 to P2... ...but tortoise moves on to P3

But note what happens if we focus on *time* rather than distance. The cars will collide with each other at a speed of 60 mi/hr (because one car is traveling at 20 mi/hr and the other is traveling in the opposite direction at 40 mi/hr). Because they initially are 120 miles apart, they will collide in precisely 2 hours:

$$120 \text{ mi} \div 60\frac{\text{mi}}{\text{hr}} = 120 \text{ mi} \times \frac{1 \text{ hr}}{60 \text{ mi}} = 2 \text{ hr}$$

The canary is flying at a speed of 150 miles per hour, so in 2 hours it will fly

$$2 \text{ hr} \times 150\frac{\text{mi}}{\text{hr}} = 300 \text{ mi}$$

The canary will fly 300 miles before the cars collide. We could have found the answer with calculus, but why bother when we were able to do it with just multiplication and division? *Now try Exercises 11–12.* ◄

HINT 4: CONSIDER SIMPLER, SIMILAR PROBLEMS

Sometimes you are confronted with a problem that at first seems daunting. Our fourth hint is to consider a simpler, but similar, problem. The insight gained from solving the easier problem may help you understand the original problem.

►**EXAMPLE 4** Coffee and Milk

Suppose you have two cups in front of you: One holds coffee and one holds milk (Figure 2.8). You take a teaspoon of milk from the milk cup and stir it into the coffee cup. Next, you take a teaspoon of the mixture in the coffee cup and put it back into the milk cup. After the two transfers, there will be either (1) more coffee in the milk cup than milk in the coffee cup, (2) less coffee in the milk cup than milk in the coffee cup, or (3) equal amounts of coffee in the milk cup and milk in the coffee cup. Which of these three possibilities is correct?

FIGURE 2.8 Set-up for the coffee and milk problem.

Solution A cup of either milk or coffee contains something like a *trillion trillion* molecules of liquid. Clearly, it would be difficult to visualize how such enormous numbers of molecules mix together, let alone to calculate the result. However, the essence of this problem is the *mixing* of two things. So one approach is to try a similar mixing problem that is much easier: mixing two piles of marbles.

Suppose that the black pile in Figure 2.9 has ten black marbles and represents the coffee. The white pile has ten white marbles and represents the milk. In this simpler problem, we can represent the first transfer—one teaspoon of milk into the coffee cup—by moving two white marbles to the black pile. This leaves the white pile with just eight white marbles, while the black pile now has ten black and two white marbles.

Two white marbles
are moved to the black pile.

FIGURE 2.9 The white and black piles of marbles represent the milk and coffee, respectively. Moving two white marbles to the black pile represents putting a teaspoon of milk in the coffee.

Representing the second transfer—of one teaspoon from the coffee cup into the milk cup—involves taking two marbles from the black pile and putting them in the white pile. We can then ask a question analogous to the original question: Are there more black marbles in the white pile or white marbles in the black pile?

Because the marbles represent molecules that mix thoroughly, the two marbles for the second transfer must be drawn at random. This opens three possible cases: The two marbles in the second transfer can be either both black, both white, or one of each. But, as shown in Figure 2.10, in all three cases we end up with the same number of white marbles in the black pile as black marbles in the white pile. By analogy, we have the answer to our original question: After the two transfers, the amounts of coffee in the milk cup and milk in the coffee cup are equal.

Two black marbles transferred Two white marbles transferred One of each transferred

FIGURE 2.10 When we transfer two marbles back to the white pile, the two marbles may be both black (left), both white (center), or one black and one white (right). In all three cases, we end up with the same number of white marbles in the black pile as black marbles in the white pile.

The only remaining step is to confirm that the simpler problem is a reasonable representation of the real problem. Our choice of using two marbles to represent a teaspoon was arbitrary. If we redo this example with transfers of one, three, or any other number of marbles, we will find the same result: We always end up with the same number of black marbles in the white pile as white marbles in the black pile. Starting with ten marbles in each pile also was arbitrary; the conclusion remains the same if we start with twenty, fifty, or a trillion trillion marbles. Because molecules can be thought of as tiny marbles for the purpose of this problem, the real problem has no essential differences from the marble problem.

Now try Exercises 13–14.

Time out to think _____

Most people are surprised by the result of the coffee and milk problem. Are you? Now that you know the solution, can you give a simple explanation of the real problem that would satisfy surprised friends?

HINT 5: CONSIDER EQUIVALENT PROBLEMS WITH SIMPLER SOLUTIONS

Replacing a problem with a similar, simpler problem can reveal essential insights about a problem, as we've just seen. However, "similar" is not good enough when we need a numerical answer. In that case a useful approach to a difficult problem is to look for an *equivalent* problem. An equivalent problem will have the same numerical answer but may be easier to solve.

▶**EXAMPLE 5** Party Decorations

Juan is decorating for a party in a room that has ten large cylindrical posts. The posts are 8 feet high and have a circumference of 6 feet. His plan is to wrap eight turns of ribbon around each post. How much ribbon does Juan need?

Solution The problem is difficult because it involves three-dimensional geometry. However, we can convert it to a simpler equivalent problem. Assume that each wrapped post is a hollow cylinder, and imagine cutting it down its length and unfolding it into a flat rectangle (Figure 2.11). The width of the rectangle is the 6-foot circumference of the post, and its length is the 8-foot length of the post.

Now, instead of dealing with ribbon wrapped around a three-dimensional post, we have a simple rectangle with eight diagonal segments of ribbon. The total length of the eight segments will be the length of the ribbon required for one of the posts in

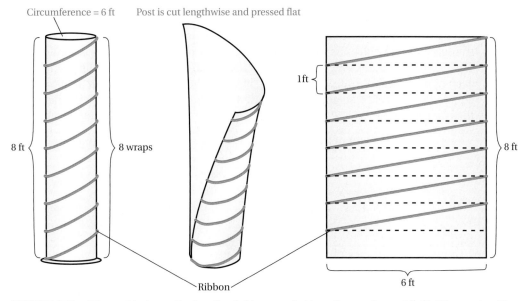

FIGURE 2.11 We want to know the length of ribbon needed to coil around a post (left). We can simplify the problem by imagining that we cut the wrapped post lengthwise (center). When the post is pressed flat, the ribbon lies in straight segments, each of which is the hypotenuse of a right triangle.

the original problem. The height of each triangle is $\frac{1}{8}$ of the length of the rectangle, or 8 ft ÷ 8 = 1 ft. The base of each triangle is the 6-foot width of the rectangle. The Pythagorean theorem tells us that

$$\text{base}^2 + \text{height}^2 = \text{hypotenuse}^2 \quad \text{or} \quad \text{hypotenuse} = \sqrt{\text{base}^2 + \text{height}^2}$$

Substituting the 6-foot base and 1-foot height yields

$$\text{hypotenuse} = \sqrt{(6 \text{ ft})^2 + (1 \text{ ft})^2} = 6.1 \text{ ft}$$

Thus, the length of each ribbon segment is 6.1 feet, and the total length of the ribbon is 8 × 6.1 ft = 48.8 ft. To decorate 10 posts, Juan needs 10 × 48.8 ft = 488 feet of ribbon. Note how much easier it was to solve this equivalent problem than the original one. *Now try Exercises 15–16.* ◄

HINT 6: APPROXIMATIONS CAN BE USEFUL

Another useful strategy is to make problems easier by using approximations. Most real problems involve approximate numbers to begin with, so an approximation often is good enough for a final answer. In other cases, an approximation will reveal the essential character of a problem, making it easier to reach an exact solution. Approximations also provide a useful check: If you come up with an "exact solution" that isn't close to the approximate one, something may have gone wrong.

➤**EXAMPLE 6** A Bowed Rail

Imagine a mile-long bar of metal such as a railroad rail. Suppose that the rail was anchored on both ends (a mile apart) and that, on a hot day, its length expanded by 1 foot. If the added length caused the rail to bow upward in a circular arc as shown in Figure 2.12a, about how high would the center of the rail rise above the ground?

Solution Because the added length is short compared to the original length, we can approximate the curved rail with two straight lines (Figure 2.12b). We now have two right triangles and the Pythagorean theorem applies. The bases of the two right triangles together give the original rail length of 1 mile, so each base is $\frac{1}{2}$ mile long. The two hypotenuses together represent the expanded length of 1 mile + 1 foot, so each hypotenuse is $\frac{1}{2}$ mile + $\frac{1}{2}$ foot long. Because there are 5280 feet in a mile, $\frac{1}{2}$ mile equals 2640 feet. The height of the rail off the ground is approximately the height of the triangles:

$$\text{height of triangle} = \sqrt{(2640.5 \text{ ft})^2 - (2640 \text{ ft})^2} \approx 51.4 \text{ ft}$$

Not drawn to scale!

1 mile + 1 foot

1 mile

(a)

$\frac{1}{2}$ mile + $\frac{1}{2}$ foot

$\frac{1}{2}$ mile

(b)

FIGURE 2.12 (a) If a rail lengthens while remaining anchored at its ends, it will bow upward as shown. (b) We can find the approximate height of the bow by ignoring its curvature and assuming that each side makes a triangle.

According to our approximation, the center of the rail would rise more than 50 feet off the ground! Because a triangle will stick up higher than a curve with the same base and length, the actual height would be less than what we found with the approximation. An exact solution shows that the top of the curved rail would be about 48 feet off the ground. *Now try Exercises 17–18.* ◄

Time out to think _____

Are you surprised by the answer to Example 6? How does the use of the approximation tell you that the original question contained at least some unrealistic assumptions? Which assumptions do you think were unrealistic?

HINT 7: TRY ALTERNATIVE PATTERNS OF THOUGHT

Try to avoid rigid patterns of thought that tend to suggest the same ideas and methods over and over again. Instead, approach every problem with an openness that allows innovative ideas to percolate. In its most wondrous form, this approach is typified by what Martin Gardner, a well-known popularizer of mathematics, calls "aha!" problems. These are problems whose best solution involves a penetrating insight that reduces the problem to its essential parts.

➤ **EXAMPLE 7** China's Population Policy

In an effort to reduce population growth, in 1978 China instituted a policy that allows only one child per family. One unintended consequence has been that, because of a cultural bias toward sons, China now has many more young boys than girls (presumably because of selective abortions or, in some cases, infanticide). To solve this problem, some people have suggested replacing the one-child policy with a *one-son* policy. That is, if a family's first child is a boy, the family has met its limit of children. But if the first child is a girl, the family can have additional children until one is a boy. Suppose that the one-son policy were implemented and birth rates returned to their natural levels (half boys and half girls). Compared to the one-child policy, how would this affect the overall birth rate and the numbers of boys and girls?

Solution Most people initially approach this problem by considering the families. Half the families would have a boy as their first and only child. Of the remaining families, half would have a boy as their second child (reaching their one-son limit), while the other half would have a girl and go on to a third child. Half of these families would have a boy on this third try, while the other half would continue on after a third girl. And so on.

However, a moment of insight makes the problem much easier. Neither policy, by itself, affects the natural probability that any child will be a boy or a girl. If the one-son policy restores natural birth rates, then equal numbers of boys and girls must be born. Moreover, nearly all families would have exactly one boy under this policy. Because there must be equal numbers of boys and girls overall, there must also be one girl per family, on average. Thus, the average number of children per family must be two (one boy and one girl) under a one-son policy. In summary, a one-son policy would lead to equal numbers of boys and girls but, compared to a one-child policy, would double the average number of children per family from one to two.

Now try Exercises 19–20. ◄

by the way

Hospital birth records show that boys and girls are born in nearly, but not precisely, equal numbers: Roughly 106 boys are born for every 100 girls. However, males have higher mortality rates than females at every age, so the numbers even out in adulthood and women outnumber men in old age.

Time out to think _____

China occupies roughly the same amount of land as the United States, but has more than four times as many people. Given these circumstances, do you think that the one-child policy is a good idea? Would a one-son policy be better? Defend your opinions.

HINT 8: DO NOT SPIN YOUR WHEELS

Finally, everyone has had the experience of getting "bogged down" with a problem. When your wheels are spinning, let up on the gas! Often the best strategy in problem solving is to put a problem aside for a few hours or days. You may be amazed at what you see (and what you overlooked) when you return to it.

EXERCISES 2C

Review Questions

1. Describe the four basic steps of problem solving.

2. Summarize the strategic hints for problem solving given in this unit, with an example of the meaning of each one.

Does It Make Sense?

Decide whether each of the following statements makes sense (or is clearly true) or does not make sense (or is clearly false). Explain your reasoning.

3. If you follow my simple problem-solving recipe, you will be able to solve any mathematical problem easily and with a minimal investment of time.

4. Whether it's a problem in mathematics or something else, it's best to start by taking time to make sure you understand the nature of the problem.

5. If your answer doesn't make sense, then you must not have solved the problem correctly.

6. Kids are often taught to draw pictures for simple math problems, but grownups should be able to solve problems without any pictures.

Basic Skills & Concepts

7. **Box Office.** You work at a box office selling tickets that cost $15 for children and $30 for adults. You sold a total of $150 worth of tickets, but did not keep track of child and adult tickets separately. Can you determine how many child and how many adult tickets you sold? Find all the possible solutions to the problem.

8. **Party Punch.** You're making up some party punch using cranberry juice and ginger ale. The cranberry juice comes in 4-quart bottles, and the ginger ale comes in 2-quart bottles. You mix a huge 48-quart (12-gallon) batch of punch, but forget how many bottles of cranberry juice and ginger ale you poured in. Can you find the exact recipe for the punch? Find all the possible solutions to the problem.

More on Jack and Jill. Refer to Example 2 for Exercises 9–10.

9. Suppose that Jill's time in the first race was 10 seconds. Using strategy 1, determine by how much she would win the second race.

10. Hack and Quill race 200 meters and Hack wins by 10 meters. They race a second time, with Hack starting 10 meters behind the starting line. Who wins the second race? Explain.

11. **Cars and Canary.** Refer to Example 3. Two cars, 150 kilometers apart, begin driving toward each other on a long, straight highway. One car travels 80 kilometers per hour and the other 100 kilometers per hour. At the same time, a canary, starting on one car, flies back and forth between the two cars as they approach each other. If the canary flies 120 kilometers per hour and spends no time to turn around at each car, how far has it flown when the cars collide?

12. **Cars and Canary.** Repeat Exercise 11, but assume the canary flies at only 60 kilometers per hour.

13. **Mixing Marbles.** Refer to Example 4. Consider the case in which each pile initially has fifteen marbles. Suppose that on the first transfer three black marbles are moved to the white pile. On the second transfer, any three marbles are taken from the white pile and put into the black pile. Demonstrate, with diagrams and words, that you will always end up with as many white marbles in the black pile as black marbles in the white pile.

14. **Mixing Marbles.** Repeat Exercise 13, but for the case of piles of twenty marbles and transfers of four marbles at a time.

15. **Wedding Decorations.** You are responsible for wrapping ribbon around the cylindrical columns in a large banquet room that will be the site of a wedding reception. The room has 20 columns that are each 12 feet high and have a circumference of 3 feet. You plan to make ten wraps of ribbon around each column. How much ribbon will you need?

16. **Coiled Wire.** Eight turns of a wire are wrapped around a pipe with a length of 20 centimeters and a circumference of 6 centimeters. What is the length of the wire?

Bowed Rail. Exercises 17–18 refer to Example 6 in the text.

17. In a few sentences, briefly explain why the triangle makes a good approximation to the actual shape of the bowed rail and why a circle with radius $\frac{1}{2}$ mile would *not* have made a good approximation.

18. Suppose the rail is 1 kilometer long, and it expands on a hot day by 10 centimeters in length. Approximately how high would the center of the rail rise above the ground?

19. **China's One-Child Policy.** To convince yourself that a one-son policy would lead to an average of two children per family, with equal numbers of boys and girls, do the following. Suppose that 10,000 families are having children according to the one-son policy. Describe the general makeup of all of the families (that is, start with the fact that 5000 families have a boy as their first and therefore only child, and continue on). Use this process to show that the average number of children is two and that boys and girls are equal in number.

20. **Demographic Imbalance.** Suppose that, as the current generation matures, China's population of young adults has more men than women by a ratio of 118 to 100. With 400 million young adults in China, how many men will be unable to find a spouse?

Further Applications

21. **Gardening Supplies.** Topsoil comes in 3-cubic-foot bags, and fertilizer comes in 1-cubic-foot bags. Assuming you use entire bags (so that you do not waste any soil or fertilizer), how many different ways can you make a 25-cubic-foot mixture of soil and fertilizer? Find all the possible solutions to the problem.

22. **Traffic Counter.** A traffic counter is a device designed to count the number of vehicles passing along a street. It usually is a thin black tube stretched across a street or high-way, connected to a "brain box" at the side of the road. The device registers one "count" each time a set of wheels (that is, wheels on a single axle) rolls over the tube. A normal automobile registers two counts: one for the front wheels and one for the rear wheels. A truck with three axles (front wheels plus a double set of rear wheels) registers three counts. A large semitrailer truck might have four or five axles and hence register four or five counts. Suppose that, during a 1-hour period, a traffic counter registered 35 counts on a residential street on which only two-axle vehicles (cars) and three-axle vehicles (light trucks) are allowed. How many cars and light trucks passed over the traffic counter? Find all the possible solutions to the problem.

23. **Stereo Wire 1.** A stereo system is being installed in a room with a rectangular floor measuring 12 feet by 10 feet and an 8-foot ceiling. The stereo amplifier is near one corner of the room, 1 foot from the floor and 1 foot from a side wall. A speaker is near the opposite corner of the room, 1 foot from the ceiling and 1 foot from the opposite wall. You must run a wire from the amplifier to the speaker, and the wire must run along the floor or walls (not through the air). What is the shortest length of wire you can use for the connection? (Hint: Turn the problem into an equivalent simpler problem by imagining cutting the room along its vertical corners and unfolding it so that it is flat. You will be able to apply the Pythagorean theorem.)

24. **Stereo Wire 2.** You are connecting the same amplifier and speaker described in Exercise 23. This time, however, the wire must run along the edges of the room for as much of the distance as possible. (That is, you cannot run it diagonally across the floor.) What is the minimum amount of wire needed for the connection?

25. **Bridge Height.** A curved bridge rises over a river. The two endpoints of the bridge are 100 meters apart (horizontally). You walk across the bridge with a device to measure its length, and you discover that the walking distance is 102 meters. Approximately how high does the bridge rise over the river? (Hint: See Example 6.)

26. **Bridge Length.** A curved bridge rises over a small creek. The two endpoints of the bridge are 25 meters apart (horizontally). The bridge rises to a height of 3 meters (above the horizontal). Approximately what is the walking distance along the bridge? (Hint: See Example 6, and note that this time you are given the height and must determine the hypotenuse.)

Puzzle Problems. **Exercises 27–41 are puzzle problems that require careful thinking and can help your problem-solving skills. Hint: Look for "aha!" solutions.**

27. It takes you 30 seconds to walk from the first (ground) floor of a building to the third floor. How long will it take to walk from the first floor to the sixth floor (at the same pace, assuming that all floors have the same height)?

28. Reuben says, "Two days ago I was 20 years old. Later next year I will be 23 years old." How is this possible?

29. There are three kinds of apples all mixed up in a basket. How many apples must you draw (without looking) from the basket to be sure of getting at least two of one kind?

30. "Brothers and sisters I have none, but that man's father is my father's son." Who is "that man"?

31. "I am the brother of the blind fiddler, but brothers I have none." How can this be?

32. A woman bought a horse for $500 and then sold it for $600. She bought it back for $700 and then sold it again for $800. How much did she gain or lose on these transactions?

33. Three boxes of fruit are labeled *Apples, Oranges,* and *Apples and Oranges.* Each label is wrong. By selecting just one fruit from just one box, how can you determine the correct labeling of the boxes?

34. Each of ten large barrels is filled with golf balls that all look alike. The balls in nine of the barrels weigh 1 ounce and the balls in one of the barrels weigh 2 ounces. With only *one* weighing on a scale, how can you determine which barrel contains the heavy golf balls?

35. A woman is traveling with a wolf, a goose, and a mouse. She must cross a river in a boat that will hold only herself and one other animal. If left to themselves, the wolf will eat the goose and the goose will eat the mouse. How many crossings are required to get all four creatures across the river alive?

36. How do you measure 9 minutes with a 7-minute and a 4-minute hourglass? Assume the hourglasses can only measure 7-minute and 4-minute intervals and cannot be used to measure other time intervals (for example, you can't tell when 2 minutes have gone by).

37. A 150-foot rope is suspended at its two ends from the tops of two 100-foot flagpoles. The lowest point of the rope is 25 feet from the ground. What is the distance between the flagpoles?

38. You are considering buying 12 gold coins that look alike, but you have been told that one of them is a heavy counterfeit. How can you find the heavy coin in three weighings on a balance scale?

39. Suppose you have 40 blue socks and 40 brown socks in a drawer. How many socks must you take from the drawer (without looking) to be sure of getting a pair of the same color?

40. Alma, Bess, Cleo, and Dina visited Elf on Saturday. Alma visited at 8:00, Bess visited at 9:00, Cleo visited at 10:00, and Dina visited at 11:00. Some clues: (1) The times may be either A.M. or P.M. (2) At least one woman visited Elf between Alma and Bess. (3) Alma did not visit Elf before both Cleo and Dina. (4) Cleo did not visit Elf between Bess and Dina. Who visited Elf last?

41. Three prisoners know that the jailer has three white hats and two red hats. The jailer gives one hat to each prisoner and says, "If you can deduce the color of your own hat, you will be freed." Each prisoner can see the hats of the other two prisoners, but not his own. The first prisoner says, "I cannot tell the color of my hat." The second prisoner says, "I cannot tell the color of my hat." The third prisoner is blind, but he is freed. What color hat does he have and how did he know?

Projects: Real Problems in the Real World. Exercises 42–50 describe complex problems that do not have single, straightforward solutions. For each, describe how you would apply the four-step problem-solving process described in the text (without actually carrying out the process to obtain a solution). You may choose either to enumerate the four steps or to describe your process in essay form. Either way, be sure that you list as many relevant factors as possible and discuss sources of uncertainty associated with each factor. Also, describe how you would work with these factors to find a solution. Conclude by describing your overall impression of whether the problem can be solved and whether any solution would likely generate controversy.

42. You are asked to calculate the cost of installing enough bike racks on campus to solve a bicycle parking problem.

43. You want to know how many new faculty members would be needed, and the total cost to the university, to make sure that all classes have twenty or fewer students (to replace large lecture classes with smaller classes).

44. You want to know whether having a top-quality football program means more money for academics at a university.

45. You want to figure out how much taxes would have to be increased to provide public school teachers with twice their present salaries.

46. You decide that, in the interest of the environment, you will convert your home heating and hot water system to

solar power. How much will this conversion cost or save over the next ten years?

47. Suppose that China and India decide to use their extensive coal reserves to supply energy to their populations at the same per capita level as in the United States. How much carbon dioxide would be added to the atmosphere?

48. Are automobile insurance companies gouging drivers? Suppose that you want to figure out whether they are justified in raising insurance rates as rapidly as they have during the past few years.

49. A large city is facing a severe shortage of water. How should water be rationed?

50. Suppose that a city added new bus routes and handed out free bus passes. How many people would give up driving in favor of the bus? How much money, overall, would this cost or save the city?

51. **Project: Textbook Analysis.** Although research shows that most adults today have difficulty with "story problems," we might hope that the next generation will have less difficulty. Find a current textbook in mathematics that is used at the upper elementary school level (grades 4–6). Read through the "story problems" in the textbook. Write a critical analysis of the problems and conclude

with an opinion as to whether the problems make mathematics meaningful.

Web Projects

Find useful links for Web Projects on the text Web site: www.aw.com/bennett-briggs

52. **China's Population.** Find current statistics regarding China's population, such as its total population and average number of children per family, and projections of its future population. Based on what you find, does it appear that China is slowing its rate of population growth? Explain.

53. **Polya and Problem Solving.** Many Web sites discuss George Polya's work and other problem-solving strategies. Explore a few such sites. Write a one- or two-paragraph summary of something you learn about problem solving that you think would be useful to many people.

In the News

54. **Multiple Solutions.** Find an example of a real problem for which, because of insufficient data, we cannot distinguish between two or more possible solutions. The problem might come from a news report or from your own experiences. What additional data would be useful?

55. **Multiple Strategies.** Find an example of a real problem that could potentially be solved by two or more competing strategies. The problem might come from a news report or from your own experiences. Describe each strategy. Which one do you think is better? Why?

56. **Novel Solution.** Find a news report concerning a problem in business or science that was solved by a surprising method. Describe the method and why it was useful.

57. **China's Population Policy.** Look for a recent news report concerning consequences of China's population policies. Does the news report affect your overall opinion of China's policy? Explain.

THE FAR SIDE® BY GARY LARSON

Hell's library

CHAPTER 2 SUMMARY

UNIT	KEY TERMS	KEY IDEAS AND SKILLS
2A	units conversion factor currency conversions	**Identify** units with key words *per, square,* or *cube* or with hyphens. **Write** conversion factors in three equivalent forms. **Work** with units to check answers and help solve problems.
2B	U.S. customary system (USCS) metric system (SI) Celsius, Fahrenheit, Kelvin absolute zero energy units: Calorie, joule, kilowatt-hour power units: 1 watt = 1 joule/s density concentration blood alcohol content	**Understand and convert** basic USCS and metric units. **Know** metric prefixes. **Know** temperature units and convert between °F, °C, and Kelvin. **Understand** energy units and power units. Energy is what makes matter move or heat up. Power is the *rate* at which energy is used. **Apply** the concept of density to materials, population, and information. **Apply** the concept of concentration to pollution and blood alcohol content.
2C	four-step problem- solving process	**Apply** eight strategic hints for problem solving.

The concept of number is the obvious distinction between the beast and man. Thanks to number, the cry becomes song, noise acquires rhythm, the spring is transformed into a dance, force becomes dynamic, and outlines figures.

—JOSEPH DE MAISTRE,
19TH-CENTURY FRENCH PHILOSOPHER

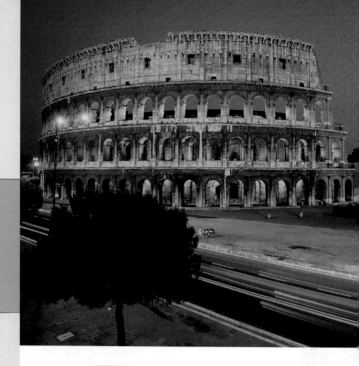

CHAPTER 3

Numbers in the Real World

Life is filled with numbers that may at first seem incomprehensible: a population measured in billions of people, national budgets measured in trillions of dollars, and distances ranging from nanometers to light-years. Moreover, numbers come in many forms, including percentages and index numbers (such as the Consumer Price Index), and nearly always involve uncertainties. In this chapter, we will discuss the use and interpretation of numbers we encounter in our daily lives.

UNIT **3A** **Uses and Abuses of Percentages:** Percentages can be surprisingly subtle and difficult to interpret. We will study both uses and abuses of percentages.

UNIT **3B** **Putting Numbers in Perspective:** Many of the numbers we hear daily are extremely large or small. We will look at several techniques for giving meaning to such numbers.

UNIT **3C** **Dealing with Uncertainty:** We will discuss ways of dealing with the inevitable uncertainty in numbers heard daily in the news. We will also consider the types of errors that affect measured numbers.

UNIT **3D** **Index Numbers: The CPI and Beyond:** We will study the important role of index numbers in modern life, paying special attention to the Consumer Price Index (CPI).

UNIT **3E** **How Numbers Deceive: Polygraphs, Mammograms, and More:** Numbers may seem straightforward, but they can be deceiving unless we interpret them carefully. We will study several interesting cases that relate to everyday issues.

USES AND ABUSES OF PERCENTAGES

*n*ews reports frequently express quantitative information with percentages. Unfortunately, while percentages themselves are rather basic—they are just an alternative form of fractions—they are often used in very subtle ways. For example, consider the following quote from a front-page article in the *New York Times:*

> The rate [of smoking] among tenth graders jumped 45 percent, to 18.3 percent, and the rate for eighth graders is up 44 percent, to 10.4 percent.

All the percentages in this statement are used correctly, but they are quite confusing. For instance, what does it mean to say "up 44%, to 10.4%"? In this unit, we will investigate some of the uses and abuses of percentages.

THREE WAYS OF USING PERCENTAGES

Consider the following statements from news reports:

- A total of 13,000 newspaper employees, 2.6% of the newspaper work force, lost their jobs.
- Cisco stock rose 5.7% last week, to $17.93.
- High definition television sets (HDTVs) have 125% more resolution than conventional TV, but cost 400% more.

On close examination, each statement uses a percentage in a different way. The first uses a percentage to express a *fraction* of the total work force. The second uses a percentage to describe a *change* in stock price. The third uses percentages to *compare* the resolutions and costs of televisions.

Using Percentages as Fractions

Percent is just a fancy way of saying "divided by 100," so *P*% simply means $P/100$. For example, 10.4% means $10.4/100$, or 0.104. Therefore, if 10.4% of eighth-graders smoke and there are 100,000 eighth-graders, then the number who smoke is 10.4% *of* 100,000, or

$$10.4\% \times 100,000 = 0.104 \times 100,000 = 10,400$$

Note that the word *of* implies multiplication. We've found that if 10.4% of 100,000 eighth-graders smoke, there are 10,400 smokers.

➤EXAMPLE 1 Newspaper Survey

A newspaper reports that 44% of 1069 people surveyed said that the president is doing a good job. How many said the president is doing a good job?

Solution The 44% represents the fraction of respondents who said the president is doing a good job. Because *of* indicates multiplication, we find that 44% of 1069 is

$$44\% \times 1069 = 0.44 \times 1069 = 470.36 \approx 470$$

PERCENTAGES

The words *per cent* simply mean "per 100" or "divided by 100." We use the symbol % as a shorthand for the words *per cent*. For example, we read 50% as "50 percent" and its meaning is

$$50\% = \frac{50}{100} = 0.5$$

More generally, for a number *P*,

$$P\% = \frac{P}{100}$$

For example:

$$100\% = \frac{100}{100} = 1 \qquad 200\% = \frac{200}{100} = 2$$

$$350\% = \frac{350}{100} = 3.5$$

Note that multiplying a number by 100% does not change its meaning, because 100% is just another way of writing 1. For example, multiplying 1.25 by 100% gives

$$1.25 \times 100\% = 125\%$$

This shows that 125% is just another way of writing 1.25. These basic ideas lead to the following rules for converting between percentages and decimals or common fractions.

- **To convert a percentage to a common fraction:** Replace the % symbol with division by 100; simplify the fraction if necessary.

$$\text{\textit{Example:}} \quad 25\% = \frac{25}{100} = \frac{1}{4}$$

- **To convert a percentage to a decimal:** Drop the % symbol and move the decimal point two places to the left (that is, divide by 100).

$$\text{\textit{Example:}} \quad 25\% = 0.25$$

- **To convert a decimal to a percentage:** Move the decimal point two places to the right (that is, multiply by 100) and add the % symbol.

$$\text{\textit{Example:}} \quad 0.43 = 43\%$$

- **To convert a common fraction to a percentage:** First convert the common fraction to decimal form, using a calculator if necessary; then convert the decimal to a percentage.

$$\text{\textit{Example:}} \quad \frac{1}{5} = 0.2 = 20\%$$

Now try Exercises 17–32.

About 470 people said the president is doing a good job. We rounded the answer to 470 to obtain a whole number of people. *Now try Exercises 45–50.* ◄

Using Percentages to Describe Change

Percentages are often used to describe how a quantity changes with time. As an example, suppose the population of a town was 10,000 in 1970 and 15,000 in 2000. We can express the change in population in two basic ways:

- Because the population rose by 5000 people (from 10,000 to 15,000), we say that the **absolute change** in the population was 5000 people.
- Because the increase of 5000 people was 50% of the starting population of 10,000, we say that the **relative change** in the population was 50%, or 0.5.

In general, calculating an absolute or relative change always involves two numbers: a starting number, or **reference value,** and a **new value.** Once we identify these

two values, we can calculate the absolute and relative change with the formulas in the following box. Note that the absolute and relative changes are *positive* if the new value is greater than the reference value and *negative* if the new value is less than the reference value.

Absolute and Relative Change

The **absolute change** describes the actual increase or decrease from a reference value to a new value:

$$\text{absolute change} = \text{new value} - \text{reference value}$$

The **relative change** is a fraction that describes the size of the absolute change in comparison to the reference value:

$$\text{relative change} = \frac{\text{absolute change}}{\text{reference value}} = \frac{\text{new value} - \text{reference value}}{\text{reference value}}$$

The relative change can be converted from a fraction to a percentage by multiplying by 100%.

As the next example shows, the relative change formula leads to the following important rules:

- When a quantity *doubles* in value, its relative change is 1, or 100%.
- When a quantity *triples* in value, its relative change is 2, or 200%.
- When a quantity *quadruples* in value, its relative change is 3, or 300%. And so on.

➤EXAMPLE 2 Stock Price Rise

During a 6-month period, Nokia's stock doubled in price from $10 to $20. What were the absolute and relative changes in the stock price?

Solution The reference value is the starting stock price of $10 and the new value is the later stock price of $20. The absolute change is the difference between the new and starting stock prices:

$$\text{absolute change} = \text{new value} - \text{reference value}$$
$$= \$20 - \$10 = \$10$$

The absolute change divided by the reference value gives the relative change:

$$\text{relative change} = \frac{\text{absolute change}}{\text{reference value}} = \frac{\$10}{\$10} = 1 = 100\%$$

That is, the doubling of the stock price means a relative increase in value of 100%.

Now try Exercises 51–52. ◄

Time out to think

In your own words, explain why doubling in value means a relative increase of 100% (as opposed to, say, 200%). Extend this thinking to explain why tripling means a 200% increase, quadrupling means a 300% increase, and so on.

▶ **EXAMPLE 3** World Population Growth

World population was 2.6 billion in 1950 and 6.0 billion in 2000. Describe the absolute and relative change in world population from 1950 to 2000.

Solution The reference value is the 1950 population of 2.6 billion and the new value is the 2000 population of 6.0 billion.

$$\text{absolute change} = \text{new value} - \text{reference value}$$
$$= 6.0 \text{ billion} - 2.6 \text{ billion} = 3.4 \text{ billion}$$

$$\text{relative change} = \frac{\text{new value} - \text{reference value}}{\text{reference value}}$$

$$= \frac{6.0 \text{ billion} - 2.6 \text{ billion}}{2.6 \text{ billion}} \approx 1.3 = 130\%$$

World population increased by 3.4 billion people, or by 130%, from 1950 to 2000.

Now try Exercises 53–54. ◀

▶ **EXAMPLE 4** Depreciating a Computer

You bought a computer three years ago for $1000. Today, it is worth only $300. Describe the absolute and relative change in the computer's value.

Solution The reference value is the original price of $1000 and the new value is its current worth of $300. The absolute change in the computer's value is

$$\text{absolute change} = \text{new value} - \text{reference value}$$
$$= \$300 - \$1000 = -\$700$$

The negative sign tells us that the computer's current worth is $700 *less than* the price you paid three years ago. The relative change is

$$\text{relative change} = \frac{\text{new value} - \text{reference value}}{\text{reference value}}$$

$$= \frac{\$300 - \$1000}{\$1000}$$

$$= -0.7 = -70\%$$

Again, the negative sign tells us that the computer is now worth 70% *less than* it was three years ago.

Now try Exercises 55–56. ◀

Using Percentages for Comparisons

Percentages are also commonly used to compare two numbers. Suppose we want to compare the price of a $50,000 Mercedes to the price of a $40,000 Lexus. The difference between the Mercedes and Lexus prices is

$$\$50,000 - \$40,000 = \$10,000$$

That is, the Mercedes costs $10,000 *more than* the Lexus. We can also express this difference as a percentage of the Lexus price:

$$\frac{\$10,000}{\$40,000} = 0.25 = 25\%$$

Thus, in relative terms, the Mercedes costs 25% *more than* the Lexus.

Because we are *comparing to* the Lexus price, we say that the Lexus price is the **reference value.** Notice that the reference value follows the word *than.* The Mercedes price is the **compared value.** We can now define the absolute and relative *difference* between the quantities, much as we defined absolute and relative change. The absolute and relative differences are *positive* if the compared value is greater than the reference value and *negative* if the compared value is less than the reference value.

Absolute and Relative Difference

The **absolute difference** is the actual difference between the compared value and the reference value:

$$\text{absolute difference} = \text{compared value} - \text{reference value}$$

The **relative difference** describes the size of the absolute difference as a fraction of the reference value:

$$\text{relative difference} = \frac{\text{absolute difference}}{\text{reference value}}$$

$$= \frac{\text{compared value} - \text{reference value}}{\text{reference value}}$$

Like the relative change formula, the relative difference formula above gives a fraction. We can convert the answer to a **percent difference** by multiplying it by 100%.

That's essentially all there is to comparisons, except for one subtlety: There's no particular reason why we chose the Lexus price as the reference value. We can just as easily use the Mercedes price as our reference value. In that case, the Lexus price is the compared value and the absolute difference is

$$\text{Lexus price} - \text{Mercedes price} = \$40,000 - \$50,000 = -\$10,000$$

The negative sign tells us that the Lexus costs $10,000 *less than* the Mercedes. The relative difference is

$$\frac{\text{absolute difference}}{\text{Mercedes price}} = \frac{-\$10,000}{\$50,000} = -0.20 = -20\%$$

The negative sign indicates that the Lexus costs 20% *less than* the Mercedes.

We now have two ways to express the relative difference:

- The Mercedes costs 25% more than the Lexus; or
- the Lexus costs 20% less than the Mercedes.

$50,000 Mercedes $40,000 Lexus

Both statements are correct, yet they contain *different* percentages. Thus, we see the importance of keeping careful track of the reference and compared values.

For those who like formulas, we can write the following rule: If quantity A is $P\%$ more than quantity B, then quantity B is $\dfrac{P}{100 + P}$ percent less than quantity A.

►EXAMPLE 5 Pay Comparison

Average pay for full-time wage earners varies from state to state. As of 2001, Connecticut ranked first in average pay, at $46,963 per person. Montana had the lowest average pay, at $25,194 per person. Compare average pay in Montana to that in Connecticut in both absolute and relative terms.

Solution Based on the wording of the statement, we use Montana pay as the compared value and Connecticut pay as the reference value.

$$\text{absolute difference} = \text{compared value} - \text{reference value}$$
$$= \$25,194 - \$46,963$$
$$= -\$21,769$$

$$\text{relative difference} = \frac{\text{compared value} - \text{reference value}}{\text{reference value}}$$
$$= \frac{\$25,194 - \$46,963}{\$46,963}$$
$$\approx -0.464 = -46.4\%$$

In absolute terms, average pay in Montana is $21,769 *less than* average pay in Connecticut. In relative terms, average pay in Montana is about 46% less than average pay in Connecticut. (You might want to verify that this also means that average pay in Connecticut is 86.4% more than average pay in Montana.) *Now try Exercises 57–62.* ◄

OF VERSUS MORE THAN

Consider a population that triples in size from 200 to 600. There are two equivalent ways to state this change with percentages:

- Using *more than:* The new population is 200% *more than* the original population. Here we are stating the *relative change* in the population. We find the increase of 200% by using the relative change formula.

by the way

Although Montana has the lowest average pay, Mississippi ranks lowest in per capita (per person) income. Average pay is the total amount of wages earned in the state divided by the total number of wage earners. Per capita income is all income earned, divided by the total number of people, employed or not. Per capita income is a better measure of poverty or wealth, which is why Mississippi is generally considered the poorest state. But jobs there pay better, on average, than those in Montana.

- Using *of*: The new population is 300% *of* the original population, which means it is three times the size of the original population. Here, we are looking at the *ratio* of the new population to the original population, which is 3 to 1.

Notice that the percentages in the *of* and *more than* statements are related by 300% = 100% + 200%. Generalizing, we find the relationships in the following box.

Of versus *More Than* (or *Less Than*)

- If the compared value is *P%* *more than* the reference value, it is (100 + *P*)% *of* the reference value.
- If the compared value is *P%* *less than* the reference value, it is (100 − *P*)% *of* the reference value.

Time out to think

Use the relative change formula to confirm that a population that grows from 200 to 600 increases in size by 200%.

WHAT IS A RATIO?

Suppose we want to compare two quantities, such as the prices of a BMW that costs $80,000 and a Honda that costs $20,000. We can, of course, find the absolute or relative difference in the quantities. But another way to make the comparison is by dividing, which tells us the *ratio* of the two quantities. In this case, the two quantities are $80,000 and $20,000, so their ratio is

$$\frac{\$80,000}{\$20,000} = \frac{4}{1} = 4$$

We can state this result in several equivalent ways:

- The ratio of the price of the BMW to that of the Honda is 4 to 1. (The ratio can also be stated as 4, but "4 to 1" is more common.)
- The price of the BMW is 4 times the price of the Honda.
- The ratio of the price of the Honda to that of the BMW is 1 to 4, which we can also state as $\frac{1}{4}$, 0.25, or 25%.

Notice how the units of *dollars* canceled on the top and bottom of the fraction. This is a general characteristic of ratios: Because comparisons make sense only when we compare quantities with the same units, ratios always end up without units.

Example: Earth's average density is about 5.5 grams per cubic centimeter and the average density of Saturn is about 0.7 gram per cubic centimeter. What is the ratio of their densities?

Solution: We divide Earth's density by Saturn's density:

$$\frac{\text{average density of Earth}}{\text{average density of Saturn}} = \frac{5.5 \text{ g/cm}^3}{0.7 \text{ g/cm}^3} \approx 8$$

The ratio of Earth's average density to Saturn's average density is about 8 to 1. Alternatively, we can say that Earth's density is about 8 times that of Saturn, or that Saturn's density is about $\frac{1}{8}$ that of Earth. Again, notice how the units canceled, leaving the ratio without units.

Now try Exercises 33–44.

➤ **EXAMPLE 6** Salary Difference

Carol earns 50% more than William. How many times larger is her income than his?

Solution We use the rule that *P%* *more than* means $(100 + P)\%$ *of*. Because Carol's income is 50% more than William's, we set $P = 50$. Thus, Carol's income is 150% of William's income. Remembering that 150% = 1.5, we know that Carol earns 1.5 times as much as William. *Now try Exercises 63–66.* ◄

➤ **EXAMPLE 7** Sale!

A store is having a "25% off" sale. How does an item's sale price compare to its original price?

Solution The "25% off" means that an item's sale price is 25% *less than* its original price. Thus, the sale price is $(100 - 25)\% = 75\%$ *of* the original price. For example, if the original price is $100, the sale price is $75. *Now try Exercises 67–70.* ◄

Time out to think

One store advertises "30% off everything!" Another store advertises "Sale prices are 30% of original prices!" Which store is having the bigger sale? Explain.

PERCENTAGES OF PERCENTAGES

Percentage changes or comparisons can be particularly confusing when the values *themselves* are percentages. Suppose your bank increases the interest rate on your savings account from 3% to 4%. It's tempting to say that the interest rate increased by 1%, but this statement is ambiguous at best. The interest rate increased by 1 *percentage point,* but the relative change in the interest rate was

$$\frac{\text{new value} - \text{reference value}}{\text{reference value}} = \frac{4\% - 3\%}{3\%} = 0.33 = 33\%$$

Thus, you can say that the bank raised your interest rate by 33%, even though the actual rate increased by only 1 percentage point. More specifically, the *absolute change* in the interest rate was 1 percentage point, while the *relative change* in the interest rate was 33%.

> **Percentage Points versus %**
>
> When you see a change or difference expressed in *percentage points,* you can assume it is an *absolute* change or difference. If it is expressed with the % sign or the word *percent,* it should be a *relative* change or difference.

➤ **EXAMPLE 8** Newspaper Readership Declines

According to a *Time Magazine* story,

The percentage of adults reading daily newspapers fell [over a 25-year period] from 78% to 64%.

TECHNICAL NOTE
Be cautious with the *of* versus *more than* rule, which may fail if we convert the percentages to ordinary fractions or numbers. For example, the statement "50 is 100% more than 25" makes sense, but the statement "50 is 1 more than 25" does not—even though 100% = 1. Similarly, most people would say that "50 is two times larger than 25," but it is not correct to say that "50 is 200% larger than 25" (it is 200% *of* 25, or 100% more than 25). Because ordinary language can be ambiguous, it's important to examine context.

Describe the change in newspaper readership.

Solution The drop in readership from 78% to 64% represents a decline of 14 percentage points. This is the *absolute change* in the percentage of adults who read a daily newspaper. The *relative change* in readership is

$$\frac{\text{new value} - \text{reference value}}{\text{reference value}} = \frac{64\% - 78\%}{78\%} = -0.18 = -18\%$$

The negative sign indicates a *decrease* in readership. Thus, we can say that readership dropped by 18% in relative terms or by 14 percentage points in absolute terms.

Now try Exercises 71–74. ◄

▶**EXAMPLE 9** Care in Wording

If you can't convince them, confuse them.

—HARRY S TRUMAN

Assume that 40% of the registered voters in Carson City are Republicans. Read the following questions carefully, and give the most appropriate answers.

a. The percentage of voters registered as Republicans is 25% higher in Freetown than in Carson City. What percentage of the registered voters in Freetown are Republicans?

b. The percentage of voters registered as Republicans is 25 percentage points higher in Freetown than in Carson City. What percentage of the registered voters in Freetown are Republicans?

Solution

a. We interpret the 25% as a relative difference, and 25% of 40% is 10% (because $0.25 \times 0.40 = 0.10$). We add this value to the Carson City percentage to find that the percentage of registered Republicans in Freetown is $40\% + 10\% = 50\%$.

b. We interpret 25 *percentage points* as an absolute difference, so we add this value to the percentage of Republicans in Carson City. The percentage of registered Republicans in Freetown is $40\% + 25\% = 65\%$.

Now try Exercises 75–76. ◄

SOLVING PERCENTAGE PROBLEMS

Consider this statement:

Retail prices are 25% more than wholesale prices.

If you knew a wholesale price, how would you calculate the retail price? One way to begin is by translating the *more than* statement into an *of* statement. The statement becomes:

Retail prices are $(100 + 25)\% = 125\%$ *of wholesale prices.*

Replacing *of* with multiplication, we write this statement mathematically as

$$\text{retail price} = 125\% \times \text{wholesale price}$$

With this equation, we can find a retail price from a wholesale price. For example, if the wholesale price is $10, the retail price is $125\% \times \$10 = 1.25 \times \$10 = \$12.50$.

Now suppose we want to find a wholesale price from a retail price. In this case, we must divide both sides of the equation by 125%:

$$\frac{\text{retail price}}{125\%} = \frac{125\% \times \text{wholesale price}}{125\%}$$

We then interchange the left and right sides to write the result:

$$\text{wholesale price} = \frac{\text{retail price}}{125\%}$$

For example, if the retail price is $15, the wholesale price is $15/1.25 = $12. We can now generalize our results.

by the way

Percentages are often called *rates*. For example, a sales tax of 6% is called a *tax rate* of 6%. If 15% of high school students smoke, we say that the *smoking rate* among high school students is 15%.

Solving Percentage Problems

If the compared value is *P% more than* the reference value, then

$$\text{compared value} = (100 + P)\% \times \text{reference value}$$

and

$$\text{reference value} = \frac{\text{compared value}}{(100 + P)\%}$$

If the compared value is *less than* the reference value, then use $(100 - P)$ instead of $(100 + P)$ in the above equations.

> **EXAMPLE 10** Tax Calculations

 a. You purchase a shirt with a labeled (pre-tax) price of $17. The local sales tax rate is 5%. What is your final cost (including tax)?

 b. Your receipt shows that you paid $19.26 for a DVD, tax included. The local sales tax rate is 7%. What was the labeled (pre-tax) price of the DVD?

Solution

 a. The final cost of an item is its labeled price plus the sales tax. For a tax rate of $P\% = 5\%$, the final cost is

$$\text{final cost} = \text{labeled price} + (5\% \times \text{labeled price})$$

In other words, the final cost is 5% more than the labeled price. Changing from a *more than* statement to an *of* statement, we have

$$\text{final cost} = (100 + 5)\% \times \text{labeled price} = 105\% \times \text{labeled price}$$

When we substitute the given labeled price of $17, the final cost is

$$\text{final cost} = 105\% \times \$17 = 1.05 \times \$17 = \$17.85$$

The final cost for the shirt is $17.85.

b. This time the tax rate is $P\% = 7\%$, so the relation between labeled price and final cost is

$$\text{final cost} = (100 + 7)\% \times \text{labeled price} = 107\% \times \text{labeled price}$$

We are given the final cost of $19.26. We find the labeled price by dividing both sides by 107%:

$$\text{labeled price} = \frac{\text{final cost}}{107\%} = \frac{\$19.26}{107\%} = \frac{\$19.26}{1.07} = \$18.00$$

The labeled price of the DVD was $18. *Now try Exercises 77–82.* ◀

►EXAMPLE 11 Up 44%, to 10.4%

Consider the following statement from the introduction to this unit:

The rate [of smoking] for eighth graders is up 44 percent, to 10.4 percent.

What was the previous smoking rate for eighth-graders?

Solution The "10.4 percent" means that 10.4% of all eighth-graders now smoke. The "44 percent" expresses the relative change from the previous smoking rate to the new rate. It tells us that the new rate is 44% *more than* the previous rate, which means it is $(100 + 44)\% = 144\%$ *of* the previous rate:

$$\text{new rate} = 144\% \times \text{previous rate}$$

We solve for the previous rate by dividing both sides by $144\% = 1.44$:

$$\text{previous rate} = \frac{\text{new rate}}{144\%} = \frac{\text{new rate}}{1.44}$$

Now we find the previous rate by substituting 10.4% for the new rate:

$$\text{previous rate} = \frac{10.4\%}{1.44} \approx 7.2\%$$

The previous smoking rate for eighth-graders was 7.2%. Note that we left the numerator as a percentage in the above equation (rather than converting it to decimal form) so that our answer would be a percentage.

Now try Exercises 83–84. ◀

ABUSES OF PERCENTAGES

Because percentages can be so subtle, many people misuse percentages—sometimes inadvertently and sometimes deliberately. In the rest of this unit, we'll investigate a few common abuses of percentages.

Beware of Shifting Reference Values

Consider the following situation: Because of losses by your employer, you agree to accept a temporary 10% pay cut. Your employer promises to give you a 10% pay raise after six months. Will the pay raise restore your original salary?

DILBERT reprinted by permission of United Feature Syndicate, Inc.

We can answer this question by assuming some arbitrary number for your original weekly pay, such as $200. A 10% pay cut means that your pay will decrease by 10% of $200, or $20, so your weekly pay after the cut is

$$\$200 - \$20 = \$180$$

The subsequent raise increases your pay by 10% of $180, or $18, making your weekly pay

$$\$180 + \$18 = \$198$$

Thus, the 10% pay cut followed by the 10% pay raise leaves you with *less* money than you started with. This result arises because the reference value for the calculations shifted during the problem: It was $200 in the first calculation and $180 in the second.

➤ EXAMPLE 12 Shifting Investment Value

A stockbroker offers the following defense to angry investors: "I admit that the value of your investments fell 60% during my first year on the job. This year, however, their value has increased by 75%, so you are now 15% ahead!" Evaluate the stockbroker's defense.

Solution Imagine that you began with an investment of $1000. During the first year, your investment lost 60% of its value, or $600, leaving you with $400. During the second year, your investment gained 75% of $400, or $0.75 \times \$400 = \300. Thus, at the end of the second year, your investment was worth $400 + $300 = $700. This is *less* than your original investment of $1000 and certainly not a 15% gain overall. We can trace the problem with the stockbroker's defense to a shifting reference value: It was $1000 for the first calculation and $400 for the second.

Now try Exercises 85–86. ◄

➤ EXAMPLE 13 Tax Cuts

A politician promises, "If elected, I will cut your taxes by 20% for each of the first three years of my term, for a total cut of 60%." Evaluate the promise.

Solution The politician neglected the effects of the shifting reference value. A cut of 20% in each of three years will *not* make an overall cut of 60%. To see what really happens, suppose that you currently pay $1000 in taxes. The following table shows how your taxes change over the three years.

Year	Tax paid in previous year	20% of previous year tax	New tax this year
1	$1000	$200	$800
2	$800	$160	$640
3	$640	$128	$512

Over three years, your taxes decline by $1000 − $512 = $488. But this is only 48.8% of $1000, so over the three years your tax bill declines by 48.8%, not by 60%.

Now try Exercises 87–88. ◄

Less than Nothing

We often see numbers that represent large "more than" percentages. For example, a price of $40 is 300% more than a price of $10. However, in most cases it is not possible to have a "less than" percentage that is greater than 100%. To see why, consider an advertisement claiming that replacing standard light bulbs with energy-efficient ones would use "200 percent less energy." If you think about it, you'll realize that such a savings is impossible. If the new light bulbs used 100% less energy, they'd be using no energy at all. Thus, the only way they could use 200% less would be if they actually *produced* energy. Clearly, whoever wrote the advertisement made a mistake.

by the way

The light bulb example related here was used as the title story in the book *200% of Nothing* by A. K. Dewdney (Wiley, 1993). The book contains many other interesting stories of misuses of numbers.

► **EXAMPLE 14** Impossible Sale

A store advertises that it will take "150% off" the price of all merchandise. What should happen when you go to the counter to buy a $500 item?

Solution If the price were 100% off, the item would be free. Thus, if the price is 150% off, the store should *pay you* half the item's cost, or $250. More likely, the store manager did not understand percentages. *Now try Exercises 89–94.* ◄

Time out to think _____

Can an athlete give a 110% effort? Can a glass of juice have 110% of the minimum daily requirement for vitamin C? Explain.

Don't Average Percentages

Suppose you got 70% of the questions correct on a midterm exam and 90% correct on the final exam. Can we conclude that you answered 80% of all the questions correctly? It might be tempting to say yes—after all, 80% is the average (mean) of 70% and 90%. But it would be wrong, unless both tests happened to have the same number of questions.

To see why, let's say that the midterm had 10 questions and the final had 100 questions, for a total of 110 questions. Your 70% score on the midterm means you answered

7 questions correctly, while your 90% score on the final means you answered 90 questions correctly. Thus, on the two exams combined, you answered 97 out of 110 questions correctly, which is 88.2% (because $97/110 = 0.882$). This is much higher than the 80% "average" of the two individual exam percentages.

This example carries a very important lesson: As a general rule, *you should never average percentages.*

> **EXAMPLE 15** Batting Average

In baseball, a player's batting average represents the percentage of at-bats in which he got a hit. For example, a batting average of .350 means the player got a hit 35% of the times he batted. Suppose a player had a batting average of .200 during the first half of the season and .400 during the second half of the season. Can we conclude that his batting average for the entire season was .300 (the average of .200 and .400)? Why or why not? Give an example that illustrates your reasoning.

Solution No. For example, suppose he had 300 at-bats during the first half of the season and 200 at-bats during the second half, for a total of 500 at-bats. His first-half batting average of .200 means he got hits on 20% of his 300 at-bats, or $0.2 \times 300 = 60$ hits. His second-half batting average of .400 means he got hits on 40% of his 200 at-bats, or $0.4 \times 200 = 80$ hits. For the season, he got a total of $60 + 80 = 140$ hits in his 500 at-bats. Thus, his season batting average was $140/500 = 28\%$, or .280—not the .300 found by averaging his first-half and second-half batting percentages. (In fact, the only case in which his season average would be .300 is if he had precisely the same number of at-bats in each half of the season.) *Now try Exercises 95–96.* ◄

> **by the way**
> We usually think of the average of two numbers as their sum divided by 2. Technically, this type of average is called a *mean*. We'll discuss other definitions of *average* in Unit 6A.

EXERCISES 3A

Review Questions

1. Describe the three basic uses of percentages. Give a sample statement that uses percentages in each of the three ways.

2. Distinguish between absolute and relative change. Give an example that illustrates how we calculate a relative change.

3. Distinguish between absolute and relative difference. Give an example that illustrates how we calculate a relative difference.

4. Explain the difference between the key words *of* and *more than* when dealing with percentages. How are their meanings related?

5. Explain the difference between the terms *percent* (%) and *percentage points*. Give an example of how they can differ for the same situation.

6. Give an example to explain why, in general, it is not legitimate to average percentages.

Does It Make Sense?

Decide whether each of the following statements makes sense (or is clearly true) or does not make sense (or is clearly false). Explain your reasoning.

7. In many European countries, the percentage change in population has been negative in recent decades.

8. The value of our home has tripled since we bought it— that's a 200% increase in value!

9. My older child weighs 25% more than my younger child.

10. I've decreased my caloric intake 125%, which has helped me lose weight.

11. If you earn 20% more than I do, then I must earn 20% less than you do.

12. If they keep cutting taxes by 10% every year, in a decade we won't be paying any tax at all!

13. We found that these rare cancers were 700% more common in children living near the toxic landfill than in the general population.

14. I have a 60% average on our assignments going into the final exam, but I still hope to raise my course average to 70% by getting an 80% on the final. (The final is worth 25% of the final grade.)

15. The rate of return on our fund increased by 50%, to 15%.

16. My bank increased the interest rate on my savings account 100%, from 2% to 4%, so my bank balance is now doubling every month.

Basic Skills & Concepts

Review of Fractions. Exercises 17–32 use skills covered in the Brief Review on p. 135. For each exercise, express the given number in three forms: fraction, decimal, and percentage.

17. $\frac{1}{4}$

18. 120%

19. 0.45

20. 0.652

21. 0.33333. . .

22. 2.34

23. 23%

24. 0.98

25. $\frac{1}{8}$

26. 45%

27. 1.34

28. 1.01

29. 0.65

30. $\frac{3}{8}$

31. 99%

32. 125%

Review of Ratios. Exercises 33–44 use skills covered in the Brief Review on p. 140. Each exercise gives a pair of quantities, A and B. Compare them in three ways. First, find the ratio of A to B. Then, find the ratio of B to A. Finally, complete this statement: A is _____ percent of B.

33. $A = 10$ and $B = 2$

34. $A = 150$ and $B = 25$

35. $A = 550$ and $B = 50$

36. $A = 45$ and $B = 550$

37. $A = 75$ and $B = 480$

38. $A = 33$ and $B = 123$

39. A is the 7.5 million population of North Carolina and B is the 16.5 million population of Texas.

40. A is the 2.8 million population of Chicago and B is the 0.75 million population of Baltimore.

41. A is the 359,000 births recorded in the United States in a recent July and B is the 290,000 births recorded in the United States in a recent January.

42. A is the average U.S. household size of 3.1 people in 1970 and B is the average U.S. household size of 2.6 people in 2003.

43. A is the average size of a new single-family home in 1970, which was 1500 square feet. B is the average size of such a new home in 2003, which was 2200 square feet.

44. A is the median price of $23,000 for a single-family house in the United States in 1970 and B is the median price of $163,000 for a single-family house in the United States in 2003.

Percentages as Fractions. Each statement in Exercises 45–50 gives two numbers. Express the first number as a percentage of the second.

45. 25 women in a room of 113 people

46. 345 blooming tulips in a field of 398 tulips

47. 1234 people who voted for the losing candidate out of 3009 voters

48. 38 men at a convention of 236 people

49. 23 purple jelly beans in a jar of 123 jelly beans

50. 2345 people who voted for the winning candidate out of 4023 voters

51. Salary Comparison. John's salary grew from $20,000 in 2000 to $28,000 in 2003. Mary's salary grew from $25,000 to $35,000 in the same period. Whose salary grew more in absolute terms? in relative terms? Explain.

52. Population Comparison. The population of Clarkville grew from 1400 to 2200 in the past decade, while the population of Centerville grew from 27,000 to 29,000. Which town had the greater absolute growth in population? Which town had the greater relative growth? Explain.

Percentage Change. Exercises 53–56 each describe measurements made at two different times. Find the absolute and relative change (as a percentage) in each case.

53. The population of the United States grew from 76 million in 1900 to 281 million in 2000.

54. Consumer debt in the United States increased from $130 billion in 1974 to $1.75 trillion ($1750 billion) in 2003.

55. The number of daily newspapers in the United States was 2226 in 1900 and 1483 in 2000.

56. The annual number of deaths from cardiovascular disease in the United States decreased from 1,008,000 in 1970 to 950,000 in 2000.

Percentage Difference. Exercises 57–62 each state two measurements. Find the absolute and relative difference (as a percentage) in each case. Assume that the first quantity is the compared value and the second quantity is the reference value.

57. The daily circulation of the *Wall Street Journal* is about 1.77 million (the largest in the country). The daily circulation of the *New York Times* is about 1.07 million (the third largest in the country).

58. In a recent year, Ford sold 401,049 Tauruses and Honda sold 382,298 Accords.

59. In a recent year, France ranked as the number one tourist destination, with about 66 million international arrivals. The United States ranked second, with about 48 million international arrivals.

60. In a recent year, Saudi Arabia produced 8.2 million barrels of crude oil per day, while the United States produced 6.4 million barrels per day.

61. In a recent year, 504,000 males had open-heart surgery in the United States, while 209,000 females had the operation.

62. The life expectancy in Canada is 79.1 years, while the life expectancy in the United States is 76.0 years.

Of **vs.** *More Than.* Fill in the blanks in Exercises 63–66. Briefly explain your reasoning in each case.

63. Jack weighs 40% more than Jill, so his weight is _____% of Jill's weight.

64. The area of Norway is 24% more than the area of Colorado, so Norway's area is _____% of Colorado's area.

65. The population of Montana is 20% less than the population of New Hampshire, so Montana's population is ____% of New Hampshire's population.

66. Henry earns 45% less than Ingrid, so his salary is ____% of Ingrid's salary.

Prices and Sales. Fill in the blanks in Exercises 67–70.

67. The wholesale cost of a car is 50% less than its retail cost. Therefore, the retail cost is ____ times the wholesale cost.

68. A store is having a 15% off sale. Therefore, a sale price is ____ times an item's original price.

69. The retail cost of a TV is 40% more than its wholesale cost. Therefore, the retail cost is ____ times the wholesale cost.

70. A store is having a 40% off sale. The sale price for an item with a regular cost of $80 is ____.

Percentages of Percentages. Exercises 71–74 describe changes in which the measurements themselves are percentages. Express each change in two ways: (1) as an absolute difference in terms of *percentage points* and (2) as a relative difference in terms of a percentage.

71. In a 22-year period, the percentage of high school seniors using alcohol decreased from 68.2% to 52.7%.

72. In a 25-year period, the percentage of the world's population living in developed countries decreased from 27% to 20%.

73. In the past four decades, the five-year survival rate for all forms of cancer increased from 39% to 60%.

74. The percentage of all bachelor's degrees awarded to women rose from 44% in 1972 to 58% in 2000.

75. Care in Wording. Assume that 30% of city employees in Carson City ride the bus to work. Consider the following two statements:

• The percentage of city employees who ride the bus to work is 10% higher in Freetown than in Carson City.

• The percentage of city employees who ride the bus to work is 10 percentage points higher in Freetown than in Carson City.

For each case, state the percentage of city employees in Freetown who ride the bus to work. Briefly explain why the two statements have different meanings.

76. Ambiguous News. The average annual precipitation on Mt. Washington, New Hampshire, is 90 inches. During

one particularly wet year, different news reports carried the following statements.

- The precipitation this year is 200% of normal.
- The precipitation this year is 200% above normal.

What does each of these statements imply about the precipitation during this year? Do the two statements have the same meaning? Explain.

Solving Percentage Problems. Solve the percentage problems in Exercises 77–84.

77. You purchase a bicycle with a labeled (pre-tax) price of $699. The local sales tax rate is 7.6%. What is the final cost?

78. The final cost of your new shoes is $107.69. The local sales tax rate is 6.2%. What was the labeled (pre-tax) price?

79. You purchase a car with a labeled (pre-tax) price of $17,600. The local sales tax rate is 6.5%. What is the final cost?

80. The final cost of your new HDTV is $3706.30. The local sales tax rate is 8.1%. What was the labeled (pre-tax) price?

81. Your dinner bill is $18.75. You leave $22. What percent tip did you leave? (Ignore taxes in this question.)

82. Your dinner bill is $42.50. You leave $50. What percent tip did you leave? (Ignore taxes in this question.)

83. Consider the statement "The smoking rate among tenth-graders jumped 45 percent, to 18.3 percent." What percentage of tenth-graders smoked before the increase?

84. Consider the statement: "The smoking rate among twelfth-graders jumped 20 percent, to 22 percent." What percentage of twelfth-graders smoked before the increase?

Shifting Reference Value. In each of Exercises 85–88, determine whether the concluding statement is true or false. If it is false, state the true change. Explain clearly.

85. Sue went on a weight loss program and lost 10% of her weight. However, she then gained 10% (from her lowest weight). In the end, her weight was unchanged.

86. You receive a pay raise of 5%, then receive a pay cut of 5%. After the two changes in pay, your salary is unchanged.

87. Your property taxes increase 5% in one year and increase 6% the next year. The total increase in taxes over the two years is 11%.

88. A high school reports that its students' SAT scores were down by 20% for one year. The next year, however, they rose by 30%. The high school principal announces, "Overall, test scores have improved by 10% over the past two years."

Is It Possible? Decide whether each claim in Exercises 89–94 is possible. If it is, explain how. If not, explain why not.

89. The price of cellular service has decreased 150% in the last ten years.

90. Mary is 100% shorter than Vivian.

91. The price of real estate in this city has increased 450% in the last ten years.

92. Super Oats have 100% more calcium than Regular Oats.

93. Your computer is 250% faster than mine.

94. Your computer is 250% slower than mine.

95. **Average Percentages.** You are a teacher. Your first-period class, with 25 students, had a mean score of 86% on the midterm exam. Your second-period class, with 30 students, had a mean score of 84% on the same exam. Does it follow that the mean score for both classes is 85%? Explain.

96. **Average Percentages.** A player has a batting average over many games of .400. In his next game, he goes 2 for 4, which is a batting average of .500 for the game. Does it follow that his new batting average is $(.400 + .500)/2 = .450$? Explain.

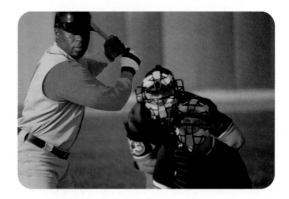

Further Applications

Analyzing Percentage Statements. For Exercises 97–100, decide if the statement is true or false. Explain.

97. The class is 60% women and 10% of the women have blond hair, so blond-haired women comprise $60\% \times 10\% = 6\%$ of the class.

98. The class is 40% men and 20% of the men are bald, so bald men comprise 40% × 20% = 80% of the class.

99. 70% of the hotels have a restaurant and 20% have a swimming pool, so 90% of the hotels in town have a restaurant or a pool.

100. 40% of the class drives to school and 15% takes a bus there, so 45% of the class neither drives nor takes a bus.

Solving News Percentages. Each of Exercises 101–110 has a quote from a *Time Magazine* article. Answer the question that follows the quote.

101. "Almost 75% of the $100 billion in U.S. trade with Mexico is delivered by truck and most of that cargo travels through Texas."
 Question: What is the value of U.S. trade with Mexico (in dollars) that is delivered by truck?

102. "In 1991 just 12% of the U.S. population was considered obese; by 1998 it was 20%."
 Question: What is the percentage increase in the proportion of the U.S. population considered obese?

103. "Bagel sales have reached $3 billion and are growing 20% annually according to Lehman Brothers."
 Question: At this growth rate, how much will be spent on bagels next year?

104. "Consumer debt has soared 39% in the last five years and now exceeds $1 trillion."
 Question: What (approximately) was consumer debt five years ago?

105. "In a new TIME/CNN poll, 80% of working people say they're not worried about losing their jobs, up 16% since last winter."
 Question: What percentage of people said they were not worried last winter?

106. "Iraq's oil industry, which in its heyday produced 3.5 million barrels a day, now produces little more than 5% of that."
 Question: Approximately how much oil does Iraq produce today?

107. "In 2003, 625,000 U.S. prisoners were released. Within three years, 60% of those former inmates will be re-arrested."
 Question: Approximately how many U.S. prisoners will be rearrested within three years?

108. "Factory-to-dealer sales of mobile video and navigation devices amounted to more than $450 million last year, a 54% jump over the previous year."
 Question: What were the sales of mobile video and navigation devices in the previous year?

109. "Oracle's stock rose $13.48 at the end of the week, up 3% over the previous week's close."
 Question: What was the price of Oracle's stock at the previous week's close?

110. "The average cost of bringing a barrel of oil out of the ground in the U.S. is $10. In Saudi Arabia, it's about $2.50. And in Iraq, it's less than $1."
 Questions: How do the costs of bringing a barrel of oil out of the ground in Iraq and in the United States compare, in terms of percentages? How do the costs of bringing a barrel of oil out of the ground in Saudi Arabia and in the United States compare, in terms of percentages?

Web Project

Find useful links for Web Projects on the text Web site: www.aw.com/bennett-briggs

111. **Web News Percentages.** Go to the Web site for a newspaper and look for at least three stories with percentages. Explain the percentages in each case.

In the News

112. **Percentages.** Find three recent news reports that quote percentages. In each case, describe the use of the percentage (as a fraction, to describe change, or for comparison) and explain its context.

113. **Percentage Change.** Find a recent news report that quotes a percentage change. Describe the meaning of the change.

114. **Abuse of Percentages.** Find a news article or report in which the use of a percentage is either suspicious or wrong. If possible, clarify or correct the statement.

UNIT 3B

PUTTING NUMBERS IN PERSPECTIVE

What is a billion dollars or how much is a trillion dollars? For many people, these numbers are just words—so large that they hardly seem to mean anything. But we hear numbers like these every day, and we cannot truly understand the major issues of our time unless we have some understanding of the numbers that are involved. In this unit, we will study several techniques for putting large or small numbers into a perspective that gives them real meaning.

WRITING LARGE AND SMALL NUMBERS

Working with large and small numbers is much easier when we write them in a special format known as **scientific notation.** We express numbers in this format by writing a number *between* 1 and 10 multiplied by a power of 10. (See the Brief Review on p. 105 for a review of powers of 10.) For example, a billion is ten to the ninth power, or 10^9, so we write 6 billion in scientific notation as 6×10^9. Similarly, we write 420 in scientific notation as 4.2×10^2, and 0.67 as 6.7×10^{-1}.

DEFINITION

Scientific notation is a format in which a number is expressed as a number *between* 1 and 10 multiplied by a power of 10.

Scientific notation makes it easy to write numbers no matter how large or small they might be. We must be careful, however, not to let this ease of writing deceive us. For example, it's so easy to write the number 10^{80} that we might think it's not all that big—but, in fact, it is a number larger than the total number of atoms in the known universe.

> ➤ **EXAMPLE 1** Numbers in Scientific Notation
>
> Rewrite each of the following statements using scientific notation.
>
> **a.** The U.S. federal debt is about $6,900,000,000,000.
> **b.** The diameter of a hydrogen nucleus is about 0.000 000 000 000 001 meter.
>
> *Solution* Notice how much easier it is to read the numbers with scientific notation.
>
> **a.** The U.S. federal debt is about 6.9×10^{12}, or $6.9 trillion.
> **b.** The diameter of a hydrogen nucleus is about 1×10^{-15} meter.
>
> *Now try Exercises 23–28.*

by the way

In the United States, a *billion* is a thousand million, or 10^9, and a trillion is a thousand billion, or 10^{12}. But in Great Britain and Germany, a billion is a million million, or 10^{12}, and a trillion is 10^{18}. We will use only the U.S. meanings in this book.

A brief review . . .

WORKING WITH SCIENTIFIC NOTATION

Converting to Scientific Notation

To convert a number from ordinary notation to scientific notation:

Step 1. Move the decimal point to come after the *first* nonzero digit.

Step 2. For the power of 10, use the number of places the decimal point moves; the power is *positive* if the decimal point moves to the left and *negative* if it moves to the right.

Examples:

$$3042 \xrightarrow[\text{3 places to left}]{\text{decimal moves}} 3.042 \times 10^3$$

$$0.00012 \xrightarrow[\text{4 places to right}]{\text{decimal moves}} 1.2 \times 10^{-4}$$

$$226 \times 10^2 \xrightarrow[\text{2 places to left}]{\text{decimal moves}} (2.26 \times 10^2) \times 10^2$$
$$= 2.26 \times 10^4$$

Converting from Scientific Notation

To convert a number from scientific notation to ordinary notation:

Step 1. The power of 10 indicates how many places to move the decimal point; move it to the *right* if the power of 10 is positive and to the *left* if it is negative.

Step 2. If moving the decimal point creates any open places, fill them with zeros.

Examples:

$$4.01 \times 10^2 \xrightarrow[\text{2 places to right}]{\text{move decimal}} 401$$

$$3.6 \times 10^6 \xrightarrow[\text{6 places to right}]{\text{move decimal}} 3,600,000$$

$$5.7 \times 10^{-3} \xrightarrow[\text{3 places to left}]{\text{move decimal}} 0.0057$$

Multiplying or Dividing with Scientific Notation

Multiplying or dividing numbers in scientific notation simply requires operating on the powers of 10 and the other parts of the number separately.

Examples:

$$(6 \times 10^2) \times (4 \times 10^5) = (6 \times 4) \times (10^2 \times 10^5)$$
$$= 24 \times 10^7$$
$$= 2.4 \times 10^8$$

$$\frac{4.2 \times 10^{-2}}{8.4 \times 10^{-5}} = \frac{4.2}{8.4} \times \frac{10^{-2}}{10^{-5}}$$
$$= 0.5 \times 10^{-2-(-5)}$$
$$= 0.5 \times 10^3$$
$$= 5 \times 10^2$$

Note that, in both examples, we first found an answer in which the number multiplied by a power of 10 was *not* between 1 and 10. We then followed the process for converting the final answer into scientific notation.

Addition and Subtraction with Scientific Notation

In general, we must write numbers in ordinary notation before adding or subtracting.

Examples:

$$(3 \times 10^6) + (5 \times 10^2) = 3,000,000 + 500$$
$$= 3,000,500$$
$$= 3.0005 \times 10^6$$

$$(4.6 \times 10^9) - (5 \times 10^8) = 4,600,000,000 - 500,000,000$$
$$= 4,100,000,000$$
$$= 4.1 \times 10^9$$

When both numbers have the *same* power of 10, we can factor out the power of 10 first.

Examples:

$$(7 \times 10^{10}) + (4 \times 10^{10}) = (7 + 4) \times 10^{10}$$
$$= 11 \times 10^{10}$$
$$= 1.1 \times 10^{11}$$

$$(2.3 \times 10^{-22}) - (1.6 \times 10^{-22}) = (2.3 - 1.6) \times 10^{-22}$$
$$= 0.7 \times 10^{-22}$$
$$= 7.0 \times 10^{-23}$$

Now try Exercises 15–22.

Approximations with Scientific Notation

Another advantage of scientific notation is that it makes it easy to *approximate* answers without a calculator. For example, we can quickly approximate the answer to 5795 × 326 by rounding 5795 to 6000 and 326 to 300. Writing the rounded numbers in scientific notation, we then see that

$$5795 \times 326 \approx (6 \times 10^3) \times (3 \times 10^2) = 18 \times 10^5 = 1,800,000$$

Because the exact answer is 1,889,170, this approximation provides a good estimate.

➤**EXAMPLE 2** Checking Answers with Approximations

You and a friend are doing a rough calculation of how much garbage New City residents produce every day. You estimate that, on average, each of the 8 million residents produces 1.8 pounds, or 0.0009 ton, of garbage each day. Thus, the total amount of garbage is

$$8,000,000 \text{ persons} \times 0.0009 \frac{\text{ton}}{\text{person}}$$

Your friend quickly presses calculator buttons and tells you that the answer is 225 tons. Without using your calculator, determine whether this answer is reasonable.

Solution You can write 8 million as 8×10^6, which is nearly 10^7. You can write 0.0009 as 9×10^{-4}, which is nearly 10^{-3}. Thus, the product should be approximately

$$10^7 \times 10^{-3} = 10^{7-3} = 10^4 = 10,000$$

Clearly, your friend's answer of 225 tons is too small. This simple approximation technique provided a useful check, even though it did not tell us the exact answer. (You should confirm that the exact answer is 7200 tons.) *Now try Exercises 29–30.* ◄

Calvin and Hobbes by Bill Watterson

GIVING MEANING TO NUMBERS

Now that we have a method for writing large and small numbers, we can move toward our goal of putting numbers in perspective. As we found for problem solving in Chapter 2, there's no single recipe for giving meaning to numbers. Nevertheless, a few techniques are particularly useful. Here, we will study three of

them: putting numbers in perspective through estimation, through comparisons, and through scaling.

Perspective Through Estimation

Making estimates is one of the best techniques for putting numbers into perspective. For example, how high is 1000 feet? For most people, the quantity 1000 feet has little meaning by itself. However, we know that each story in a building is typically about 10 feet from floor to ceiling. Therefore, we can put 1000 feet in perspective by thinking of it as the approximate height of a 100-story building.

Keep in mind that estimates are not meant to be exact. For example, the 102-story Empire State Building stands 1250 feet tall (not counting its needle), which means our estimate of 1000 feet for a 100-story building was off by about 250 feet, or 25%, in this case. But if we started with no idea about what 1000 feet looks like, this 25% difference is not very important—picturing a 100-story building still gives us a good sense of 1000 feet.

Often, estimates are useful even if they only get us within a very broad range of an exact value. For example, you can infer a lot about the character of a town or city simply by knowing whether its population is in the ten thousands or the millions— the former is the size of a small college town and the latter is a huge city. Estimates that give only a broad range of values, such as "in the millions," are called **order of magnitude** estimates.

DEFINITION

An **order of magnitude** estimate specifies only a broad range of values, such as "in the ten thousands" or "in the millions."

In ordinary language, we usually indicate an order of magnitude estimate by actually using the word *order*. For example, we might say that the population of the United States is "on the order of 300 million," by which we mean it is nearer to 300 million than to, say, 200 million or 400 million. (The U.S. population in the 2000 census was about 281 million.) Note that the context determines how we interpret an order of magnitude estimate. When astronomers say that the number of stars in a galaxy is "of order 100 billion," they mean it could be anywhere within about a factor of 10 of this number—that is, between about 10 billion and 1 trillion. This is a much wider range than the range implied when we say "on the order of 300 million" for the population of the United States, but it is appropriate to the context.

▶EXAMPLE 3 Order of Magnitude of Ice Cream Spending

Make an order of magnitude estimate of total annual spending on ice cream in the United States.

Solution We begin by devising a strategy for the problem. Clearly, we can calculate total annual ice cream spending if we know how much the average person spends each year. We simply multiply the per person spending by the population to find the total spending. We can find the per person annual spending by multiplying the average cost

by the way

To make it *convincingly* taller than all other buildings in the world at the time, the architects of the Empire State Building designed a 200-foot cap and needle for the top of the building. The original purpose of the needle was to serve as a docking station for air ships. While the needle was never used for this purpose, it has served as a good radio and TV antenna.

TECHNICAL NOTE

In science, the term *order of magnitude* refers specifically to powers of 10. For example, 10^{23} is said to be five orders of magnitude larger than 10^{18}, because it is 10^5 times larger ($10^{23} = 10^{18} \times 10^5$).

of a serving of ice cream by the average number of servings a person eats in a year. Putting these ideas together, we can find the total annual ice cream spending with the following equation:

$$\underbrace{\frac{\text{total annual}}{\text{spending}}}_{\text{units of } \frac{\$}{\text{yr}}} = \underbrace{\frac{\text{servings per}}{\text{person per year}}}_{\text{units of } \frac{\text{servings}}{\text{person} \times \text{yr}}} \times \underbrace{\frac{\text{price per}}{\text{serving}}}_{\text{units of } \frac{\$}{\text{serving}}} \times \underbrace{\text{population}}_{\text{units of persons}}$$

Notice how the units work out, telling us that our equation will indeed give us the units of dollars per year that we expect for the answer.

Now we need a value for each term in the equation. We can reasonably guess that an average person has something like 50 servings of ice cream per year (which means about one a week). The price of ice cream is of order $1 per serving (as opposed to, say, 10¢ or $10). The U.S. population is of order 300 million (3×10^8) people. Thus, a reasonable estimate is

$$\text{total annual spending} = 50 \, \frac{\text{servings}}{\text{person} \times \text{yr}} \times \frac{\$1}{\text{serving}} \times (3 \times 10^8 \, \text{persons})$$

$$= \frac{\$1.5 \times 10^{10}}{1 \, \text{yr}}$$

Total annual spending on ice cream in the United States is of order 1.5×10^{10}, or $15 billion, per year. The actual number could be up to a few times larger or smaller. For example, actual spending might be anywhere between about $5 billion and $50 billion. Nevertheless, we've learned a lot—before we started, we had no idea how much Americans spend on ice cream, and now we know that they spend billions of dollars on ice cream each year. *Now try Exercises 31–40.* ◄

Time out to think _____

Suppose you work for a company that distributes ice cream to stores throughout the United States. Market research tells you that a $5 million advertising campaign could give you an additional 5% of the total U.S. market for ice cream. Based on the estimate in Example 3, would you expect the advertising campaign to be worth its cost? Explain.

Perspective Through Comparisons

A second general way to put numbers in perspective is by making comparisons. For example, consider $100 billion, which is roughly the wealth of the world's richest individuals. It's easy to say a number like 100 billion, but how big is it? Let's think of it in terms of counting. Suppose you were asked to count $100 billion in $1 bills. How long would it take? Clearly, if we assume you can count 1 bill each second, it would take 100 billion seconds. We can put 100 billion (10^{11}) seconds in perspective by converting to years, using a chain of conversions:

$$10^{11} \, \text{s} \times \left(\frac{1 \, \text{min}}{60 \, \text{s}}\right) \times \left(\frac{1 \, \text{hr}}{60 \, \text{min}}\right) \times \left(\frac{1 \, \text{days}}{24 \, \text{hr}}\right) \times \left(\frac{1 \, \text{yr}}{365 \, \text{days}}\right) \approx 3171 \, \text{yr}$$

We see that 100 billion seconds is equivalent to 3171 years. In other words, you would need *more than three thousand years* just to count $100 billion in $1 bills. And

Using Your Calculator

Scientific Notation

Most calculators have a special key for entering scientific notation. The key, usually labeled exp (short for "exponent") or EE (short for "enter exponent"), is used to enter a power of 10. For example, enter the number 3.5×10^6 with the key sequence

3.5 (EE) 6

You should think of the EE (or exp) as "times 10 to the power entered next."

Beware of two common mistakes: First, some people try to enter 10^5 by pressing 10 (EE) 5, but this key sequence actually means 10×10^5, or 10^6. The correct way to enter 10^5 is

1 (EE) 5

Second, some people try to enter a number like 7×10^6 by pressing 7 (EE) 10 (EE) 6, which actually means $7 \times 10 \times 10^6$, or 7×10^7. The correct way to enter 7×10^6 is

7 (EE) 6

that assumes that you never take a break: no sleeping, no eating, and absolutely no dying!

Comparisons are particularly useful in dealing with units that are relatively unfamiliar, such as energy units. Table 3.1 lists various energies that you can use in comparisons. For example, we immediately see that U.S. annual energy consumption is about $\frac{1}{5}$ of world annual energy consumption.

| TABLE 3.1 | Selected Energy Comparisons | |
|---|---|
| ITEM | ENERGY (JOULES) |
| Energy released by metabolism of 1 average candy bar | 1×10^6 |
| Energy needed for 1 hour of running (adult) | 4×10^6 |
| Energy released by burning 1 liter of oil | 1.2×10^7 |
| Electrical energy used in an average home daily | 5×10^7 |
| Energy released by burning 1 kilogram of coal | 1.6×10^9 |
| Energy released by fission of 1 kilogram of uranium-235 | 5.6×10^{13} |
| Energy released by fusion of hydrogen in 1 liter of water | 7×10^{13} |
| U.S. annual energy consumption | 1×10^{20} |
| World annual energy consumption | 5×10^{20} |
| Annual energy generation of Sun | 1×10^{34} |

NOTE: *Consumption data for 2000, from the U.S. Energy Information Administration.*

> **EXAMPLE 4** U.S. versus World Energy Consumption

Compare the U.S. population to the world population and U.S. energy consumption to world energy consumption. What does this tell you about energy usage by Americans?

Solution The world population is of order 6 billion (6×10^9) and the U.S. population is of order 300 million (3×10^8). Comparing the two quantities, we find

$$\frac{\text{U.S. population}}{\text{world population}} \approx \frac{3 \times 10^8}{6 \times 10^9} = \frac{1}{2} \times 10^{-1} = \frac{1}{20}$$

The U.S. population is roughly $\frac{1}{20}$ of the world's population but, as shown in Table 3.1, the United States uses roughly $\frac{1}{5}$ of the world's energy. This tells us that Americans use much more energy per person (about four times as much) than the world average.

Now try Exercises 41–44. ◀

> **EXAMPLE 5** Fusion Power

No one has yet succeeded in creating a commercially viable way to produce energy through nuclear fusion. However, suppose that at some time in the future we are able to build fusion power plants that are safe and cost-efficient. If we could extract all the hydrogen from water and use it for fusion, how much water would we need each minute to meet U.S. energy needs?

by the way

Nuclear fission means splitting large atomic nuclei into smaller ones, while nuclear fusion means combining small nuclei into larger ones. All current nuclear reactors involve fission of uranium or plutonium. In contrast, nuclear bombs generate most of their destructive force through fusion of hydrogen into helium. The Sun also generates energy by fusion.

by the way

Scientists working on fusion power actually use a rare form of hydrogen called *deuterium*. (Ordinary hydrogen nuclei contain just a single proton; deuterium contains a proton and a neutron.) About 1 in 50,000 hydrogen atoms is deuterium, so the flow rate needed to power the United States would be 50,000 times higher than that calculated in Example 5—but still only about the flow rate of a small creek.

Solution First, we find the amount of water needed each year by dividing the U.S. annual energy consumption by the amount of energy available through fusion of 1 liter of water (from Table 3.1):

$$\frac{10^{20} \text{ joule}}{7 \times 10^{13} \dfrac{\text{joule}}{\text{L}}} = 1.4 \times 10^{6} \text{ L}$$

That is, we would need about 1.4 million liters of water *per year*. Now we convert 1.4 million liters per year to units of liters per minute (note how the units work out).

$$\frac{1.4 \times 10^{6} \text{ L}}{\text{yr}} \times \frac{1 \text{ yr}}{365 \text{ days}} \times \frac{1 \text{ day}}{24 \text{ hr}} \times \frac{1 \text{ hr}}{60 \text{ min}} \approx \frac{2.7 \text{ L}}{\text{min}}$$

In other words, it would take less than 3 liters of water per minute—less than the rate at which water flows from a typical kitchen faucet—to meet all U.S. energy needs through nuclear fusion. Thus, if we could attach a nuclear fusion reactor to your kitchen faucet, we could stop producing and importing oil, remove all the hydroelectric dams, shut down all the coal-burning power plants, and still have energy to spare.

Now try Exercises 45–48. ◀

Time out to think _____

Scientists have been working to achieve viable nuclear fusion power plants for decades, but so far without success. Nevertheless, it's conceivable that nuclear fusion will eventually become a realistic source of energy. Based on what you've learned in Example 5, how do you think this would change our world?

Perspective Through Scaling

A third general technique for giving meaning to numbers uses *scaling* or scale models. You are probably familiar with three common ways of expressing scales:

- *Verbally:* A scale can be described in words such as "One centimeter represents one kilometer" or, more simply, as "1 cm = 1 km."
- *Graphically:* A marked miniruler can show the scale visually (Figure 3.1).
- *As a ratio:* We can state the ratio of a scaled size to an actual size. For example, there are 100,000 centimeters in a kilometer (because there are 100 centimeters in 1 meter and 1000 meters in 1 kilometer). Thus, a scale where 1 centimeter represents 1 kilometer can be described as a **scale ratio** of 1 to 100,000 (or 1/100,000).

Scales have many uses other than maps. Architects and engineers build scale models to visualize their plans. *Timelines* represent the scale of time; a given distance along a timeline represents a certain number of years. *Time-lapse photography* and computer simulations allow us to represent large blocks of time in short periods. For example, in a 4-second time-lapse video clip that shows 24 hours of weather, each second of the video represents 6 hours of real time. Similarly, a computer simulation of continental drift might represent 1 billion years of geographical change in just 1 minute.

FIGURE 3.1 The miniruler acts as the map scale. In this case, 1 cm on the map represents 1 km in the city.

▶**EXAMPLE 6** Scale Ratio

A city map states, "One inch represents one mile." What is the scale ratio for this map?

Solution We find the ratio by converting 1 mile into inches:

$$1 \text{ mi} \times 5280 \frac{\text{ft}}{\text{mi}} \times 12 \frac{\text{in.}}{\text{ft}} = 63,360 \text{ in.}$$

One inch on the map represents 1 mile, or 63,360 inches. Therefore, the scale ratio for this map is 1 to 63,360, meaning that real distances are 63,360 times the distances on the map. Note that scale ratios never have units. *Now try Exercises 49–56.* ◀

▶**EXAMPLE 7** Earth and Sun

The distance from the Earth to the Sun is about 150 million kilometers. The diameter of the Sun is about 1.4 million kilometers and the diameter of the Earth is about 12,760 kilometers. Put these numbers in perspective by using a scale model of the solar system with a 1 to 10 billion scale.

Solution The scale ratio tells us that real sizes and distances are 10 billion (10^{10}) times as large as the sizes and distances in the model. Therefore, we find scaled sizes and distance by dividing the actual values by 10^{10}. For the Earth–Sun distance, we find

$$\text{scaled Earth–Sun distance} = \frac{\text{actual distance}}{10^{10}}$$
$$= \frac{1.5 \times 10^8 \text{ km}}{10^{10}}$$
$$= 1.5 \times 10^{-2} \text{ km} \times 10^3 \frac{\text{m}}{\text{km}} = 15 \text{ m}$$

Note that, in the last step, we converted the distance from kilometers to meters because it is easier to understand "15 meters" than "0.015 kilometer." We find the scaled Sun and Earth diameters similarly, this time converting the units to centimeters and millimeters, respectively:

$$\text{scaled Sun diameter} = \frac{\text{actual Sun diameter}}{10^{10}}$$

$$= \frac{1.4 \times 10^6 \text{ km}}{10^{10}}$$

$$= 1.4 \times 10^{-4} \text{ km} \times 10^5 \frac{\text{cm}}{\text{km}} = 14 \text{ cm}$$

$$\text{scaled Earth diameter} = \frac{\text{actual Earth diameter}}{10^{10}}$$

$$= \frac{1.276 \times 10^4 \text{ km}}{10^{10}}$$

$$= 1.276 \times 10^{-6} \text{ km} \times 10^6 \frac{\text{mm}}{\text{km}} = 1.276 \text{ mm}$$

The model Sun, at 14 centimeters in diameter, is roughly the size of a grapefruit. The model Earth, at about 1.3 millimeters in diameter, is about the size of the ball point in a pen, and the distance between them is 15 meters. At this scale, it is easy to visualize relative sizes and distances. *Now try Exercise 57.* ◄

Time out to think _____

Find a grapefruit or similar-sized ball and the ball point from a pen. Set them 15 meters apart to represent the Sun and Earth with a 1 to 10 billion scale. How does a scale model like this one make it easier to understand our solar system? Discuss.

►EXAMPLE 8 Distances to the Stars

The distance from the Earth to the nearest stars besides the Sun (the three stars of the *Alpha Centauri* system) is about 4.3 light-years. On the 1 to 10 billion scale of Example 7, how far are these stars from the Earth? Note: A light-year is the distance that light can travel in one year; 1 light-year = 9.5×10^{12} km.

Solution Because a light-year is about 9.5×10^{12} kilometers, 4.3 light-years is equivalent to

$$4.3 \text{ light-years} \times \frac{9.5 \times 10^{12} \text{ km}}{1 \text{ light-year}} = 4.1 \times 10^{13} \text{ km}$$

We divide this actual distance by 10 billion to find the scaled distance to the nearest stars:

$$\text{scaled distance} = \frac{\text{actual distance}}{10^{10}} = \frac{4.1 \times 10^{13} \text{ km}}{10^{10}}$$

$$= 4.1 \times 10^3 \text{ km} = 4100 \text{ km}$$

The distance to even the nearest stars on this scale is more than 4000 kilometers, or approximately the distance across the United States. *Now try Exercise 58.* ◄

Time out to think _____

Suppose that an Earth-like planet is orbiting a nearby star. Based on the results of Examples 7 and 8, discuss the challenge of trying to detect such a planet.

▶**EXAMPLE 9** Timeline

Human civilization, at least since the time of ancient Egypt, is on the order of 5000 years old. The age of the Earth is on the order of 5 billion years. Suppose we use the length of a football field, or about 100 meters, as a timeline to represent the age of the Earth. If we put the birth of the Earth at the start of the timeline, how far from the line's end would human civilization begin?

Solution First, we compare the 5000-year history of human civilization to the 5 billion-year age of the Earth:

$$\frac{5000 \text{ yr}}{5 \text{ billion yr}} = \frac{5 \times 10^3 \text{ yr}}{5 \times 10^9 \text{ yr}} = 10^{-6}$$

That is, 5000 years is about 10^{-6}, or one *millionth*, of the age of the Earth. One millionth of a 100-meter (10^2 m) timeline is

$$10^{-6} \times 10^2 \text{ m} = 10^{-6+2} \text{ m} = 10^{-4} \text{ m} \times \frac{10^3 \text{ mm}}{1 \text{ m}} = 0.1 \text{ mm}$$

In the last step, we converted from meters to millimeters. On a timeline where the Earth's history stretches the length of a football field, human civilization shows up only in the final tenth of a millimeter. *Now try Exercise 59.* ◀

PUTTING IT ALL TOGETHER: CASE STUDIES

We've studied several techniques for putting numbers in perspective, but in many cases we gain more perspective by using two or more techniques together. Sometimes we need additional creativity to think of ways to make sense of a particular number. The following case studies illustrate a few more of the many ways of putting numbers in perspective. As you study them, ask whether the numbers now have more meaning and how else you might give meaning to the numbers.

CASE STUDY How Big Is a University?

Consider a large university with 25,000 students. The number 25,000 is small compared to many of the numbers we've dealt with in this chapter, but it still takes thought to put it in perspective. Here's one way: Imagine that the new university president wants to get to know the students. She proposes to meet for lunch with groups of 5 students at a time. Is it possible for her to have lunch with *all* the students?

To answer, let's assume she holds the lunch meetings five days per week. With five students at each lunch, she will meet $5 \times 5 = 25$ students per week. Since there are 25,000 students at the school, it would take her 1000 weeks to have lunch with everyone. If we now assume that she has the lunches for 50 weeks each year (out of the total of 52 weeks in a year), it would take her 1000 weeks ÷ 50 weeks/yr = 20 years to meet all the students. But after only 4 years, most of the 25,000 students would have graduated and been replaced by a new group of 25,000 students, so she could never keep up. Thus, it would *not* be possible to meet all the students at these small group lunches.

by the way

According to modern science, the Earth and solar system formed from the collapse of a large cloud of interstellar gas. Meteorites represent some of the first rocks that formed as the cloud collapsed. Radiometric dating shows the oldest meteorites to be 4.6 billion years old, so scientists conclude that the Earth formed 4.6 billion years ago.

The lesson from this example is that while 25,000 people might not sound like much, you could not get to know all of them in 4 years—or even a lifetime.

Time out to think

Today, a typical congressional representative represents about 650,000 constituents. Is it possible for a representative to campaign by going door to door to meet everyone in his or her district? Explain.

CASE STUDY What Is a Billion Dollars?

One way to put $1 billion in perspective is to ask a question like "How many people can you employ with $1 billion per year?" Let's take $25,000 as a typical starting salary for a college graduate. Further, let's assume that it costs a business an additional $25,000 per year to have a $25,000 employee on the payroll (costs for office space, computer services, health insurance, and other benefits). At $50,000 per employee, $1 billion would allow a business to hire

A billion here, a billion there; soon you're talking real money.

—FORMER ILLINOIS SENATOR EVERETT DIRKSEN

$$\frac{\$1 \text{ billion}}{\$50,000 \text{ per employee}} = \frac{\$10^9}{\$5 \times 10^4/\text{employee}} = 2 \times 10^4 \text{ employees}$$

One billion dollars per year could support a work force of some 20,000 employees.

Another way to put $1 billion in perspective also points out how different numbers can be, even when they sound similar (like million, billion, and trillion). Suppose you become a sports star and earn a salary of $1 million per year. How long would it take you to earn a billion dollars? We simply divide $1 billion by your salary of $1 million/year:

$$\frac{\$1 \text{ billion}}{\$1 \text{ million/yr}} = \frac{\$10^9}{\$10^6/\text{yr}} = 10^3 \text{ yr} = 1000 \text{ yr}$$

Even at a salary of $1 million per year, earning a billion dollars would take a thousand years.

CASE STUDY The Scale of the Atom

Everything is made from *atoms,* which consist of a nucleus (made from *protons* and *neutrons*) surrounded by a "cloud" of *electrons.* A typical atom has a diameter of about 10^{-10} meter (as defined by its electron cloud), while its nucleus is about 10^{-15} meter in diameter. Can we put these numbers in perspective?

Let's begin with the atom itself. Because its diameter is 10^{-10} meter, or one *ten-billionth* of a meter, we could fit 10 billion (10^{10}) atoms in a line along a meter stick. A centimeter is $1/100$ (or $1/10^2$) of a meter, so we could fit $10^{10}/10^2 = 10^8$, or 100 million, atoms in a 1-centimeter line. Thus, if we could shrink people down to the size of atoms, the roughly 300 million people in the United States could fit in a line only about 3 centimeters—just over an inch—long.

Next, let's compare the diameter of the atom to the diameter of its nucleus:

$$\frac{\text{diameter of atom}}{\text{diameter of nucleus}} = \frac{10^{-10} \text{ m}}{10^{-15} \text{ m}} = 10^{-10-(-15)} = 10^5$$

by the way

The size of an atom is determined by the size of its electron cloud, but most of the mass of an atom is contained in its nucleus. Because the nucleus is so tiny compared to the atom itself, we have the surprising fact that atoms consist mostly of empty space!

The atom itself is about 10^5, or 100,000, times as large as its nucleus. Thus, if we made a scale model of an atom in which the nucleus were the size of a marble (1 cm), the atom would have a diameter of about 100,000 centimeters, which is 1 kilometer, or more than half a mile.

CASE STUDY Until the Sun Dies

We can hope that living things will flourish on Earth until the Sun dies, which astronomers estimate will be in about 5 billion (5×10^9) years. How long is 5 billion years? First, let's compare the Sun's remaining lifetime to a long human lifetime of 100 years:

$$\frac{5 \times 10^9 \text{ yr}}{100 \text{ yr}} = 5 \times 10^7$$

The Sun's remaining lifetime is equivalent to about 5×10^7, or 50 million, human lifetimes. We can add perspective by dividing a human lifetime of 100 years by 50 million and converting the result from years to minutes:

$$\frac{100 \text{ yr}}{5 \times 10^7} = 2 \times 10^{-6} \text{ yr} \times \frac{365 \text{ days}}{1 \text{ yr}} \times \frac{24 \text{ hr}}{1 \text{ day}} \times \frac{60 \text{ min}}{1 \text{ hr}} \approx 1 \text{ min}$$

Thus, a human lifetime in comparison to the Sun's remaining life is roughly the same as *one minute* in comparison to a human lifetime.

How about human creations? The Egyptian pyramids are often described as "eternal." But they are slowly eroding because of wind, rain, air pollution, and the impact of tourists. All traces of them will have vanished within a few million years. If we estimate the lifetime of the pyramids as 5 million (5×10^6) years, then

$$\frac{\text{Sun's remaining lifetime}}{\text{lifetime of pyramids}} \approx \frac{5 \times 10^9 \text{ yr}}{5 \times 10^6 \text{ yr}} = 10^3 = 1000$$

A few million years may seem like a long time, but the Sun's remaining lifetime is a thousand times as long.

On a more somber note, we can gain perspective on the Sun's remaining 5 billion years by considering evolutionary time scales. During the past century, we have acquired sufficient technology and power to destroy human life totally, if we so choose. However, even if we make that unfortunate choice, some species (including many insects) are likely to survive. Would another intelligent species ever emerge on the Earth? There is no way to know, but we can look to the past for guidance. It's been 65 million years since the dinosaurs were wiped out in what many scientists believe was a fairly sudden cataclysm caused by the impact of an asteroid. To keep the numbers simple, imagine that another intelligent species could emerge some 50 million years after a human extinction. Then, if they also destroyed themselves, another species might emerge 50 million years after that, and so on. If this pattern continued, how many

I met a traveller from an antique land
Who said: Two vast and trunkless legs of stone
Stand in the desert . . . Near them, on the sand,
Half sunk, a shattered visage lies, whose frown,
And wrinkled lip, and sneer of cold command,
Tell that its sculptor well those passions read
Which yet survive, stamped on these lifeless things,
The hand that mocked them, and the heart that fed:
And on the pedestal these words appear:
'My name is Ozymandias, king of kings:
Look on my works, ye Mighty, and despair!'
Nothing beside remains. Round the decay
Of that colossal wreck, boundless and bare,
The lone and level sands stretch far away.

—PERCY BYSSHE SHELLEY, *OZYMANDIAS*

intelligent species would eventually emerge on Earth? We divide the Sun's remaining 5 billion years by the 50 million years between successive intelligent species:

$$\frac{5 \text{ billion years}}{50 \text{ million years}} = \frac{5 \times 10^9 \text{ years}}{5 \times 10^7 \text{ years}} = 100$$

Even at 50 million years per shot, the Earth would have time for some 100 more intelligent species to emerge. Perhaps one of those species will not destroy itself, and descendants of the Earth might move on to other star systems by the time the Sun finally dies. Perhaps that species will be ours.

EXERCISES 3B

Review Questions

1. Briefly describe scientific notation. How is it useful for writing large and small numbers? How is it useful for making approximations?

2. Explain how we can use estimation to put numbers in perspective. Give an example.

3. What is an order of magnitude estimate? Explain why such an estimate can be useful even though it may be as much as 10 times too large or too small.

4. Explain how we can use comparisons to put numbers in perspective. Give an example.

5. Describe three common ways of expressing the scale of a map or model. How would you show a scale of 1 cm = 100 km graphically? How would you describe it as a ratio?

6. Explain how we can use scaling to put numbers in perspective. Give an example.

7. Suppose that the Sun were the size of a grapefruit. How big and how far away would the Earth be on this scale? How far would the nearest stars (besides the Sun) be?

8. Describe several ways of putting each of the following in perspective: the size of a large university; $1 billion; the size of an atom; the Sun's remaining lifetime.

Does It Make Sense?

Decide whether each of the following statements makes sense (or is clearly true) or does not make sense (or is clearly false). Explain your reasoning.

9. I read a book that had 10^5 words in it.

10. I've seen about 10^{50} commercials on TV.

11. I live in an apartment building that is 200 feet tall.

12. In total, Americans spend about a billion dollars per year on housing costs (rent and home mortgage payments).

13. During a recent sold-out NFL football game, the star player signed autographs for every single person in attendance.

14. The CEO of the company earned more money last year than the company's 500 lowest paid employees combined.

Basic Skills & Concepts

Review of Scientific Notation. Exercises 15–20 use skills covered in the Brief Review on p. 153.

15. Convert each of the following numbers from scientific to ordinary notation and write its name.

 Example: $2 \times 10^3 = 2000 =$ two thousand

 a. 5×10^6 b. 7×10^9

 c. 2×10^{-2} d. 8×10^{11}

 e. 1×10^{-7} f. 9×10^{-4}

16. Convert each of the following numbers from scientific to ordinary notation and write its name.

 a. 7.0×10^3 b. 1.5×10^9

 c. 3.0906×10^3 d. 2.3×10^6

 e. 2.2×10^{-4} f. 7.06×10^{-1}

17. Write each of the following numbers in scientific notation.

 a. 600 b. 0.9

 c. 50,000 d. 0.003

 e. 0.0005 f. 70,000,000,000

18. Write each of the following numbers in scientific notation.

 a. 64,598 **b.** 0.92

 c. 500.098 **d.** 0.002×10^6

 e. 250 million **f.** $-23,800$

19. Do the following operations without a calculator and show your work clearly. Be sure to express answers in scientific notation. You may round your answers to two decimal places (as in 3.2×10^5).

 a. $(3 \times 10^4) \times (8 \times 10^5)$

 b. $(6.3 \times 10^2) + (1.5 \times 10^1)$

 c. $(9 \times 10^3) \times (5 \times 10^{-7})$

 d. $(4.4 \times 10^{99}) \div (2.0 \times 10^{11})$

20. Do the following operations without a calculator and show your work clearly. Be sure to express answers in scientific notation. You may round your answers to two decimal places (as in 3.2×10^5).

 a. $(8 \times 10^{12}) \div (4 \times 10^9)$

 b. $(7.5 \times 10^{21}) \div (1.5 \times 10^{13})$

 c. $(3.2 \times 10^{22}) \div (1.6 \times 10^{-14})$

 d. $(6 \times 10^{10}) - (5 \times 10^9)$

They Don't Look That Different! In Exercises 21–22, compare each pair of numbers. By what factor do the numbers differ?

Example: 10^6 is 10^2, or 100, times as large as 10^4.

21. a. $10^{35}, 10^{26}$ **b.** $10^{17}, 10^{27}$

 c. 1 billion, 1 million **d.** 7 trillion, 7 thousand

 e. $2 \times 10^{-6}, 2 \times 10^{-9}$ **f.** $6.1 \times 10^{27}, 6.1 \times 10^{29}$

22. a. 250 million , 5 billion **b.** $9.3 \times 10^2, 3.1 \times 10^{-2}$

 c. $10^{-8}, 2 \times 10^{-13}$ **d.** $3.5 \times 10^{-2}, 7 \times 10^{-8}$

 e. 1 thousand, 1 thousandth

 f. $10^{12}, 10^{-9}$

Using Scientific Notation. Rewrite each statement in Exercises 23–28 with a number in scientific notation.

23. Total annual energy consumption in the United States is about 100,000,000,000,000,000,000 joules.

24. The United States imports about 420 million gallons of crude oil each day.

25. The hard drive on my computer has a capacity of 12.0 gigabytes. (Hint: Recall that the prefix *giga* means 1 billion.)

26. The area of the Earth's surface is 509,600,000 square kilometers.

27. The diameter of a typical bacterium is about 0.000001 meter.

28. A beam of light can travel the length of a football field in about 30 nanoseconds. (Hint: Recall that *nano* means one billionth.)

Approximation with Scientific Notation. In Exercises 29–30, make an estimate of the answer without a calculator, showing your method of estimation. Then do the exact calculation (with a calculator if necessary), and describe how well your approximation technique worked.

29. a. $20,000 \times 100$ **b.** $9642 \div 31$

 c. $-12.5 \times 11,890$ **d.** 250 million \times 40

 e. 7.453×291 **f.** $6,570,999 \div 32.7$

30. a. 5.6 billion \div 200 **b.** 4 trillion \div 260 million

 c. $9000 \times 54,986$ **d.** 3 billion \div 25,000

 e. 5987×341 **f.** $43 \div 765$

Perspective Through Estimation. Make estimates as needed to answer Exercises 31–34.

31. Which is bigger, the height of a 10-story apartment building or the length of a football field? By how much?

32. Could a person walk across the United States (New York to California) in a year? If not, about how long would it take?

33. Which is more, the number of miles Americans fly each year or the number of miles Americans drive each year?

34. Which holds more people, a football stadium or 10 movie theaters?

Orders of Magnitude Estimates. For each quantity described in Exercises 35–40, make an order of magnitude estimate. Be sure to explain your estimates clearly.

35. The total amount of money spent each year by Americans in going to the movies

36. The number of breaths you take in a week

37. The amount of water you drink in a year

38. The total number of words in this textbook

39. The amount of gasoline an average adult uses per year

40. The total amount you spend on food in a year

Energy Comparisons. Use data from Table 3.1 in Exercises 41–48.

41. How many average candy bars would you have to eat to supply the energy needed for four hours of running?

42. How many liters of oil are required to supply the electrical energy needs of an average home for a month?

43. Compare the energy released by burning 1 kilogram of coal to that released by fission of 1 kilogram of uranium-235.

44. Compare the energy released by burning 1 liter of oil to that released by fusion of the hydrogen in 1 liter of water.

45. If you could generate energy by fusing the hydrogen in ordinary tap water, how much water would you need to generate the electrical energy used daily by a typical home?

46. If you could generate energy by fusing the hydrogen in ordinary tap water, how much water would you need to meet all the world's current energy consumption?

47. How many kilograms of uranium would be required to supply the energy needs of the United States for one year using fission?

48. Suppose that we could somehow capture all the energy released by the Sun for just one second. Would this energy be enough to supply U.S. energy needs for a year? Explain.

Scale Ratio. Find the scale ratio for each of the maps described in Exercises 49–56.

49. 1 centimeter on the map represents 1 kilometer on the ground.

50. 1 inch on the map represents 1 mile on the ground.

51. 2 inches on the map represents 0.5 mile on the ground.

52. 3 inches on the map represents 2 miles on the ground.

53. 5 cm (map) = 100 km (actual)

54. 10 cm (map) = 10 km (actual)

55. 1 ft (map) = 100 m (actual)

56. 1 ft (map) = 1 km (actual)

57. **Scale Model Solar System.** The following table gives size and distance data for all the planets. Calculate the scaled size and distance for each planet using a 1 to 10 billion scale model solar system. Give your results in table form. Then write one or two paragraphs that describe your findings in words and give perspective to the size of our solar system.

Planet	Diameter	Average distance from Sun
Mercury	4880 km	57.9 million km
Venus	12,100 km	108.2 million km
Earth	12,760 km	149.6 million km
Mars	6790 km	227.9 million km
Jupiter	143,000 km	778.3 million km
Saturn	120,000 km	1427 million km
Uranus	52,000 km	2870 million km
Neptune	48,400 km	4497 million km
Pluto	2260 km	5900 million km

58. **Interstellar Travel.** The fastest spaceships launched to date are traveling away from Earth at speeds of about 50,000 kilometers per hour. How long would such a spaceship take to reach *Alpha Centauri?* (Hint: See Example 8.) Based on your answer, write one or two paragraphs discussing whether interstellar travel is a realistic possibility today.

59. **Universal Timeline.** According to modern astronomy, the universe is about 14 billion years old. Suppose you represent this age with a timeline 100 meters long. What distance along the timeline represents 1 billion years? Written human history extends back about 10,000 years. How far would that be on the timeline?

Further Applications

Making Numbers Understandable. Restate the facts in Exercises 60–67 as indicated.

60. There are approximately 4 million births in the United States in a year. Express this quantity in births per minute.

61. There are approximately 2.2 million marriages in the United States each year. Express this quantity in marriages per day.

62. There are approximately 132 million births worldwide in a year. Express this quantity in births per minute.

63. There are roughly 30,000 suicides in the United States each year. Express this quantity in suicides per day.

64. American businesses spend about $250 billion on advertising each year. Express this quantity as the height of a stack of $1 bills. (Hint: Measure the thickness of a stack of ten $1 bills.)

65. The national debt was about $6.9 trillion at the end of 2003. Express this quantity as the height of a stack of $1 bills.

66. Approximately 720,000 Americans die of a heart attack each year. Express this quantity in heart attacks per day.

67. Approximately 1.7 million college students receive bachelor's degrees each year. Express this number of students in terms of the number of football stadiums (capacity 70,000) that could be filled by these graduates.

68. CO_2 Emissions. For every gallon of gasoline burned by an automobile, approximately 10.2 kilograms of carbon dioxide are emitted into the atmosphere. Estimate the total amount of carbon dioxide added to the atmosphere by all automobile travel in the United States over the past year.

69. The Amazing Amazon. An issue of *National Geographic* contained the following statement:

> *Dropping less than two inches per mile after emerging from the Andes, the Amazon drains a sixth of the world's runoff into the ocean. One day's discharge at its mouth—4.5 trillion gallons—could supply all U.S. households for five months.*

Based on this statement, determine how much water an average U.S. household uses each month. Does this answer seem reasonable? Explain any estimates you make.

70. Wood for Energy? A total of about 180,000 terawatts of solar power reaches the Earth's surface, of which about 0.06% is used by plants for photosynthesis. Of the energy that goes to photosynthesis, about 1% ends up stored in plant matter (including wood). (Recall that 1 watt = 1 joule/s; 1 terawatt = 10^{12} watts.)

a. Calculate the total amount of energy that becomes stored in plant matter each second.

b. Suppose that power stations generated electricity by burning plant matter. If all the energy stored in plants could be converted to electricity, what average power, in terawatts, would be possible? Would it be enough to meet world electricity demand, which is of order 10 terawatts?

c. Based on your answer to b, can you draw any conclusions about why humans depend on fossil fuels, such

as oil and coal, which are the remains of plants that died long ago? Explain.

71. Stellar Corpses: White Dwarfs and Neutron Stars. A few billion years from now, after exhausting its nuclear engines, the Sun will become a type of remnant star called a *white dwarf*. It will still have nearly the same mass (about 2×10^{30} kg) as the Sun today, but its radius will be only about that of the Earth (about 6400 km).

a. Calculate the average density of the white dwarf in units of kilograms per cubic centimeter.

b. What is the mass of a teaspoon of material from the white dwarf? (Hint: A teaspoon is about 4 cubic centimeters.) Compare this mass to the mass of something familiar (for example, a person, a car, a tank).

c. A neutron star is a type of stellar remnant compressed to even greater densities than a white dwarf. Suppose that a neutron star has a mass 1.4 times the mass of the Sun but a radius of only 10 kilometers. What is its density? Compare the mass of 1 cubic centimeter of neutron star material to the total mass of Mt. Everest (about 5×10^{10} kg).

Sampling Problems. Sampling techniques can be used to estimate physical quantities. To estimate a large quantity, you might measure a representative small sample and find the total quantity by "scaling up." To estimate a small quantity, you might measure several of the small quantities together and "scale down." Carry out the necessary measurements and use sampling techniques to answer Exercises 72–75.

Example: How thick is a sheet of a paper?

Solution: One way to estimate the thickness of a sheet of paper is to measure the thickness of a ream (500 sheets) of paper. A particular ream was 7.5 centimeters thick. Thus, a sheet of paper from this ream was 7.5 cm ÷ 500 = 0.015 cm thick, or 0.15 millimeter.

72. How much does a sheet of paper weigh?

73. How thick is a penny? a nickel? a dime? a quarter? Would you rather have your height stacked in pennies, nickels, dimes, or quarters? Explain.

74. How much does a grain of sand weigh? How many grains of sand are in a typical playground sand box?

75. How many stars are visible in the sky on the clearest, darkest nights? How could astronomers estimate the total number of stars in the universe?

Web Projects

Find useful links for Web Projects on the text Web site:
www.aw.com/bennett-briggs

76. Energy Comparisons. Using data available from the Energy Information Administration Web site, choose a few measures of U.S. or world energy consumption or production. Make comparisons that put these numbers in perspective.

77. Nuclear Fusion. Learn about the current state of research into building commercially viable fusion power plants. What obstacles must still be overcome? Do you think fusion power will be a reality in your lifetime? Explain.

78. Scale Model Solar System. Visit the Web site for the Voyage scale model solar system on the National Mall in Washington, D.C. Write a brief report on what you learn.

79. Richest People. Find the net worth of the world's three richest people. Put these monetary values in perspective through any techniques you wish.

In the News

80. Large Numbers. Search today's newspaper for as many numbers larger than 100,000 as you can find. Briefly explain the context within which each large number is used.

81. Perspective in the News. Find an example in the recent news in which a reporter uses a technique to put a number in perspective. Describe the example. Do you think the technique is effective? Can you think of a better way to put the number in perspective? Explain.

82. Putting Numbers in Perspective. Find at least two examples of very large or small numbers in recent news reports. Use a technique of your choosing to put each number into perspective in a way that you believe most people would find meaningful.

DEALING WITH UNCERTAINTY

*T*he rosy outlook for the United States federal budget was big news in early 2001. Government economists projected a cumulative surplus of $5.6 *trillion* for the coming 10 years (through 2011). Politicians argued about whether this windfall should be returned to taxpayers, spent on new programs, or used to pay down the federal debt.

A mere two years later, the projected surplus had completely vanished. Instead of surpluses, most economists were predicting that the government would run huge deficits for years to come. How did the $5.6 trillion surplus—some $20,000 for every man, woman, and child in the United States—disappear?

It didn't really disappear, of course. Rather, it never existed in the first place—it was only an estimate. Like all estimates, it was only as good as the assumptions that went into it, and these assumptions included highly uncertain predictions about the future of the economy, future tax rates, and future spending. The government economists who made the projection were well aware of the uncertainties (and diligently reported these uncertainties in press releases), but the news media and politicians tended to report the surplus projection as an indisputable fact.

This story of vanishing trillions holds an important lesson. Many of the numbers we encounter in daily life are far less certain than we are told, and we can therefore be severely misled unless we learn to examine and interpret uncertainties for ourselves. In this unit, we will discuss ways of dealing with uncertainty properly.

None of us really understands what's going on with all these numbers.

—DAVID STOCKMAN, BUDGET DIRECTOR FOR PRESIDENT REAGAN, 1981

Understanding the nature of uncertainty will make you better equipped to assess the reliability of numbers as they are typically reported in the news.

SIGNIFICANT DIGITS

One key consideration in dealing with uncertainty is how we state measurements. For example, suppose you measure your weight to be 132 pounds on a scale that can be read only to the nearest pound. Saying that you weigh 132.00 pounds would be misleading, because it would incorrectly imply that you know your weight to the nearest *hundredth* of a pound, rather than to the nearest pound. In other words, 132 pounds and 132.00 pounds do *not* have the same meaning.

The digits in a number that represent actual measurements and therefore have meaning are called **significant digits.** For example, 132 pounds has 3 significant digits and implies a measurement to the nearest pound, while 132.00 pounds has 5 significant digits and implies a measurement to the nearest hundredth of a pound.

Note that zeros are significant when they represent actual measurements, but not when they serve only to locate the decimal point. We assume that the zeros in 132.00 pounds are significant because there was no reason to include them unless they represented an actual measurement. In contrast, we assume that the zeros in 600 centimeters are *not* significant, because they serve only to tell us that the decimal point comes to their right. Rewriting 600 centimeters as 6 meters makes it easier to see that only the 6 is a significant digit.

The only subtlety in counting significant digits arises when we cannot be sure whether zeros are truly significant. For example, suppose your professor states that there are 200 students in your class. Without further information, you have no way to know whether she means exactly 200 students or roughly 200. We can avoid this kind of ambiguity by writing numbers in scientific notation. In that case, zeros appear only when they are significant. For example, an enrollment of 2×10^2 implies a measurement to the nearest hundred students, while 2.00×10^2 implies exactly 200 students.

SUMMARY When Are Digits Significant?

Type of Digit	Significance
Nonzero digits	Always significant
Zeros that follow a nonzero digit *and* lie to the right of the decimal point (as in 4.20 or 3.00)	Always significant
Zeros *between* nonzero digits (as in 4002 or 3.06) or other significant zeros (such as the first zero in 30.0)	Always significant
Zeros to the *left* of the first nonzero digit (as in 0.006 or 0.00052)	Never significant
Zeros to the *right* of the last nonzero digit but before the decimal point (as in 40,000 or 210)	Not significant unless stated otherwise

➤EXAMPLE 1 Counting Significant Digits

State the number of significant digits and the implied meaning of the following numbers.

 a. a time of 11.90 seconds
 b. a length of 0.000067 meter
 c. a population reported as 240,000
 d. a population reported as 2.40×10^5

Solution

 a. The number 11.90 seconds has 4 significant digits and implies a measurement to the nearest 0.01 second.
 b. The number 0.000067 meter has 2 significant digits and implies a measurement to the nearest 0.000001 meter. Note that we can rewrite this number as 67 micrometers, showing clearly that it has only 2 significant digits.
 c. We assume that the zeros in 240,000 people are not significant. Thus, the number has 2 significant digits and implies a measurement to the nearest 10,000 people.
 d. The number 2.40×10^5 has 3 significant digits. Although this number means 240,000, the scientific notation shows that only 1 zero is significant, so it implies a measurement to the nearest 1000 people.

Now try Exercises 17–28.

ROUNDING

The basic process of rounding numbers takes just two steps.

Step 1. Decide which decimal place (e.g., tens, ones, tenths, or hundredths) is the smallest that should be kept.

Step 2. Look at the number in the next place to the *right* (for example, if rounding to tenths, look at hundredths). If the value in the next place is *less than 5*, round *down*; if it is *5 or greater*, round *up*.

For example, the number 382.2593 is given to the nearest ten thousandth. It can be rounded in the following ways:

382.2593 rounded to the nearest thousandth is 382.259.
382.2593 rounded to the nearest hundredth is 382.26.
382.2593 rounded to the nearest tenth is 382.3.
382.2593 rounded to the nearest one is 382.
382.2593 rounded to the nearest ten is 380.
382.2593 rounded to the nearest hundred is 400.

(Some statisticians use a more complex rounding rule if the value in the next column is exactly 5: They round up if the last digit being kept is odd and down if it is even. We won't use this rule in this book.)

Now try Exercises 15–16.

➤**EXAMPLE 2** Rounding with Significant Digits

For each of the following operations, give your answer with the specified number of significant digits.

 a. 7.7 mm × 9.92 mm; give your answer with 2 significant digits
 b. 240,000 × 72,106; give your answer with 4 significant digits

Solution

 a. 7.7 mm × 9.92 mm = 76.384 mm². Because we are asked to give the answer with 2 significant digits, we round to 76 mm².
 b. 240,000 × 72,106 = 1.730544 × 10¹⁰. Because we are asked to give the answer with 4 significant digits, we round to 1.731 × 10¹⁰.

Now try Exercises 29–34. ◄

UNDERSTANDING ERRORS

We are now ready to deal with errors themselves. We'll begin by describing two types of error. Then we'll discuss how to quantify the sizes of errors and how to report final results to account for errors.

Types of Error: Random and Systematic

Broadly speaking, measurement errors fall into two categories: random errors and systematic errors. The difference between them can be illustrated with an example.

Suppose you work in a pediatric office and use a digital scale to weigh babies. If you've ever worked with babies, you know that they usually aren't very happy about being put on a scale. Their thrashing and crying tends to shake the scale, making the readout jump around. For the case shown in Figure 3.2a, you could equally well record the baby's weight as anything between about 14.5 and 15.0 pounds. We say that the shaking of the scale introduces a **random error,** because any particular measurement may be either too high or too low.

Now suppose you have weighed babies all day. At the end of the day, you notice that the scale reads 1.2 pounds when there is nothing on it (Figure 3.2b). In that case, every measurement you made was too high by 1.2 pounds. This type of error is called a **systematic error,** because it is caused by an error in the measurement *system*—that is, an error that consistently (systematically) affects all measurements.

(a) (b)

FIGURE 3.2 (a) The baby's motion introduces random errors. (b) The scale reads 1.2 pounds when empty, introducing a systematic error that makes all measurements 1.2 pounds too high.

Two Types of Measurement Error

Random errors occur because of random and inherently unpredictable events in the measurement process.

Systematic errors occur when there is a problem in the measurement system that affects all measurements in the same way, such as making them all too low or too high by the same amount.

A systematic error affects all measurements in the same way, such as making them all too high or all too low. Thus, if you discover a systematic error, you can go back and adjust the affected measurements. In contrast, the unpredictable nature of random errors makes it impossible to correct for them. However, you can minimize the effects of random errors by making many measurements and averaging them. For example, if you measure the baby's weight ten times, your measurements will probably be too high in some cases and too low in others. Thus, when you average the ten measurements, you are likely to get a value that better represents the true weight.

Time out to think

Call the local phone number that gives the current time or check a Web site to get the time. How far off is your clock or watch? Describe possible sources of random and systematic errors in your time-keeping.

by the way

The fact that urban areas tend to be warmer than they would be in the absence of human activity is often called the *urban heat island effect*. Major causes of this effect include heat released by burning fuel in automobiles, homes, and industry and the fact that pavement and large masonry buildings tend to retain heat from sunlight.

►EXAMPLE 3 Errors in Global Warming Data

Scientists studying global warming need to know how the average temperature of the entire Earth, or the *global average temperature,* has changed with time. Consider two difficulties (among many others) in trying to interpret historical temperature data from the early 20th century: (1) temperatures were measured with simple thermometers and the data were recorded by hand; and (2) most temperature measurements were recorded in or near urban areas, which tend to be warmer than surrounding rural areas because of heat released by human activity. Discuss whether each of these two difficulties produces random or systematic errors, and consider the implications of these errors.

Solution The first difficulty involves *random errors* because people undoubtedly made occasional errors in reading the thermometers, in calibrating the thermometers, and in recording temperature readings. There is no way to predict whether any individual reading is correct, too high, or too low. However, if there are several readings for the same region on the same day, averaging these readings can minimize the effects of the random errors.

The second difficulty involves a *systematic error* because the excess heat in urban areas always causes the temperature reading to be higher than it would be otherwise. If the researchers could determine how much this systematic error affected the temperature readings, they could correct the data for this problem. ◄

Now try Exercises 35– 42.

▶ **EXAMPLE 4** The Census

The Constitution of the United States mandates a census of the population every 10 years. The United States Census Bureau conducts the census by distributing household surveys through the mail or through personal visits. Suggest several sources of both random and systematic error in the census.

Solution The surveys are useful if they are properly filled out and recorded. However, some people make errors when they complete the surveys, and Census Bureau employees may make errors when they enter the data into their computers. These kinds of errors are random errors because too few or too many people may end up being counted for any individual household.

Systematic errors arise from problems in the surveying process. For example, it is very difficult to deliver surveys to the homeless, and it is difficult to count undocumented aliens, who may be reluctant to reveal their presence in the United States. Thus, the homeless and undocumented aliens tend to be systematically undercounted. Other systematic errors lead to overcounts. For example, college students often get counted both by their parents and in the households where they reside at school, and children of divorced parents are sometimes counted in both parents' households.

Although there is no way to determine the overall effects of random errors, statistical studies can estimate the effects of systematic errors. For example, the 2000 census originally counted about 281.4 million people. However, follow-up statistical studies suggested that the surveys had missed about 7.6 million people and counted 4.3 million people twice. Overall, then, the census undercounted the population by over 3 million people. *Now try Exercises 43– 46.* ◀

Time out to think

The question of whether the Census Bureau should be allowed to adjust its "official" count on the basis of statistical surveys is very controversial. The Constitution calls for an "actual enumeration" of the population (Article 1, Section 2, Subsection 2). Do you believe this wording precludes or allows the use of statistical surveys in the official count? Defend your opinion. Also discuss reasons why Democrats tend to favor the use of sampling methods while Republicans tend to oppose it.

Size of Errors: Absolute versus Relative

Besides wanting to know whether an error is random or systematic, we often want to know something about the overall size of the error. For example, is the error big enough to be of concern or small enough to be unimportant? Let's explore the concept of error size.

Suppose you go to the store and ask for 6 pounds of hamburger. However, because the store's scale is poorly calibrated, you actually get only 4 pounds. You'd probably be upset by this 2-pound error. Now suppose you buy a car which the owner's manual says weighs 3132 pounds, but you find that it really weighs 3130 pounds. The error is the same 2 pounds as before, but this error probably doesn't seem very important for the car. This simple example shows that the importance of an error can depend on how you look at it.

In more technical language, the 2-pound error in both the hamburger weight and the car weight is an **absolute error**—it describes how far the measured value lies from the true value. A **relative error** compares the size of an absolute error to the true value. The relative error for the hamburger is fairly large because the absolute error of 2 pounds is half the true weight of 4 pounds. Thus, the relative error for the hamburger is 2/4, or 50%. In contrast, the relative error for the car is quite small, because 2 pounds is very small in comparison to the car's true weight of 3130 pounds. Here, the relative error is 2/3130, which is about 0.0006, or 0.06%. Note the similarity between the ideas of absolute and relative error and those of absolute and relative change or difference (see Unit 3A).

Absolute and Relative Error

The **absolute error** describes how far a measured value lies from the true value:

$$\text{absolute error} = \text{measured value} - \text{true value}$$

The **relative error** compares the size of the error to the true value:

$$\text{relative error} = \frac{\text{absolute error}}{\text{true value}} = \frac{\text{measured value} - \text{true value}}{\text{true value}}$$

The absolute and relative errors are *positive* when the measured value is greater than the true value and *negative* when the measured value is less than the true value. Note that the above formula gives the relative error as a fraction, which can be converted to a percentage.

▶ **EXAMPLE 5** Absolute and Relative Error

Find the absolute and relative error in each case.

 a. Your true weight is 125 pounds, but a scale says you weigh 130 pounds.
 b. The government claims that a program costs $49.0 billion, but an audit shows that the true cost is $50.0 billion.

Solution

 a. The measured value is the scale reading of 130 pounds, and the true value is your true weight of 125 pounds. The absolute and relative errors are

$$\text{absolute error} = \text{measured value} - \text{true value}$$
$$= 130\,\text{lb} - 125\,\text{lb}$$
$$= 5\,\text{lb}$$

$$\text{relative error} = \frac{\text{absolute error}}{\text{true value}} = \frac{5\,\text{lb}}{125\,\text{lb}} = 0.04 = 4\%$$

The measured weight is too high by 5 pounds, or 4%.

And now for some temperatures around the nation: 58, 72, 85, 49, 77.

—GEORGE CARLIN,
COMEDIAN

b. The measured value is the claimed cost of $49.0 billion, and the true value is the true cost of $50.0 billion. The absolute and relative errors are

$$\text{absolute error} = \$49.0 \text{ billion} - \$50.0 \text{ billion}$$
$$= -\$1.0 \text{ billion}$$

$$\text{relative error} = \frac{\$49.0 \text{ billion} - \$50.0 \text{ billion}}{\$50.0 \text{ billion}}$$
$$= -0.02 = -2\%$$

The claimed cost is too low by $1.0 billion, or 2%.

Now try Exercises 47–54. ◄

Describing Results: Accuracy and Precision

Once a measurement is reported, we should evaluate it to see whether it is believable in light of any potential errors. In particular, we should consider two key ideas about any reported value: its *accuracy* and its *precision*. The terms are often used interchangeably in English, but mathematically they are different.

The goal of any measurement is to obtain a value that is as close as possible to the *true value*. The closer the measured value lies to the true value, the greater is the **accuracy** of the measurement. If a census says that the population of your home town is 72,453 but the true population is 96,000, then the census report is not very accurate. In contrast, if a company projects sales of $7.30 billion and true sales turn out to be $7.32 billion, we would say that projection is quite accurate.

The **precision** of a number describes the amount of detail in a measurement. That is, it describes how *precisely* a number is measured. For example, a distance given as 2.345 kilometers is more precise than a distance given as 2.3 kilometers, because the first number gives detail to the nearest 0.001 kilometer whereas the second number gives detail only to the nearest 0.1 kilometer. Similarly, an income of $45,678.90 has greater precision than one of $46,000, because the first income is precise to the nearest penny whereas the second income is precise only to the nearest thousand dollars.

by the way

The Mars Climate Orbiter was supposed to go into orbit of Mars in late 1999. NASA engineers controlled the spacecraft from Earth by sending precise computer instructions telling it when and for how long to fire its rockets. However, the instructions were sent in English units (pounds) and the spacecraft software interpreted them in metric units (kilograms). Result: The very precise instructions were actually quite inaccurate, sending the $160 million spacecraft so close to Mars that it burned up in the Martian atmosphere.

DILBERT reprinted by permission of United Feature Syndicate, Inc.

DEFINITIONS

Accuracy describes how closely a measurement approximates a true value. An accurate measurement is very close to the true value.

Precision describes the amount of detail in a measurement.

▶ **EXAMPLE 6 Accuracy and Precision in Your Weight**

Suppose your true weight is 102.4 pounds. The scale at the doctor's office, which can be read only to the nearest quarter pound, says that you weigh $102\frac{1}{4}$ pounds. The scale at the gym, which gives a digital readout to the nearest 0.1 pound, says that you weigh 100.7 pounds. Which scale is more *precise?* Which is more *accurate?*

Solution The scale at the gym is more *precise,* because it gives your weight to the nearest tenth of a pound whereas the doctor's scale gives your weight only to the nearest quarter pound. However, the scale at the doctor's office is more *accurate,* because its value is closer to your true weight. <u>Now try Exercises 55–58.</u> ◀

Time out to think _____

In Example 6, we needed to know your *true* weight in order to determine which scale was more accurate. But how would we know your true weight? Can we ever be *sure* that we know your true weight? Explain.

CASE STUDY Does the Census Measure the True Population?

Upon completing the 2000 census, the U.S. Census Bureau reported a population of 281,421,906 (on April 1, 2000), thereby implying an exact count of everyone living in the United States. Unfortunately, such a precise count could not possibly be accurate.

Even in principle, the only way to get an exact count of the number of people living in the United States would be to count everyone *instantaneously.* Otherwise, the count would be off because the number of people changes very rapidly. An average of about eight births and four deaths occur every minute in the United States, and a new immigrant enters the United States about every three minutes on average. Moreover, random and systematic errors can easily make the census inaccurate by a few million people (see Example 4).

Thus, in reality, the Census Bureau did *not* know the exact population and only found an estimate of the population to within a few million people. Given these facts, it is dishonest to report the population as an exact count. A more honest report would use much less precision—for example, stating the population as "about 280 million." In fairness to the Census Bureau, their detailed reports did explain the uncertainties in the population count, but these uncertainties were rarely mentioned by the press.

Summary: Dealing with Errors

Mistakes are the portals of discovery.

—JAMES JOYCE

The ideas we've covered in this section are a bit technical, but very important to understanding measurements and errors. Let's briefly summarize how the ideas relate to one another.

- Errors can occur in many ways, but generally can be classified as one of two basic types: random errors or systematic errors.
- Whatever the source of an error, its size can be described in two different ways: as an absolute error or as a relative error.
- Once a measurement is reported, we can evaluate it in terms of its accuracy and its precision.

COMBINING MEASURED NUMBERS

Suppose you live in a city with a population of 300,000. One day, your best friend moves to your city to share an apartment with you. What is the population of your city now?

You might be tempted to add your friend to the city's population, making the new population 300,001. However, the number 300,000 has only 1 significant digit, implying that the population is known only to the nearest 100,000 people. The number 300,001 has 6 significant digits, which implies that you know the population *exactly*. Clearly, your friend's move cannot change the fact that the population is known only to the nearest 100,000. The population is still 300,000, despite the addition of your friend.

As this example illustrates, we must be very careful when we combine measured numbers. Otherwise, we may state answers with more certainty than they deserve. In scientific or statistical work, researchers conduct careful analyses to determine how to combine numbers properly. For most purposes, however, we can use two simple rounding rules.

Combining Measured Numbers

Rounding rule for addition or subtraction: Round your answer to the same precision as the *least precise* number in the problem.

Rounding rule for multiplication or division: Round your answer to the same number of *significant digits* as the measurement with the *fewest significant digits*.

Note: You should do the rounding only after completing all the operations, not during intermediate steps.

➤ **EXAMPLE 7** Combining Measured Numbers

a. A book written in 1962 states that the oldest Mayan ruins are 2000 years old. How old are they now?

b. The government in a town of 82,000 people plans to spend $41.5 million this year. Assuming all this money must come from taxes, what average amount must the city collect from each resident?

Solution

a. A book written in 1962 is about 40 years old, so we might be tempted to add 40 years to 2000 years to get

2040 years for the age of the ruins. However, 2000 years is the least precise of the two numbers: It is precise only to the nearest 1000 years, while 40 years is precise to the nearest 10 years. Thus, the answer also should be precise only to the nearest 1000 years:

$$\underbrace{2000 \text{ yr}}_{\text{precise to nearest 1000}} + \underbrace{40 \text{ yr}}_{\text{precise to nearest 10}} = \underbrace{2040 \text{ yr}}_{\text{must round to nearest 1000}} \approx \underbrace{2000 \text{ yr}}_{\text{correct final answer}}$$

To the precision with which we know the age of the ruins, they are still 2000 years old, despite the 40-year age of the book.

b. We find the average tax by dividing the $41.5 million, which has 3 significant digits, by the population of 82,000, which has 2 significant digits. Thus, the answer should have 2 significant digits:

$$\underbrace{\$41,500,000}_{\text{3 significant digits}} \div \underbrace{82,000 \text{ persons}}_{\text{2 significant digits}} = \underbrace{\$506.10 \text{ per person}}_{\substack{\text{must round to} \\ \text{2 significant digits}}} \approx \underbrace{\$510 \text{ per person}}_{\text{correct final answer}}$$

The average resident must pay about $510 in taxes.

Now try Exercises 59–66. ◀

EXERCISES 3C

Review Questions

1. What are significant digits? How can you tell whether zeros are significant?

2. Distinguish between random errors and systematic errors. Give an example of how each type might affect measurements of weight. How can we minimize the effects of random errors? How can we account for the effects of a systematic error?

3. Distinguish between the absolute error and the relative error in a measurement. Give an example in which the absolute error is large but the relative error is small and another example in which the absolute error is small but the relative error is large.

4. Distinguish between accuracy and precision. Give an example of a measurement that is precise but inaccurate and another example of a measurement that is accurate but imprecise.

5. Why can it be misleading to give measurements with more precision than is justified by the measurement process?

6. State the rounding rules for adding and subtracting measured numbers and for multiplying and dividing measured numbers. Give examples of their use.

Does It Make Sense?

Decide whether each of the following statements makes sense (or is clearly true) or does not make sense (or is clearly false). Explain your reasoning.

7. Next year's federal deficit will be $275.734 billion.

8. In many developing nations, official estimates of the population may be off by 10% or more.

9. I weigh 110.3627 pounds.

10. The civil rights rally on the National Mall was attended by 27,503 people.

11. The number of people who registered and paid for this year's Bolder Boulder (a popular 10-kilometer race) was 48,103.

12. More precision is useless if the measurement is inaccurate.

13. A $1 million error may sound like a lot, but when compared to our company's revenue it represents a relative error of only 0.1%.

14. Once we corrected the measurements for all known systematic errors, we were sure that our results had perfect accuracy.

Basic Skills & Concepts

Review of Rounding. Exercises 15–16 use skills covered in the Brief Review on p. 170.

15. Round the following numbers to the nearest whole number.

a. 2.4 b. 14.500 c. 779.49

d. 4.1999 e. 13.5001 f. 234.5

g. 1999.5 h. 88.71 i. −13.998

16. Round the following numbers to the nearest thousandth, tenth, ten, and hundred.

a. 2365.98521 b. 322354.09005 c. 6000

d. $\frac{34}{3}$ e. 78.555 f. 0.45232768

g. −12.1 h. −850.7654 i. −12.001

Counting Significant Digits. In Exercises 17–28, state the number of significant digits and the implied meaning of the given number.

17. 96.2 km/hr **18.** 401 people

19. 100.020 seconds **20.** 200.0 liters

21. 0.00098 mm **22.** 1.00098 mm

23. 0.0002020 meter **24.** 0.000202 meter

25. 300,000 **26.** 550,000,000

27. 3.0000×10^5 **28.** 5.500×10^8

Rounding with Significant Digits. In Exercises 29–34, carry out the indicated operation and give your answer with the specified number of significant digits.

29. 1452×9076.7; 2 significant digits

30. 1452×9076.7; 4 significant digits

31. $46,371 \div 0.00728$; 3 significant digits

32. $46,371 \div 0.00728$; 1 significant digit

33. $(3.4 \times 10^9) \div (4.1 \times 10^{-6})$; 2 significant digits

34. $(4 \times 10^{-7}) \div (9.2489 \times 10^{-2})$; 2 significant digits

Sources of Errors. For each measurement described in Exercises 35–42, briefly describe any likely sources of random errors and any likely sources of systematic errors.

35. A count of sport utility vehicles passing through a busy intersection during a 20-minute period

36. The population of a county in Mississippi, according to a census count

37. The average income of 25 people, found by checking their tax returns

38. The average income of 25 people, found by asking them their incomes during interviews conducted at a supermarket

39. Weights of sandwich meats, measured in a delicatessen using a well-calibrated scale

40. Weights of adults, measured by weighing them with their clothes on

41. The numbers of popped kernels in "large" boxes of popcorn at a movie theater

42. Times in a swimming meet

43. Tax Audit. A tax auditor reviewing a tax return looks for several kinds of problems, including these two: (1) mistakes made in entering or calculating numbers on the tax return and (2) places where the taxpayer reported income dishonestly. Discuss whether each problem involves random or systematic errors.

44. AIDS Epidemic. Researchers studying the progression of the AIDS epidemic need to know how many people are suffering from AIDS, which they can try to determine by studying medical records. Two of the many problems they face in this research are that (1) some people who are suffering from AIDS are misdiagnosed as having other diseases, and vice versa, and (2) some people with AIDS never seek medical help and therefore do not have medical records. Discuss whether each problem involves random or systematic errors.

45. Safe Air Travel. Before taking off, a pilot is supposed to set the aircraft altimeter to the elevation of the airport. A pilot leaves Denver (altitude 5280 ft) with her altimeter set to 2500 feet. Explain how this affects the altimeter readings throughout the flight. What kind of error is this?

46. **Lumber Yard.** A lumber yard is cutting boards that are supposed to be 4 feet long. Each board is cut by an employee who first measures and marks the 4-foot length. Later, careful measurements show that the average length of the boards was indeed 4 feet, but that most of the boards were actually slightly longer or shorter than 4 feet. What type of measurement error is involved in this situation? Explain.

Absolute and Relative Errors. In Exercises 47–54, find the absolute and relative errors.

47. The price of a gift you purchase by mail-order is supposed to be $18.50, but instead you are billed for $19.00.

48. The label on a bag of dog food says "20 pounds," but the true weight is only 18 pounds.

49. The bank is supposed to charge you interest of $65 on your loan, but instead charges you only $48.

50. You cut a piece of linen to a length of 48.5 inches, but the instructions call for a piece that is 48 inches long.

51. The recipe calls for $5\frac{1}{2}$ cups of flour, but you measure $5\frac{1}{4}$ cups of flour.

52. The meeting is supposed to last one hour, but it lasts one hour and five minutes.

53. The official distance to Longmont is 12 miles, but your odometer reads 11.7 miles.

54. The diameter of a ball bearing is supposed to be 0.1 centimeter, but instead is 0.2 centimeter.

Accuracy and Precision. Exercises 55–58 each describe a pair of measurements. In each case, state which measurement is more precise and which measurement is more accurate. Explain.

55. Your true height is 62.50 inches. A tape measure that can be read to the nearest $\frac{1}{8}$ inch gives your height as $62\frac{3}{8}$ inches. A new laser device at the doctor's office that gives readings to the nearest 0.05 inch gives your height as 62.90 inches.

56. Your true height is 62.50 inches. A tape measure that can be read to the nearest $\frac{1}{8}$ inch gives your height as $62\frac{5}{8}$ inches. A new laser device at the doctor's office that gives readings to the nearest 0.05 inch gives your height as 62.50 inches.

57. Your weight is 52.55 kilograms. A scale at a health clinic that gives weight measurements to the nearest half kilogram gives your weight as 53 kilograms. A digital scale at the gym that gives readings to the nearest 0.01 kilogram gives your weight as 52.88 kilograms.

58. Your weight is 52.55 kilograms. A scale at a health clinic that gives weight measurements to the nearest half kilogram gives your weight as $52\frac{1}{2}$ kilograms. A digital scale at the gym that gives readings to the nearest 0.01 kilogram gives your weight as 51.48 kilograms.

Combining Measured Numbers. Use the appropriate rounding rules to answer Exercises 59–66 with the correct precision or correct number of significant digits.

59. Subtract the volumes 140 liters and 1.09 liters.

60. Add the times 2 hours, 37 minutes and 1 hour, 22 minutes, 15 seconds.

61. Multiply the weights 9.7 kilograms and 165 kilograms.

62. Divide the distance 110 kilometers by the time 55 minutes.

63. As you drive down the freeway, a sign tells you that it is 36 miles to the downtown. Your destination lies 2.2 miles beyond city hall (which is downtown). How much farther do you have to drive?

64. At the hardware store, you buy a 50-kilogram bag of sand. You also buy 1.25 kilograms of nails. What is the total weight of your purchases?

65. The government in a city of 480,000 people plans to spend $112.4 million on a transportation project. Assuming all this money must come from taxes, what average amount must the city collect from each resident?

66. Representatives in Congress serve an average of 630,000 constituents. If a political consultant recommends campaign spending of $4.25 per constituent for an upcoming election, how much should a representative spend in total?

Further Applications

Believable Facts? Exercises 67–74 give statements of "fact" taken from official sources. For each statement, briefly discuss possible sources of error in the measurement. Then in light of the precision with which the measurement is given, discuss whether you think the fact is believable.

67. The population of the United States in 1860 was 31,443,321.

68. The number of deaths worldwide due to coronary heart disease in 1997 was 7.2 million.

69. The population of Tokyo in 2015 will be 28.7 million.

70. The Petrona Towers in Kuala Lumpur are 1483 feet tall, making them the world's tallest buildings as of 2003.

71. The average maximum temperature in January in Lagos, Nigeria, is 88 degrees.

72. The number of deaths due to AIDS in the United States as of last year was 390,242.

73. The number of cell phone users in the United States last year was 55,312,293.

74. Worldwide in 1998, there were 1694 officially declared endangered or threatened species.

Web Projects

Find useful links for Web Projects on the text Web site: www.aw.com/bennett-briggs

75. **The 2000 Census.** Go to the Web site for the Census Bureau and find the results from the 2000 census. According to the census, what was the population of the United States in 2000? How does the Census Bureau estimate the population in non-census years?

76. **Census Controversies.** Use the Library of Congress's "Thomas" Web site to find out about any pending legislation concerning the collection or use of census data. Choose one legislative bill to study in depth. Summarize the proposed legislation, and briefly discuss arguments both for and against it.

77. **Global Warming.** Find current estimates regarding how much the Earth has warmed over the past century and how much human activity contributes to that warming. Based on what you learn, write a one- to two-page essay about the uncertainty involved in understanding how human activity is contributing to global warming.

In the News

78. **Random and Systematic Errors.** Find a recent news report that gives a measured quantity (for example, a report of population, average income, or the number of homeless people). Write a short description of how the quantity was measured, and briefly describe any likely sources of either random or systematic errors. Overall, do you think that the reported measurement was accurate? Why or why not?

79. **Absolute and Relative Errors.** Find a recent news report that describes some mistake in a measured, estimated, or projected number (for example, a budget projection that turned out to be incorrect). In words, describe the size of the error in terms of both absolute error and relative error.

80. **Accuracy and Precision.** Find a recent news article that causes you to question accuracy or precision. For example, the article might report a figure with more precision than you think is justified, or it may cite a figure that you know is inaccurate. Write a one-page summary of the report and explain why you question its accuracy or precision (or both).

81. **Uncertainty in the News.** Look through the newspapers from the past week for articles that mention numbers in any context. Find at least two numbers from each of the following sections: national/international news, local news, sports, and business. In each case, describe the number and its context, and discuss any uncertainty that you think is associated with the number.

82. **Real-World Rounding.** Find at least three different examples of numbers that were rounded in news stories. Explain how and why each number was rounded. (Hint: You might consider baseball batting averages, times in athletic events, or the stock market.)

UNIT **3D**

INDEX NUMBERS: THE CPI AND BEYOND

*I*f you listen to the nightly business report, you've probably heard about **index numbers** such as the Consumer Price Index, the Producer Price Index, and the Consumer Confidence Index. Index numbers are common because they provide a simple way to compare measurements made at different times or in different places. In this unit, we'll investigate the meaning and use of index numbers. We'll focus special attention on the Consumer Price Index (CPI), because it plays such an important role in modern economics.

WHAT IS AN INDEX NUMBER?

Let's start with an example using gasoline prices. Table 3.2 shows the average prices of gasoline in the United States for selected years from 1955 to 2000. Suppose that, instead of the prices themselves, we want to find out how the price of gasoline in different years compared to the 1975 price. One way to do this would be to express each year's price as a percentage of the 1975 price. For example, by dividing the 1965 price by the 1975 price, we find that the 1965 price was 55.0% of the 1975 price:

$$\frac{1965 \text{ price}}{1975 \text{ price}} = \frac{31.2¢}{56.7¢} = 0.550 = 55.0\%$$

Proceeding similarly for each of the other years, we can calculate all the prices as percentages of the 1975 price. The third column of Table 3.2 shows the results. Note that the percentage for 1975 is 100%, because we chose the 1975 price as the reference value.

TABLE 3.2	Average Gasoline Prices (per gallon)		
YEAR	PRICE	PRICE AS A PERCENTAGE OF 1975 PRICE	PRICE INDEX 1975 = 100
1955	29.1¢	51.3%	51.3
1965	31.2¢	55.0%	55.0
1975	56.7¢	100.0%	100.0
1985	119.6¢	210.9%	210.9
1995	120.5¢	212.5%	212.5
2000	155.0¢	273.4%	273.4

SOURCE: U.S. Department of Energy.

Now look at the last column of Table 3.2. It is identical to the third column, except we dropped the % signs. This simple change converts the numbers from percentages to a *price index*, which is one type of index number. The statement "1975 = 100" in the column heading shows that the reference value is the 1975 price. In this case, there's really no difference between stating the comparison in terms of

percentages and in terms of index numbers—it's a matter of choice and convenience. However, as we'll see shortly, it's traditional to use index numbers rather than percentages in cases where many factors are considered simultaneously.

Index Numbers

An **index number** provides a simple way to compare measurements made at different times or in different places. The value at one particular time (or place) must be chosen as the *reference value*. The index number for any other time (or place) is

$$\text{index number} = \frac{\text{value}}{\text{reference value}} \times 100$$

➤ **EXAMPLE 1** Finding an Index Number

Suppose the current price of gasoline is $1.60 per gallon. Using the 1975 price as the reference value, find the price index number for gasoline today.

Solution Table 3.2 shows that the price of gas was 56.7¢, or $0.567, per gallon in 1975. If we use the 1975 price as the reference value, the index number for the current price is

$$\frac{\text{current price}}{\text{1975 price}} \times 100 = \frac{\$1.60}{\$0.567} \times 100 = 282.2$$

The index number for the current gasoline price is 282.2, meaning that the current price is 282.2% of the 1975 price. *Now try Exercises 11–12.* ◄

Time out to think _____

What is the actual price of gasoline today? What is the gasoline price index for today's price, with the 1975 price as the reference value?

Making Comparisons with Index Numbers

The primary purpose of index numbers is to facilitate comparisons. For example, suppose we want to know how much more expensive gas was in 2000 than in 1975. We can get the answer easily from Table 3.2, which uses the 1975 price as the reference value. This table shows that the price index for 2000 was 273.4, which means that the price of gasoline in 2000 was 273.4% of the 1975 price. Equivalently, we can say that the 2000 price was 2.734 times the 1975 price.

 We can also do comparisons when neither value is the reference value. For example, suppose we want to know how much more expensive gas was in 1995 than in 1965. We find the answer by dividing the index numbers for the two years:

$$\frac{\text{index number for 1995}}{\text{index number for 1965}} = \frac{212.5}{55.0} = 3.86$$

That is, the 1995 price was 3.86 times the 1965 price, or 386% of the 1965 price. The same amount of gas that cost $1.00 in 1965 would have cost $3.86 in 1995.

➤ **EXAMPLE 2** Using the Gas Price Index

Use Table 3.2 to answer the following questions:

 a. Suppose it cost $7.00 to fill a gas tank in 1975. How much would it have cost to fill the same tank in 1995?

 b. Suppose it cost $20.00 to fill a gas tank in 1995. How much would it have cost to fill the same tank in 1955?

Solution

 a. Table 3.2 shows that the price index for 1995 was 212.5, which means that the price of gasoline in 1995 was 212.5% of the 1975 price. Thus, the 1995 price of gas that cost $7.00 in 1975 would have been

$$212.5\% \times \$7.00 = 2.125 \times \$7.00 = \$14.88$$

 b. Table 3.2 shows that the price index for 1995 was 212.5 and the index for 1955 was 51.3. Thus, the cost of gasoline in 1955 compared to the cost in 1995 was

$$\frac{\text{index number for 1955}}{\text{index number for 1995}} = \frac{51.3}{212.5} = 0.2414$$

A tank of gas that cost $20.00 in 1995 would have cost only $0.2414 \times \$20.00 = \4.83 in 1955. *Now try Exercises 13–14.* ◄

Changing the Reference Value

There's no particular reason why we chose the 1975 price as the reference value in Table 3.2, and it is easy to recast the table with a different reference value. For example, let's compute the gasoline price index with the 1985 price as the reference value.

The first two columns of the table do not change because they give actual years and prices of gasoline. But this time, the reference value is the 1985 price of 119.6¢. We use this value to find the index numbers for the other years. For example, the index number for the 1955 price of 29.1¢ is

$$\frac{\text{1955 price}}{\text{1985 price}} \times 100 = \frac{29.1¢}{119.6¢} \times 100 = 24.3$$

As we should expect, the index number is *not* the same as it was when we used 1975 as the reference year. Table 3.3 shows the index numbers with both 1975 and 1985 as the reference year. Both sets of numbers are equally valid, but meaningful only when we know the reference value. Thus, including the reference year is crucial.

TABLE 3.3		The Gasoline Price Index with Two Different Reference Years	
YEAR	PRICE	PRICE INDEX (1975 = 100)	PRICE INDEX (1985 = 100)
1955	29.1¢	51.3	24.3
1965	31.2¢	55.0	26.1
1975	56.7¢	100.0	47.4
1985	119.6¢	210.9	100.0
1995	120.5¢	212.5	100.8
2000	155.0¢	273.4	129.6

Time out to think _____

For practice in finding index numbers, use the index number formula to confirm the rest of the values in the last column of Table 3.3.

► **EXAMPLE 3** 1985 Index

Suppose the current price of gasoline is $1.60 per gallon. Using the 1985 price as the reference value, find its price index number. Compare this number to the answer in Example 2, where 1975 was the reference year.

Solution Table 3.3 shows that the price of gas was 119.6¢, or $1.196, per gallon in 1985. With the 1985 price as the reference value, the index number for the current price is

$$\frac{\text{current price}}{\text{1985 price}} \times 100 = \frac{\$1.60}{\$1.196} \times 100 = 133.8$$

As we should expect, this index number is different from the value of 282.2 found in Example 2, when we used 1975 as the index year. Both numbers are valid, but meaningful only when we know the reference value. Specifically, the 1985 index number tells us that the current price is 133.8% of the 1985 price, while the 1975 index number tells us that the current price is 282.2% of the 1975 price. *Now try Exercises 15–16.* ◄

THE CONSUMER PRICE INDEX

We've seen that the price of gas has risen substantially with time. Most other prices and wages have also risen, a phenomenon we call **inflation.** (Prices and wages occasionally decline with time, which is *deflation.*) Thus, changes in the actual price of gasoline are not very meaningful unless we compare them to the overall rate of inflation, which is measured by the **Consumer Price Index (CPI).**

 The Consumer Price Index is computed and reported monthly by the U.S. Bureau of Labor Statistics. It represents an average of prices for a sample of goods, services, and housing. The monthly sample consists of more than 60,000 items. The details of the data collection and index calculation are fairly complex, but the CPI itself is a simple index number. Table 3.4 shows the average annual CPI over a 30-year

TABLE 3.4	Average Annual Consumer Price Index (1982–1984 = 100)				
YEAR	CPI	YEAR	CPI	YEAR	CPI
1973	44.4	1983	99.6	1993	144.5
1974	49.3	1984	103.9	1994	148.2
1975	53.8	1985	107.6	1995	152.4
1976	56.9	1986	109.6	1996	156.9
1977	60.6	1987	113.6	1997	160.5
1978	65.2	1988	118.3	1998	163.0
1979	72.6	1989	124.0	1999	166.6
1980	82.4	1990	130.7	2000	172.2
1981	90.9	1991	136.2	2001	177.1
1982	96.5	1992	140.3	2002	179.9

TECHNICAL NOTE
The government measures two consumer price indices. The *CPI-U* is based on products thought to reflect the purchasing habits of all urban consumers. The *CPI-W* is based on the purchasing habits of only wage earners. (Table 3.4 shows the CPI-U.)

period. Currently, the reference value for the CPI is an average of prices during the period 1982–1984, which is why the table says "1982–1984 = 100."

The Consumer Price Index

The **Consumer Price Index (CPI)** is based on an average of prices for a sample of more than 60,000 goods, services, and housing costs. It is computed and reported monthly.

The CPI allows us to see how prices have changed with time. For example, to find out how much higher typical prices were in 2000 than in 1975, we compute the ratio of the CPIs for the two years, using the shorthand CPI_{2000} to represent the CPI for 2000 (and similar notation for other years):

$$\frac{CPI_{2000}}{CPI_{1975}} = \frac{172.2}{53.8} = 3.20$$

Based on the CPI, typical prices in 2000 were 3.20 times those in 1975. For example, a typical item that cost $1.00 in 1975 would have cost $3.20 in 2000.

►EXAMPLE 4 CPI Changes

Suppose you needed $30,000 to maintain a particular standard of living in 1995. How much would you have needed in 2002 to maintain the same standard of living? Assume that the average price of your typical purchases has risen at the same rate as the Consumer Price Index (CPI).

Solution We compare prices for two different years by comparing the CPIs for those years:

$$\frac{CPI_{2002}}{CPI_{1995}} = \frac{179.9}{152.4} = 1.180$$

That is, typical prices in 2002 were 1.180 times those in 1995. If you needed $30,000 for your standard of living in 1995, then in 2002 you would have needed $1.180 \times \$30,000 = \$35,400$ for the same standard of living. *Now try Exercises 17–18.* ◄

The Rate of Inflation

The **rate of inflation** refers to the relative change in the CPI from one year to the next. For example, the inflation rate from 1980 to 1981 was the relative increase in the CPI between those two years:

$$\text{inflation rate 1980 to 1981} = \frac{CPI_{1981} - CPI_{1980}}{CPI_{1980}}$$

$$= \frac{90.9 - 82.4}{82.4} = 0.103 = 10.3\%$$

As measured by the CPI, the rate of inflation from 1980 to 1981 was 10.3%.

The Rate of Inflation

The **rate of inflation** from one year to the next is usually defined as the relative change in the Consumer Price Index.

▶**EXAMPLE 5** Inflation Comparison

Find the inflation rate from 1998 to 1999. How does it compare to the inflation rate in the late 1970s?

Solution The inflation rate from 1998 to 1999 was

$$\frac{\text{CPI}_{1999} - \text{CPI}_{1998}}{\text{CPI}_{1998}} = \frac{166.6 - 163.0}{163.0} = 0.022 = 2.2\%$$

This rate of 2.2% is much lower than the rate of inflation a couple decades earlier.

Now try Exercises 19–20. ◀

Adjusting Prices for Inflation

During the spring and summer of 2000, a sudden jump in gasoline prices became a hot consumer and political issue. Average pump prices in the Midwest peaked at $1.87 per gallon—easily shattering the previous record price of $1.47 per gallon from 1981. News reports trumpeted the "highest gasoline prices in history."

But were prices *really* at a record high? Not if you take inflation into account. If we want to compare prices fairly, we must first adjust the prices for the effects of inflation. We can see how typical prices in 2000 compared to typical prices in 1981 by finding the ratio of the CPIs for those years:

$$\frac{\text{CPI}_{2000}}{\text{CPI}_{1981}} = \frac{172.2}{90.9} \approx 1.89$$

Thus, the 1981 price of $1.47 per gallon for gasoline was equivalent to a 2000 price of 1.89 × $1.47 = $2.78. In the language of economics, we say that $1.47 in "1981 dollars" is equivalent to $2.78 in "2000 dollars." Because the actual price of gasoline in 2000 was much less than $2.78, the 2000 prices were *not* as high as the 1981 prices adjusted for inflation. Figure 3.3 shows gasoline prices going back more than 80 years. Note that, adjusted for inflation, recent gasoline prices have been much lower than prices at most other times.

If I had to populate an asylum with people certified insane, I'd just pick 'em from all those who claim to understand inflation.

—WILL ROGERS

Time out to think _____

Suppose that adjusted prices in Figure 3.3 were adjusted to 1982 dollars rather than 2002 dollars. Where would the two curves cross in that case? Would the general trends look the same or different? Explain.

Cost of Gasoline

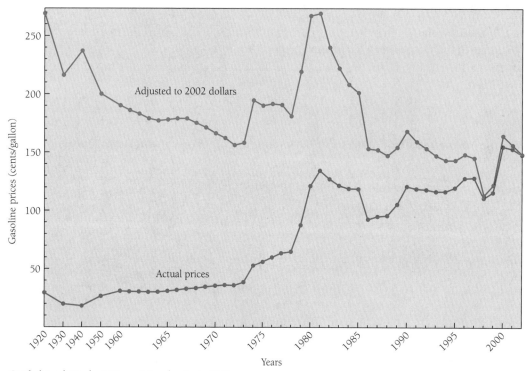

Graph shows data only at 10-year intervals prior to 1960.

FIGURE 3.3 Gasoline prices, 1918–2002. While actual prices have generally risen with time, recent prices adjusted for inflation are low by historical standards. Note that, because we use 2002 dollars for the adjusted prices, the actual and adjusted prices are the same for 2002. That is why the two curves converge at 2002. *SOURCE: American Petroleum Institute.*

Adjusting Prices for Inflation

Given a price in dollars for year X ($\$_X$), the equivalent price in dollars for year Y ($\$_Y$) is

$$\text{price in } \$_Y = (\text{price in } \$_X) \times \frac{\text{CPI}_Y}{\text{CPI}_X}$$

where X and Y represent years, such as 1992 and 2002.

➤ EXAMPLE 6 Baseball Salaries

In 1987, the mean (average) salary for major league baseball players was $412,000. In 2002, it was $2,295,000. Compare the two salaries in "2002 dollars" ($\$_{2002}$). How does the rise in baseball salaries compare to the overall rise in prices with inflation?

Solution We convert the 1987 salary of $412,000 into 2002 dollars:

$$\text{salary in } \$_{2002} = (\text{salary in } \$_{1987}) \times \frac{\text{CPI}_{2002}}{\text{CPI}_{1987}}$$

$$= \$412,000 \times \frac{179.9}{113.6}$$

$$= \$652,000$$

That is, a salary of $412,000 in 1987 was equivalent to a salary of $652,000 in 2002. However, the actual 2002 average salary of $2,295,000 was more than *triple* the adjusted salary from 1987, showing that baseball salaries have far outpaced the rate of inflation. More specifically, we say that average baseball salaries more than tripled in "real terms" (terms adjusted for inflation) from 1987 to 2002. In other words, the average baseball salary in 2002 has more than three times the buying power of the average baseball salary in 1987. *Now try Exercises 21–22.* ◄

►**EXAMPLE 7** Computer Prices

According to tests by scientists at NASA's Ames Research Center in California, the Macintosh G4 computers that sold for $3000 in mid-2000 had computing power equivalent to that of a Cray supercomputer that sold for $30 million in 1985. If computer prices had risen with inflation, how much would the computing power of the 1985 Cray have cost in 2000? What does this tell us about the cost of computers?

Solution We convert the 1985 price of $30 million into 2000 dollars:

$$\text{price in } \$_{2000} = (\text{price in } \$_{1985}) \times \frac{\text{CPI}_{2000}}{\text{CPI}_{1985}}$$

$$= \$30 \text{ million} \times \frac{172.2}{107.6} = \$48 \text{ million}$$

If computer prices had risen with inflation, the computing power of the 1985 Cray would have cost $48 million in 2000. Instead, it cost only $3000—less than 1/1000 as much. Thus, in real terms, the price of computing power fell by a factor of more than 1000 from 1985 to 2000. In other words, in a mere 15 years, computing power that had been affordable only by government agencies and large corporations became cheaper than a typical used car. *Now try Exercises 23–26.* ◄

OTHER INDEX NUMBERS

The Consumer Price Index is only one of many index numbers you'll see in news reports. Some are also price indices, such as the Producer Price Index (PPI), which measures the prices that producers (manufacturers) pay for the goods they purchase (rather than the prices that consumers pay). Other indices attempt to measure more qualitative variables. For example, the Consumer Confidence Index is based on a survey designed to measure consumer attitudes so that businesses can gauge whether people are likely to be spending or saving. New indices are created frequently by groups attempting to provide simple comparisons.

by the way

Salaries of professional athletes were once kept low because the players were not allowed to offer their skills in the free market ("free agency"). That changed after star baseball player Curt Flood filed suit against Major League Baseball in 1970. Flood ultimately lost when the Supreme Court ruled in favor of baseball management in 1972, but the process he set in motion (toward free agency) was unstoppable.

Copyright © 1998, 2004 Sidney Harris.

►EXAMPLE 8 Health Care Quality Index

Statisticians at Health Risk Management, Inc. invented the Health Care Quality Index to compare the quality of health care in different states. The index represents an average of 46 different measures of health care quality, such as spending on public health, the fraction of children receiving immunizations, and the quality of public drinking water. They chose the average score among all the states as the reference value. Thus, the average score is defined to be 100 on this index. Minnesota had the highest index value of 118, while Louisiana had the lowest value at 83. Discuss the meaning of these index values.

Solution We can start by comparing the Health Care Quality Indices for Minnesota and Louisiana:

$$\frac{\text{Health Care Quality Index Minnesota}}{\text{Health Care Quality Index Louisiana}} = \frac{118}{83} \approx 1.4$$

The Health Care Quality Index for Minnesota is 1.4 times, or 140% of, the index for Louisiana. Remembering how the word *of* relates to the words *more than*, we can also say that the index for Minnesota is 40% more than that for Louisiana. Of course, not everyone will agree that the 46 measures used in this index are the best set of measures on which to rate health care. Thus, while the index provides a useful starting point for discussions of health care quality, it is by no means definitive. Without much more discussion, it is unfair to conclude that health care is 40% better in Minnesota than in Louisiana. *Now try Exercises 27–30.* ◄

Practical matters

DOES THE CPI REALLY MEASURE INFLATION?

The Consumer Price Index affects everyone because it plays such a huge role in modern economics. For example:

• Your tax rate depends on the CPI. Each year, the government raises the levels of income at which different income tax rates take effect (*tax brackets*) by the amount that the CPI changes. This annual adjustment is supposed to protect you from the effects of inflation. If the brackets did not change, then people with no change in their living standards would gradually move to higher tax brackets.

• If you receive government payments from Social Security or other programs, your monthly benefits are raised each year to keep up with the CPI. These increases are supposed to make sure that benefits increase enough that recipients can maintain the same standard of living.

• Spending on all government programs, from education to the military, is largely determined by the CPI. Just as maintaining your standard of living requires that you receive pay increases to keep up with inflation, government programs must keep up with inflation if they are to maintain their same level of commitment. Thus, for example, increasing education spending by *less* than the increase in the CPI is considered a "cut" in real spending. One result is that total spending generally rises each year just to keep up with the CPI, even before decisions about new programs are made.

The CPI also influences how we perceive nearly all economic data. For example, over the past two to three decades, the rise in wages for lower-income people has been less than the rise in the CPI. This is taken to mean that the "real" spending power of lower-income people has decreased—that is, the poor are getting poorer.

But what if the CPI does not really measure the true rate of inflation? In that case, all the decisions and beliefs that are pegged to the CPI could be wrong. In fact, many economists argue that the CPI significantly overstates the true rate of inflation.

The thrust of the argument is that the standard calculation of the CPI contains systematic errors that tend to overstate the increase in the cost of living. One problem is that, from one month to the next, the CPI is based on changes in the prices of particular items at particular stores. However, consumers may not experience an increase in their cost of living if they can find the same item for a lower price at a different store or if they can substitute a similar but lower-priced item. Another problem arises because the CPI tracks prices of items purchased by "typical" consumers. Thus, today's CPI includes items that were not available just a couple of decades ago, such as DVD players, Internet access, and computers. The CPI does not factor in any improvement in the standard of living that these items represent. For example, consider the fact that the average person can now afford computing power that would have cost $30 million in 1985 (see Example 7). Economists would say that this represents a significant improvement in our standard of living, but the CPI does not take it into account.

The implications of an overstated CPI are immense. It might mean that the poor are *not* getting poorer after all. It would also mean that Social Security benefits have been rising more than necessary to keep pace with inflation, thereby affecting the amount of money available for all other government programs. In fact, if the CPI overstates the rate of inflation by 1 percentage point per year, the total cost for Social Security over the next decade will be more than *$1 trillion* higher than it would be if the CPI accurately tracked inflation.

Clearly, knowing the truth about the CPI and inflation is very important to all of us. Unfortunately, the CPI remains a topic of hot debate. Given the huge economic and political implications of changes to the CPI, it's unlikely that the debate will end soon.

■ EXERCISES 3D

Review Questions

1. What is an index number? Briefly describe how index numbers are calculated and what they mean.

2. What is the Consumer Price Index (CPI)? How is it supposed to be related to inflation?

3. In making price comparisons, why is it important to adjust prices for the effects of inflation? Briefly describe how we use the CPI to adjust prices.

4. List a few other uses of index numbers besides the CPI. Why is it important to understand an index before deciding whether to trust it?

Does It Make Sense?

Decide whether each of the following statements makes sense (or is clearly true) or does not make sense (or is clearly false). Explain your reasoning.

5. The price per gallon of gasoline has risen from only a quarter in 1918 to nearly $2 today, thereby making it much more difficult for the poor to afford fuel for their cars.

6. Even though my salary has remained the same for the past 7 years, my standard of living has fallen.

7. Benjamin Franklin said, "A penny saved is a penny earned," but if he were alive today, he would be talking about a dollar rather than a penny.

8. The prices of cars have risen steadily, but when the prices are adjusted for inflation, cars are actually cheaper today than they were a couple of decades ago.

9. When we chart the price of milk in 1995 dollars we find that it has become slightly more expensive, but when we chart it in 1975 dollars we find that it has become cheaper.

10. The Consumer Price Index is a nice theoretical idea, but it has no impact on me, as a student on financial aid.

Basic Skills & Concepts

Gasoline Price Index. In Exercises 11–16, specify a particular price for gasoline (see Table 3.2).

11. Suppose the current cost of gasoline is $1.80 per gallon. Find the current price index number, using the 1975 price as the reference value.

12. Suppose the current cost of gasoline is $1.45 per gallon. Find the current price index number, using the 1975 price as the reference value.

13. Suppose it cost $10 to fill a gas tank in 1975. How much would it have cost to fill the same tank in 1995?

14. Suppose it cost $15 to fill a gas tank in 1985. How much would it have cost to fill the same tank in 1965?

15. Recast the gasoline price index in Table 3.2 using the 1965 price as the reference value.

16. Recast the gasoline price index in Table 3.2 using the 2000 price as the reference value.

Understanding the CPI. Use Table 3.4 to answer Exercises 17–26. Assume that all prices have risen at the same rate as the Consumer Price Index (CPI).

17. Suppose you needed $20,000 to maintain a particular standard of living in 1975. How much would you have needed in 2002 to maintain the same standard of living?

18. Suppose you needed $45,000 to maintain a particular standard of living in 1981. How much would you have needed in 2002 to maintain the same standard of living?

19. Find the inflation rate from 1977 to 1978.

20. Find the inflation rate from 2001 to 2002.

21. A loaf of bread cost $0.75 in 1979. What was its price in 2000 dollars?

22. A car cost $4500 in 1975. What was its price in 2000 dollars?

23. Admission to a movie cost $7.00 in 2000. What was its price in 1975 dollars?

24. A haircut cost $15 in 2000. What was its price in 1980 dollars?

25. What was the purchasing power of $1 in 1973, in terms of 2002 dollars?

26. What was the purchasing power of $1 in 1976, in terms of 1996 dollars?

Housing Price Index. The following table shows a housing index that can be used to compare housing prices in different cities. If you know the price of a particular home in your town, you can use the index to find the price of a comparable house in another town with the following formula:

$$\begin{array}{c} \text{price} \\ \text{(other town)} \end{array} = \begin{array}{c} \text{price} \\ \text{(your town)} \end{array} \times \frac{\text{index (other town)}}{\text{index (your town)}}$$

Use this index for Exercises 27–30.

City	Index	City	Index
Juneau, AK	100	Manhattan, NY	495
Palo Alto, CA	365	Tulsa, OK	52
Denver, CO	87	Providence, RI	91
Sioux City, IA	47	Spokane, WA	78
Boston, MA	182	Cheyenne, WY	75

27. Suppose you see a house valued at $250,000 in Denver. Find the price of a comparable house in Palo Alto, Sioux City, and Boston.

28. Suppose you see a house valued at $380,000 in Boston. Find the price of a comparable house in Juneau, Manhattan, and Tulsa.

29. Suppose you see a house valued at $250,000 in Cheyenne. Find the price of a comparable house in Spokane, Denver, and Juneau.

30. Suppose you see a house valued at $1,000,000 in Manhattan. Find the price of a comparable house in Providence, Spokane, and Tulsa.

Further Applications

31. **Health Care Spending.** Total spending on health care in the United States rose from $80 billion in 1973 to $1.5 trillion in 2002. Compare this rise in health care spending to the overall rate of inflation as measured by the Consumer Price Index.

32. **Airfare.** The average price for a low-fare airline ticket between New York and Los Angeles rose from $230 in 1980 to $310 in 2002. Calculate the relative change in cost from 1980 to 2002, and compare this value to the overall rate of inflation as measured by the Consumer Price Index.

33. **Private College Costs.** The average cost of college (tuition, fees, and room and board) at four-year private universities rose from $5900 in 1980 to $25,100 in 2002. Calculate the relative change in cost from 1980 to 2002, and compare it to the overall rate of inflation as measured by the Consumer Price Index.

34. **Public College Costs.** The average cost of college (tuition, fees, and room and board) at four-year public universities rose from $2490 in 1980 to $9700 in 2002. Calculate the relative change in cost from 1980 to 2002, and compare it to the overall rate of inflation as measured by the Consumer Price Index.

35. **Home Prices—South.** The typical (median) price of a single-family home in the South (United States) rose from

$75,300 in 1990 to $123,600 in 2000. Calculate the relative change in price from 1990 to 2000, and compare it to the overall rate of inflation as measured by the Consumer Price Index.

36. **Home Prices—West.** The typical (median) price of a single-family home in the West (United States) rose from $129,600 in 1990 to $183,000 in 2000. Calculate the relative change in price from 1990 to 2000, and compare it to the overall rate of inflation as measured by the Consumer Price Index.

Federal Minimum Wage. Exercises 37–44 refer to the following table showing the federal hourly minimum wage over the past 60 years, in both actual dollars at the time and 1996 dollars.

Year	Actual dollars	1996 dollars
1938	$0.25	$2.78
1939	$0.30	$3.39
1945	$0.40	$3.49
1950	$0.75	$4.88
1956	$1.00	$5.77
1961	$1.25	$6.41
1967	$1.40	$6.58
1968	$1.60	$7.21
1974	$2.00	$6.37
1976	$2.30	$6.34
1978	$2.65	$6.38
1979	$2.90	$6.27
1981	$3.35	$5.78
1990	$3.50	$4.56
1991	$4.25	$4.90
1996	$4.75	$4.75
1997	$5.15	$5.03
2003	$5.15	$4.48

NOTE: 1996 dollars based on CPI-U; entries in table are years in which the minimum wage changed.
SOURCE: Department of Labor.

37. According to this table, how much is a quarter ($0.25) in 1938 dollars worth in 1996 dollars?

38. According to this table, how much is $1.00 in 1956 dollars worth in 1996 dollars?

39. Explain why the minimum wage for 1996 is the same in actual and 1996 dollars.

40. Explain why the minimum wage for 2003 in actual dollars is greater than the minimum wage for 2003 in 1996 dollars.

41. Use Table 3.4 to convert the 1979 minimum wage from actual dollars to 1996 dollars. Is the result consistent with the entry in the minimum wage table above?

42. Use Table 3.4 to convert the 1981 minimum wage from actual dollars to 1996 dollars. Is the result consistent with the entry in the minimum wage table above?

43. In what year was the purchasing power of the minimum wage the highest? Explain.

44. You are listening to an argument in which Paul claims that the minimum wage has never been higher, because it has been rising for the past 60 years. Paula counters that the minimum wage actually needs to be increased, because it has been decreasing almost consistently since 1968. Based on the data in the minimum wage table above, write a one-paragraph explanation of each argument. Which argument do you think is stronger? Why?

45. Project: Convenience Store Index. Go to a local supermarket and find the prices of a few staples, such as bread, milk, juice, and coffee. Compute the total cost of those items. Then go to a few smaller convenience stores and find the prices of the same items. Using the supermarket total as the reference value, compute the index numbers for the convenience stores.

Web Projects

Find useful links for Web Projects on the text Web site: www.aw.com/bennett-briggs

46. Consumer Price Index. Go to the Consumer Price Index home page and find the latest news release with updated figures for the CPI. Summarize the news release and any important ongoing trends in the CPI.

47. Producer Price Index. Go to the Producer Price Index (PPI) home page. Read the overview and recent news releases. Write a short summary describing the purpose of the PPI and how it is different from the CPI. Also summarize any important recent trends in the PPI.

48. Consumer Confidence Index. Use a search engine to find recent news about the Consumer Confidence Index. After studying the news, write a short summary of what the Consumer Confidence Index is trying to measure and describe any recent trends in the Consumer Confidence Index.

49. Human Development Index. The United Nations Development Programme regularly releases its Human Development Report. A closely watched finding of this report is the Human Development Index (HDI), which measures the overall achievements in a country in three basic dimensions of human development: life expectancy, educational attainment, and adjusted income. Find the most recent copy of this report and investigate exactly how the HDI is defined and computed.

50. Does the CPI Overstate Inflation? Find arguments on both sides of the question of whether the CPI overstates inflation. Write a short summary of the arguments. Then state and defend your own opinion as to whether the CPI overstates inflation.

In the News

51. Consumer Price Index. Find a recent news report that includes a reference to the Consumer Price Index. Briefly describe how the Consumer Price Index is important in the story.

52. Index Numbers. Find a recent news report that includes an index number *other than* the Consumer Price Index. Describe the index number and its meaning, and discuss how the index is important in the story.

53. Accurate CPI? Find a news story that talks about some long-term economic change that is tied to inflation, such as change in personal income over the past few decades. Briefly explain the report. How would the story be different if it turned out that the Consumer Price Index overstated inflation by, say, 1 percentage point per year?

HOW NUMBERS DECEIVE: POLYGRAPHS, MAMMOGRAMS, AND MORE

*T*he government administers polygraph tests ("lie detectors") to new applicants for sensitive security jobs. The polygraph tests are reputed to be 90% accurate. That is, they supposedly catch 90% of the people who are lying and validate 90% of the people who are truthful. Most people therefore guess that only 10% of the people who fail a polygraph test have been falsely identified as lying. In fact, the actual percentage of false accusations can be *much* higher—in some cases, more than 90% of those who fail a test have been falsely accused! How can this be?

We'll discuss the answer soon, but first let's draw the important moral of this story: Numbers may not lie, but they can be deceiving if we do not interpret them carefully. In this unit, we'll discuss several common ways in which numbers can deceive.

BETTER IN EACH CASE, BUT WORSE OVERALL

Suppose a pharmaceutical company creates a new treatment for acne. To decide whether the new treatment is better than an older treatment, the company conducts a test. The old treatment is given to 90 patients, and the new treatment is given to 110 patients. Here are the results after four weeks of treatment (summarized in Table 3.5):

- Among patients with mild acne, 10 received the old treatment and 90 received the new treatment. Among the 10 receiving the old treatment, 2 were cured, for a 20% cure rate. Among the 90 receiving the new treatment, 30 were cured, for a 33% cure rate. Thus, the new treatment had a higher cure rate among patients with mild acne.

- Among patients with severe acne, 80 received the old treatment and 20 received the new treatment. Among the 80 receiving the old treatment, 40 were cured, for a 50% cure rate. Among the 20 receiving the new treatment, 12 were cured, for a 60% cure rate. Thus, the new treatment had a higher cure rate among patients with severe acne.

TABLE 3.5	Results of Acne Treatments			
	MILD ACNE		SEVERE ACNE	
	Cured	Not Cured	Cured	Not Cured
Old Treatment	2	8	40	40
New Treatment	30	60	12	8

Notice that the new treatment had a higher cure rate both for patients with mild acne and for patients with severe acne. Is it fair for the company to claim that its new treatment is better than the old treatment?

At first, this might seem to make sense. But let's look at the overall results:

- Among the total of 90 patients receiving the old treatment, 42 were cured—2 out of 10 with mild acne and 40 out of 80 with severe acne. Thus, the overall cure rate for the old treatment was $42/90 = 46.7\%$.
- Among the total of 110 patients receiving the new treatment, 42 were cured—30 out of 90 with mild acne and 12 out of 20 with severe acne. Thus, the overall cure rate for the new treatment was $42/110 = 38.2\%$.

Overall, the old treatment had the higher cure rate, despite the fact that the new treatment had a higher rate for both mild and severe acne cases.

This example illustrates that it is possible for something to appear better in each of two or more group comparisons, but actually be worse overall. This situation is an example of **Simpson's paradox.** If you look carefully, you'll see that it occurs because of the way in which the overall results were divided into unequally sized groups (in this case, the groups were mild acne patients and severe acne patients). Simpson's paradox arises surprisingly often and is one way in which numbers can deceive unless they are examined with great care.

➤ **EXAMPLE 1** Who Played Better?

Table 3.6 gives the shooting performance of two players in a basketball game during the two halves. Shaq had a higher shooting percentage both in the first half (40% to 25%) and in the second half (75% to 70%). Does this mean that Shaq had the better shooting percentage for the game?

TABLE 3.6 Basketball Shots

Player	FIRST HALF			SECOND HALF		
	Baskets	Attempts	Percent	Baskets	Attempts	Percent
Shaq	4	10	40%	3	4	75%
Vince	1	4	25%	7	10	70%

Solution No, and we can see why by looking at the overall game statistics. Shaq made a total of 7 baskets (4 in the first half and 3 in the second half) on 14 shots (10 in the first half and 4 in the second half), for an overall shooting percentage of $7/14 = 50\%$. Vince made a total of 8 baskets on 14 shots, for an overall shooting percentage of $8/14 = 57.1\%$. Thus, even though Shaq had a higher shooting percentage in both halves, Vince had a better overall shooting percentage for the game.

Now try Exercises 11–12. ◄

IS IT DISCRIMINATION?—AND OTHER CASES OF SIMPSON'S PARADOX

Variations on Simpson's paradox arise in many situations. One famous case involved an investigation of possible gender discrimination in graduate programs at the University of California, Berkeley. Examining graduate admissions for 1973, university officials found that male applicants were being admitted at a significantly higher

| TABLE 3.7 | Applicants and Admissions for Six Graduate Programs at UC Berkeley, 1973 |

	MEN			WOMEN		
Department	Applied	Admitted	Percent	Applied	Admitted	Percent
A	825	512	62%	108	89	82%
B	560	353	63%	25	17	68%
C	325	120	37%	593	202	34%
D	417	138	33%	375	131	35%
E	191	53	28%	393	94	24%
F	373	22	6%	341	24	7%
Total	**2691**	**1198**	**44.5%**	**1835**	**557**	**30.4%**

Source: P. Bickel and J. W. O'Connell, Science, Vol. 187, 1975, pp. 398–404.

rate than female applicants. Assuming that male and female applicants were equally qualified, the university appeared to be engaging in discrimination against women.

The university began investigating individual departments to see which ones were responsible for the discrimination. Table 3.7 shows the results for the six departments with the largest numbers of graduate applicants. (University policy requires that departments not be named, so we refer to them as Departments A through F.)

Note that the admission rates (percentages) for women were actually higher than those for men in all but Departments C and E, and the rates were quite close in those cases. Nevertheless, the totals confirm a much higher overall admission rate for men than for women. Thus, we see Simpson's paradox in action again: While women were admitted at a significantly lower rate overall, no individual department was guilty of this practice.

Time out to think
Look closely at the data in Table 3.7. Would it be reasonable to conclude that Department A discriminates against men? Why or why not?

►EXAMPLE 2 Does Smoking Make You Live Longer?

In the early 1970s, a medical study in England involved many adult residents from a district called Wickham. Twenty years later, a follow-up study looked at the survival rates of the people from the original study. The follow-up study found the following surprising results (D. R. Appleton, J. M. French, and M. P. Vanderpump, *American Statistician*, Vol. 50, 1996, pp. 340–341):

- Among the adults who smoked, 24% died during the 20 years since the original study.
- Among the adults who did not smoke, 31% died during the 20 years since the original study.

Do these results suggest that smoking can make you live longer?

Solution Not necessarily, because the given results don't tell us the ages of the smokers and nonsmokers. It turned out that, in the original study, the nonsmokers

by the way

A National Science Foundation study once found that, on average, women in science and engineering earned only 77% as much as men. But when the data were examined within each field, in every case the average salary for women was at least 92% of the average salary for men. The explanation for this case of Simpson's paradox: Women were more highly represented in life sciences and social sciences, which tend to be lower paying than physical sciences and engineering.

were older on average than the smokers. Thus, the higher death rate among non-smokers simply reflected the fact that death rates tend to increase with age. When the results were broken into age groups, they clearly showed that for any given age group, nonsmokers had a higher 20-year survival rate than smokers. That is, 55-year-old smokers were less likely to reach age 75 than 55-year-old nonsmokers, and so on. Rather than suggesting that smoking prolongs life, a careful study of the data showed just the opposite. *Now try Exercises 13–16.* ◄

DOES A POSITIVE MAMMOGRAM MEAN CANCER?

We often associate tumors with cancers, but most tumors are not cancers. Medically, any kind of abnormal swelling or tissue growth is considered a tumor. A tumor caused by cancer is said to be *malignant* (or *cancerous*), while others are said to be *benign*.

Imagine you are a doctor or nurse treating a patient who has a breast tumor. The patient will be understandably nervous, but you can give her some comfort by telling her that only about 1 in 100 breast tumors turn out to be malignant. But, just to be safe, you order a mammogram to determine whether her tumor is one of the 1% that are malignant.

Now, suppose the mammogram comes back positive, suggesting the tumor is malignant. Mammograms are not perfect, so even the positive result does not necessarily mean that your patient has breast cancer. More specifically, let's assume that the mammogram screening is 85% accurate. It will correctly identify 85% of malignant tumors as malignant and 85% of benign tumors as benign. (Most real cancer tests have different accuracy rates for malignant and benign tumors.) When you tell your patient that her mammogram was positive, what should you tell her about the chance that she actually has cancer?

Because the mammogram screening is 85% accurate, most people guess that the positive result means that the patient probably has cancer. Studies have shown that in this situation most doctors also believe this to be the case, and would tell the patient to be prepared for cancer treatment. But a more careful analysis shows otherwise. In fact, the chance that the patient has cancer is still quite small—about 5%. We can see why by analyzing some numbers.

Consider a study in which mammograms are given to 10,000 women with breast tumors. Assuming that 1% of tumors are malignant, $1\% \times 10,000 = 100$ of the women actually have cancer; the remaining 9900 women have benign tumors. Table 3.8 summarizes the mammogram results. Note the following:

- The mammogram screening correctly identifies 85% of the 100 malignant tumors as malignant. Thus, it gives positive (malignant) results for 85 of the malignant tumors. These cases are called **true positives.** In the other 15 malignant cases, the result is negative, even though the women actually have cancer. These cases are **false negatives.**
- The mammogram screening correctly identifies 85% of the 9900 benign tumors as benign. Thus, it gives negative (benign) results for $85\% \times 9900 = 8415$ of the benign tumors. These cases are **true negatives.** The rest of the 9900 women with benign tumors ($9900 - 8415 = 1485$ women) get positive results in which the mammogram incorrectly suggests their tumors are malignant. These cases are **false positives.**

Overall, the mammogram screening gives positive results to 85 women who actually have cancer and to 1485 women who do *not* have cancer. The total number

by the way

The mammogram example and the polygraph example that follow illustrate cases in which *conditional probabilities* (discussed in Unit 7B) lead to confusion. The proper way of handling conditional probabilities was discovered by the Reverend Thomas Bayes (1702–1761) and is often called *Bayes rule*.

TABLE 3.8	Summary of Results for 10,000 Mammograms (in which 100 of the women have malignant tumors and 9900 have benign tumors)		
	TUMOR IS MALIGNANT	TUMOR IS BENIGN	TOTAL
POSITIVE MAMMOGRAM	85 true positives	1485 false positives	1570
NEGATIVE MAMMOGRAM	15 false negatives	8415 true negatives	8430
TOTAL	100	9900	10,000

of positive results is 85 + 1485 = 1570. Because only 85 of these are true positives (the rest are false positives), the chance that a positive result really means cancer is only 85/1570 = 0.054, or 5.4%. Therefore, when your patient's mammogram comes back positive, you should reassure her that there's still only a small chance that she has cancer.

> **EXAMPLE 3** False Negatives

Suppose you are a doctor seeing a patient with a breast tumor. Her mammogram comes back negative. Based on the numbers in Table 3.8, what is the chance that she has cancer?

Solution For the 10,000 cases summarized in Table 3.8, the mammograms are negative for 15 women with cancer and for 8415 women with benign tumors. The total number of negative results is 15 + 8415 = 8430. Thus, the fraction of women with false negatives is 15/8430 = 0.0018, or slightly less than 2 in 1000. In other words, the chance that a woman with a negative mammogram has cancer is only about 2 in 1000.

Now try Exercises 17–18. ◄

Time out to think _____
While the chance of cancer with a negative mammogram is small, it is not zero. Therefore, it might seem like a good idea to biopsy all tumors, just to be sure. However, biopsies involve surgery, which means they can be painful and expensive, among other things. Given these facts, do you think that biopsies should be routine for all tumors? Should they be routine for cases of positive mammograms? Defend your opinion.

POLYGRAPHS AND DRUG TESTS

We're now ready to return to the question asked at the beginning of this unit, about how a 90% accurate polygraph test can lead to a surprising number of false accusations. The explanation is very similar to that found in the case of the mammograms.

Suppose the government gives the polygraph test to 1000 applicants for sensitive security jobs. Further suppose that 990 of these 1000 people tell the truth on their polygraph test, while only 10 people lie. For a test that is 90% accurate, we find the following results:

by the way
The accuracy of mammograms is improving with technology. In 2000, studies showed the overall accuracy rate to be over 91%, and some experimental technologies appeared to achieve accuracy rates of over 98%. The accuracy is lower, however, for women receiving hormone replacement therapy, which is quite common among women over 40.

- Of the 10 people who lie, the polygraph correctly identifies 90%, meaning that 9 fail the test (they are identified as liars) and 1 passes.
- Of the 990 people who tell the truth, the polygraph correctly identifies 90%, meaning that 90% × 990 = 891 truthful people pass the test and the other 10% × 990 = 99 people fail the test.

Figure 3.4 summarizes these results. The total number of people who fail the test is 9 + 99 = 108. Of these, only 9 were actually liars; so the other 99 were falsely accused of lying. That is, 99 out of 108, or 99/108 = 91.7%, of the people who fail the test were actually telling the truth.

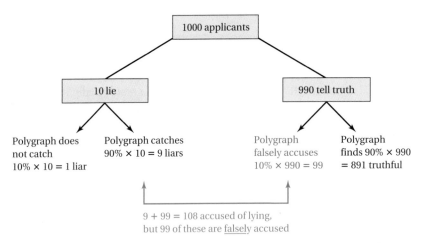

FIGURE 3.4 A tree diagram summarizes results of a 90% accurate polygraph test for 1000 people, of whom only 10 are lying.

The percentage of people who are falsely accused in any real situation depends on both the accuracy of the test and the proportion of people who are lying. Nevertheless, for the numbers given here, we have an astounding result: Assuming the government rejects applicants who fail the polygraph test, then almost 92% of the rejected applicants were actually being truthful and may have been highly qualified for the jobs.

Time out to think _____

Imagine that you are falsely accused of a crime. The police suggest that, if you are truly innocent, you should agree to take a polygraph test. Would you do it? Why or why not?

➤**EXAMPLE 4** High School Drug Testing

All athletes participating in a regional high school track and field championship must provide a urine sample for a drug test. Those who fail are eliminated from the meet and suspended from competition for the following year. Studies show that, at the laboratory selected, the drug tests are 95% accurate. Assume that 4% of the athletes actually use drugs. What fraction of the athletes who fail the test are falsely accused and therefore suspended without cause?

Solution The easiest way to answer this question is by using some sample numbers. Suppose there are 1000 athletes in the meet. We are told to assume that 4%, or 40 athletes, actually use drugs. The remaining 960 athletes do not use drugs. In that case, the 95% accurate drug test should return the following results:

- 95% of the 40 athletes who use drugs, or $0.95 \times 40 = 38$ athletes, fail the test. The other 2 athletes who use drugs pass the test.
- 95% of the 960 athletes who do not use drugs pass the test, but 5% of these 960 athletes fail. Thus, the number of athletes who fail despite not using drugs is $0.05 \times 960 = 48$.

The total number of athletes who fail the test is $38 + 48 = 86$. But 48 of these athletes, or $48/86 = 56\%$, are actually nonusers. Despite the 95% accuracy of the drug test, more than half of the suspended students are innocent of drug use.

Now try Exercises 19–20. ◄

POLITICAL MATH

Another type of deception occurs when two sides in a debate argue with different sets of numbers. Some of the classic cases arise in the politics of taxes and tax cuts.

Both of the charts in Figure 3.5 purport to show the effects of the *same* proposed tax cut. Figure 3.5a reflects numbers supplied by Republicans, who supported the tax cut. It suggests that the tax cut would benefit families of all incomes similarly in percentage terms, with slightly greater benefits going to middle-income families than to the poor or the rich. Figure 3.5b reflects numbers supplied by Democrats, who opposed the cut. This chart suggests that the benefits would go disproportionately to the rich. If we assume that neither side is lying outright, how can they make

Two Views of a Tax Cut

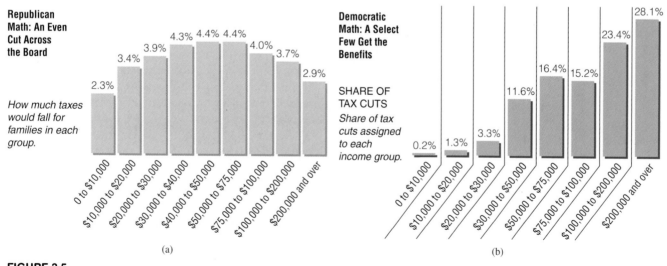

FIGURE 3.5

Source: Adapted from the New York Times.

such remarkably different claims? The answer lies in how each side chose to calculate "benefits" of the tax cut.

The Republicans calculated the *average* tax cut that would be received by families in each group. For example, the last bar in Figure 3.5a shows that families with income over $200,000 would get an average tax cut of 2.9%. Of course, a 2.9% tax cut means many more actual dollars to someone paying a lot in taxes than to someone paying a little. For example, someone who pays $100,000 in taxes would save $2900 with a 2.9% tax cut, while someone paying $1000 in taxes would save only $29.

The Democrats calculated the percentage of *total* benefits that would go to each group. For example, the last bar in Figure 3.5b shows that families with an income over $200,000 would receive 28.1% of the total benefits from the tax cut. But this does *not* mean that these families get a 28.1% tax cut. To see why, consider the effects of an across-the-board tax cut of the same percentage for everyone. Because families with incomes above $200,000 pay more than one-fourth of the total income taxes collected by the U.S. government, these families would receive about one-fourth of the total benefit of *any* across-the-board cut. For example, if the government collected $1 trillion in taxes, a 10% across-the-board cut would mean savings of $100 billion for taxpayers as a whole. In that case, because one-fourth of the $1 trillion was paid by families with income over $200,000, one-fourth of the $100 billion in savings would go to these same families. Thus, these families would get 25% of the *benefits* of the tax cut, even though their actual tax cut would be only 10%.

Which side was being more fair? Neither, really. The Republicans chose to neglect the fact that most of the total tax savings would go to the wealthy, while the Democrats chose to neglect the fact that the wealthy already pay most of the taxes. Unfortunately, this type of "selective truth" is very common when it comes to numbers, bringing home the necessity of listening carefully to quantitative arguments.

►EXAMPLE 5 A Cut or an Increase?

Government spending for a popular education program was $100 million last year. When Congress prepares its budget for next year, spending for the program is slated to rise to $102 million. Lobbyists immediately go into action. Those who support the program complain about the planned cut in the program. Those who oppose the program complain that it is being increased when it should be cut. Is one side lying? Explain. Assume that the Consumer Price Index is expected to rise by 3% over the next year.

Solution Neither side is lying. It's easiest to see for those who oppose the program: Clearly, spending *is* rising, since it is going from $100 million to $102 million. However, the opponents are neglecting the effects of inflation. If the Consumer Price Index rises by 3%, then this year's $100 million is equivalent to $103 million next year. Those who support the program say it is being cut because the increase to $102 million is not enough to keep up with inflation. In other words, $102 million next year will not buy as much as $100 million this year, so the program is being cut in the sense that it will not be able to do as much next year. *Now try Exercise 21.* ◄

EXERCISES 3E

Review Questions

1. What is Simpson's paradox? Give an example to show how it can make something look better in individual cases yet be worse overall.

2. Briefly explain why a positive result on a cancer test such as a mammogram does not necessarily mean that a patient has cancer.

3. Explain how it is possible for a very accurate polygraph or drug test to result in a large proportion of false accusations.

4. Give an example explaining how politicians on both sides of an issue can use numbers to support their case without actually lying.

Does It Make Sense?

Decide whether each of the following statements makes sense (or is clearly true) or does not make sense (or is clearly false). Explain your reasoning.

5. Despite the fact that the new drug lowered blood pressure more than the old drug did in both the men and the women in the study, an overall analysis shows that the old drug was actually more effective.

6. I clearly played a better basketball game than you, because I had a higher shooting percentage in both the first and the second half.

7. Baggage screening machines are 98% accurate in identifying bags that contain banned materials. Therefore, if the screening shows a bag contains banned materials, then it almost certainly does.

8. The polygraph test showed that the suspect was lying when he claimed to be innocent, so he must be guilty.

9. The Republicans claim the tax cut benefits everyone equally, but the Democrats say it favors the rich. Clearly, one side must be lying.

10. The agency suffered a real cut in its annual budget, even though it got a higher dollar amount this year than last.

Basic Skills & Concepts

11. **Batting Percentages.** The table below shows the batting records of two baseball players in the first half (first 81 games) and last half of a season.

Player	Hits	At-bats	First Half Batting average
Josh	50	150	.333
Jude	10	50	.200

Player	Hits	At-bats	Second Half Batting average
Josh	35	70	.500
Jude	70	150	.467

Who had the higher batting average in the first half of the season? Who had the higher batting average in the second half? Who had the higher overall batting average? Explain how these results illustrate Simpson's paradox.

12. **Passing Percentages.** The table below shows the passing records of two rival quarterbacks in the first half and second half of a football game.

Player	Completions	Attempts	First Half Percent
Allan	8	20	40%
Abner	2	6	33%

Player	Completions	Attempts	Second Half Percent
Allan	3	6	50%
Abner	12	25	48%

Who had the higher completion percentage in the first half? Who had the higher completion percentage in the second half? Who had the higher overall completion percentage? Explain how these results illustrate Simpson's paradox.

13. **Test Scores.** The table below shows eighth-grade mathematics test scores in Nebraska and New Jersey. The

scores are separated according to the race of the students. Also shown are the state averages for all races.

	White	Nonwhite	Average for all races
Nebraska	281	250	277
New Jersey	283	252	272

National Assessment of Educational Progress scores for 1992, from Chance, *Spring 1999.*

a. Which state had the higher scores in both racial categories? Which state had the higher overall average across both racial categories?

b. Explain how a state could score lower in both categories and still have a higher overall average.

c. Now consider the table below that gives the percentages of whites and nonwhites in each state. Use these percentages to verify that the overall average test score in Nebraska is 277, as claimed in the first table.

	White	Nonwhite
Nebraska	87%	13%
New Jersey	66%	34%

d. Use the racial percentages to verify that the overall average test score in New Jersey is 272, as claimed in the first table.

e. Explain briefly, in your own words, how Simpson's paradox appeared in this case.

14. Test Scores. Consider the following table comparing the grade point average (GPA) and mathematics SAT scores of high school students in 1988 and 1998.

GPA	% students 1988	% students 1998	SAT score 1988	SAT score 1998	Change
A+	4	7	632	629	−3
A	11	15	586	582	−4
A−	13	16	556	554	−2
B	53	48	490	487	−3
C	19	14	431	428	−3
Overall average			504	514	+10

Cited in Chance, *Vol. 12, No. 2, 1999, from data in* New York Times, *September 2, 1999.*

a. In general terms, how did the SAT scores of the students in the five grade categories change between 1988 and 1998?

b. How did the overall average SAT score change between 1988 and 1998?

c. How is this an example of Simpson's paradox?

15. Tuberculosis Deaths. The following table shows deaths due to tuberculosis (TB) in New York City and Richmond, Virginia, in 1910.

Race	New York Population	New York TB deaths
White	4,675,000	8400
Nonwhite	92,000	500
Total	4,767,000	8900

Race	Richmond Population	Richmond TB deaths
White	81,000	130
Nonwhite	47,000	160
Total	128,000	290

SOURCE: *Cohen and Nagel,* An Introduction to Logic and Scientific Method, *Harcourt, Brace and World, 1934.*

a. Compute the death rates for whites, nonwhites, and all residents in New York City.

b. Compute the death rates for whites, nonwhites, and all residents in Richmond.

c. Explain why this is an example of Simpson's paradox and explain how the paradox arises.

16. Weight Training. Two cross-country running teams participated in a (hypothetical) study in which a fraction of each team used weight training to supplement a running workout. The remaining runners did not use weight training. At the end of the season, the mean improvement in race times (in seconds) was recorded in the table below.

	Time improvement with weight training	Time improvement without weight training	Team average time improvement
Gazelles	10 s	2 s	6.0 s
Cheetahs	9 s	1 s	6.2 s

Describe how Simpson's paradox arises in this table. Resolve the paradox by finding the percentage of each team that used weight training.

More Accurate Mammograms. Like Table 3.8, the following table is based on the assumption that 1% of breast tumors are malignant. However, this table assumes that

mammogram screening is 95% accurate (versus 85% accurate in Table 3.8). Use these data in Exercises 17–18.

	Tumor is malignant	Tumor is benign	Total
Positive mammogram	95 true positives	495 false positives	590
Negative mammogram	5 false negatives	9405 true negatives	9410
Total	100	9900	10,000

17. Suppose a patient has a positive mammogram. What is the chance that she really has cancer?

18. Suppose a patient has a negative mammogram. What is the chance that she actually does have cancer?

19. **Polygraph Test.** Suppose that a polygraph is 90% accurate (it will correctly detect 90% of people who are lying and it will correctly detect 90% of people who are telling the truth). The 2000 employees of a company are given a polygraph test during which they are asked whether they use drugs. All of them deny drug use, when, in fact, 1% of the employees actually use drugs. Assume that anyone whom the polygraph operator finds untruthful is accused of lying.

 a. Verify that the entries in the table below follow from the given information. Explain each entry.

	Users	Non-users	Total
Test finds employee lying	18	198	216
Test finds employee truthful	2	1782	1784
Total	20	1980	2000

 b. How many employees are accused of lying? Of these, how many were actually lying and how many were telling the truth? What percentage of those accused of lying were falsely accused?

 c. How many employees are found truthful? Of these, how many were actually truthful? What percentage of those found truthful really were truthful?

20. **Disease Test.** Suppose a test for a disease is 80% accurate for those who have the disease (true positives) and 80% accurate for those who do not have the disease (true negatives). Within a sample of 4000 patients, the incidence rate of the disease is the national average, which is 1.5%.

 a. Verify that the entries in the table below follow from the information given and that the overall incidence rate is 1.5%. Explain.

	Disease	No disease	Total
Test positive	48	788	836
Test negative	12	3152	3164
Total	60	3940	4000

 b. Of those with the disease, what percentage test positive?

 c. Of those who test positive, what percentage have the disease? Compare this result to the one in part b and explain why they are different.

 d. Suppose a patient tests positive for the disease. As a doctor using this table, how would you describe the patient's chance of actually having the disease? Compare this figure to the overall incidence rate of the disease.

21. **Political Math.** Government spending for a popular housing program was $1 billion last year. When Congress prepares its budget for next year, spending for the program is slated to rise to $1.01 billion. Assume that the Consumer Price Index is expected to rise by 3% over the next year. Those who support the program complain that the program is being cut. Those who oppose the program complain that the program is being increased. Explain each position.

Further Applications

22. **Basketball Records.** Consider the following hypothetical basketball records for Spelman and Morehouse Colleges.

	Spelman College	Morehouse College
Home games	10 wins, 19 losses	9 wins, 19 losses
Away games	12 wins, 4 losses	56 wins, 20 losses

 a. Give numerical evidence to support the claim that Spelman College has a better team than Morehouse College.

 b. Give numerical evidence to support the claim that Morehouse College has a better team than Spelman College.

 c. Which claim do you think makes more sense? Why?

23. **Better Drug.** Two drugs, A and B, were tested on a total of 2000 patients, half of whom were women and half of whom were men. Drug A was given to 900 patients, and Drug B to 1100 patients. The results appear in the table below.

	Women	Men
Drug A	5 of 100 cured	400 of 800 cured
Drug B	101 of 900 cured	196 of 200 cured

 a. Give numerical evidence to support the claim that Drug B is more effective than Drug A.

 b. Give numerical evidence to support the claim that Drug A is more effective than Drug B.

 c. Which claim do you think makes more sense? Why?

24. **HIV Risks.** The New York State Department of Health estimates a 10% rate of HIV for the "at risk" population and a 0.3% rate for the general population. Tests for HIV are 95% accurate in detecting both true negatives and true positives. Random selection of 5000 "at risk" people and 20,000 people from the general population results in the following table.

	"At risk" population	
	Test positive	Test negative
Infected	475	25
Not infected	225	4275

	General population	
	Test positive	Test negative
Infected	57	3
Not infected	997	18,943

 a. Verify that incidence rates for the general and "at risk" populations are 0.3% and 10%, respectively. Also verify that detection rates for the general and "at risk" populations are 95%.

 b. Consider a patient in the "at risk" category. Of those with HIV, what percentage test positive? Of those who test positive, what percentage have HIV? Explain why these two percentages are different.

 c. Suppose a patient in the "at risk" category tests positive for the disease. As a doctor using this table, how would you describe the patient's chance of actually having the disease? Compare this figure to the overall rate of the disease in the "at risk" population.

 d. Consider a patient in the general population. Of those with HIV, what percentage test positive? Of those who test positive, what percentage have HIV? Explain why these two percentages are different.

 e. Suppose a patient in the general population tests positive for the disease. As a doctor using this table, how would you describe the patient's chance of actually having the disease? Compare this figure to the overall incidence rate of the disease.

25. **Hiring Statistics.** (This problem is based on an example in "Ask Marilyn," *Parade Magazine,* April 28, 1996.) A company decided to expand, so it opened a factory, generating 455 jobs. For the 70 white-collar positions, 200 males and 200 females applied. Of the females who applied, 20% were hired, while only 15% of the males were hired. Of the 400 males applying for the blue-collar positions, 75% were hired, while 85% of the 100 females who applied were hired. How does looking at the white-collar and blue-collar positions separately suggest a hiring preference for women? Do the overall data support the idea that the company hires women preferentially? Explain why this is an example of Simpson's paradox and how the paradox can be resolved.

26. **Drug Trials.** (This problem is based on an example in "Ask Marilyn," *Parade Magazine,* April 28, 1996.) A company runs two trials of two treatments for an illness. In the first trial, Treatment A cures 20% of the cases (40 out of 200) and Treatment B cures 15% of the cases (30 out of 200). In the second trial, Treatment A cures 85% of the cases (85 out of 100) and Treatment B cures 75% of the cases (300 out of 400). Which treatment had the better cure rate in the two trials individually? Which treatment had the better overall cure rate? Explain why this is an example of Simpson's paradox and how the paradox can be resolved.

Exercises 27–30 refer to the graph in Figure 3.6 (p. 207), which shows changes in income share and tax share for the 400 Americans who earned the most in each year.

27. Use the data in the graph to support the argument that the highest-income taxpayers are paying more than their fair share of taxes and therefore their taxes should be cut.

28. Use the data in the graph to support the argument that the highest-income taxpayers have seen their tax rates fall sharply in recent years and therefore their taxes should be raised.

29. Use the data in the graph to support the argument that lower tax rates on the rich lead to greater tax revenue for the federal government.

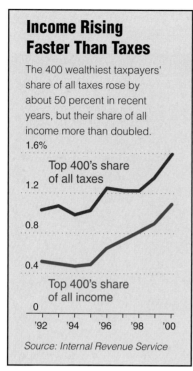

Income Rising Faster Than Taxes

The 400 wealthiest taxpayers' share of all taxes rose by about 50 percent in recent years, but their share of all income more than doubled.

Top 400's share of all taxes

Top 400's share of all income

Source: Internal Revenue Service

FIGURE 3.6

SOURCE: New York Times, *June 26, 2003*.

30. Use the data in the graph to support the argument that the rich earn too much money compared to the average person.

31. **2003 Tax Cut.** According to an analysis of the 2003 federal tax cut by the accounting firm Deloitte and Touche, a single person with a household income of $41,000 would save $211 in income taxes, while a single person with a household income of $530,000 would save $12,838 in taxes. A married couple with two children and a household income of $41,000 would save $1208 in income taxes, while a married couple with two children and a household income of $530,000 would save $13,442 in taxes.

 a. Compare, in absolute and relative terms, the effects of the tax cut on a single person with a household income of $41,000 and a single person with a household income of $530,000.

 b. Compare, in absolute and relative terms, the effects of the tax cut on a married couple with a household income of $41,000 and a married couple with a household income of $530,000.

 c. Write a paragraph either defending or denying the position that the tax cut helps lower income people.

Web Projects

Find useful links for Web Projects on the text Web site: www.aw.com/bennett-briggs

32. **Polygraph Arguments.** Visit Web sites devoted to either opposing or supporting the use of polygraph tests. Summarize the arguments on both sides, specifically noting the role that false negative rates play in the discussion.

33. **Drug Testing.** Explore the issue of drug testing either in the workplace or for athletic competitions. Discuss the legality of drug testing in these settings and the accuracy of the tests that are commonly conducted.

34. **Cancer Screening.** Investigate recommendations concerning routine screening for some type of cancer (for example, breast cancer, prostate cancer, colon cancer). Explain how the accuracy of the screening test is measured. How is the test useful? How can its results be misleading?

In the News

35. **Simpson's Paradox.** Find a recent news article that gives data whose interpretation may be complicated by Simpson's paradox (such as an article about possible discrimination). Explain what additional information you would need to reach a definitive conclusion.

36. **Polygraphs.** Find a recent article in which someone or some group proposes a polygraph test to determine whether a person is being truthful. In light of what you know about polygraph tests, do you think the results will be meaningful? Why or why not?

37. **Drug Testing and Athletes.** Find a news report concerning drug testing of athletes. Summarize how the testing is being used, and discuss whether the testing is reliable.

38. **Tax Cut.** Find information about a recently proposed tax cut. According to its proponents, who will benefit most from the cut? Who will benefit most according to its opponents? Discuss the numerical data cited by each side. Which side do you think has the stronger argument? Why?

CHAPTER 3 SUMMARY

UNIT	KEY TERMS	KEY IDEAS AND SKILLS
3A	absolute change relative change absolute difference relative difference	**Absolute change** = new value − reference value **Relative change** $= \dfrac{\text{absolute change}}{\text{reference value}} = \dfrac{\text{new value} - \text{reference value}}{\text{reference value}}$ **Absolute difference** = compared value − reference value **Relative difference** $= \dfrac{\text{absolute difference}}{\text{reference value}} = \dfrac{\text{compared value} - \text{reference value}}{\text{reference value}}$ **Understand** the *of* versus *more than* rule. **Understand** percentage points versus %. **Solve** percentage problems. **Identify** common abuses of percentages.
3B	scientific notation order of magnitude	**Write** and interpret numbers in scientific notation. **Put numbers in perspective** through estimation, including order of magnitude estimates comparisons scaling
3C	significant digits systematic error relative error accuracy precision	**Distinguish significant digits** from nonsignificant zeros. **Identify and distinguish** between random and systematic errors. **Absolute error** = measured value − true value **Relative error** $= \dfrac{\text{absolute error}}{\text{true value}} = \dfrac{\text{measured value} - \text{true value}}{\text{true value}}$ **Distinguish** between accuracy and precision. **Apply rounding rules** for combining approximate numbers. Addition/subtraction: Round to the precision of the *least precise* number in the problem. Multiplication/division: Round to the same number of significant digits as in the number in the problem with the fewest significant digits.
3D	index number reference value Consumer Price Index (CPI) inflation	**Understand index numbers** and how they are useful for comparisons. **Index number** $= \dfrac{\text{value}}{\text{reference value}} \times 100$ **Understand** how the CPI is used to measure inflation. **Adjust** prices for inflation with the CPI: price in \$$_Y$ = (price in \$$_X$) $\times \dfrac{\text{CPI}_Y}{\text{CPI}_X}$
3E	Simpson's paradox test results: false positive true positive false negative true negative	**Understand and give examples** of Simpson's paradox. **Understand and give examples** of how mammograms and polygraphs can lead to surprising results. **Understand and give examples** of how Democrats and Republicans can make different claims about the same data, even if neither side is lying.

A fool and his money are soon parted.

—ENGLISH PROVERB

CHAPTER

4

Financial Management

Managing your personal finances is a complex task in the modern world. If you are like most Americans, you already have a bank account and at least one credit card. You may also have student loans, a home mortgage, and various investment plans. In this chapter, we discuss key issues in personal financial management, including savings, loans, taxes, and investments.

UNIT **4A** **The Power of Compounding:** If you put $100 in a piggy bank, many years from now it will still be $100. But if you invest $100 well, it may grow to much more. We'll study how money grows with compound interest.

UNIT **4B** **Savings Plans and Investments:** One way to save is to make regular deposits of, say, $100 per month. We'll calculate the future value of such savings plans and study investments such as stocks and bonds.

UNIT **4C** **Loan Payments, Credit Cards, and Mortgages:** Nearly everyone has some type of loan. We'll calculate monthly payments and explore loan issues.

UNIT **4D** **Income Taxes:** We'll calculate income taxes and discuss a few of the hot political issues that surround them.

UNIT **4E** **Understanding the Federal Budget:** Everyone's personal finances are ultimately tied to government finances. We'll look at how the federal budget works and study some of the political issues involved.

UNIT **4A**

THE POWER OF COMPOUNDING

O n July 18, 1461, King Edward IV of England borrowed the modern equivalent of $384 from New College of Oxford. The King soon paid back $160, but never repaid the remaining $224. The debt was forgotten for 535 years. Upon its rediscovery in 1996, a New College administrator wrote to the Queen of England asking for repayment, with interest. Assuming an interest rate of 4% per year, he calculated that the college was owed $290 billion.

This example illustrates what is sometimes called the "power of compounding": the remarkable way that money grows when interest continues to accumulate year after year. In the New College case, there is no clear record of a promise to repay the debt with interest, and even if there were, the Queen might not feel obliged to pay a debt that had been forgotten for more than 500 years. But anyone can take advantage of compound interest simply by opening a savings account. With patience, the results may be truly astonishing.

SIMPLE VERSUS COMPOUND INTEREST

Imagine that you deposit $1000 in Honest John's Money Holding Service, which promises to pay 5% interest each year. At the end of the first year, Honest John's sends you a check for

$$5\% \times \$1000 = 0.05 \times \$1000 = \$50$$

You also get $50 at the end of the second and third years. Over the 3 years, you receive total interest of

$$3 \times \$50 = \$150$$

Your original $1000 has grown in value to $1150. Honest John's method of payment represents **simple interest,** in which interest is paid only on your actual investment, or **principal.**

Now, suppose that you place the $1000 in a bank account that pays the same 5% interest once a year. But instead of paying you the interest directly, the bank adds the interest to your account. At the end of the first year, the bank deposits $50 interest into your account, raising your balance to $1050. At the end of the second year, the bank again pays you 5% interest. This time, however, the 5% interest is paid on the balance of $1050, so it amounts to

$$5\% \times \$1050 = 0.05 \times \$1050 = \$52.50$$

Adding this $52.50 raises your balance to

$$\$1050 + \$52.50 = \$1102.50$$

This is the new balance on which your 5% interest is computed at the end of the third year. So your third interest payment is

$$5\% \times \$1102.50 = 0.05 \times \$1102.50 = \$55.13$$

by the way

The New College administrator did not seriously believe that the Queen would pay $290 billion. However, he suggested a compromise of assuming a 2% per year interest rate, in which case the college was owed only $8.9 million. This, he said, would be enough to pay for a modernization project at the College. The Queen has not yet paid.

Therefore, your balance at the end of the third year is

$$\$1102.50 + \$55.13 = \$1157.63$$

Despite identical interest rates, you end up with $7.63 more if you use the bank instead of Honest John's. The difference comes about because the bank pays you *interest on the interest* as well as on the original principal. This type of interest payment is called **compound interest.**

DEFINITIONS

The **principal** in financial formulas is the balance upon which interest is paid.

Simple interest is interest paid only on the original principal, and not on any interest added at later dates.

Compound interest is interest paid both on the original principal and on all interest that has been added to the original principal.

▶**EXAMPLE 1** Savings Bond

While banks almost always pay compound interest, bonds usually pay simple interest. Suppose you invest $1000 in a savings bond that pays simple interest of 10% per year. How much total interest will you receive in 5 years? If the bond paid compound interest, would you receive more or less total interest? Explain.

Solution With simple interest, every year you receive the same interest payment of 10% × $1000 = $100. Thus, you receive a total of $500 in interest over 5 years. With compound interest, you receive *more* than $500 in interest because the interest each year is calculated on your growing balance rather than on your original principal. For example, because your first interest payment of $100 raises your balance to $1100, your next compound interest payment is 10% × $1100 = $110, which is more than the simple interest payment of $100. For the same interest rate, compound interest always raises your balance faster than simple interest. *Now try Exercises 31–34.* ◀

THE COMPOUND INTEREST FORMULA

Let's return to King Edward's debt to the New College. We can calculate the amount owed to the College by pretending that the $224 he borrowed was deposited into an interest-bearing account for 535 years. Let's assume, as did the New College administrator, that the interest rate was 4% per year. For each year, we can calculate the interest and the new balance with interest. The first three columns of Table 4.1 show these calculations for 4 years.

To find the total balance, we could continue the calculations to 535 years. Fortunately, there's a much easier way. The 4% annual interest rate means that each end-of-year balance is 104% of, or 1.04 times, the previous year's balance. Thus, as shown in the last column of Table 4.1, we can get each balance as follows:

- The balance at the end of 1 year is the original principal times 1.04:

$$\$224 \times 1.04 = \$232.96$$

TABLE 4.1	Calculating Compound Interest		
AFTER *N* YEARS	INTEREST	BALANCE	OR EQUIVALENTLY
1 year	4% × $224　　= $8.96	$224　　+ $8.96　= $232.96	$224 × 1.04　　= $232.96
2 years	4% × $232.96 = $9.32	$232.96 + $9.32　= $242.28	$224 × $(1.04)^2$ = $242.28
3 years	4% × $242.28 = $9.69	$242.28 + $9.69　= $251.97	$224 × $(1.04)^3$ = $251.97
4 years	4% × $251.97 = $10.08	$251.97 + $10.08 = $262.05	$224 × $(1.04)^4$ = $262.05

by the way

Financial planners often call the principal, *P,* the *present value* (PV) of the money and the accumulated amount, *A,* the *future value* (FV). This terminology is also used on many financial calculators and software packages.

- The balance at the end of 2 years is the 1-year balance times 1.04:

$$\$224 \times 1.04 \times 1.04 = \$224 \times (1.04)^2 = \$242.28$$

- The balance at the end of 3 years is the 2-year balance times 1.04:

$$\$224 \times 1.04 \times 1.04 \times 1.04 = \$224 \times (1.04)^3 = \$251.97$$

Continuing the pattern, we find that the balance after *N* years is the original principal times 1.04 raised to the *N*th power. For example, the balance after $N = 10$ years is

$$\$224 \times (1.04)^{10} = \$331.57$$

We can generalize this result as the **compound interest formula.**

The Compound Interest Formula (General Form)

Suppose you invest a principal *P* in an account that earns compound interest. Each time interest is paid, your balance increases by an interest rate *i* (note that *i* should be expressed as a decimal rather than a percentage). After interest has been paid *N* times, your accumulated balance *A* is

$$A = P \times (1 + i)^N$$

Using Your Calculator

Most calculators have a key for raising to powers, labeled (y^x) or (\wedge). For example, calculate 1.04^{535} by pressing 1.04 (y^x) 535 $(=)$ or 1.04 (\wedge) 535.

In the New College case, the interest is paid at a rate of $i = 4\% = 0.04$ per year, and it is paid a total of $N = 535$ times (once a year for 535 years). Thus, the accumulated balance over the 535 years is

$$A = P \times (1 + i)^N = \$224 \times (1 + 0.04)^{535} = \$224 \times (1.04)^{535}$$
$$= \$224 \times 1{,}296{,}691{,}085 \approx \$2.9 \times 10^{11} = \$290 \text{ billion}$$

As the administrator claimed, a 4% interest rate means the Queen owes about $290 billion.

▶ **EXAMPLE 2**　Simple and Compound Interest

You invest $100 in two accounts that each pay an interest rate of 10% per year. However, one account pays simple interest and one account pays compound interest. Make a table that shows the growth of each account over a 5-year period. Use the compound interest formula to verify the result in the table for the compound interest case.

Solution　The simple interest is the same absolute amount each year. The compound interest grows from year to year, because it is paid on the accumulated interest as well as on the starting balance. Table 4.2 summarizes the calculations.

TABLE 4.2	Calculations for Example 2				
	SIMPLE INTEREST ACCOUNT		COMPOUND INTEREST ACCOUNT		
End of year	Interest paid	Old balance + interest = new balance	Interest paid	Old balance + interest = new balance	
1	10% × $100 = $10	$100 + $10 = **$110**	10% × $100 = $10	$100 + $10 = **$110**	
2	10% × $100 = $10	$110 + $10 = **$120**	10% × $110 = $11	$110 + $11 = **$121**	
3	10% × $100 = $10	$120 + $10 = **$130**	10% × $121 = $12.10	$121 + $12.10 = **$133.10**	
4	10% × $100 = $10	$130 + $10 = **$140**	10% × $133.10 = $13.31	$133.10 + $13.31 = **$146.41**	
5	10% × $100 = $10	$140 + $10 = **$150**	10% × $146.41 = $14.64	$146.41 + $14.64 = **$161.05**	

To verify the final entry in the table with the compound interest formula, we use a starting principal $P = \$100$ and an interest rate $i = 10\% = 0.1$ with interest paid $N = 5$ times (once a year for each of the 5 years). The accumulated balance A is

$$A = P \times (1 + i)^N = \$100 \times (1 + 0.1)^5 = \$100 \times 1.1^5 = \$100 \times 1.6105 = \$161.05$$

This result agrees with the one in the table. Overall, the account paying compound interest builds to $161.05 while the simple interest account reaches only $150, even though both pay at the same 10% rate. This is a significant difference, especially when you consider that the difference will continue to grow with time. Clearly, compound interest is much better for an investor than simple interest at the same rate.

Now try Exercises 35–36. ◄

COMPOUND INTEREST AS EXPONENTIAL GROWTH

The New College case demonstrates the remarkable way in which money can grow with compound interest. Figure 4.1 shows how the value of the New College debt rises during the first 100 years, assuming a starting value of $224 and an interest

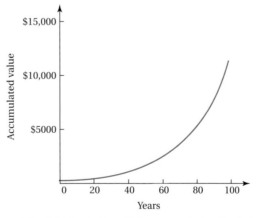

FIGURE 4.1 The value of the debt in the New College case during the first 100 years. The original value in year 0 is $224, and the interest rate is 4% per year. Note that the value rises much more rapidly in later years than in earlier years—a hallmark of exponential growth.

rate of 4% per year. Note that while the value rises slowly at first, it rapidly accelerates, so in later years the value grows by much more each year than it did during earlier years.

This rapid growth is a hallmark of what we generally call *exponential growth.* You can see how exponential growth gets its name by looking again at the general compound interest formula:

$$A = P \times (1 + i)^N$$

Because the principal P and the interest rate i have fixed values for any particular compound interest calculation, the growth of the accumulated value A depends only on N (the number of times interest has been paid) and N appears in the *exponent* of the calculation.

Exponential growth is one of the most important topics in mathematics, with applications that include population growth, resource depletion, and radioactivity. We will study exponential growth in much more detail in Chapter 8. In this chapter, we focus only on its applications in finance.

CALCULATIONS FOR DIFFERENT COMPOUNDING PERIODS

We can use the general form of the compound interest formula for any compound interest problem. However, working with the formula is often easier if we rewrite it to fit the problem at hand. In this section, we'll restate and use the compound interest formula for three common cases: interest paid once a year (annually), interest paid two or more times per year, and interest paid continuously at all times.

Compound Interest Paid Once a Year

We can make the compound interest formula specific to any particular case by rewriting it with appropriate replacements for the interest rate i and the number of payments N. For interest paid once a year, the interest rate is the **annual percentage rate,** or **APR.** The total number of interest payments is simply the number of years, which we call Y. We can therefore rewrite the compound interest formula in the following form.

Compound Interest Formula for Interest Paid Once a Year

$$A = P \times (1 + \text{APR})^Y$$

where

$$A = \text{acumulated balance after } Y \text{ years}$$
$$P = \text{starting principal}$$
$$\text{APR} = \text{annual percentage rate (as a decimal)}$$
$$Y = \text{number of years}$$

As in the general compound interest formula, we must write the interest rate APR as a decimal. For example, an annual interest rate of 4% means APR $= 0.04$.

►EXAMPLE 3 New College Debt at 2%

If the interest rate is 2%, calculate the amount due to New College using

 a. simple interest **b.** compound interest

Solution

 a. At a rate of 2%, the simple interest due each year is

$$2\% \times \$224 = 0.02 \times \$224 = \$4.48$$

Over 535 years, the total interest due would be

$$535 \times \$4.48 = \$2396.80$$

Adding this to the original loan principal of $224 gives the payoff amount of

$$\$224 + \$2396.80 = \$2620.80$$

 b. To find the amount due with compound interest, we set the annual interest rate to APR = 2% = 0.02 and the number of years to $Y = 535$. Then we use the formula for compound interest paid once a year:

$$A = P \times (1 + \text{APR})^Y = \$224 \times (1 + 0.02)^{535}$$
$$= \$224 \times (1.02)^{535}$$
$$= \$224 \times 39{,}911$$
$$\approx \$8.94 \times 10^6$$

The amount due with compound interest is about $8.94 million—far higher than the amount due with simple interest. You should note the remarkable effects of small changes in the compound interest rate. Here, we found that a 2% compound interest rate leads to $8.94 million after 535 years. Earlier, we found that a 4% interest rate for the same 535 years leads to $290 billion—which is more than 30,000 times as large as $8.94 million. Figure 4.2 contrasts the values of the New College debt during the first 100 years at interest rates of 2% and 4%. Note that the rates don't make much difference for the first few years, but over time the higher rate becomes far more valuable. *Now try Exercises 37–38.* ◄

Time out to think

Suppose the interest rate for the New College debt were 3%. Without calculating, do you think the value after 535 years would be halfway between the values at 2% and 4% or closer to one or the other of these values? Now, check your guess by calculating the value at 3%. What happens at an interest rate of 6%? Briefly discuss why small changes in the interest rate can lead to large changes in the accumulated value.

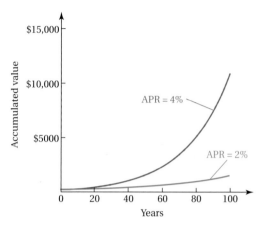

FIGURE 4.2 This figure contrasts the debt in the New College case during the first 100 years at interest rates of 2% and 4%. The original value in year 0 is $224. Note that the higher interest rate has an ever-greater impact at later times.

> **EXAMPLE 4** Mattress Investments

Your grandfather put $100 under his mattress 50 years ago. If he had instead invested it in a bank account paying 3.5% interest compounded yearly (roughly the average U.S. rate of inflation during that period), how much would it be worth now?

Solution The starting principal is $P = \$100$. The annual percentage rate is APR = 3.5% = 0.035. The number of years is $Y = 50$. So the accumulated balance is

$$A = P \times (1 + \text{APR})^Y = \$100 \times (1 + 0.035)^{50}$$
$$= \$100 \times (1.035)^{50}$$
$$= \$558.49$$

Invested at a rate of 3.5%, the $100 would be worth over $550 today. Unfortunately, the $100 was put under a mattress, so it still has a face value of only $100.

Now try Exercises 39–42. ◄

Compound Interest Paid More Than Once a Year

Suppose you deposit $1000 into a bank that pays compound interest at an annual percentage rate of APR = 8%. If the interest is paid all at once at the end of a year, you'll receive interest of

$$8\% \times \$1000 = 0.08 \times \$1000 = \$80$$

Thus, your year-end balance will be $1000 + $80 = $1080.

Now, assume instead that the bank pays the interest *quarterly,* or four times a year (that is, once every 3 months). The quarterly interest rate is one-fourth of the annual interest rate:

$$\text{quarterly interest rate} = \frac{\text{APR}}{4} = \frac{8\%}{4} = 2\% = 0.02$$

Table 4.3 shows how quarterly compounding affects the $1000 starting principal during the first year.

TABLE 4.3	Quarterly Interest Payments		
AFTER *N* QUARTERS		INTEREST PAID	NEW BALANCE
1st quarter (3 months)	2% × $1000	= $20	$1000 + $20 = $1020
2nd quarter (6 months)	2% × $1020	= $20.40	$1020 + $20.40 = $1040.40
3rd quarter (9 months)	2% × $1040.40	= $20.81	$1040.40 + $20.81 = $1061.21
4th quarter (1 full year)	2% × $1061.21	= $21.22	$1061.21 + $21.22 = $1082.43

Note that the year-end balance with quarterly compounding ($1082.43) is *greater* than the year-end balance with interest paid all at once ($1080). That is, when interest is compounded more than once a year, the balance increases by *more* than the APR in 1 year.

We can find the same results with the general compound interest formula. In this case, the starting principal is $P = \$1000$. The quarterly payments have an interest rate of $i = \text{APR}/4 = 0.02$. In 1 year, interest is paid $N = 4$ times. Thus, the accumulated balance at the end of 1 year is

$$A = P \times (1 + i)^N = \$1000 \times (1 + 0.02)^4 = \$1000 \times (1.02)^4 = \$1082.43$$

We see that if interest is paid quarterly, the interest rate at each payment is $i = \text{APR}/4$. Generalizing, if interest is paid n times per year, the interest rate at each payment is $i = \text{APR}/n$. The total number of times that interest is paid after Y years is $N = nY$. (Note that n and N have different meanings: Lowercase n is the number of times interest is paid *each year*, while uppercase N is the total number of interest payments.) Substituting these expressions for i and N into the general compound interest formula, we get a formula for interest paid more than once each year.

Compound Interest Formula for Interest Paid *n* Times per Year

$$A = P\left(1 + \frac{\text{APR}}{n}\right)^{(nY)}$$

where

A = accumulated balance after Y years

P = starting principal

APR = annual percentage rate (as a decimal)

n = number of compounding periods per year

Y = number of years (may be a fraction)

Time out to think _____

Confirm that substituting $n = 1$ into the formula for interest paid n times per year gives you the formula for interest paid once a year. Explain why this should be true.

➤ **EXAMPLE 5** Monthly Compounding at 3%

You deposit $5000 in a bank account that pays an APR of 3% and compounds interest monthly. How much money will you have after 5 years? Compare this amount to the amount you'd have if interest were paid only once each year.

Solution The starting principal is $P = \$5000$ and the interest rate is $APR = 0.03$. Monthly compounding means that interest is paid $n = 12$ times a year, and we are considering a period of $Y = 5$ years. We put these values into the compound interest formula to find the accumulated balance, A.

$$A = P \times \left(1 + \frac{APR}{n}\right)^{(nY)} = \$5000 \times \left(1 + \frac{0.03}{12}\right)^{(12 \times 5)}$$

$$= \$5000 \times (1.0025)^{60}$$

$$= \$5808.08$$

For interest paid only once each year, we find the balance after 5 years by using the formula for compound interest paid once a year:

$$A = P \times (1 + APR)^{Y} = \$5000 \times (1 + 0.03)^5$$

$$= \$5000 \times (1.03)^5$$

$$= \$5796.37$$

After 5 years, monthly compounding gives you a balance of $5808.08 while annual compounding gives you a balance of $5796.37. That is, monthly compounding

Using Your Calculator

The Compound Interest Formula (for Interest Paid More Than Once per Year)

You can do compound interest calculations on any calculator that has a $\boxed{y^x}$ or $\boxed{\wedge}$ key for raising to powers. Here's a five-step procedure that will work on most scientific calculators, along with an example in which $P = \$1000$, APR $= 8\% = 0.08$, $Y = 10$ years, and $n = 12$ (monthly compounding). With a graphing calculator, you may be able to do the calculation more directly by using the parentheses keys. Some business calculators have built-in functions for calculating compound interest in a single step. Note: It is very important that you not round any answers until you have completed all the calculations.

	In general	Example	Display
Starting Formula:	$A = P \times \left(1 + \dfrac{APR}{n}\right)^{(nY)}$	$\$1000 \times \left(1 + \dfrac{0.08}{12}\right)^{(12 \times 10)}$	____
Step 1. Multiply factors in exponent.	$n \boxed{\times} Y \boxed{=}$	$12 \boxed{\times} 10 \boxed{=}$	120.
Step 2. Store product in memory (or write down).	$\boxed{\text{Store}}$	$\boxed{\text{Store}}$	120.
Step 3. Add terms 1 and APR/n.	$1 \boxed{+} APR \boxed{\div} n \boxed{=}$	$1 \boxed{+} 0.08 \boxed{\div} 12 \boxed{=}$	1.0066666667
Step 4. Raise result to power in memory.	$\boxed{y^x} \boxed{\text{Recall}} \boxed{=}$	$\boxed{y^x} \boxed{\text{Recall}} \boxed{=}$	2.219640235
Step 5. Multiply result by P.	$\boxed{\times} P \boxed{=}$	$\boxed{\times} \$1000 \boxed{=}$	2219.640235

With the calculation complete, you can round to the nearest cent, writing the answer as $2219.64. Finally, because it's easy to push the wrong buttons by accident, you should always check the calculation (preferably twice) and make sure your answer makes sense.

earns $5808.08 − $5796.37 = $11.71 more, even though the APR is the same in both cases. *Now try Exercises 43–50.* ◄

Annual Percentage Yield (APY)

We've seen that in one year, money grows by *more* than the APR when interest is compounded more than once a year. For example, we found that with quarterly compounding and an 8% APR, a $1000 principal increases to $1082.43 in one year. This represents a relative increase of 8.243%:

$$\text{relative increase} = \frac{\text{absolute increase}}{\text{starting principal}} = \frac{\$82.43}{\$1000} = 0.08243 = 8.243\%$$

This relative increase over one year is called the **annual percentage yield (APY).** Note that the APY of 8.243% is greater than the APR of 8%.

> ### DEFINITION
>
> The **annual percentage yield (APY)** is the actual percentage by which a balance increases in one year. It is equal to the APR if interest is compounded annually. It is greater than the APR if interest is compounded more than once a year. The APY does not depend on the starting principal.

► **EXAMPLE 6** More Compounding Means a Higher Yield

You deposit $1000 into an account with APR = 8%. Find the annual percentage yield with *monthly* compounding and with *daily* compounding.

Solution The easiest way to find the annual percentage yield is by finding the balance at the end of one year. We have $P = \$1000$, APR = 8% = 0.08, $Y = 1$ year. For monthly compounding, we set $n = 12$. Thus, at the end of one year, the accumulated balance with monthly compounding is

$$A = P \times \left(1 + \frac{\text{APR}}{n}\right)^{(nY)} = \$1000 \times \left(1 + \frac{0.08}{12}\right)^{(12 \times 1)}$$
$$= \$1000 \times (1.006666667)^{12} = \$1083.00$$

Your balance increases by $83.00, so the annual percentage yield is

$$\text{APY} = \text{relative increase in 1 year} = \frac{\$83.00}{\$1000} = 0.083 = 8.3\%$$

With monthly compounding, the annual percentage yield is 8.3%.

Daily compounding means that interest is paid $n = 365$ times per year. At the end of one year, your accumulated balance with daily compounding is

$$A = P \times \left(1 + \frac{\text{APR}}{n}\right)^{(nY)} = \$1000 \times \left(1 + \frac{0.08}{365}\right)^{(365 \times 1)}$$
$$= \$1000 \times (1.000219178)^{365} = \$1083.28$$

by the way

Banks usually list both the annual percentage rate (APR) and the annual percentage yield (APY). The APY is what your money really earns and is the more important number when you are comparing interest rates. Banks are required by law to state the APY on interest-bearing accounts. The APY is sometimes called the *effective yield,* or simply the *yield.*

TECHNICAL NOTE

Most banks divide the APR by 360, rather than 365, when calculating the interest rate and APY for daily compounding. Thus, the results found here may not agree exactly with actual bank results.

Your balance increases by $83.28, so the annual percentage yield is

$$\text{APY} = \text{relative increase in 1 year} = \frac{\$83.28}{\$1000} = 0.08328 = 8.328\%$$

With an APR of 8% and daily compounding, the annual percentage yield is 8.328%, slightly higher than the APY for monthly compounding. The same APY would have been found using any starting principal. *Now try Exercises 51–54.* ◄

Continuous Compounding

Suppose that interest were compounded more often than daily—say, every second or every trillionth of a second. How would this affect the annual percentage yield?

Let's examine what we've found so far for APR = 8%. If interest is compounded annually (once a year), the annual yield is simply APY = APR = 8%. With quarterly compounding, we found APY = 8.243%. With monthly compounding, we found APY = 8.300%. With daily compounding, we found APY = 8.328%. Clearly, more frequent compounding means a higher APY (for a given APR).

However, notice that the change gets smaller as the frequency of compounding increases. For example, changing from annual compounding ($n = 1$) to quarterly compounding ($n = 4$) increases the APY quite a bit, from 8% to 8.243%. In contrast, going from monthly ($n = 12$) to daily ($n = 365$) compounding increases the APY only slightly, from 8.300% to 8.328%. You probably won't be surprised to learn that the APY can't get much bigger than it already is for daily compounding.

Table 4.4 shows the APY for various compounding periods and Figure 4.3 is a graph of the results. As expected, the annual yield does *not* grow indefinitely. Instead, it approaches a *limit* that is very close to the APY of 8.3287068% found for $n = 1$ billion. In other words, even if we could compound *infinitely many times* per year, the annual yield would not go much above 8.3287068%. Compounding infinitely many times per year is called **continuous compounding.** It represents the best possible compounding for a particular APR. With continuous compounding, the compound interest formula takes the following special form.

Compound Interest Formula for Continuous Compounding

$$A = P \times e^{(\text{APR} \times Y)}$$

where

A = accumulated balance after Y years

P = starting principal

APR = annual percentage rate (as a decimal)

Y = number of years

The number e is a special irrational number with a value of $e \approx 2.71828$. You can compute e to a power with the $\boxed{e^x}$ key on your calculator.

Time out to think _____

Look for the $\boxed{e^x}$ key on your calculator. Use it to enter e^1 and thereby verify that $e \approx 2.71828$.

TABLE 4.4	Annual Yield (APY) for APR = 8% with Various Numbers of Compounding Periods (n)		
n	APY	n	APY
1	8.0000000%	1000	8.3283601%
4	8.2432160%	10,000	8.3286721%
12	8.2999507%	1,000,000	8.3287064%
365	8.3277572%	10,000,000	8.3287067%
500	8.3280135%	1,000,000,000	8.3287068%

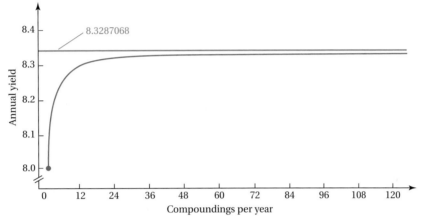

FIGURE 4.3 The annual percentage yield (APY) for APR = 8% depends on the number of times interest is compounded per year.

▶**EXAMPLE 7** Continuous Compounding

You deposit $100 in an account with an APR of 8% and continuous compounding. How much will you have after 10 years?

Solution We have $P = \$100$, APR = 8% = 0.08, and $Y = 10$ years of continuous compounding. The accumulated balance after 10 years is

$$A = P \times e^{(\text{APR} \times Y)} = \$100 \times e^{(0.08 \times 10)}$$
$$= \$100 \times e^{0.8}$$
$$= \$222.55$$

Your balance will be $222.55 after 10 years. Note: Be sure you can get the above answer by using the $\boxed{e^x}$ on your calculator. *Now try Exercises 55–60.* ◀

PLANNING AHEAD WITH COMPOUND INTEREST

Suppose you have a new baby and want to make sure that you'll have $100,000 for his or her college education. Assuming your baby will start college in 18 years, how much money should you deposit now?

If we know the interest rate, this problem is simply a "backwards" compound interest problem. We start with the amount A needed after 18 years and

HISTORICAL NOTE

Like the number π that arises so frequently in mathematics, the number e is one of the universal constants of mathematics. It appears in countless applications, most importantly to describe exponential growth and decay processes. The notation e was proposed by the Swiss mathematician Leonhard Euler in 1727. Like π, the number e is not only an irrational number, but also a transcendental number.

then calculate the necessary starting principal, P. The following two examples illustrate the calculations.

▶ **EXAMPLE 8** College Fund at 7%

Suppose you could make a single deposit in an investment with an interest rate of APR = 7%, compounded annually, and leave it there for the next 18 years. How much would you have to deposit now to realize $100,000 after 18 years?

Solution We know the interest rate (APR = 0.07), the number of years of compounding ($Y = 18$), and the amount desired after 18 years ($A = \$100,000$). We want to find the starting principal, P, that must be deposited now. We therefore solve the compound interest formula (for interest paid once a year) for P, by dividing both sides by $(1 + \text{APR})^Y$.

$$\underbrace{A = P \times (1 + \text{APR})^Y}_{\substack{\text{compound interest formula} \\ \text{(interest paid once a year)}}} \xrightarrow[\text{then interchange left and right sides}]{\text{divide both sides by } (1 + \text{APR})^Y;} P = \frac{A}{(1 + \text{APR})^Y}$$

Now we substitute the given values for A, APR, and Y. The original starting principal is

$$P = \frac{A}{(1 + \text{APR})^Y} = \frac{\$100,000}{(1 + 0.07)^{18}} = \frac{\$100,000}{(1.07)^{18}} = \$29,586.39$$

Depositing $29,586 now will yield the desired $100,000 in 18 years—assuming that the 7% APR doesn't change and that you make no withdrawals or additional deposits.
Now try Exercises 61–64. ◀

▶ **EXAMPLE 9** College Fund at 10%, Compounded Monthly

Repeat Example 8, but with an interest rate of APR = 10% and monthly compounding. Compare the results.

Solution This time we must solve for P in the compound interest formula for interest paid more than once a year.

$$\underbrace{A = P \times \left(1 + \frac{\text{APR}}{n}\right)^{(nY)}}_{\substack{\text{compound interest formula} \\ \text{(interest paid } n \text{ times per year)}}} \xrightarrow[\text{then interchange left and right sides}]{\text{divide both sides by } \left(1 + \frac{\text{APR}}{n}\right)^{(nY)}} P = \frac{A}{\left(1 + \frac{\text{APR}}{n}\right)^{nY}}$$

We substitute the given interest rate (APR = 0.10), the number of years ($Y = 18$), and the balance after 18 years ($A = \$100,000$). With monthly compounding, we have $n = 12$. The required starting principal is

$$P = \frac{A}{\left(1 + \dfrac{APR}{n}\right)^{(nY)}} = \frac{\$100,000}{\left(1 + \dfrac{0.10}{12}\right)^{(12 \times 18)}} = \frac{\$100,000}{(1.008333333)^{216}} = \$16,653.64$$

With a 10% APR and monthly compounding, you can reach $100,000 in 18 years by depositing about $16,650 today. This is just over half the $29,586 that you must deposit to reach the same goal with an interest rate of 7% (compounded annually).
Now try Exercises 65–68. ◀

A brief review . . .

THREE BASIC RULES OF ALGEBRA

In prior units, we've already encountered several instances in which we needed to solve an equation by adding, subtracting, multiplying, or dividing both sides by the same quantity. These operations will be useful in this unit and the rest of the book, so it is important to review the basic rules.

Three Basic Rules

The following three rules can always be used:

1. We can add or subtract the same quantity on both sides of an equation.
2. We can multiply or divide both sides of an equation by the same quantity, as long as we do not multiply or divide by zero.
3. We can interchange the left and right sides of an equation. That is, if $x = y$, it is also true that $y = x$.

Adding and Subtracting

The following examples show how adding to or subtracting from both sides can help solve equations that involve unknowns.

Example: Solve the equation $x - 9 = 3$ for x.
Solution: We isolate x by adding 9 to both sides:

$$x - 9 + 9 = 3 + 9 \quad \rightarrow \quad x = 12$$

Example: Solve the equation $y + 6 = 2y$ for y.
Solution: Because we have y on the left side and $2y$ on the right, we can isolate y on the right by subtracting y from both sides:

$$y + 6 - y = 2y - y \quad \rightarrow \quad 6 = y$$

We interchange the left and right sides, writing the answer as $y = 6$.

Example: Solve the equation $8q - 17 = p + 4q - 2$ for p.

Solution: We isolate p by subtracting $4q$ from both sides while also adding 2 to both sides:

$$8q - 17 - 4q + 2 = p + 4q - 2 - 4q + 2$$
$$\downarrow$$
$$4q - 15 = p$$

We interchange the left and right sides, writing the answer more simply as $p = 4q - 15$. Note that the equation is solved for p, but we cannot state a numerical value for p until we know the value of q.

Multiplying and Dividing

When we cannot isolate a variable by addition or subtraction alone, we may need to multiply or divide both sides of an equation by the same quantity.

Example: Solve the equation $4x = 24$ for x.
Solution: We isolate x by dividing both sides by 4:

$$\frac{\cancel{4}x}{\cancel{4}} = \frac{24}{4} \quad \rightarrow \quad x = 6$$

Example: Solve the equation $\frac{3z}{4} - 2 = 10$ for z.

Solution: First, we isolate the term containing z by adding 2 to both sides:

$$\frac{3z}{4} - 2 + 2 = 10 + 2 \quad \rightarrow \quad \frac{3z}{4} = 12$$

Now we multiply both sides by $\frac{4}{3}$:

$$\frac{3z}{\cancel{4}} \times \frac{\cancel{4}}{\cancel{3}} = \cancel{12}^{4} \times \frac{4}{\cancel{3}_{1}} \quad \rightarrow \quad z = 16$$

Example: Solve the equation $7w = 3s + 5$ for s.
Solution: We isolate the term containing s by subtracting 5 from both sides:

$$7w - 5 = 3s + 5 - 5 \quad \rightarrow \quad 7w - 5 = 3s$$

Next we divide both sides by 3 to isolate s. To write the final answer more simply, we also switch the left and right sides after dividing.

$$\frac{7w - 5}{3} = \frac{\cancel{3}s}{\cancel{3}} \quad \rightarrow \quad s = \frac{7w - 5}{3}$$

Remember that you should always check that your solution satisfies the original equation.

Now try Exercises 15–30.

Time out to think _____

Aside from long-term government bonds, it is extremely difficult to find investments with a constant interest rate for 18 years. Nevertheless, financial planners often make such assumptions when exploring investment options. Explain why such calculations can be useful, despite the fact that you can't be sure of a steady interest rate.

■ EXERCISES 4A

Review Questions

1. What is the difference between simple interest and compound interest? Why do you end up with more money with compound interest?

2. Explain how New College could claim that a debt of $224 from 535 years ago is worth $290 billion today. How does this show the "power of compounding"?

3. State the general form of the compound interest formula, and define A, P, i, and N. Give an example that uses this formula.

4. State the compound interest formula for interest paid once a year. Define APR and Y. Briefly explain how this formula is related to the general compound interest formula.

5. State the compound interest formula for interest paid more than once a year. Define n, and explain how this formula is related to the general compound interest formula.

6. What is an annual percentage yield (APY)? Explain why, for a given APR, the APY is higher if the interest is compounded more frequently.

7. What is continuous compounding? How does the APY for continuous compounding compare to the APY for, say, daily compounding? Explain the use of the formula for continuous compounding.

8. Give an example of a situation in which you might want to solve the compound interest formula to find the principal P that must be invested now to yield a particular amount A in the future.

Does It Make Sense?

Decide whether each of the following statements makes sense (or is clearly true) or does not make sense (or is clearly false). Explain your reasoning.

9. Simple Bank was offering simple interest at 5% per year, which was clearly a better deal than the 4.5% compound interest rate at Complex Bank.

10. Both banks were paying the same annual percentage rate (APR), but one had a higher annual percentage yield than the other (APY).

11. The bank that pays the highest annual percentage rate (APR) is always the best deal.

12. No bank could afford to pay interest every trillionth of a second because, with compounding, it'd soon owe everyone infinite dollars.

13. My bank paid an annual interest rate (APR) of 5.0%, but at the end of the year my account balance had grown by 5.1%.

14. If you deposit $10,000 in an investment account today, it can double in value to $20,000 in just a couple decades even at a relatively low interest rate (say, 5–7%).

Basic Skills & Concepts

Algebra Review. Exercises 15–30 use skills covered in the Brief Review on p. 223. Solve the equations for the unknown quantity.

15. $x - 4 = 6$

16. $x - 8 = 2$

17. $y + 5 = 10$

18. $x + 7 = 4$

19. $2z = 12$

20. $4y = 28$

21. $3x = 15$

22. $5n = 100$

23. $2x - 5 = 13$

24. $5z - 5 = 35$

25. $3n + 4 = 13$

26. $3x + 4 = 13 - 6x$

27. $4x + 8 = 24$

28. $s + 4 = 24 - 4s$

29. $5 - 2w = 9$

30. $6 - 2y = 9 - y$

Simple Interest. In Exercises 31–34, calculate the amount of money you'll have at the end of the indicated period of time.

31. You invest $500 in an account that pays simple interest of 5% for 5 years.

32. You invest $1000 in an account that pays simple interest of 7% for 5 years.

33. You invest $2000 in an account that pays simple interest of 3% for 20 years.

34. You invest $5000 in an account that pays simple interest of 6.5% for 20 years.

Simple vs. Compound Interest. Exercises 35–36 describe two similar, but not identical, investment accounts. Make a table that shows the performance of both accounts for 5 years. The table should list the amount of interest earned each year and the balance in each account. Compare the balances after 5 years.

35. Yancy invests $500 in an account that earns simple interest at an annual rate of 5% per year. Samantha invests $500 in a savings account with annual compounding at a rate of 5% per year.

36. Trevor invests $1000 in an account that earns simple interest at an annual rate of 6% per year. Kendra invests $1000 in a savings account with annual compounding at a rate of 6% per year.

Compound Interest. In Exercises 37–42, use the compound interest formula to determine the accumulated balance after the stated period. Assume that interest is compounded annually.

37. $2000 is invested at an APR of 3% for 10 years.

38. $10,000 is invested at an APR of 5% for 20 years.

39. $30,000 is invested at an APR of 7% for 25 years.

40. $3000 is invested at an APR of 4% for 12 years.

41. $10,000 is invested at an APR of 6% for 25 years.

42. $40,000 is invested at an APR of 8.5% for 30 years.

Compounding More Than Once a Year. In Exercises 43–50, use the compound interest formula for compounding more than once a year to determine the accumulated balance after the stated period.

43. A $1000 deposit at an APR of 3.5% with monthly compounding for 10 years

44. A $2000 deposit at an APR of 3% with daily compounding for 5 years

45. A $5000 deposit at an APR of 5.6% with quarterly compounding for 20 years

46. A $10,000 deposit at an APR of 2.75% with monthly compounding for 5 years

47. A $1000 deposit at an APR of 7% with monthly compounding for 15 years

48. A $3000 deposit at an APR of 5% with daily compounding for 10 years

49. A $5000 deposit at an APR of 6.2% with quarterly compounding for 30 years

50. A $15,000 deposit at an APR of 7.8% with monthly compounding for 15 years

Annual Percentage Yield (APY). Find the annual percentage yield (APY) for the banks described in Exercises 51–54.

51. A bank offers an APR of 2.5% compounded daily.

52. A bank offers an APR of 4.5% compounded monthly.

53. A bank offers an APR of 3.25% compounded monthly.

54. A bank offers an APR of 2.25% compounded quarterly.

Continuous Compounding. In Exercises 55–60, use the compound interest formula for continuous compounding to determine the accumulated balance after 1 year, 5 years, and 20 years. Also find the APY for each account.

55. A $1000 deposit in an account with an APR of 4%

56. A $2000 deposit in an account with an APR of 5%

57. A $10,000 deposit in an account with an APR of 6%

58. A $3000 deposit in an account with an APR of 4%

59. A $5000 deposit in an account with an APR of 6.5%

60. A $500 deposit in an account with an APR of 7%

Planning Ahead with Compounding. For Exercises 61–64, suppose you start saving today for a $10,000 down payment that you plan to make on a house in 10 years. Assume that you make no deposits into the account after your initial deposit. For each account described, how much would you have to deposit now to reach your $10,000 goal in 10 years?

61. An account with annual compounding and an APR of 5%

62. An account with quarterly compounding and an APR of 4.5%

63. An account with monthly compounding and an APR of 6%

64. An account with daily compounding and an APR of 4%

College Fund. You want to have a $100,000 college fund in 18 years. How much will you have to deposit now under each of the scenarios in Exercises 65–68?

65. An APR of 4%, compounded daily

66. An APR of 5.5%, compounded daily

67. An APR of 9%, compounded monthly

68. An APR of 3.5% compounded monthly

Further Applications

Small Rate Differences. Exercises 69–70 describe two similar, but not identical, investment accounts. In each case, compare the balances after 10 years and after 30 years. Briefly discuss the effects of the small difference in interest rates.

69. Chang invests $500 in a savings account that earns 3.5% compounded annually. Kio invests $500 in a different savings account that earns 3.75% compounded annually.

70. José invests $1500 in a savings account that earns 5.6% compounded annually. Marta invests $1500 in a different savings account that earns 5.7% compounded annually.

71. **Comparing Annual Yields.** Consider an account with an APR of 6.6%. Find the APY with quarterly compounding, monthly compounding, and daily compounding. Comment on how changing the compounding period affects the annual yield.

72. **Comparing Annual Yields.** Consider an account with an APR of 5%. Find the APY with quarterly compounding, monthly compounding, and daily compounding. Comment on how changing the compounding period affects the annual yield.

73. **Rates of Compounding.** Compare the accumulated balance in two accounts that both start with an initial deposit of $1000. Both accounts have an APR of 5.5%, but one account compounds interest annually while the other account compounds interest daily. Make a table that shows the interest earned each year and the accumulated balance in both accounts for the first 10 years. Compare the balance in the accounts, in percentage terms, after 10 years. Round all figures to the nearest dollar.

74. **Understanding Annual Percentage Yield (APY).**

 a. Explain why APR and APY are the same with annual compounding.

 b. Explain why APR and APY are different with daily compounding.

 c. Does APY depend on the starting principal, *P*? Why or why not?

 d. How does APY depend on the number of compoundings during a year, *n*? Explain.

75. **Comparing Investment Plans.** Bernard deposits $1600 in a savings account that compounds interest annually at an APR of 4%. Carla deposits $1400 in a savings account that compounds interest daily at an APR of 5%. Who will have the higher accumulated balance after 5 years and after 20 years? Explain.

76. **Comparing Investment Plans.** Brian invests $1600 in an account with annual compounding and an APR of 5.5%. Celeste invests $1400 in an account with continuous compounding and an APR of 5.2%. Determine who has the higher accumulated balance after 5 years and after 20 years. Discuss the effect of the APR and the compounding period.

77. **Retirement Fund.** You want to accumulate $75,000 for your retirement in 35 years. You have two choices. Plan A is an account with annual compounding and an APR of 5%. Plan B is an account with continuous compounding and an APR of 4.5%. How much of an investment does each plan require to reach your goal?

78. **Your Bank Account.** Find the current APR, the compounding period, and the claimed APY for your personal savings account. (Choose just one account if you have more than one, or pick a rate from a nearby bank if you don't have an account.)

 a. Calculate the APY on your account. Does your calculation agree with the APY claimed by the bank? Explain.

 b. Suppose you receive a gift of $10,000 and place it in your account. If the interest rate never changes, how much will you have in 10 years?

 c. Suppose you find another account that offers interest at an APR that is 2 percentage points higher than yours, with the same compounding period. For the $10,000 deposit, how much will you have after 10 years? Briefly discuss how this result compares to the result from part b.

Finding Time Periods. Use a calculator and possibly some trial and error to answer Exercises 79–81.

79. How long will it take your money to triple at an APR of 8% compounded annually?

80. How long will it take your money to grow by 50% at an APR of 7% compounded annually?

81. You deposit $1000 in an account that pays an APR of 7% compounded annually. How long will it take for your balance to reach $100,000?

82. Continuous Compounding. Explore continuous compounding by answering the following questions.

 a. For an APR of 12%, make a table similar to Table 4.4 in which you display the APY for $n = 1, 4, 12, 365, 500, 1000$.

 b. Find the APY for continuous compounding at an APR of 12%.

 c. Show the results of parts a and b on a graph similar to Figure 4.3.

 d. In words, compare the APY with continuous compounding to the APY with other types of compounding.

 e. You deposit $500 in an account with an APR of 12%. With continuous compounding, how much money will you have at the end of 1 year? at the end of 5 years?

Web Projects

Find useful links for Web Projects on the text Web site: www.aw.com/bennett-briggs

83. Compound Interest Calculators. Although you know how to calculate balances with the compound interest formula, the Web has many compound interest calculators. Find such a calculator on the Web. Experiment with various APRs, initial deposits, and compounding periods to determine if the Web calculator is accurate. Note and discuss any terms that are new or different from those you encountered in this unit.

84. Money Stretcher. The Money Stretcher is a Web-based tutorial on compound interest. Read this short article and comment on its accuracy, given what you have learned in this unit.

85. Rate Comparisons. Find a Web site that compares interest rates available for ordinary savings accounts at different banks. What is the range of rates currently being offered? What is the best deal? How does your own bank account compare?

In the News

86. Bank Advertisement. Find two bank advertisements that refer to compound interest rates. Explain the terms in each advertisement. Which bank offers the better deal? Explain.

87. Power of Compounding. In an advertisement or article about an investment, find a description of how money has grown (or will grow) over a period of many years. What is the annual yield listed? How does the value of the account change?

UNIT 4B

SAVINGS PLANS AND INVESTMENTS

Suppose you want to save money for retirement, for your child's college expenses, or for some other reason. You could deposit a lump sum of money today and let it grow through the power of compound interest. But what if you don't have a large lump sum to start such an account?

For most people, a more realistic way to save is by depositing smaller amounts on a regular basis. For example, you might put $50 a month into savings. Such long-term **savings plans** are so popular that many have special names—and some even get special tax treatment (see Unit 4D). Popular types of savings plans include Individual Retirement Accounts (IRAs), 401(k) plans, Keogh plans, and employee pension plans.

THE SAVINGS PLAN FORMULA

We can study savings plans with a simple example. Suppose you deposit $100 into your savings plan at the end of each month. Further suppose that your plan pays interest monthly at an annual rate of APR = 12%, or 1% per month.

by the way

Financial planners call any series of equal, regular payments an *annuity*. Thus, savings plans are a type of annuity, as are loans that you pay with equal monthly payments.

TABLE 4.5	Savings Plan Calculations			
END OF . . .	PRIOR BALANCE	INTEREST ON PRIOR BALANCE	END-OF-MONTH DEPOSIT	NEW BALANCE
Month 1	$0	$0	$100	$100
Month 2	$100	1% × $100 = $1	$100	$201
Month 3	$201	1% × $201 = $2.01	$100	$303.01
Month 4	$303.01	1% × $303.01 = $3.03	$100	$406.04
Month 5	$406.04	1% × $406.04 = $4.06	$100	$510.10
Month 6	$510.10	1% × $510.10 = $5.10	$100	$615.20

NOTE: *The last column shows the new balance at the end of each month, which is the sum of the prior balance, the interest, and the end-of-month deposit.*

- You begin with $0 in the account. At the end of month 1, you make the first deposit of $100.
- At the end of month 2, you receive the monthly interest on the $100 already in the account, which is 1% × $100 = $1. In addition, you make your monthly deposit of $100. Thus, your balance at the end of month 2 is

$$\underbrace{\$100}_{\text{prior balance}} + \underbrace{\$1.00}_{\text{interest}} + \underbrace{\$100}_{\text{new deposit}} = \$201.00$$

- At the end of month 3, you receive 1% interest on the $201 already in the account, or 1% × $201 = $2.01. Adding your monthly deposit of $100, you have a balance at the end of month 3 of

$$\underbrace{\$201.00}_{\text{prior balance}} + \underbrace{\$2.01}_{\text{interest}} + \underbrace{\$100}_{\text{new deposit}} = \$303.01$$

Table 4.5 continues these calculations through 6 months.

In principle, we could extend this table indefinitely—but it would take a lot of work! Fortunately, there's a much easier way: the **savings plan formula.**

TECHNICAL NOTE
This formula assumes the same payment and compounding periods. For example, if payments are made monthly, interest also is calculated and paid monthly. If the compounding period is different from the payment period, replace the term APR/*n* by the effective yield for each payment period.

Savings Plan Formula (Regular Payments)

$$A = \text{PMT} \times \frac{\left[\left(1 + \dfrac{\text{APR}}{n}\right)^{(nY)} - 1\right]}{\left(\dfrac{\text{APR}}{n}\right)}$$

where

$$A = \text{accumulated savings plan balance}$$
$$\text{PMT} = \text{regular payment (deposit) amount}$$
$$\text{APR} = \text{annual percentage rate (as a decimal)}$$
$$n = \text{number of payment periods per year}$$
$$Y = \text{number of years}$$

Using Your Calculator

The Savings Plan Formula

There are many ways to do the savings plan formula on your calculator. On a graphing calculator, you may be able to do the calculations directly if you use the parentheses keys. Some business calculators have built-in functions that allow you to make savings plan calculations in a single step. However, the following procedure will work on most scientific calculators. The example uses numbers from Example 2 (pp. 230–231): $n = 12$ payments (monthly payments), PMT $= \$100$, APR $= 8\% = 0.08$, and $Y = 35$ years. *It is very important that you not round any number until the end of the calculation.*

		In general	Example	Display
Starting Formula:		$A = \text{PMT} \times \left[\dfrac{\left(1 + \dfrac{\text{APR}}{n}\right)^{(nY)} - 1}{\dfrac{\text{APR}}{n}} \right]$	$\$100 \times \left[\dfrac{\left(1 + \dfrac{0.08}{12}\right)^{(12 \times 35)} - 1}{\dfrac{0.08}{12}} \right]$	——
Step 1.	Multiply factors in exponent.	n ⊗ Y ⊜	12 ⊗ 35 ⊜	420.
Step 2.	Store product in memory (or write down).	⌇Store⌇	⌇Store⌇	420.
Step 3.	Add terms 1 and APR$/n$.	1 ⊕ APR ⊘ n ⊜	1 ⊕ 0.08 ⊘ 12 ⊜	1.0066666667
Step 4.	Raise result to power in memory.	⌇y^x⌇ ⌇Recall⌇ ⊜	⌇y^x⌇ ⌇Recall⌇ ⊜	16.2925499
Step 5.	Subtract 1 from result.	⊖ 1 ⊜	⊖ 1 ⊜	15.2925499
Step 6.	Store result in memory (or write down).	⌇Store⌇	⌇Store⌇	15.2925499
Step 7.	Compute denominator, then take its reciprocal.	APR ⊘ n ⊜ ⌇$^1/_x$⌇	0.08 ⊘ 12 ⊜ ⌇$^1/_x$⌇	150.
Step 8.	Multiply by result in memory and payment.	⊗ ⌇Recall⌇ ⊗ PMT ⊜	⊗ ⌇Recall⌇ ⊗ 100 ⊜	229,388.249

With the calculation complete, you can round to the nearest cent, writing the answer as $229,388.25. Be sure to check the calculation.

▶ **E X A M P L E 1** Using the Savings Plan Formula

Use the savings plan formula to calculate the balance after 6 months for an APR of 12% and monthly payments of $100.

Solution We have monthly payments of PMT = $100, annual interest rate of APR = 0.12, $n = 12$ because the payments are made monthly, and $Y = \frac{1}{2}$ because 6 months is a half year. Using the savings plan formula, we can find the balance after 6 months:

$$A = \text{PMT} \times \frac{\left[\left(1 + \dfrac{\text{APR}}{n}\right)^{(nY)} - 1\right]}{\left(\dfrac{\text{APR}}{n}\right)} = \$100 \times \frac{\left[\left(1 + \dfrac{0.12}{12}\right)^{(12 \times \frac{1}{2})} - 1\right]}{\left(\dfrac{0.12}{12}\right)}$$

$$= \$100 \times \frac{\left[(1.01)^6 - 1\right]}{0.01} = \$615.20$$

Note that this answer agrees with Table 4.5.

Now try Exercises 35–38. ◀

> **by the way**
>
> A savings plan in which payments are made at the end of each month is called an *ordinary annuity*. A plan in which payments are made at the beginning of each period is called an *annuity due*. In both cases, the accumulated amount, *A,* at some future date is called the *future value* of the annuity. The formulas in this unit apply only to ordinary annuities.

𝒯hinking about . . .

DERIVATION OF THE SAVINGS PLAN FORMULA

We can derive the savings plan formula by looking at the example in Table 4.5 in a different way. Instead of calculating the balance at the end of each month (as in Table 4.5), let's calculate the value of each individual payment (deposit) and its interest at the end of month 6.

The first $100 payment was made at the *end* of month 1. Therefore, by the end of $N = 6$ months, it has collected interest for $6 - 1 = 5$ months (at the end of months 2, 3, 4, 5, and 6). We can find its value at the end of month 6 with the general compound interest formula (Unit 4A). The payment amount is PMT = $100 and the monthly interest rate is $i = 0.01$. Thus, after the collection of 5 interest payments, its value is

$$\text{PMT} \times (1 + i)^5 = \$100 \times 1.01^5$$

Similarly, the second $100 payment has collected interest for $6 - 2 = 4$ months, so its value at the end of month 6 is

$$\text{PMT} \times (1 + i)^4 = \$100 \times 1.01^4$$

The first two columns of the following table continue the calculations for the remaining months. The last column shows how the compound interest formula applies in general to each individual payment. The accumulated balance A after $N = 6$ months is the sum of the values of the individual payments. Note that the second column sum agrees with the result found in Table 4.5.

End-of-month payment	Value after month 6	Value generalized for N months
1	$100 × 1.01^5$	$\text{PMT} \times (1 + i)^{N-1}$
2	$100 × 1.01^4$	$\text{PMT} \times (1 + i)^{N-2}$
3	$100 × 1.01^3$	⋮
4	$100 × 1.01^2$	
5	$100 × 1.01$	$\text{PMT} \times (1 + i)^1$
6	$100	PMT
Total (accumulated balance, A)	**$615.20**	(sum of terms above)

➤EXAMPLE 2 Retirement Plan

At age 30, Michelle starts an IRA to save for retirement. She deposits $100 at the end of each month. If she can count on an APR of 8%, how much will she have when she retires 35 years later at age 65? Compare the IRA's value to her total deposits over this time period.

Solution We use the savings plan formula for payments of PMT = $100, an interest rate of APR = 0.08, and $n = 12$ for monthly deposits. The balance after $Y = 35$ years is

$$A = \text{PMT} \times \frac{\left[\left(1 + \dfrac{\text{APR}}{n}\right)^{(nY)} - 1\right]}{\left(\dfrac{\text{APR}}{n}\right)} = \$100 \times \frac{\left[\left(1 + \dfrac{0.08}{12}\right)^{(12 \times 35)} - 1\right]}{\left(\dfrac{0.08}{12}\right)}$$

$$= \$100 \times \frac{\left[(1.006666667)^{420} - 1\right]}{0.006666667}$$

$$= \$229{,}388.25$$

Because the last column contains only general formulas, we can use the sum of its terms as the accumulated balance A for any savings plan after N months:

$$A = PMT + PMT \times (1 + i)^1 + \cdots + PMT \times (1 + i)^{N-1}$$

(We've used "\cdots" to indicate a continuing pattern.)

We could use this formula for A, but we can simplify it further with the algebra shown in Equation 1 below. As shown, we multiply both sides by $(1 + i)^1$, and then subtract the original equation from the result. Note how all but two terms cancel on the right. The last line in Equation 1 contains the formula we seek, but we need to solve it for A.

We do this by rewriting the term on the left as

$$\underbrace{A(1 + i)}_{=A+Ai} - A = A + Ai - A = Ai$$

Thus, the equation becomes

$$Ai = PMT(1 + i)^N - PMT = PMT \times [(1 + i)^N - 1]$$

(The last step above comes from factoring out PMT from both terms on the right.) Now, we divide both sides by i to get the savings plan formula:

$$A = PMT \times \frac{[(1 + i)^N - 1]}{i}$$

To put the savings plan formula in the form given in the text, we simply substitute $i = APR/n$ for the interest rate per period and $N = nY$ for the total number of payments (where n is the number of payments per year and Y is the number of years).

Equation 1:

$$
\begin{aligned}
A(1 + i) &= \cancel{PMT(1 + i)^1} + \cdots + \cancel{PMT(1 + i)^{N-1}} + PMT(1 + i)^N \\
-A &= PMT + \cancel{PMT(1 + i)^1} + \cdots + \cancel{PMT(1 + i)^{N-1}} \\
\hline
A(1 + i) - A &= -PMT + PMT(1 + i)^N \\
&= PMT(1 + i)^N - PMT
\end{aligned}
$$

Because 35 years is 420 months ($35 \times 12 = 420$), the total amount of her deposits over 35 years is

$$420 \, \cancel{months} \times \frac{\$100}{\cancel{month}} = \$42,000$$

She will deposit a total of $42,000 over 35 years. However, thanks to compounding, her IRA will have a balance of almost $230,000—*more than five times* the amount of her contributions. *Now try Exercises 39–42.* ◄

PLANNING AHEAD WITH SAVINGS PLANS

Most people start savings plans with a particular goal in mind, such as saving enough for retirement or enough to buy a new car in a couple of years. For planning ahead, the important question is this: Given a financial goal (the total amount, A, desired after a certain number of years), what regular payments are needed to reach the goal? The following two examples show how the calculations work.

by the way

The lump sum deposit that would give you the same end result as regular payments into a savings plan is called the *present value* of the savings plan.

►**EXAMPLE 3** College Savings Plan at 7%

You want to build a $100,000 college fund in 18 years by making regular, end-of-month deposits. Assuming an APR of 7%, calculate how much you should deposit monthly. How much of the final value comes from actual deposits and how much from interest?

Solution The goal is to accumulate $A = \$100,000$ over $Y = 18$ years. The interest rate is APR $= 0.07$ and monthly payments mean $n = 12$. The goal is to calculate the required monthly payments, PMT. We therefore need to solve the savings plan formula for PMT. The savings plan formula is

$$A = \text{PMT} \times \frac{\left[\left(1 + \dfrac{\text{APR}}{n}\right)^{(nY)} - 1\right]}{\left(\dfrac{\text{APR}}{n}\right)}$$

To isolate PMT, we multiply both sides by $\left(\dfrac{\text{APR}}{n}\right)$ and divide both sides by $\left[\left(1 + \dfrac{\text{APR}}{n}\right)^{(nY)} - 1\right]$. You should confirm the following result:

$$\text{PMT} = \frac{A \times \dfrac{\text{APR}}{n}}{\left[\left(1 + \dfrac{\text{APR}}{n}\right)^{(nY)} - 1\right]}$$

Using Your Calculator

On most calculators, it is easiest to calculate the denominator first and then take its reciprocal and multiply by the other terms.

Now we substitute the given values for A, APR, n, and Y.

$$\text{PMT} = \frac{A \times \dfrac{\text{APR}}{n}}{\left[\left(1 + \dfrac{\text{APR}}{n}\right)^{(nY)} - 1\right]} = \frac{\$100,000 \times \dfrac{0.07}{12}}{\left[\left(1 + \dfrac{0.07}{12}\right)^{(12 \times 18)} - 1\right]}$$

$$= \frac{\$100,000 \times 0.005833333}{\left[(1.005833333)^{216} - 1\right]}$$

$$= \$232.17$$

Assuming the APR remains 7%, monthly payments of $232.17 will give you $100,000 after 18 years. During that time, you deposit a total of

$$18 \text{ yr} \times \frac{12 \text{ mo}}{\text{yr}} \times \frac{\$232.17}{\text{mo}} = \$50,148.72$$

Just over half of the $100,000 comes from your actual deposits; the rest is the result of compound interest. *Now try Exercises 43–46.* ◄

Time out to think _____

Suppose you want a $100,000 college fund in 18 years and you are counting on an APR of 7%. In Example 3 above, we found that you could reach your goal with monthly deposits of about $232. In Unit 4A (Example 8), we found that you could reach the same goal with a lump sum deposit of $29,586 today. Discuss the circumstances under which you might choose either the lump sum or the savings plan.

> **EXAMPLE 4** A Comfortable Retirement

You would like to retire 25 years from now, and you would like to have a retirement fund from which you can draw an income of $50,000 per year—forever! How can you do it? Assume a constant APR of 9%.

Solution You can achieve your goal by building a retirement fund that is large enough to earn $50,000 per year *from interest alone*. In that case, you can withdraw the interest for your living expenses while leaving the principal untouched (for your heirs!). The principal will then continue to earn the same $50,000 interest year after year (assuming there is no change in interest rates).

What balance do you need to earn $50,000 annually from interest? Since we are assuming an APR of 9%, the $50,000 must be 9% = 0.09 of the total balance. That is,

$$\$50{,}000 = 0.09 \times (\text{total balance})$$

Dividing both sides by 0.09, we find

$$\text{total balance} = \frac{\$50{,}000}{0.09} = \$555{,}556$$

In other words, with a 9% APR, a balance of about $556,000 allows you to withdraw $50,000 per year without ever reducing the principal.

Let's assume you will try to accumulate this balance of $A = \$556{,}000$ by making regular, monthly deposits into a savings plan. We have APR = 0.09, $n = 12$ (for monthly deposits), and $Y = 25$ years. As in Example 3, we calculate the required monthly deposits by using the savings plan formula solved for PMT.

$$\text{PMT} = \frac{A \times \dfrac{\text{APR}}{n}}{\left[\left(1 + \dfrac{\text{APR}}{n}\right)^{(nY)} - 1\right]} = \frac{\$556{,}000 \times \dfrac{0.09}{12}}{\left[\left(1 + \dfrac{0.09}{12}\right)^{(12 \times 25)} - 1\right]}$$

$$= \frac{\$556{,}000 \times 0.0075}{\left[(1.0075)^{300} - 1\right]}$$

$$= \$495.93$$

If you deposit about $500 per month over the next 25 years, you will achieve your retirement goal—assuming you can count on a 9% APR. Although saving $500 per month may seem like a lot, it can be easier than it sounds thanks to special tax treatment for retirement plans (see Unit 4D). *Now try Exercises 47–48.* ◄

TOTAL AND ANNUAL RETURN

In the examples so far, we've assumed that you get a constant interest rate for a long period of time. In reality, interest rates usually vary over time. Consider a case in which you invest a starting principal of $1000 and it grows to $1500 in 5 years.

by the way

An account that provides a permanent source of income without reducing its principal is called an *endowment*. Many charitable foundations are endowments. They spend each year's interest (or a portion of the interest) on charitable activities, while leaving the principal untouched to earn interest again in future years. Of course, the *value* of a particular dollar amount tends to decline with time, because inflation reduces the value of a dollar.

Although the interest rate may have varied during the 5 years, we can still describe the change in both total and annual terms.

Your **total return** is the relative change in the investment value over the 5-year period (see Unit 3A for a discussion of relative change):

$$\text{total return} = \frac{\text{new value} - \text{starting principal}}{\text{starting principal}} = \frac{\$1500 - \$1000}{\$1000} = 0.5$$

The total return on this investment is 50% over 5 years.

Your **annual return** is the *average annual rate* at which your money grew over the 5 years. That is, it is the constant annual percentage yield (APY) that would give the same result in 5 years. In this case, the annual return is about 8.5%. We can see why by using the compound interest formula for interest paid once a year. We set the interest rate (APR) to the annual yield of APY = 8.5% = 0.085 and the number of years to $Y = 5$. The compound interest formula confirms that a starting principal $P = \$1000$ grows to about $A = \$1500$ in 5 years:

$$A = P \times (1 + \text{APY})^Y = \$1000 \times (1 + 0.085)^5 = \$1503.66$$

In this case, we "guessed" the APY, then confirmed our guess with the compound interest formula. We can calculate the annual return more directly by solving the compound interest formula above for APY. The algebra is shown in the Brief Review on pp. 236–237. The result is summarized in the following box.

Total and Annual Return

Consider an investment that grows from an original principal P to a later accumulated balance A.

The **total return** is the relative change in the investment value:

$$\text{total return} = \frac{(A - P)}{P}$$

The **annual return** is the annual percentage yield (APY) that would give the same overall growth. The formula is

$$\text{annual return} = \left(\frac{A}{P}\right)^{(1/Y)} - 1$$

where Y is the investment period in years.

➤**EXAMPLE 5** Mutual Fund Gain

You invest $3000 in the Clearwater mutual fund. Over 4 years, your investment grows in value to $8400. What are your total and annual returns for the 4-year period?

Solution You have a starting principal $P =$ \$3000 and an accumulated value of $A =$ \$8400 after $Y = 4$ years. Thus, your total and annual returns are

$$\text{total return} = \frac{(A - P)}{P} = \frac{(\$8400 - \$3000)}{\$3000}$$
$$= 1.8 = 180\%$$

and

$$\text{annual return} = \left(\frac{A}{P}\right)^{(1/Y)} - 1 = \left(\frac{\$8400}{\$3000}\right)^{(1/4)} - 1$$
$$= \sqrt[4]{2.8} - 1 = 0.294 = 29.4\%$$

Your total return is 1.8, or 180%, meaning that the value of your investment after 4 years is 1.8 times its original value. Your annual return is 0.294, or 29.4%, meaning that your investment has grown by an average of 29.4% each year.

You should check your answer for the annual return by using the compound interest formula. If you use the annual return as the APY, the compound interest formula should give you the correct accumulated value. In this case,

$$A = P \times (1 + \text{APY})^Y = \$3000 \times (1 + 0.294)^4 = \$8411.21$$

This is very close to the correct value of \$8400. The slight difference is due to rounding when we calculated the annual return. *Now try Exercises 49–52.* ◄

▶ **EXAMPLE 6** Investment Loss

You purchased shares in NewWeb.com for \$2000. Three years later, you sold them for \$1100. What were your total return and annual return on this investment?

Solution You had a starting principal $P =$ \$2000 and an accumulated value of $A =$ \$1100 after $Y = 3$ years. Thus, your total and annual returns were

$$\text{total return} = \frac{(A - P)}{P} = \frac{(\$1100 - \$2000)}{\$2000} = -0.45$$

and

$$\text{annual return} = \left(\frac{A}{P}\right)^{(1/Y)} - 1 = \left(\frac{\$1100}{\$2000}\right)^{(1/3)} - 1 = \sqrt[3]{0.55} - 1 = -0.18$$

The returns are negative because you *lost* money on this investment. Your total return was -0.45, or -45%, meaning that your investment lost 45% of its original value. Your annual return was -0.18, or -18%, meaning that your investment lost an average of 18% of its value each year. *Now try Exercises 53–56.* ◄

TYPES OF INVESTMENTS

Savings plans can involve many types of investments. By combining what we've covered about savings plans with the ideas of total and annual return, we can now study investment options. Most investments fall into one of the three basic categories described in the following box.

Using Your Calculator

Raising a number to the $1/Y$ power is the same as taking the Yth root. The key for taking a root will be labeled something like $\boxed{\sqrt[x]{\ }}$ or $\boxed{x^{1/y}}$. For example, on many calculators you would calculate $\sqrt[4]{2.8}$ by pressing 2.8 $\boxed{\sqrt[x]{\ }}$ 4 $\boxed{=}$.

by the way

There are many other types of investments besides the basic three, such as rental properties, precious metals, commodities futures, and derivatives. These investments are generally more complex and often higher risk than the basic three.

Three Basic Types of Investments

Stock (or *equity*) gives you a share of ownership in a company. You invest some principal amount to purchase the stock, and the only way to get your money out is to sell the stock. Because stock prices change with time, the sale may give you either a gain or a loss on your original investment.

A **bond** (or *debt*) represents a promise of future cash. You buy a bond by paying some principal amount to the issuing government or corporation. The issuer pays you simple interest (as opposed to compound interest) and promises to pay back your principal at some later date.

Cash investments include money you deposit into bank accounts, certificates of deposit (CD), and U.S. Treasury bills. Cash investments generally earn interest.

A brief review . . .

ALGEBRA WITH POWERS AND ROOTS

As we have seen, powers and roots are often used in financial calculations. A review of these operations may be helpful.

Basics of Powers

A number raised to the nth power is that number multiplied by itself n times (n is called an *exponent*). For example:

$$2^1 = 2 \qquad 2^2 = 2 \times 2 = 4 \qquad 2^3 = 2 \times 2 \times 2 = 8$$

A number to the *zero* power is defined to be 1. For example:

$$2^0 = 1$$

Negative powers are the reciprocals of the corresponding positive powers. For example:

$$5^{-2} = \frac{1}{5^2} = \frac{1}{5 \times 5} = \frac{1}{25} \qquad 2^{-3} = \frac{1}{2^3} = \frac{1}{2 \times 2 \times 2} = \frac{1}{8}$$

Power Rules

In the following rules, x represents a number being raised to a power, and n and m are exponents. Note that these rules work only when all powers involve the same number x. (See also A Brief Review on p. 105 [Unit 2B].)

- To multiply powers of the same number, *add* the exponents:

$$x^n \times x^m = x^{n+m}$$

Example: $2^3 \times 2^2 = 2^{3+2} = 2^5 = 32$

- To divide powers of the same number, *subtract* the exponents:

$$\frac{x^n}{x^m} = x^{n-m}$$

Example: $\dfrac{5^3}{5^2} = 5^{3-2} = 5^1 = 5$

- When a power is raised to another power, *multiply* the exponents:

$$(x^n)^m = x^{n \times m}$$

Example: $(2^2)^3 = 2^{2 \times 3} = 2^6 = 64$

Basics of Roots

Finding a root is the reverse of raising a number to a power. Second roots, or square roots, are written with a number under the root symbol $\sqrt{\ }$. More generally, we indicate an nth root by writing a number under the symbol $\sqrt[n]{\ }$. For example:

$$\sqrt{4} = 2 \text{ because } 2^2 = 2 \times 2 = 4$$
$$\sqrt[3]{27} = 3 \text{ because } 3^3 = 3 \times 3 \times 3 = 27$$
$$\sqrt[4]{16} = 2 \text{ because } 2^4 = 2 \times 2 \times 2 \times 2 = 16$$
$$\sqrt[6]{1,000,000} = 10 \text{ because } 10^6 = 1,000,000$$

There are two basic ways to invest in any of these categories. First, you can invest directly, which means buying individual investments yourself. For example, you can directly purchase individual stocks through a stockbroker and you can buy bonds directly from the government. In general, the only costs associated with direct investments are commissions that you pay to brokers.

Alternatively, you can invest indirectly by purchasing shares in a **mutual fund,** where a professional fund manager invests your money (and the money of others participating in the fund). *Stock mutual funds* invest primarily in stocks, *bond mutual funds* invest primarily in bonds, *money market funds* invest only in cash, and *diversified funds* invest in a mixture of stocks, bonds, and cash.

Investment Considerations: Liquidity, Risk, and Return

No matter what type of investment you make, you should evaluate the investment in terms of three general considerations.

Roots as Fractional Powers

The nth root of a number is the same as the number raised to the $1/n$ power. That is,

$$\sqrt[n]{x} = x^{1/n}$$

For example:

$$64^{1/3} = \sqrt[3]{64} = 4$$
$$1{,}000{,}000^{1/6} = \sqrt[6]{1{,}000{,}000} = 10$$

Power and Root Algebra

The following two rules hold true for working with equations:

1. We can raise both sides of an equation to the same power.
2. We can take the same root of both sides of an equation, which is equivalent to raising both sides to the same fractional power. (Note: This process may produce both positive and negative roots; we consider only positive roots here.)

Example: Find the positive solution of the equation $x^4 = 16$.

Solution: We isolate x by raising both sides to the $1/4$ power:

$$(x^4)^{1/4} = 16^{1/4} \quad \rightarrow \quad x^{4 \times \frac{1}{4}} = 16^{1/4} \quad \rightarrow \quad x = 16^{1/4} = 2$$

Therefore, one solution of the equation is $x = 2$. Note that, in the last step, we recognized that $x^{4 \times 1/4} = x^1 = x$.

Example: Solve the equation $A = P \times (1 + \text{APY})^Y$ for APY.

Solution: First, we isolate the term containing APY by dividing both sides of the equation by P:

$$\frac{A}{P} = (1 + \text{APY})^Y$$

Next, we raise both sides of the equation to the $1/Y$ power, then simplify the right side:

$$\left(\frac{A}{P}\right)^{1/Y} = [(1 + \text{APY})^Y]^{1/Y}$$

$$= \underbrace{(1 + \text{APY})^{Y \times 1/Y}}_{\substack{\text{This step} \\ \text{from the rule} \\ (x^n)^m = x^{n \times m}}} = \underbrace{1 + \text{APY}}_{\substack{\text{This step} \\ \text{from} \\ \text{the rule} \\ x^{n/n} = x^1 = x}}$$

Finally, we isolate APY by subtracting 1 from both sides:

$$\left(\frac{A}{P}\right)^{1/Y} - 1 = 1 + \text{APY} - 1 \quad \rightarrow \quad \text{APY} = \left(\frac{A}{P}\right)^{1/Y} - 1$$

(In the last step, we interchanged the left and right sides.) To get the formula in the box on p. 234, we replace APY by *annual return.* This replacement is valid because the annual return is the same as the constant APY that would lead to an accumulated balance A.

Now try Exercises 15–34.

by the way

The U.S. Treasury issues bills, notes, and bonds. *Treasury bills* are essentially cash investments that are highly liquid and very safe. *Treasury notes* are essentially bonds with 2- to 10-year terms. *Treasury bonds* have 20- to 30-year terms.

- How difficult is it to take out your money? An investment from which you can withdraw money easily, such as an ordinary bank account, is said to be **liquid.** The liquidity of an investment like real estate is much lower because real estate can be difficult to sell.
- Is your investment principal at **risk?** The safest investments are federally insured bank accounts and U.S. Treasury bills—there's virtually no risk of losing the principal you've invested. Stocks and bonds are much riskier because they can drop in value, in which case you may lose part or all of your principal.
- How much **return** (total or annual) can you expect on your investment? A higher return means you earn more money. In general, low-risk investments offer relatively low returns, while high-risk investments offer the prospects of higher returns—along with the possibility of losing your principal.

Historical Returns

One of the most difficult tasks of investing is trying to balance risk and return. Although there are no guarantees for the future, historical trends offer at least some guidance. Table 4.6 shows historical average annual returns for several different types of investments.

BUILDING A PORTFOLIO

Before you bought a new television for a few hundred dollars, you'd probably do a fair amount of research to make sure that you were getting a good buy. You should be even more diligent when making investments that may determine your entire financial future.

The best way to plan your savings is to learn about investments by reading financial pages of newspapers and some of the many books and magazines devoted to finance. You may also want to consult a professional financial planner. With this background, you will be prepared to create a personal financial *portfolio* (set of investments) that meets your needs.

Most financial advisors recommend that you create a *diversified* portfolio—that is, a portfolio with a mixture of low-risk and high-risk investments. No single mixture is right for everyone. Your portfolio should balance risk and return in a way that is appropriate for your situation. For example, if you are young and retirement is far in the future, you may be willing to have a relatively risky portfolio that offers the hope of high returns. In contrast, if you are already retired, you may want a low-risk portfolio that promises a safe and steady stream of income.

However you structure your portfolio, the most important step in meeting your financial goals is making sure that you save enough money. You can use the tools in this unit to help you determine what is "enough." Make a reasonable estimate of the annual return you can expect from your overall portfolio. Use this annual return as the interest rate in the savings plan formula, and calculate how much you must invest each month or each year to meet your goals (see Examples 3 and 4). Then make sure you actually put this money in your investment plan. If you need further motivation, consider this: Every $100 you spend today is gone, but even at a fairly low (by historical standards) annual return of 4%, every $100 you invest today will be worth $148 in 10 years, $219 in 20 years, and $711 in 50 years.

TABLE 4.6 Returns on Different Investment Categories, 1926–2000

INVESTMENT TYPE	AVERAGE ANNUAL RETURN*	BEST YEAR	WORST YEAR
Small-company stocks	12.4%	142.9% (1933)	−58.0% (1937)
Large-company stocks	11.1%	54.0% (1933)	−43.3% (1931)
Long-term corporate bonds	5.6%	42.6% (1982)	−8.1% (1969)
Cash (U.S. Treasury bills)	3.8%	14.7% (1981)	−0.02% (1938)

*Includes both increases in price and any dividends or interest.
SOURCE: Stocks, Bonds, Bills & Inflation Yearbook™, Ibbotson Associates, Chicago.

Time out to think

How do the data in Table 4.6 confirm that higher returns tend to involve higher risk? Explain.

A common way of tracking an investment category over time is to use an *index* that describes the average return for some category of investments. The best-known index is the *Dow Jones Industrial Average* (DJIA), which reflects the average stock prices of 30 of the largest and most stable companies. Different investment categories are tracked by different indices. Figure 4.4 shows the historical performance of the DJIA.

by the way

The DJIA is the most famous financial index, but many others are important. *Standard and Poor's 500* (S&P 500) tracks 500 large-company stocks; the *Russell 2000* tracks 2000 small-company stocks; the *Nasdaq composite* tracks 100 large-company stocks listed on the Nasdaq exchange; the *Lehman Brothers T-Bond Index* tracks the performance of U.S. Treasury bonds; and the *Federal Funds Index* tracks short-term interest rates.

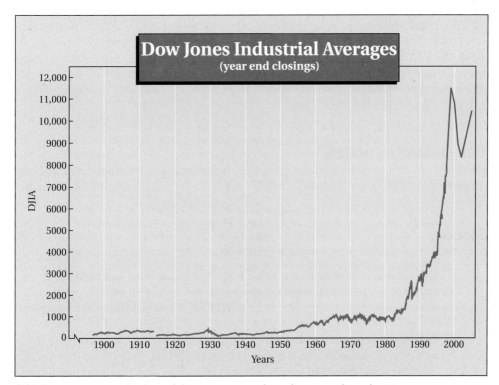

FIGURE 4.4 Historical values of the Dow Jones Industrial Average through 2003.

➤ **EXAMPLE 7** Historical Returns

Suppose your great-grandmother invested $1000 at the beginning of 1926 in each of the following: small-company stocks, large-company stocks, long-term corporate bonds, and U.S. Treasury bills. Assuming her investments grew at the rates given in Table 4.6, approximately how much was each investment worth at the end of 2000?

Solution We find the value of each of the four investments with the compound interest formula, setting the interest rate (APR) to the annual return. In all four cases, the starting principal is $P = \$1000$ and $Y = 75$ years from the beginning of 1926 to the end of 2000. Table 4.7 shows the calculations.

TABLE 4.7	Calculations for Example 7	
INVESTMENT TYPE	ANNUAL RETURN	INVESTMENT VALUE: $A = P \times (1 + APR)^Y$
Small-company stocks	12.4% = 0.124	$A = \$1000 \times (1 + 0.124)^{75} = \$6,419,100$
Large-company stocks	11.1% = 0.111	$A = \$1000 \times (1 + 0.111)^{75} = \$2,682,600$
Corporate bonds	5.6% = 0.056	$A = \$1000 \times (1 + 0.056)^{75} = \$59,500$
Treasury bills	3.8% = 0.038	$A = \$1000 \times (1 + 0.038)^{75} = \$16,400$

Note the enormous difference between the categories. The $1000 investment in cash grew to $16,400 in 75 years, while the same investment in small-company stocks grew to over $6 million! *Now try Exercises 57–58.* ◄

Time out to think _____

Although stocks have outperformed other investments over the long term, Figure 4.4 shows that there have been some periods during which stocks gained little or lost value. For example, what happened to typical stock portfolios during 2000–2002? What do *you* think will happen to the stock market over the next 5 years? the next 50 years? Why?

THE FINANCIAL PAGES

If you are investing money, you can track your investments in the financial pages or through many Web sites. Let's look briefly at what you must know to understand commonly published data about stocks, bonds, and mutual funds.

Stocks

In general, there are two ways to make money on stocks:

- You can make money if you sell a stock for more than you paid for it, in which case you have a **capital gain** on the sale of the stock. Of course, you also can lose money on a stock (a capital loss) if you sell shares for less than you paid for them or if the company goes into bankruptcy.
- You can make money while you own the stock if the corporation distributes part or all of its profits to stockholders as **dividends.** Each share of stock is paid the same dividend, so the amount of money you receive depends on the number of shares you own. Not all companies distribute profits as dividends. Some reinvest all profits within the corporation.

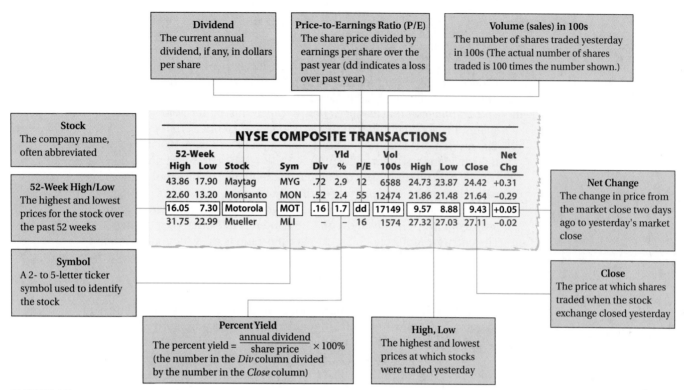

FIGURE 4.5

Daily stock tables provide a wealth of information about stocks, summarized in Figure 4.5. Nevertheless, it pays to get even more information if you are buying stocks. For example, you can learn a lot by studying a company's annual report. Many companies have Web sites with information for investors. You can also get independent research reports from many investment services (usually for a fee) or by working with a stockbroker (to whom you pay commissions when you buy or sell stock).

► **EXAMPLE 8** Motorola Stock

Suppose that Figure 4.5 comes from today's paper.

a. What is the ticker symbol for the Motorola Corporation?

b. What was the range of selling prices for Motorola shares yesterday? How do these prices compare to prices over the past year?

c. What was the closing price of Motorola shares yesterday and 2 days ago?

d. How many shares of Motorola were traded yesterday?

e. Suppose you own 100 shares of Motorola. What total dividend payment should you expect this year?

f. Compare what you can expect to earn from dividends to what you would earn from a bank account offering a 1.5% annual interest rate.

g. Has Motorola made a profit in the past year?

Solution

a. The *symbol* column shows that Motorola's ticker symbol is MOT.

b. The *high* and *low* columns show that, yesterday, Motorola stock traded in the range from $8.88 to $9.57 per share. The middle of this range is close to $2, or about 25%, above the 52-week low of $7.30.

c. The *close* column shows that Motorola closed at $9.43 per share. The *change* column shows that the share price rose $0.05 from the previous day. Thus, the closing price 2 days ago was $9.43 − $0.05 = $9.38 per share.

d. The *volume* column reads 17,149. Because this figure is in *hundreds* of shares, the actual number of shares traded was 17,149 × 100 = 1,714,900.

e. The annual dividend rate is $0.16 per share. If you own 100 shares, your total dividend payment will be 100 × $0.16 = $16. (However, dividends are usually paid quarterly, so your actual dividend may be different if the company changes its dividend rate during the year.)

f. The *percent yield* column shows that dividends alone represent an annual return of 1.7%—slightly above the 1.5% interest rate offered by the bank.

g. The entry for Motorola's price-to-earnings ratio is dd, which means that it had a loss, not a profit, in the past year. *Now try Exercises 59–66.* ◄

▶ **EXAMPLE 9** P/E Ratio

Using the data from Figure 4.5, compare Monsanto's share price to its profit per share in the past year. How much profit per share did Monsanto earn in the past year? Historically, stocks trade at an average P/E ratio of about 12–14. Based on this historical average and its current P/E ratio, does Monsanto's stock price seem cheap or expensive right now? Given this P/E ratio, what might explain the current stock price?

Solution Monsanto's price-to-earnings ratio is 55, which means that its current (closing) stock price is 55 times its earnings (profit) per share in the past year. Thus, its earnings per share over the past year must have been $\frac{1}{55}$ of its current stock price:

$$\text{earnings per share} = \frac{\text{stock price}}{\text{P/E ratio}} = \frac{\$21.64}{55} = \$0.393$$

Monsanto earned a profit of about 39¢ per share over the past year. The P/E ratio of 55 is far above the historical average P/E at which stocks trade, which makes the stock seem quite expensive on this basis alone. One possible explanation for the high P/E ratio is that investors expect Monsanto's profits to grow substantially in

Frank and Ernest

1997 Thaves/Reprinted with permission. Newspaper distribution by NEA, Inc.

the near future, since higher earnings would bring the P/E ratio closer to historical norms. *Now try Exercises 67–72.* ◄

Bonds

Most bonds are issued with three main characteristics:

- The **face value** (or *par value*) of the bond is the price you must pay the issuer to buy it at the time it is issued.
- The **coupon rate** of the bond is the *simple interest rate* that the issuer promises to pay. For example, a coupon rate of 8% on a bond with a face value of $1000 means that the issuer will pay you interest of 8% × $1000 = $80 each year.
- The **maturity date** of the bond is the date on which the issuer promises to repay the face value of the bond.

Bonds would be simple if that were the end of the story. However, bonds can also be bought and sold after they are issued, in what is called the *secondary bond market*. For example, suppose you own a bond with a $1000 face value and a coupon rate of 8%. Further suppose that new bonds with the same level of risk and same time to maturity are issued with a coupon rate of 9%. In that case, no one would pay $1000 for your bond because the new bonds offer a higher interest rate. However, you may be able to sell your bond at a *discount*—that is, for less than its face value. In contrast, suppose that new bonds are issued with a coupon rate of 7%. In that case, buyers will prefer your 8% bond to the new bonds and therefore may pay a *premium* for your bond—a price greater than its face value.

Consider a case in which you buy a bond with a face value of $1000 and a coupon rate of 8% for only $800. The bond issuer will still pay simple interest of 8% of $1000, or $80 per year. However, because you paid only $800 for the bond, *your* return for each year is

$$\frac{\text{amount you earn}}{\text{amount you paid}} = \frac{\$80}{\$800} = 0.1 = 10\%$$

More generally, the **current yield** of a bond is defined as the amount of interest it pays each year divided by the bond's current price (*not* its face value).

Current Yield of a Bond

$$\text{current yield} = \frac{\text{annual interest payment}}{\text{current price of bond}}$$

A bond selling at a discount from its face value has a current yield that is higher than its coupon rate. The reverse is also true: A bond selling at a premium over its face value has a current yield that is lower than its coupon rate. Thus, we have the rule that *bond prices and yields move in opposite directions.*

Bond prices are usually quoted in *points*, which means percentage of face value. Most bonds have a face value of $1000. Thus, for example, a bond that closes at 102 points is selling for 102% × $1000 = $1020.

➤ **EXAMPLE 10** Bond Interest

The closing price of a U.S. Treasury bond with a face value of $1000 is quoted as 105.97 points, for a current yield of 3.7%. If you buy this bond, how much annual interest will you receive?

Solution The 105.97 points means the bond is selling for 105.97% of its face value or

$$105.97\% \times \$1000 = \$1059.70$$

This is the current price of the bond. We are also given its current yield of 3.7%, so we can solve the current yield formula to find the annual interest payment:

$$\text{current yield} = \frac{\text{annual interest}}{\text{current price}}$$

↓ multiply both sides by current price

$$\text{annual interest} = \text{current yield} \times \text{current price}$$

Substituting the price and yield, we find

$$\text{annual interest} = 3.7\% \times \$1059.70 = 0.037 \times \$1059.70 = \$39.21$$

The annual interest payments on this bond are $39.21. *Now try Exercises 73–80.* ◄

Mutual Funds

When you buy shares in a mutual fund, the fund manager takes care of the day-to-day decisions about when to buy and sell individual stocks or bonds. Thus, in comparing mutual funds, the most important factors are the fees charged for investing and measures of how well the manager is doing with the fund's money. Figure 4.6 shows a sample mutual fund table. The table makes it easy to compare the past performance of funds. Of course, as stated in every mutual fund prospectus, *past performance is no guarantee of future results.*

Most mutual fund tables do not show the fees charged. For that, you must call or check the Web site of the company offering the mutual fund. Because fees are generally withdrawn automatically from your mutual fund account, they can have a big impact on your long-term gains. For example, if you invest $100 in a fund that charges a 5% annual fee, only $95 is actually invested. Over many years, this can significantly reduce your total return.

➤ **EXAMPLE 11** Mutual Fund Growth

Suppose that Figure 4.6 represents a table from today's paper. If you invested $500 in the Calvert Income fund 3 years ago, what is your investment worth now? (Assume you reinvested all dividends and gains, and do not count fees.)

Solution Figure 4.6 shows that the annual return for the past 3 years was 10.3%, or 0.103. Therefore, we can use this as the APR in the compound interest formula, with a term of $Y = 3$ years and principal of $P = \$500$.

$$A = P \times (1 + \text{APR})^Y = \$500(1 + 0.103)^3 = \$670.96$$

Your $500 investment is now worth about $671. *Now try Exercises 81–82.* ◄

FIGURE 4.6

Text boxes from figure:

Rating
A system for comparing fund performance, with 1 as the worst and 5 as the best (The first number is performance compared to a broad group of similar funds, such as all stock funds, and the second number is performance compared only to funds of the same type.)

NAV
The net asset value of the fund's shares—that is, the amount that each fund share is currently worth

Weekly % Return
The total return for the week, including capital gains from sales and any dividends

Fund Family
A group of funds from the same company

Fund Name
The name of an individual mutual fund

Type
An abbreviation describing the type of investments the fund manages (*The New York Times* categorizes funds into about 50 different types and includes an index each Sunday.)

YTD % Return
The total return for the year-to-date (since Jan. 1)

1-year % Return
The total return for the past 1-year period

3-year % Return
The annual return over the past 3 years, calculated by assuming that any dividends and gains are reinvested into that fund

MUTUAL FUND QUOTATIONS

Fund Family Fund Name	Type	Rating	NAV	Wkly. % Ret.	YTD % Ret.	1-Yr. % Ret.	3-Yr. % Ret.
USGovSec	GI	3/3	11.00	– 1.1	+ 3.4	+ 8.9	+ 8.4
Calvert 800-368-2748							
CapAccmA m	MG	2/3	19.09	+ 0.9	+ 13.2	– 0.7	– 11.99
Income m	CI	3/3	17.43	– 0.6	+ 8.5	+ 10.5	+ 10.3
NatMulntA m	MI	2/3	11.39	– 1.4	+ 4.4	+ 9.0	+ 7.9
SocInvBdA m	CI	3/3	16.39	– 0.9	+ 6.4	+ 9.4	+ 10.0
SocInvEqA m	LB	3/5	29.55	+ 1.1	+ 11.9	+ 2.5	– 0.9
SocIvBalA m	MA	3/2	24.31	+ 0.1	+ 9.3	+ 2.8	– 4.9
TaxFVTA f	SI	3/3	16.57	– 1.4	+ 4.1	+ 8.8	+ 7.7
WdVaIntlA m	FS	3/3	14.00	+ 0.9	+ 9.6	+ 0.3	+ 13.6
Capstone 800-262-6631							
Growth b	LB	3/2	11.01	+ 1.1	+ 10.1	– 6.1	– 12.9

■ EXERCISES 4B

Review Questions

1. What is a savings plan? Explain the savings plan formula.

2. Give an example of a situation in which you might want to solve the savings plan formula to find the payments, PMT, required to achieve some goal.

3. Distinguish between the total return and the annual return on an investment. How do we calculate the annual return? Give an example.

4. Briefly describe the three basic types of investments: stocks, bonds, and cash. How can you invest in these types directly? How can you invest in them indirectly through a mutual fund?

5. Explain what we mean by an investment's liquidity, risk, and return. How are risk and return usually related?

6. Contrast the historical returns for different types of investments. How do financial indices, such as the DJIA, help keep track of historical returns?

7. Define the face value, coupon rate, and maturity date of a bond. What does it mean to buy a bond at a premium? at a discount? How can you calculate the current yield of a bond?

8. Briefly describe the meaning of each column in typical financial tables for stocks and mutual funds.

Does It Make Sense?

Decide whether each of the following statements makes sense (or is clearly true) or does not make sense (or is clearly false). Explain your reasoning.

9. If interest rates stay where they are and I continue to make my monthly $200 deposits into my retirement plan, I should be able to retire in 25 years with a comfortable income.

10. My financial advisor showed me that I could reach my retirement goal with deposits of $200 per month and an average annual return of 7%. But I don't want to deposit that much of my paycheck, so I'm going to reach the same goal by getting an average annual return of 15% instead.

11. I'm putting all my savings into stocks because stocks always outperform other types of investment over the long term.

12. I'm hoping to withdraw money to buy my first house soon, so I need to put it into an investment that is fairly liquid.

13. I bought a fund advertised on the Web that says it uses a secret investment strategy to get an annual return twice that of stocks, with no risk at all.

14. I'm already retired, so I need low-risk investments. That's why I put most of my money in U.S. Treasury bills, notes, and bonds.

Basic Skills & Concepts

Review of Powers and Roots. Exercises 15–26 use skills covered in the Brief Review on pp. 236–237. Evaluate the expressions and express the answer in simplest terms.

15. 4^3

16. $25^{1/2}$

17. $2^3 \times 2^5$

18. $27^{1/3}$

19. $3^6 \div 3^2$

20. $81^{1/2} \times 3^2$

21. 4^{-2}

22. $16^{-1/2}$

23. $4^5 \times 4^{-2}$

24. $64^{-1/3}$

25. $5^3 \div 5^{-4}$

26. $36^{1/2} \div 16^{1/4}$

Solving with Powers and Roots. Solve the equations in Exercises 27–34 for the unknown, x.

27. $x^2 = 100$

28. $(x - 2)^2 = 36$

29. $x^3 = 27$

30. $x^3 + 3 = 30$

31. $x^{1/3} = 2$

32. $(x/2)^2 = 25$

33. $x^{1/5} = 2$

34. $x^2 - 4 = 32$

Savings Plan Formula. In Exercises 35–38, calculate the balance under the given assumptions.

35. Find the savings plan balance after 9 months with an APR of 12% and monthly payments of $100.

36. Find the savings plan balance after 1 year with an APR of 12% and monthly payments of $100.

37. Find the savings plan balance after 18 months with an APR of 6% and monthly payments of $200.

38. Find the savings plan balance after 24 months with an APR of 5% and monthly payments of $250.

Investment Plans. Use the savings plan formula in Exercises 39–42.

39. You set up an IRA (individual retirement account) with an APR of 5% at age 25. At the end of each month, you deposit $50 in the account. How much will the IRA contain when you retire at age 65? Compare that amount to the total amount of deposits made over the time period.

40. A friend creates an IRA with an APR of 6.25%. She starts the IRA at age 25 and deposits $50 per month. How much will her IRA contain when she retires at age 65? Compare that amount to the total amount of deposits made over the time period.

41. You put $200 per month in an investment plan that pays an APR of 7%. How much money will you have after 18 years? Compare this amount to the total amount of deposits made over the time period.

42. You put $200 per month in an investment plan that pays an APR of 4.5%. How much money will you have after 18 years? Compare this amount to the total amount of deposits made over the time period.

Investment Planning. Use the savings plan formula in Exercises 43–46.

43. You intend to create a college fund for your baby. If you can get an APR of 7.5% and want the fund to have a value of $150,000 after 18 years, how much should you deposit monthly?

44. At age 35 you start saving for retirement. If your investment plan pays an APR of 6% and you want to have $2 million when you retire in 30 years, how much should you deposit monthly?

45. You want to purchase a new car in 3 years and expect the car to cost $10,000. Your bank offers a plan with a guaran-

teed interest rate of $APR = 5.5\%$ if you make regular monthly deposits. How much should you deposit each month to end up with $10,000 in 3 years?

46. At age 20 when you graduate, you start saving for retirement. If your investment plan pays an APR of 8% and you want to have $5 million when you retire in 45 years, how much should you deposit monthly?

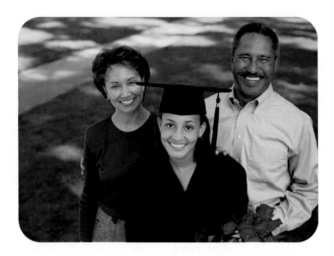

47. **Comfortable Retirement.** Suppose you are 30 years old and would like to retire at age 60. Furthermore, you would like to have a retirement fund from which you can draw an income of $50,000 per year—forever! How can you do it? Assume a constant APR of 6%.

48. **Very Comfortable Retirement.** Suppose you are 25 years old and would like to retire at age 65. Furthermore, you would like to have a retirement fund from which you can draw an income of $200,000 per year—forever! How can you do it? Assume a constant APR of 6%.

Total and Annual Returns. In Exercises 49–56, compute the total and annual returns on the described investment.

49. Five years after buying 100 shares of XYZ stock for $55 per share, you sell the stock for $10,300.

50. You pay $8000 for a municipal bond. When it matures after 20 years, you receive $12,500.

51. Ten years after purchasing shares in a mutual fund for $5500, you sell them for $11,300.

52. Three years after buying 200 shares of XYZ stock for $25 per share, you sell the stock for $8500.

53. Three years after paying $4500 for shares in a startup company, you sell the shares for $2500 (at a loss).

54. Five years after paying $5000 for shares in a new company, you sell the shares for $3000 (at a loss).

55. Eight years after purchasing shares in a mutual fund for $7500, you sell them for $12,600.

56. Ten years after purchasing shares in a mutual fund for $10,000, you sell them for $2200 (at a loss).

57. **Historical Returns.** Suppose your great-uncle invested $500 at the beginning of 1931 in each of the following: small-company stocks, large-company stocks, long-term corporate bonds, and U.S. Treasury bills. Assuming his investments grew at the long-term average annual returns in Table 4.6, approximately how much would each investment have been worth at the end of 2000?

58. **Best and Worst Years.** Suppose you invest $2000 in each of the following: small-company stocks, large-company stocks, long-term corporate bonds, and U.S. Treasury bills. Using the returns shown in Table 4.6, calculate how much your investments would be worth a year later if it was the *best* of years? How much would your investments be worth a year later if it was the *worst* of years?

Reading Stock Tables. Use the data in Figure 4.5 to answer the questions in Exercises 59–66. Assume the data come from today's newspaper.

59. Of the four stocks shown in Figure 4.5, which one had the biggest gain in price yesterday? State the company's name and symbol and the amount it gained per share. Based on the closing price and the gain, what was its closing price two days ago?

60. Of the four stocks shown in Figure 4.5, which one had the biggest decline in price yesterday? State the company's name and symbol and the amount it lost per share. Based on the closing price and the loss, what was its closing price two days ago?

61. Of the four stocks shown in Figure 4.5, which one is currently trading at prices nearest to its highest price over the past year? Explain.

62. Of the four stocks shown in Figure 4.5, which one is currently trading at prices nearest to its lowest price over the past year? Explain.

63. Suppose you own 1000 shares of Monsanto. What total dividend payment can you expect this year?

64. Suppose you own 100 shares of each of the four stocks shown in Figure 4.5. Which one will pay you the highest dividend, in absolute dollars? How much will your dividend payment be?

65. Suppose your primary investment goal is to receive income from dividends. Assuming the stock prices and dividends in Figure 4.5 continue to hold, which of the four stocks would be the best investment for you? Explain.

66. Suppose your primary investment goal is to receive income from dividends. Which stock(s) in Figure 4.5 would it make no sense for you to invest in? Explain.

Price-to-Earning Ratio. For each stock listed in Exercises 67–72, answer the following questions:

 a. Did the company earn a profit in the past year? If so, how does its share price compare to the profit per share that it earned in the past year?

 b. How much profit per share did the company earn in the past year?

 c. Based on the fact that stocks historically trade at an average P/E ratio of about 12–14, does the stock price seem cheap, about right, or expensive right now? If it seems cheap or expensive, what might explain the current stock price?

67. Maytag, assuming Figure 4.5 comes from today's newspaper

68. Maytag, based on yesterday's actual closing stock price (from a newspaper or Web site)

69. Motorola, assuming Figure 4.5 comes from today's newspaper

70. Motorola, based on yesterday's actual closing stock price (from a newspaper or Web site)

71. Mueller Industries, assuming Figure 4.5 comes from today's newspaper

72. Mueller Industries, based on yesterday's actual closing stock price (from a newspaper or Web site)

Bond Yields. In Exercises 73–76, calculate the current yield on the described bond.

73. A $1000 Treasury bond with a coupon rate of 2.0% that has a market value of $950

74. A $1000 Treasury bond with a coupon rate of 2.5% that has a market value of $1050

75. A $1000 Treasury bond with a coupon rate of 5.5% that has a market value of $1100

76. A $10,000 Treasury bond with a coupon rate of 3.0% that has a market value of $9500

Bond Interest. In Exercises 77–80, calculate the annual interest that you will receive on the described bond.

77. A $1000 Treasury bond with a current yield of 3.9% that is quoted at 105 points

78. A $1000 Treasury bond with a current yield of 1.5% that is quoted at 98 points

79. A $1000 Treasury bond with a current yield of 6.2% that is quoted at 114.3 points

80. A $10,000 Treasury bond with a current yield of 3.6% that is quoted at 102.5 points

81. **Mutual Fund Growth.** Assume that Figure 4.6 comes from today's paper. Suppose you invested $500 in the Calvert Social Investment Bond fund (SocInvBdA) three years ago and reinvested all dividends and gains. What is your investment worth now?

82. **Mutual Fund Growth.** Assume that Figure 4.6 comes from today's paper. Suppose you invested $500 in the Calvert Social Investment Equity fund (SocInvEqA) three years ago and reinvested all dividends and gains. What is your investment worth now?

Further Applications

Who Comes Out Ahead? Exercises 83–86 each describe two savings plans. Compare the balances in the two plans after 10 years. Who deposits more money in each case? Who comes out ahead in each case? Comment on any lessons about savings plans that you find in the results. (Assume that, for each plan, the payment and compounding periods are the same, so the savings plan formula is valid.)

83. Yolanda deposits $100 per month in an account with an APR of 5%, while Zach deposits $1200 at the end of each year in an account with an APR of 5%.

84. Polly deposits $50 per month in an account with an APR of 6%, while Quint deposits $40 per month in an account with an APR of 6.5%.

85. Juan deposits $200 per month in an account with an APR of 6%, while Maria deposits $2500 at the end of each year in an account with an APR of 6.5%.

86. George deposits $40 per month in an account with an APR of 7%, while Harvey deposits $150 per quarter in an account with an APR of 7.5%.

Comparing Investment Plans. Suppose you want to accumulate $50,000 for your child's college fund within the next

15 years. Explain fully whether the investment plans in Exercises 87–90 will allow you to reach your goal.

87. You deposit $50 per month into an account with an APR of 7%.

88. You deposit $75 per month into an account with an APR of 7%.

89. You deposit $100 per month into an account with an APR of 6%.

90. You deposit $200 per month into an account with an APR of 5%.

91. **Total Return on Stock.** Suppose you bought XYZ stock 1 year ago for $5.80 per share and sell it at $8.25. You also pay a commission of $0.25 per share on your sale. What is the total return on your investment?

92. **Total Return on Stock.** Suppose you bought XYZ stock 1 year ago for $46.00 per share and sell it at $8.25. You also pay a commission of $0.25 per share on your sale. What is the total return on your investment?

93. **Death and the Maven (A True Story).** In December 1995, 101-year-old Anne Scheiber (whose passport photo is shown below) died and left $22 million to Yeshiva University. This fortune was accumulated through shrewd and patient investment of a $5000 nest egg over the course of 50 years. In turning $5000 into $22 million, what were her total and annual returns? How did her annual return compare to the average annual return for large-company stocks (see Table 4.6)?

94. **Personal Savings Plan.** Describe something for which you would like to save money right now. How much do you need to save? How long do you have to save it? Based on these needs, calculate how much you should deposit each month in a savings plan to meet your goal. For the interest rate, use the highest rate currently available at local banks.

Web Projects

Find useful links for Web Projects on the text Web site: www.aw.com/bennett-briggs

95. **Investment Tracking.** Choose three stocks, three bonds, and three mutual funds that you think would make good investments. Imagine that you invest $100 in each of these nine investments. Use the Web to track the value of your investment portfolio over the next 5 weeks. Based on the portfolio value at the end, find your return for the 5-week period. Which investments fared the best, and which did most poorly?

96. **Dow Jones Industrial Average.** The Dow Jones Company has an extensive Web site that includes its history and functions, as well as information on the Dow Jones Industrial Average (DJIA) and links to the companies that make up the DJIA. Visit the Web site and choose a specific topic related to the DJIA (for example, the history of the DJIA, the original companies in the DJIA, the best and worst days for the DJIA, how the DJIA is computed). Using the Web site and any other resources, write a two-page paper on your topic.

97. **Company Research.** Go to the Web site of a specific company (links to the 30 DJIA companies are on the Dow Jones Web site) and carry out research on that company as if you were a prospective investor. You should consider the following questions: How has the company performed over the last year? 5 years? 10 years? Does the company offer dividends? How do you interpret its P/E ratio? Overall, do you think the company is a good investment? Why or why not?

98. **Financial Web Sites.** Visit one of the many financial news and advising Web sites. Describe the services offered by the Web site. Explain whether, as an active or prospective investor, you find the Web site useful.

99. **Other Averages.** Investigate one of several other stock averages, such as Standard and Poor's or the Russell 2500. How do these averages differ from the Dow Jones

Industrial Average? What services do they offer on their Web pages?

100. **Online Brokers.** It is possible to buy and sell stocks on the Internet through online brokers. Visit the Web sites of at least two online brokers. How do their services differ? Compare the commissions charged by the brokers.

In the News

101. **Advertised Investment.** Find an advertisement for an investment plan. Describe some of the cited benefits of the plan. Using what you learned in this unit, identify at least one possible drawback of the plan.

102. **Financial Pages.** Choose a major newspaper and study its financial pages. Can you identify all the investment data described in this unit? If not, what data are missing? If so, what additional financial data are offered? Explain how to read the pages.

103. **Personal Investment Options.** Does your employer offer you the option of enrolling in a savings or retirement plan? If so, describe the available options and discuss the advantages and disadvantages of each.

UNIT 4C

LOAN PAYMENTS, CREDIT CARDS, AND MORTGAGES

D o you have a credit card? Do you have a loan for your car? Do you have student loans? Do you own a house? Chances are that you owe money for at least one of these purposes. If so, you not only have to pay back the money you borrowed but also have to pay interest on the money that you owe. In this unit, we will begin by studying the basic ideas of loans and then apply these ideas to common loans, including credit cards and mortgages.

LOAN BASICS

Suppose you borrow $1200 at an annual interest rate of APR = 12%, or 1% per month. At the end of the first month, you owe interest in the amount of

$$1\% \times \$1200 = \$12$$

If you paid *only* this $12 in interest, you'd still owe $1200. That is, the total amount of the loan, called the loan **principal,** would still be $1200. In that case, you'd owe the same $12 in interest the next month. In fact, if you paid only the interest from one month to the next, the loan would never be paid off and you'd have to pay $12 per month forever.

If you hope to make progress in paying off the loan, you need to pay part of the principal as well as interest. For example, suppose that you pay $200 toward your loan principal each month, plus the current interest. At the end of the first month, you'd pay $200 toward principal *plus* $12 for the 1% interest you owe, making a total payment of $212. Because you've paid $200 toward principal, your new loan principal would be $1200 − $200 = $1000.

At the end of the second month, you'd again pay $200 toward principal and 1% interest. But this time the interest is on the $1000 that you still owe. Thus, your interest payment would be $1\% \times \$1000 = \10, making your total payment $210. Table 4.8 shows how the calculations continue until the loan is paid off after 6 months.

TABLE 4.8	Payments and Loan Principal for a $1200 Loan with Principal Paid Off at a Constant $200 per Month				
END OF . . .	PRIOR PRINCIPAL	INTEREST ON PRIOR PRINCIPAL	PAYMENT TOWARD PRINCIPAL	TOTAL PAYMENT	NEW PRINCIPAL
Month 1	$1200	$1\% \times \$1200 = \12	$200	$212	$1000
Month 2	$1000	$1\% \times \$1000 = \10	$200	$210	$800
Month 3	$800	$1\% \times \$800\ \ = \8	$200	$208	$600
Month 4	$600	$1\% \times \$600\ \ = \6	$200	$206	$400
Month 5	$400	$1\% \times \$400\ \ = \4	$200	$204	$200
Month 6	$200	$1\% \times \$200\ \ = \2	$200	$202	$0

Loan Basics

For any loan, the **principal** is the amount of money owed at any particular time. Interest is charged on the loan principal. To pay off a loan, you must gradually pay down the principal. Thus, in general, every payment should include all the interest you owe plus some amount that goes toward paying off the principal.

Installment Loans

For the case illustrated in Table 4.8, your total payment decreases from month to month because of the declining amount of interest that you owe. There's nothing inherently wrong with this method of paying off a loan, but most people prefer to pay the *same* total amount each month because it makes planning a budget easier. A loan that you pay off with equal regular payments is called an **installment loan** (or *amortized loan*).

Suppose you wanted to pay off your $1200 loan with 6 equal monthly payments. How much should you pay each month? Because the payments in Table 4.8 vary between $202 and $212, it's clear that the equal monthly payments must lie somewhere in this range. The exact amount is not obvious, but we can calculate it with the **loan payment formula**.

Loan Payment Formula (Installment Loans)

$$\text{PMT} = \frac{P \times \left(\dfrac{\text{APR}}{n}\right)}{\left[1 - \left(1 + \dfrac{\text{APR}}{n}\right)^{(-nY)}\right]}$$

where

$$\text{PMT} = \text{regular payment amount}$$
$$P = \text{starting loan principal (amount borrowed)}$$
$$\text{APR} = \text{annual percentage rate}$$
$$n = \text{number of payment periods per year}$$
$$Y = \text{loan term in years}$$

In our current example, the starting loan principal is $P = \$1200$, the annual interest rate is APR $= 12\%$, the loan term is $Y = \frac{1}{2}$ year (6 months), and monthly payments mean $n = 12$. The loan payment formula gives

$$\text{PMT} = \frac{P \times \left(\dfrac{\text{APR}}{n}\right)}{\left[1 - \left(1 + \dfrac{\text{APR}}{n}\right)^{(-nY)}\right]} = \frac{\$1200 \times \left(\dfrac{0.12}{12}\right)}{\left[1 - \left(1 + \dfrac{0.12}{12}\right)^{(-12 \times 1/2)}\right]}$$

$$= \frac{\$1200 \times (0.01)}{\left[1 - (1 + 0.01)^{-6}\right]}$$

$$= \frac{\$12}{1 - 0.942045235}$$

$$= \$207.06$$

The monthly payments would be $207.06, which, as we expected, is between $202 and $212.

Because the loan principal is gradually paid down with the installment payments, the interest due each month must also decline gradually. Thus, because the payments remain the same, the amount paid toward principal each month gradually rises. We therefore have the general relationship between principal and interest summarized in the following box.

Principal and Interest for Installment Loans

The portions of installment loan payments going toward principal and toward interest vary as the loan is paid down. Early in the loan term, the portion going toward interest is relatively high and the portion going toward principal is relatively low. As the term proceeds, the portion going toward interest gradually decreases and the portion going toward principal gradually increases.

Using Your Calculator

The Loan Payment Formula

As with other formulas in this chapter, there are many ways to do loan calculations on your calculator. Graphing or business calculators may make the calculations easier. Here is a procedure that will work on most scientific calculators. The example uses $P = \$1200$, APR $= 12\%$, $n = 12$ (monthly payments), and $Y = \frac{1}{2}$ year (6 months). It is important that you not round any numbers until the last step.

	In general	Example	Display
Starting Formula:	$$\text{PMT} = \dfrac{P \times \left(\dfrac{\text{APR}}{n}\right)}{\left[1 - \left(1 + \dfrac{\text{APR}}{n}\right)^{(-nY)}\right]}$$	$$\dfrac{\$1200 \times \left(\dfrac{0.12}{12}\right)}{\left[1 - \left(1 + \dfrac{0.12}{12}\right)^{(-12 \times 1/2)}\right]}$$	—
Step 1. Multiply factors in exponent.	$n \; \boxed{+/-}^{*} \; \boxed{\times} \; Y \; \boxed{=}$	$12 \; \boxed{+/-} \; \boxed{\times} \; 1 \; \boxed{\div} \; 2 \; \boxed{=}$	$-6.$
Step 2. Store product in memory (or write down).	$\boxed{\text{Store}}$	$\boxed{\text{Store}}$	$-6.$
Step 3. Add denominator terms 1 and APR/n.	$1 \; \boxed{+} \; \text{APR} \; \boxed{\div} \; n \; \boxed{=}$	$1 \; \boxed{+} \; 0.12 \; \boxed{\div} \; 12 \; \boxed{=}$	1.01
Step 4. Raise result to power in memory.	$\boxed{y^x} \; \boxed{\text{Recall}} \; \boxed{=}$	$\boxed{y^x} \; \boxed{\text{Recall}} \; \boxed{=}$	0.942045235
Step 5. Subtract result from 1 by making result negative and adding 1.	$\boxed{+/-} \; \boxed{+} \; 1 \; \boxed{=}$	$\boxed{+/-} \; \boxed{+} \; 1 \; \boxed{=}$	0.057954765
Step 6. Denominator is now complete; take its reciprocal.	$\boxed{1/x}$	$\boxed{1/x}$	17.25483667
Step 7. Multiply result by factors P and APR/n.	$\boxed{\times} \; P \; \boxed{\times} \; \text{APR} \; \boxed{\div} \; n \; \boxed{=}$	$\boxed{\times} \; 1200 \; \boxed{\times} \; 0.12 \; \boxed{\div} \; 12 \; \boxed{=}$	207.0580401

With the calculation complete, you can round to the nearest cent, writing the answer as $207.06. Be sure to check the calculation.
*The $\boxed{+/-}$ key is used on scientific calculators to change the sign of a number.

▶ **EXAMPLE 1** Student Loan

Suppose you have student loans totaling $7500 when you graduate from college. The interest rate is APR $= 9\%$ and the loan term is 10 years. What are your monthly payments? How much will you pay over the lifetime of the loan? What is the total interest you will pay on the loan?

Solution The starting loan principal is $P = \$7500$, the interest rate is APR $= 0.09$, the loan term is $Y = 10$ years, and $n = 12$ for monthly payments. We use the loan payment formula to find the monthly payments:

$$\text{PMT} = \frac{P \times \left(\dfrac{\text{APR}}{n}\right)}{\left[1 - \left(1 + \dfrac{\text{APR}}{n}\right)^{(-nY)}\right]} = \frac{\$7500 \times \left(\dfrac{0.09}{12}\right)}{\left[1 - \left(1 + \dfrac{0.09}{12}\right)^{(-12 \times 10)}\right]}$$

TECHNICAL NOTE
Because we assume the compounding period is the same as the payment period and because we round payments to the nearest cent, the calculated payments may differ slightly from actual payments.

<table>
<tr><td>

by the way

A table of principal and interest payments over the life of a loan is called an *amortization schedule.* Most banks will provide an amortization schedule for any loan you are considering.

</td></tr>
</table>

$$= \frac{\$7500 \times (0.0075)}{[1 - (1.0075)^{-120}]}$$

$$= \frac{\$56.25}{[1 - 0.407937305]}$$

$$= \$95.01$$

Your monthly payments are $95.01. Over the 10-year term, your total payments will be

$$10 \text{ yr} \times 12\,\frac{\text{mo}}{\text{yr}} \times \frac{\$95.01}{\text{mo}} = \$11,401.20$$

Of this amount, $7500 pays off the principal. The rest, or $11,401 − $7500 = $3901, represents interest payments. *Now try Exercises 13–24.* ◄

▶ **EXAMPLE 2 Principal and Interest Payments**

For the loan in Example 1, calculate the portions of your payments that go to principal and to interest during the first 3 months.

Solution The monthly interest rate is APR/12 = 0.09/12 = 0.0075. For a $7500 starting loan principal, the interest due at the end of the first month is

$$0.0075 \times \$7500 = \$56.25$$

Your monthly payment (calculated in Example 1) is $95.01. We've found that the interest due is $56.25, so the rest, or $95.01 − $56.25 = $38.76, goes to principal. Thus, after your first payment, your new loan principal is

$$\$7500 - \$38.76 = \$7461.24$$

Table 4.9 continues the same calculations for months 2 and 3. Note that, as expected, the interest payment gradually decreases and the payment toward principal gradually increases. But also note that, for these first 3 months of a 10-year loan, more than half of each payment goes toward interest. We could continue this table through the life of the loan, but it's generally easier to use software that finds principal and interest payments with built-in functions.

TABLE 4.9	Interest and Principal Portions of Payments on a $7500 Loan (10-year term, APR = 9%)		
END OF . . .	INTEREST = 0.0075 × BALANCE	PAYMENT TOWARD PRINCIPAL	NEW PRINCIPAL
Month 1	0.0075 × $7500 = $56.25	$95.01 − $56.25 = $38.76	$7500 − $38.76 = $7461.24
Month 2	0.0075 × $7461.24 = $55.96	$95.01 − $55.96 = $39.05	$7461.24 − $39.05 = $7422.19
Month 3	0.0075 × $7422.19 = $55.67	$95.01 − $55.67 = $39.34	$7422.19 − $39.34 = $7382.85

Now try Exercises 25–26. ◄

*T*hinking about . . .

DERIVATION OF THE LOAN PAYMENT FORMULA

Suppose you borrow a principal P for a loan term of N months at a monthly interest rate i. In most real cases, you would make monthly payments on this loan. However, suppose the lender did not want monthly payments, but instead wanted you to pay back the principal with compound interest in a lump sum at the end of the loan term. We can find this lump sum amount with the compound interest formula:

$$A = P \times (1 + i)^N$$

In financial terms, this lump sum amount, A, is called the *future value* of your loan. (The *present value* is the original loan principal, P.) From the lender's point of view, allowing you to spread your payments out over time should not affect this future value. Thus, in the end, your monthly payments should represent the same future value, A. We already have a formula for determining the future value with monthly payments—it is the savings plan formula from Unit 4B:

$$A = \text{PMT} \times \frac{[(1 + i)^N - 1]}{i}$$

We now have two different expressions for A, so we set them equal:

$$\text{PMT} \times \frac{[(1 + i)^N - 1]}{i} = P \times (1 + i)^N$$

To find the loan payment formula, we need to solve this equation for PMT. We first multiply both sides by the reciprocal of the fraction on the left:

$$\text{PMT} \times \frac{\cancel{[(1 + i)^N - 1]}}{\cancel{i}} \times \frac{\cancel{i}}{\cancel{[(1 + i)^N - 1]}}$$

$$= P \times (1 + i)^N \times \frac{i}{[(1 + i)^N - 1]}$$

Simplified, this becomes

$$\text{PMT} = \frac{P \times (1 + i)^N \times i}{[(1 + i)^N - 1]}$$

Next, we divide both the numerator and the denominator of the fraction on the right by $(1 + i)^N$:

$$\text{PMT} = \frac{\dfrac{P \times \cancel{(1 + i)^N} \times i}{\cancel{(1 + i)^N}}}{\dfrac{[(1 + i)^N - 1]}{(1 + i)^N}}$$

The numerator simplifies easily to $P \times i$. To simplify the denominator, note that

$$\frac{[(1 + i)^N - 1]}{(1 + i)^N} = \frac{\cancel{(1 + i)^N}}{\cancel{(1 + i)^N}} - \underbrace{\frac{1}{(1 + i)^N}}_{\substack{\text{apply rule} \\ \frac{1}{x^N} = x^{-N}}}$$

$$= 1 - (1 + i)^{-N}$$

Substituting the simplified terms for the numerator and the denominator, we find the loan payment formula:

$$\text{PMT} = \frac{P \times i}{1 - (1 + i)^{-N}}$$

To put the loan payment formula in the form given in the text, we substitute $i = \text{APR}/n$ for the interest rate per period and $N = nY$ for the total number of payments (where n is the number of payments per year and Y is the number of years).

Time out to think _____

In a case such as the student loan in Examples 1 and 2, many people are surprised to find that more than half of their early loan payments goes to interest when the annual interest rate is only 9%. By referring to Table 4.9, explain why this is the case. How will the payments toward principal and interest compare toward the end of the loan?

Choices of Rate and Term

You'll usually have several choices of interest rate and loan term when seeking a loan. For example, a bank might offer a 3-year car loan at 8%, a 4-year loan at 9%, and a 5-year loan at 10%. You'll pay less total interest with the shortest-term, lowest-rate loan, but this loan will have the highest monthly payments. Thus, you'll have to evaluate your choices and make the decision that is best for your personal situation.

➤ **EXAMPLE 3** Choice of Auto Loans

You need a $6000 loan to buy a used car. Your bank offers a 3-year loan at 8%, a 4-year loan at 9%, and a 5-year loan at 10%. Calculate your monthly payments and total interest over the loan term with each option.

Solution Let's begin with the 3-year loan at 8%. The starting loan principal is $P = \$6000$, the interest rate is APR $= 0.08$, the loan term is $Y = 3$ years, and $n = 12$ for monthly payments. Your monthly payments would be

$$\text{PMT} = \frac{P \times \left(\dfrac{\text{APR}}{n}\right)}{\left[1 - \left(1 + \dfrac{\text{APR}}{n}\right)^{(-nY)}\right]} = \frac{\$6000 \times \left(\dfrac{0.08}{12}\right)}{\left[1 - \left(1 + \dfrac{0.08}{12}\right)^{(-12 \times 3)}\right]}$$

$$= \frac{\$40}{\left[1 - (1.006666667)^{-36}\right]}$$

$$= \$188.02$$

Three years is 36 months, so your payments would total $36 \times \$188.02 = \6768.72. Of this total, $6000 pays off your principal, so the total interest is the remaining $768.72.

For the 4-year loan at 9%, we repeat the calculations with APR $= 0.09$ and $Y = 4$ years:

$$\text{PMT} = \frac{P \times \left(\dfrac{\text{APR}}{n}\right)}{\left[1 - \left(1 + \dfrac{\text{APR}}{n}\right)^{(-nY)}\right]} = \frac{\$6000 \times \left(\dfrac{0.09}{12}\right)}{\left[1 - \left(1 + \dfrac{0.09}{12}\right)^{(-12 \times 4)}\right]} = \$149.31$$

Your total payments over 4 years, or 48 months, would be $48 \times \$149.31 = \7166.88. After we subtract the $6000 that goes to principal, the total interest is $1166.88.

For the 5-year loan, we set APR $= 0.1$ and $Y = 5$ years:

$$\text{PMT} = \frac{P \times \left(\dfrac{\text{APR}}{n}\right)}{\left[1 - \left(1 + \dfrac{\text{APR}}{n}\right)^{(-nY)}\right]} = \frac{\$6000 \times \left(\dfrac{0.1}{12}\right)}{\left[1 - \left(1 + \dfrac{0.1}{12}\right)^{(-12 \times 5)}\right]} = \$127.48$$

by the way

You should always watch out for financial scams, especially when borrowing money. Keep in mind what is sometimes called the first rule of finance: *If it sounds too good to be true, it probably is!*

Your total payments over 5 years, or 60 months, would be 60 × $127.48 = $7648.80. After we subtract the $6000 that goes to principal, the total interest is $1648.80. As we expected, the monthly payments are lower with the longer-term loans, but the total interest is higher. *Now try Exercises 27–28.* ◄

Time out to think _____

Consider your own current financial situation. If you needed a $6000 car loan, which option from Example 3 would you choose? Why?

CREDIT CARDS

Credit card loans differ from installment loans in that you are not required to pay off your balance in any set period of time. Instead, you are required to make only a minimum monthly payment that generally covers all the interest but very little principal. As a result, it takes a very long time to pay off your credit card loan if you make only the minimum payments. If you wish to pay off your loan in a particular amount of time, you should use the loan payment formula to calculate the necessary payments.

A word of caution: Most credit cards have very high interest rates compared to other types of loans. As a result, it is easy to get into financial trouble if you get overextended with credit cards. The trouble is particularly bad if you miss your payments. In that case, you will probably be charged a late fee that is added to your principal, thereby increasing the amount of interest due the next month. With the interest charges operating like compound interest in reverse, failure to pay on time can put a person into an ever-deepening financial hole.

► EXAMPLE 4 Credit Card Debt

You have a credit card balance of $2300 with an annual interest rate of 21%. You decide to pay off your balance over 1 year. How much will you need to pay each month? Assume you make no further credit card purchases.

Solution Your starting loan principal is P = $2300, the interest rate is APR = 0.21, and monthly payments mean n = 12. Because you want to pay off the loan in 1 year, we set Y = 1. The required payments are

$$\text{PMT} = \frac{P \times \left(\dfrac{\text{APR}}{n}\right)}{\left[1 - \left(1 + \dfrac{\text{APR}}{n}\right)^{(-nY)}\right]} = \frac{\$2300 \times \left(\dfrac{0.21}{12}\right)}{\left[1 - \left(1 + \dfrac{0.21}{12}\right)^{(-12 \times 1)}\right]} = \$214.16$$

You must pay $214.16 per month to pay off the balance in 1 year.
Now try Exercises 29–32. ◄

Time out to think _____

Continuing Example 4, suppose you can get a personal loan at a bank at an annual interest rate of 10%. Should you take this loan and use it to pay off your $2300 credit card debt? Why or why not?

by the way

About three-fourths of American households have at least one credit card, and their average credit card balance is about $8000. The average credit card interest rate is about 17%—far higher than the interest rate on most other consumer loans.

Practical matters

AVOIDING CREDIT CARD TROUBLE

Most adults have credit cards for good reason. Used properly, credit cards offer many conveniences: They are safer and easier to carry than cash, they offer monthly statements that list everything charged to the card, and they can be used as ID to rent a car. But credit card trouble can compound quickly, and many people get into financial trouble as a result. A few simple guidelines can help you avoid credit card trouble.

• Use only one credit card. People who accumulate balances on several cards often lose track of their overall debt. A lost wallet or purse means more credit cards that must be canceled.
• If possible, pay off your balance in full each month. Then there's no chance of getting into a financial hole.
• If you plan to pay off your balance in full each month, be sure that your credit card offers an interest-free "grace period" on purchases (usually of about 1 month) so that you will not have to pay any interest.
• Compare the interest rate and annual fee (if any) of your credit card and others. Fees and rates differ greatly among credit cards, so be sure you are getting

a good deal. In particular, if you carry a balance, look for a card with a relatively low interest rate.
• When choosing a credit card, watch out for teaser rates. These are low interest rates that are offered for a short period, such as 6 months, after which the card reverts to very high rates.
• Never use your credit card for a cash advance except in an emergency, because nearly all credit cards charge both fees and high interest rates for cash advances. In addition, most credit cards charge interest immediately on cash advances, even if there is a grace period on purchases. When you need cash, get it directly from your own bank account by cashing a check or using an ATM card.
• If you own a home, consider replacing a common credit card with a home equity credit line. You'll generally get a lower interest rate, and the interest may be tax deductible.
• If you find yourself in a deepening financial hole, consult a financial advisor right away. A good place to start is with the National Foundation for Credit Counseling (www.nfcc.org). The longer you wait, the worse off you'll be in the long run.

►EXAMPLE 5 A Deepening Hole

Paul has gotten into credit card trouble. He has a balance of $9500 and just lost his job. His credit card company charges interest of APR = 21%, compounded daily. Suppose the credit card company allows him to suspend his payments until he finds a new job—but continues to charge interest. If it takes him a year to find a new job, how much will he owe when he starts his new job?

Solution Because Paul is not making payments during the year, this is not a loan payment problem. Instead, it is a compound interest problem, in which Paul's balance of $9500 grows at an annual rate of 21%, compounded daily. We use the compound interest formula with a starting balance of $P = \$9500$, APR = 0.21, $Y = 1$ year, and $n = 365$ (for daily compounding). At the end of the year, his loan balance will be

$$A = P \times \left(1 + \frac{\text{APR}}{n}\right)^{(nY)} = \$9500 \times \left(1 + \frac{0.21}{365}\right)^{(365 \times 1)} = \$11{,}719.23$$

During his year of unemployment, interest alone will make Paul's credit card balance grow from $9500 to over $11,700, an increase of more than $2200. Clearly, this increase will only make it more difficult for Paul to get back on his financial feet.

Now try Exercises 33–36. ◄

MORTGAGES

One of the most popular types of installment loans is designed specifically to help you buy a home. It's called a home **mortgage.** Mortgage interest rates generally are lower than interest rates on other types of loans because your home itself serves as a payment guarantee. If you fail to make your payments, the lender (usually a bank or mortgage company) can take possession of your home and sell it to recover the amount loaned to you.

There are several considerations in getting a home mortgage. First, the lender will probably require a **down payment,** typically 10% to 20% of the purchase price. Then the lender will loan you the rest of the money needed to purchase the home. Most lenders also charge fees, or **closing costs,** at the time you take out a loan. Closing costs can be substantial and may vary significantly between lenders, so you should be sure that you understand them. In general, there are two types of closing costs:

- Direct fees, such as fees for getting the home appraised and checking your credit history, for which the lender charges a fixed dollar amount. These fees typically range from a few hundred dollars to a couple thousand dollars.
- Fees charged as **points,** where each point is 1% of the loan amount. Many lenders divide points into two categories: an "origination fee" that is charged on all loans and "discount points" that vary for loans with different rates. For example, a lender might charge an origination fee of 1 point (1%) on all loans, then offer you a choice of adding 1 discount point for a loan at 8% or 2 discount points for a loan at 7.75%. Despite their different names, there is no essential difference between an origination fee and discount points.

As always, you should watch out for any fine print that may affect the cost of your loan. For example, you should check to make sure that there are no *prepayment penalties* if you decide to pay off your loan early. Most people pay off mortgages early, either because they sell the home or because they decide to refinance the loan to get a better interest rate or to change their monthly payments.

by the way

The idea of a mortgage contract originated in early British real estate law. The curious word *mortgage* comes from Latin and old French. It literally means "dead pledge."

Mortgage Basics

If you are seeking a home mortgage, be sure to keep the following considerations in mind as you compare lenders:

- What interest rate and down payment are required for the loan?
- What closing costs will be charged? Be sure you identify all closing costs, including origination fees and discount points, since different lenders may quote their fees differently.
- Watch out for fine print, such as prepayment penalties, that may make the loan more expensive than it seems on the surface.

Fixed Rate Mortgages

The simplest type of home loan is a **fixed rate mortgage,** in which you are guaranteed that the interest rate will not change over the life of the loan. Most fixed rate loans have a term of either 15 or 30 years, with lower interest rates on the shorter-term loans. We can calculate payments on fixed rate loans with the loan payment formula.

▶ **EXAMPLE 6** Fixed Rate Payment Options

You need a loan of $100,000 to buy your new home. The bank offers a choice of a 30-year loan at an APR of 8% or a 15-year loan at 7.5%. Compare your monthly payments and total loan cost under the two options. Assume that the closing costs are the same in both cases and therefore do not affect the choice.

Solution The starting loan principal is $P = \$100,000$ and we set $n = 12$ for monthly payments. For the 30-year loan, we have APR = 0.08 and $Y = 30$. The monthly payments are

$$\text{PMT} = \frac{P \times \left(\dfrac{\text{APR}}{n}\right)}{\left[1 - \left(1 + \dfrac{\text{APR}}{n}\right)^{(-nY)}\right]} = \frac{\$100,000 \times \left(\dfrac{0.08}{12}\right)}{\left[1 - \left(1 + \dfrac{0.08}{12}\right)^{(-12 \times 30)}\right]} = \$733.76$$

Over the 30-year life of the loan, your total payments are

$$30 \text{ yr} \times \frac{12 \text{ mo}}{\text{yr}} \times \frac{\$733.76}{\text{mo}} \approx \$264,150$$

For the 15-year loan, we have APR = 0.075 and $Y = 15$. The monthly payments are

$$\text{PMT} = \frac{P \times \left(\dfrac{\text{APR}}{n}\right)}{\left[1 - \left(1 + \dfrac{\text{APR}}{n}\right)^{(-nY)}\right]} = \frac{\$100,000 \times \left(\dfrac{0.075}{12}\right)}{\left[1 - \left(1 + \dfrac{0.075}{12}\right)^{(-12 \times 15)}\right]} = \$927.01$$

Over the 15-year life of the loan, your total payments are

$$15 \text{ yr} \times \frac{12 \text{ mo}}{\text{yr}} \times \frac{\$927.01}{\text{mo}} \approx \$166,860$$

Note that the payments of $927.01 on the 15-year loan are almost $200 higher than the payments of $733.76 on the 30-year loan. However, the 15-year loan saves you almost $100,000 in total payments. Thus, the 15-year loan saves you a lot in the long run, but it's a good plan only if you are confident that you can afford the additional $200 per month that it will cost you for the next 15 years. (See Example 9 for an alternative payment strategy.) *Now try Exercises 37–40.* ◀

▶ **EXAMPLE 7** Closing Costs

Great Bank offers a $100,000, 30-year, 8% fixed rate loan with closing costs of $500 plus 2 points. Big Bank offers a lower rate of 7.9%, but with closing costs of $1000 plus 2 points. Evaluate the two options.

Solution In Example 6, we calculated the payments on the 8% loan to be $733.76. At the lower 7.9% rate, the payments are

$$\text{PMT} = \frac{P \times \left(\dfrac{\text{APR}}{n}\right)}{\left[1 - \left(1 + \dfrac{\text{APR}}{n}\right)^{(-nY)}\right]} = \frac{\$100{,}000 \times \left(\dfrac{0.079}{12}\right)}{\left[1 - \left(1 + \dfrac{0.079}{12}\right)^{(-12 \times 30)}\right]} = \$726.81$$

Thus, you'll save about $7 per month with Big Bank's lower interest rate. Now we must consider the difference in closing costs. Both banks charge the same 2 points, so this portion of the closing costs won't affect your decision. (Note, however, that 2 points means 2% of the $100,000 loan, which is a $2000 fee!) But you must consider Big Bank's extra $500 in direct fees.

The choice comes down to this: Big Bank costs you an extra $500 now, but saves you $7 per month in payments. Dividing $500 by $7 per month, we find the time it will take to recoup the extra $500:

$$\frac{\$500}{\$7/\text{mo}} = 71.4 \text{ mo} \approx 6 \text{ yr}$$

Thus, it will take you about 6 years to save the extra $500 that Big Bank charges up front. Unless you are sure that you will be staying in your house (and keeping the same loan) for much more than 6 years, you probably should go with the lower closing costs at Great Bank, even though your monthly payments will be slightly higher.

Now try Exercises 41–42. ◀

► EXAMPLE 8 Points Decision

Continuing Example 7, suppose you've decided to go with Great Bank's lower closing costs. You learn that Great Bank actually offers two options for 30-year loans: an 8% interest rate with 2 points or a 7.5% rate with 4 points. Evaluate your options.

Solution We already know that the monthly payments for the 8% loan are $733.76. For the 7.5% loan, we have $P = \$100{,}000$, $n = 12$, and $Y = 30$. Setting APR = 7.5% = 0.075, we find that the monthly payments are

$$\text{PMT} = \frac{P \times \left(\dfrac{\text{APR}}{n}\right)}{\left[1 - \left(1 + \dfrac{\text{APR}}{n}\right)^{(-nY)}\right]} = \frac{\$100{,}000 \times \left(\dfrac{0.075}{12}\right)}{\left[1 - \left(1 + \dfrac{0.075}{12}\right)^{(-12 \times 30)}\right]} = \$699.21$$

The 7.5% loan lowers the monthly payments by $733.76 − $699.21 = $34.55. However, this loan has 2 additional points in closing costs, which means 2% of your $100,000 loan, or $2000. Thus, you must decide whether it is worth an extra $2000 up front for a monthly savings of just under $35. Let's calculate how long it will take to make up the added up-front costs:

$$\frac{\$2000}{\$34.55/\text{mo}} = 57.9 \text{ mo}$$

This is not quite 5 years. If you think it's likely that you will sell or refinance within 5 years, you should not pay the extra points. However, if you expect to keep the

by the way

When you pay points, most lenders give you a choice between paying them up front and folding them into the loan. For example, if you pay 2 points on a $100,000 loan, your choice is to pay $2000 up front or to make the loan amount $102,000 rather than $100,000. For the examples in this book, we assume that you pay the points up front.

by the way

Mortgage rates vary substantially with time. In the 1980s, average U.S. rates for new mortgages (30-year fixed) were almost always above 10%, peaking at more than 18% in 1981. During 2003, average rates fell to near 5%—the lowest rates in over 40 years.

loan for a long time, the added points might be worth it. For example, if you keep the loan for the full 30 years (360 months), you'll save 360 × $34.55 = $12,438 in monthly payments over the life of the loan—far more than the extra $2000 you pay for the lower rate today. *Now try Exercises 43–44.* ◄

Prepayment Strategies

Because of the long loan term, the early payments on a mortgage tend to be almost entirely interest. For example, Figure 4.7 shows the portion of each payment going to principal and interest for a 30-year, $100,000 loan at 8%. In addition, the total interest paid on mortgages is often much more than the principal. In Example 6, we found that the total payments for this $100,000 loan would be about $264,000—more than $2\frac{1}{2}$ times the starting principal!

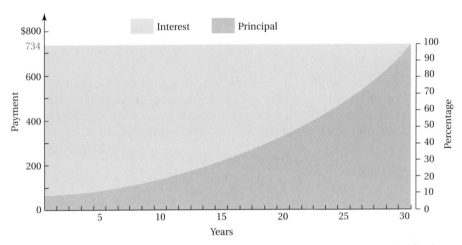

FIGURE 4.7 Portions of monthly payments going to principal and interest over the life of a 30-year, $100,000 loan at 8%.

Clearly, you can save a lot if you can reduce your interest payments. One way to do this is to pay extra toward the principal, particularly early in the term. For example, suppose you pay an extra $100 toward principal in the first monthly payment of your $100,000 loan. That is, instead of paying the required $734 (see Example 6), you pay $834. Because you've reduced your loan balance by $100, you will save the compounded value of this $100 over the rest of the 30-year loan term—which is nearly $1100. In other words, paying an extra $100 in the first month saves you about $1100 in interest over the 30 years.

►EXAMPLE 9 An Alternative Strategy

An alternative strategy to the mortgage options in Example 6 is to take the 30-year loan at 8%, but to try to pay it off in 15 years by making larger payments than are required. How much would you have to pay each month? Discuss the pros and cons of this strategy.

Solution To reflect paying off an 8% loan in 15 years, we set APR = 0.08 and $Y = 15$; we still have $P = \$100{,}000$ and $n = 12$. The monthly payments are

$$\text{PMT} = \frac{P \times \left(\dfrac{\text{APR}}{n}\right)}{\left[1 - \left(1 + \dfrac{\text{APR}}{n}\right)^{(-nY)}\right]} = \frac{\$100{,}000 \times \left(\dfrac{0.08}{12}\right)}{\left[1 - \left(1 + \dfrac{0.08}{12}\right)^{(-12\times15)}\right]} = \$955.65$$

In Example 6, we found that the 30-year loan requires payments of $733.76. Thus, to pay off the loan in 15 years, you must make payments that are more than the minimum required by $955.65 − $733.76 = $221.89 per month.

Note that this payment is about $30 per month *more* than the payment of $927.01 required with the 15-year loan (see Example 6), because the 15-year loan had a lower interest rate. Clearly, if you *know* you're going to pay off the loan in 15 years, you should take the lower-interest 15-year loan. However, taking the 30-year loan has one advantage: Because your required payments are only $733.76, you can always drop back to this level if you find it difficult to afford the extra needed to pay off the loan in 15 years. *Now try Exercises 45–46.* ◀

Time out to think

Consider two options for paying off a loan in 15 years: taking out a 15-year loan or taking out a 30-year loan and making an extra principal payment each month. Assuming that you would like to pay off the loan in 15 years, how would you decide which strategy is better for you?

Adjustable Rate Mortgages

A fixed rate mortgage is advantageous for you because your monthly payments never change. However, it poses a risk to the lender. Imagine that you take out a fixed, 30-year loan of $100,000 from Great Bank at a 6% interest rate. Initially, the loan may seem like a good deal for Great Bank. But suppose that, 2 years later, prevailing interest rates have risen to 8%. If Great Bank still had the $100,000 that it lent to you, it could lend it out to someone else at this higher 8% rate. Instead, it's stuck with the 6% rate that you are paying. In effect, Great Bank loses potential future income if prevailing rates rise substantially and you have a fixed rate loan.

Lenders can lessen the risk of rising interest rates by charging higher rates for longer-term loans. That is why rates generally are higher for 30-year loans than for 15-year loans. But an even lower-risk strategy for the lender is an **adjustable rate mortgage (ARM),** in which the interest rate you pay changes whenever prevailing rates change. Because of the reduced long-term risk to lenders, ARMs generally have much lower initial interest rates than fixed rate loans. For example, a bank offering a 6% rate on a fixed 30-year loan might offer an ARM that begins at 4%. Most ARMs guarantee their starting interest rate for the first 6 months or 1 year, but interest rates in subsequent years move up or down to reflect prevailing rates. Most ARMs also include a *rate cap* that cannot be exceeded. For example, if your ARM begins at an interest rate of 4%, you may be promised that your interest rate can never go higher than a rate cap of 10%. Making a decision between a fixed rate loan and an ARM can be one of the most important financial decisions of your life.

by the way

Watch out for "teaser" rates on adjustable rate mortgages. Under normal circumstances, your rate on an ARM rises only if prevailing interest rates rise. However, some lenders offer low teaser rates—rates below the prevailing rates—for the first few months of an ARM. Teaser rates are certain to rise as soon as the teaser period is over. Thus, while teaser rates may be attractive, the longer-term policies of the ARM are far more important.

➤ **EXAMPLE 10** Rate Approximations for ARMs

You have a choice between a 30-year fixed rate loan at 8% and an ARM with a first-year rate of 5%. Neglecting compounding and changes in principal, estimate your monthly savings with the ARM during the first year on a $100,000 loan. Suppose that the ARM rate rises to 11% by the fourth year. How will your payments be affected?

Solution Because mortgage payments are mostly interest in the early years of a loan, we can make approximations by pretending that the principal remains unchanged. For the 8% fixed rate loan, the interest on the $100,000 loan for the first year will be approximately $8\% \times \$100,000 = \8000. With the 5% loan, your first-year interest will be approximately $5\% \times \$100,000 = \5000. Thus, the ARM will save you about $3000 in interest during the first year, which means a monthly savings of about $\$3000 \div 12 = \250.

By the fourth year, when rates reach 11%, the situation is reversed. The rate on the ARM is now 3 percentage points above the rate on the fixed rate loan. Instead of saving $250 per month, you'd be paying $250 per month *more* on the ARM than on the 8% fixed rate loan. Moreover, if interest rates remain high on the ARM, you will continue to make these high payments for many years to follow. Thus, while ARMs reduce risk for the lender, they add risk for the borrower.

Now try Exercises 47–48.

■ EXERCISES 4C

Review Questions

1. Suppose you pay only the interest on a loan. Will the loan ever be paid off? Why not?

2. What is an installment loan? Explain the meaning and use of the loan payment formula.

3. Explain, in general terms, how the portions of installment loan payments going to principal and interest change over the life of the loan.

4. Suppose that you need a loan of $10,000 and are offered a choice of a 3-year loan at 7% interest or a 5-year loan at 8% interest. Discuss the pros and cons of each choice.

5. How do credit card loans differ from ordinary installment loans? Why are credit card loans particularly dangerous?

6. What is a mortgage? What is a down payment on a mortgage? Explain how closing costs, including points, can affect loan decisions.

Does It Make Sense?

Decide whether each of the following statements makes sense (or is clearly true) or does not make sense (or is clearly false). Explain your reasoning.

7. The interest rate on my student loan is only 7%, yet more than half of my payments are currently going toward interest rather than principal.

8. My student loans were all 20-year loans at interest rates of 8% or above, so when my bank offered me a 20-year loan at 7%, I took it and used it to pay off the student loans.

9. I make only the minimum required payments on my credit card balance each month, because that way I'll have more of my own money to keep.

10. I carry a large credit card balance, and I had a credit card that charged an annual interest rate of 12%. So when I found another credit card that promised a 3% interest rate for the first 3 months, it was obvious that I should switch to this new card.

11. I had a choice between a fixed rate mortgage at 6% and an adjustable rate mortgage that started at 3% for the first year with a maximum increase of 1.5 percentage points a year. I took the adjustable rate, because I'm planning to move within three years.

12. Fixed rate loans with 15-year terms have lower interest rates than loans with 30-year terms, so it always makes sense to take the 15-year loan.

Basic Skills & Concepts

Loan Terminology. For the loans described in Exercises 13–14, do the following:

a. Clearly identify the starting loan principal, the annual interest rate, the number of payments per year, the loan term, and the payment amount.

b. How many payments will you make in total? What total amount will you pay over the full term of the loan?

c. Of the total amount you pay, how much will go toward principal and how much toward interest?

13. You borrowed $40,000 at an APR of 7%, which you are paying off with monthly payments of $310 for 20 years.

14. You borrowed $7500 at an APR of 9%, which you are paying off with monthly payments of $95 for 10 years.

Loan Payments. For the loans described in Exercises 15–24, do the following:

a. Calculate the monthly payment.

b. Determine the total payment over the term of the loan.

c. Determine how much of the total payment over the loan term goes to principal and how much to interest.

15. A student loan of $25,000 at a fixed APR of 10% for 20 years

16. A student loan of $12,000 at a fixed APR of 8% for 10 years

17. A home mortgage of $150,000 with a fixed APR of 7.5% for 30 years

18. A home mortgage of $150,000 with a fixed APR of 7.5% for 15 years

19. A home mortgage of $100,000 with a fixed APR of 9% for 15 years

20. A home mortgage of $100,000 with a fixed APR of 8.5% for 15 years

21. You borrow $5000 over a period of 3 years at a fixed APR of 12%.

22. You borrow $10,000 over a period of 5 years at a fixed APR of 10%.

23. You borrow $50,000 over a period of 15 years at a fixed APR of 8%.

24. You borrow $100,000 over a period of 30 years at a fixed APR of 7%.

Principal and Interest Payments. For the loans described in Exercises 25–26, calculate the monthly payment and the portions of the payments that go to principal and to interest during the first 3 months. (Hint: Use a table as in Example 2.)

25. A home mortgage of $100,000 with a fixed APR of 8.5% for 30 years

26. A student loan of $12,000 at a fixed APR of 8% for 15 years

27. **Choosing an Auto Loan.** You need to borrow $10,000 to buy a car and you determine that you can afford monthly payments of $220. The bank offers three choices: a 3-year loan at 7% APR, a 4-year loan at 7.5% APR, or a 5-year loan at 8% APR. Which loan best meets your needs? Explain your reasoning.

28. **Choosing a Personal Loan.** You need to borrow $4000 to pay off your credit cards and you can afford monthly payments of $150. The bank offers three choices: a 2-year loan at 8% APR, a 3-year loan at 9% APR, or a 4-year loan at 10% APR. Which loan best meets your needs? Explain your reasoning.

Credit Card Debt. For Exercises 29–32, assume you have a balance of $2500 on your credit card that you want to pay off. Calculate your monthly payment and total payment under the conditions listed. Assume you make no additional charges to the card.

29. The credit card APR is 18% and you want to pay off the balance in 1 year.

30. The credit card APR is 20% and you want to pay off the balance in 2 years.

31. The credit card APR is 21% and you want to pay off the balance in 3 years.

32. The credit card APR is 22% and you want to pay off the balance in 1 year.

33. **Credit Card Debt.** Assume you have a balance of $1200 on a credit card with an APR of 18%, or 1.5% per month. You start making monthly payments of $200, but at the same time you charge an additional $75 per month to the credit card. Assume that interest for a given month is

based on the balance for the previous month. The following table shows how you can calculate your monthly balance.

Month	Payment	Expenses	Interest	New Balance
0	—	—	—	$1200
1	$200	$75	1.5% × $1200 = $18	$1200 − $200 + $75 + 18 = $1093
2	$200	$75		
3	$200	$75		

Complete and extend the table to show your balance at the end of each month until the debt is paid off. How long does it take to pay off the credit card debt?

34. **Credit Card Debt.** Repeat the table of Exercise 33, but this time assume that you make monthly payments of $300. Extend the table as long as necessary until your debt is paid off. How long does it take to pay off your debt?

35. **Credit Card Woes.** The following table shows the expenses and payments for 8 months on a credit card account with an initial balance of $300. Assume that the interest rate is 1.5% per month (18% APR) and that interest for a given month is charged on the balance for the previous month. Complete the table. After 8 months, what is the balance on the credit card? Comment on the effect of the interest and the initial balance, in light of the fact that for 7 of the 8 months expenses never exceeded payments.

Month	Payment	Expenses	Interest	Balance
0	—	—	—	$300
1	$300	$175	1.5% × $300 = $4.50	$179.50
2	$150	$150		
3	$400	$350		
4	$500	$450		
5	0	$100		
6	$100	$100		
7	$200	$150		
8	$100	$80		

36. **Teaser Rate.** You have a total credit card debt of $4000. You receive an offer to transfer this debt to a new card with an introductory APR of 6% for the first 6 months. After that, the rate becomes 24%.

a. What is the monthly interest payment on $4000 during the first 6 months? (Assume you pay nothing toward principal and don't charge any further debts.)

b. What is the monthly interest payment on $4000 *after* the first 6 months? Comment on the change from the teaser rate.

Fixed Rate Options. Compare your monthly payments and total loan cost under the two options listed in each of Exercises 37–40. Assume that the loans are fixed rate and that closing costs are the same in both cases. Briefly discuss the pros and cons of each option.

37. You need a $100,000 loan.

Option 1: a 30-year loan at an APR of 8%

Option 2: a 15-year loan at 7.5%

38. You need a $75,000 loan.

Option 1: a 30-year loan at an APR of 8%

Option 2: a 15-year loan at 7%

39. You need a $120,000 loan.

Option 1: a 30-year loan at an APR of 7.15%

Option 2: a 15-year loan at 6.75%

40. You need a $180,000 loan.

Option 1: a 30-year loan at an APR of 7.25%

Option 2: a 15-year loan at 6.8%

Closing Costs. You need a loan of $80,000 to buy a home. Each of Exercises 41–44 offers two choices. Calculate your monthly payments and total closing costs in each case. Briefly discuss how you would decide between the two choices.

41. Choice 1: 30-year fixed rate at 8% with closing costs of $1200 and no points

Choice 2: 30-year fixed rate at 7.5% with closing costs of $1200 and 2 points

42. Choice 1: 30-year fixed rate at 8.5% with no closing costs and no points

Choice 2: 30-year fixed rate at 7.5% with closing costs of $1200 and 4 points

43. Choice 1: 30-year fixed rate at 7.25% with closing costs of $1200 and 1 point

Choice 2: 30-year fixed rate at 6.75% with closing costs of $1200 and 3 points

44. Choice 1: 30-year fixed rate at 7.5% with closing costs of $1000 and no points

Choice 2: 30-year fixed rate at 6.5% with closing costs of $1500 and 4 points

45. Accelerated Loan Payment. Suppose you have a student loan of $25,000 with an APR of 9% for 20 years.

 a. What are your required monthly payments?

 b. Suppose you would like to pay the loan off in 10 years instead of 20. What monthly payments will you need to make?

 c. Compare the total amounts you'll pay over the loan term if you pay the loan off in 20 years versus 10 years.

46. Accelerated Loan Payment. Suppose you have a student loan of $60,000 with an APR of 8% for 25 years.

 a. What are your required monthly payments?

 b. Suppose you would like to pay the loan off in 15 years instead of 25. What monthly payments will you need to make?

 c. Compare the total amounts you'll pay over the loan term if you pay the loan off in 25 years versus 15 years.

47. ARM Rate Approximations. You have a choice between a 30-year fixed rate loan at 7% and an ARM with a first-year rate of 5%. Neglecting compounding and changes in principal, estimate your monthly savings with the ARM during the first year on a $150,000 loan. Suppose that the ARM rate rises to 8.5% at the start of the third year. Approximately how much extra will you then be paying over what you would have paid if you had taken the fixed rate loan?

48. ARM Rate Approximations. You have a choice between a 30-year fixed rate loan at 8.5% and an ARM with a first-year rate of 5.5%. Neglecting compounding and changes in principal, estimate your monthly savings with the ARM during the first year on a $125,000 loan. Suppose that the ARM rate rises to 10% at the start of the second year. Approximately how much extra will you then be paying over what you would have paid if you had taken the fixed rate loan?

Further Applications

49. How Much House Can You Afford? You can afford monthly payments of $500. If current mortgage rates are 9% for a 30-year fixed rate loan, what loan principal can you afford? If you are required to make a 20% down payment and you have the cash on hand to do it, what price

home can you afford? (Hint: You will need to solve the loan payment formula for P.)

50. How Much House Can You Afford? You can afford monthly payments of $1200. If current mortgage rates are 7.5% for a 30-year fixed rate loan, what loan principal can you afford? If you are required to make a 20% down payment and you have the cash on hand to do it, what price home can you afford? (Hint: You will need to solve the loan payment formula for P.)

51. Student Loan Consolidation. Suppose you have the following three student loans: $10,000 with an APR of 8% for 15 years, $15,000 with an APR of 8.5% for 20 years, and $12,500 with an APR of 9% for 10 years.

 a. Calculate the monthly payment for each loan individually.

 b. Calculate the total you'll pay in payments during the life of all three loans.

 c. A bank offers to consolidate your three loans into a single loan with an APR of 8.5% and a loan term of 20 years. What will your monthly payments be in that case? What will your total payments be over the 20 years? Discuss the pros and cons of accepting this loan consolidation.

52. Bad Deals: Car-Title Lenders. Some "car-title lenders" offer quick cash loans in exchange for being allowed to hold the title to your car as collateral (you lose your car if you fail to pay off the loan). In many states, these lenders operate under pawnbroker laws that allow them to charge fees as a percentage of the unpaid balance. Suppose you need $2000 in cash, and a car-title company offers you a loan at an interest rate of 2% per month *plus* a monthly fee of 20% of the unpaid balance.

a. How much will you owe in interest and fees on your $2000 loan at the end of the first month?

b. Suppose that you pay only the interest and fees each month. How much will you pay over the course of a full year?

c. Suppose instead that you obtain a loan from a bank with a term of 3 years and an APR of 10%. What are your monthly payments in that case? Compare these to the payments to the car-title lender.

53. **Project: Choosing a Mortgage.** Imagine that you work for an accounting firm and a client has told you that he is buying a house and needs a loan of $120,000. His monthly income is $4000 and he is single with no children. He has $14,000 in savings that can be used for a down payment. Find the current rates available from local banks for both fixed rate mortgages and adjustable rate mortgages (ARMs). Analyze the offerings and summarize orally or in writing the best options for your client, along with the pros and cons of each option.

Web Projects

Find useful links for Web Projects on the text Web site: www.aw.com/bennett-briggs

54. **Credit Card Comparisons.** Visit a Web site that gives comparisons between credit cards. Briefly explain the factors that are considered in the comparisons. How does your own credit card compare to other credit cards? Based on this comparison, do you think you would be better off with a different credit card?

55. **Home Financing.** Visit a Web site that offers online home financing. Describe the terms of a particular home mortgage. Discuss the advantages and disadvantages of financing a home online rather than at a local bank.

56. **Online Car Purchase.** Find a car online that you might want to buy. Find a loan that you would qualify for, and calculate your monthly payments and total payments over the life of the loan. Next, suppose that you started a savings plan instead of buying the car, depositing the same amounts that would have gone to car payments. Estimate how much you would have in your savings plan by the time you graduate from college. Explain your assumptions.

57. **Student Financial Aid.** There are many Web sites that offer student loans. Visit a Web site that offers student loans and describe the terms of a particular loan. Discuss the advantages and disadvantages of financing a student loan online rather than through a bank or through your university or college.

58. **Scholarship Scams.** The Federal Trade Commission keeps track of many financial scams related to college scholarships. Read about two different types of scams, and report on how they work and how they hurt people who are taken in by them.

59. **Financial Scams.** Many Web sites keep track of current financial scams. Visit some of these sites and report on one scam that has already hurt a lot of people. Describe how the scam works and how it hurts those who are taken in by it.

In the News

60. **Mortgage Rates.** Find advertisements in the newspaper for two different home mortgages companies. Using the ideas of this unit, evaluate the terms of loans from each company and decide which company you would use for a home mortgage.

61. **Credit Card Statement.** Look carefully at the terms of financing explained on your most recent credit card statement. Explain all the important terms, including the interest rates that apply, annual fees, and grace periods.

62. **Bank Rates.** Find the interest rates that your bank (or another local bank) charges for different types of loans, such as auto loans, personal loans, and home mortgages. Why do you think the rates are different in the different cases?

UNIT 4D

INCOME TAXES

*T*here are many different types of taxes, including sales tax, gasoline tax, and property tax. But for most Americans, the largest tax burden comes from taxes on wages and other income. In this unit, we explore a few of the many aspects of federal income taxes.

In this world, nothing is certain but death and taxes.

—BENJAMIN FRANKLIN

INCOME TAX BASICS

It's quite possible that *no one* fully understands federal income taxes. The complete tax code consists of thousands of pages of detailed regulations. Many of the regulations are difficult to interpret, and disputes about their meaning are often taken to court. Congress frequently tinkers with tax laws and occasionally undertakes major reforms. For example, tax laws were greatly simplified by Congress in 1986. Unfortunately, politicians were unable to resist making modifications to the simplified tax code, so it gradually became more complex once again.

Nevertheless, understanding a few basic concepts is all that is required to calculate most people's taxes. Figure 4.8 summarizes the steps in a typical tax calculation. We'll follow the flow of the steps, defining terms as we go along.

The hardest thing in the world to understand is the income tax.

—ALBERT EINSTEIN

- The process begins with your **gross income,** which is *all* your income for the year. Thus, gross income includes wages, tips, profits from a business, interest or dividends from investments, and any other income you receive.
- Some gross income is not taxed (at least not in the year it is received), such as contributions to IRAs and other tax-deferred savings plans. These untaxed portions of gross income are called *adjustments*. Subtracting adjustments from your gross income gives your **adjusted gross income.**
- Everyone is entitled to certain **exemptions** and **deductions**—amounts that you subtract from your adjusted gross income before calculating your taxes. (The amounts you can subtract depend on factors that we'll discuss shortly.)

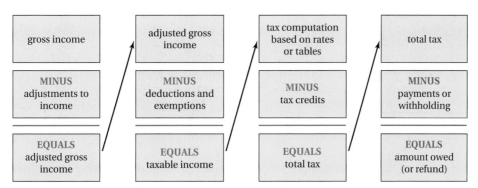

FIGURE 4.8 Flow chart showing the basic steps in calculating income tax.

HISTORICAL NOTE

An income tax was first levied in the United States in 1862 (during the Civil War), but was abandoned a few years later. The 16th Amendment to the Constitution, ratified in 1913, gave the federal government full authority to levy an income tax.

Once you subtract the exemptions and deductions, you are left with your **taxable income.**

- A tax table or tax rate computation allows you to determine how much tax you owe on your taxable income. However, you may not actually have to pay this much tax if you are entitled to any tax credits. For example, you may be entitled to a tax credit of $1000 per child. From your tax rate computation, you subtract the amount of any credits to find your **total tax.**
- Finally, most people have already paid part or all of their tax bill during the year, either through *withholdings* (by your employer) or through paying quarterly *estimated taxes* (if you are self-employed). You subtract the taxes that you've already paid to determine how much you still owe. In many cases, you may have paid more than you owe, in which case you should receive a **tax refund.**

▶ **EXAMPLE 1 Income on Tax Forms**

Karen earned wages of $34,200, received $750 in interest from a savings account, and contributed $1200 to a tax-deferred retirement plan. She was entitled to a personal exemption of $3050 and to deductions totaling $5400. Find her gross income, adjusted gross income, and taxable income.

Solution Karen's gross income is the sum of all her income, which means the sum of her wages and her interest:

$$\text{gross income} = \$34{,}200 + \$750 = \$34{,}950$$

Her $1200 contribution to a tax-deferred retirement plan counts as an adjustment to her gross income, so her adjusted gross income (AGI) is

$$\text{AGI} = \text{gross income} - \text{adjustments} = \$34{,}950 - \$1200 = \$33{,}750$$

To find her taxable income, we subtract her exemptions and deductions:

$$\text{taxable income} = \text{AGI} - \text{exemptions} - \text{deductions}$$
$$= \$33{,}750 - \$3050 - \$5400 = \$25{,}300$$

Her taxable income is $25,300. *Now try Exercises 19–22.* ◀

by the way

United States federal income taxes are collected by the *Internal Revenue Service* (IRS), which is part of the United States Department of the Treasury. Most people file federal taxes by completing a tax form, such as Form 1040, 1040A, or 1040EZ.

Filing Status

Tax calculations depend on your **filing status,** such as single or married. Most people fall into one of four filing status categories:

- *Single* applies if you are unmarried, divorced, or legally separated.
- *Married filing jointly* applies if you are married *and* you and your spouse file a single tax return. (In some cases, this category also applies to widows or widowers.)
- *Married filing separately* applies if you are married *and* you and your spouse file two separate tax returns.
- *Head of household* applies if you are unmarried and are paying more than half the cost of supporting a dependent child or parent.

We will use these four categories in the rest of our discussion.

Exemptions and Deductions

Both exemptions and deductions are subtracted from your adjusted gross income. However, they are calculated differently, which is why they have different names.

Exemptions are a fixed amount per person ($3050 in 2003). You can claim the amount of an exemption for yourself and each of your *dependents* (for example, children whom you support).

Deductions vary from one person to another. The most common deductions include interest on home mortgages, contributions to charity, and taxes you've paid to other agencies (such as state income taxes or local property taxes). However, you don't necessarily have to add up all your deductions. When you file your taxes, you have two options for deductions:

- You can choose a **standard deduction,** the amount of which depends on your filing status.
- You can choose **itemized deductions,** in which case you add up all the individual deductions to which you are entitled.

Note that you get *either* the standard deduction *or* itemized deductions, not both. Because deductions lower your tax bill, you should choose whichever option is larger.

▶EXAMPLE 2 Should You Itemize?

Suppose you have the following deductible expenditures: $2500 for interest on a home mortgage, $900 for contributions to charity, and $250 for state income taxes. Your filing status entitles you to a standard deduction of $4750. Should you itemize your deductions or claim the standard deduction?

Solution The total of your deductible expenditures is

$$\$2500 + \$900 + \$250 = \$3650$$

If you itemize your deductions, you can subtract $3650 when finding your taxable income. But if you take the standard deduction, you can subtract $4750. Thus, you are better off with the standard deduction. *Now try Exercises 23–28.* ◀

Tax Rates

The United States has a **progressive income tax,** meaning that people with higher taxable incomes pay at a higher tax *rate*. The system works by assigning different **marginal tax rates** to different income ranges (or *margins*). For example, suppose you are single and your taxable income is $20,000. Under 2003 tax rates, you would pay 10% tax on the first $7000 and 15% tax on the remaining $13,000. In this case, we say that your *marginal rate* is 15%, or that you are in the 15% *tax bracket*. For each major filing status, Table 4.10 shows the marginal tax rate, standard deduction, and exemptions for 2003.

The income levels for the tax brackets, the standard deductions, and the exemption amounts all rise each year to keep pace with inflation. In addition, Congress and the President tend to change the rates for the tax brackets every few years. Thus, if you are calculating taxes for a year other than 2003, you must get an updated tax rate table. You can find current tax rates on the IRS Web site.

by the way

Special tax rules come into play for people with very high incomes or large amounts of itemized deductions. As a result, not all taxpayers get the full advantage of exemptions and deductions. In particular, many middle- to high-income tax payers are subject to the alternative minimum tax (AMT), which disallows most or all deductions.

TABLE 4.10	2003 Marginal Tax Rates, Standard Deductions, and Exemptions			
TAX RATE	SINGLE	MARRIED FILING JOINTLY	MARRIED FILING SEPARATELY	HEAD OF HOUSEHOLD
10%	up to $7000	up to $14,000	up to $7000	up to 10,000
15%	$7001 to $28,400	$14,001 to $56,800	$7001 to $28,400	$10,001 to $38,050
25%	$28,401 to $68,800	$56,801 to $114,650	$28,401 to $57,325	$38,051 to $98,250
28%	$68,801 to $143,500	$114,651 to $174,700	$57,326 to $87,350	$98,251 to $159,100
33%	$143,501 to $311,950	$174,701 to $311,950	$87,351 to $155,975	$159,101 to $311,950
35%	$311,951 or more	$311,951 or more	$155,976 or more	$311,951 or more
standard deduction	$4750	$9500	$4750	$7000
exemptions (per person)	$3050	$3050	$3050	$3050

➤ EXAMPLE 3 Marginal Tax Computations

Using 2003 rates, calculate the tax owed by each of the following people. Assume that they all claim the standard deduction and neglect any tax credits.

a. Deirdre is single with no dependents. Her adjusted gross income is $75,000.

b. Robert is a head of household taking care of two dependent children. His adjusted gross income also is $75,000.

c. Jessica and Frank are married with no dependents. They file jointly. They each have $75,000 in adjusted gross income, making a combined income of $150,000.

Solution

a. First, we must find Deirdre's taxable income. She is entitled to a personal exemption of $3050 and a standard deduction of $4750. We subtract these amounts from her adjusted gross income to find her taxable income:

$$\text{taxable income} = \$75,000 - \$3050 - \$4750 = \$67,200$$

Now we calculate her taxes using the *single* rates in Table 4.10. She is in the 25% tax bracket because her taxable income is above $28,400 but below the 28% threshold of $68,801. Thus, she owes 10% on the first $7000 of her taxable income, 15% on her taxable income between $7000 and $28,400, and 25% on her taxable income above $28,400.

$$\underbrace{(10\% \times \$7000)}_{\substack{\text{10\% marginal rate on first} \\ \text{\$7000 of taxable income}}} + \underbrace{(15\% \times [\$28,400 - \$7000])}_{\substack{\text{15\% marginal rate on taxable income} \\ \text{between \$7001 and \$28,400}}} + \underbrace{(25\% \times [\$67,200 - \$28,400])}_{\substack{\text{25\% marginal rate on taxable} \\ \text{income above \$28,400}}}$$

$$= \$700 + \$3210 + \$9700$$
$$= \$13,610$$

Deirdre's tax is $13,610.

b. Robert is entitled to *three* exemptions of $3050 each—one for himself and one for each of his two children. As a head of household, he is also entitled to a standard deduction of $7000. We subtract these amounts from his adjusted gross income to find his taxable income:

$$\text{taxable income} = \$75,000 - (3 \times \$3050) - \$7000 = \$58,850$$

We calculate Robert's taxes using the *head of household* rates. His taxable income of $58,850 puts him in the 25% tax bracket, so his tax is

$$\underbrace{(10\% \times \$10,000)}_{\substack{\text{10\% marginal rate on first} \\ \text{\$10,000 of taxable income}}} + \underbrace{(15\% \times [\$38,050 - \$10,000])}_{\substack{\text{15\% marginal rate on taxable} \\ \text{income between \$10,001 and \$38,050}}} + \underbrace{(25\% \times [\$58,850 - \$38,050])}_{\substack{\text{25\% marginal rate on taxable} \\ \text{income above \$38,050}}}$$

$$= \$1000 + \$4207.50 + \$5200$$
$$= \$10,407.50$$

Robert's tax is $10,407.50.

c. Jessica and Frank are each entitled to one exemption of $3050. Because they are married filing jointly, their standard deduction is $9500. We subtract these amounts from their adjusted gross income to find their taxable income:

$$\text{taxable income} = \$150,000 - (2 \times \$3050) - \$9500 = \$134,400$$

We calculate their taxes using the *married filing jointly* rates. Their taxable income of $134,400 puts them in the 28% tax bracket, so their tax is

$$\underbrace{(10\% \times \$14,000)}_{\substack{\text{10\% marginal rate on first} \\ \text{\$14,000 of taxable income}}} + \underbrace{(15\% \times [\$56,800 - \$14,000])}_{\substack{\text{15\% marginal rate on taxable income} \\ \text{between \$14,001 and \$56,800}}}$$

$$+ \underbrace{(25\% \times [\$114,650 - \$56,800])}_{\substack{\text{25\% marginal rate on taxable income} \\ \text{between \$56,801 and \$114,650}}} + \underbrace{(28\% \times [\$134,400 - \$114,650])}_{\substack{\text{28\% marginal rate on taxable} \\ \text{income above \$114,650}}}$$

$$= \$1400 + \$6420 + \$14,462.50 + \$5530$$
$$= \$27,812.50$$

Jessica and Frank's combined tax is $27,812.50, equivalent to $13,906.25 each. *Now try Exercises 29–36.* ◄

by the way

In Example 3, Jessica and Frank each earned the same amount as Deirdre, but together they paid more than twice as much tax (by almost $600). This feature of the tax code, whereby people pay more when they are married than they would if they were single, is called the *marriage penalty*. Not all couples are affected the same way by the marriage penalty. Some couples even get a marriage bonus instead, especially if one spouse earns much more than the other (or only one is employed).

Time out to think

Note that all four individuals in Example 3 have the same $75,000 in adjusted gross income, yet they each pay a different amount in taxes. Explain why this is the case. Do you believe the outcomes are fair? Why or why not? (Bonus: Could their gross incomes have differed even though their adjusted gross incomes were the same? Explain.)

Tax Credits and Deductions

Tax credits and tax deductions may sound similar, but they are very different. Suppose you are in the 15% tax bracket. A tax *credit* of $500 reduces your total tax bill by the full $500. In contrast, a tax *deduction* of $500 reduces your taxable income by

$500, which means it saves you only 15% × $500 = $75 in taxes. Thus, as a rule, tax credits are more valuable than tax deductions.

Of course, Congress has authorized tax credits for only a few specific situations, such as a (maximum) $1000 tax credit for each child. In contrast, your spending determines how much you claim in deductions, at least if you are itemizing. The most valuable deduction for most people is the **mortgage interest tax deduction,** which allows you deduct all the *interest* (but not the principal) you pay on a home mortgage. Many people also get substantial deductions from donating money to charities.

▶ **EXAMPLE 4** Tax Credits vs. Tax Deductions

Suppose you are in the 28% tax bracket. How much does a $1000 tax credit save you? How much does a $1000 charitable contribution (which is tax deductible) save you? Answer these questions both for the case in which you itemize deductions and for the case in which you take the standard deduction.

Solution The entire $1000 tax credit is deducted from your tax bill and therefore saves you a full $1000, whether you itemize deductions or take the standard deduction. In contrast, a $1000 deduction reduces your taxable income, not your total tax bill, by $1000. Thus, for the 28% tax bracket, *at best* your $1000 deduction will save you 28% × $1000 = $280. However, you will save this $280 only if you are itemizing deductions. If your total itemized deductions are less than the standard deduction (see Example 2), you will still be better off with the standard deduction. In that case, the $1000 contribution will save you nothing at all. *Now try Exercises 37–42.* ◀

▶ **EXAMPLE 5** Rent or Own?

Suppose you are in the 28% tax bracket and you itemize your deductions. You are trying to decide whether to rent an apartment or buy a house. The apartment rents for $1400 per month. You've investigated your loan options, and you've determined that if you buy the house, your monthly mortgage payments will be $1600, of which an average of $1400 goes toward interest during the first year. Compare the monthly rent to the mortgage payment. Is it cheaper to rent the apartment or buy the house?

Solution The monthly cost of the apartment is $1400 in rent. For the house, however, we must take into account the value of the mortgage deduction. The monthly interest of $1400 is tax deductible. Because you are in the 28% tax bracket, this deduction saves you 28% × $1400 = $392. Thus, the true monthly cost of the mortgage is the payment minus the tax savings, or

$$\$1600 - \$392 = \$1208$$

Despite the fact that the mortgage payment is $200 higher than the rent, its true cost is almost $200 per month less because of the tax savings from the mortgage interest deduction. (This example assumes you would be itemizing deductions regardless of whether you rent or buy.) *Now try Exercises 43–44.* ◀

Time out to think _____
Aside from the lower monthly cost, what other factors would affect your decision about whether to rent or buy in Example 5?

➤EXAMPLE 6 Varying Value of Deductions

Drew is in the 15% marginal tax bracket. Marian is in the 35% marginal tax bracket. They each itemize their deductions. They each donate $5000 to charity. Compare their true costs for the charitable donation.

Solution The $5000 contribution to charity is tax deductible. Because Drew is in the 15% tax bracket, this contribution saves him $15\% \times \$5000 = \750 in taxes. Thus, its true cost to him is the contributed amount of $5000 minus his tax savings of $750, or $4250. For Marian, who is in the 35% tax bracket, the contribution saves $35\% \times \$5000 = \1750 in taxes. Thus, its true cost to her is $\$5000 - \$1750 = \$3250$. The true cost of the donation is considerably lower for Marian because she is in a higher tax bracket. *Now try Exercises 45–46.* ◄

Time out to think _____

As shown in Example 6, tax deductions are more valuable to people in higher tax brackets. Some people argue that this is unfair because it means that tax deductions save more money for richer people than for poorer people. Others argue that it is fair, because richer people pay a higher tax rate in the first place. What do you think? Defend your opinion.

SOCIAL SECURITY AND MEDICARE TAXES

In addition to being subject to taxes computed with the marginal rates, some income is subject to Social Security and Medicare taxes, which are collected under the obscure name of **FICA** (Federal Insurance Contribution Act) taxes. Taxes collected under FICA are used to pay Social Security and Medicare benefits, primarily to people who are retired.

FICA applies only to income from wages (including tips) and self-employment. It does not apply to income from such things as interest, dividends, or profits from sales of stock. In 2003, the FICA tax rates for individuals who were *not* self-employed were

- 7.65% on the first $87,000 of income from wages
- 1.45% on any income from wages in excess of $87,000

In addition, the individual's employer is required to pay matching amounts of FICA taxes.

Individuals who are self-employed must pay both the employee and the employer shares of FICA. Thus, the rates for self-employed individuals are *double* the rates paid by individuals who are not self-employed.

FICA is calculated on *all* wages, tips, and self-employment income. You may not subtract any adjustments, exemptions, or deductions when calculating FICA taxes.

➤EXAMPLE 7 FICA Taxes

In 2003, Jude earned $22,000 in wages and tips from her job waiting tables. Calculate her FICA taxes and her total tax bill including marginal taxes. What is her overall tax rate on her gross income, including both FICA and income taxes? Assume she is single and takes the standard deduction.

by the way

The portion of FICA going to Social Security is called OASDI (Old Age, Survivors, and Disability Insurance). The portion going to Medicare is called HI (Hospital Insurance).

Solution Jude's entire income of $22,000 is subject to the 7.65% FICA tax:

$$\text{FICA tax} = 7.65\% \times \$22,000 = \$1683$$

Now we must find her income tax. We get her taxable income by subtracting her $3050 personal exemption and $4750 standard deduction:

$$\text{taxable income} = \$22,000 - \$3050 - \$4750 = \$14,200$$

From Table 4.10, her income tax is 10% on the first $7000 of her taxable income and 15% on the remaining amount of $14,200 - $7000 = $7200. Thus, her income tax is $(10\% \times \$7000) + (15\% \times \$7200) = \$1780$. Her total tax, including both FICA and income tax, is

$$\text{total tax} = \$1683 + \$1780 = \$3463$$

Her overall tax rate is

$$\frac{\text{total tax}}{\text{gross income}} = \frac{\$3463}{\$22,000} = 0.157$$

Jude's overall tax rate is 15.7%. Note that she pays nearly as much in FICA tax as in income tax. *Now try Exercises 47–52.* ◄

DIVIDENDS AND CAPITAL GAINS

Not all income is created equal, at least not in the eyes of the tax collector! In particular, dividends (on stocks) and **capital gains**—profits from the sale of stock or other property—get special tax treatment. Capital gains are divided into two subcategories. **Short-term capital gains** are profits on items sold within 12 months of their purchase, and **long-term capital gains** are profits on items held for more than 12 months before being sold.

Long-term capital gains and most dividends are taxed at lower rates than other income such as wages and interest earnings. As of 2003, the rates were

- a maximum of 5% for income in the 10% and 15% tax brackets
- a maximum of 15% for income in all higher tax brackets

In a few cases, capital gains get even better tax treatment. For example, capital gains on the sale of your home are often tax exempt.

►**EXAMPLE 8** Dividend and Capital Gains Income

In 2003, Serena was single and lived off an inheritance. Her gross income consisted solely of $85,000 in dividends and long-term capital gains. She had no adjustments to her gross income, but had $10,000 in itemized deductions and a personal exemption of $3050. How much tax does she owe? What is her overall tax rate?

Solution She owes no FICA tax because her income is not from wages. She had no adjustments to her gross income, so we find her taxable income by subtracting her itemized deductions and personal exemption:

$$\text{taxable income} = \$85,000 - \$10,000 - \$3050 = \$71,950$$

Because her income is all dividends and long-term capital gains, she pays tax at the special rates for these types of income. The special 5% rate for dividends and long-term capital gains applies to the income on which she would have been taxed at 10% or 15% if it had been ordinary income. From Table 4.10, therefore, we see that this 5% rate applies to her first $28,400 of income. The rest of her income is taxed at the special 15% rate. Thus, her total tax is

$$\underbrace{(5\% \times \$28{,}400)}_{\text{5\% capital gains rate}} + \underbrace{(15\% \times [\$71{,}950 - \$28{,}400])}_{\text{15\% capital gains rate}} = \$1420 + \$6532.50 = \$7952.50$$

Her overall tax rate is

$$\frac{\text{total tax}}{\text{gross income}} = \frac{\$7952.50}{\$85{,}000} = 0.094$$

Serena's overall tax rate is 9.4%. *Now try Exercises 53–54.* ◄

Time out to think

Note that Serena in Example 8 had a gross income almost quadruple that of Jude in Example 7. Compare their tax payments and overall tax rates. Who pays more tax? Who pays at a higher tax rate? Explain.

TAX-DEFERRED INCOME

The tax code tries to encourage long-term savings by allowing you to *defer* income taxes on contributions to certain types of savings plans, called **tax-deferred savings plans.** Money that you deposit into such savings plans is not taxed now. Instead, it will be taxed in the future when you withdraw the money.

Tax-deferred savings plans go by a variety of names, such as *individual retirement accounts* (IRAs), *qualified retirement plans* (QRPs), *401(k) plans,* and more. All are subject to strict rules. For example, you generally are not allowed to withdraw money from any of these plans until you reach age $59\frac{1}{2}$. Anyone can set up a tax-deferred savings plan, and *you should,* regardless of your current age. Why? Because they offer two key advantages in saving for your long-term future.

First, contributions to tax-deferred savings plans count as *adjustments* to your present gross income and are not part of your taxable income. As a result, the contributions cost you less than contributions to savings plans without special tax treatment. For example, suppose you are in the 28% marginal tax bracket. If you deposit $100 in an ordinary savings account, your tax bill is unchanged and you have $100 less to spend on other things. But if you deposit $100 in a tax-deferred savings account, you do not have to pay tax on that $100. With your 28% marginal rate, you therefore save $28 in taxes. Thus, the amount you have to spend on other things decreases by only $100 − $28 = $72.

The second advantage of tax-deferred savings plans is that their earnings are also tax deferred. With an ordinary savings plan, you must pay taxes on the earnings each year, which effectively reduces your earnings. With a tax-deferred savings plan, *all* of the earnings accumulate from one year to the next. Over many years, this tax saving makes the value of tax-deferred savings accounts rise much more quickly than that of ordinary savings accounts (Figure 4.9).

by the way

With tax-deferred savings, you will eventually pay tax on the money when you withdraw it. With *tax-exempt* investments, you never have to pay tax on the earnings. Some government bonds are tax-exempt. A Roth IRA is a special type of individual retirement account in which you pay taxes on money you deposit now, but all earnings on the account are tax-exempt when you withdraw them.

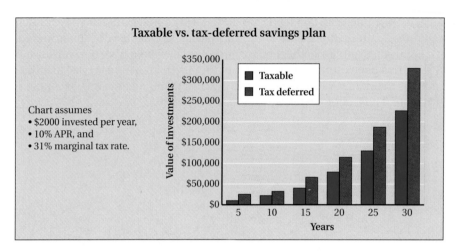

FIGURE 4.9 This graph compares the values of a tax-deferred savings plan and an ordinary savings plan, assuming that tax on the interest is paid from the plan in the latter case. After 30 years, the tax-deferred savings plan is worth over $100,000 more than the ordinary plan.

➤ **EXAMPLE 9** Tax-Deferred Savings Plan

Suppose you are single, have a taxable income of $65,000, and make monthly payments of $500 to a tax-deferred savings plan. How do the tax-deferred contributions affect your monthly take-home pay?

Solution Table 4.10 shows that your marginal tax rate is 25%. Each $500 contribution to a tax-deferred savings plan therefore reduces your tax bill by

$$25\% \times \$500 = \$125$$

In other words, $500 goes into your tax-deferred savings account each month, but your monthly paychecks go down by only $500 − $125 = $375. The special tax treatment makes it significantly easier for you to afford the monthly contributions needed to build your retirement fund. (The monthly saving found here is your *average* monthly saving for the year, after any tax refund or tax bill. It will be your precise monthly saving only if your withholding is computed so that you have zero tax due at year end.) *Now try Exercises 55–58.* ◄

■ EXERCISES 4D

Review Questions

1. Explain the basic process of calculating income taxes, as shown in Figure 4.8. What is the difference between gross income, adjusted gross income, and taxable income?

2. What is meant by filing status? How does it affect tax calculations?

3. What are exemptions and deductions? How should you choose between taking the standard deduction and itemizing deductions?

4. What do we mean by a progressive income tax? Explain the use of marginal tax rates in calculating taxes. What do we mean by a tax bracket?

5. What is the difference between a tax deduction and a tax credit? Why is a tax credit more valuable?

6. Explain how a deduction, such as the mortgage interest tax deduction, can save you money. Why do deductions benefit people in different tax brackets differently?

7. What are FICA taxes? What type of income is subject to FICA taxes?

8. How are dividends and capital gains treated differently than other income by the tax code?

9. Explain how you can benefit from a tax-deferred savings plan.

10. Why do tax-deferred savings plans tend to grow faster than ordinary savings plans?

Does It Make Sense?

Decide whether each of the following statements makes sense (or is clearly true) or does not make sense (or is clearly false). Explain your reasoning.

11. We're both single with no children and we both have the same total (gross) income, so we must both pay the same amount in taxes.

12. The $1000 child tax credit sounds like a good idea, but it doesn't help me because I take the standard deduction rather than itemized deductions.

13. When I calculated carefully, I found that it was cheaper for me to buy a house than to continue renting, even though my rent payments were lower than my new mortgage payments.

14. My husband and I paid $7000 in mortgage interest this year, but we didn't get any tax benefit from it.

15. Bob and Sue were planning to get married in December of this year, but they postponed their wedding until January when they found it would save them money in taxes.

16. The top marginal tax rate may be 35%, but I never pay more than 15% because I live off the dividends from my inheritance.

17. I didn't owe any ordinary income tax because my business (self-employed) made only a $7000 profit, but my total tax bill still came to 15.3% of my income.

18. I started contributing $400 each month to my tax-deferred savings plan, but my take-home pay declined by only $300.

Basic Skills & Concepts

Income on Tax Forms. For each situation described in Exercises 19–22, find the person's gross income, adjusted gross income, and taxable income.

19. Antonio earned wages of $47,200, received $2400 in interest from a savings account, and contributed $3500 to a tax-deferred retirement plan. He was entitled to a personal exemption of $3050 and had deductions totaling $7400.

20. Marie earned wages of $28,400, received $95 in interest from a savings account, and was entitled to a personal exemption of $3050 and a standard deduction of $4750.

21. Isabella earned wages of $88,750, received $4900 in interest from a savings account, and contributed $6200 to a tax-deferred retirement plan. She was entitled to a personal exemption of $3050 and had deductions totaling $9050.

22. Lebron earned wages of $3,452,000, received $54,200 in interest from savings, and contributed $30,000 to a tax-deferred retirement plan. He was not allowed to claim a personal exemption (because of his high income) but was allowed deductions totaling $674,500.

Should You Itemize? In Exercises 23–24, decide whether you should itemize your deductions or claim the standard deduction. Explain your reasoning.

23. Your deductible expenditures are $8600 for interest on a home mortgage, $2700 for contributions to charity, and $645 for state income taxes. Your filing status entitles you to a standard deduction of $9500.

24. Your deductible expenditures are $3700 for contributions to charity and $760 for state income taxes. Your filing status entitles you to a standard deduction of $4750.

Income Calculations. In Exercises 25–28, compute the individual's (or couple's) gross income, adjusted gross income, and taxable income. Use the 2003 values for exemptions and standard deductions in Table 4.10. Be sure to explain how you decide to claim standard or itemized deductions. (Note: Do your calculations based only on the given data, which may not include all credits and deductions.)

25. Suzanne is single and earned wages of $33,200. She received $350 in interest from a savings account. She contributed $500 to a tax-deferred retirement plan. She had $450 in itemized deductions from charitable contributions.

26. Malcolm is single and earned wages of $23,700. He had $4500 in itemized deductions from interest on a house mortgage.

27. Wanda is married, but she and her husband filed separately. Her gross salary was $35,400, and she earned $500 in interest. She had $1500 in itemized deductions and claimed three exemptions for herself and two children.

28. Emily and Juan are married and filed jointly. Their combined wages were $75,300. They earned $2000 from a rental property they own, and they received $1650 in interest. They claimed four exemptions for themselves and two children. They contributed $3240 to their tax-deferred retirement plans, and their itemized deductions totaled $9610.

Marginal Tax Calculations. In Exercises 29–36, use the 2003 marginal tax rates in Table 4.10 to compute the tax owed.

29. Gene is single and had a taxable income of $35,400.

30. Sarah and Marco are married filing jointly with a taxable income of $87,500.

31. Bobbi is married filing separately with a taxable income of $77,300.

32. Abraham is single with a taxable income of $23,800.

33. Paul is a head of household with a dependent child and a taxable income of $89,300. He also is entitled to a $1000 tax credit.

34. Pat is a head of household with a dependent child and a taxable income of $57,000. He also is entitled to a $1000 tax credit.

35. Winona and Jim are married filing jointly with a taxable income of $105,500. They also are entitled to a $2000 tax credit.

36. Chris is married filing separately with a taxable income of $127,300.

Tax Credits and Tax Deductions. In Exercises 37–42, state how much each individual or couple will save in taxes with the tax credit or tax deduction specified.

37. Midori and Tremaine are in the 28% tax bracket and claim the standard deduction. How much will their tax bill be reduced if they qualify for a $500 tax credit?

38. Vanessa is in the 35% tax bracket and itemizes her deductions. How much will her tax bill be reduced if she qualifies for a $500 tax credit?

39. Rosa is in the 15% tax bracket and claims the standard deduction. How much will her tax bill be reduced if she makes a $1000 contribution to charity?

40. Shiro is in the 15% tax bracket and itemizes his deductions. How much will his tax bill be reduced if he makes a $1000 contribution to charity?

41. Sebastian is in the 28% tax bracket and itemizes his deductions. How much will his tax bill be reduced if he makes a $1000 contribution to charity?

42. Santana is in the 35% tax bracket and itemizes her deductions. How much will her tax bill be reduced if she makes a $1000 contribution to charity?

Rent or Own? Exercises 43–44 state a tax bracket, an apartment rent, and a house payment, along with the average amount going toward interest in the first year. Including savings through the mortgage interest deduction, determine whether renting or buying is cheaper (in terms of monthly payments) during the first year. Assume you are itemizing deductions in all cases.

43. You are in the 33% tax bracket. The apartment rents for $1600 per month. Your monthly mortgage payments would be $2000, of which an average of $1800 per month goes toward interest during the first year.

44. You are in the 15% tax bracket. The apartment rents for $600 per month. Your monthly mortgage payments would be $675, of which an average of $600 per month goes toward interest during the first year.

45. **Varying Value of Deductions.** Maria is in the 33% tax bracket. Steve is in the 15% tax bracket. They each itemize their deductions and pay $10,000 in mortgage interest during the year. Compare their true costs for mortgage interest.

46. **Varying Value of Deductions.** Yolanna is in the 35% tax bracket. Alia is in the 10% tax bracket. They each itemize their deductions, and they each donate $4000 to charity. Compare their true costs for charitable donations.

FICA Taxes. Exercises 47–52 each describe a person's income. In each case, calculate the person's FICA taxes and total tax bill, including marginal income taxes. Then find the person's overall tax rate on his or her gross income, including both FICA and income taxes. Assume all individuals are single and take the standard deduction. Use the 2003 tax rates in Table 4.10. (Round tax calculations to the nearest dollar.)

47. Luis earned $28,000 from wages as a computer programmer and made $2500 in tax-deferred contributions to a retirement fund.

48. Carla earned $34,500 in salary and $750 in interest and made $3000 in tax-deferred contributions to a retirement fund.

49. Jack earned $44,800 in salary and $1250 in interest and made $2000 in tax-deferred contributions to a retirement fund.

50. Alejandro earned $102,400 in salary and $4450 in interest and made $9500 in tax-deferred contributions to a retirement fund.

51. Brittany earned $48,200 in wages and tips. She had no other income and made no contributions to retirement plans.

52. Larae earned $21,200 in wages and tips. She had no other income and made no contributions to retirement plans.

Dividends and Capital Gains. In Exercises 53–54, calculate the total tax owed by each of the two people, including both FICA and income taxes. Compare their overall tax rates, including both FICA and income taxes. Assume all individuals are single and take the standard deduction. Use the 2003 tax rates in Table 4.10 for ordinary income and the special rates for dividends and capital gains listed in the text.

53. Pierre earned $120,000 in wages. Katarina earned $120,000, all in dividends and long-term capital gains.

54. Deion earned $60,000 in wages. Josephina earned $60,000, all in dividends and long-term capital gains.

Tax-Deferred Savings Plans. In Exercises 55–58, calculate the effect on annual take-home pay of the tax-deferred contributions described. Use the 2003 tax rates in Table 4.10.

55. You are single and have a taxable income of $18,000. You make monthly contributions of $400 to a tax-deferred savings plan.

56. You are single and have a taxable income of $45,000. You make monthly contributions of $600 to a tax-deferred savings plan.

57. You are married filing jointly and have a taxable income of $90,000. You make monthly contributions of $800 to a tax-deferred savings plan.

58. You are married filing jointly and have a taxable income of $200,000. You make monthly contributions of $800 to a tax-deferred savings plan.

Further Applications

Marriage Penalty. Exercises 59–62 give the adjusted gross incomes of a couple that is engaged to be married. Calculate the tax owed by the couple in two ways: (1) if they delay their marriage until next year so that they can each file a tax return at the single tax rate this year and (2) if they marry before the end of the year and file a joint return. Assume that each person takes one exemption and the standard deduction. Use the 2003 tax rates in Table 4.10. Does the couple face a "marriage penalty" if they marry before the end of the year? Explain. (Note: Married rates apply for the entire year, no matter when during a year you are married.)

59. Gabriella and Roberto have adjusted gross incomes of $44,500 and $33,400, respectively.

60. Joan and Paul have adjusted gross incomes of $32,500 and $29,400, respectively.

61. Mia and Steve each have an adjusted gross income of $185,000.

62. Lisa has an adjusted gross income of $85,000, and Patrick is a student with no income.

63. **Estimating Your Taxes.** List all the gross income you expect for the coming year, along with any expenses you are entitled to deduct from gross income. Then calculate your adjusted gross income and taxable income.

 a. Based on your estimates, how much tax will you owe this year? Use the 2003 tax rates in Table 4.10, or find updated rates on the Web.

 b. How much (if any) tax is being withheld from your paychecks each month? Should you expect a tax refund next year? Explain.

 c. Suppose you begin making a $100 monthly contribution to a tax-deferred retirement plan. How will it affect your take-home pay? Explain.

 d. Suppose you make a $1000 contribution to charity. By how much, if at all, will this contribution reduce your tax bill? Explain.

Web Projects

Find useful links for Web Projects on the text Web site: www.aw.com/bennett-briggs

64. **Tax Simplification Plans.** Use the Web to investigate a recent proposal to simplify federal tax laws and filing procedures. What are the advantages and disadvantages of the simplification plan, and who supports it?

65. **Fairness Issues.** Choose a tax question that has issues of fairness associated with it (for example, capital gains rates, the marriage penalty, or the alternative minimum tax [AMT]). Use the Web to investigate the current status of this question. Have new laws been passed that affect it? What are the advantages and disadvantages of recent or proposed changes, and who supports the changes? Summarize your own opinion about whether current tax law is unfair and, if so, what should be done about it.

66. The Digital Daily. The electronic news publication of the Internal Revenue Service is the Digital Daily. Visit the Web site for the publication. Choose a current "front page" issue and report on it in terms of how it affects you as a taxpayer.

67. Current Tax Rates. Use the IRS Web site to find the current tax rates. Recast Table 4.10 with these tax rates.

In the News

68. Alternative Minimum Tax (AMT). The calculations described in this unit all assume that a person pays taxes according to the "normal" tax code. However, some people will be subject to the alternative minimum tax (AMT) in coming years. The AMT calculates taxes in a very different way, making taxes higher for people who pay it. The AMT is becoming a hot political issue, because more and more people are expected to owe it over the next few years. Find a recent article about the impact of the AMT. Write a short report on what you learn.

69. Tax Changes. Find a recent news article about proposed changes to federal tax laws. Briefly describe the proposed changes and their impact. What parties support and oppose the changes?

70. Your Tax Return. Briefly describe your own experiences with filing a federal income tax return. Do you file your own returns? If so, do you use a computer software package or a professional tax advisor? Will you change your filing method in the future? Why or why not?

UNIT 4E

UNDERSTANDING THE FEDERAL BUDGET

So far in this chapter, we have discussed issues of financial management that affect us directly as individuals. But we are also affected by the way our government manages its finances. In this unit, we will discuss a few of the basic concepts needed to understand the federal budget.

BUDGET BASICS

In theory, the federal budget works much like your personal budget or the budget of a small business. All have **receipts,** or income, and **outlays,** or expenses. **Net income** is the difference between receipts and outlays. When receipts exceed outlays, net income is positive and the budget has a **surplus** (profits). When outlays exceed receipts, net income is negative and the budget has a **deficit** (losses).

DEFINITIONS

Receipts, or income, represent money that has been collected.

Outlays, or expenses, represent money that has been spent.

Net income = receipts − outlays

If net income is positive, the budget has a **surplus.**

If net income is negative, the budget has a **deficit.**

Note that a deficit means spending more money than was collected. The only way you (or a business or government) can survive a deficit is by spending savings or borrowing money. When you borrow, you accumulate a **debt.** Every year that you borrow to cover a deficit, your debt grows. In addition, the lender will surely charge interest on your debt. Thus, in addition to accumulating a debt, you will also face growing interest payments as your debt rises. In contrast, a surplus means collecting more money than was spent. If you have a surplus, you can use it either to add to your savings or to reduce your debt.

There can be no freedom or beauty about a home life that depends on borrowing and debt.

—HENRIK IBSEN, 1879

Debt versus Deficit

A **deficit** represents money that is borrowed (or taken from savings) during a *single* year.

The **debt** is the total amount of money owed to lenders, which may result from accumulating deficits over many years.

A national debt, if it is not excessive, will be to us a national blessing.

—ALEXANDER HAMILTON, 1781

For over half a century, the U.S. government has run a deficit almost every year, with the notable exception of the years 1998–2001. Figure 4.10a shows the deficits and surpluses since 1955. Figure 4.10b shows the national debt that has accumulated during this period.

➤ EXAMPLE 1 Personal Budget

Suppose your gross income last year was $40,000. Your expenditures were as follows: $20,000 for rent and food, $2000 for interest on your credit cards and student loans, $6000 for car expenses, and $9000 for entertainment and miscellaneous expenses. You also paid $8000 in taxes. Did you have a deficit or a surplus?

Solution The total of your outlays, including tax, was

$$\$20,000 + \$2000 + \$6000 + \$9000 + \$8000 = \$45,000$$

Because your outlays were greater than your $40,000 income by $5000, your personal budget had a $5000 deficit. Therefore, you must have either withdrawn $5000 from savings or borrowed $5000 to cover your expenditures. *Now try Exercises 15–16.* ◄

➤ EXAMPLE 2 The Federal Debt

The federal debt at the end of 2003 was roughly $7 trillion. If this debt were divided evenly among the roughly 300 million citizens of the United States, how much would *you* owe?

Solution This question is easiest to answer by putting the numbers in scientific notation. We divide the debt of $7 trillion ($7 \times 10^{12}$) by the 300 million ($3 \times 10^8$) population:

$$\frac{\$7 \times 10^{12}}{3 \times 10^8 \text{ persons}} = \$2.3 \times 10^4/\text{person}$$

Your personal share of the total debt is roughly 2.3×10^4, or $23,000.

Now try Exercises 17–18. ◄

Surplus or Deficit (millions of dollars)

(a)

Gross Federal Debt (millions of dollars)

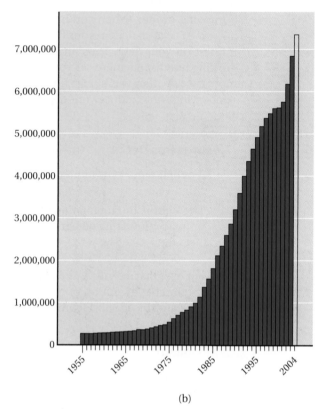

(b)

FIGURE 4.10 (a) Annual deficits or surpluses since 1955. (b) Accumulated gross federal debt since 1955. In both cases, the value for 2004 is a projection (unshaded bar).

SOURCE: Budget of the United States Government, 2003.

Time out to think _____

How does your share of the national debt compare to personal debts that you owe? Explain.

A SMALL-BUSINESS ANALOGY

Before we focus on the federal budget, let's investigate the simpler books of an imaginary company with not-so-imaginary problems. Table 4.11 summarizes four years of budgets for the Wonderful Widget Company, which started with a clean slate at the beginning of 2001.

The first column shows that, during 2001, the company had receipts of $854,000 and total outlays of $1,000,000. Thus, the company's net income was

$$\$854{,}000 - \$1{,}000{,}000 = -\$146{,}000$$

TABLE 4.11	Budget Summary for the Wonderful Widget Company (all amounts in thousands of dollars)			
	2001	2002	2003	2004
Total Receipts	$854	$908	$950	$990
Outlays				
Operating	525	550	600	600
Employee Benefits	200	220	250	250
Security	275	300	320	300
Interest on Debt	0	12	26	47
Total Outlays	1000	1082	1196	1197
Surplus/Deficit	−146	−174	−246	−207
Debt (accumulated)	**−146**	**−320**	**−566**	**−773**

The negative sign tells us that the company had a *deficit* of $146,000. The company had to borrow money to cover this deficit and ended the year with a debt of $146,000. The debt is shown as a negative number because it represents money owed to someone else.

In 2002, receipts increased to $908,000, while outlays increased to $1,082,000. These outlays included a $12,000 interest payment on the debt from the first year. Thus, the deficit for 2002 was

$$\$908,000 - \$1,082,000 = -\$174,000$$

The company had to borrow $174,000 to cover this deficit. Further, it had no money with which to pay off the debt from 2001. Thus, the total debt at the end of 2002 was

$$\$146,000 + \$174,000 = \$320,000$$

Here is the key point: Because the company again failed to balance its budget in its second year, its total debt continued to grow. As a result, its interest payment in 2003 increased to $26,000.

In 2004, the company's owners decided to change strategy. They froze operating expenses and employee benefits (relative to 2003) and actually *cut* security expenses. However, the interest payment rose substantially because of the rising debt. Despite the attempts to curtail outlays and despite another increase in receipts, the company still ran a deficit in 2004 and the total debt continued to grow.

Time out to think

Suppose you were a loan officer for a bank in 2005, when the Wonderful Widget Company came asking for further loans to cover its increasing debt. Would you lend it the money? If so, would you attach any special conditions to the loan? Explain.

➤ **EXAMPLE 3** Growing Interest Payments

Consider Table 4.11 for the Wonderful Widget Company. Assume that the $47,000 interest payment in 2004 was for the prior debt of $566,000. What was the annual interest rate? If the interest rate remains the same, what will the payment be on the debt at the end of 2004? What will the payment be if the interest rate rises by 2 percentage points?

Solution Paying $47,000 interest on a debt of $566,000 means an interest rate of

$$\frac{\$47{,}000}{\$566{,}000} = 0.083$$

The interest rate was 8.3%. At the end of 2004, the debt stands at $773,000. At the same interest rate, the next interest payment will be

$$0.083 \times \$773{,}000 = \$64{,}159$$

If the interest rate rises by 2 percentage points, to 10.3%, the next interest payment will be

$$0.103 \times \$773{,}000 = \$79{,}619$$

A 2-percentage-point change in the interest rate increases the interest payment by more than $15,000. *Now try Exercises 19–20.* ◄

THE FEDERAL BUDGET

The Widget Company example shows that a succession of deficits leads to a rising debt. The increasing interest payments on that debt, in turn, make it even easier to run deficits in the future. The Widget Company story is a mild version of what happened to the U.S. budget. Table 4.12 shows a summary of the federal budget in recent years. Moreover, as the debt has risen, interest payments have generally increased, consuming a larger and larger proportion of outlays. By 1997, interest on the debt made up more than 15% of total outlays. Fortunately, low interest rates have helped ease this burden, with interest on the debt falling below 10% of the total outlays in 2003. Still, this interest payment represented close to $200 billion—or about $700 per person in the United States.

TABLE 4.12	**U.S. Federal Budget Summary, 1996–2003 (all amounts in billions of dollars)**							
	1996	1997	1998	1999	2000	2001	2002	2003
Total Receipts	$1453	$1579	$1722	$1827	$2025	$1991	$1853	$1782
Total Outlays	1560	1601	1653	1703	1789	1863	2011	2156
Net Income	−107	−22	69	124	236	127	−158	−374

SOURCE: United States Office of Management and Budget.

The future of the federal budget is notoriously difficult to predict. As recently as 2001, projections suggested that the government would run a cumulative surplus of $5.6 trillion over the next 10 years (see Unit 3C). A mere two years later, the projections had changed from a multi-trillion-dollar surplus to a multi-trillion-dollar deficit. Over the past couple decades, each year's budget projections for the following year have been off by an average of about 11%. Projections more than one year out have been even further off.

►EXAMPLE 4 Budget Projections

As of 2003, the government projected total receipts of $2135 billion ($2.135 trillion) and a deficit of $208 billion in 2005. How would net income change if the projection of receipts turned out to be too high by 11%? How would it change if the projection of receipts were too low by 11%? Assume that outlays are unchanged.

Solution An error of 11% of the projected receipts of $2135 billion is

$$0.11 \times \$2135 \text{ billion} = \$235 \text{ billion (rounded to nearest \$1 billion)}$$

Thus, if receipts were 11% lower than expected, net income would be $235 billion less than projected, thereby increasing the deficit from its projected $208 billion to $208 billion + $235 billion = $442 billion. That is, the deficit would be more than double the projection. On the other hand, if receipts were 11% higher than expected, net income would be $235 billion higher than projected, turning the projected $208 billion deficit into a $27 billion surplus. *Now try Exercises 21–22.* ◄

Time out to think

Do you think it is wise to base long-term spending or taxing plans on long-term budget projections? Why or why not?

Federal Government Receipts

To understand the federal budget more deeply, we need to understand how the government gets its receipts and how it spends its outlays. Figure 4.11 on the next page shows the basic makeup of government receipts as of 2003. The categories for individual and corporate income taxes simply represent income taxes (see Unit 4D) paid by individuals and corporations, respectively. The other categories are

- Social insurance taxes, which primarily represent FICA taxes (see Unit 4D) for Social Security and Medicare but also include payments into retirement plans by federal employees and taxes for unemployment insurance
- Excise taxes, which include taxes on alcohol, tobacco, gasoline, and other products
- Borrowing from the public, which adds to the government's publicly held debt (see p. 289)
- Other, which includes such things as gift taxes and fines collected by the government

Note that most of the receipts currently come from income taxes. However, social insurance taxes are expected to represent a rising share of total receipts in the future.

by the way

The government also collects revenues from a few "business-like" activities, such as charging entrance fees at national parks. However, for historical reasons, these revenues are subtracted from outlays instead of being added to receipts when the government publishes its budget. Although this method of accounting may seem odd, it does not affect overall calculations of the surplus or deficit.

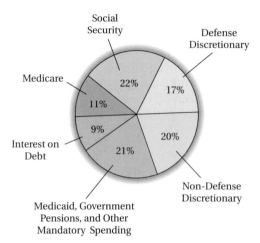

FIGURE 4.11 Makeup of federal government receipts, 2003.
SOURCE: United States Office of Management and Budget.

FIGURE 4.12 Makeup of federal government outlays, 2003.
SOURCE: United States Office of Management and Budget.

Federal Government Outlays

Figure 4.12 shows the basic makeup of government outlays as of 2003. For purposes of projecting budgets, the government generally groups spending into two major areas.

- **Mandatory outlays** are expenses that will be paid automatically unless Congress acts to change them. Most of the mandatory outlays are for "entitlements" such as Social Security, Medicare, and other payments to individuals. (They are called *entitlements* because the law specifically states the conditions under which individuals are entitled to them.) Interest on the debt is also a mandatory outlay, because it must be paid to prevent the government from being in default on its loans.

- **Discretionary outlays** are decided on a year-to-year basis. The amounts for discretionary programs must be approved by Congress in authorization bills, which then must be signed by the President to become law. Discretionary outlays are subdivided into programs for defense (military) and non-defense. *Non-defense discretionary outlays* include everything except the mandatory outlays and defense. For example, non-defense discretionary outlays include education, transportation, housing, international aid, the space program, and scientific research.

Note that Social Security is by far the single largest item in the budget. Moreover, its portion of the budget is expected to continue to grow as more and more people retire in coming decades.

▶ **EXAMPLE 5** Discretionary Squeeze

Suppose that Social Security rises to 30% of total outlays while all other programs except non-defense discretionary spending hold steady at the proportions shown in Figure 4.12. As a percentage of total outlays, how much would non-defense discre-

tionary spending have to decrease to cover the increase in Social Security? Comment on how this scenario would affect Congress's power to control the surplus or deficit.

Solution Figure 4.12 shows that 22% of outlays currently go to Social Security, so a rise to 30% would be a rise of 8 percentage points. Thus, the proportion of spending for all other programs would have to drop by 8 percentage points for the total to remain 100%. If this drop came entirely from non-defense discretionary spending, non-defense discretionary spending would fall from 20% to 12% of total outlays.

If non-defense discretionary spending were only 12% of total outlays, Congress would lose much of its power to control surpluses or deficits. Here's why: First, remember that Congress authorizes only discretionary spending (as opposed to mandatory spending) on a year-to-year basis. In essence, this is the only portion of the budget that Congress can easily control. Second, 12% turns out to be roughly the average error in budget projections (see Example 4). Thus, the proportion of the budget that Congress can easily control would be roughly the same as the uncertainty that Congress must deal with in making a budget. Clearly, this would make it nearly impossible for Congress to predict a surplus or deficit accurately.

Now try Exercises 23–28. ◄

STRANGE NUMBERS: PUBLICLY HELD AND GROSS DEBT

Take another look at Figure 4.10 and you may notice something rather strange: Even in the years when the government ran a surplus (1998–2001), the debt still continued to increase. Why did the debt keep rising, even when the government collected more money than it spent? More generally, why does the debt tend to rise from one year to the next by more than the amount of the deficit for the year? To answer these questions, we must investigate government accounting in a little more detail.

Financing the Debt

Remember that whenever you run a deficit, you must cover it either by withdrawing from savings or by borrowing money. The federal government does both. It withdraws money from its "savings," and it borrows money from people and institutions willing to lend to it.

Let's consider borrowing first. The government borrows money by selling Treasury bills, notes, and bonds (see Unit 4B) to the public. If you buy one of these Treasury issues, you are effectively lending the government money that it promises to pay back with interest. Because Treasury issues are considered to be very safe investments, the government has never had trouble finding people willing to buy them. By the end of 2003, the government had borrowed a total of about $4 trillion through the sale of Treasury issues. This debt, which the government must eventually pay back to the people who hold the Treasury issues, is called the **publicly held debt** (sometimes called the *net debt* or the *marketable debt*).

The government's "savings" consist of special accounts designed to meet future obligations. These accounts are called *trust funds*. The biggest trust fund by far is for Social Security, which is primarily a retirement program. People "invest" by paying Social Security taxes (most of the FICA taxes; see Unit 4D) and then collect Social Security benefits after they retire.

Currently, the government is collecting much more in Social Security taxes than it is paying out in Social Security benefits (see Figures 4.11 and 4.12). This reflects the fact that, today, many more people are working and paying Social Security taxes than are collecting benefits. However, as today's workers retire, the government will have to pay more and more in Social Security benefits. Therefore, to make sure there is enough money to pay future Social Security benefits, the government should invest the excess Social Security taxes that it collects today.

Legally, the government must invest the excess Social Security money in the Social Security trust fund. It does the same for several other trust funds, including those for the pensions of government workers. In a sense, these trust funds are like the government's savings accounts. But there's a catch: Before the government borrows from the public to finance a deficit, it first tries to cover the deficit by borrowing from its own trust funds.

In fact, the government has to date borrowed every penny it ever deposited into these trust funds. Thus, *there is no money in any of the trust funds, including Social Security.* Instead, the trust fund is filled with the equivalent of a stack of IOUs (more technically, with Treasury bills), in which the government has promised to return the money it borrowed, with interest.

As of the end of 2003, the government's debt to its own trust funds stood at about $3 trillion. Adding this amount to the publicly held debt of $4 trillion, we get a **gross debt** of $7 trillion. This is the total debt shown in Figure 4.10b, and it represents the total amount that the government is eventually obligated to repay from other government receipts (that is, receipts besides those collected for Social Security and other trust funds).

Two Kinds of National Debt

The **publicly held debt** (or *net debt*) represents money the government must repay to individuals and institutions that bought Treasury issues.

The **gross debt** includes both the publicly held debt and money that the government owes to its own trust funds, such as the Social Security trust fund.

On-Budget and Off-Budget: Effects of Social Security

As an example of how trust funds affect the two kinds of debt, consider 1998—when the federal government ran a $69 billion surplus (see Table 4.12). The government used this surplus money to buy back some of the Treasury notes and bonds it had sold to the public, which reduced the nearly $4 trillion publicly held debt.

However, remember that the government also collected excess Social Security taxes, which legally had to be deposited in the Social Security trust fund. In addition, the government owed the trust fund interest for all the money it had borrowed from the trust fund in the past. When we add both the excess Social Security taxes and the

owed interest, it turns out that the government was supposed to deposit $99 billion in the Social Security trust fund in 1998. But the government had already spent the $99 billion, leaving no cash available to deposit in the trust fund. The government therefore "deposited" $99 billion worth of IOUs in the trust fund, adding to the stack of IOUs already there from the past. Because IOUs represent loans, the government effectively borrowed $99 billion from the Social Security trust fund. When we subtract this borrowed amount from the $69 billion surplus, the government's income for 1998 becomes

$$\underbrace{\$69\ \text{billion}}_{\text{unified surplus}} - \underbrace{\$99\ \text{billion}}_{\text{off-budget surplus}} = \underbrace{-\$30\ \text{billion}}_{\text{on-budget deficit}}$$

With Social Security counted, the $69 billion surplus turns into a $30 billion deficit!

In government-speak, Social Security is said to be **off-budget.** Because the government really did collect $69 billion more than it spent, this number is called the unified surplus. The **on-budget** deficit is what remains after we subtract the portion of the surplus that came from Social Security. It represents the amount by which the government overspent its revenue when Social Security is included.

Although Social Security is the only major expenditure that is legally considered off-budget, other trust funds also represent future repayment obligations. Because the government borrowed from all these other trust funds as well, the gross debt rose by considerably more than the $30 billion on-budget deficit. In fact, when all was said and done, the gross debt rose by $109 billion during 1998. Despite the surplus, the debt to be repaid in the future grew substantially.

Unified Budget, On Budget, and Off Budget

The U.S. government's **unified budget** represents all federal revenues and spending. For accounting purposes, the government divides this unified budget into two parts:

- The portion of the unified budget that is involved in Social Security (that is, revenue from Social Security tax and spending on Social Security benefits) is considered **off-budget.**
- The rest of the unified budget (that is, everything that is not involved in Social Security) is considered **on-budget.**

Thus, the following relationships hold for any surplus or deficit in the budget:

$$\begin{matrix}\text{unified budget} \\ \text{(surplus or deficit)}\end{matrix} = \begin{matrix}\text{on-budget} \\ \text{(surplus or deficit)}\end{matrix} + \begin{matrix}\text{off-budget} \\ \text{(surplus or deficit)}\end{matrix}$$

Or, equivalently,

$$\begin{matrix}\text{unified budget} \\ \text{(surplus or deficit)}\end{matrix} - \begin{matrix}\text{on-budget} \\ \text{(surplus or deficit)}\end{matrix} = \begin{matrix}\text{off-budget} \\ \text{(surplus or deficit)}\end{matrix}$$

➤ **EXAMPLE 6** On- and Off-Budget

The federal government ran a $158 billion deficit (unified deficit) in 2002. However, this number does not separate the effects of excess Social Security taxes. For 2002, the government collected $160 billion more in Social Security revenue than it paid out in Social Security benefits. What do we call this excess $160 billion of Social Security revenue, and what happened to it? What was the government's on-budget deficit for 2002? Explain.

Solution The $160 billion excess Social Security revenue represents what we call the off-budget surplus for 2002, because it is counted separately from the rest of the budget (that's what makes it "off" budget). By law, this $160 billion had to be added to the Social Security trust fund. Unfortunately, it had already been spent (on programs other than Social Security), so the government instead added $160 billion worth of IOUs (Treasury bills) to the trust fund. Since this $160 billion worth of IOUs will have to be repaid eventually, it should be included in the calculation of what we call the on-budget deficit—the amount by which the government actually overspent in 2002. That is, the on-budget deficit for 2002 was

$$\underbrace{-\$158 \text{ billion}}_{\text{unified deficit}} - \underbrace{\$160 \text{ billion}}_{\text{off-budget surplus}} = \underbrace{-\$318 \text{ billion}}_{\text{on-budget deficit}}$$

In summary, the unified deficit of $158 billion means the government's total revenue fell $158 billion short of its outlays. But because the government added $160 billion to its long-term obligation to repay its own Social Security trust fund, the government's future repayment obligations really rose by the on-budget deficit of $318 billion. (In fact, because of the government's other trust funds and other accounting details, the debt rose by even more than this amount.) *Now try Exercises 29–30.* ◄

THE FUTURE OF SOCIAL SECURITY

Imagine that you decide to set up a retirement savings plan that will allow you to retire comfortably at age 65. Using the savings plan formula (see Unit 4B), you determine that you can achieve your retirement goal by making monthly deposits of $250 into your retirement plan. So you start the plan today by making your first $250 deposit.

However, in the first month you see a new music system that looks really cool. Being short on cash, you decide to buy it by withdrawing the $250 that you had put into your retirement plan. Because you don't want to fall behind on your retirement savings, you write yourself an IOU stating that you "owe" your retirement plan $250.

Next month, you again deposit $250 in your retirement plan—but soon withdraw it for a nice weekend getaway. As before, you write an IOU to remind yourself that you owe $250 to your retirement plan. Moreover, recognizing that you would have earned interest on the previous month's deposit if you hadn't withdrawn it, you write yourself an IOU for the lost interest.

Month after month and year after year, you continue in the same way. Because you always spend the money you had planned to put in your retirement plan, you keep writing yourself IOUs for the payments plus interest. When you finally reach age 65, your retirement plan contains IOUs that say you owe yourself enough money to retire on—*but your retirement account contains no actual money.* Obviously, it will be difficult to live off the IOUs you wrote to yourself.

This method of "saving" for retirement may sound silly, but it essentially describes the Social Security trust fund. Officially, the Social Security trust fund is growing larger and larger because of excess Social Security taxes and interest on past IOUs. According to recent projections, its balance will grow to over $3 trillion by 2015. But, in reality, the trust fund contains no cash today and will contain no cash in 2015. It will just be filled with $3 trillion worth of IOUs from the government to itself.

Now comes the bad news. Sometime around or after 2015, the increasing number of retirees will mean that Social Security payments will exceed the receipts from Social Security taxes. In order to pay benefits, the government will have to begin withdrawing money from the trust fund. This means the government will somehow have to start redeeming the IOUs that it has written to itself.

To see the problem vividly, consider the year 2035—shortly before current projections say the Social Security trust fund will go bankrupt. By then, projected Social Security payments will be about $700 billion more than collections from Social Security taxes. The government will therefore have to redeem $700 billion in IOUs from the Social Security trust fund, which means it will somehow have to find $700 billion in cold, hard cash. Generally speaking, the government can do this through some combination of the following three options:

1. It can cut spending on discretionary programs, such as the military or education, in order to free up money to redeem the IOUs. Unfortunately, $700 billion exceeds the amount currently spent on *all* discretionary spending. Thus, even if the government eliminated discretionary spending entirely, it would not have enough to cover the Social Security payments.
2. It can borrow the money from the public by issuing more Treasury notes and bonds. But the needed $700 billion would be larger than any single-year deficit in history.
3. It can raise taxes to collect extra cash.

You may notice that any of these three possibilities could have a dramatic impact on *you*. Programs such as education for your kids may be cut, or the economy will be hurt by huge deficits, or you'll be taxed much more heavily than you are today. Clearly, something must be done to solve this problem before it arrives. Unfortunately, the politics of Social Security makes a solution hard to come by.

Time out to think

Some proposals for solving the Social Security problem call for converting part or all of the program to private savings accounts. This would have the advantage of making sure that the government couldn't keep borrowing from Social Security. In addition, private investments have historically grown at a faster rate than the trust fund interest the government pays itself. However, the fact that private accounts can also lose value makes them at least somewhat risky. What's your opinion of privatizing Social Security? Explain.

►EXAMPLE 7 Tax Increase

In 2001, individual income taxes made up about half of total government receipts of nearly $2.0 trillion. Suppose that the government needed to raise an additional $700 billion through individual income taxes. How much would taxes have to increase? Neglect any economic problems that the tax increase might cause.

TECHNICAL NOTE
Projections are made in current dollars, so it is not necessary to adjust projected numbers for future inflation.

by the way
Social Security is sometimes called the "third rail" of politics. The term comes from the New York City subways, where the trains run on two rails and the third rail carries electricity at very high voltage. Touching the third rail generally causes instant death.

Solution To bring in half the $2.0 trillion in receipts, individual income taxes had to account for $1.0 trillion in government revenue. Thus, raising an additional $700 billion ($0.7 trillion) would require an additional 70% in revenue from individual income taxes. To generate an extra $700 billion, overall income taxes would have to rise by 70%. *Now try Exercises 31–32.* ◄

EXERCISES 4E

Review Questions

1. Define receipts, outlays, net income, surplus, and deficit as they apply to annual budgets.

2. What is the difference between a deficit and a debt? How large is the federal debt?

3. Explain why years of running deficits makes it increasingly difficult to get a budget into balance.

4. How large is the deficit at present? Should we assume future deficit projections are correct? Explain.

5. Briefly summarize the makeup of federal receipts and federal outlays. Distinguish between mandatory outlays and discretionary outlays.

6. How does the federal government finance its debt? Distinguish between the publicly held debt and the gross debt.

7. Briefly describe the Social Security trust fund. What's in it? What problems may this cause in the future?

8. Distinguish between an off-budget deficit (or surplus) and an on-budget deficit (or surplus). What is the unified deficit (or surplus)?

Does It Make Sense?

Decide whether each of the following statements makes sense (or is clearly true) or does not make sense (or is clearly false). Explain your reasoning.

9. My share of the federal government's debt is larger than all of my personal debt combined.

10. My share of the federal government's annual interest payments on the federal debt is greater than the payments I make on my car and credit card combined.

11. Since Social Security is off-budget, we could cut Social Security taxes with no impact on the rest of the federal government.

12. The government collected more money than it spent, but its total debt still increased.

13. I read today that in 10 years the government will return to surpluses (from deficits), so we should start planning how we'll use the surplus.

14. The Social Security trust fund will have a positive balance for at least 40 years to come, so there's no need to be concerned about how the government will pay Social Security benefits.

Basic Skills & Concepts

15. **Personal Budget Basics.** Suppose your after-tax income is $34,000. Your annual expenses are $10,000 for rent, $5500 for food and household expenses, $1200 for interest on credit cards, and $8200 for entertainment, travel, and other.

 a. Do you have a surplus or a deficit? Explain.

 b. Next year, you expect to get a 3% raise. You think you can keep your expenses unchanged, with one exception: You plan to spend $8500 on a car. Explain the effect of this purchase on your budget.

 c. As in part b, assume you get a 3% raise for next year. If you can limit your expenses to a 1% increase (over the prior year), could you afford $7500 in tuition and fees without going into debt?

16. **Personal Budget Basics.** Suppose your after-tax income is $28,000. Your annual expenses are $8000 for rent, $4500 for food and household expenses, $1600 for interest on credit cards, and $10,400 for entertainment, travel, and other.

 a. Do you have a surplus or a deficit? Explain.

 b. Next year, you expect to get a 2% raise, but plan to keep your expenses unchanged. Will you be able pay off $5200 in credit card debt? Explain.

c. As in part b, assume you get a 2% raise for next year. If you can limit your expenses to a 1% increase, could you afford $3500 for a wedding and honeymoon without going into debt?

17. **Per Worker Debt.** Suppose the government decided to pay off the $7 trillion debt with a one-time charge distributed equally among all workers. Assuming the total U.S. work force is 150 million people, how much would each worker be charged?

18. **Per Family Debt.** Suppose the government decided to pay off the $7 trillion debt with a one-time charge distributed equally among all families. Assuming there are 100 million families in the United States, how much would each family be charged?

19. **The Wonderful Widget Company Future.** Extending the budget summary of the Widget Company (Table 4.11), assume that, for 2005, total receipts are $1,050,000, operating expenses are $600,000, employee benefits are $200,000, and security costs are $250,000.

 a. Based on the accumulated debt at the end of 2004, calculate the 2005 interest payment. Assume an interest rate of 8.2%.

 b. Calculate the total outlays for 2005, the year-end surplus or deficit, and the year-end accumulated debt.

 c. Based on the accumulated debt at the end of 2005, calculate the 2006 interest payment, again assuming an 8.2% interest rate.

 d. Assume that in 2006 the Widget Company has receipts of $1,100,000, holds operating costs and employee benefits to their 2005 levels, and spends no money on security. Calculate the total outlays for 2006, the year-end surplus or deficit, and the year-end accumulated debt.

 e. Imagine that you are the CFO (Chief Financial Officer) of the Wonderful Widget Company at the end of 2006. Write a three-paragraph statement to shareholders about the company's future prospects.

20. **The Wonderful Widget Company Future.** Extending the budget summary of the Widget Company (Table 4.11), assume that, for 2005, total receipts are $975,000, operating expenses are $850,000, employee benefits are $290,000, and security costs are $210,000.

 a. Based on the accumulated debt at the end of 2004, calculate the 2005 interest payment. Assume an interest rate of 8.2%.

 b. Calculate the total outlays for 2005, the year-end surplus or deficit, and the year-end accumulated debt.

 c. Based on the accumulated debt at the end of 2005, calculate the 2006 interest payment, again assuming an 8.2% interest rate.

 d. Assume that in 2006 the Widget Company has receipts of $1,050,000, holds operating costs and employee benefits to their 2005 levels, and spends no money on security. Calculate the total outlays for 2006, the year-end surplus or deficit, and the year-end accumulated debt.

 e. Imagine that you are the CFO (Chief Financial Officer) of the Wonderful Widget Company at the end of 2006. Write a three-paragraph statement to shareholders about the company's future prospects.

21. **Budget Projections.** Refer to the 2002 data in Table 4.12. How would the deficit have been affected by a 1% decrease in total receipts? How would it have been affected by a 0.5% increase in total outlays?

22. **Budget Projections.** Refer to the 2002 data in Table 4.12. How would the deficit have been affected by a 0.5% decrease in total receipts? How would it have been affected by a 1% increase in total outlays?

Budget Analysis. Consider the 2002 total receipts and outlays shown in Table 4.12. Based on Figures 4.11 and 4.12, answer the questions in Exercises 23–28.

23. How much income came from individual income taxes?

24. How much income came from social insurance taxes?

25. How much income came from excise taxes?

26. How much was spent on Social Security?

27. How much was spent on Medicare?

28. How much was spent on defense?

29. **On- and Off-Budget.** Suppose the government has a unified surplus of $50 billion, but was supposed to deposit $160 billion in the Social Security trust fund. What was the on-budget surplus or deficit? Explain.

30. **On- and Off-Budget.** Suppose the government has a unified deficit of $215 billion, but was supposed to deposit $195 billion in the Social Security trust fund. What was the on-budget deficit? Explain.

31. **Social Security Finances.** Suppose the year is 2020, and the government needs to pay out $350 billion more in Social Security benefits than it collects in Social Security taxes. Briefly discuss the options for finding this money.

32. Social Security Finances. Suppose the year is 2025, and the government needs to pay out $525 billion more in Social Security benefits than it collects in Social Security taxes. Briefly discuss the options for finding this money.

Further Applications

33. Counting the Federal Debt. Suppose you began counting the $7 trillion federal debt, $1 at a time. If you could count $1 each second, how long would it take to complete the count?

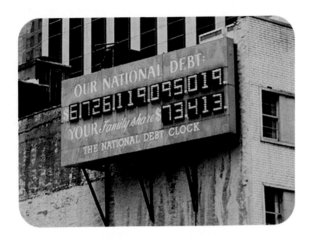

34. Paving with the Federal Debt. Suppose you began covering the ground with $1 bills. If you had the $7 trillion federal debt in $1 bills, how much total area could you cover? Compare this area to the total land area of the United States, which is about 10 million square kilometers. (Hint: Measure the length and width of a $1 bill in centimeters. Then compute its area in square centimeters and convert the area to square kilometers.)

35. Rising Debt. Suppose the federal debt rises at an annual rate of 1% per year. Use the compound interest formula to determine the size of the debt in 10 years and in 50 years. Assume that the current size of the debt (the principal for the compound interest formula) is $7 trillion.

36. Rising Debt. Suppose the federal debt rises at an annual rate of 2% per year. Use the compound interest formula to determine the size of the debt in 10 years and in 50 years. Assume that the current size of the debt (the principal for the compound interest formula) is $7 trillion.

37. Budget 2005. Consider the 2005 projection described in Example 4. What are the projected outlays? Suppose that

outlays turn out to be higher than projected by 5%, while receipts are lower by 5%. In that case, what is the 2005 surplus or deficit?

38. Budget 2005. Consider the 2005 projection described in Example 4. What are the projected outlays? Suppose that outlays turn out to be higher than projected by 10%, while receipts are lower by 11%. In that case, what is the 2005 surplus or deficit?

39. Retiring the Public Debt. Consider the publicly held debt of $4.0 trillion in 2003. Use the loan payment formula to determine the annual payments needed to pay this debt off in 10 years. Assume an annual interest rate of 4%.

40. Retiring the Public Debt. Consider the publicly held debt of $4.0 trillion in 2003. Use the loan payment formula to determine the annual payments needed to pay this debt off in 15 years. Assume an annual interest rate of 2%.

41. National Debt Lottery. Imagine that, through some political or economic miracle, the gross debt stopped rising. To retire the gross debt, the government decided to have a national lottery. Suppose that every U.S. citizen bought a $1 lottery ticket every week, thereby generating about $300 million in weekly lottery revenue. Because lotteries typically use half their revenue for prizes and lottery operations, half the $300 million, or $150 million, would go toward debt reduction each week. How long would it take to retire the debt through this lottery? Use the 2003 gross debt of $7 trillion.

42. National Debt Lottery. Suppose the government hopes to pay off the gross debt of $7 trillion with a national lottery. For the debt to be paid off in 50 years, how much would each citizen have to spend on lottery tickets each year? Assume that half of the lottery revenue goes toward debt reduction and that there are 300 million citizens.

Web Projects

Find useful links for Web Projects on the text Web site: www.aw.com/bennett-briggs

43. Federal Budget Deficit/Surplus. Use the Web to find the most recent projections of the federal deficit/surplus for the next 10 years.

44. Debt Problem. How serious of a problem is the gross debt? Use the Web to find arguments on both sides of this question. Summarize the arguments and state your own opinion.

45. **Social Security Problems.** Using information available on the Web, research the current status of the Social Security trust fund and potential future problems in paying out benefits. For example, when is the fund projected to start paying out more than it takes in each year? Write a one- to two-page report that summarizes your findings.

46. **Social Security Solutions.** Research various proposals for solving the problems with Social Security. Choose one proposal that you think is worthwhile and write a one- to two-page report summarizing it and describing why you think it is a good idea.

47. **Privatizing Social Security.** One proposal for saving the Social Security program is privatization—removing it from the government and running it like a for-profit business. Find an argument for and an argument against privatization of Social Security. Summarize each argument and discuss which case you think is stronger.

In the News

48. **Federal Budget.** Choose one of the many current news stories concerning federal finances. Summarize the story and the issues involved.

49. **Social Security.** Find a news article that concerns either the present or the future state of the Social Security system. Briefly summarize the article and interpret it in light of what you learned in this unit.

50. **Relying on Projections.** Find a news story in which Congress or the President is relying on projections several years into the future to make a budget today. Report on how the uncertainty in the projections is being dealt with, and discuss whether the decisions are being made wisely.

CHAPTER 4 SUMMARY

UNIT	KEY TERMS	KEY IDEAS AND SKILLS
4A	principal simple interest compound interest annual percentage rate (APR) annual percentage yield (APY) variable definitions: A, P, i, N, n, Y	**General form of the compound interest formula:** $A = P \times (1 + i)^N$ **Compound interest formula for interest paid once a year:** $A = P \times (1 + \text{APR})^Y$ **Compound interest formula for interest paid n times a year:** $A = P\left(1 + \dfrac{\text{APR}}{n}\right)^{(nY)}$ **Compound interest formula for continuous compounding:** $A = P \times e^{(\text{APR} \times Y)}$ **Know** when and how to apply these formulas.
4B	savings plan total return annual return mutual fund investment considerations liquidity risk return bond characteristics face value coupon rate maturity date current yield	**Savings plan formula:** $A = \text{PMT} \times \dfrac{\left[\left(1 + \dfrac{\text{APR}}{n}\right)^{(nY)} - 1\right]}{\left(\dfrac{\text{APR}}{n}\right)}$ **Return on investments:** total return $= \dfrac{(A - P)}{P}$ annual return $= \left(\dfrac{A}{P}\right)^{(1/Y)} - 1$ **Understand** investment types: stock, bond, cash. **Read** financial tables for stocks, bonds, and mutual funds. **Remember** important principles of investing, such as Higher returns usually involve higher risk. High commissions and fees can dramatically lower returns. Build an appropriately diversified portfolio.
4C	installment loan mortgages down payment closing costs points fixed rate mortgage adjustable rate mortgage	**Loan payment formula:** $\text{PMT} = \dfrac{P \times \left(\dfrac{\text{APR}}{n}\right)}{\left[1 - \left(1 + \dfrac{\text{APR}}{n}\right)^{(-nY)}\right]}$ **Understand** the uses and dangers of credit cards. **Understand** strategies for early payment of loans. **Understand** considerations in choosing a mortgage.

(Continues on the next page)

4D	gross income adjusted gross income exemptions, deductions, credits taxable income filing status progressive income tax marginal tax rates Social Security, FICA, self-employment tax capital gains	**Define** different types of income as they apply to taxes. **Use** tax rate tables to calculate taxes. **Distinguish between** tax credits and tax deductions. **Calculate** FICA taxes. **Be aware of** special tax rates for dividends and capital gains. **Understand** the benefits of tax-deferred savings plans.
4E	receipts, outlays net income surplus deficit debt mandatory outlays discretionary outlays publicly held debt gross debt on budget, off budget unified budget	**Distinguish between** a deficit and a debt. **Understand** basic principles of the federal budget. **Distinguish between** publicly held debt and gross debt. **Be familiar with** major issues concerning the future of Social Security.

Statistical thinking will one day be as necessary for efficient citizenship as the ability to read and write.

—H. G. WELLS

CHAPTER **5**

Statistical Reasoning

Is your drinking water safe? Do most people approve of the President's tax plan? How much is the cost of health care rising? These questions and thousands more like them can be answered only through statistical studies. Indeed, statistical information appears in the news every day, making the ability to understand and reason with statistics crucial to modern life. In this chapter, you will study several key concepts necessary to understanding the statistics that appear daily in the news.

FUNDAMENTALS OF STATISTICS

The subject of statistics plays a major role in modern society. It's used to determine whether a new drug is effective in treating cancer. It's involved when agricultural inspectors check the safety of the food supply. It's used in every opinion poll and survey. In business, it's used for market research. Sports statistics are part of daily conversation for millions of people. Indeed, you'll be hard-pressed to think of a topic that is not linked in some way to statistics.

But what *is* (or *are*) statistics? There are two answers, because the term *statistics* can be either singular or plural. When it is singular, *statistics* refers to the *science* of statistics. The science of statistics helps us collect, organize, and interpret *data,* which are numbers or other pieces of information about some topic. When it is plural, the word *statistics* refers to the data themselves, especially those that describe or summarize something. For example, if there are 30 students in your class and they range in age from 17 to 64, the numbers "30 students," "17 years," and "64 years" are statistics that describe your class.

TWO DEFINITIONS OF STATISTICS

- Statistics is the *science* of collecting, organizing, and interpreting data.
- Statistics are the *data* that describe or summarize something.

HOW STATISTICS WORKS

Statistical studies are conducted in many different ways and for many different purposes, but they all share a few characteristics. To get the basic ideas, consider the *Nielsen ratings,* which are used to estimate the numbers of people watching various television shows. These ratings are used, for example, to determine the most popular television show of the week.

Suppose the Nielsen ratings tell you that *Survivor* was last week's most popular show, with 22 million viewers. You probably know that no one actually counted all 22 million people. But you may be surprised to learn that the Nielsen ratings are based on the television-viewing habits of people in only 5000 homes. To understand how Nielsen can draw a conclusion about millions of Americans from only 5000 homes, we need to investigate the principles behind statistical research.

Nielsen's goal is to draw conclusions about the viewing habits of all Americans. In the language of statistics, we say that Nielsen is interested in the **population** of all Americans. The characteristics of this population that Nielsen seeks to learn—such as the number of people watching each television show—are called **population parameters.** Note that, although we usually think of a population as a group of people, in statistics a population can be any kind of group—people, animals, or things. For example, in a study of college costs, the population might be *all colleges and uni-*

versities, and the population parameters might include prices for tuition, fees, and housing.

Nielsen seeks to learn about the population of all Americans by studying a much smaller **sample** of Americans in depth. More specifically, Nielsen has devices (called "people meters") attached to televisions in 5000 homes, so the people who live in these homes make up the sample of Americans that Nielsen studies. The individual measurements that Nielsen collects from the sample, such as who is watching each show at each time, constitute the *raw data*. Nielsen then consolidates these raw data into a set of numbers that characterize the sample, such as the percentage of young male viewers watching *Survivor*. These numbers are called **sample statistics.**

➤ EXAMPLE 1 Population and Sample

For each of the following cases, describe the population, sample, population parameters, and sample statistics.

a. Agricultural inspectors for Jefferson County measure the levels of residue from three common pesticides on 25 ears of corn from each of the 104 corn-producing farms in the county.

b. Anthropologists determine the average brain size of early Neanderthals in Europe by studying skulls found at three sites in southern Europe.

Solution

a. The inspectors seek to learn about the population of all ears of corn grown in the county. They do this by studying a sample that consists of 25 ears from each farm. The population parameters are the average levels of residue from the three pesticides on all corn grown in the county. The sample statistics describe the average levels of residue that are actually measured on the corn in the sample.

b. The anthropologists seek to learn about the population of all early Neanderthals in Europe. Specifically, they seek to determine the average brain size of all Neanderthals, which is the population parameter in this case. The sample consists of the relatively few individual Neanderthals whose skulls are found at the three sites. The sample statistic is the average brain size (skull size) of the individuals in the sample. *Now try Exercises 15–20.* ◄

The Process of a Statistical Study

Because Nielsen does not study the entire population of all Americans, it cannot actually measure any population parameters. Instead, the company tries to infer reasonable values for population parameters from the sample statistics (which it *did* measure).

The process of inference is simple in principle, though it must be carried out with great care. For example, suppose Nielsen finds that 7% of the people in its sample watched *Survivor*. If this sample accurately represents the entire population of all Americans, then Nielsen can infer that *approximately* 7% of all Americans watched the show. In other words, the sample statistic of 7% is used as an estimate for the population parameter. (By using statistical techniques that we'll discuss in Unit 6D, Nielsen can also estimate the uncertainty in the inferred population parameters.)

Once Nielsen has estimates of the population parameters, it can draw general conclusions about what Americans were watching. The process used by Nielsen Media Research is similar to that used in many statistical studies. Figure 5.1 summarizes the general relationships among a population, a sample, the sample statistics, and the population parameters.

Basic Steps in a Statistical Study

1. State the goal of your study precisely. That is, determine the population you want to study and exactly what you'd like to learn about it.
2. Choose a representative sample from the population.
3. Collect raw data from the sample and summarize these data by finding sample statistics of interest.
4. Use the sample statistics to infer the population parameters.
5. Draw conclusions: Determine what you learned and whether you achieved your goal.

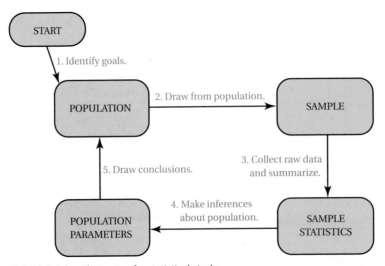

FIGURE 5.1 Elements of a statistical study.

► **EXAMPLE 2** Unemployment Survey

Each month, the U.S. Labor Department surveys 60,000 households to determine characteristics of the U.S. work force. One population parameter of interest is the U.S. unemployment rate, defined as the percentage of people who are unemployed among all those who are either employed or actively seeking employment. Describe how the five basic steps of a statistical study apply to this research.

Solution The steps apply as follows.

Step 1: The goal of the research is to learn about the employment (or unemployment) within the population of all Americans who are either employed or actively seeking employment.

Step 2: The Labor Department chooses a sample consisting of people employed or seeking employment in 60,000 households.

Step 3: The Labor Department asks questions of the people in the sample, and their responses constitute the raw data for the research. The Department then consolidates these data into sample statistics, such as the percentage of people in the sample who are unemployed.

Step 4: Based on the sample statistics, the Labor Department makes estimates of the corresponding population parameters, such as the unemployment rate for the entire United States.

Step 5: The Labor Department draws conclusions based on the population parameters and other information. For example, it might use the current and past unemployment rates to draw conclusions about whether jobs have been created or lost.

Now try Exercises 21–26. ◄

by the way

According to the Labor Department, someone who is not working is not necessarily unemployed. For example, stay-at-home moms and dads are not counted among the unemployed unless they are actively trying to find a job, and people who had been trying to find work but gave up in frustration are not counted as unemployed.

Choosing a Sample

Choosing a sample may be the most important step in any statistical study. If the sample fairly represents the population as a whole, then it's reasonable to make inferences from the sample to the population. But if the sample is *not* representative, then there's little hope of drawing accurate conclusions about the population.

Suppose you want to determine the average height and weight of students at a large university by measuring the heights and weights of a sample of 100 students. A sample consisting only of members of the football and basketball teams would not be reliable, because these athletes tend to be larger than most students. In contrast, suppose you select your sample with a computer program that randomly draws student numbers from the entire university population. In this case, the 100 students in your sample are likely to be representative of the entire student body. You can therefore expect that the average height and weight of students in the sample are reasonable estimates of the averages for all students.

DEFINITION

A **representative sample** is a sample in which the relevant characteristics of the sample members match those of the population.

Now try Exercises 27–28.

A sample drawn with a computer program that selects students at random is an example of a **simple random sample.** More technically, simple random sampling means that every sample of a particular size has the same chance of being selected. In the case of the student sample, every set of 100 students has an equal chance of being selected by the computer program.

Simple random sampling is usually the best way to choose a representative sample. However, it is not always practical or necessary, so other sampling techniques are sometimes used. The following box summarizes four of the most common sampling techniques, and Figure 5.2 illustrates the ideas.

Common Sampling Methods

Simple random sampling: We choose a sample of items in such a way that every sample of a given size has an equal chance of being selected.

Systematic sampling: We use a simple system to choose the sample, such as selecting every 10th or every 50th member of the population.

Convenience sampling: We use a sample that is convenient to select, such as people who happen to be in the same classroom.

Stratified sampling: We use this method when we are concerned about differences among subgroups, or *strata*, within a population. We first identify the subgroups and then draw a simple random sample within each subgroup. The total sample consists of all the samples from the individual subgroups.

Simple Random Sampling:
Every sample of the same size has an equal chance of being selected. Computers are often used to generate random telephone numbers.

Convenience Sampling:
Use results that are readily available.

Systematic Sampling:
Select every *k*th member.

Stratified Sampling:
Partition the population into at least two strata, then draw a sample from each.

FIGURE 5.2 Common sampling techniques.

Regardless of what type of sampling is used, always keep the following two key ideas in mind:

- No matter how a sample is chosen, the study can be successful only if the sample is representative of the population.
- Even if a sample is chosen in the best possible way, it is still just a sample (as opposed to the entire population). Thus, we can never be *sure* that a sample is representative of the population. In general, a larger sample is more likely to be representative of the population, as long as it is chosen well.

►EXAMPLE 3 Sampling Methods

Identify the type of sampling used in each of the following cases, and comment on whether the sample is likely to be representative of the population.

 a. You are conducting a survey of students in a dormitory. You choose your sample by knocking on the door of every 10th room.

 b. To survey opinions on a possible property tax increase, a research firm randomly draws the addresses of 150 homeowners from a public list of all homeowners.

 c. Agricultural inspectors for Jefferson County check the levels of residue from three common pesticides on 25 ears of corn from each of the 104 corn-producing farms in the county.

 d. Anthropologists determine the average brain size of early Neanderthals in Europe by studying skulls found at three sites in southern Europe.

Solution

 a. Choosing every 10th room makes this a systematic sample. The sample may be representative, as long as students were randomly assigned to rooms.

 b. The records presumably list all homeowners, so drawing randomly from this list produces a simple random sample. It has a good chance of being representative of the population.

 c. Each farm may have different pesticide use, so the inspectors consider corn from each farm as a subgroup (stratum) of the full population. By checking 25 ears of corn from each of the 104 farms, the inspectors are using stratified sampling. If the ears are collected randomly on each farm, each set of 25 is likely to be representative of its farm.

 d. By studying skulls found at selected sites, the anthropologists are using a convenience sample. They have little choice, because only a few skulls remain from the many Neanderthals who once lived in Europe. However, it seems reasonable to assume that these skulls are representative of the larger population.
 Now try Exercises 29–34. ◄

Watching Out for Bias

Consider a study designed to estimate the average weight of all men at a college. As we discussed earlier, a sample consisting only of football players would *not* be representative of the population with respect to weight. We say that this sample is *biased* because the men in the sample differ in a critical way from "typical" men at

by the way

Neanderthals lived between about 100,000 and 30,000 years ago in Eurasia and northern Africa. They were physiologically distinct from modern humans, but scientists are not yet sure whether they represented a separate species or could interbreed with Homo sapiens. Neanderthals developed many aspects of culture, including caring for the sick and burying their dead. Skull measurements suggest that Neanderthals had larger brains than modern humans.

the college. More generally, the term **bias** refers to any problem in the design or conduct of a statistical study that tends to favor certain results.

Besides occurring in a poorly chosen sample, bias can arise in many other ways. For example, a researcher may be biased if he or she has a personal stake in the outcome of the study. In that case, the researcher might distort (intentionally or unintentionally) the true meaning of the data. You should always be on the lookout for any type of bias that may affect the results or interpretation of a statistical study. We'll discuss sources of bias further in Unit 5B.

DEFINITION

A statistical study suffers from **bias** if its design or conduct tends to favor certain results.

Time out to think _____

Thinking about issues of bias, explain why television networks use Nielsen to measure ratings rather than doing it themselves.

TYPES OF STATISTICAL STUDY

Broadly speaking, most statistical studies fall into one of two categories: observational studies and experiments. Nielsen's studies of television viewing are **observational** because they are designed to *observe* the television-viewing behavior of the people in its 5000 sample homes. Note that observational studies may still involve some interaction. For example, an opinion poll is observational, even though researchers may conduct in-depth interviews, because the poll's goal is to learn (observe) people's opinions, not to change them. Similarly, a study in which individuals in the sample are weighed is also observational, because the measurement process records (observes) but does not change a person's weight.

In contrast, consider a medical study designed to test whether large doses of vitamin C can help prevent colds. To conduct this study, the researchers must ask some people in the sample to take large doses of vitamin C. This type of statistical study is called an **experiment,** because some participants receive a treatment (in this case, vitamin C) that they would not otherwise receive.

Two Basic Types of Statistical Study

1. In an **observational study,** researchers observe or measure characteristics of the sample members but do not attempt to influence or modify these characteristics.
2. In an **experiment,** researchers apply a treatment to some or all of the sample members and then look to see whether the treatment has any effects.

It is difficult to determine whether an experimental treatment works unless you compare groups that receive the treatment to groups that don't. In the vitamin C study, for example, researchers might create two groups of people: a **treatment**

group that takes large doses of vitamin C and a **control group** that does not take vitamin C. The researchers can then look for differences in the numbers of colds among people in the two groups. Having a control group is usually crucial to interpreting the results of experiments.

In an experiment, it is very important for the treatment and control groups to be alike in all respects except for the treatment. For example, if the treatment group consisted of active people with good diets and the control group consisted of sedentary people with poor diets, we could not attribute any differences in colds to vitamin C alone. To avoid this type of problem, assignments to the control and treatment groups must be done randomly.

With proper treatment, a cold can be cured in a week. Left to itself, it may linger for seven days.
—A MEDICAL FOLK SAYING

Treatment and Control Groups

The **treatment group** in an experiment is the group of sample members who receive the treatment being tested.

The **control group** in an experiment is the group of sample members who do *not* receive the treatment being tested.

It is important for the treatment and control groups to be selected randomly and to be alike in all respects except for the treatment.

The Placebo Effect and Blinding

For experiments involving people, using a treatment and a control group might not be enough to get reliable results. The problem is that people can be affected by their beliefs as well as by real treatments. For example, stress and other psychological factors have been shown to affect resistance to colds. If people taking vitamin C get fewer colds than people who don't, we can't conclude that the vitamin C was responsible. It might be that people stayed healthier because they believed that vitamin C works. Therefore, people in the control group should be given a **placebo**—in this case, pills that look like vitamin C pills but don't actually contain vitamin C. As long as the participants don't know whether they are in the treatment or control group (that is, whether they got the real pills or the placebo), any effect arising from psychological factors—known as a **placebo effect**—should affect both groups equally. Then, if people in the vitamin C group get fewer colds than people in the control group, we have evidence that vitamin C really works.

by the way

The placebo effect can be surprisingly powerful. Consider a drug now used to combat balding, which was tested on balding men. The drug maker was pleased to learn that 86% of the men receiving the drug either stopped balding or grew new hair. But remarkably, so did 42% of the men who received the placebo! In other studies, as many as 75% of participants receiving a placebo have actually improved.

DEFINITIONS

A **placebo** lacks the active ingredients of a treatment being tested in a study, but is identical in appearance to the treatment. Thus, study participants cannot distinguish the placebo from the real treatment.

The **placebo effect** refers to the situation in which patients improve simply because they believe they are receiving a useful treatment.

In statistical terminology, the practice of keeping people in the dark about who is in the treatment group and who is in the control group is called **blinding.** A **single-blind** experiment is one in which the participants don't know which group they belong to, but the experimenters (the people administering the treatment) do know. Using a placebo is one way to create a single-blind experiment. Sometimes, a single-blind experiment can still be unreliable if the experimenters can subtly influence outcomes. For example, in an experiment that involves interviews, the experimenters might speak differently to people who received the real treatment than to those who received the placebo. This type of problem can be avoided by making the experiment **double-blind,** which means neither the participants nor the experimenters know who belongs to each group. (Of course, someone must keep track of the two groups in order to evaluate the results at the end. In typical double-blind experiments, researchers hire experimenters to make any necessary contact with the participants.)

Blinding in Experiments

An experiment is **single-blind** if the participants do not know whether they are members of the treatment group or members of the control group, but the experimenters do know.

An experiment is **double-blind** if neither the participants nor the experimenters (people administering the treatment) know who belongs to the treatment group and who belongs to the control group.

➤ **EXAMPLE 4** What's Wrong with This Experiment?

For each of the experiments described below, identify any problems and explain how the problems could have been avoided.

 a. A chiropractor wants to know if his adjustments relieve back pain. He performs adjustments on 25 patients with back pain. Afterward, 18 of the patients say they feel better. He concludes that the adjustments are an effective treatment.

 b. A new drug for attention deficit disorder (ADD) is supposed to make the affected children more polite. Randomly selected children suffering from ADD are divided into treatment and control groups. Those in the control group receive a placebo that looks just like the real drug. The experiment is single-blind. Experimenters interview the children one-on-one to decide whether they became more polite.

Solution

 a. The 25 patients who receive adjustments represent a treatment group, but this study lacks a control group. The patients may be feeling better because of a placebo effect rather than any real effect of the adjustments. The chiropractor might have improved his study by hiring an actor to do a fake adjustment (one that feels like a real manipulation, but doesn't actually conform to

DILBERT reprinted by permission of United Feature Syndicate, Inc.

chiropractic guidelines) on a control group. Then he could have compared the results in the two groups to see whether a placebo effect was involved.

b. Because the experimenters know which children received the real drug, during the interviews they may inadvertently speak differently or interpret behavior differently with these children. In that case, their conclusions might not be valid. The experiment should have been double-blind, so that the experimenters conducting the interviews would not have known which children received the real drug and which children received the placebo.

Now try Exercises 35–40. ◀

Case-Control Studies

Sometimes it may be impractical or unethical to conduct an experiment. For example, suppose we want to study how alcohol consumed during pregnancy affects newborn babies. Because it is already known that alcohol can be harmful during pregnancy, it would be unethical to divide a sample of pregnant mothers randomly into two groups and then force the members of one group to consume alcohol. However, we may be able to conduct a **case-control study,** in which the participants naturally form groups by choice. In this example, the **cases** consist of mothers who consume alcohol during pregnancy *by choice,* and the **controls** consist of mothers who choose *not* to consume alcohol.

A case control study is observational because the researchers do not change the behavior of the participants. But it also resembles an experiment because the cases effectively represent a treatment group and the controls represent a control group.

DEFINITIONS

A **case-control study** is an observational study that resembles an experiment because the sample naturally divides into two (or more) groups. The participants who engage in the behavior under study form the **cases,** which makes them like a treatment group in an experiment. The participants who do not engage in the behavior are the **controls,** making them like a control group in an experiment.

➤ **E X A M P L E 5** Which Type of Study?

For each of the following questions, what type of statistical study is most likely to lead to an answer? Why?

 a. What is the average income of stock brokers?
 b. Do seat belts save lives?
 c. Can lifting weights improve runners' times in a 10-kilometer race?
 d. Can a new herbal remedy reduce the severity of colds?

Solution

 a. An *observational study* can tell us the average income of stock brokers. We need only survey (observe) the brokers.

 b. It would be unethical to do an experiment in which some people were told to wear seat belts and others were told *not* to wear them. Instead, we can conduct an observational *case-control study.* Some people choose to wear seat belts (the cases) and others choose not to wear them (the controls). By comparing the death rates in accidents between cases and controls, we can learn whether seat belts save lives. (They do.)

 c. We need an *experiment* to determine whether lifting weights can improve runners' 10K times. One group of runners will be put on a weight-lifting program, and a control group will be asked to stay away from weights. We must try to ensure that all other aspects of their training are similar. Then we can see whether the runners in the lifting group improve their times more than those in the control group. Note that we cannot use blinding in this experiment because there is no way to prevent participants from knowing whether they are lifting weights.

 d. We should use a *double-blind experiment,* in which some participants get the actual remedy while others get a placebo. We need double-blind conditions because the severity of a cold may be affected by mood or other factors that experimenters might inadvertently influence.

Now try Exercises 41–46. ◀

by the way

Politicians and marketers often pretend they are trying to conduct a true opinion poll or survey when, in fact, they are deliberately trying to get particular results. These types of surveys are called *push polls* because they try to "push" people's opinions.

SURVEYS AND OPINION POLLS

Surveys and opinion polls may be the most common types of statistical study, and we must be very careful when we interpret them. Fortunately, survey and poll results usually include something called the **margin of error.**

 Suppose a poll finds that 76% of the public supports the President, with a margin of error of 3 percentage points. The 76% is a sample statistic; that is, 76% of the people in a sample said they support the President. The margin of error helps us understand how well this sample statistic is likely to approximate the true population parameter (in this case, the percentage of *all* Americans who support the President). By adding and subtracting the margin of error from the sample statistic, we find a range of values, or a **confidence interval,** *likely* to contain the population parameter. In this case, we add and subtract 3 percentage points to find a confidence interval from 73% to 79%.

DEFINITION

The **margin of error** in a statistical study is used to describe a **confidence interval** that is likely to contain the true population parameter. We find this interval by subtracting and adding the margin of error from the sample statistic obtained in the study. That is, the confidence interval is

from (sample statistic − margin of error)
to (sample statistic + margin of error)

How confident can we be in a poll result? Unless we are told otherwise, we assume that the margin of error is defined to give us 95% confidence that the confidence interval contains the population parameter. We'll discuss the precise meaning of "95% confidence" in Unit 6D, but for now you can think of it as follows: If the poll were repeated 20 times with 20 different samples, 19 of the 20 polls (that is, 95% of the polls) would have a confidence interval that contains the true population parameter.

►**EXAMPLE 6** Close Election

An election eve poll finds that 52% of surveyed voters plan to vote for Smith, and she needs a majority (more than 50%) to win without a runoff. The margin of error in the poll is 3 percentage points. Will she win?

Solution We subtract and add the margin of error of 3 percentage points to find a confidence interval

from 52% − 3% = 49% to 52% + 3% = 55%

We can be 95% confident that the actual percentage of people planning to vote for her is between 49% and 55%. Because this confidence interval leaves open the possibility of both a majority and less than a majority, this election is too close to call.

Now try Exercises 47–50. ◄

Time out to think _____

In Example 6, suppose the poll found the candidate had 55% of the vote. Should she be confident of a win?

■ EXERCISES 5A

Review Questions

1. Why do we say that the term *statistics* has two meanings? Describe both meanings.

2. Define the terms *population, sample, population parameter,* and *sample statistics* as they apply to statistical studies.

3. Describe the five basic steps in a statistical study, and give an example of their application.

4. Why is it so important that a statistical study use a representative sample? Briefly describe four common sampling methods.

5. What is bias? How can it affect a statistical study? Give examples of several forms of bias.

6. Describe and contrast observational studies and experiments. What do we mean by the treatment group and control group in an experiment? What do we mean by the cases and controls in an observational case-control study?

7. What is a placebo? Describe the placebo effect and how it can make experiments difficult to interpret. How can making an experiment single-blind or double-blind help?

8. What is meant by the margin of error in a survey or opinion poll? How is it used to identify a confidence interval?

Does It Make Sense?

Decide whether each of the following statements makes sense (or is clearly true) or does not make sense (or is clearly false). Explain your reasoning.

9. In my experimental study, I used a sample that was larger than the population.

10. I followed all the guidelines for sample selection carefully, yet my sample still did not reflect the characteristics of the population.

11. I wanted to test the effects of vitamin C on colds, so I gave the treatment group vitamin C and gave the control group vitamin D.

12. I don't believe the results of the experiment, because the results were based on interviews but the study was not double-blind.

13. The pre-election poll found that Kennedy would get 58% of the vote, with a margin of error of 4%, but he ended up losing the election.

14. By choosing my sample carefully, I can make a good estimate of the average height of Americans by measuring the heights of only 500 people.

Basic Skills & Concepts

Population and Sample. For the studies described in Exercises 15–20, describe the population, sample, population parameters, and sample statistics.

15. In order to gauge public opinion about the candidates in a recent election for Governor of California, CBS News conducted telephone interviews with 1026 people who were registered to vote in California.

16. Astronomers typically determine the distance to a galaxy (a galaxy is a huge collection of billions of stars) by measuring the distances to just a few stars within it and taking the mean (average) of these distance measurements.

17. To rate the speed of new computers, *Byte* magazine editors purchased five new computers and measured the time required by each to run a standard program.

18. The Gallup Organization conducted a poll of 1200 adults to determine how Americans respond to the question "Do you think college coaches use physical force with their athletes?"

19. A nationwide Harris poll of 1010 adults surveyed by telephone revealed that the professions viewed by the public as most prestigious are doctor, scientist, teacher, and minister/clergy.

20. The American Institute of Education conducts an annual study of attitudes of incoming college students by surveying approximately 261,000 first-year students at 462 colleges and universities. There are approximately 1.6 million first-year college students in this country.

Steps in a Study. Describe how you would apply the five basic steps of a statistical study to the issues in Exercises 21–26.

21. You want to determine the average amount of pizza consumed each year by students at your school.

22. As an airline marketing executive, you want to know if there has been an increase in frustration with air travel among business travelers.

23. You want to know the mean (average) height of adult American women.

24. You want to know the typical percentage of the bill that is left as a tip in restaurants.

25. You want to know how long alkaline batteries last in digital cameras.

26. You want to know the percentage of high school students who are vegetarians.

27. **Representative Sample?** You want to determine the mean (average) number of calories eaten daily by girls on high school track teams. Which of the following samples is most likely to be representative, and why? Also explain why each of the other choices is *not* likely to make a representative sample for this study.

 • The seniors on your high school's girls' track team

 • The shot putters and javelin throwers on your high school's girls' track team

 • The sprinters on your high school's girls' track team

 • The entire girls' track team at your high school

28. **Representative Sample?** You want to determine the typical dietary habits of students at a college. Which of the following would make the best sample, and why? Also explain why each of the other choices would *not* make a good sample for this study.

 • Students in a single dormitory

 • Students majoring in public health

 • Students who participate in intercollegiate sports

 • Students enrolled in a required mathematics class

Identify the Sampling Method. Exercises 29–34 each describe a sample. Identify the sampling method as simple random sampling, systematic sampling, convenience sampling, or stratified sampling. Briefly explain why you think this sampling method was chosen.

29. An IRS (Internal Revenue Service) auditor randomly selects for audits 30 taxpayers whose gross income is less than $30,000 and 30 taxpayers whose gross income is more than $30,000.

30. *People* magazine chooses its "best-dressed celebrities" by looking at responses from readers who choose to mail in a survey printed in the magazine.

31. A study of sleep habits uses 50 participants whose ages are between 20 and 29, 50 participants whose ages are between 30 and 39, and 50 participants whose ages are between 40 and 49.

32. Every 100th roll of rope that is produced is given a strength test.

33. A computer randomly selects 400 names from a list of all registered voters. Those selected are surveyed to predict who will win the election for Mayor.

34. At a football game, a computer randomly selects 50 ticket numbers. Those whose tickets are selected are invited to participate in a contest.

Type of Study. For Exercises 35–40, state whether the study is an observational study or an experiment. If it is an experiment, describe the treatment and control groups and discuss whether single- or double-blinding is needed. If it is observational, state whether it is a case-control study and, if it is, distinguish between the cases and the controls.

35. A Michigan State University study of starting salaries for college graduates found that computer science majors had the highest salaries ($42,500), while communications majors had the lowest ($25,600).

36. A National Cancer Institute study of 716 melanoma patients and 1014 cancer-free patients matched by age, sex, and race found that those having a single large mole had twice the risk of melanoma. Having 10 or more moles was associated with a 12 times greater risk of melanoma (*Journal of the American Medical Association*).

37. A Gallup survey done for CNN/*USA Today* found that 80% of Americans think we're less civil than we were ten years ago and 67% think that we are more likely to use vulgar language than we were ten years ago.

38. A breast cancer study began by asking 25,624 women questions about how they spent their leisure time. The health of these women was tracked over the next 15 years. Those women who said they exercise regularly were found to have a lower incidence of breast cancer (*New England Journal of Medicine*).

39. A (hypothetical) study found that tomato plants treated with a new fertilizer had a 23% greater yield than plants with no fertilizer.

40. A survey of 275,811 first-year college students revealed that 32.4% of these students had an A average in high school (Higher Education Research Institute).

Which Type of Study? For each of the questions in Exercises 41–46, what type of statistical study is most likely to lead to an answer? Why?

41. What is the average income of public school teachers?

42. Can listening to jazz music while studying improve students' grades?

43. Are high levels of artificial flavoring harmful to humans when eaten in foods?

44. Does playing soccer help swimmers improve their times?

45. Does an aspirin a day reduce the incidence of heart attacks?

46. Does a self-proclaimed mind-reader really have supernatural abilities?

Margin of Error. Each of Exercises 47–50 states both a sample statistic and a margin of error. Find the confidence interval in each case, and answer any additional questions asked. Be sure to explain your answers clearly.

47. A poll is conducted the day before a state election for Senator. There are only two candidates running. The poll shows that 53% of the voters surveyed favor the Republican candidate, with a margin of error of 2.5 percentage points. Should the Republican plan a victory party? Why or why not?

48. A poll is conducted the day before an election for U.S. Representative. There are only two candidates running. The poll shows that 48.5% of the voters surveyed favor the Democratic candidate, with a margin of error of 2.0 percentage points. Based on this poll, should the Democratic candidate expect to lose the election? Why or why not?

49. The U.S. Labor Department reports that the January unemployment rate was 4.2% and the February unemployment rate was 4.3%. The margin of error for each report is 0.2 percentage point. Can we conclude that unemployment increased from January to February?

50. In a survey of 1002 people, 701 (which is 70%) said that they voted in the most recent presidential election (based on data from ICR Research Group). The margin of error for the survey was 3 percentage points. However, actual voting records show that only 61% of all eligible voters actually did vote. Does this necessarily imply that people lied when they answered the survey?

Further Applications

Experiment Results. Consider an experiment designed to determine the effectiveness of a new drug. The drug is given to participants in the treatment group, while participants in the control group receive a placebo. For each set of results described in Exercises 51–54, discuss whether there appears to be evidence that the treatment is effective.

51. 60% of those in the treatment group showed improvement; 20% of those in the placebo group showed improvement.

52. 40% of those in the treatment group showed improvement; 40% of those in the placebo group showed improvement.

53. 90% of those in the treatment group showed improvement; 50% of those in the placebo group showed improvement.

54. 30% of those in the treatment group showed improvement; 40% of those in the placebo group showed improvement.

Interpreting Real Studies. For each of Exercises 55–60, do the following:

 a. Identify the population and the population parameter of interest.

 b. Briefly describe the sample and sample statistic for the study.

 c. Find the confidence interval likely to contain the population parameter of interest.

55. In a TIME/CNN poll, 748 adults were asked whether they believed their children would have a higher standard of living than they have; 63% of those polled said "yes." The margin of error was 3.6 percentage points.

56. A Yankelovich poll determined that 70% of 4000 people surveyed agreed with the statement "People have to realize that they can only count on their own skills and abilities if they're going to win in this world." The margin of error in the poll was 1.6 percentage points.

57. Based on its survey of 60,000 households (see Example 2), the U.S. Labor Department reported an unemployment rate of 6.4% in June 2003. The margin of error for the report was 0.2 percentage point.

58. The Pew Research Center asked 1546 adult Americans whether humans would land on Mars within the next 50 years; 76% of these people said either definitely yes or probably yes. The margin of error for the poll was 3 percentage points.

59. The Pew Research Center asked 1546 adult Americans whether they believed that humans would be cloned within the next 50 years; 51% answered either definitely or probably yes. The margin of error was 3 percentage points.

60. A Roper Organization survey of 2000 adults revealed that 64% of those surveyed kept money in a regular savings account. The margin of error for the survey was 2.0 percentage points.

Web Projects

Find useful links for Web Projects on the text Web site: www.aw.com/bennett-briggs

61. Current Nielsen Ratings. Find the Nielsen ratings for the past week. What were the three most popular television shows? Explain both the "rating" and the "share" for each show.

62. Nielsen Sample. Use information available on the Nielsen Media Research Web site to answer each of the following questions.

 a. How does Nielsen select the sample of homes to be included in a viewer survey?

 b. Describe a few ways by which Nielsen attempts to check that the results from its people meter surveys are accurate.

 c. Based on what you have learned, do you think the Nielsen ratings are reliable? If so, why? If not, why not?

63. Attitude Update. The Pew Research Center for the People and the Press studies public attitudes toward the press, politics, and public policy issues. Go to its Web site and find the latest survey about attitudes. Write a one-page summary of what Pew surveyed, how it conducted the survey, and what it found.

64. Labor Statistics. Use the Bureau of Labor Statistics Web page to learn about its monthly survey. Choose one aspect of the survey, such as how the sample is chosen or how it is used to compare unemployment rates over time. Write a short summary of what you learn.

65. Professional Polling. Visit the Web site of a national polling organization and report on a recent poll. Write a short description of the poll and its results, commenting on features such as sampling technique, sample size, and margin of error.

In the News

66. Statistics in the News. Select three news stories from the past week that involve statistics in some way. In each case, write one or two paragraphs describing the role of statistics in the story.

67. Statistics in Your Major. Write two to three paragraphs describing the ways in which you think the science of statistics is important in your major field of study. (If you have not chosen a major, answer this question for a major that you are considering.)

68. Statistics in Sports. Choose a sport and describe three different statistics commonly tracked by participants in or spectators of the sport. In each case, briefly describe the importance of the statistic to the sport.

69. Sample and Population. Find a report in today's news concerning any type of statistical study. What is the population being studied? What is the sample? Why do you think the sample was chosen as it was?

70. Poor Sampling. In a recent newspaper or magazine, find an article about a study that attempts to describe some characteristic of a population, but that you believe involved poor sampling (for example, a sample that was too small or unrepresentative of the population under study). Describe the population, the sample, and what you think was wrong with the sample. Briefly discuss how you think the poor sampling affected the study results.

71. Good Sampling. In a recent newspaper or magazine, find an article that describes a statistical study in which the sample was well chosen. Describe the population, the sample, and why you think the sample was a good one.

72. Margin of Error. Find a report of a recent survey or poll. Interpret the sample statistic and margin of error quoted for the survey or poll.

SHOULD YOU BELIEVE A STATISTICAL STUDY?

Most statistical research is carried out with integrity and care. Nevertheless, statistical research is sufficiently complex that bias can arise in many different ways. Thus, we must always examine reports of statistical research carefully, looking for anything that might make us question the results. In this unit, we discuss eight guidelines that can help you answer the question "Should I believe a statistical study?"

GUIDELINE 1: IDENTIFY THE GOAL, POPULATION, AND TYPE OF STUDY

Before evaluating the details of a statistical study, we must know what it is about. Based on what you hear or read, try to answer basic questions such as these:

- What was the goal of the study?
- What was the population under study? Was the population clearly and appropriately defined?
- What type of study was used? Was the type appropriate for the goal?

If you can't find reasonable answers to these questions, it will be difficult to evaluate other aspects of the study.

▶ **EXAMPLE 1 Appropriate Type of Study?**

A newspaper reports: "Researchers gave each of the 100 participants their astrological horoscopes, and asked them whether the horoscopes appeared to be accurate. Eighty-five percent of the participants reported that the horoscopes were accurate. The researchers concluded that horoscopes are valid most of the time." Analyze this study according to Guideline 1.

Solution The goal of the study was to determine the validity of horoscopes. Based on the news report, it appears that the study was *observational:* The researchers simply asked the participants about the accuracy of the horoscopes. However, because the accuracy of a horoscope is somewhat subjective, this study should have been a controlled experiment in which some people were given their actual horoscope and others were given a fake horoscope. Then the researchers could have looked for differences between the two groups. Moreover, because researchers could easily influence the results by how they questioned the participants, the experiment should have been double-blind. In summary, the type of study was inappropriate to the goal and its results are meaningless.

Now try Exercises 9–10. ◀

by the way

Surveys show that nearly half of Americans believe their horoscopes. However, in controlled experiments, the predictions of horoscopes come true no more often than would be expected by chance.

Copyright © 1998, 2004 by Sidney Harris.

GUIDELINE 2: CONSIDER THE SOURCE

Statistical studies are supposed to be objective, but the people who carry them out and fund them may be biased. Thus, it is important to consider the source of a study and evaluate the potential for biases that might invalidate its conclusions.

> **EXAMPLE 2** Is Smoking Healthy?

By 1963, enough research on the health dangers of smoking had accumulated that the Surgeon General of the United States publicly announced that smoking is bad for health. Research done since that time has built further support for this claim. However, while the vast majority of studies show that smoking is unhealthy, a few studies found no dangers from smoking, and perhaps even health *benefits*. These studies generally were carried out by the Tobacco Research Institute, funded by the tobacco companies. Analyze the Tobacco Research Institute studies according to Guideline 2.

Solution Tobacco companies had a financial interest in minimizing the dangers of smoking. Because the studies carried out at the Tobacco Research Institute were funded by the tobacco companies, there may have been pressure on the researchers to produce results to the companies' liking. This *potential* for bias does not mean their research *was* biased, but the fact that it contradicts virtually all other research on the subject should be cause for concern. *Now try Exercises 11–12.* ◀

GUIDELINE 3: LOOK FOR BIAS IN THE SAMPLE

Look for bias that may prevent the sample from being representative of the population. There are two particularly common forms of bias that can affect sample selection.

by the way

After decades of arguing to the contrary, in 1999 the Philip Morris Company—the world's largest seller of tobacco products—publicly acknowledged that smoking causes lung cancer, heart disease, emphysema, and other serious diseases. Shortly thereafter, Philip Morris changed its name to Altria.

Bias in Choosing a Sample

Selection bias occurs whenever researchers *select* their sample in a way that tends to make it unrepresentative of the population. For example, a pre-election poll that surveys only registered Republicans has selection bias because it is unlikely to reflect the opinions of all voters.

Participation bias occurs primarily with surveys and polls; it arises whenever people *choose* whether to participate. Because people who feel strongly about an issue are more likely to participate, their opinions may not represent the larger population that is less emotionally attached to the issue.

CASE STUDY The 1936 Literary Digest Poll

The *Literary Digest,* a popular magazine of the 1930s, successfully predicted the outcomes of several elections using large polls. In 1936, editors of the *Literary Digest* conducted a particularly large poll in advance of the presidential election. They randomly chose a sample of 10 million people from various lists, including names in telephone books and rosters of country clubs. They mailed a postcard "ballot" to each of these 10 million people. About 2.4 million people returned the postcard ballots. Based on the returned ballots, the editors of the *Literary Digest* predicted that Alf Landon would win the presidency by a margin of 57% to 43% over Franklin Roosevelt. Instead, Roosevelt won with 62% of the popular vote. How did such a large survey go so wrong?

The sample suffered from both selection bias and participation bias. The selection bias arose because the *Literary Digest* chose its 10 million names in ways that favored affluent people. For example, selecting names from telephone books meant choosing only from those who could afford telephones back in 1936. Similarly, country club members are usually quite wealthy. The selection bias favored Landon because he was the Republican, and affluent voters of the 1930s tended to vote for Republican candidates.

The Franklin Delano Roosevelt Memorial in Washington D.C.

The participation bias arose because return of the postcard ballots was voluntary. People who felt most strongly about the election were more likely to be among those who returned their postcard ballots. This bias also tended to favor Landon because he was the challenger—people who did not like President Roosevelt could express their desire for change by returning the postcards. Together, the two forms of bias made the sample results useless, despite the large number of people surveyed.

➤ **EXAMPLE 3** Self-Selected Poll

The television show *Nightline* conducted a poll in which viewers were asked whether the United Nations headquarters should be kept in the United States. Viewers could respond to the poll by paying 50 cents to call a "900" phone number with their opinions. The poll drew 186,000 responses, of which 67% favored moving the United Nations out of the United States. Around the same time, a poll using simple random sampling of 500 people found that 72% wanted the United Nations to *stay* in the United States. Which poll is more likely to be representative of the general opinions of Americans?

Solution The *Nightline* sample suffered from severe participation bias. Not only did viewers choose whether to call in for the survey, but they had to *pay* to participate. This cost made it even more likely that respondents would be those who felt a need for change. Thus, despite its large number of respondents, the *Nightline* survey was too biased to be trusted. In contrast, a simple random sample of 500 people is quite likely to be representative, so the finding of this small survey has a better chance of representing the true opinions of all Americans. *Now try Exercises 13–14.* ◄

GUIDELINE 4: LOOK FOR PROBLEMS IN DEFINING OR MEASURING THE VARIABLES OF INTEREST

Statistical studies usually attempt to measure *something*, and we call the things being measured the **variables** of interest in the study. The term *variable* simply refers to an item or quantity that can vary or take on different values. For example, variables in the Nielsen ratings include *show being watched* and *number of viewers*.

DEFINITION

A **variable** is any item or quantity that can vary or take on different values. The *variables of interest* in a statistical study are the items or quantities that the study seeks to measure.

Results of a statistical study may be especially difficult to interpret if the variables under study are difficult to define or measure. For example, imagine trying to conduct a study of how exercise affects resting heart rates. The variables of interest would be *amount of exercise* and *resting heart rate*. However, both variables are difficult to define and measure. In the case of *amount of exercise*, it's not clear what the definition covers: Does it include walking to class? Even if we specify the definition, how can we measure *amount of exercise* given that some forms of exercise are more vigorous than others? The following two examples describe real cases in which defining or measuring variables caused problems in statistical studies.

by the way

More than a third of all Americans routinely shut the door or hang up the phone when contacted for a survey, thereby making self-selection a problem for legitimate pollsters. One reason people hang up may be the proliferation of selling under the guise of market research (often called "sugging"), in which a telemarketer pretends you are part of a survey in order to get you to buy something.

Time out to think _____

How would you measure your resting heart rate? Describe some difficulties in defining and measuring *resting heart rate.*

►**EXAMPLE 4** Can Money Buy Love?

A Roper poll reported in *USA Today* involved a survey of the wealthiest 1% of Americans. The survey found that these people would pay an average of $487,000 for "true love," $407,000 for "great intellect," $285,000 for "talent," and $259,000 for "eternal youth." Analyze this result according to Guideline 4.

Solution The variables in this study are very difficult to define. How, for example, do you define "true love"? And does it mean true love for a day, a lifetime, or something else? Similarly, does the ability to balance a spoon on your nose constitute "talent"? Because the variables are so poorly defined, it's likely that different people interpreted them differently, making the results very difficult to interpret. *Now try Exercise 15.* ◄

►**EXAMPLE 5** Illegal Drug Supply

Law enforcement authorities try to stop illegal drugs from entering the country. A commonly quoted statistic is that they succeed in stopping only about 10% to 20% of the drugs entering the United States. Should you believe this statistic?

Solution There are essentially two variables in the study: *quantity of illegal drugs intercepted* and *quantity of illegal drugs NOT intercepted.* It should be relatively easy to measure the quantity of illegal drugs that law enforcement officials intercept. However, because the drugs are illegal, it's unlikely that anyone is reporting the quantity of drugs that are *not* intercepted. How, then, can anyone know that the intercepted drugs are 10% to 20% of the total? In a *New York Times* analysis, a police officer was quoted as saying that his colleagues refer to this type of statistic as "P.F.A.," for "pulled from the air." *Now try Exercise 16.* ◄

GUIDELINE 5: WATCH OUT FOR CONFOUNDING VARIABLES

Variables that are *not intended* to be part of the study can sometimes make it difficult to interpret results properly. Such variables are often called *confounding* variables, because they confound (confuse) a study's results.

It's not always easy to discover confounding variables. Sometimes they are discovered years after a study was completed, and sometimes they are not discovered at all. Fortunately, confounding variables are sometimes more obvious and can be discovered simply by thinking hard about factors that may have influenced a study's results.

►**EXAMPLE 6** Radon and Lung Cancer

Radon is a radioactive gas produced by natural processes (the decay of uranium) in the ground. The gas can leach into buildings through the foundation and can accumulate in relatively high concentrations if doors and windows are closed. Imagine a

by the way

Many hardware stores sell simple kits that you can use to test whether radon gas is accumulating in your home. If it is, the problem can be eliminated by installing an appropriate "radon mitigation" system, which usually consists of a fan that blows the radon out from under the house before it can get in.

study that seeks to determine whether radon gas causes lung cancer by comparing the lung cancer rate in Colorado, where radon gas is fairly common, with the lung cancer rate in Hong Kong, where radon gas is less common. Suppose the study finds that the lung cancer rates are nearly the same. Is it fair to conclude that radon is *not* a significant cause of lung cancer?

Solution The variables under study are *amount of radon* and *lung cancer rate*. However, because smoking can also cause lung cancer, *smoking rate* may be a confounding variable in this study. In particular, the smoking rate in Hong Kong is much higher than the smoking rate in Colorado, so any conclusions about radon and lung cancer must take the smoking rate into account. In fact, careful studies have shown that radon gas *can* cause lung cancer, and the U.S. Environmental Protection Agency (EPA) recommends taking steps to prevent radon from building up indoors.

Now try Exercises 17–18. ◄

GUIDELINE 6: CONSIDER THE SETTING AND WORDING IN SURVEYS

Even when a survey is conducted with proper sampling and with clearly defined terms and questions, it's important to watch out for problems in the setting or wording that might produce inaccurate or dishonest responses. Dishonest responses are particularly likely when the survey concerns sensitive subjects, such as personal habits or income. For example, the question "Do you cheat on your income taxes?" is unlikely to elicit honest answers from those who cheat, especially if the setting does not guarantee complete confidentiality.

In other cases, even honest answers may not be accurate if the wording of questions invites bias. Sometimes just the order of the words in a question can affect the outcome. A poll conducted in Germany asked the following two questions:

- *Would you say that traffic contributes more or less to air pollution than industry?*
- *Would you say that industry contributes more or less to air pollution than traffic?*

With the first question, 45% answered traffic and 32% answered industry. With the second question, only 24% answered traffic while 57% answered industry. Thus, simply changing the order of the words *traffic* and *industry* dramatically changed the survey results.

►EXAMPLE 7 Do You Want a Tax Cut?

The Republican National Committee commissioned a poll to find out whether Americans supported a tax-cut proposal. Asked whether they favored the tax cut, 67% of respondents answered yes. Should we conclude that Americans supported the proposal?

Solution A question like "Do you favor a tax cut?" is biased because it does not give other options (much like the fallacy of *limited choice* discussed in Unit 1A). In fact, an independent poll conducted at the same time gave respondents a list of options for using surplus revenues. This poll found that 31% wanted the money de-

by the way

People are more likely to choose the item that comes first in a survey because of what psychologists call the *availability error*—the tendency to make judgments based on what is *available* in the mind. Professional polling organizations must be very careful to avoid this problem, sometimes by posing the question to some people in one order and to others in the opposite order.

voted to Social Security, 26% wanted it used to reduce the national debt, and only 18% favored using it for a tax cut. (The remaining 25% of respondents chose a variety of other options.) *Now try Exercises 19–20.* ◄

GUIDELINE 7: CHECK THAT RESULTS ARE PRESENTED FAIRLY

Even when a statistical study is done well, it may be misrepresented in graphs or concluding statements. Researchers may occasionally misinterpret the results of their own studies or jump to conclusions that are not supported by the results, particularly when they have personal biases toward certain interpretations. In other cases, news reporters or others may misinterpret a survey or jump to unwarranted conclusions that make a story seem more spectacular. Misleading graphs are an especially common problem (see Unit 5D). In general, you should look for inconsistencies between the interpretation of a study (in pictures and words) and any actual data given with it.

► **EXAMPLE 8** Does the School Board Need a Statistics Lesson?

The school board in Boulder, Colorado, created a hubbub when it announced that 28% of Boulder school children were reading "below grade level," and hence concluded that methods of teaching reading needed to be changed. The announcement was based on reading tests on which 28% of Boulder school children scored below the national average for their grade. Do these data support the board's conclusion?

Solution The fact that 28% of Boulder children scored below the national average for their grade implies that 72% scored at or above the national average. Thus, the school board's ominous statement about students reading "below grade level" makes sense only if "grade level" means the national average score for a particular grade. This interpretation of "grade level" is curious because it means that half the students in the nation are always below grade level—no matter how high the scores. The conclusion that teaching methods needed to be changed was not justified by these data. *Now try Exercises 21–22.* ◄

GUIDELINE 8: STAND BACK AND CONSIDER THE CONCLUSIONS

Finally, even if a study seems reasonable according to all the previous guidelines, you should stand back and consider the conclusions. Ask yourself questions such as these:

- Did the study achieve its goals?
- Do the conclusions make sense?
- Can you rule out alternative explanations for the results?
- If the conclusions do make sense, do they have any practical significance?

► **EXAMPLE 9** Practical Significance

An experiment is conducted in which the weight losses of people who try a new "Fast Diet Supplement" are compared to the weight losses of a control group of people who try to lose weight in other ways. After eight weeks, the results show that the

Extraordinary claims require extraordinary evidence.

—CARL SAGAN (1934–1996)

treatment group lost an average of $\frac{1}{2}$ pound more than the control group. Assuming that it has no dangerous side effects, does this study suggest that the Fast Diet Supplement is a good treatment for people wanting to lose weight?

Solution Compared to the average person's body weight, the difference of $\frac{1}{2}$ pound hardly matters at all. Thus, while the statistics in this case may be interesting, they don't seem to have much practical significance. *Now try Exercises 23–24.* ◄

SUMMARY Eight Guidelines for Evaluating a Statistical Study

1. Identify the goal of the study, the population considered, and the type of study.

2. Consider the source, particularly with regard to whether the researchers may be biased.

3. Look for bias that may prevent the sample from being representative of the population.

4. Look for problems in defining or measuring the variables of interest, which can make it difficult to interpret results.

5. Watch out for confounding variables that can invalidate the conclusions of a study.

6. Consider the setting and the wording of questions in any survey, looking for anything that might tend to produce inaccurate or dishonest responses.

7. Check that results are presented fairly in graphs and concluding statements, since both researchers and media often create misleading graphics or jump to conclusions that the results do not support.

8. Stand back and consider the conclusions. Did the study achieve its goals? Do the conclusions make sense? Do the results have any practical significance?

■ EXERCISES 5B

Review Questions

1. Briefly describe each of the eight guidelines for evaluating statistical studies. Give an example to which each guideline applies.

2. Describe and contrast selection bias and participation bias in sampling. Give an example of each.

3. What do we mean by variables of interest in a study?

4. What are confounding variables, and what problems can they cause?

Does It Make Sense?

Decide whether each of the following statements makes sense (or is clearly true) or does not make sense (or is clearly false). Explain your reasoning.

5. The TV survey got more than 1 million phone-in responses, so it is clearly more valid than the survey by the professional pollsters, which involved interviews with only a few hundred people.

6. The survey of religious beliefs suffered from selection bias because the questionnaires were handed out only at Catholic churches.

7. My experiment proved beyond a doubt that vitamin C can reduce the severity of colds, because I controlled the experiment carefully for every possible confounding variable.

8. Everyone who jogs for exercise should try the new training regimen, because careful studies suggest it can increase your speed by 1%.

Basic Skills & Concepts

Would You Believe This Study? Exercises 9–20 each describe some aspect of a statistical study. Based solely on the information given in each case, decide whether you have any reason to doubt the results of the study. Explain your reasoning.

9. A study designed to determine whether homeopathic medicines can prevent cancer asked visitors to a natural foods store whether they had ever had cancer and whether they took homeopathic medicines.

10. An experimental, double-blind study finds that people who eat more fast food are more likely to feel tired throughout the day.

11. The staff at the Democratic Party headquarters conducted a study to find out what people think of the Republican tax-cut plan.

12. A study financed by a major pharmaceutical company finds that its new drug is no more effective against high blood pressure than older, less expensive drugs.

13. A magazine tested its new format by including a postcard that readers could return to say whether or not they preferred the new format over the old format.

14. In trying to determine whether their candidate for Governor has a chance of defeating the incumbent Democrat, the Republican Party conducts a survey of 1000 of its members, selected at random.

15. Researchers claim to have learned that rich people are more likely to be happy than poor people.

16. A government study finds, based on people who had their tax returns audited, that 15% of taxpayers understate their income.

17. A study compares accident rates among people who drive at different speeds, but does not consider driver ages or whether drivers were intoxicated.

18. A study seeks to learn about obesity among children. The researchers monitor the eating and exercise habits of the children in the study, carefully recording everything they eat and all their activity.

19. A survey was conducted that asked people whether they cheat on their income taxes. The survey required respondents to provide their names and addresses, but the letter that came with the survey promised full confidentiality.

20. To gauge public opinion on whether there should be a constitutional amendment to ban flag burning, a survey asked people, "Do you support the American flag?"

Would You Believe This Claim? Exercises 21–24 each describe a claim based on a statistical study. Based solely on the information given in each case, decide whether you have any reason to doubt the claim. Explain your reasoning.

21. A group of economists conducts a study of unemployment data and claims that their results "have overturned virtually all of standard economic theory."

22. Citing statistical data indicating that half the children in the school district are of above average weight, the School Board claims to have proved that new exercise classes should be mandated for everyone.

23. A researcher claims to have statistical proof showing that following his nutritional and exercise regimen will allow most people to live to 150 years of age.

24. Based on data showing that a new cold treatment can shorten the average duration of a cold from 7 days to 6.8 days, the company that sells the treatment claims that everyone should use it.

Further Applications

Bias. Exercises 25–32 present situations in which bias may be an issue. Describe one potential source of bias in the situation, and briefly discuss whether the bias should affect your view of the situation.

25. A film critic on ABC gives her opinion of the latest movie from Disney, which happens to own ABC.

26. Market researchers conduct a survey at a supermarket on a weekday between 10:00 a.m. and noon to determine what fraction of customers use coupons.

27. An exit poll designed to predict the winner of a local election uses interviews with everyone who votes between 7:00 and 7:30 a.m.

28. An exit poll designed to predict the winner of a national election uses interviews with randomly selected voters in New York.

29. In order to determine the opinions of people in the 18- to 24-year age group on capital punishment, researchers survey a random sample of 500 Marines in this age group.

30. A college mails survey forms to all current seniors, asking for the students' choice of their all-time best and worst professor. Students are asked to return the survey in the campus mail.

31. Monsanto conducts a study to determine whether its new, genetically engineered soybean poses any threat to the environment.

32. Scientists working for Greenpeace (which opposes genetically engineered crops) conduct a study to determine whether Monsanto's new, genetically engineered soybean poses any threat to the environment.

33. **It's All in the Wording.** Princeton Survey Research Associates did a study for *Newsweek* magazine illustrating the effects of wording in a survey. Two questions were asked:

 • Do you personally believe that abortion is wrong?

 • Whatever your own personal view of abortion, do you favor or oppose a woman in this country having the choice to have an abortion with the advice of her doctor?

To the first question, 57% of the respondents replied yes, while 36% responded no. In response to the second question, 69% of the respondents favored the choice, while 24% opposed the choice. Discuss why the two questions produced seemingly contradictory results. How could the results of the questions be used selectively by various groups?

34. **Tax or Spend?** A Gallup poll asked the following two questions:

 • Do you favor a tax cut or "increased spending on other government programs"? *Result:* 75% for tax cut.

 • Do you favor a tax cut or "spending to fund new retirement savings accounts, as well as increased spending on education, defense, Medicare and other programs"? *Result:* 60% for the spending.

Discuss why the two questions produced seemingly contradictory results. How could the results of the questions be used selectively by various groups?

Stat-Bytes. Politicians must make their political statements (often called sound-bytes) very short because the attention span of listeners is so short. A similar effect occurs in reporting statistical news. Major statistical studies are often reduced to one or two sentences. The summaries of statistical reports in Exercises 35–40 are taken from various news sources. Discuss what crucial information is missing and what more you would want to know before you acted on the report.

35. *USA Today* reports on a Harris poll claiming that the percentage of adults with a "great deal of confidence" in military leaders stands at 54%.

36. CNN reports on a Zagat Survey of America's Top Restaurants which found that "only nine restaurants achieved a rare 29 out of a possible 30 rating and none of those restaurants is in the Big Apple."

37. A University of Michigan study concluded that drunken driving by high school seniors had declined. The report was based on two studies. In 1984, 31.2% of high school seniors reported driving after drinking sometime during the previous two weeks. By 1997, only 18.3% of high school seniors reported driving after drinking sometime during the previous two weeks.

38. Only 2% of the estates of Americans who died in the past year paid estate taxes, while 60% of Americans favor repealing estate taxes.

39. The size of network TV audiences has decreased by 24% since 1991, while the annual advertising revenue of the networks increased by 42%.

40. Thirty percent of newborns in India would qualify for intensive care if they were born in the United States.

Accurate Headlines? Exercises 41–43 give a headline and a brief description of the statistical news story that accompanied the headline. In each case, discuss whether the headline accurately represents the story.

41. Headline: "Drugs shown in 98 percent of movies"

 Story summary: A "government study" claims that drug use, drinking, or smoking was depicted in 98% of the top movie rentals (Associated Press).

42. Headline: "Sex more important than jobs"

 Story summary: A survey found that 82% of 500 people interviewed by phone ranked a satisfying sex life as important or very important, while 79% ranked job satisfaction as important or very important (Associated Press).

43. Headline: "Grape juice may fight disease"

 Story summary: A study of 15 people, partially funded by Welch Foods, found that grape juice helps to expand blood vessels and increase the levels of HDL cholesterol. Both constricted blood vessels and low HDL levels are risk factors for heart disease (*Milwaukee Journal Sentinel*).

Web Projects

Find useful links for Web Projects on the text Web site: www.aw.com/bennett-briggs

44. **Polling Organization.** Go to the Web site for a major professional polling organization. Study results from a recent poll, and evaluate the poll according to the guidelines in this section.

45. **Harper's Index.** Go to the Web site for the Harper's Index and study a few of the recently quoted statistics. Be sure to select the option on the page that allows you to see the sources for the statistics. Choose three statistics that you find particularly interesting, and discuss whether, in accord with the guidelines given in this section, you believe them.

In the News

46. **Applying the Guidelines.** Find a recent newspaper article or television report about a statistical study on a topic that you find interesting. Write a short report applying each of the eight guidelines given in this section. (Some of the guidelines may not apply to the particular study you are analyzing. In that case, explain why the guideline is not applicable.)

47. **Believable Results.** Find a recent news report about a statistical study whose results you believe are meaningful and important. In one page or less, summarize the study and explain why you find it believable.

48. **Unbelievable Results.** Find a recent news report about a statistical study whose results you *don't* believe are meaningful or important. In one page or less, summarize the study and why you don't believe its claims.

49. **Legal Experts.** Find a news report concerning a major ongoing trial. Find out whether any of the "expert witnesses" are being paid by either side. Based on what you learn, describe whether you think the experts are giving biased testimony.

50. **Biased Questioning?** Find a recent news report of responses to a single question in an opinion poll. State the exact words of the question and the results of the poll. Analyze the question and the reported results for potential biases. At the end of your analysis, state whether you believe the results, and defend your opinion.

UNIT **5C**

STATISTICAL TABLES AND GRAPHS

W hether you look at a newspaper, a corporate annual report, or a government study, you are almost sure to see tables and graphs of statistical data. Some of these tables and graphs are simple; others can be quite complex. Some make it easy to understand the data; others may be confusing or even misleading. In this unit, we'll investigate some of the basic principles behind tables and graphs, preparing for more complex graphics in Unit 5D.

FREQUENCY TABLES

A teacher makes the following list of the grades she gave to her 25 students on an essay:

A C C B C D C C F D C C C B B A B D B A A B F C B

This list contains all the raw data, but it isn't easy to read. A better way to display these data is with a **frequency table**—a table showing the number of times, or **frequency,** that each grade appears (Table 5.1). The five possible grades are called the **categories** for the table.

There are two common variations on the idea of frequency. The **relative frequency** for a category expresses its frequency as a fraction or percentage of the total. For example, 4 of the 25 students received A grades, so the relative frequency for A grades is 4/25, or 16%. Note that the total relative frequency must always be 1, or 100%.

The **cumulative frequency** is the number of responses in a particular category and all preceding categories. For example, the cumulative frequency for grades of C and above is 20, because 20 students received grades of either A, B, or C.

TABLE 5.1

GRADE	FREQUENCY
A	4
B	7
C	9
D	3
F	2
Total	**25**

DEFINITION

A basic **frequency table** has two columns:

- The first column lists all the **categories** of data.
- The second column lists the **frequency** of each category, which is the number of times each category appears in the data set.

Additional columns may include **relative frequency** (frequency expressed as a fraction or percentage of the total) or **cumulative frequency** (total of frequencies for the given category and all previous categories).

►**EXAMPLE 1** Relative and Cumulative Frequency

Add to Table 5.1 columns showing the relative and cumulative frequencies.

Solution Table 5.2 shows columns and calculations for relative and cumulative frequency.

TABLE 5.2

GRADE	FREQUENCY	RELATIVE FREQUENCY	CUMULATIVE FREQUENCY
A	4	4/25 = 16%	4
B	7	7/25 = 28%	7 + 4 = 11
C	9	9/25 = 36%	9 + 7 + 4 = 20
D	3	3/25 = 12%	3 + 9 + 7 + 4 = 23
F	2	2/25 = 8%	2 + 3 + 9 + 7 + 4 = 25
Total	25	1 = 100%	25

Now try Exercises 15–16. ◀

Time out to think _____
Briefly explain why the total relative frequency must always be 1, or 100%.

Data Types

Essay grades represent subjective ratings, not actual measurements or counts. We say that the grade categories are **qualitative,** because they represent *qualities* such as bad or good. In contrast, scores on a multiple-choice exam are **quantitative,** because they represent an actual count (or measurement) of the number of correct answers. As we'll see shortly, distinguishing between qualitative and quantitative data can be useful in creating tables or graphs.

Data Types

Qualitative data describe *qualities* or nonnumerical categories.

Quantitative data represent counts or measurements.

➤ EXAMPLE 2 Data Types

Classify each of the following types of data as either qualitative or quantitative.

 a. Brand names of shoes in a consumer survey
 b. Heights of students
 c. Audience ratings of a film on a scale of 1 to 5, where 5 means excellent

Solution

 a. Brand names are nonnumerical categories, so they are qualitative data.
 b. Heights are measurements, so they are quantitative data.

c. Although the film rating categories involve numbers, the numbers represent subjective opinions about a film, not counts or measurements. Thus, they are qualitative data, despite being stated as numbers.

Now try Exercises 17–24. ◄

Time out to think
Give another example in which numbers are used to represent qualitative data rather than quantitative data.

Binning Data

When we deal with quantitative data categories, it's often useful to group, or *bin*, the data into categories that cover a range of possible values. For example, in a table of income levels, it might be useful to create bins of $0 to $20,000, $20,001 to $40,000, and so on. In this case, the frequency of each bin is simply the number of people with incomes in that bin.

► **EXAMPLE 3** Binned Exam Scores

Consider the following set of 20 scores from a 100-point exam:

76 80 78 76 94 75 98 77 84 88 81 72 91 72 74 86 79 88 72 75

Determine appropriate bins and make a frequency table. Include columns for relative and cumulative frequency, and interpret the cumulative frequency for this case.

Solution The scores range from 72 to 98. One way to group the data is with 5-point bins. The first bin represents scores from 95 to 99, the second bin represents scores from 90 to 94, and so on. Note that there is no overlap between bins. We then count the frequency (the number of scores) in each bin. For example, only 1 score is in bin 95 to 99 (the high score of 98) and 2 scores are in bin 90 to 94 (the scores of 91 and 94). Table 5.3 shows the complete frequency table. In this case, we interpret the cumulative frequency of any bin to be the total number of scores in *or above* that bin.

TABLE 5.3 Frequency Table for Binned Exam Scores

SCORES	FREQUENCY	RELATIVE FREQUENCY	CUMULATIVE FREQUENCY
95 to 99	1	0.05 = 5%	1
90 to 94	2	0.10 = 10%	3
85 to 89	3	0.15 = 15%	6
80 to 84	3	0.15 = 15%	9
75 to 79	7	0.35 = 35%	16
70 to 74	4	0.20 = 20%	20
Total	20	1.00 = 100%	20

For example, the cumulative frequency of 6 for the bin 85 to 89 means that 6 scores are either between 85 and 89 or higher than 89. *Now try Exercises 25–26.* ◀

BAR GRAPHS AND PIE CHARTS

Bar graphs and pie charts are commonly used to show data when the categories are qualitative. You are probably familiar with both, but let's review the basic ideas.

Consider the essay grade data in Table 5.1. A bar graph would show each category with a bar whose length corresponded to its frequency. If you make a bar graph by hand (as opposed to with a computer), you should measure the bar lengths carefully to make sure they correctly correspond to the frequencies. In Figure 5.3, for example, the vertical axis is marked with frequencies $\frac{1}{2}$ centimeter apart. Thus, the bar for A grades is 2 centimeters long, because the frequency of A grades is 4. Note that the left side of the bar graph in Figure 5.3 is marked with frequency, while the right side is marked with relative frequency. As you can see, bar graphs make it easy to display both frequencies simultaneously.

In contrast, pie charts are used primarily for relative frequencies, because the total pie must always represent the total relative frequency of 100%. The size of each wedge is proportional to the relative frequency of the category it represents. Figure 5.4 shows a pie chart for the essay grade data. To make comparisons easier, relative frequencies are often written on pie chart wedges.

Nowadays, most people make graphs with the aid of computers that measure bar lengths or wedge sizes automatically. However, you must still specify any labels or axis marks you want on a graph. This labeling is extremely important: Without proper labels, a graph is meaningless. The following summary lists the important labels for graphs. Of course, not all labels are necessary in all cases. For

FIGURE 5.3 Bar graph for the essay grade data in Table 5.1.

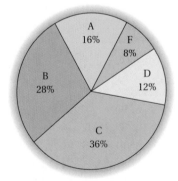

FIGURE 5.4 Pie chart for the essay grade data in Table 5.1.

HISTORICAL NOTE
A bar graph with the bars in descending order is often called a *Pareto chart,* after Italian economist Vilfredo Pareto (1848–1923).

example, pie charts do not require a vertical or horizontal scale. Notice how these rules were applied in Figure 5.3.

Important Labels for Graphs

Title/caption: The graph should have a title or caption (or both) that explains what is being shown and, if applicable, lists the source of the data.

Vertical scale and title: Numbers along the vertical axis should clearly indicate the scale. The numbers should line up with the tick marks—the marks along the axis that precisely locate the numerical values. Include a label that describes the variable shown on the vertical axis.

Horizontal scale and title: The categories should be clearly indicated along the horizontal axis. (Tick marks may not be necessary for qualitative data, but should be included for quantitative data.) Include a label that describes the variable shown on the horizontal axis.

Legend: If multiple data sets are displayed on a single graph, include a legend or key to identify the individual data sets.

▶**EXAMPLE 4** Carbon Dioxide Emissions

Carbon dioxide is released into the atmosphere by the combustion of fossil fuels (oil, coal, natural gas). Because carbon dioxide in the atmosphere can potentially lead to global warming, the U.S. Department of Energy tracks carbon dioxide emissions of countries around the world. Table 5.4 lists the eight countries that emit the most carbon dioxide each year. Make bar graphs for the total emissions and the emissions per person. Put the bars in descending order of size.

TABLE 5.4 The World's Eight Leading Emitters of Carbon Dioxide

COUNTRY	TOTAL CARBON DIOXIDE EMISSIONS (MILLIONS OF METRIC TONS OF CARBON)	POPULATION (MILLIONS)	PER PERSON CARBON DIOXIDE EMISSIONS (METRIC TONS OF CARBON)
United States	1495	271	5.5
China	740	1244	0.6
Russia	405	147	2.8
Japan	288	126	2.3
India	253	955	0.3
Germany	227	82	2.8
United Kingdom	147	59	2.5
Canada	138	30	4.6

SOURCE: *U.S. Department of Energy, based on 1998 emissions.*

Solution The categories are the countries. Because country names are qualitative data, a bar graph is appropriate.

The values for total carbon dioxide emissions go from 138 to 1495 (millions of tons), so a range of 0 to 1500 makes a good choice for the vertical scale. Each bar's height corresponds to its data value, and we label the category (country) under the bar. Figure 5.5a shows the bar graph for total emissions, with bars in order of decreasing height.

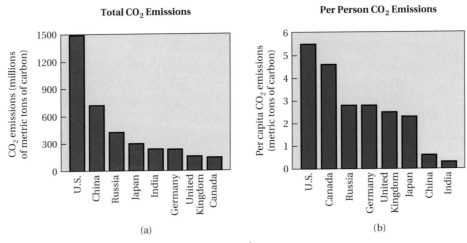

FIGURE 5.5 Bar graphs for (a) total carbon dioxide emissions by country and (b) per person carbon dioxide emissions by country.

The data values for per person emissions range from 0.3 to 5.5 (tons), so a range of 0 to 6 will work for this vertical scale. Figure 5.5b shows the bar graph, again with bars placed in order of descending height. *Now try Exercises 27–28.* ◄

**Registered Voters
in Rochester County**

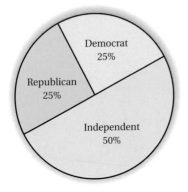

FIGURE 5.6 Party affiliations of registered voters in Rochester County.

Time out to think _____

Note that the two bar graphs in Figure 5.5 do not show the countries in the same order. Why not? What can we learn by comparing the two graphs? Explain.

► **EXAMPLE 5** Simple Pie Chart

Among the registered voters in Rochester County, 25% are Democrats, 25% are Republicans, and 50% are Independents. Make a pie chart showing the breakdown of party affiliations in Rochester County.

Solution The wedge sizes should correspond to the relative frequencies. Thus, the wedges for Republicans and Democrats each occupy one-fourth of the pie, while the wedge for Independents occupies the remaining half of the pie (Figure 5.6). Note the importance of clear labeling.
 Now try Exercises 29–30. ◄

➤EXAMPLE 6 Student Majors

Figure 5.7 is a pie chart showing planned major areas for first-year college students. Make a bar graph showing the same data, with the bars in order of decreasing size. What are the three most popular major areas? Comment on the relative ease with which this question can be answered with the pie chart and the bar graph.

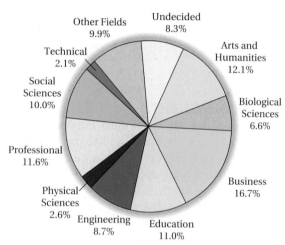

What Students Expect to Major In

FIGURE 5.7 Planned major areas for first-year college students, 2001.

SOURCE: The Chronicle of Higher Education.

Solution Figure 5.8 shows the bar graph for the data.

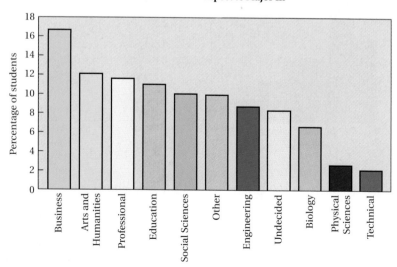

What Students Expect to Major In

FIGURE 5.8 Bar graph for the data in Figure 5.7.

TECHNICAL NOTE
Different books define
the terms *histogram*
and *bar graph*
differently. In this
book, a bar graph is
any graph that uses
bars, and histograms
are bar graphs used
for quantitative data
categories.

Note that, because we have only relative frequency data from the pie chart, we can show only relative frequencies on the bar graph. This bar graph makes it immediately obvious that the three most popular major areas are business (16.7%), arts and humanities (12.1%), and professional (11.6%). ("Professional" includes fields with professional licensing, such as architecture, nursing, and pharmacy.) In contrast, it takes a fair amount of study of the pie chart before we can easily list the three most popular major areas. *Now try Exercises 31–32.* ◄

Time out to think _____

Example 6 discussed an advantage of a bar graph over a pie chart for showing the data concerning major areas. Do you think the pie chart has any advantages over the bar graph? If so, what?

HISTOGRAMS AND LINE CHARTS

For quantitative data categories, the two most common types of graphics are *histograms* and *line charts*. Figure 5.9a shows a histogram for the binned exam data of Table 5.3. Figure 5.9b shows a line chart for the same data.

A **histogram** is essentially a bar graph in which the data categories are quantitative. Thus, the bars on a histogram must follow the natural order of the numerical categories. In addition, the widths of histogram bars have a specific meaning. For example, the width of each bar in Figure 5.9a represents 5 points on the exam. Because there are no gaps between the categories, the bars on a histogram touch each other.

(a) (b)

FIGURE 5.9 (a) Histogram for the data in Table 5.3. (b) Line chart for the same data.

A **line chart** serves the same basic purpose as a histogram, but instead of using bars, a line chart connects a series of dots. When data are binned, the dot is placed at the center of each bin. Histograms and line charts are often used to show how some variable changes with time. For example, the line chart in Figure 5.10 shows how the U.S. homi-

U.S. Homicide Rate

FIGURE 5.10 U.S. homicide rate per 100,000 people.
SOURCE: FBI Uniform Crime Reports.

cide rate has changed with time. The categories are time intervals. In this case, each bin represents a year in the data. Histograms and line charts with time on the horizontal axis are often called **time-series diagrams.**

DEFINITIONS

A **histogram** is a bar graph for quantitative data categories. The bars have a natural order and the bar widths have specific meaning.

A **line chart** shows the data value for each category as a dot, and the dots are connected with lines. For each dot, the horizontal position is the *center* of the bin it represents and the vertical position is the data value for the bin.

A **time-series diagram** is a histogram or line chart in which the horizontal axis represents time.

➤ **EXAMPLE 7** Oscar-Winning Actresses

Table 5.5 shows the ages of 34 recent Academy Award–winning actresses at the time when they won their award. Make a histogram and a line chart to display these data. Discuss the results.

Solution The fact that the categories are 10-year bins makes the data quantitative. Thus, a histogram is appropriate. Figure 5.11a shows the histogram. The bars touch one another because there are no gaps between the categories.

Figure 5.11b shows the same data as a line chart. The histogram is also included to show how it relates to the line chart. In looking at these data, we see that actresses are most likely to win Oscars when they are fairly young.

TABLE 5.5

AGE	NUMBER OF ACTRESSES
20–29	7
30–39	15
40–49	6
50–59	1
60–69	3
70–79	1
80–89	1

Ages of 34 Academy Award–Winning Actresses

(a)

Ages of 34 Academy Award–Winning Actresses

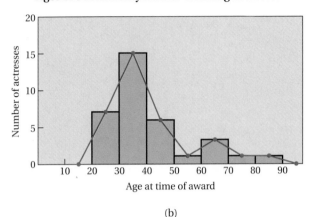

(b)

FIGURE 5.11 (a) Histogram for ages of 34 recent Academy Award–winning actresses. (b) Line chart for the same data, with histogram overlaid for comparison.

Now try Exercises 33–34. ◄

> **EXAMPLE 8** Reading a Time-Series Diagram

Figure 5.12 shows a time-series line chart of stock, bond, and gold prices over a 12-week period. Suppose that, on July 7, you invested $100 in a stock fund that tracks the S&P 500, $100 in a bond fund that follows the Lehman Index, and $100 in gold. If you sold all three funds on August 4, how much did you gain or lose?

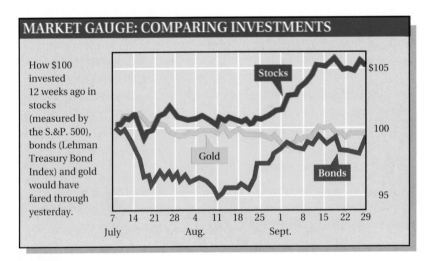

FIGURE 5.12

<label>HISTORICAL NOTE</label>

HISTORICAL NOTE

Gold was once considered to be a solid investment and an important part of any investment portfolio. However, gold prices have languished in recent decades. In 2003, gold was worth only about $350 per ounce—much less than half its peak price of $850 per ounce in 1980.

Solution The graph shows that the $100 in the stock fund would have been worth about $101 on August 4. The $100 bond investment would have declined in value to about $96. The gold investment would have held its initial value of $100. Thus, on August 4, your complete portfolio would have been worth $101 + $96 + $100 = $297. You would have lost $3 on your total investment of $300. *Now try Exercises 35–36.* ◄

■ EXERCISES 5C

Review Questions

1. What is a frequency table? Explain what we mean by the categories and frequencies. What do we mean by relative frequency? What do we mean by cumulative frequency?

2. What is the distinction between qualitative data and quantitative data? Give a few examples of each.

3. What is the purpose of binning? Give an example in which binning is useful.

4. What two types of graphs are most common when the categories are qualitative data? Describe the construction of each.

5. Describe the importance of labeling on a graph, and briefly discuss the kinds of labels that should be included on graphs.

6. What two types of graphs are most common when the categories are quantitative data? Describe the construction of each.

Does It Make Sense?

Decide whether each of the following statements makes sense (or is clearly true) or does not make sense (or is clearly false). Explain your reasoning.

7. I made a frequency table with two columns, one labeled *State* and one labeled *State Capitol*.

8. The relative frequency of B grades in our class was 0.3.

9. Your bar graph must be wrong, because your bars are wider than the ones shown on the teacher's answer key.

10. Your bar graph must be wrong, because it shows different frequencies than the ones shown on the teacher's answer key.

11. Your pie chart must be wrong, because you have the 45% frequency wedge near the upper left and the answer key shows it near the lower right.

12. Your pie chart must be wrong, because when I added the percentages on your wedges, they totaled 124%.

13. I was unable to make a bar chart, because the data categories were qualitative rather than quantitative.

14. I rearranged the bars on my histogram so that the tallest bar would come first.

Basic Skills & Concepts

Frequency Tables. Make a frequency table for the data in each of Exercises 15–16. Include columns for relative frequency and cumulative frequency. Briefly explain the meaning of each column.

15. Final grades of 24 students in a math class:

A	A	A	A	B	B	B	B	B	B	B	C
C	C	C	C	C	C	C	D	D	D	F	F

16. A guide book for New York City lists 5 five-star restaurants (the highest rating), 10 four-star restaurants, 20 three-star restaurants, 15 two-star restaurants, and 5 one-star restaurants.

Qualitative vs. Quantitative. In Exercises 17–24, determine whether the variable described is qualitative or quantitative, and explain why.

17. The blood types of individuals

18. The average waiting time in check-out lines at a grocery store

19. The responses of people to potato chips in a taste test where 0 = inedible up to 5 = outstanding

20. The high temperature in Anchorage, Alaska, on each day of the year

21. The responses of people to the question "Will you vote to reelect your mayor?"

22. The federal income tax paid by individuals

23. The choice of TV shows to watch on a given night

24. The number of people watching various TV shows on a given night

Binned Frequency Tables. In Exercises 25–26, use the indicated bin size to make a frequency table for the following set of exam scores:

67	54	89	67	98	68	88	78	87
82	91	80	73	59	85	86	76	99
38	75	84	87	77	69	90		

Include columns for relative frequency and cumulative frequency. Briefly explain the meaning of each column.

25. Use 5-point bins (95 to 99, 90 to 94, etc.).

26. Use 10-point bins (90 to 99, 80 to 89, etc.).

27. Largest Cities. The following table shows the five most populous U.S. cities at the time of the 2000 census. Make a bar graph for these data, with the bars in descending order.

City	Population
Chicago	2.9 million
Houston	2.0 million
Los Angeles	3.7 million
New York	8.0 million
Philadelphia	1.5 million

28. Vehicle Sales. The following table shows the five vehicles with the largest number of sales in 2002. Make a bar graph for these data, with the bars in descending order.

Vehicle	Number sold
Chevy Silverado	652,646
Ford Explorer	433,847
Ford F-series	813,701
Honda Accord	398,980
Toyota Camry	434,145

Constructing Pie Charts. Exercises 29–30 each give a data set. Construct a pie chart for the data.

29. In a survey of favorite ice cream flavors, two-thirds of the respondents preferred chocolate, one-sixth preferred vanilla, and the rest chose other flavors.

30. A survey of students at a high school found that 50% planned to attend a 4-year college, 33% planned to attend a 2-year college, and 17% did not plan to attend college.

31. Government Income. The pie chart in Figure 4.11 on p. 288 shows the makeup of federal government receipts. Make a bar graph for these data.

32. Government Spending. The pie chart in Figure 4.12 on p. 288 shows the makeup of federal government spending. Make a bar graph for these data.

33. Oscar-Winning Actors. The following data show the ages of 34 recent Academy Award–winning actors at the time they won their award. Make a frequency table for these data, using bins of 20–29, 30–39, and so on. Then draw both a histogram and a line chart to display the binned data.

32	37	36	32	51	53	33	61	35	45	55	39
76	37	42	40	32	60	38	56	48	48	40	43
62	43	42	44	41	56	39	46	31	47		

34. Oscar Winners. In words, contrast the bar graphs in Example 7 with those you drew in Exercise 33. Do actors appear to be more likely to win Oscars when they are younger, older, or neither? Do you think these graphs indicate any difference in how movie makers treat male and female performers? Defend your opinion.

35. Homicide Rates. Study Figure 5.10. Write one to two paragraphs summarizing how the homicide rate has changed with time since 1950.

36. Death Rates. Figure 5.13 shows overall death rates in the United States during the 20th century. Note that the spike in 1919 was due to a worldwide epidemic of influenza. Write a few sentences summarizing the overall trend, describing how much the death rate changed during the century, and putting the 1919 spike into context in terms of its impact on the population.

FIGURE 5.13 *SOURCE: National Center for Health Statistics.*

Further Applications

Statistical Graphs. Each of Exercises 37–45 gives a table of data. For each exercise, do the following:

a. Explain whether the data categories are qualitative or quantitative.

b. If the data categories are qualitative, draw either a bar graph or pie chart for the data. If the data categories are quantitative, draw either a histogram or a line chart for the data.

c. Write a one-paragraph summary of any interesting information revealed by the graphic.

37. The following table gives colleges and universities that have won more than one NCAA men's basketball championship from 1960 to 2003.

Team	Number of championships
Cincinnati	2
Duke	3
Indiana	3
Kentucky	3
Louisville	2
Michigan State	2
North Carolina	2
North Carolina State	2
UCLA	11

38. The following table lists the top eight retail companies in the United States, by total sales volume.

Company	Sales (billions of dollars)
Albertson's	36.8
Home Depot	45.7
JC Penney	33.0
Kmart	37.0
Kroger	49.0
Sears, Roebuck	40.9
Target	36.9
Wal-Mart	193.3

SOURCE: Wall Street Journal Almanac.

39. The following table shows high school graduation rates for various ethnic groups in the United States in 1997.

Ethnic group	Graduation rate
white	83.0%
black	74.9%
Asian/Pacific Islander	84.9%
Hispanic	54.7%

40. The following table gives world population by region.

Region	Population (billions)
Asia	3.6
Africa	0.8
North America	0.3
Latin America	0.5
Europe	0.7
Oceania	0.03

41. The following table lists areas of the world's major land masses.

Land mass	Area (millions of sq. miles)
Asia	17.2
Africa	11.6
North America	9.3
South America	6.9
Australia	3.0
Europe	3.8
Antarctica	5.1
All others	2.1

42. The following table gives the ages of first-year college students (2001). (Hint: For the "16 or less" and "21 or more" categories, position the bars or dots at precisely 16 and 21, respectively, but label the bins clearly with their actual meanings.)

Age	Percent of sample
16 or less	0.1
17	1.9
18	68.9
19	27.4
20	1.0
21 or more	0.7

SOURCE: UCLA Higher Education Research Institute.

43. The following table gives the stated religions of first-year college students. (Note: The "other religions" category consists of religions that were stated by less than 1% of the students in the sample.)

Religion	Percent of sample
Baptist	11.6
Catholic	30.5
Episcopal	1.7
Jewish	2.8
Lutheran	5.8
Methodist	6.4
Mormon	1.5
Presbyterian	4.0
United Church of Christ	1.5
Other religions	19.3
No religion	14.9

SOURCE: *UCLA Higher Education Research Institute.*

44. The following table gives the number of "other colleges" (besides the one they attend) to which first-year college students applied. (Hint: For the "11 or more" category, position the bar or dot at precisely 11, but label the bin clearly.)

Other colleges	Percent of sample
0	34.0
1	14.2
2	15.2
3	14.5
4	8.9
5	5.4
6	3.3
7–10	3.8
11 or more	0.7

SOURCE: *UCLA Higher Education Research Institute.*

45. The following table gives average family size in the United States since 1940.

Year	Family size	Year	Family size
1940	3.76	1980	3.29
1950	3.54	1985	3.23
1960	3.67	1990	3.17
1965	3.70	1995	3.19
1970	3.58	2000	3.17
1975	3.42		

SOURCE: *U.S. Bureau of Census.*

46. **Drunk Driving Deaths.** Figure 5.14 shows the number of automobile fatalities in the United States in which alcohol was involved for each year from 1982 to 2000.

 a. How many alcohol-related fatalities were there in 1982? in 2000? Comment on the overall trend over this period.

 b. What is the percent change in alcohol-related fatalities over this period?

 c. The total numbers of automobile fatalities in 1982 and 2000 were 43,945 and 41,821, respectively. What percentage of all fatalities in these two years involved alcohol?

 d. In view of your answer to part c, can you offer explanations for the trend in these data? Explain.

Total Fatalities in Alcohol-Related Crashes

FIGURE 5.14 SOURCE: *National Highway Traffic Safety Administration.*

Web Projects

Find useful links for Web Projects on the text Web site: www.aw.com/bennett-briggs

47. **CO_2 Emissions.** Look for updated data concerning international carbon dioxide emissions at the Web site for the *International Energy Annual,* published by the U.S. Energy Information Administration (EIA). Create an updated or expanded version of Figure 5.5. Discuss any new features of your updated graphs.

48. **Energy Table.** Explore some of the many energy tables at the U.S. Energy Information Administration (EIA) Web site. Choose a table that you find interesting, and make a graph of its data. You may choose any of the graph types discussed in this section. Explain how you made your graph, and briefly discuss what can be learned from it.

49. **Statistical Abstract.** Go to the Web site for the *Statistical Abstract of the United States.* Explore the selection of "frequently requested tables." Choose one table of interest to

you, and make a graph from its data. You may choose any of the graph types discussed in this section. Explain how you made your graph, and briefly discuss what can be learned from it.

In the News

50. Frequency Tables. Find a recent news article that includes some type of frequency table. Briefly describe the table and how it is useful to the news report. Do you think the table was constructed in the best possible way for the article? If so, why? If not, what would you have done differently?

51. Bar Graph. Find a recent news article that includes a bar graph with qualitative data categories. Briefly explain what the graph shows, and discuss whether it helps make the point of the news article.

52. Pie Chart. Find a recent news article that includes a pie chart. Briefly discuss the effectiveness of the pie chart. For example, would it be better if the data were displayed in a bar graph rather than a pie chart? Could the pie chart be improved in other ways?

53. Histogram. Find a recent news article that includes a histogram. Briefly explain what the histogram shows, and discuss whether it helps make the point of the news article. Are the labels clear? Is the histogram a time-series diagram? Explain.

54. Line Chart. Find a recent news article that includes a line chart. Briefly explain what the line chart shows, and discuss whether it helps make the point of the news article. Are the labels clear? Is the line chart a time-series diagram? Explain.

UNIT 5D

GRAPHICS IN THE MEDIA

ow that we've discussed basic types of statistical graphs, we are ready to explore some of the fancier graphics that appear daily in the news. We will also discuss several cautions to keep in mind when interpreting media graphics.

GRAPHICS BEYOND THE BASICS

Many graphical displays of data go beyond the basic types discussed in Unit 5C. Here, we explore a few of the types that are most common in the news media.

Multiple Bar Graphs

A **multiple bar graph** is a simple extension of a regular bar graph. It has two or more sets of bars that allow comparison between two or more data sets. All the data sets must involve the same categories so that they can be displayed on the same graph. For example, Figure 5.15 is a multiple bar graph showing trends in home computing. The categories are years. The two sets of bars represent two different measures of home computing: ownership of personal computers and connection to the Internet. Note that a legend clearly identifies the two sets of bars.

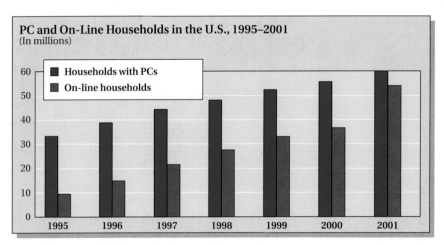

FIGURE 5.15 Trends in home computing.
SOURCE: Statistical Abstract of the United States.

> **EXAMPLE 1** Computing Trends

Summarize two major trends shown in Figure 5.15.

Solution The most obvious trend is that both data sets show an increase with time. That is, the number of homes with computers and the number of online homes both increased with time. We see a second trend by comparing the bars within each year. In 1995, the number of online homes (about 10 million) was less than one-third the number of homes with computers (about 33 million). By 2001, the number of online homes (about 54 million) was about 90% of the number of homes with computers (about 60 million). This tells us that a higher percentage of computer users are going online. *Now try Exercises 13–14.* ◄

Stack Plots

Another common type of graph, called a **stack plot,** shows different data sets in a vertical stack. Figure 5.16 uses a stack plot to show trends in death rates (deaths per 100,000 people) for four diseases over the period from 1900 to 2000. Each disease has its own color-coded region, or wedge; note the importance of the legend. The *thickness* of a wedge at a particular time tells you its value at that time: When a wedge is thick it has a large value, and when it is thin it has a small value.

> **EXAMPLE 2** Stack Plot

Based on Figure 5.16, what was the death rate for pneumonia in 1980? Discuss the general trends visible on this graph.

Solution For 1980, the pneumonia wedge extends from about 620 to 640 on the vertical axis (see dashed lines in Figure 5.16). Thus, the thickness of this wedge is 20, meaning that the death rate in 1980 for pneumonia was 20 deaths per 100,000 people.

Death Rates for Various Diseases: 1900–2000

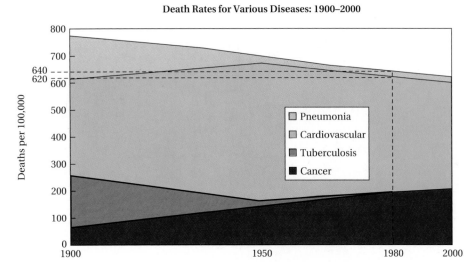

FIGURE 5.16 A stack plot showing trends in death rates from four diseases.

by the way

Since the mid-1980s, there has been a small but noticeable resurgence of tuberculosis in the United States. Part of the resurgence is due to new strains of the disease that resist most common drug treatments.

The graph shows several important trends. First, the downward slope of the top wedge shows that the overall death rate from these four diseases decreased substantially, from nearly 800 deaths per 100,000 in 1900 to about 600 in 2000. The drastic decline in the thickness of the tuberculosis wedge shows that this disease was once a major killer, but has been nearly wiped out since 1950. Meanwhile, the thickening of the cancer wedge shows that the death rate from cancer has gradually increased.

Now try Exercises 15–18. ◀

Graphs of Geographical Data

We are often interested in geographical patterns in data. Figure 5.17 shows one common way of displaying geographical data. In this case, the data on per capita (per person) income are shown state by state. The legend explains that different colors represent different income levels. Similar colors are used for similar income levels. Thus, it is easy to see that income levels tend to be highest in the northeast and lowest in the south.

The display in Figure 5.17 works well because each state is associated with a unique income level. For data that vary continuously across geographical areas, a **contour map** is more convenient. Figure 5.18 shows a contour map of temperature over the United States at a particular time. Each of the *contours* connects locations with the same temperature. For example, the temperature is 50°F everywhere along the contour labeled 50° and 60°F everywhere along the contour labeled 60°F. Between these two contours, the temperature is between 50°F and 60°F. Note that in regions where contours are tightly spaced, there are greater temperature changes. For example, the closely packed contours in the northeast indicate that the temperature varies substantially over small distances. To make the graph easier to read, the regions between adjacent contours are color-coded.

State Per Capita Income

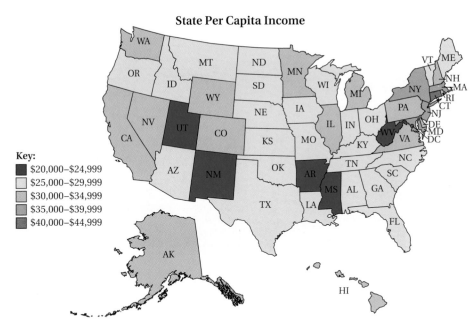

Key:
- $20,000–$24,999
- $25,000–$29,999
- $30,000–$34,999
- $35,000–$39,999
- $40,000–$44,999

FIGURE 5.17 Per capita income in the 50 states (2002).
SOURCE: U.S. Department of Commerce.

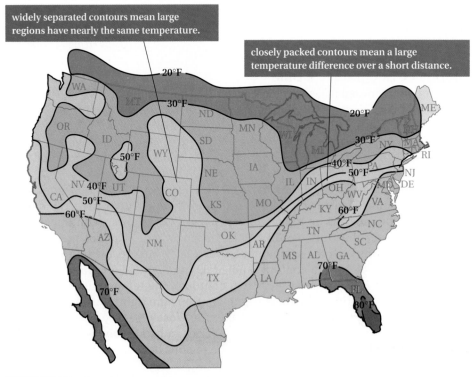

widely separated contours mean large regions have nearly the same temperature.

closely packed contours mean a large temperature difference over a short distance.

FIGURE 5.18 A contour map of temperature.

➤ **EXAMPLE 3** Interpreting Geographical Data

Study Figures 5.17 and 5.18, using them to answer the following questions.

a. Which state(s) had the highest per capita income in 2002?
b. Were there any temperatures above 80°F in the United States on the date shown in Figure 5.18? If so, where?

Solution

a. Connecticut was the only state with a per capita income in the highest category shown on the graph ($40,000–$44,999), so it had the highest per capita income. (The District of Columbia was also in this category, but it is not a state.)
b. The 80° contour passes through southern Florida, so the parts of Florida south of this contour had a high temperature above 80°.

Now try Exercises 19–20. ◄

The greatest value of a picture is when it forces us to notice what we never expected to see.

—JOHN TUKEY

Time out to think _____

Look for a weather map in today's news. How are the temperature contours shown? Interpret the temperature data.

Three-Dimensional Graphics

Today, computer software makes it easy to give almost any graph a three-dimensional appearance. For example, Figure 5.19 shows the bar graph of Figure 5.3, but "dressed up" with a three-dimensional look. It may look nice, but the three-dimensional effects are purely cosmetic. They don't provide any information that wasn't already in the two-dimensional graph in Figure 5.3. As this example shows, many "three-dimensional" graphics really only make two-dimensional data look a little fancier.

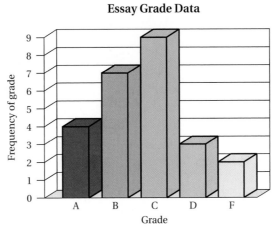

FIGURE 5.19 This graph has a three-dimensional appearance, but shows only two-dimensional data.

In contrast, each of the three axes in Figure 5.20 carries distinct information, making it a true three-dimensional graph. Researchers studying migration patterns of a bird species (the *Bobolink*) counted the number of birds flying over seven New York cities throughout the night. As shown on the inset map, the cities were aligned east-west so that the researchers would learn what parts of the state the birds flew over, and at what times of night, as they headed south for the winter. Thus, the three axes measure *number of birds, time of night,* and *east-west location.*

SONIC MAPPING TRACES BIRD MIGRATION

Sensors across New York State counted each occurrence of the nocturnal flight call of the bobolink to trace the fall migration on the night of Aug. 28–29, 1993. The data showed the heaviest swath passing over the eastern part of the state.

Source: Bill Evans/Cornell Laboratory of Ornithology

FIGURE 5.20 This graph shows true three-dimensional data.
SOURCE: New York Times.

▶ **EXAMPLE 4** Three-Dimensional Bird Migration

Based on Figure 5.20, at about what time was the largest number of birds flying over the east-west line marked by the seven cities? Over what part of New York did most of the birds fly? Approximately how many birds passed over Oneonta at about 12:30 a.m.?

Solution The number of birds detected in all the cities peaked between 4 and 6 hours after 8:30 p.m., or between about 12:30 and 2:30 a.m. More birds flew over the two easternmost cities of Oneonta and Jefferson than over cities farther west. Thus, most of the birds were flying over the eastern part of the state. To answer the specific question about Oneonta, note that 12:30 a.m. is 4 hours after 8:30 p.m. On the graph, this time aligns with the dip between peaks on the line at Oneonta. Looking across to the *number of birds* axis, we see that about 25 to 30 birds were flying over Oneonta at that time.

Now try Exercises 21–22. ◄

Combination Graphics

All of the graphic types we have studied so far are common and fairly easy to create. But the media today are often filled with many varieties of even more complex graphics. For example, Figure 5.21 shows a graphic concerning the participation of women in the summer Olympics. This single graphic combines a line chart, many pie charts, and numerical data. It is certainly a case of a picture being worth far more than a thousand words.

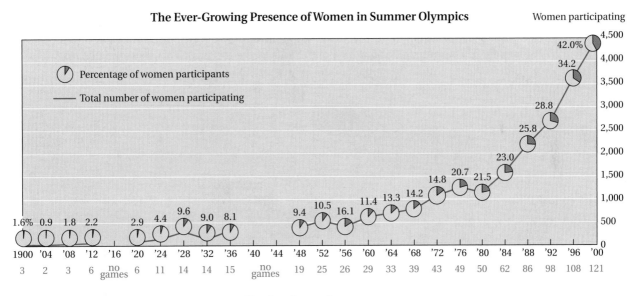

FIGURE 5.21 *SOURCE:* New York Times.

► **EXAMPLE 5** Olympic Women

Describe three trends shown in Figure 5.21.

Solution The line chart shows that the total number of women competing in the summer Olympics has risen fairly steadily, especially since the 1960s, reaching nearly 4500 in the 2000 games. The pie charts show that the percentage of women among all competitors has also increased, reaching 42% in the 2000 games. The bold red numbers at the bottom show that the number of events for women has also increased dramatically, reaching 121 in the 2000 games. *Now try Exercises 23–24.* ◄

Time out to think _____

Do you think the upward trend of the pie charts in Figure 5.21 will continue over the next few Olympic games? Why or why not?

A FEW CAUTIONS ABOUT GRAPHICS

As we have seen, graphics can offer clear and meaningful summaries of statistical data. However, even well-made graphics can be misleading if we are not careful in interpreting them, and poorly made graphics are almost always misleading. Moreover, some people use graphics in deliberately misleading ways. Here, we discuss a few of the more common ways in which graphics can lead us astray.

Perceptual Distortions

Many graphics are drawn in a way that distorts our perception of them. Figure 5.22 shows one of the most common types of distortion. The dollar-shaped bars are used to represent the declining value of the dollar over time. The *lengths* of the bars represent the data, but our eyes tend to focus on the *areas* of the bars. For example, the bottom bar is supposed to show that a dollar in 2000 was worth only 48% as much as a dollar in 1980. Its length is indeed 48% that of the top bar, but its area is much smaller in comparison (about 23% of the area of the top bar). This gives the perception that the value of the dollar shrank even more than it really did.

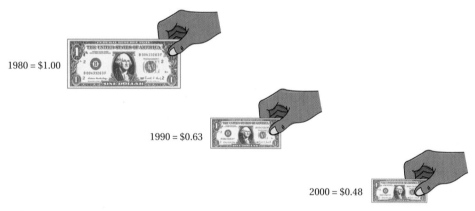

1980 = \$1.00

1990 = \$0.63

2000 = \$0.48

FIGURE 5.22 The lengths of the dollars are proportional to their spending power, but our eyes are drawn to the areas, which decline more than the lengths.

Now try Exercises 25–26.

Watch the Scales

Figure 5.23a shows the percentage of college students between 1910 and 2000 who were women. At first glance, it appears that there was a huge increase in this percentage between 1950 and 2000. But note that the vertical axis scale does not begin at zero and does not end at 100%. The increase looks far less dramatic if we redraw the graph with the vertical axis covering the full range of 0 to 100% (Figure 5.23b). From a mathematical point of view, leaving out the zero point on a scale is perfectly honest

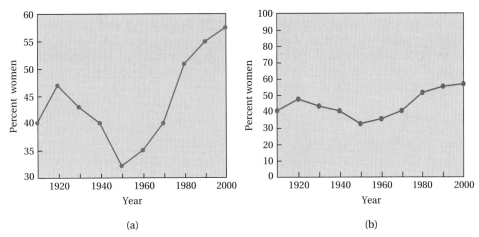

Women as a Percentage of All College Students

(a)

(b)

FIGURE 5.23 Both graphs show the same data, but they look very different because their vertical scales have different ranges.
SOURCE: National Center for Education Statistics.

and can make it easier to see small-scale trends in the data. Nevertheless, as this example shows, it can be visually deceptive if you don't study the scale carefully.

Now try Exercises 27–28.

Sometimes the scale may not be deceptive, but still requires care to avoid misinterpretation. Consider Figure 5.24a, which shows maximum computer speeds from 1950 to 2000. At first glance, it appears that speeds have been increasing linearly. For example, it might look as if the speed increased by the same amount from 1990 to 2000 as it did from 1950 to 1960. However, if we look closely, we see that each tick mark on the vertical scale represents a *tenfold* increase in speed. Now we see that computer speed grew from about 1 to 100 calculations per second between 1950 and

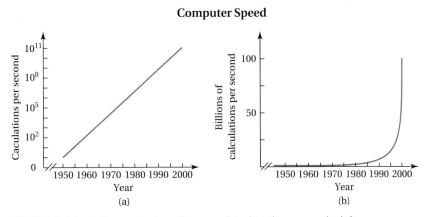

Computer Speed

(a)

(b)

FIGURE 5.24 Both graphs show the same data, but the one on the left uses an exponential scale.

HISTORICAL NOTE
In 1965, Intel founder Gordon E. Moore predicted that advances in technology would allow computer chips to double in power roughly every two years. This idea is now called *Moore's law,* and it has held fairly true ever since Moore first stated it.

1960, and from about 100 million to 10 billion calculations per second between 1990 and 2000. This type of scale is called an *exponential scale* (or *logarithmic scale*), because each unit corresponds to a power of 10. In general, exponential scales are useful for displaying data that vary over a huge range of values. You can see this usefulness by looking at Figure 5.24b, where the computer data have been recast with an ordinary scale. Because the speeds have grown so rapidly, the ordinary scale makes it impossible to see any detail in the early years shown on the graph.

Now try Exercise 29.

Time out to think _____

Based on Figure 5.24a, can you predict the speed of the fastest computers in 2010? Could you make the same prediction with Figure 5.24b? Explain.

Percentage Change Graphs

Is college getting more or less expensive? A quick look at the downward-sloping lines in the lower graph in Figure 5.25 might give the impression that college costs fell steadily in the 1990s. But on closer examination, we find that this is not the case at all. The vertical axis on the lower graph represents the *percentage increase* in costs. Thus, the downward-sloping lines show only that the rate of increase in college costs decreased, *not* that the actual costs decreased. In fact, the actual costs have risen substantially, as

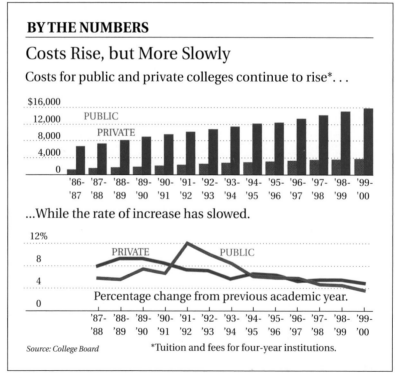

FIGURE 5.25 *SOURCE:* New York Times.

shown in the upper graph. Moreover, because the rate of inflation (as measured by the Consumer Price Index; see Unit 3D) has been less than the rate of increase in college costs, the *real cost* of public colleges has steadily risen. Graphs that show percentage change are very common, particularly with economic data. Although they are perfectly honest, you can be easily misled unless you interpret them with great care.

Now try Exercise 30.

Pictographs

Pictographs are graphs embellished with additional artwork. The artwork may make the graph more appealing, but it can also distract or mislead. Figure 5.26 is a pictograph showing the rise in world population from 1804 to 2054 (numbers for future years are based on United Nations projections). The lengths of the bars correspond correctly to world population for the different years listed. However, the artistic embellishments of this graph are deceptive in several ways. For example, your eye may be drawn to the figures of people lining the globe. Because this line of people rises from the left side of the pictograph to the center and then falls, it might give the impression that future world population will be declining. In fact, the line of people is purely decorative and carries no information.

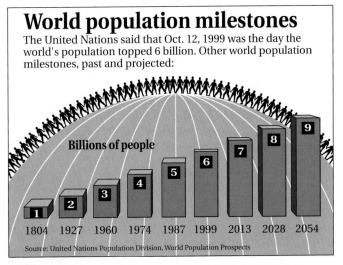

World population milestones
The United Nations said that Oct. 12, 1999 was the day the world's population topped 6 billion. Other world population milestones, past and projected:

Billions of people

1 — 1804 2 — 1927 3 — 1960 4 — 1974 5 — 1987 6 — 1999 7 — 2013 8 — 2028 9 — 2054

Source: United Nations Population Division, World Population Prospects

FIGURE 5.26 *SOURCE: James Abundis and Bob Laird,* USA Today, *October 18, 1999.*

Perhaps the most serious problem with this pictograph is that it makes it appear that world population has been rising linearly. However, notice that the time intervals on the horizontal axis are not uniform in size. For example, the interval between the bars for 1 billion and 2 billion people is 123 years (from 1804 to 1927), but the interval between the bars for 5 billion and 6 billion people is only 12 years (from 1987 to 1999).

Pictographs are very common, but as this example shows, you have to study them carefully to extract the essential information and not be distracted by the cosmetic effects.

Now try Exercise 31.

by the way

If world population continues to double at the same rate as in the late 20th century, it will reach 34 billion by 2100 and 192 billion by 2200. By about 2650, human population would be so large that it would not fit on the Earth, even if everyone stood elbow-to-elbow everywhere.

■ EXERCISES 5D

Review Questions

1. Briefly describe the construction and use of multiple bar charts and stack plots.

2. What are geographical data? Briefly describe at least two ways to display geographical data. Be sure to explain the meaning of contours on a contour map.

3. What are three-dimensional graphics? Explain the difference between graphics that only appear three-dimensional and those that show truly three-dimensional data.

4. Describe how perceptual distortions can arise in graphics and how they can be misleading.

5. How can graphics be misleading when the scales do not go all the way to zero? Why are such graphics sometimes useful?

6. What is an exponential scale? When is an exponential scale useful?

7. Explain how a graph that shows percentage change can show descending bars (or a descending line) even when the variable of interest is increasing.

8. What is a pictograph? How can a pictograph enhance a graph? How can it make a graph misleading?

Does It Make Sense?

Decide whether each of the following statements makes sense (or is clearly true) or does not make sense (or is clearly false). Explain your reasoning.

9. My bar chart contains more information than yours, because I made my bars three-dimensional.

10. I used an exponential scale because the data values for my categories ranged from 7 to 450,000.

11. There's been only a very slight rise in our stock price over the past few months, but I wanted to make it look dramatic so I started the vertical scale from the lowest price rather than from zero.

12. A graph showing the yearly rate of increase in the number of computer users has a slight downward trend, even though the actual number of users is rising.

Basic Skills & Concepts

13. **Net Grain Production.** Net grain production is the difference between the amount of grain a country produces and the amount of grain its citizens consume. It is positive if the country produces more than it consumes, and negative if the country consumes more than it produces. Figure 5.27 shows the net grain production of four countries in 1990 and projected for 2030.

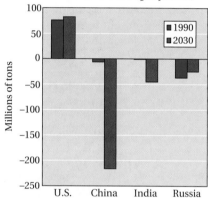

FIGURE 5.27

a. Which of the four countries had to import grain to meet its needs in 1990?

b. Which of the four countries are expected to need to import grain to meet needs in 2030?

c. Given that India and China are the world's two most populous countries, what does this graph tell you about how world agriculture will have to change between now and 2030?

14. **Education and Earnings.** Figure 5.28 shows mean (average) earnings in three different years according to level of education.

a. Briefly explain the meaning of each of the three sets of bars on the graph.

b. Compare the change in earnings between 1975 and 1995 for people with bachelor's degrees to the change for people who did not graduate from high school. What do these data say about the value of a college education?

c. The graph has a three-dimensional appearance. Is it showing true three-dimensional data, or is the appearance purely cosmetic? Do you think the three-dimensional appearance helps or hinders the display?

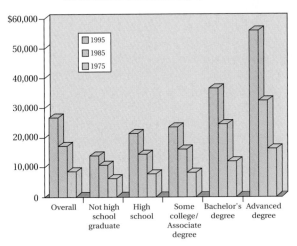

Mean Earnings of Workers 18 Years and Over by Educational Attainment, 1975 to 1995

Source: Census Bureau, Current Population Survey

FIGURE 5.28 *SOURCE:* TIME Almanac, *1999, p. 886.*

15. **Stack Plot.** Answer the following based on Figure 5.16.

 a. State whether the death rate for each of the four diseases individually decreased or increased between 1900 and 2000.

 b. When was the death rate due to cardiovascular diseases the greatest, and what was it?

 c. What was the death rate due to cancer in 2000?

 d. Based on the trends in the graph, which of these four diseases will be responsible for the most deaths in 2050? Explain.

16. **College Degrees.** Figure 5.29 shows the numbers of college degrees awarded to men and women over time.

 a. Estimate the numbers of college degrees awarded to men and to women (separately) in 1930 and in 2000.

 b. Compare the numbers of degrees awarded to men and to women (separately) in 1980 and 2000.

 c. During what decade did the *total* number of degrees awarded increase the most?

 d. Compare the *total* numbers of degrees awarded in 1950 to 2000.

 e. Do you think the stack plot is an effective way to display these data? Briefly discuss other ways that might have been used instead.

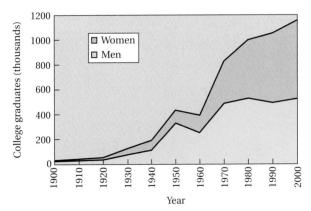

College Degrees Awarded

FIGURE 5.29

17. **Federal Spending.** Figure 5.30 shows the changes in major spending categories of the federal budget. (*Payments to individuals* includes Social Security and Medicare; *net interest* represents interest payments on the national debt; *all other* represents non-defense discretionary spending.) Interpret the stack plot and discuss some of the trends it reveals.

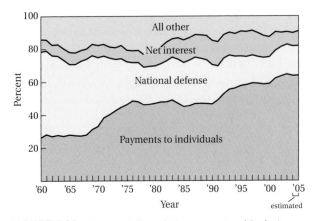

Percentage Composition of Federal Government Outlays

FIGURE 5.30 *SOURCE: Office of Management and Budget.*

 a. Find the percentage of the budget that went to net interest in 1980, 1990, 1995, and 2000.

 b. Find the percentage of the budget that went to defense in 1960, 1980, and 2000.

 c. Find the percentage of the budget that went to payments to individuals in 1960, 1980, and 2000.

18. **Federal Trends.** Consider Figure 5.30. Summarize at least three trends shown in the figure.

Probability That a Black Student Would Have White Classmates

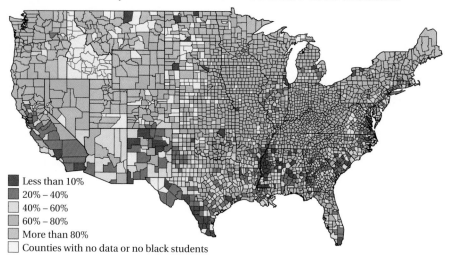

- ■ Less than 10%
- ■ 20% – 40%
- ☐ 40% – 60%
- ▨ 60% – 80%
- ▨ More than 80%
- ☐ Counties with no data or no black students

FIGURE 5.31 SOURCE: New York Times, *April 2, 2000.*

19. School Segregation. One way of measuring segregation is to determine the likelihood that a black student will have white classmates. A *New York Times* study found that, by this measure, segregation increased significantly in the 1990s. Figure 5.31 shows the probability that a black student had white classmates, by county, during the 1997–1998 academic year. Do there appear to be any significant regional differences? Can you pick out any differences between urban and rural areas? Discuss possible explanations for a few of the trends that you see in the figure.

20. Contour Elevations. Contour maps are often used to show geographical elevations. Figure 5.32 shows elevation

contours around Boulder, Colorado. Discuss a few key features shown on the map.

U.S. Age Distribution. Exercises 21–22 refer to Figure 5.33.

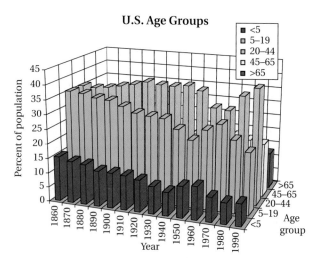

FIGURE 5.33

21. a. Briefly describe the meaning of each bar.

b. Based on this graph, how has the percentage of the youngest Americans changed since 1860?

c. Does this graph show true three-dimensional data, or is the three-dimensional look purely cosmetic? Explain.

FIGURE 5.32

22. **a.** Estimate the percentages of 5- to 19-year-olds in 1860 and 1990.

 b. Estimate the percentages of 20- to 44-year-olds in 1860 and 1990.

 c. Describe and explain the change in the percentage of under-5-year-olds from 1940 to 1960.

23. **Extending the Olympic Graph.** Make a list of all the data you would need in order to extend the graph in Figure 5.21 to the 2004 Olympics and beyond.

24. **Data for 2004 Olympics.** By the time you read this book, the 2004 Olympics should be over. Use the Web to find the data you need to extend Figure 5.21 (see Exercise 23). Then photocopy the graph and add the 2004 data on the same graph.

25. **Volume Distortion.** Figure 5.34 uses television sets to represent the numbers of homes with cable in 1980 and 2000. Note that the heights of the TVs represent the numbers of homes. Briefly explain how the graph creates a perceptual distortion that exaggerates the true change in the number of homes with cable.

Homes with Cable TV

18 million homes 72 million homes

FIGURE 5.34

26. **Three-Dimensional Pies.** The pie charts in Figure 5.35 represent the percentage of Americans in three age categories in 1990 and 2050 (projected). Briefly explain how the three-dimensional effects create a perceptual distortion in this case. Why would flat pies (without the three-dimensional effects) give a more accurate representation of the data?

1990 Age Distribution 2050 Age Distribution

Others Others

FIGURE 5.35 SOURCE: U.S. Census Bureau.

27. **Comparing Earnings.** Figure 5.36 compares the average weekly earnings of men and women. Identify any misleading aspects of the display. Draw the display in a fairer way.

FIGURE 5.36 SOURCE: U.S. Census Bureau.

28. **Braking Distances.** Figure 5.37 shows the braking distance for four different cars. Discuss the ways in which it might be deceptive. How much greater is the braking distance of Lincolns than the braking distance of Oldsmobiles? Draw the display in a fairer way.

FIGURE 5.37 SOURCE: Car and Driver.

29. Computer Sales. The following table shows the number of worldwide shipments of personal computers from 1981 through 1999. Draw the graph both with an ordinary vertical scale and with an exponential vertical scale. (Hint: For the exponential scale, use ticks at 1 million, 10 million, and 100 million.) Which graph is more useful? Why?

Year	Worldwide shipments
1981	1,630,000
1983	11,120,000
1985	14,700,000
1987	16,680,000
1989	21,330,000
1991	26,970,000
1993	38,850,000
1995	60,170,000
1997	82,400,000
1999	112,000,000

30. Rising College Costs. Refer to Figure 5.25 to answer the following questions.

a. In what academic year did public college costs rise by the largest percentage? What was the percentage increase?

b. In the same year, what was the percentage increase in private college costs?

c. In the same year, which had the larger increase in actual cost (in dollars): public or private colleges? Explain.

31. World Population. Recast Figure 5.26 with a proper horizontal axis. What trends are clear in your new graph that are not clear in the original? Explain.

Further Applications

Creating Graphics. Exercises 32–35 give tables of real data. For each table, make a graphical display of the data. You may choose any graphic type that you feel is appropriate to the data set. In addition to making the display, write a few sentences explaining why you chose this type of display and a few sentences describing interesting patterns in the data.

32. Percent Never Married. The following table shows the percentages, for 1970 and 2000, of men and women in various age categories who were never married.

Women	1970	2000	Men	1970	2000
20–24	35.8	72.8	20–24	54.7	83.7
25–29	10.5	38.9	25–29	19.1	51.7
30–34	6.2	21.9	30–34	9.4	30.0
35–39	5.4	14.3	35–39	7.2	20.3
40–44	4.9	11.8	40–44	6.3	15.7

SOURCE: U.S. Census Bureau.

33. Alcohol on the Road. The following table gives the number of automobile fatalities in which (i) no alcohol was involved, (ii) moderate alcohol was involved (blood alcohol content between 0.01 and 0.09), and (iii) a high level of alcohol was involved (blood alcohol content above 0.1). All figures are in thousands of deaths.

Year	No alcohol	Moderate alcohol	High alcohol
1982	18.7	4.8	20.4
1984	20.5	4.8	19.0
1986	22.0	5.1	18.9
1988	23.5	4.9	18.7
1990	22.5	4.4	17.7
1992	21.4	3.6	14.2
1994	24.1	3.5	13.1
1996	24.8	3.8	13.4
1998	25.5	3.7	12.2
2000	25.1	3.9	12.8

SOURCE: National Highway Traffic Safety Administration.

34. Daily Newspapers. The following table gives the number of daily newspapers and their total circulation (in millions) for selected years since 1920.

Year	Number of daily newspapers	Circulation (millions)
1920	2042	27.8
1930	1942	39.6
1940	1878	41.1
1950	1772	53.9
1960	1763	58.8
1970	1748	62.1
1980	1747	62.2
1990	1611	62.3
2000	1485	56.1

SOURCE: Editor & Publisher.

35. Firearm Fatalities. The following table summarizes deaths due to firearms in different nations in a recent year.

Country	Total	Homicides	Suicides	Fatal accidents
U.S.	35,563	15,835	18,503	1225
Germany	1197	168	1004	25
Canada	1189	176	975	38
Australia	536	96	420	20
Spain	396	76	219	101
U.K.	277	72	193	12
Sweden	200	27	169	4
Vietnam	131	85	16	30
Japan	93	34	49	10

SOURCE: *Coalition to Stop Gun Violence.*

36. Seasonal Effects on Schizophrenia? The graph in Figure 5.38 shows data regarding the relative risk of schizophrenia among people born in different months.

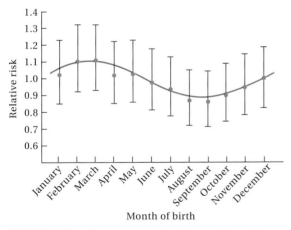

FIGURE 5.38 SOURCE: *New England Journal of Medicine.*

a. Note that the scale of the vertical axis does not include zero. Sketch the same risk curve using an axis that includes zero. Comment on the effect of this change.

b. Each value of the relative risk is shown with a dot at its most likely value and with an "error bar" indicating the range in which the data value probably lies. The study concludes that "the risk was also significantly associated with the season of birth." Given the size of the error bars, does this claim appear justified? (Is it possible to draw a flat line that passes through all of the error bars?)

Web Projects

Find useful links for Web Projects on the text Web site: www.aw.com/bennett-briggs

37. Weather Maps. Many Web sites offer contour maps with current weather data. For example, you can use the Yahoo Weather site to generate many different contour weather maps. Generate at least two contour weather maps and discuss what they show.

38. Cancer Cure. As shown in Figure 5.16, cancer is one of the leading causes of death today. Nevertheless, scientists have made great progress in treating many forms of cancer. Go to the American Cancer Society Web site and investigate research into cancer cures. Read about one or two recent studies, and write a short report on what you learn. Be sure to include graphics in your report.

39. USA Snapshot. The *USA Today* Web site offers a daily pictograph for its "USA Snapshot." Study today's snapshot. Briefly discuss its purpose and effectiveness.

In the News

40. News Graphics. Find a recent news report that shows a multiple bar chart or stack plot. Comment on the effectiveness of the display. Could another display have been used to depict the same data?

41. Geographical Data. Find an example of a graph of geographical data in a recent news report. Comment on the effectiveness of the display. Could another display have been used to depict the same data?

42. Three-Dimensional Effects. Find an example of a three-dimensional display in a recent news report. Are the data three-dimensional or are the three-dimensional effects cosmetic? Comment on the effectiveness of the display. Could another display have been used to depict the same data?

43. Graphic Confusion. Find an example in a recent news report of a graph that is misleading in one of the ways discussed in this unit. Explain what makes the graph misleading, and describe how it could have been drawn more honestly.

44. Outstanding News Graph. Find a graph from a recent news report that, in your opinion, is truly outstanding in displaying data visually. Discuss what the graph shows, and explain why you think it is so outstanding.

Smoking is one of the leading causes of statistics.

—FLETCHER KNEBEL

CORRELATION AND CAUSALITY

A major goal of many statistical studies is to determine whether one factor *causes* another. For example, does smoking cause lung cancer? In this unit, we will discuss how statistics can be used to search for *correlations* that might suggest a cause-and-effect relationship. Then we'll explore the more difficult task of establishing causality.

SEEKING CORRELATION

What does it mean when we say that smoking *causes* lung cancer? It certainly does *not* mean that you'll get lung cancer if you smoke a single cigarette. It does not even mean that you'll definitely get lung cancer if you smoke heavily for many years, since some heavy smokers do not get lung cancer. Rather, it is a *statistical* statement meaning that you are *much more likely* to get lung cancer if you smoke than if you don't smoke.

Let's try to understand how researchers learned that smoking causes lung cancer. Before they could investigate *cause*, researchers first needed to establish correlations between smoking and cancer. The process of establishing correlations began with observations. The early observations were informal. Doctors noticed that smokers made up a surprisingly high proportion of their patients with lung cancer. This suggestion of a linkage led to carefully conducted studies in which researchers compared lung cancer rates among smokers and nonsmokers. These studies showed clearly that heavier smokers were more likely to get lung cancer. In more formal terms, we say that there is a correlation between the variables *amount of smoking* and *incidence of lung cancer*. A correlation is a special type of relationship between variables, in which a rise or fall in one goes along with a corresponding rise or fall in the other.

DEFINITION

A **correlation** exists between two variables when higher values of one variable consistently go with higher values of another or when higher values of one variable consistently go with lower values of another.

Here are a few other examples of correlations:

- There is a correlation between the variables *height* and *weight* for people. That is, taller people tend to weigh more than shorter people.
- There is a correlation between the variables *demand for apples* and *price of apples*. That is, demand tends to decrease as prices increase.
- There is a correlation between *practice time* and *skill* among piano players. That is, those who practice more tend to be more skilled.

Establishing a correlation between two variables does *not* mean that a change in one variable *causes* a change in the other. Thus, finding the correlation between smoking and lung cancer did not by itself prove that smoking causes lung cancer. We could imagine, for example, that some gene predisposes a person both to smoking and to lung cancer. Nevertheless, identifying the correlation was the crucial first step in learning that smoking causes lung cancer.

Time out to think _____

Suppose there really were a gene that made people prone to both smoking and lung cancer. Explain why we would still find a strong correlation between smoking and lung cancer in that case, but would not be able to say that smoking caused lung cancer.

by the way

Smoking is linked to many serious diseases besides lung cancer, including heart disease and emphysema. Smoking is also linked with less lethal health conditions such as premature skin wrinkling and sexual impotence.

Scatter Diagrams

Table 5.6 shows the production cost and gross receipts (total revenue from ticket sales) for 15 of the biggest-budget science fiction and fantasy movies of all time. Movie executives presumably hope there is a favorable correlation between the production budget and the receipts. That is, they hope that spending more to produce a movie will result in higher box office receipts. But is there such a correlation? We can look for a correlation by making a **scatter diagram** showing the relationship between the variables *production cost* and *gross receipts*.

TABLE 5.6 Movie Data

MOVIE	PRODUCTION COST (MILLIONS OF DOLLARS)	GROSS RECEIPTS (MILLIONS OF DOLLARS)
Adventures of Pluto Nash (2002)	100	5
Armageddon (1998)	140	200
Battlefield Earth (2000)	80	21
Final Fantasy: The Spirits Within (2001)	137	32
Harry Potter and the Chamber of Secrets (2002)	100	300
Harry Potter and the Sorcerer's Stone (2001)	130	320
Lord of the Rings: The Fellowship of the Ring (2001)	109	315
Lord of the Rings: The Two Towers (2002)	94	350
Red Planet (2000)	80	17
Spider-Man (2002)	140	404
Star Wars: Attack of the Clones (2002)	115	302
Terminator 3: Rise of the Machines (2003)	170	350
The Matrix Reloaded (2003)	127	300
X2: X-Men United (2003)	125	215
Waterworld (1995)	175	88

NOTE: *Gross receipts are for United States only.*

DEFINITION

A **scatter diagram** is a graph in which each point represents the values of two variables.

The following procedure describes how we make the scatter diagram, which is shown in Figure 5.39:

1. We assign one variable to each axis, and we label each axis with values that comfortably fit the data. Here, we assign *production cost* to the horizontal axis and *gross receipts* to the vertical axis. We choose a range of $50 to $200 million for the production cost axis and $0 to $450 million for the gross receipts axis.

2. For each movie in Table 5.6, we plot a *single point* at the horizontal position corresponding to its production cost and the vertical position corresponding to its gross receipts. For example, the point for the movie *X2* goes at a position of $125 million on the horizontal axis and $215 million on the vertical axis. The dashed lines on Figure 5.39 show how we locate this point.

3. (Optional) If we wish, we can label data points, as is done for selected points in Figure 5.39.

Time out to think _____

By studying Table 5.6, associate each of the unlabeled data points in Figure 5.39 with a particular movie.

Types of Correlation

Look carefully at the scatter diagram for movies in Figure 5.39. The dots seem to be scattered about with no apparent pattern. In other words, at least for the movies shown, there appears to be little or no correlation between the amount of money spent producing the movie and the amount of money it earned in gross receipts.

Now consider the scatter diagram in Figure 5.40, which shows the weights (in carats) and retail prices of 23 diamonds. Here, the dots show a clear upward trend,

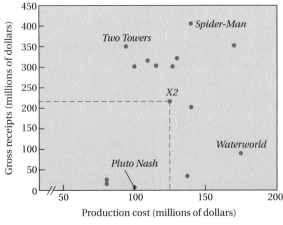

FIGURE 5.39 Scatter diagram for the data in Table 5.6.

FIGURE 5.40 A scatter diagram for diamond weights and prices.

indicating that larger diamonds generally cost more. The correlation is not perfect. For example, the heaviest diamond is not the most expensive. But the overall trend seems fairly clear. Because the prices tend to increase with the weights, we say that Figure 5.40 shows a **positive correlation.**

In contrast, Figure 5.41 shows a scatter diagram for the variables *life expectancy* and *infant mortality* in 16 countries. We again see a clear trend, but this time it is a **negative correlation:** Countries with *higher* life expectancy tend to have *lower* infant mortality.

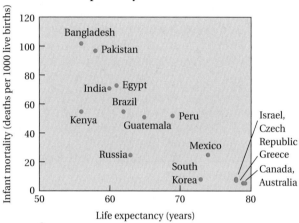

FIGURE 5.41 A scatter diagram for life expectancy and infant mortality.

Besides stating whether a correlation exists, we can also discuss its strength. The more closely the data follow the general trend, the stronger is the correlation.

Relationships Between Two Data Variables

No correlation: There is no apparent relationship between the two variables.

Positive correlation: Both variables tend to increase (or decrease) together.

Negative correlation: The two variables tend to change in opposite directions, with one increasing while the other decreases.

Strength of a correlation: The more closely two variables follow the general trend, the stronger the correlation (which may be either positive or negative). In a perfect correlation, all data points lie on a straight line.

> *by the way*
> In statistics, the *correlation coefficient* provides a quantitative measure of the strength of a correlation. It is defined to be 1 for a perfect positive correlation (i.e., all data points lie on a single, straight line), −1 for a perfect negative correlation, and 0 for no correlation.

> **EXAMPLE 1** Inflation and Unemployment

Prior to the 1990s, most economists assumed that the unemployment rate and the inflation rate were negatively correlated. That is, when unemployment goes down, inflation goes up, and vice versa. Table 5.7 shows unemployment and inflation data for the period 1990–2002. Make a scatter diagram for these data. Based on

TABLE 5.7	U.S. Inflation and Unemployment	
YEAR	UNEMPLOYMENT RATE (%)	INFLATION RATE (%)
1990	5.6	5.4
1991	6.8	4.2
1992	7.5	3.0
1993	6.9	3.0
1994	6.1	2.6
1995	5.6	2.8
1996	5.4	3.0
1997	4.9	2.3
1998	4.6	2.3
1999	4.3	2.2
2000	4.0	3.4
2001	4.2	1.8
2002	5.8	1.6

SOURCE: U.S. Bureau of Labor Statistics.

your diagram, does it appear that the data support the historical claim of a link between the unemployment and inflation rates?

Solution We make the scatter diagram by plotting the variable *unemployment rate* on the horizontal axis and the variable *inflation rate* on the vertical axis. To make the graph easy to read, we use values ranging from 3.5% to 8% for the unemployment rate and from 0 to 6% for the inflation rate. Figure 5.42 shows the result. To the eye, there does not appear to be any obvious correlation between the two variables. (A calculation confirms that there is no appreciable correlation.) Thus, these data do *not* support the historical claim of a negative correlation between the unemployment and inflation rates.

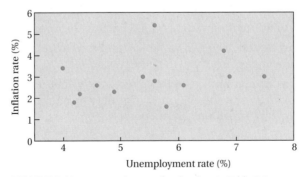

FIGURE 5.42 Scatter diagram for the data in Table 5.7.

Now try Exercises 13–14. ◄

➤**EXAMPLE 2** Accuracy of Weather Forecasts

The scatter diagrams in Figure 5.43 show two weeks of data comparing the actual high temperature for the day with the same-day forecast (left diagram) and the three-day forecast (right diagram). Discuss the types of correlation on each diagram.

FIGURE 5.43 Comparison of actual high temperatures with same-day and three-day forecasts.

Solution Both scatter diagrams show a general trend in which higher predicted temperatures mean higher actual temperatures. Thus, both show positive correlations. However, the points in the left diagram lie more nearly on a straight line, indicating a stronger correlation than in the right diagram. This makes sense, because we expect weather forecasts to be more accurate on the same day than three days in advance. *Now try Exercises 15–16.* ◄

Possible Explanations for a Correlation

We began by stating that correlations can help us search for cause-and-effect relationships. But we've already seen that causality is not the only possible explanation for a correlation. For example, the predicted temperatures on the horizontal axis of Figure 5.43 certainly do not *cause* the actual temperatures on the vertical axis. The following box summarizes three possible explanations for a correlation.

Possible Explanations for a Correlation

1. The correlation may be a *coincidence*.
2. Both variables might be directly influenced by some *common underlying cause.*
3. One of the correlated variables may actually be a *cause* of the other. Note that, even in this case, we may have identified only one of several causes.

➤**EXAMPLE 3** Explanation for a Correlation

Consider the correlation between infant mortality and life expectancy in Figure 5.41. Which of the three possible explanations for a correlation applies? Explain.

Solution The negative correlation between infant mortality and life expectancy is probably an example of common underlying cause. Both variables respond to an underlying variable that we might call *quality of health care*. In countries where health care is better in general, infant mortality is lower and life expectancy is higher.

Now try Exercises 17–18. ◄

➤ **EXAMPLE 4** How to Get Rich in the Stock Market (Maybe)

Every financial advisor has a strategy for predicting the direction of the stock market. Most focus on fundamental economic data, such as interest rates and corporate profits. But an alternative strategy relies on a remarkable correlation between the Super Bowl winner in January and the direction of the stock market for the rest of the year: The stock market tends to rise when a team from the old, pre-1970 NFL wins the Super Bowl, and tends to fall otherwise. This correlation successfully matched 28 of the first 32 Super Bowls to the stock market. Suppose that the Super Bowl just ended and the winner was the Detroit Lions, an old NFL team. Should you invest all your spare cash (and maybe even some that you borrow) in the stock market?

Solution Based on the reported correlation, you might be tempted to invest, since the old-NFL winner suggests a rising stock market over the rest of the year. However, this investment would make sense only if you believed that the Super Bowl result actually *causes* the stock market to move in a particular direction. This belief is clearly preposterous, and the correlation is undoubtedly a coincidence. If you are going to invest, don't base your investment on this correlation. *Now try Exercises 19–24.* ◄

ESTABLISHING CAUSALITY

Suppose you have discovered a correlation and suspect causality. How can you test your suspicion? Let's return to the issue of smoking and lung cancer. The strong correlation between smoking and lung cancer did not by itself prove that smoking causes lung cancer. In principle, we could have looked for proof with a controlled experiment. But such an experiment would be unethical, since it would require forcing a group of randomly selected people to smoke cigarettes. So how was smoking established as a cause of lung cancer?

The answer involves several lines of evidence. First, researchers found correlations between smoking and lung cancer among many groups of people: women, men, and people of different races and cultures. Second, among groups of people that seemed otherwise identical, lung cancer was found to be rarer in nonsmokers. Third, people who smoked more and for longer periods of time were found to have higher rates of lung cancer. Fourth, when researchers accounted for other potential causes of lung cancer (such as exposure to radon gas or asbestos), they found that almost all the remaining lung cancer cases occurred among smokers.

These four lines of evidence made a strong case, but still did not rule out the possibility that some other factor, such as genetics, predisposes people both to smoking and to lung cancer. However, two additional lines of evidence made this possibility highly unlikely. One line of evidence came from animal experiments. In controlled experiments, animals were divided into randomly chosen treatment and control groups. The experiments still found a correlation between inhalation of ciga-

The truth is rarely pure and never simple.

—OSCAR WILDE

rette smoke and lung cancer, which seems to rule out a genetic factor, at least in the animals. The final line of evidence came from biologists studying cell cultures (that is, small samples of human lung tissue). The biologists discovered the basic process by which ingredients in cigarette smoke can create cancer-causing mutations. This process does not appear to depend in any way on specific genetic factors, making it all but certain that lung cancer is caused by smoking and not by any preexisting genetic factor.

The following box summarizes these ideas about establishing causality. Generally speaking, the case for causality is stronger when more of these guidelines are met.

Guidelines for Establishing Causality

To investigate whether a suspected cause actually causes an effect:

1. Look for situations in which the effect is correlated with the suspected cause even while other factors vary.

2. Among groups that differ only in the presence or absence of the suspected cause, check that the effect is similarly present or absent.

3. Look for evidence that larger amounts of the suspected cause produce larger amounts of the effect.

4. If the effect might be produced by other potential causes (besides the suspected cause), make sure that the effect still remains after accounting for these other potential causes.

5. If possible, test the suspected cause with an experiment. If the experiment cannot be performed with humans for ethical reasons, consider doing the experiment with animals, cell cultures, or computer models.

6. Try to determine the physical mechanism by which the suspected cause produces the effect.

Time out to think

There's a great deal of controversy concerning whether animal experiments are ethical. What is your opinion of animal experiments? Defend your opinion.

CASE STUDY Air Bags and Children

By the mid-1990s, passenger-side air bags had become commonplace in cars. Statistical studies showed that the air bags saved many lives in moderate- to high-speed collisions. But a disturbing pattern also appeared. In at least some cases, young children, especially infants and toddlers in child car seats, were killed by air bags in low-speed collisions.

At first, many safety advocates found it difficult to believe that air bags could be the cause of the deaths. But the observational evidence became stronger, meeting the first four guidelines for establishing causality. For example, the greater risk to infants in child car seats fit Guideline 3, because it indicated that being closer to the

by the way

The first four guidelines for establishing causality are called *Mill's methods*, after the English philosopher and economist John Stuart Mill (1806–1873). Mill was a leading scholar of his time and an early advocate of women's right to vote.

by the way

Based on these studies, the government now recommends that child car seats *never* be used on the front seat, and that children under age 12 sit in the back seat if possible.

air bags increased the risk of death. (A child car seat sits on top of the built-in seat, thereby putting a child closer to the air bags than the child would be otherwise.)

To seal the case, safety experts undertook experiments using dummies. They found that children, because of their small size, often sit where they could be easily hurt by the explosive opening of an air bag. The experiments also showed that an air bag could impact a child car seat hard enough to cause death, thereby revealing the physical mechanism by which the deaths occurred.

CASE STUDY What Causes Global Warming?

Statistical measurements suggest that the average temperature of the entire Earth, or the *global average temperature,* has risen slightly (by about 1°F) in the past 50 years. But what is causing this so-called *global warming?*

Most scientists attribute the rise to the increasing concentration of carbon dioxide and other greenhouse gases in our atmosphere (Figure 5.44). Comparative studies of the Earth and other planets, particularly Venus and Mars, show that greenhouse gases meet the first three guidelines for establishing causality. Indeed, there is virtually no doubt that the average temperature of each planet is strongly influenced by its atmospheric concentration of greenhouse gases. The case for greenhouse gases also meets Guideline 6, as scientists have clearly established the physical mechanism by which these gases can cause global warming. (Basically, greenhouse gases slow the escape of heat from a planet's surface, thereby causing the surface to warm up.) Thus, greenhouse gases, the suspected cause of global warming, satisfy at least four of the six guidelines for establishing causality.

<div style="float:left; width:22%;">

by the way

Some greenhouse gases are present naturally in the Earth's atmosphere. Without them, the global average temperature would be a frigid −10°F. With them, the global average temperature is about 59°F. Thus, the greenhouse effect is a good thing for life on Earth. On Venus, where the temperature would be pleasantly warm with an Earth-like atmosphere, a very strong greenhouse effect produces a global average temperature of about 870°F—demonstrating that it's possible to have too much of a good thing.

</div>

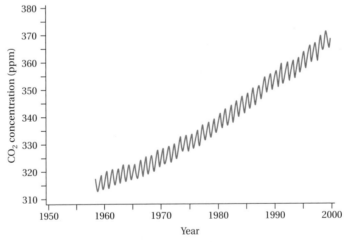

FIGURE 5.44 This graph shows data for carbon dioxide concentration (in parts per million; see Unit 2B) collected on Mauna Loa in Hawaii. The yearly wiggles show seasonal changes in the concentration, but the long-term trend is clearly upward.
SOURCE: Dave Keeling and Tim Whorf, Scripps Institution of Oceanography.

Nevertheless, we cannot be certain that the observed global warming of the past 50 years has been caused by greenhouse gases. The trouble comes mainly from

Guideline 4, which tells us to account for other potential causes. One such potential cause is an increase in heating of the Earth by the Sun. Some evidence suggests that the Sun has become slightly warmer during the past century, thereby increasing the amount of heat the Earth receives by a few hundredths of 1%. If true, this could account for some or all of the observed global warming.

The ideal way to sort out the causes would be with an experiment (Guideline 5), but it is impossible to perform experiments with entire planets. Instead, researchers run experiments with *computer models* of planets that account for the way weather and climate work. In this way, researchers can compare results among models with varying amounts of heat from the Sun and varying amounts of greenhouse gases. Unfortunately, the models made by different research groups tend to give different results. Unless everyone can agree on what constitutes a valid computer model, the controversy about the cause of global warming will persist.

CASE STUDY Ozone and CFCs

While the human connection to global warming remains controversial, evidence is much more definitive for another environmental issue: the depletion of the Earth's atmospheric ozone. Ozone is a gas that exists naturally at high altitudes, where it absorbs harmful ultraviolet radiation from the Sun. Ozone depletion (a reduction in the total amount of ozone) could pose a serious problem by allowing more of this ultraviolet radiation to reach the ground, where it can cause mutations, cancers, and deaths of both plants and animals.

By the early 1970s, experiments had shown that human-made chemicals called *chlorofluorocarbons* (CFCs) could destroy ozone. CFCs were invented in the 1930s and have been widely used in air conditioners and refrigerators, in the manufacture of packaging foam, as propellants in spray cans, in industrial solvents used in the computer industry, and for many other purposes. Because of their many everyday uses, it seemed unwise to ban CFCs on the basis of a few experiments, especially since there was no evidence at the time that any ozone depletion was actually occurring.

The situation changed in the mid-1980s with the discovery of the so-called *ozone hole* over Antarctica (Figure 5.45). Researchers found that the formation of this hole (generally lasting from August through November) was strongly correlated with the presence of chlorine and fluorine—both components of CFCs—in the atmosphere. They soon identified the physical mechanism that led to the fairly extreme ozone depletion over Antarctica compared to other places. (The mechanism involves chemical reactions that occur on the surfaces of ice crystals that form in the Antarctic clouds.)

With CFCs confirmed as a cause of damage to the Earth's ozone, negotiations soon began on international treaties that now ban the manufacture of CFCs. Unfortunately, these treaties have not by themselves put the problem of ozone depletion behind us. Most of the CFCs made during the 20th century are either still in use or in discarded appliances and materials in landfills, and probably will leak into the atmosphere over the next few decades. In addition, because substitute chemicals tend to be more expensive than CFCs, there is now a large international black market for CFCs manufactured in violation of the international treaties.

by the way

Paul Crutzen, Mario Molina, and F. Sherwood Rowland received the 1995 Nobel Prize in chemistry for their work in discovering the linkage between CFCs and ozone destruction.

EP/TOMS Total Ozone Oct 1, 2000

GSFC/916

Dobson Units
GEN:277/2000 Dark Gray < 100, Red > 500 DU

FIGURE 5.45 This map shows ozone distribution on October 1, 2000, as measured from a satellite. The key explains how colors represent ozone concentrations. The regions of gray, pink, and purple around Antarctica represent the ozone hole, where ozone concentrations are far below normal. The white strips on the map are regions where the satellite did not collect data.
SOURCE: NASA/NOAA Earth Probe TOMS data.

Time out to think

Suppose your car air conditioner is broken and will cost $200 to fix. You hear about a shop that can fix it for only $100—undoubtedly by using black market CFCs. Given that your single air conditioner won't make much difference to the ozone layer, is it worth the $100 difference to avoid the black market CFCs? Defend your opinion.

CONFIDENCE IN CAUSALITY

If human activity is *causing* global warming, we'd be wise to change our activities so as to stop it. But we cannot yet be sure that this is the case. Moreover, the changes needed to slow global warming might be very expensive. Given the huge cost and the uncertainty as to whether the changes would even make a difference, how can we set sensible policies?

In an ideal world, we would continue to study the issue until we could establish whether human activity is the cause of global warming. However, we have seen that it is difficult to establish causality and often impossible to *prove* causality beyond all doubt. We are therefore forced to make decisions about global warming, and many other important issues, despite great uncertainty about cause and effect.

In other areas of mathematics, accepted techniques help us deal with uncertainty by allowing us to calculate numerical measures of possible errors. But there are no accepted ways to assign such numbers to the uncertainty that comes with questions of causality. Fortunately, another area of study has dealt with practical problems of causality for hundreds of years: our legal system. You may be familiar with the following three broad ways of expressing a legal level of confidence.

Broad Levels of Confidence in Causality

Possible cause: We have discovered a correlation, but cannot yet determine whether the correlation implies causality. In the legal system, possible cause (such as thinking that a particular suspect possibly caused a particular crime) is often the reason for starting an investigation.

Probable cause: We have good reason to suspect that the correlation involves cause, perhaps because some of the guidelines for establishing causality are satisfied. In the legal system, probable cause is the general standard for getting a judge to grant a warrant for a search or wiretap.

Cause beyond reasonable doubt: We have found a physical model that is so successful in explaining how one thing causes another that it seems unreasonable to doubt the causality. In the legal system, cause beyond reasonable doubt is the usual standard for conviction. It generally demands that the prosecution show how and why (essentially the physical model) the suspect committed the crime. Note that beyond *reasonable* doubt does *not* mean beyond *all* doubt.

While these broad levels remain fairly vague, they give us at least some common language for discussing confidence in causality. If you study law, you will learn much more about the subtleties of interpreting these terms. However, because statistics has little to say about them, we will not discuss them much further in this book.

Time out to think
Given what you know about global warming, do you think that human activity is a possible cause, probable cause, or cause beyond reasonable doubt? Defend your opinion. Based on your level of confidence in the causality, how would you recommend setting policies with regard to global warming?

■ EXERCISES 5E

Review Questions

1. What is a correlation? Give three examples of pairs of variables that are correlated.

2. What is a scatter diagram, and how do you make one? How can we use a scatter diagram to look for a correlation?

3. Define and distinguish among positive correlation, negative correlation, and no correlation. How do we determine the strength of a correlation?

4. Describe the three general categories of explanation for a correlation. Give an example of each.

5. Briefly describe each of the six guidelines presented in this unit for establishing causality. Give an example of the application of each guideline.

6. Briefly describe three levels of confidence in causality and how they can be useful when we do not have absolute proof of causality.

Does It Make Sense?

Decide whether each of the following statements makes sense (or is clearly true) or does not make sense (or is clearly false). Explain your reasoning.

7. There is a strong negative correlation between the price of tickets and the number of tickets sold. This suggests that if we want to sell a lot of tickets, we should lower the price.

8. There is a strong positive correlation between the amount of time spent studying and grades in mathematics classes. This suggests that if you want to get a good grade, you should spend more time studying.

9. I found a nearly perfect positive correlation between variable A and variable B, and therefore was able to conclude that an increase in variable A causes an increase in variable B.

10. I found a nearly perfect negative correlation between variable C and variable D, and therefore was able to conclude that an increase in variable C causes a decrease in variable D.

11. I had originally suspected that an increase in variable E would cause a decrease in variable F, but I no longer believe this because I found no correlation between the two variables.

12. I agree that we should require kids to wear helmets if helmets really lower injury rates, but it makes no sense to start this requirement until we have absolute proof that helmets *cause* the lower injury rate.

Basic Skills & Concepts

Interpreting Scatter Diagrams. Exercises 13–16 each show a scatter diagram with its axes labeled. For each exercise, do the following:

a. Indicate the variables for which we can seek a correlation with this diagram.

b. State whether the diagram shows a positive correlation, a negative correlation, or no correlation. If there is a positive or negative correlation, state whether it is strong or weak.

c. In words, summarize any conclusions you can draw from the diagram.

13.

14.

15.

16.

Children Ages 14–16

Types of Correlation. Exercises 17–24 list pairs of variables. State the units you would use to measure each of the two variables (for example, pounds, years, or miles per hour). Then state whether you believe the two variables are correlated. If you believe they are correlated, state whether the correlation is positive or negative and strong or weak. Explain your reasoning.

17. Body weight and intake of calories per day

18. Height and shoe size

19. Rate of pedaling and speed of the bicycle

20. Altitude on a mountain hike and surrounding air temperature

21. Blood alcohol content and reaction time (for an individual)

22. Weight of car and gas mileage

23. Weight of car and price of car

24. Retail price of an item and the number of items that can be sold in a day

Further Applications

Making Scatter Diagrams. Exercises 25–30 each give a table of data. In each case, do the following:

a. Make a scatter diagram for the data.

b. State whether the two variables appear to be correlated and, if so, whether the correlation is positive or negative and strong or weak.

c. Suggest a reason for the correlation (or lack of correlation). If you suspect causality, briefly discuss what further evidence you would need to establish it.

25. The following table gives speed limits and death rates for automobile accidents in selected countries in the 1980s.

Country	Death rate (per 100 million vehicle miles)	Speed limit (miles per hour)
Norway	3.0	55
United States	3.3	55
Finland	3.4	55
Britain	3.5	70
Denmark	4.1	55
Canada	4.3	60
Japan	4.7	55
Australia	4.9	65
Netherlands	5.1	60
Italy	6.1	75

SOURCE: D. J. Rivkin, New York Times.

26. The following table gives number of home runs and batting averages for baseball's Most Valuable Players, 1996–2002 (NL = National League and AL = American League).

Player	Home runs	Batting average
Ken Caminiti (1996 NL)	40	.326
Juan Gonzalez (1996 AL)	47	.314
Larry Walker (1997 NL)	49	.366
Ken Griffey Jr. (1997 AL)	56	.304
Sammy Sosa (1998 NL)	66	.308
Juan Gonzalez (1998 AL)	45	.318
Chipper Jones (1999 NL)	45	.319
Ivan Rodriguez (1999 AL)	35	.332
Jeff Kent (2000 NL)	33	.334
Jason Giambi (2000 AL)	43	.333
Barry Bonds (2001 NL)	73	.328
Ichiro Suzuki (2001 AL)	8	.350
Barry Bonds (2002 NL)	46	.370
Miguel Tejada (2002 AL)	34	.308

27. The following table gives per capita personal income and percent of the population below the poverty level for ten states between 1980 and 1990.

State	Per capita personal income (dollars)	Percent of population below poverty level
Nevada	30,529	8.2
California	32,275	12.9
New Hampshire	33,332	4.8
Utah	23,907	9.6
Colorado	32,949	7.9
Minnesota	32,101	5.9
Montana	22,569	15.9
Illinois	32,259	11.6
Iowa	26,723	7.1
West Virginia	21,915	13.9

SOURCE: U.S. Census Bureau; U.S. Bureau of Economic Analysis.

28. The following table gives the average hours of television watched in households in five categories of annual income. (Hint: For the first and last categories of the household income data, place the dot at the position corresponding to $25,000 and $65,000, respectively. For other categories, place the dot at the center of each bin.)

Household income	Weekly TV hours
Less than $30,000	56.3
$30,000–$40,000	51.0
$40,000–$50,000	50.5
$50,000–$60,000	49.7
More than $60,000	48.7

SOURCE: Nielsen Media Research.

29. The following table gives the average teacher salary and the expenditure on public education per pupil for ten states.

State	Average teacher salary (dollars)	Per pupil expenditure (dollars)
Alabama	36,689	4946
Alaska	47,262	8834
Arizona	35,650	4505
Connecticut	51,780	9792
Massachusetts	46,250	8750
North Dakota	29,863	4512
Oregon	40,919	8605
Texas	37,567	6092
Utah	34,946	4036
Wyoming	34,140	6911

SOURCE: U.S. Department of Education.

30. The following table gives mean daily Caloric intake (all residents) and infant mortality rate (per 1000 births) for ten countries.

Country	Mean daily Calories	Infant mortality rate (per 1000 births)
Afghanistan	1523	154
Austria	3495	6
Burundi	1941	114
Colombia	2678	24
Ethiopia	1610	107
Germany	3443	6
Liberia	1640	153
New Zealand	3362	7
Turkey	3429	44
United States	3671	7

Correlation and Causality. Exercises 31–36 make statements about a correlation. In each case, state the correlation clearly (for example, there is a positive correlation between variable A and variable B). Then state whether the correlation is most likely due to coincidence, a common underlying cause, or a direct cause. Explain your answer.

31. In one U.S. city, the crime rate increased at the same time that the number of people in prison increased.

32. Over the past three decades, the number of miles of free-ways in Los Angeles has grown, and traffic congestion has worsened.

33. When gasoline prices rise, the number of airline passengers increases.

34. Sales of ice tea in a local restaurant are positively correlated with ticket sales at the local swimming pool.

35. When deer hunting permit costs were gradually increased in one region, the incidence of deer sightings increased.

36. Over a period of twenty years, the number of ministers and priests increased, as did attendance at movies.

37. **Identifying Causes: Headaches.** You are trying to identify the cause of late-afternoon headaches that plague you several days each week. For each of the following tests and observations, explain which of the six guidelines for establishing causality you used and what you concluded.

 • The headaches occur only on days that you go to work.

 • If you stop drinking Coke at lunch, the headaches persist.

 • In the summer, the headaches occur less frequently if you open the windows of your office slightly. They occur even less often if you open the windows of your office fully.

 Having made all these observations, what reasonable conclusion can you reach about the cause of the headaches?

38. **Smoking and Lung Cancer.** There is a strong correlation between tobacco smoking and incidence of lung cancer, and most physicians believe that tobacco smoking causes lung cancer. Yet, not everyone who smokes gets lung cancer. Briefly describe how smoking could cause cancer when not all smokers get cancer.

39. **Longevity of Orchestra Conductors.** A famous study in *Forum on Medicine* (1978) concluded that the mean lifetime of conductors of major orchestras was 73.4 years, about 5 years longer than that of all American males at the time. The author claimed that a life of music *causes* a longer life. Evaluate the claim of causality and propose other explanations for the longer life expectancy of conductors.

40. **High-Voltage Power Lines.** Suppose that people living near a particular high-voltage power line have a higher incidence of cancer than people living farther from the power line. Can you conclude that the high-voltage power line is the cause of the elevated cancer rate? If not, what other explanations might there be for it? What other types of research would you like to see before you conclude that high-voltage power lines cause cancer?

Web Projects

Find useful links for Web Projects on the text Web site: www.aw.com/bennett-briggs

41. **Success in the NFL.** Use the Web to find last season's NFL team statistics. Make a table showing the following for each team: number of wins, average yards gained on offense per game, and average yards allowed on defense per game. Make scatter diagrams to explore the correlations between offense and wins and between defense and wins. Discuss your findings. Do you think that there are other team statistics that would yield stronger correlations with the number of wins?

42. **Statistical Abstract.** Explore the "frequently requested tables" at the Web site for the *Statistical Abstract of the United States*. Choose data that are of interest to you and explore at least two correlations. Briefly discuss what you learn from the correlations.

43. **Air Bags and Children.** Starting from the Web site of the National Highway Traffic Safety Administration, research the latest studies on the safety of air bags, especially with regard to children. Write a short report summarizing your findings and offering recommendations for improving child safety in cars.

44. **Global Warming.** Use the Web to find recent information about global warming and its potential consequences. Discuss the evidence linking human activity to global warming. In light of your findings, suggest how we should deal with the issue of global warming.

45. **Ozone Depletion.** Starting with information available at the Environmental Protection Administration (EPA) Web site, research the latest information about ozone depletion and efforts to combat it. Summarize your findings and suggest what should be done in the future to help prevent ozone depletion.

46. **Tobacco Lawsuits.** Tobacco companies have been the subject of many lawsuits relating to the dangers of smoking. Research one recent lawsuit. What were the plaintiffs

trying to prove? What statistical evidence did they use? How well do you think they established causality? Did they win? Summarize your findings in one to two pages.

In the News

47. **Correlations in the News.** Find a recent news report that describes some type of correlation. Describe the correlation. Does the article give any sense of the strength of the correlation? Does it suggest that the correlation reflects any underlying causality? Briefly discuss whether you believe the implications the article makes with respect to the correlation.

48. **Causation in the News.** Find a recent news report in which a statistical study has led to a conclusion of causation. Describe the study and the claimed causation. Do you think the claim of causation is legitimate? Explain.

49. **Legal Causation.** Find a news report concerning an ongoing legal case, either civil or criminal, in which establishing causality is important to the outcome. Briefly describe the issue of causation in the case and how the ability to establish or refute causality will influence the outcome of the case.

CHAPTER 5 SUMMARY

UNIT	KEY TERMS	KEY IDEAS AND SKILLS
5A	statistics *is* a science *are* data population, sample population parameters, sample statistics bias observational study case-control study experiment placebo, placebo effect blinding single-blind double-blind margin of error confidence interval	**Understand and interpret** the five basic steps in a statistical study. **Understand** the importance of a representative sample. **Be familiar with** four common sampling methods: simple random sampling systematic sampling convenience sampling stratified sampling **Distinguish between** observational studies and experiments; also recognize observational case-control studies. **Understand** the placebo effect and the importance of blinding in experiments. **Find** a confidence interval from a margin of error: from (sample statistic − margin of error) to (sample statistic + margin of error)
5B	selection bias participation bias variable (in a statistical study)	**Understand and apply** eight guidelines for evaluating a statistical study.
5C	frequency table categories frequency relative frequency cumulative frequency data types qualitative quantitative bar chart pie chart histogram line chart time-series diagram	**Interpret and create** frequency tables. **Interpret and create** bar graphs and pie charts. **Interpret and create** histograms and line charts.

(Continues on the next page)

5D	multiple bar graph stack plot geographical data contour map	**Interpret** multiple bar graphs, stack plots, contour maps, and other media graphs. **Distinguish between** true three-dimensional data and graphs that have a three-dimensional look for cosmetic reasons only. **Be aware of** common cautions about graphs.
5E	correlation cause scatter diagram	**Distinguish between** correlation and causality. **Create and interpret** scatter diagrams and use them to identify correlations: positive, negative, or no correlation strength of correlation **Know** three possible explanations for a correlation: coincidence common underlying cause true cause **Understand and apply** six guidelines for establishing causality.

Some people hate the very name statistics, but I find them full of beauty and interest. Whenever they are not brutalized, but delicately handled by the higher methods, and are warily interpreted, their power of dealing with complicated phenomena is extraordinary.

—SIR FRANCIS GALTON (1822–1911)

CHAPTER 6

Putting Statistics to Work

In Chapter 5, we explored statistics as they are presented in the news: reports on major statistical studies, opinion polls, graphics, and attempts to establish cause and effect. Now, we are ready to do some actual statistical calculations. With the tools in this chapter, you will have a strong introduction to the powerful science of statistics.

UNIT **6A** **Characterizing Data:** We often describe data by their average and their distribution. We'll study such descriptions and discover the multiple meanings of "average."

UNIT **6B** **Measures of Variation:** We'll discuss common measures of the variation, or spread, in a data set: the range, the five-number summary, and the standard deviation.

UNIT **6C** **The Normal Distribution:** The normal distribution is the formal name for the famous bell-shaped curve. We'll study the characteristics of the normal distribution and learn why it is so important.

UNIT **6D** **Statistical Inference:** The heart of most statistical studies lies with *inferring* a conclusion about a population from results for a sample. We'll explore three important ideas about inference: statistical significance, the margin of error, and hypothesis testing.

UNIT 6A

CHARACTERIZING A DATA DISTRIBUTION

Frequency tables and graphs (see Units 5C and 5D) show us how data are distributed over various categories. More formally, we say that a table or graph gives the **distribution** of a data set. For example, a distribution of essay grades shows how many students received grades of A, B, C, D, and F. In this case, the categories are the grades.

DEFINITION

The **distribution** of a variable (or data set) describes the values taken on by the variable and the frequency (or relative frequency) of these values.

In many cases, however, we are less interested in the complete distribution than in a few descriptive terms that summarize it. In this unit, we'll study how we can characterize a data distribution by its center, or average, and by its shape.

WHAT IS AVERAGE?

The term *average* comes up frequently in the news, but you may be surprised to learn that it does not always have the same meaning. Let's begin by studying three common ways of characterizing the center of a data distribution, all of which are sometimes called the "average."

Mean, Median, and Mode

Table 6.1 shows the number of movies (original and sequels or prequels) in each of five popular science fiction series.

What is the average number of films in these series? One way to answer this question is to compute the **mean.** We find the mean by dividing the total number of movies by five (because there are five series):

$$\text{mean} = \frac{4 + 3 + 5 + 10 + 5}{5} = \frac{27}{5} = 5.4$$

In other words, these five series have a mean of 5.4 movies. More generally, we find the mean of any data set by summing all the data values and then dividing by the number of data values. The mean is what most people think of as the average. In essence, it represents the balance point for a data distribution, as shown in Figure 6.1.

We could also describe the average number of films in these five series by computing the **median,** or middle value, of the data set. To find a median, we arrange the data values in ascending (or descending) order, repeating data values that

TABLE 6.1	Five Science Fiction Movie Series	
TITLE	NUMBER OF MOVIES IN SERIES (AS OF 2003)	
Alien	4	
Back to the Future	3	
Planet of the Apes*	5	
Star Trek	10	
Star Wars	5	

*Not including remake.

appear more than once. If the number of values is odd, there is exactly one value in the middle of the list, and this value is the median. If the number of values is even, there are two values in the middle of the list, and the median is the number that lies halfway between them. Putting the data in Table 6.1 in ascending order gives the list 3, 4, 5, 5, 10. The median number of movies is 5 because 5 is the middle number in the list.

Another value sometimes used as a measure of average is the **mode,** which is the most common value (or group of values) in a data set. In the case of the movies, the mode is 5 because this value occurs twice in the data set, while the other values occur once. A data set may have one mode, more than one mode, or no mode. Sometimes the mode refers to a group of closely spaced values rather than a single value.

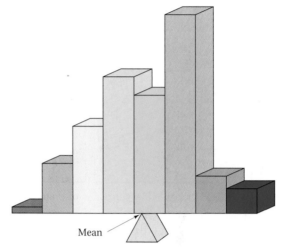

FIGURE 6.1 A histogram made from blocks would balance at the position of its mean.

Measures of Center in a Distribution

The **mean** is what we most commonly call the average value. It is defined as follows:

$$\text{mean} = \frac{\text{sum of all values}}{\text{total number of values}}$$

The **median** is the middle value in the sorted data set (or halfway between the two middle values if the number of values is even).

The **mode** is the most common value (or group of values) in a distribution.

➤ **EXAMPLE 1** Price Data

Eight grocery stores sell the PR energy bar for the following prices:

$1.09 $1.29 $1.29 $1.35 $1.39 $1.49 $1.59 $1.79

Find the mean, median, and mode for these prices.

Solution The mean price is $1.41:

$$\text{mean} = \frac{\left(\begin{array}{c}\$1.09 + \$1.29 + \$1.29 + \$1.35 \\ + \ \$1.39 + \$1.49 + \$1.59 + \$1.79\end{array}\right)}{8} = \$1.41$$

To find the median, we first sort the data in ascending order:

$1.09 $1.29 $1.29 $1.35 $1.39 $1.49 $1.59 $1.79

 3 values below 2 middle values 3 values above

by the way

A University of Chicago study found that the median allowance for teenagers in this country is about $50 per week. The study estimated that teenagers receive over $1 billion per week in allowances.

FRAZZ reprinted by permission of United Feature Syndicate, Inc.

Because there are eight prices (an even number), there are two values in the middle of the list: $1.35 and $1.39. Therefore the median lies halfway between these two values, which we calculate by adding them and dividing by 2:

$$\text{median} = \frac{\$1.35 + \$1.39}{2} = \$1.37$$

The mode is $1.29, because this is the only price that occurs more than once.

Now try Exercises 13–18. ◄

Effects of Outliers

To explore the differences among the mean, median, and mode, imagine that the five graduating seniors on a college basketball team receive the following first-year contract offers to play in the National Basketball Association (zero indicates that the player did not receive a contract offer):

$$\$0 \quad \$0 \quad \$0 \quad \$0 \quad \$3{,}500{,}000$$

The mean contract offer is

$$\text{mean} = \frac{\$0 + \$0 + \$0 + \$0 + \$3{,}500{,}000}{5} = \$700{,}000$$

Is it therefore fair to say that the *average* contract offer was $700,000?

Not really. The problem is that the single player receiving the large offer makes the mean much larger than it would be otherwise. In fact, if we ignore this one player and look only at the other four, the mean contract offer is zero! Because this one value of $3,500,000 is so extreme compared to the others, we call it an **outlier** (or *outlying value*). As our example shows, an outlier can pull the mean significantly upward (or downward), thereby making the mean unrepresentative of the data set as a whole.

DEFINITION

An **outlier** is a data value that is much higher or much lower than almost all other values.

While the outlier pulls the mean contract offer upward, it has no effect on the median contract offer, which remains zero for the five players. In fact, the value of an

outlier never has an effect on the median, because outliers don't lie in the middle of a data set. Outliers do not affect the mode either.

Time out to think _____
Is it fair to use the median as the *average* contract offer for the five players? Why or why not?

➤ **EXAMPLE 2** Mistake?

A track coach wants to determine an appropriate heart rate for her athletes during their workouts. She chooses five of her best runners and asks them to wear heart rate monitors during a workout. In the middle of the workout, she reads the following heart rates for the five athletes: 130, 135, 140, 145, 325. Which is a better measure of the average in this case: the mean or the median? Why?

Solution Four of the five values are fairly close together and seem reasonable for mid-workout heart rates. The high value of 325 is an outlier. This outlier seems likely to be a mistake (perhaps caused by a faulty heart monitor), because anyone with such a high heart rate should be in cardiac arrest. If the coach uses the mean as the average, she will be including this outlier—which means she will be including any mistake made when it was recorded. If she uses the median as the average, she'll have a more reasonable value, because the median won't be affected by the outlier.

Now try Exercises 19–20. ◄

Confusion About "Average"

The different meanings of "average" can lead to confusion. Sometimes this confusion arises because we are not told whether the "average" is the mean or the median, and other times because we are not given enough information about how the average was computed. The following two examples illustrate such situations.

➤ **EXAMPLE 3** Wage Dispute

A newspaper surveys wages for assembly workers in regional high-tech companies and reports an average of $22 per hour. The workers at one large firm immediately request a pay raise, claiming that they work as hard as employees at other companies but their average wage is only $19. The management rejects their request, telling them that they are *overpaid* because their average wage, in fact, is $23. Can both sides be right? Explain.

Solution Both sides can be right if they are using different definitions of average. In this case, the workers may be using the median while management uses the mean. For example, imagine that there are only five workers at the firm and their wages are $19, $19, $19, $19, and $39. The median of these five wages is $19 (as the workers claimed), but the mean is $23 (as management claimed).

Now try Exercises 21–22. ◄

Figures won't lie, but liars will figure.

—CHARLES H. GROSVENOR

➤ **EXAMPLE 4** Which Mean?

All 100 first-year students at a small college take three courses in the Core Studies Program. The first two courses are taught in large lectures, with all 100 students in a single class. The third course is taught in ten classes of 10 students each.

Students and administrators get into an argument about whether classes are too large. The students claim that the mean size of their Core Studies classes is 70. The administrators claim that the mean class size is only 25 students. Can both sides be right? Explain.

Solution Both sides are right, but they are using different means. The students have calculated the mean size of the classes in which each student is personally enrolled. Each student is taking two classes with enrollments of 100 each and one class with an enrollment of 10, so the mean size of each student's classes is

$$\frac{\text{total enrollment in student's classes}}{\text{number of classes student is taking}} = \frac{100 + 100 + 10}{3} = 70$$

The administrators have calculated the mean enrollment in all classes. There are two classes with 100 students each and ten classes with 10 students each, making a total enrollment of 300 students in 12 classes. Thus, the mean enrollment per class is

$$\frac{\text{total enrollment}}{\text{number of classes}} = \frac{300}{12} = 25$$

The two claims about the mean are both correct, but very different, because the students and administrators used different means. The students calculated the mean *class size per student*, while the administrators calculated the mean number of *students per class*. *Now try Exercises 23–26.* ◄

Time out to think _____

In Example 4, could the administrators redistribute faculty assignments so that all classes had 25 students each? How? Discuss the advantages and disadvantages of such a change.

SHAPES OF DISTRIBUTIONS

We next turn our attention to describing the overall shape of a distribution. We can *see* the complete shape of a distribution on a graph. In essence, our goal is to describe the general shape in words. Although such descriptions carry less information than a table or graph, they are still useful. We will focus on three characteristics of a distribution: its number of peaks, its symmetry or skewness, and its variation.

Number of Peaks

Because we are interested primarily in the *general* shapes of distributions, it's often easier to work with smooth curves than actual data. Figure 6.2 shows three examples of this idea, two in which the actual distributions are shown as histograms and one in which the actual distribution is shown as a line chart. The smooth curves approximate the actual distributions, but do not show all of their details.

The first way we might characterize the shape is by its number of peaks. The distributions in Figures 6.2a and 6.2c are single-peaked (or *unimodal*). Note that the peak represents the mode of each distribution. The distribution in Figure 6.2b has two peaks, so we say that it is double-peaked (or *bimodal*). Other distributions may have no peaks (in which case they are called *uniform distributions*) or more than two peaks.

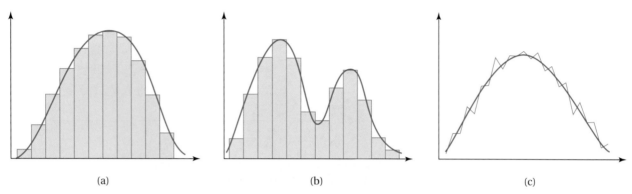

FIGURE 6.2 The smooth curves approximate the actual shapes of the distributions. Note that (a) and (c) are single-peaked (or unimodal), while (b) is double-peaked (or bimodal).

> **EXAMPLE 5** Number of Peaks

How many peaks would you expect for each of the following distributions? Why?

 a. Heights of all women at a college
 b. Heights of all students at a college
 c. The numbers of people with particular last digits (0 through 9) in their Social Security numbers

Solution

 a. The distribution of heights of women is single-peaked because many women are at or near the mean height, with fewer and fewer women at heights much taller or shorter than the mean.
 b. The distribution of heights of all students (men and women) will likely have two peaks, one corresponding to the mode of men's heights and one corresponding to the mode of women's heights.
 c. The last digits of Social Security numbers are essentially random, so the frequencies of the different digits (0 through 9) should be nearly equal. That is, about 10% of all Social Security numbers end in 0, 10% end in 1, and so on. The result is a uniform distribution with no peak.

Now try Exercises 27–34, part a. ◄

Symmetry or Skewness

A second way to describe shape is in terms of symmetry or skewness. A distribution is **symmetric** if its left half is a mirror image of its right half. The distributions in Figure 6.3 are all symmetric.

A distribution that is not symmetric must have values that tend to be more spread out on one side than the other. In this case, we say that the distribution is **skewed.** Figure 6.4a shows a distribution in which the values are more spread out on the left, making it **left-skewed.** Sometimes it is helpful to think of such a distribution as having a tail that has been pulled toward the left. Figure 6.4b shows a distribution in which the values are more spread out on the right, making it **right-skewed.** It looks as if it has a tail pulled toward the right.

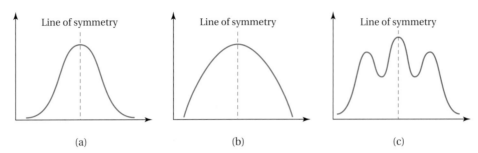

FIGURE 6.3 These distributions are all symmetric because their left halves are mirror images of their right halves. Note that (a) and (b) are single-peaked while (c) is triple-peaked (or trimodal).

Figure 6.4 also shows how skewness affects the relative positions of the mean, median, and mode. By definition, the mode is the peak in a single-peaked distribution. In a left-skewed distribution, the tail pulls the mean and median to the left of the mode. Moreover, the low values in the data set tend to make the mean less than the median. Similarly, a right-skewed distribution pulls the mean and median to the right—that is, to values greater than the mode. In such cases, the large values in the data set tend to make the mean greater than the median. When the distribution is symmetric and single-peaked, both the mean and the median are equal to the mode.

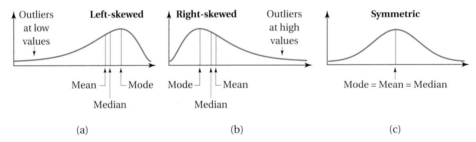

FIGURE 6.4 (a) Skewed to the left (left-skewed): The mean and median are less than the mode. (b) Skewed to the right (right-skewed): The mean and median are greater than the mode. (c) Symmetric distribution: The mean, median, and mode are the same.

Symmetry and Skewness

A distribution is **symmetric** if its left half is a mirror image of its right half.

A distribution is **left-skewed** if its values are more spread out on the left side.

A distribution is **right-skewed** if its values are more spread out on the right side.

Time out to think _____

In a skewed distribution, the median is a better measure of the center of the distribution than the mean. Why do you think this is true?

➤ **EXAMPLE 6** Skewness

For each of the following situations, state whether you expect the distribution to be symmetric, left-skewed, or right-skewed. Explain.

 a. Heights of a sample of 100 women
 b. Family income in the United States
 c. Speeds of cars on a road where a visible patrol car is using radar to detect speeders

Solution

 a. The distribution of heights of women is symmetric, because roughly equal numbers of women are shorter and taller than the mean and extremes of height are rare on either side of the mean.
 b. The distribution of family income is right-skewed. Most families are middle-class, so the mode of this distribution is a middle-class income (somewhere around \$40,000 in the United States). But a few very high-income families pull the mean to a considerably higher value, stretching the distribution to the right (high-income) side.
 c. Drivers usually slow down when they are aware of a patrol car looking for speeders. Few if any drivers will be exceeding the speed limit, but some drivers tend to slow to well below the speed limit. Thus, the distribution of speeds will be left-skewed, with a mode near the speed limit but a few cars going well below the speed limit. *Now try Exercises 27–34, part b.* ◄

by the way
Speed kills. On average in the United States, someone is killed in an auto accident about every 12 minutes. About one-third of these fatalities involve a speeding driver.

Time out to think _____

In ordinary English, the term *skewed* is often used to mean something that is distorted or depicted in an unfair way. How is this use of *skew* related to its meaning in statistics?

Variation

A third way to describe a distribution is by its **variation,** which is a measure of how much the data values are spread out. A distribution in which most data values are clustered together has a low variation. As shown in Figure 6.5a, such a distribution has a fairly sharp peak. The variation is higher when the data are distributed more widely around the center, which makes the peak broader. Figure 6.5b shows a distribution with moderate variation and Figure 6.5c shows a distribution with high variation. We'll discuss methods for quantitatively describing variation in Unit 6B.

 (a) (b) (c)

FIGURE 6.5 From left to right, these three distributions have increasing variation.

DEFINITION

Variation describes how widely data values are spread out about the center of a distribution.

▶**EXAMPLE 7** Variation in Marathon Times

How would you expect the variation to differ between times in the Olympic marathon and times in the New York marathon? Explain.

Solution The Olympic marathon invites only elite runners, whose times are likely to be clustered relatively near world record times. The New York marathon allows runners of all abilities, whose times are spread over a very wide range (from near the world record to many hours). Therefore, the variation among the times should be greater in the New York marathon than in the Olympic marathon.

Now try Exercises 27–34, part c.

EXERCISES 6A

Review Questions

1. Define and distinguish among mean, median, and mode.

2. What are outliers? Describe the effects of outliers on the mean, median, and mode.

3. Briefly describe at least two possible sources of confusion about the "average."

4. Give simple examples of a single-peaked distribution and a double-peaked distribution.

5. What do we mean when we say that a distribution is symmetric? Give simple examples of a symmetric distribution, a left-skewed distribution, and a right-skewed distribution.

6. What do we mean by the variation of a distribution? Give simple examples of distributions with different amounts of variation.

Does It Make Sense?

Decide whether each of the following statements makes sense (or is clearly true) or does not make sense (or is clearly false). Explain your reasoning.

7. In my data set of 10 exam scores, the mean turned out to be the score of the person with the third highest grade. No two people got the same score.

8. In my data set of 10 exam scores, the median turned out to be the score of the person with the third highest grade. No two people got the same score.

9. I made a distribution of 15 apartment rents in my neighborhood. One apartment had a much higher rent than all of the others, and this outlier caused the mean rent to be higher than the median rent.

10. If management and employees use the same data and do the calculations properly, they will always agree on the average wage.

11. The distribution of grades was left-skewed, but the mean, median, and mode were all the same.

12. There's much more variation in the ages of the general population than in the ages of students in my college extension course, but both turn out to have the same mean.

Basic Skills & Concepts

Mean, Median, and Mode. Exercises 13–18 each list a set of numbers. In each case, find the mean, median, and mode of the listed numbers.

13. Weights (in milligrams) of Bufferin aspirin tablets:

 672.2 679.2 669.8 672.6 672.2 662.2

14. Body temperatures (in degrees Fahrenheit) of randomly selected normal and healthy adults:

 98.6 98.6 98.0 98.0 99.0

 98.4 98.4 98.4 98.4 98.6

15. Blood alcohol content of drivers involved in fatal crashes and then convicted and given jail sentences (based on data from the U.S. Department of Justice):

 0.27 0.17 0.17 0.16 0.13 0.24

 0.29 0.24 0.14 0.16 0.12 0.16

16. Golf scores from a recent WPGA tournament:

 69 70 70 71 73 73 75 76 76 78

 80 81 81 82 83 84 85 86 87 88

17. High temperatures (in °F) during a 15-day period in Alaska in March:

 15 11 10 9 0 2 4 5

 5 7 10 12 15 18 19

18. Weights (in grams) of randomly selected M&M plain candies:

 0.957 0.912 0.842 0.925 0.939 0.886

 0.914 0.913 0.958 0.947 0.920

19. Outlier Coke. Cans of Coca-Cola vary slightly in weight. Here are the measured weights of seven cans, in pounds:

 0.8161 0.8194 0.8165 0.8176

 0.7901 0.8143 0.8126

Find the mean and median of these weights. Which, if any, of these weights would you consider to be an outlier? What are the mean and median weights if the outlier is excluded?

20. Alphabetic States. The following table gives the total area in square miles (land and water) of the seven states whose names begin with the letters A through C.

State	Area
Alabama	52,200
Alaska	615,200
Arizona	114,000
Arkansas	53,200
California	158,900
Colorado	104,100
Connecticut	5,500

Find the mean area and median area for these states. Which state is an outlier on the high end? If you eliminate

this state, what are the new mean and median areas for this data set? Which state is an outlier on the low end? If you eliminate this state, what are the new mean and median areas for this data set?

Appropriate Average. Exercises 21–26 list "averages" that someone might want to know. In each case, state whether the mean, median, or some other measure would give a better description of the "average." Explain your reasoning.

21. The average income of all adults in a large city

22. The average weight of the oranges in a large box

23. The average number of times that people change jobs during their careers

24. The average number of pieces of lost luggage per flight for each airline company

25. The average daily high temperature over a period of a month

26. The average waiting time in the checkout lines of a supermarket

Describing Distributions. For each distribution described in Exercises 27–34, answer the following:

 a. **How many peaks would you expect for the distribution? Explain.**

 b. **Would you expect the distribution to be symmetric, left-skewed, or right-skewed? Explain.**

 c. **Would you expect the variation of the distribution to be small or large? Explain.**

27. The exam scores on a 100-point exam where 50 students got an A, 20 got a B, and 5 got a C

28. The exam scores on an exam where 5 students got an A, 20 got a B, 40 got a C, 20 got a D, and 5 got an F

29. The weights of new cars at a dealership where about half the inventory consists of compact cars and the other half consists of sport utility vehicles

30. The numbers of people with particular last digits (0 through 9) in their phone numbers

31. The monthly sales of swimming suits over a one-year period at a store in San Diego

32. The monthly sales of milk over a one-year period at a large grocery store

33. The salaries of major league baseball players

34. The batting averages of major league baseball players

Further Applications

Smooth Distributions. For each of the histograms in Exercises 35–37, draw a smooth curve that captures its general features. Then classify the distribution according to its number of peaks, symmetry or skewness, and variation. In words, summarize the meaning of your results.

35. Times between 300 eruptions of Old Faithful geyser in Yellowstone National Park, shown in Figure 6.6.

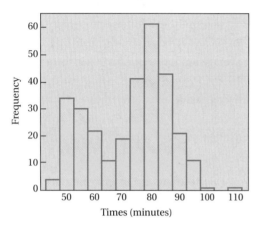

Times Between Eruptions of Old Faithful

FIGURE 6.6
SOURCE: *Hand et al.,* Handbook of Small Data Sets, *1994.*

36. Time until failure for a sample of 108 computer chips that failed, shown in Figure 6.7.

Failure Time of Computer Chips

FIGURE 6.7
SOURCE: *Hand et al.,* Handbook of Small Data Sets, *1994.*

37. Weights of a sample of 391 rugby players, shown in Figure 6.8

Weight of Rugby Players

FIGURE 6.8
SOURCE: *Hand et al.,* Handbook of Small Data Sets, *1994.*

38. Family Income. Suppose you study family income in a random sample of 300 families. You find that the mean family income is $41,000; the median is $35,000; and the highest and lowest incomes are $250,000 and $2400, respectively.

 a. Draw a rough sketch of the income distribution, with clearly labeled axes. Describe the distribution as symmetric, left-skewed, or right-skewed.

 b. How many families in the sample earned less than $35,000? Explain how you know.

 c. Based on the given information, can you determine how many families earned more than $41,000? Why or why not?

Web Projects

Find useful links for Web Projects on the text Web site: **www.aw.com/bennett-briggs**

39. Salary Data. Many Web sites offer data on salaries in different careers. Find salary data for a career you are considering. What are the mean and median salaries for this career? How do these salaries compare to those of other careers that interest you?

40. New York Marathon. The Web site for the New York marathon gives frequency data for finish times in the most recent marathon. Study the data, make a rough sketch of the distribution, and describe the shape of the distribution in words.

41. Tax Statistics. The IRS Web site provides statistics collected from tax returns on income, refunds, and much more. Choose a set of statistics from this Web site and study the distribution. Describe the distribution in words, and discuss anything you learn that is relevant to national tax policies.

42. Education Statistics. Visit the Web site of the National Center for Education Statistics and view the most recent copy of the *Digest of Educational Statistics*. Choose one of the many tables or graphs in the report that display a distribution. Describe the variables that are presented, discuss the effectiveness of the presentation, and interpret the results.

In the News

43. Averages in the News. Find three recent news articles that refer to some type of average. In each case, explain whether the average is a mean, a median, or some other type of average.

44. Daily Averages. Cite three examples of averages that you deal with in your own life (such as grade point average or batting average). In each case, explain whether the average is a mean, a median, or some other type of average. Briefly describe how the average is useful to you.

45. Distributions in the News. Find three recent examples in the news of distributions shown as histograms or line charts. Over each distribution, draw a smooth curve that captures its general features. Then classify the distribution according to its number of peaks, symmetry or skewness, and variation.

MEASURES OF VARIATION

n the previous unit, we saw how to describe variation in a qualitative sense. But the idea of variation is so important that we should also be able to describe it quantitatively. In this unit, we'll study several common ways of measuring variation.

WHY VARIATION MATTERS

Imagine customers waiting in line for tellers at two different banks. Customers at Big Bank can enter any one of three different lines leading to three different tellers. Best Bank also has three tellers, but all customers wait in a single line and are called to the next available teller. The following values are waiting times, in minutes, for eleven customers at each bank. The times are arranged in ascending order.

Big Bank (three lines):

| 4.1 | 5.2 | 5.6 | 6.2 | 6.7 | 7.2 | 7.7 | 7.7 | 8.5 | 9.3 | 11.0 |

Best Bank (one line):

| 6.6 | 6.7 | 6.7 | 6.9 | 7.1 | 7.2 | 7.3 | 7.4 | 7.7 | 7.8 | 7.8 |

You'll probably find more unhappy customers at Big Bank than at Best Bank, but it's *not* because the average wait is any longer. In fact, you can verify that the mean and median waiting times are 7.2 minutes at both banks. The difference in customer satisfaction comes from the *variation* in waiting times at the two banks. The

waiting times at Big Bank vary over a fairly wide range, so a few customers have long waits and are likely to become annoyed. In contrast, the variation of the waiting times at Best Bank is small, so all customers feel they are being treated roughly equally. The histograms in Figure 6.9 show the differences in variation visually.

FIGURE 6.9 Histograms for the waiting times at Big Bank and Best Bank, shown with data binned to the nearest minute.

Time out to think
Explain *why* Big Bank, with three separate lines, should have a greater variation in waiting times than Best Bank. Then consider several places where you commonly wait in lines, such as the grocery store, a bank, a concert ticket outlet, or a fast food restaurant. Does each place use a single customer line or multiple lines? If the place uses multiple lines, do you think a single line would be better? Explain.

Now try Exercises 13–14.

RANGE

The simplest way to describe the variation of a data set is to compute its **range,** defined as the difference between the highest (maximum) and lowest (minimum) values. For example, the waiting times for Big Bank vary from 4.1 to 11.0 minutes, so the range is $11.0 - 4.1 = 6.9$ minutes. The waiting times for Best Bank vary from 6.6 to 7.8 minutes, so the range is $7.8 - 6.2 = 1.2$ minutes. The range is much larger for Big Bank, reflecting its greater variation.

DEFINITION

The **range** of a data set is the difference between its highest and lowest data values:

$$\text{range} = \text{highest value (max)} - \text{lowest value (min)}$$

Although the range is easy to compute and can be useful, it occasionally can be misleading, as the next example shows.

➤ **EXAMPLE 1** Misleading Range

Consider the following two sets of quiz scores for nine students. Which set has the greater range? Would you also say that the scores in this set are more varied?

Quiz 1: 1 10 10 10 10 10 10 10 10

Quiz 2: 2 3 4 5 6 7 8 9 10

Solution The range for Quiz 1 is $10 - 1 = 9$ points and the range for Quiz 2 is $10 - 2 = 8$ points. Thus, the range is greater for Quiz 1. However, aside from a single low score (an outlier), Quiz 1 has no variation at all because every other student got a 10. In contrast, no two students got the same score on Quiz 2, and the scores are spread throughout the list of possible scores. Thus, the scores on Quiz 2 are more varied even though Quiz 1 has the greater range.

Now try Exercises 15–18, part a. ◄

QUARTILES AND THE FIVE-NUMBER SUMMARY

A better way to describe variation is to consider a few intermediate data values in addition to the high and low values. A common way involves looking at the quartiles, or values that divide the data distribution into quarters. The following list repeats the waiting times at the two banks, with the **quartiles** shown in bold. Note that the middle quartile, which divides the data set in half, is simply the median.

			lower quartile			median			upper quartile		
			↓			↓			↓		
Big Bank:	4.1	5.2	**5.6**	6.2	6.7	**7.2**	7.7	7.7	**8.5**	9.3	11.0
Best Bank:	6.6	6.7	**6.7**	6.9	7.1	**7.2**	7.3	7.4	**7.7**	7.8	7.8

DEFINITIONS

The **lower quartile** (or *first quartile*) divides the lowest fourth of a data set from the upper three-fourths. It is the median of the data values in the *lower half* of a data set.

The **middle quartile** (or *second* quartile) is the overall median.

The **upper quartile** (or *third quartile*) divides the lowest three-fourths of a data set from the upper fourth. It is the median of the data values in the *upper half* of a data set.

TECHNICAL NOTE
Statisticians do not universally agree on the precise methods for calculating quartiles, and different procedures can result in slightly different values.

Once we know the quartiles, we can describe a data set with a **five-number summary** consisting of the lowest value, the lower quartile, the median, the upper

DILBERT reprinted by permission of United Feature Syndicate, Inc.

quartile, and the highest value. For the waiting times at the two banks, the five-number summaries are

Big Bank:	Best Bank:
low = 4.1	low = 6.6
lower quartile = 5.6	lower quartile = 6.7
median = 7.2	median = 7.2
upper quartile = 8.5	upper quartile = 7.7
high = 11.0	high = 7.8

We can display the five-number summary with a **boxplot** (sometimes called a *box-and-whisker plot*). Using a number line for reference, we enclose the values from the lower to the upper quartile in a box. We then draw a line through the box at the median and add two "whiskers" extending from the box to the lowest and highest values. Figure 6.10 shows boxplots for the bank waiting times. Both the box and the whiskers are broader for Big Bank than for Best Bank, indicating that the waiting times have greater variation at Big Bank.

HISTORICAL NOTE

The boxplot was invented by American statistician John Tukey, widely regarded as one of the most accomplished statisticians of the 20th century. Tukey invented many powerful statistical methods and was a proponent of a thriving area of statistics called exploratory data analysis (EDA). Tukey died in July 2000 at the age of 85.

FIGURE 6.10 Boxplots for the waiting times at Big Bank and Best Bank.

The Five-Number Summary

The **five-number summary** for a data set consists of the following five numbers:

low value lower quartile median upper quartile high value

A **boxplot** shows the five-number summary visually, with a rectangular box enclosing the lower and upper quartiles, a line marking the median, and whiskers extending to the low and high values.

▶ **EXAMPLE 2** Race Times

Consider the following two sets of twenty 100-meter running times (in seconds):

Set 1:

| 9.92 | 9.97 | 9.99 | 10.01 | 10.06 | 10.07 | 10.08 | 10.10 | 10.13 | 10.13 |
| 10.14 | 10.15 | 10.17 | 10.17 | 10.18 | 10.21 | 10.24 | 10.26 | 10.31 | 10.38 |

Set 2:

| 9.89 | 9.90 | 9.98 | 10.05 | 10.35 | 10.41 | 10.54 | 10.76 | 10.93 | 10.98 |
| 11.05 | 11.21 | 11.30 | 11.46 | 11.55 | 11.76 | 11.81 | 11.85 | 11.87 | 12.00 |

Compare the variation in the two data sets with five-number summaries and boxplots.

Solution Each set has 20 data points, so the median lies halfway between the 10th and 11th times. The lower quartile is the median of the *lower half* of the times, which lies halfway between the 5th and 6th times. The upper quartile is the median of the *upper half* of the times, which lies halfway between the 15th and 16th times. You should confirm that the five-number summaries are as follows.

First Set of Times:	*Second Set of Times:*
low = 9.92	**low = 9.89**
lower quartile = 10.065	**lower quartile = 10.38**
median = 10.135	**median = 11.015**
upper quartile = 10.195	**upper quartile = 11.655**
high = 10.38	**high = 12.00**

Figure 6.11 shows the corresponding boxplots on the same scale, clearly illustrating the greater variation in the second set of times.

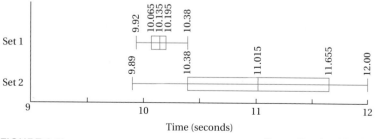

FIGURE 6.11 *Now try Exercises 15–18, part b.* ◀

STANDARD DEVIATION

The five-number summary characterizes variation well, but statisticians often prefer to describe variation with a single number. The single number most commonly used to describe variation is called the **standard deviation.**

The standard deviation is a measure of how far data values are spread around the mean of a data set. Thus, to calculate a standard deviation, we first find the mean and then find how much each data value "deviates" from the mean. Consider our earlier data sets, in which the mean waiting time was 7.2 minutes for both Big Bank and Best Bank. A waiting time of 8.2 minutes has a **deviation** from the mean of 8.2 minutes − 7.2 minutes = 1.0 minute, because it is 1.0 minute greater than the mean. A waiting time of 5.2 minutes has a deviation from the mean of 5.2 minutes − 7.2 minutes = −2.0 minutes (*negative* 2 minutes), because it is 2.0 minutes *less* than the mean.

In essence, the standard deviation is a measure of the average of all the deviations from the mean. However, for technical reasons, this average is not a simple mean. Instead, it is computed as described in the box below.

Although the standard deviation formula may look complex, it is easy to use for small data sets. Note that, because we square the deviations in Step 3 and take the square root in Step 5, the units of the standard deviation are the same as the units of the data values. For example, if the data values have units of minutes, the standard deviation also has units of minutes. The only difficulty with the standard deviation formula is that it is tedious to use for large data sets. Fortunately, many calculators and computers automate these calculations.

TECHNICAL NOTE

The formula given here, in which we divide the sum of the squared deviations by the total number of data values *minus* 1, is technically valid only for data from *samples*. When dealing with populations, we do not subtract 1. In this book, we will use only the sample formulas. Also note that the result of Step 4, which is the square of the standard deviation, is called the *variance*.

Calculating the Standard Deviation

Step 1: Compute the mean of the data set. Then find the deviation from the mean for every data value by subtracting the mean from the data value. That is, for every data value:

$$\text{deviation from mean} = \text{data value} - \text{mean}$$

Step 2: Find the squares (second power) of all the deviations from the mean.

Step 3: Add all the squares of the deviations from the mean.

Step 4: Divide this sum by the total number of data values *minus* 1.

Step 5: The standard deviation is the square root of this quotient.

All of these steps are included in the standard deviation formula:

$$\text{standard deviation} = \sqrt{\frac{\text{sum of (deviations from the mean)}^2}{\text{total number of data values} - 1}}$$

➤ **EXAMPLE 3** Calculating Standard Deviation

Calculate the standard deviations for the waiting times at Big Bank and Best Bank.

Solution We follow the five steps to calculate the standard deviations. Table 6.2 shows how to organize the work in the first three steps. The first column for each bank lists the waiting times (in minutes), the second column lists the deviations from the mean (Step 1), and the third column lists the squares of the deviations (Step 2). We add all the squared deviations to find the sum at the bottom of the third column (Step 3).

TABLE 6.2 Calculating Standard Deviation

BIG BANK			BEST BANK		
Time	Deviation (Time − Mean)	(Deviation)²	Time	Deviation (Time − Mean)	(Deviation)²
4.1	$4.1 - 7.2 = -3.1$	$(-3.1)^2 = 9.61$	6.6	$6.6 - 7.2 = -0.6$	$(0.6)^2 = 0.36$
5.2	$5.2 - 7.2 = -2.0$	$(-2.0)^2 = 4.00$	6.7	$6.7 - 7.2 = -0.5$	$(0.5)^2 = 0.25$
5.6	$5.6 - 7.2 = -1.6$	$(-1.6)^2 = 2.56$	6.7	$6.7 - 7.2 = -0.5$	$(0.5)^2 = 0.25$
6.2	$6.2 - 7.2 = -1.0$	$(-1.0)^2 = 1.00$	6.9	$6.9 - 7.2 = -0.3$	$(0.3)^2 = 0.09$
6.7	$6.7 - 7.2 = -0.5$	$(-0.5)^2 = 0.25$	7.1	$7.1 - 7.2 = -0.1$	$(0.1)^2 = 0.01$
7.2	$7.2 - 7.2 = 0.0$	$(0.0)^2 = 0.0$	7.2	$7.2 - 7.2 = 0.0$	$(0.0)^2 = 0.0$
7.7	$7.7 - 7.2 = 0.5$	$(0.5)^2 = 0.25$	7.3	$7.3 - 7.2 = 0.1$	$(0.1)^2 = 0.01$
7.7	$7.7 - 7.2 = 0.5$	$(0.5)^2 = 0.25$	7.4	$7.4 - 7.2 = 0.2$	$(0.2)^2 = 0.04$
8.5	$8.5 - 7.2 = 1.3$	$(1.3)^2 = 1.69$	7.7	$7.7 - 7.2 = 0.5$	$(0.5)^2 = 0.25$
9.3	$9.3 - 7.2 = 2.1$	$(2.1)^2 = 4.41$	7.8	$7.8 - 7.2 = 0.6$	$(0.6)^2 = 0.36$
11.0	$11.0 - 7.2 = 3.8$	$(3.8)^2 = 14.44$	7.8	$7.8 - 7.2 = 0.6$	$(0.6)^2 = 0.36$
Mean = 7.2		Sum = 38.46	Mean = 7.2		Sum = 1.98

For Step 4, we divide the sums from Step 3 by the total number of data values *minus* 1. Because there are 11 data values, we divide by 10:

Big Bank: $\dfrac{38.46}{10} = 3.846$

Best Bank: $\dfrac{1.98}{10} = 0.198$

Finally, Step 5 tells us that the standard deviation is the square root of the numbers from Step 4:

Big Bank: standard deviation $= \sqrt{3.846} \approx 1.96$ minutes

Best Bank: standard deviation $= \sqrt{0.198} \approx 0.44$ minute

We conclude that the standard deviation of the waiting times is 1.96 minutes at Big Bank and 0.44 minute at Best Bank. As expected, the standard deviation is larger for the bank with the greater variation in waiting times.

Now try Exercises 15–18, part c. ◄

Time out to think _____

Look closely at the individual deviations in Table 6.2 in Example 3. Do the standard deviations for the two data sets seem like reasonable "averages" for the deviations? Explain.

by the way

Another way of interpreting the standard deviation uses a mathematical rule called *Chebyshev's theorem*. It states that, for any data set, at least 75% of all data values lie within 2 standard deviations of the mean, and at least 89% of all data values lie within 3 standard deviations of the mean.

INTERPRETING THE STANDARD DEVIATION

A good way to develop a deeper understanding of the standard deviation is to consider an approximation called the **range rule of thumb,** summarized in the following box.

The Range Rule of Thumb

The standard deviation is *approximately* related to the range of a data set by the **range rule of thumb:**

$$\text{standard deviation} \approx \frac{\text{range}}{4}$$

If we know the range of a data set (range = high − low), we can use this rule to estimate the standard deviation. Alternatively, if we know the standard deviation for a data set, we estimate the low and high values as follows:

$$\text{low value} \approx \text{mean} - 2 \times \text{standard deviation}$$
$$\text{high value} \approx \text{mean} + 2 \times \text{standard deviation}$$

The range rule of thumb works reasonably well for data sets in which values are distributed fairly evenly. It does not work well when the high or low values are extreme outliers. Thus, you must use judgment in deciding whether the range rule of thumb is applicable in a particular case.

➤ **EXAMPLE 4 Using the Range Rule of Thumb**

Use the range rule of thumb to estimate the standard deviations for the waiting times at Big Bank and Best Bank. Compare the estimates to the actual standard deviations found in Example 3.

Solution The waiting times for Big Bank vary from 4.1 to 11.0 minutes, which means a range of 11.0 − 04.1 = 6.9 minutes. The waiting times for Best Bank vary from 6.6 to 7.8 minutes, for a range of 7.8 − 6.6 = 1.2 minutes. Thus, the range rule of thumb gives the following estimates for the standard deviations:

Big Bank: $\text{standard deviation} \approx \dfrac{6.9}{4} \approx 1.7$

Best Bank: $\text{standard deviation} \approx \dfrac{1.2}{4} = 0.3$

The actual standard deviations calculated in Example 3 are 1.96 and 0.44, respectively. For these two cases, the estimates from the range rule of thumb slightly underestimate the actual standard deviations. Nevertheless, the estimates put us in the right ballpark, showing that the rule is useful. *Now try Exercises 15–18, part d.* ◀

➤EXAMPLE 5 Estimating a Range

Studies of the gas mileage of a BMW under varying driving conditions show that it gets a mean of 22 miles per gallon with a standard deviation of 3 miles per gallon. Estimate the minimum and maximum gas mileage that you can expect under ordinary driving conditions.

Solution From the range rule of thumb, the low and high values for gas mileage are approximately

$$\text{low value} \approx \text{mean} - 2 \times \text{standard deviation}$$
$$= 22 - (2 \times 3)$$
$$= 16$$
$$\text{high value} \approx \text{mean} + 2 \times \text{standard deviation}$$
$$= 22 + (2 \times 3)$$
$$= 28$$

The range of gas mileage for the car is roughly from a minimum of 16 miles per gallon to a maximum of 28 miles per gallon. *Now try Exercises 15–18, part e.* ◄

by the way

Technologies such as catalytic converters have helped reduce the amounts of many pollutants emitted by cars (per mile driven), but increasing gas mileage is the only way to reduce the carbon dioxide emissions implicated in global warming. This is a major reason why auto manufacturers are developing high-mileage "hybrid" vehicles and zero-emission vehicles that run on electricity or fuel cells.

EXERCISES 6B

Review Questions

1. Consider two grocery stores at which the mean time in line is the same but the variation is different. At which store would you expect the customers to have more complaints about the waiting time? Explain.

2. Describe how we define and calculate the range of a distribution.

3. What are the quartiles of a distribution? How do we find them?

4. Define the five-number summary, and explain how to depict it visually with a boxplot.

5. Describe the process of calculating a standard deviation. Give a simple example of its calculation (such as calculating the standard deviation of the numbers 2, 3, 4, 4, and 6). What is the standard deviation if all of the sample values are the same?

6. Briefly describe the use of the range rule of thumb for interpreting the standard deviation. What are its limitations?

Does It Make Sense?

Decide whether each of the following statements makes sense (or is clearly true) or does not make sense (or is clearly false). Explain your reasoning.

7. Both exams had the same range, so they must have had the same median.

8. The highest exam score was in the upper quartile of the distribution.

9. For the 30 students who took the test, the high score was 80, the median was 74, and the low score was 40.

10. I examined the data carefully, and the range was greater than the standard deviation.

11. The standard deviation for the heights of a group of 5-year-old children is smaller than the standard deviation for the heights of a group of children who range in age from 3 to 15.

12. The mean gas mileage of the compact cars we tested was 34 miles per gallon, with a standard deviation of 5 gallons.

Basic Skills & Concepts

13. **Big Bank Verification.** Find the mean and median for the waiting times at Big Bank given in the beginning of this unit. Show your work clearly, and verify that both are 7.2 minutes.

14. **Best Bank Verification.** Find the mean and median for the waiting times at Best Bank given in the beginning of this unit. Show your work clearly, and verify that both are 7.2 minutes.

Comparing Variations. For each of Exercises 15–18, do the following:

 a. Find the mean, median, and range for each of the two data sets.

 b. Give the five-number summary and draw a boxplot for each of the two data sets.

 c. Find the standard deviation for each of the two data sets.

 d. Apply the range rule of thumb to estimate the standard deviation of each of the two data sets. How well does the rule work in each case? Briefly discuss why it does or does not work well.

 e. Based on all your results, compare and discuss the two data sets in terms of their center and variation.

15. The following data sets give waiting times in minutes of 11 customers at bus terminals in Atlanta and Boston.

Atlanta:	5.5	6.0	4.5	5.0	7.0	6.5
	5.0	7.5	5.5	4.0	8.0	
Boston:	5.5	8.0	2.0	5.0	8.5	12.0
	1.5	6.5	9.5	10.0	6.0	

16. The following data sets give the numbers of games played in the ten Stanley Cup finals (hockey) from 1990 to 1999 and in the NBA finals from 1990 to 1999. Seven games is the maximum and four games is the minimum in each final series.

Stanley Cup:	5	5	4	5	7	4	4	4	4	6
NBA Finals:	5	5	6	6	7	4	6	6	6	5

17. The following data sets show the ages of the first seven U.S. Presidents (Washington through Jackson) and the last seven U.S. Presidents (Nixon through Bush) at the time of inauguration.

First 7:	57	61	57	57	58	57	61
Last 7:	56	61	52	69	64	46	54

18. The following data sets give the approximate lengths of Beethoven's nine symphonies and Mahler's nine symphonies (in minutes).

Beethoven:	28	36	50	33	30	40	38	26	68
Mahler:	52	85	94	50	72	72	80	90	80

Further Applications

Understanding Variation. Exercises 19–20 each give four sets of seven numbers. In each case, do the following:

 a. Make a histogram for each set.

 b. Give the five-number summary and draw a boxplot for each set.

 c. Compute the standard deviation for each set.

 d. Based on your results, briefly explain how the standard deviation provides a useful single-number summary of the variation in these data sets.

19. The following sets of numbers all have a mean of 9:

$$\{9, 9, 9, 9, 9, 9, 9\}, \{8, 8, 9, 9, 9, 10, 10\},$$

$$\{8, 8, 8, 9, 10, 10, 10\}, \{6, 6, 6, 9, 12, 12, 12\}$$

20. The following sets of numbers all have a mean of 6:

$$\{6, 6, 6, 6, 6, 6, 6\}, \{5, 5, 6, 6, 6, 7, 7\},$$
$$\{5, 5, 5, 6, 7, 7, 7\}, \{3, 3, 3, 6, 9, 9, 9\}$$

21. Pizza Deliveries. After recording the pizza delivery times for two different pizza shops, you conclude that one pizza shop has a mean delivery time of 45 minutes with a standard deviation of 3 minutes. The other shop has a mean delivery time of 42 minutes with a standard deviation of 20 minutes. Interpret these figures. If you like the pizzas from both shops equally well, which one would you order from? Why?

22. Managing Complaints. You manage a small ice cream shop in which your employees scoop the ice cream by hand. Each night, you record your total sales and total volume of ice cream sold. You find that on nights when an employee named Sam is working, the mean price of the ice cream sold is $1.75 per pint with a standard deviation of $0.05. On nights when an employee named Kevin is working, the mean price of the ice cream sold is $1.75 per pint with a standard deviation of $0.35. Which employee is more likely to be generating complaints of "too small" servings? Explain.

23. Portfolio Standard Deviation. The book *Investments* by Zvi Bodie, Alex Kane, and Alan Marcus claims that the returns for investment portfolios with a single stock have a standard deviation of 0.55, while the returns for portfolios with 32 stocks have a standard deviation of 0.325. Explain how the standard deviation measures the risk in these two types of portfolios.

24. Choosing a Fund. Your job allows you to contribute to a retirement plan in which you have a choice of two investment funds. In the past, Fund A has grown by a mean of 5% per year with a standard deviation of 1%. Fund B has grown by a mean of 9% per year with a standard deviation of 7%. Which fund is a better choice for long-term investments? Which fund is a better choice if you will be withdrawing your money next year? Explain.

25. Batting Standard Deviation. For the last 100 years, the mean batting average in the major leagues has remained fairly constant at about .260. However, the standard deviation of batting averages has decreased from about .049 in the 1870s to .031 in the present. What does this tell us about the batting averages of players? Based on these facts, would you expect batting averages above .350 to be more or less common today than in the past? Explain.

Web Projects

Find useful links for Web Projects on the text Web site: www.aw.com/bennett-briggs

26. Web Data Sets. Go to any Web site that gives data sets, such as the Census Bureau, the U.S. Energy Information Administration, or the Statistical Abstract of the United States. Choose three data sets that you find interesting. In each case, describe the data distribution and discuss the variation in the distribution. How is the variation important in understanding the data?

27. Kids and the Media. A 1999 study by the Kaiser Family Foundation looked at the role of media (for example, television, books, computers) in the lives of children. The report, which is on the Kaiser Family Foundation Web site, gives many data distributions concerning, for example, how much time children spend daily with each medium. Study at least three distributions that you find interesting. Describe the data distribution and discuss the variation in the distribution. How is the variation important in understanding the data? Discuss your opinions of the social consequences of the findings.

In the News

28. Ranges in the News. Find two examples of data distributions in recent news reports; they may be given either as tables or as graphs. In each case, state the range of the distribution and explain its meaning in the context of the news report.

29. Summarizing a News Data Set. Find an example of a data distribution given in the form of a table in a recent news report. Make a five-number summary and a boxplot for the distribution.

30. Range Rule in the News. For each of the two data distributions you found in Exercise 28 above, estimate the standard deviation by applying the range rule of thumb. Discuss whether you think the estimate is valid in each case.

UNIT 6C

THE NORMAL DISTRIBUTION

A very special and common type of distribution is represented by a *bell-shaped curve;* it is formally called a **normal distribution.** The normal distribution is extremely important, in part because it describes many data sets concerning human characteristics. For example, distributions of men's or women's heights and weights tend to be normal, as do distributions of scores on standardized tests.

All normal distributions have the same characteristic bell shape, but they can differ from one another in two ways. First, they may differ in the location of their central peaks. Because a normal distribution is symmetric, the peak represents the mean, median, and mode of the data set. Second, they can differ in their variation, or the spread of data values around the peak. Figure 6.12 shows two normal distributions. Both have the same mean of 75, but distribution (a) has a lower peak and greater variation.

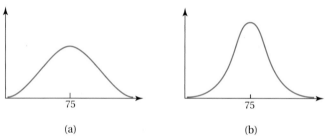

(a) (b)

FIGURE 6.12 Both distributions are normal and have the same mean of 75, but distribution (a) has a lower peak and greater variation (a larger standard deviation).

DEFINITION

The **normal distribution** is a symmetric, bell-shaped distribution with a single peak. Its peak corresponds to the mean, median, and mode of the distribution.

➤ **EXAMPLE 1** The Normal Shape

Figure 6.13a shows a famous data set of the chest sizes of 5738 Scottish militiamen collected in about 1846. Figure 6.13b shows the distribution of the population densities of the 50 states. Which distribution appears to be normal? Explain.

Solution The distribution of chest sizes in Figure 6.13a is nearly symmetric with a mean between 39 and 40 inches. Values far from the mean are less common, giving it the bell shape of a normal distribution. The distribution in Figure 6.13b shows that most states have low population densities, but a few have much higher densities. This fact makes the distribution right-skewed, so it is not a normal distribution.

Chest circumference of
Scottish militiamen (inches)

(a)

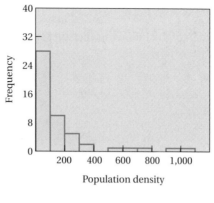

Population density

(b)

FIGURE 6.13 *Now try Exercises 11–12.* ◄

WHAT IS NORMAL?

We can better appreciate the importance of the normal distribution if we understand why it is so common. The bell shape tells us why the normal distribution fits human characteristics such as height: Most men or women have heights clustered near the mean height (for their sex), so a data set of heights has a peak at the mean height. But as we consider heights increasingly far from the mean on either side, we find fewer and fewer people. This "tailing off" of heights far from the mean produces the two tails of the normal distribution.

On a deeper level, any quantity that is the result of *many* factors is likely to follow a normal distribution. Adult heights are the result of many genetic and environmental factors. Scores on SAT tests or IQ tests tend to be normally distributed because each test score is determined from many individual test questions. Sports statistics, such as batting averages, tend to be normally distributed because they involve many people with many different levels of skill. More generally, we can expect a data set to have a nearly normal distribution if it meets the following conditions.

Conditions for a Normal Distribution

A data set that satisfies the following four criteria is likely to have a nearly normal distribution:

1. Most data values are clustered near the mean, giving the distribution a well-defined single peak.
2. Data values are spread evenly around the mean, making the distribution symmetric.
3. Larger deviations from the mean become increasingly rare, producing the tapering tails of the distribution.
4. Individual data values result from a combination of many different factors, such as genetic and environmental factors.

➤**EXAMPLE 2** Is It a Normal Distribution?

Which of the following variables would you expect to have a normal or nearly normal distribution?

 a. Scores on a very easy test

 b. Shoe sizes of a random sample of adult women

Solution

 a. Tests have a maximum score (100%) that limits the size of the data values. If a test is easy, the mean will be high and many scores will be near the maximum. The fewer lower scores can be spread out well below the mean. We therefore expect the distribution of scores to be left-skewed and not normal.

 b. Shoe size is a human trait determined by many genetic and environmental factors. We expect women's shoe sizes to cluster near a mean and become less common farther from the mean. Therefore, we expect shoe sizes to have a normal distribution. *Now try Exercises 13–18.* ◄

THE STANDARD DEVIATION IN NORMAL DISTRIBUTIONS

Because all normal distributions have the same bell shape, we can characterize the variation of any normal distribution with just a single number: the standard deviation. Recall from Unit 6B that, for any distribution, the standard deviation gives a measure of the spread of data values around the mean. For a normal distribution, the standard deviation gives a very precise description of the variation, based on a simple rule often called the **68-95-99.7 rule.**

The 68-95-99.7 Rule for a Normal Distribution

- About 68% (actually 68.3%), or just over two-thirds, of the data points fall within 1 standard deviation of the mean.
- About 95% (actually 95.4%) of the data points fall within 2 standard deviations of the mean.
- About 99.7% of the data points fall within 3 standard deviations of the mean.

FIGURE 6.14 The 68-95-99.7 rule for normal distribution.

Figure 6.14 shows the 68-95-99.7 rule graphically. The horizontal axis is marked with the position of the mean and the positions corresponding to 1, 2, and 3 standard deviations above and below the mean. The standard deviation is abbreviated by the Greek letter σ (*sigma*). The vertical axis may represent the frequency or relative frequency of data values.

▶ **EXAMPLE 3** SAT Scores

Each test that makes up the SAT is designed so that its scores are normally distributed with a mean of 500 and a standard deviation of 100. Interpret this statement according to the 68-95-99.7 rule.

Solution From the 68-95-99.7 rule, about 68% of students have scores within 1 standard deviation (100 points) of the mean of 500 points; that is, about 68% of the students score between 400 and 600. About 95% of the students score within 2 standard deviations (200 points) of the mean, or between 300 and 700. About 99.7% of the students score within 3 standard deviations (300 points) of the mean, or between 200 and 800. Figure 6.15 illustrates this interpretation graphically. Note that the horizontal axis shows both actual scores and the distance from the mean in standard deviations.

Now try Exercise 19. ◀

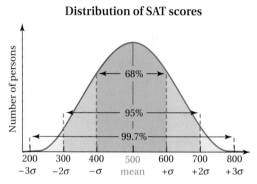

Distribution of SAT scores

FIGURE 6.15 The distribution of SAT scores.

▶ **EXAMPLE 4** Detecting Counterfeits

Vending machines can be adjusted to reject coins above and below certain weights. The weights of legal U.S. quarters are normally distributed with a mean of 5.67 grams and a standard deviation of 0.0700 gram. If a vending machine is adjusted to reject quarters that weigh more than 5.81 grams and less than 5.53 grams, what percentage of legal quarters will be rejected by the machine?

Solution A weight of 5.81 is 0.14 gram, or 2 standard deviations, above the mean. A weight of 5.53 is 0.14 gram, or 2 standard deviations, below the mean. Thus, the machine accepts quarters that are within 2 standard deviations of the mean. Conversely, the machine rejects quarters that are more than 2 standard deviations from the mean. By the 68-95-99.7 rule, 95% of legal quarters will be accepted and 5% of legal quarters will be rejected.

Now try Exercise 20. ◀

Applying the 68-95-99.7 Rule

We can use the 68-95-99.7 rule to answer many questions about the frequencies (or relative frequencies) of data values in a normal distribution. For example, suppose we want to know the percentage of values that lie *more* than 2 standard deviations above the mean. The 68-95-99.7 rule tells us that 95% of the data values lie *within* 2 standard deviations of the mean, so 5% of the data values lie *more than* 2 standard deviations from the mean. Because the normal distribution is symmetric, half of this 5%, or 2.5%, must lie more than 2 standard deviations below the mean. Thus, the remaining 2.5% of data values must lie more than 2 standard deviations above the mean. Figure 6.16 shows this idea graphically.

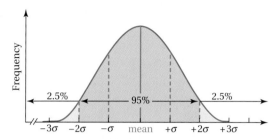

FIGURE 6.16 Because 95% of the data values lie within 2 standard deviations of the mean, 5% must lie farther than 2 standard deviations from the mean. Of this 5%, half lie more than 2 standard deviations below the mean and half lie more than 2 standard deviations above the mean.

➤**EXAMPLE 5** Normal Auto Prices

A survey finds that the prices paid for two-year-old Ford Explorers are normally distributed with a mean of $16,500 and a standard deviation of $500. Consider a sample of 10,000 people who bought two-year-old Ford Explorers.

 a. How many people paid between $16,000 and $17,000?
 b. How many paid less than $16,000?
 c. How many paid more than $18,000?

Solution

 a. Given the mean of $16,500 and standard deviation of $500, prices between $16,000 and $17,000 are within 1 standard deviation of the mean. From the 68-95-99.7 rule, about 68% of the 10,000 car buyers, or 6800 people, paid prices in this range.

 b. The remaining 32%, or 3200 people, paid prices more than 1 standard deviation from the mean, or outside the $16,000 to $17,000 range. Half of these people, or 1600 people, paid prices below $16,000 (while the other half paid prices above $17,000).

 c. A price of $18,000 is $1500 above the mean of $16,500. Because the standard deviation is $500, this price is 3 standard deviations above the mean. From the 68-95-99.7 rule, about 99.7% of the car buyers paid prices *within* 3 standard deviations of the mean, leaving 0.3% paying either more or less. Half of the 0.3%, or 0.15%, paid a price *more than* 3 standard deviations above the mean. Out of 10,000 people, 0.15% represents 15 people. Thus, approximately 15 people paid more than $18,000. *Now try Exercises 21–24.* ◄

Standard Scores

The 68-95-99.7 rule applies to data values that are exactly 1, 2, or 3 standard deviations from the mean. For other cases, we can generalize this rule if we know precisely how many standard deviations from the mean a particular data value lies. The number of standard deviations a data value lies above or below the mean is called its **standard score** (or *z-score*), which is often abbreviated by the letter *z*. For example:

- The standard score of the mean is $z = 0$, because it is 0 standard deviations from the mean.
- The standard score of a data value 1.5 standard deviations above the mean is $z = 1.5$.
- The standard score of a data value 2.4 standard deviations below the mean is $z = -2.4$.

The following box explains the computation of standard scores.

Computing Standard Scores

The number of standard deviations a data value lies above or below the mean is called its **standard score** (or *z-score*), defined by

$$z = \text{standard score} = \frac{\text{data value} - \text{mean}}{\text{standard deviation}}$$

The standard score is positive for data values above the mean and negative for data values below the mean.

▶ **EXAMPLE 6 Standard IQ Scores**

The Stanford-Binet IQ test is designed so that scores are normally distributed with a mean of 100 and a standard deviation of 16. Find the standard scores for IQ scores of 95 and 125.

Solution The standard score for an IQ of 95 is

$$z = \frac{\text{data value} - \text{mean}}{\text{standard deviation}} = \frac{95 - 100}{16} = -0.31$$

That is, an IQ of 95 lies 0.31 standard deviation *below* the mean.
The standard score for an IQ of 125 is

$$z = \frac{\text{data value} - \text{mean}}{\text{standard deviation}} = \frac{125 - 100}{16} = 1.56$$

That is, an IQ of 125 lies 1.56 standard deviations *above* the mean.

Now try Exercises 25–28. ◀

STANDARD SCORES AND PERCENTILES

Once we know the standard score of a data value, the properties of the normal distribution allow us to find its **percentile** in the distribution. You are probably familiar with the idea of percentiles. For example, if you scored in the 45th percentile on the SAT, 45% of the SAT scores were lower than yours.

Percentiles

The ***n*th percentile** of a data set is the smallest value in the set with the property that *n*% of the data values are *less than or equal to* it. A data value that lies between two percentiles is said to lie *in* the lower percentile.

by the way

By comparing IQ tests and questions that have stayed the same over the years, Dr. James Flynn discovered a startling fact: Raw scores on IQ tests (before they are set to a mean of 100) rose steadily during the 20th century, so that someone scoring an "intellectually deficient" IQ of 70 today would have rated an "intellectually superior" IQ of 130 a century ago. It seems unlikely that people today are really so much smarter than their parents and grandparents, but no one has yet found a clear explanation for Flynn's discovery.

Time out to think _____

Is it possible for someone to score *above* the 100th percentile on a standardized test? Why or why not?

A *standard score table,* such as Table 6.3, allows us to find percentiles from standard scores. For each of many standard scores in a normal distribution, the table gives the percentage of values in the distribution *less than or equal to* that value. For example, Table 6.3 shows that 55.96% of the values in a normal distribution have a standard score less than or equal to 0.15. In other words, a data value with a standard score of 0.15 lies in the 55th percentile.

As Table 6.3 shows, we sometimes use fractions of percentiles instead of whole percentiles. For example, because 55.96% of the values in a normal distribution have a standard score less than 0.15, we might say that this standard score lies in the 55th percentile, the 55.9th percentile, or the 55.96th percentile, depending on how much precision we desire.

TABLE 6.3 Standard Scores and Percentiles

Z-SCORE	PERCENTILE	Z-SCORE	PERCENTILE	Z-SCORE	PERCENTILE	Z-SCORE	PERCENTILE
−3.5	0.02	−1.0	15.87	0.0	50.00	1.1	86.43
−3.0	0.13	−0.95	17.11	0.05	51.99	1.2	88.49
−2.9	0.19	−0.90	18.41	0.10	53.98	1.3	90.32
−2.8	0.26	−0.85	19.77	0.15	55.96	1.4	91.92
−2.7	0.35	−0.80	21.19	0.20	57.93	1.5	93.32
−2.6	0.47	−0.75	22.66	0.25	59.87	1.6	94.52
−2.5	0.62	−0.70	24.20	0.30	61.79	1.7	95.54
−2.4	0.82	−0.65	25.78	0.35	63.68	1.8	96.41
−2.3	1.07	−0.60	27.43	0.40	65.54	1.9	97.13
−2.2	1.39	−0.55	29.12	0.45	67.36	2.0	97.72
−2.1	1.79	−0.50	30.85	0.50	69.15	2.1	98.21
−2.0	2.28	−0.45	32.64	0.55	70.88	2.2	98.61
−1.9	2.87	−0.40	34.46	0.60	72.57	2.3	98.93
−1.8	3.59	−0.35	36.32	0.65	74.22	2.4	99.18
−1.7	4.46	−0.30	38.21	0.70	75.80	2.5	99.38
−1.6	5.48	−0.25	40.13	0.75	77.34	2.6	99.53
−1.5	6.68	−0.20	42.07	0.80	78.81	2.7	99.65
−1.4	8.08	−0.15	44.04	0.85	80.23	2.8	99.74
−1.3	9.68	−0.10	46.02	0.90	81.59	2.9	99.81
−1.2	11.51	−0.05	48.01	0.95	82.89	3.0	99.87
−1.1	13.57	0.0	50.00	1.0	84.13	3.5	99.98

➤ **E X A M P L E 7** Cholesterol Levels

Cholesterol levels in men 18 to 24 years of age are normally distributed with a mean of 178 and a standard deviation of 41.

 a. In what percentile is a man with a cholesterol level of 190?
 b. What cholesterol level corresponds to the 90th percentile, the level at which treatment may be necessary?

Solution
 a. The standard score for a cholesterol level of 190 is

$$z = \frac{\text{data value} - \text{mean}}{\text{standard deviation}} = \frac{190 - 178}{41} \approx 0.29$$

 Table 6.3 shows that a standard score of 0.29 corresponds to about the 61st percentile.
 b. Table 6.3 shows that 90.32% of all data values have a standard score less than 1.3. Thus, the 90th percentile is about 1.3 standard deviations above the mean. The standard deviation is 41, so 1.3 standard deviations is $1.3 \times 41 = 53.3$. Adding this value to the mean, we find that the 90th percentile begins at about $178 + 53 = 231$. A person with a cholesterol level above 231 may need treatment. *Now try Exercises 29–30.* ◄

➤ **E X A M P L E 8** Women in the Army

The heights of American women aged 18 to 24 are normally distributed with a mean of 65 inches and a standard deviation of 2.5 inches. In order to serve in the U.S. Army, women must be between 58 inches and 80 inches tall. What percentage of women are ineligible to serve based on their height?

Solution The standard scores for the army's minimum and maximum heights of 58 inches and 80 inches are

For 58 inches: $z = \dfrac{\text{data value} - \text{mean}}{\text{standard devation}} = \dfrac{58 - 65}{2.5} = -2.8$

For 80 inches: $z = \dfrac{\text{data value} - \text{mean}}{\text{standard deviation}} = \dfrac{80 - 65}{2.5} = 6.0$

Table 6.3 shows that a standard score of -2.8 corresponds to the 0.26 percentile. A standard score of 6.0 does not appear in the table, which means it is above the 99.98th percentile (since that is the highest percentile shown in the table). Thus, 0.26% of all women are too short to serve in the army and fewer than 0.02% of all women are too tall to serve in the army. Altogether, this means that less than about 0.28% of women, or about one out of every 400 women, are ineligible to serve in the army based on their height. *Now try Exercises 31–32.* ◄

Time out to think _____

Note that the army standards reject a larger percentage of short women than tall women. Why do you think this is the case? Is it fair?

■ EXERCISES 6C

Review Questions

1. What is a normal distribution? Briefly describe the conditions that make a normal distribution. Give an example of a data set that is likely to have a normal distribution, and explain why.

2. What is the 68-95-99.7 rule for normal distributions? Explain how it can be used to answer questions about frequencies of data values in a normal distribution.

3. What is a standard score? How do you find the standard score for a particular data value?

4. What is a percentile? Describe how Table 6.3 allows you to relate standard scores and percentiles.

Does It Make Sense?

Decide whether each of the following statements makes sense (or is clearly true) or does not make sense (or is clearly false). Explain your reasoning.

5. The heights of male basketball players at Kentucky College are normally distributed with a mean of 6 feet 3 inches and a standard deviation of 3 inches.

6. The weights of babies born at Belmont Hospital are normally distributed with a mean of 6.8 pounds and a standard deviation of 1.2 pounds.

7. The weights of babies born at Belmont Hospital are normally distributed with a mean of 6.8 pounds and a standard deviation of 7 pounds.

8. On yesterday's mathematics exam, the standard score was 75.

9. My professor graded the final on a curve, and she gave a grade of A to anyone who had a standard score of 2 or more.

10. Jack is in the 50th percentile for height, so he is of median height.

Basic Skills & Concepts

Normal Shape. Exercises 11–12 each show three distributions, of which two are normal. Identify the two that are normal. Of the two, which has the larger standard deviation?

11.

(a) (b) (c)

12.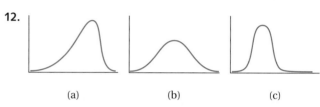

(a) (b) (c)

Normal Distribution? For each of the data sets in Exercises 13–18, state whether you would expect it to be normally distributed. Explain.

13. Rental rates for office space in large cities

14. The weights of candy bars that say "4 ounces" on the wrapper

15. The time it takes you to drive to school using the same route over the course of a school year

16. The weights of bags of a particular brand of potato chips

17. The waiting times of people who arrive randomly at a bus stop where the bus arrives on the hour

18. The times of swimmers in the 100-meter butterfly at the Olympics

19. **The 68-95-99.7 Rule.** A set of test scores is normally distributed with a mean of 100 and a standard deviation of 20. Use the 68-95-99.7 rule to find the percentage of scores in each of the following categories.

 a. less than 120

 b. less than 140

 c. less than 80

 d. less than 60

 e. greater than 120

 f. greater than 140

 g. greater than 80

 h. between 80 and 120

20. **The 68-95-99.7 Rule.** The resting heart rates for a sample of individuals are normally distributed with a mean of 70 and a standard deviation of 15. Use the 68-95-99.7 rule to find the percentage of heart rates in each of the following categories.

 a. less than 55

 b. less than 40

 c. less than 85

 d. less than 100

 e. greater than 85

 f. greater than 55

 g. greater than 40

 h. between 55 and 85

Standard Scores. Scores on a chemistry exam were normally distributed with a mean of 67 and a standard deviation of 8. Use this information in Exercises 21–28.

21. About what percentage of the exam scores were above 75?

22. About what percentage of the exam scores were below 83?

23. A failing grade on the exam was anything 2 or more standard deviations below the mean. What was the cutoff for a failing score? Approximately what percentage of the students failed?

24. Suppose that 1000 students took the test. About how many students scored above 94?

25. What is the standard score for an exam score of 67?

26. What is the standard score for an exam score of 75?

27. What is the standard score for an exam score of 58?

28. What is the standard score for an exam score of 86?

Standard Scores and Percentiles. For each of the data values in Exercises 29–30, find the standard score and use Table 6.3 to find the percentile.

29. a. A data value 1 standard deviation above the mean

 b. A data value 1.5 standard deviations below the mean

 c. A data value 2 standard deviations below the mean

30. a. A data value 0.5 standard deviation above the mean

 b. A data value 1 standard deviation below the mean

 c. A data value 1.9 standard deviations above the mean

Percentile Scores. For each of the data values in Exercises 31–32, use Table 6.3 to find the approximate standard score.

Then state where the data value lies in terms of number of standard deviations above or below the mean.

31. a. A data value in the 94th percentile

 b. A data value in the 6th percentile

 c. A data value in the 14th percentile

32. a. A data value in the 85th percentile

 b. A data value in the 10th percentile

 c. A data value in the 54th percentile

Further Applications

Pregnancy Due Dates. When a woman learns she is pregnant, her doctor usually calculates a "due date" for the baby. The due date represents the date on which the baby will be born if her pregnancy lasts the mean length for all pregnancies. Actual lengths of pregnancy terms turn out to be nearly normally distributed about this due date (the mean) with a standard deviation of 15 days. Use these facts to answer Exercises 33–36.

33. About what percentage of births occur within 15 days of the due date?

34. About what percentage of births occur within 1 month of the due date?

35. About what percentage of births occur more than 15 days before the due date? Based on your answer, do you think it is a good idea for a pregnant woman to fly to an important meeting 15 days before the due date? Explain.

36. About what percentage of births occur more than 45 days before the due date? Based on your answer, do you think it is a good idea for a pregnant woman to schedule a family trip 45 days before the due date? Explain.

37. **Heights.** According to data from the National Health Survey, the heights of all adult women are normally distributed with a mean of 63.6 inches and a standard deviation of 2.5 inches. Give the standard score and approximate percentile for women with each of the following heights.

 a. 65 inches

 b. 63 inches

 c. 62.5 inches

 d. 63.8 inches

38. **Pregnancy Lengths.** Lengths of pregnancies are normally distributed with a mean of 268 days and a standard

deviation of 15 days. Give the standard score and approximate percentile of pregnancies of the following lengths.

a. 260 days

b. 270 days

c. 255 days

d. 265 days

39. **Is It Likely?** Suppose you are told that the weights of a group of 150 first-graders are normally distributed with a mean of 38 pounds and a standard deviation of 30 pounds. Is this likely? Explain.

40. **Is It Likely?** Suppose you are told that the test scores for a statistics class are normally distributed with a mean of 85 and a standard deviation of 100. Is this likely? Explain.

GRE Scores. For Exercises 41–42, assume that the scores on the Graduate Record Exam (GRE) are normally distributed with a mean of 497 and a standard deviation of 115.

41. A graduate school requires a GRE score of 650 for admission. What percentage of all students taking the GRE meet this standard?

42. A graduate school requires a GRE score in the 95th percentile for admission. What actual score is required?

Web Projects

Find useful links for Web Projects on the text Web site: www.aw.com/bennett-briggs

43. **SAT Scores.** The College Board Web site gives the distribution of scores for the verbal and mathematics SAT (in 50-point bins). Collect these data and make histograms for both tests. Discuss the truth of the claim that SAT scores are normally distributed.

44. **Data and Story Library.** Visit the Data and Story Library Web site, and find one example (of many) of a normal distribution. Write a short description of the variable under consideration, details of the distribution, and an explanation of why you expect this variable to have an approximately normal distribution.

45. **Using FERRET.** The Federal Electronic Research and Review Extraction Tool (FERRET) provides access to many economic, demographic, and health data sets. Following the step-by-step instructions on the FERRET site, view a data set concerning a variable of interest to you (such as adjusted gross income or computer ownership). Download all or part (if the data set is very large) of the data set and make a histogram. Discuss whether the data set is normally distributed, and why or why not.

46. **Normal Distribution Demonstration.** Do a Web search on the keywords "normal distribution" and find an animated demonstration of the normal distribution. Describe how the demonstration works and the useful features that you observed.

In the News

47. **Normal Distributions.** Many data sets described in the news have nearly normal distributions, though the article may not say so explicitly. In news reports, find two data sets that you suspect have nearly normal distributions. Explain your reasoning.

48. **Non-Normal Distributions.** In news reports, find two data sets that you suspect *do not* have nearly normal distributions. Explain your reasoning.

STATISTICAL INFERENCE

*T*he ideas we have discussed so far in this chapter apply to describing actual data collected from a sample. However, as discussed in Unit 5A, the goal of most statistical studies is to learn something about an entire population. Thus, the heart of most statistical studies lies with *inferring* a conclusion about a population from results for a sample.

The process of statistical inference can be complex, but many of the ideas are fairly easy to understand. In this unit, we explore three important ideas: *statistical significance*, the *margin of error* (already introduced in Unit 5A), and *hypothesis testing*. Although we will not work with these ideas in great detail, understanding them is important for interpreting statistical studies in the news.

STATISTICAL SIGNIFICANCE

Suppose you are trying to test whether a coin is fair—that is, whether it is equally likely to land heads or tails. You toss the coin 100 times, finding 52 heads and 48 tails. Should you conclude that the coin is unfair? No. While you should expect to see *roughly* 50 heads and 50 tails in every 100 tosses, you should also expect some variation from one sample of 100 tosses to another. Thus, a small deviation from a perfect 50-50 split between heads and tails does not necessarily mean that the coin is unfair. That is, we expect small deviations to occur *by chance*.

Now suppose you toss the coin 100 times and get 20 heads and 80 tails. It's difficult to attribute such a large deviation from 50-50 to chance. So while it is *possible* that you observed a very rare set of 100 tosses, it's more likely that the coin is unfair. When the difference between what is observed and what is expected is too great to be explained by chance alone, we say the difference is **statistically significant.**

DEFINITION

A set of measurements or observations in a statistical study is said to be **statistically significant** if it is unlikely to have occurred by chance.

> **EXAMPLE 1** Significant Events?

Are the following events statistically significant? Explain.

 a. The major league baseball team with the worst win-loss record defeats the team with the best record.

 b. In terms of the global average temperature, five of the years between 1990 and 1999 were the five hottest years in the 20th century.

Solution

a. We expect a team with a poor win-loss record to lose most of its games. However, we also expect it to win occasionally, even against the team with the best record. Thus, this event is not statistically significant.

b. By chance alone, any particular year in a century would have a 5 in 100, or 1 in 20, chance of being one of the five hottest years. Having five of those years come in the same decade is unlikely to have occurred by chance alone. Thus, this event is statistically significant and suggests that the world is warming up.

Now try Exercises 15–18. ◄

From Sample to Population

We can now apply the idea of statistical significance to making an inference from a sample to a population. Suppose that in a poll of 1000 randomly selected people, 51% support the President. A week later, in another poll with a different random sample of 1000 people, only 49% support the President. Should we conclude that the opinions of all Americans changed during the one week between the polls?

You can probably guess that the answer is no. Because each poll involved a *sample* of only 1000 people, we shouldn't expect its result to be an *exact* match for the entire population of Americans. Because the two polls found only slightly different percentages (51% versus 49%), it's quite possible that the real percentage of Americans supporting the President did not change at all.

In terms of statistical significance, the change in the poll results from 51% to 49% is *not* statistically significant. Therefore, we should not infer any real change in the population based on these polls. In contrast, if the poll results had fallen from 70% one week to 30% the next, we would conclude that the change *was* statistically significant and that the opinions of Americans really did change in just one week.

► **EXAMPLE 2** Statistical Significance in Experiments

A researcher conducts a double-blind experiment that tests whether a new herbal formula is effective in preventing colds. During a three-month period, the 100 randomly selected people in a treatment group take the herbal formula while the 100 randomly selected people in a control group take a placebo. The results show that 30% of the people in the treatment group get colds, compared to 32% of the people in the control group. Can we conclude that the herbal formula is effective in preventing colds?

Solution The difference between 30% getting colds in the treatment group and 32% getting colds in the control group seems small enough to be the result of chance alone. Thus, the difference is not statistically significant, and we should not conclude that the treatment made any difference at all.

Now try Exercises 19–20. ◄

Quantifying Statistical Significance

In Example 2, it's fairly obvious that the difference between 30% colds in the treatment group and 32% colds in the control group is not statistically significant. But suppose the findings are less obvious. For example, suppose 22% of the people in the

treatment group had colds compared to 34% of the people in the control group. Is this difference enough to be statistically significant?

To answer this question, we need a quantitative definition for statistical significance. Quantitatively, statistical significance is defined as follows. (Note: We use the term *probability* in its everyday sense. We'll discuss probability in more detail in Chapter 7.)

Quantifying Statistical Significance

- If the probability of an observed difference occurring by chance is 1 in 20 (0.05) or less, the difference is statistically significant *at the 0.05 level*.
- If the probability of an observed difference occurring by chance is 1 in 100 (0.01) or less, the difference is statistically significant *at the 0.01 level*.

For the case of 22% colds in the treatment group and 34% in the control group, the difference turns out to be significant at the 0.05 level, but not at the 0.01 level. This means that if the treatment had no effect at all, we could still expect to see these results in roughly 1 out of 20 similar experiments *just by chance alone*. You can probably see that some caution is in order when working with statistical significance. While significance at the 0.05 level gives at least some evidence that the treatment works, it is by no means conclusive.

Time out to think

Suppose an experiment finds that people taking a new herbal remedy get fewer colds than people taking a placebo. The results are statistically significant at the 0.01 level. Has the experiment proven that the herbal remedy works? Explain.

➤ EXAMPLE 3 Polio Vaccine Significance

In 1954, a large experiment was conducted to test the effectiveness of a new vaccine for polio created by Dr. Jonas Salk (1914–1995). The experiment involved a sample of 400,000 children chosen from the population of all children in the United States. Half of these 400,000 children received an injection of the Salk vaccine. The other half received an injection of a placebo containing only salt water. Among the children receiving the Salk vaccine, only 33 contracted polio. In contrast, there were 115 cases of polio among the children who did not get the Salk vaccine. The probability of this difference between the groups occurring by chance turns out to be much less than 0.01. Describe the implications of this result.

Solution We say that the results of the polio vaccine test are statistically significant at much less than the 0.01 level. That is, the probability that the difference in the control and treatment groups occurred by chance is less than 1 in 100. Therefore, we can be confident that the vaccine really was responsible for the fewer cases of polio in the treatment group. As a result of this conclusion, the vaccine was immediately put into wide use, ending the polio epidemic that scarred so many people for life.

Now try Exercises 21–24. ◄

TECHNICAL NOTE
The method used to calculate the statistical significance for this example involves something called the *chi-square statistic.*

by the way

If you had been a parent in the 1940s or 1950s, polio would have been one of your greatest fears. Each year during this long polio epidemic, thousands of young children were paralyzed by the disease. Today, thanks to the polio vaccine, the disease is rare. A global effort to vaccinate children against polio, begun at the urging of the World Health Organization, seeks to eradicate polio entirely within the next few years.

MARGIN OF ERROR AND CONFIDENCE INTERVALS

Suppose an opinion poll finds that 74% of Americans are happy with their home life, with a margin of error of 3 percentage points. Recall that the margin of error (see Unit 5A) defines an interval in which the *population parameter*—in this case, the percentage of Americans who are happy with their home life—is likely to fall. Thus, the true percentage of all Americans who are happy with their home life is likely to be within 3 percentage points of 74%, or between 71% and 77%. Now we are ready to explore in more detail what we mean by an interval in which the population parameter is "likely" to fall.

Imagine that you conduct a survey of cell phone usage among *all* 400 students at a small high school. Each Y (for yes) or N (for no) below is one student's answer to the question "Do you own a cell phone?"

```
YNYYNYNNYNYYNNNYYYNYNYNYYYNYYNYYNNNYYYNYNYNYNYYYY
YYNYYNYYNNNYYYNYNYNYNYYYYYNYYNYNYNYYNNNYYYYNYNYNYN
YYYYNYNYNYNYNYNYYNNNYYYNYNYNYNYYYYYNYYNYNNNYNYYNNNN
NYYYNYNYNYNYYYYYNYYNYNNNYNYYNNNYYYYNYNYNYYYNYYNYYNY
YNYNNYNYYNNNYYYNNNYYYNYNYNYNYYYYYYYNYNYNYYYNYNYNYYNY
YYNYYNYYYYNNNYYYNNNYYYNYNYNYNYNNNYNYNYNYNYYYYYYYNYN
YNYYYYYYNYNYNYNYNYNNNYNYYNNNYYYNNNYYYNYNYNNYYYNYYNYYYN
YYNYNYNYYYNYYNYNNNYNYYNNNYYYNNNYYYNYNYNYNYYYNYYNYYY
```

If you counted carefully, you would find that 240 of the 400 responses, or 60%, are Y. Because this survey involved the complete population of students at the school, the proportion of cell phone owners, 60%, is a *population parameter*.

Now suppose that, as in most surveys, it's impractical to question every individual in the population. So instead of asking all 400 students about cell phones, you choose a random sample of only 25 students. The responses of this sample are

<div style="text-align:center">YNYYNYYNYNYYNYYNNYYYNNYY</div>

Counting carefully, you find that 16 of the 25 responses, or 64%, are Y. As we should expect, this sample statistic is *not* exactly equal to the true population parameter of 60%.

Moreover, because the sample was chosen randomly, we should expect to get a different result if we choose a different sample of 25 students. For example, suppose another sample produces the following responses:

<div style="text-align:center">YYNNNYYNYYYNNYNYYNYNYNYYN</div>

This time, 14 of the 25 responses, or 56%, are Y.

Now, imagine that you could repeat the process of drawing a random sample of 25 students hundreds or thousands of times. Because the true proportion of Y responses in the population is 60%, most of the samples would have a proportion of Ys that is close to 60%. Samples with proportions far from 60% would be relatively rare, and samples with proportions higher than 60% would occur about as often as samples with proportions lower than 60%. If you look back at Unit 6C, you'll see that we have just described three of the four conditions for a *normal distribution*.

TECHNICAL NOTE

In this discussion, we deal with percentages, or *proportions*. In other situations (such as test scores or height measurements), we may be interested in a *mean* rather than a proportion. The general ideas are the same, but the specific details and formulas are different for means and proportions.

In fact, a mathematical theorem, called the *central limit theorem,* states that the distribution of proportions from many samples of the same size is approximately a normal distribution. The theorem also says that the mean of this distribution is the true proportion for the entire population. Figure 6.17 shows this distribution. It is called a **sampling distribution** because it consists of proportions from many individual samples.

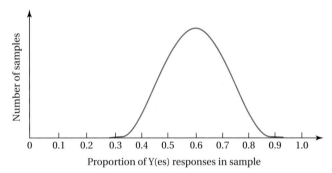

FIGURE 6.17 Suppose a true population proportion is 0.6, or 60%. If you record the proportions you find in hundreds or thousands of individual samples of the same size, the resulting distribution approximates a normal distribution. The mean of this distribution is the population proportion of 0.6.

The central limit theorem tells us not only that the sampling distribution is approximately normal with the true population proportion as its mean, but also that its standard deviation is approximately $1/(2\sqrt{n})$, where n is the size of the samples. Two standard deviations is twice this value, or $1/\sqrt{n}$. From the 68-95-99.7 rule (see Unit 6C), about 95% of all possible samples should have proportions within 2 standard deviations, or $1/\sqrt{n}$, of the true population proportion. Thus, if we take just a single sample, there is a 95% chance that its sample proportion falls within $1/\sqrt{n}$ of the true population proportion. In other words, if we add and subtract $1/\sqrt{n}$ from the sample result, we can be 95% confident that this **confidence interval** contains the true population proportion. We define $1/\sqrt{n}$ to be the **margin of error** for 95% confidence. It is important to note that the margin of error decreases as the sample size increases.

Margin of Error and Confidence Interval

Suppose you draw a single sample of size n from a large population and measure its *sample proportion.* The **margin of error** for 95% confidence is

$$\text{margin of error} \approx \frac{1}{\sqrt{n}}$$

The 95% **confidence interval** is found by subtracting and adding the margin of error from the sample proportion. You can be 95% confident that the true population proportion lies within this interval. The margin of error decreases as the sample size increases.

TECHNICAL NOTE
The precise formula for the margin of error is

$$E = 1.96\sqrt{\frac{\hat{p}(1 - \hat{p})}{n}},$$

where \hat{p} is the proportion found for the sample. This formula requires $n\hat{p} \geq 5$ and $n(1 - p) \geq 5$, conditions that are usually met in practice.

➤ EXAMPLE 4 Poll Margins

Find the margin of error and 95% confidence interval for the following surveys.

 a. A survey of 500 people finds that 52% plan to vote for Smith for Governor.
 b. A survey of 1500 people finds that 87% support stricter penalties for child abuse.

Solution

a. The proportion measured for the sample is 52%, or 0.52. For a poll of $n = 500$ people, the margin of error is approximately

$$\frac{1}{\sqrt{n}} = \frac{1}{\sqrt{500}} \approx 0.045$$

Adding and subtracting 0.045, or 4.5%, from the sample proportion of 52% produces a 95% confidence interval from 47.5% to 56.5%. We can be 95% confident that the true proportion of people who plan to vote for Smith lies in this interval.

b. The proportion measured for the sample is 87%, or 0.87. For a poll of $n = 1500$ people, the margin of error is approximately

$$\frac{1}{\sqrt{n}} = \frac{1}{\sqrt{1500}} \approx 0.026$$

Adding and subtracting 0.026, or 2.6%, from the sample proportion of 87% produces a 95% confidence interval from 84.4% to 89.6%. We can be 95% confident that the true proportion of people who support the stricter penalties lies in this interval. *Now try Exercises 25–28.* ◄

➤ EXAMPLE 5 Unemployment Rate

Suppose the Bureau of Labor Statistics finds 2160 unemployed people in a sample of $n = 60{,}000$ people. Estimate the population unemployment rate and give a 95% confidence interval.

Solution The sample proportion is the unemployment rate for the sample:

$$\frac{2160}{60{,}000} = 0.036$$

This proportion is likely to be close to the true population unemployment rate. The margin of error is approximately

$$\frac{1}{\sqrt{60{,}000}} \approx 0.004$$

We add and subtract the margin of error of 0.004 from the sample proportion of 0.036, yielding a 95% confidence interval from 0.032 to 0.040, or 3.2% to 4.0%. We can be 95% confident that the interval from 3.2% to 4.0% contains the true unemployment rate for the population. *Now try Exercises 29–32.* ◄

by the way

The actual margin of error for the unemployment survey is smaller than that found here because the Labor Department surveys 60,000 *households,* which contain more than 60,000 people.

Time out to think

Note that the margin of error is generally defined for a 95% confidence interval, as opposed to some higher level of confidence such as 99% or 99.99%. Do you think this level of confidence is enough for most surveys? Can you think of cases in which it is *not* enough?

HYPOTHESIS TESTING

Advertisers make claims about their products. Schools make claims about the quality of education they offer. Lawyers make claims about a suspect's guilt or innocence. Medical diagnoses are claims about the presence or absence of disease. Pharmaceutical companies make claims about the effectiveness of their drugs. But how do we know whether any of these claims are true? Statistics offers an answer, through a powerful set of techniques for testing the validity of claims. The techniques go by the name of **hypothesis testing,** and they play a fundamental role in almost every aspect of modern life.

Let's look at an example that illustrates the basic ideas behind hypothesis testing. A company once claimed that its product, called Gender Choice, could increase a woman's chances of giving birth to a baby girl. The company actually claimed that the chance of a baby girl could be increased "up to 80%," but let's focus simply on the claim that the chance of having a baby girl is greater than the expected 50%. How could you test whether this claim is true?

One way would be to study a random sample of, say, 100 babies born to women who used the Gender Choice product. If the product does *not* work, we would expect about half of these babies to be girls. If it does work, we would expect significantly more than half of the babies to be girls.

First, we need to define a few terms used in hypothesis testing as they relate to the Gender Choice study. Notice that the study compares the actual proportion of baby girls in the sample to the proportion expected if the product does not work. This means we are dealing with two claims, or *hypotheses,* about the product:

- The **null hypothesis** is the claim that Gender Choice does *not* work and the proportion of baby girls is 50%.
- The **alternative hypothesis** is the claim that Gender Choice does work and the proportion of girls is greater than 50%.

More generally, the null hypothesis should make a specific claim about the value of a population parameter (generally either a population mean or a population proportion). In this case, the null hypothesis makes the specific claim that the proportion of baby girls (the population parameter) equals 50%, as expected under normal conditions.

by the way

The Gender Choice product consisted of printed instructions and a thermometer. It was based on the idea that the gender of a baby could be determined through careful timing of conception. There was no evidence that the product actually worked, and it is no longer available.

Null and Alternative Hypotheses

The **null hypothesis** claims a specific value for a population parameter. (It is often the value expected in the case of no special effect.) It takes the form

null hypothesis: population parameter = claimed value

The **alternative hypothesis** is the claim that is accepted if the null hypothesis is rejected.

by the way

The word *null* comes from the Latin *nullus*, meaning "nothing." A null hypothesis often states that there is no special effect or difference. The null hypothesis is often denoted by H_0 (pronounced "H-naught"). The alternative hypothesis is often denoted by H_a.

➤ **EXAMPLE 6** Formulating Hypotheses

The manufacturer of a new car model advertises that the car averages 38 miles per gallon on the highway. A consumer group says that this claim is overstated and that the actual average (mean) is less than 38 miles per gallon. State the null and alternative hypotheses for a hypothesis test.

Solution The null hypothesis must make a specific claim about the value of a population parameter, which in this case is the average (mean) gas mileage of all cars of this model. Thus, the null hypothesis is the manufacturer's claim that the gas mileage is 38 miles per gallon on the highway. The alternative hypothesis is the consumer group's claim that the actual gas mileage is less than 38 miles per gallon. In summary:

$$\text{null hypothesis:} \qquad \text{mean gas mileage} = 38 \text{ miles per gallon}$$
$$\text{alternative hypothesis:} \qquad \text{mean gas mileage} < 38 \text{ miles per gallon}$$

Now try Exercises 33–38, part a. ◄

Outcomes of a Hypothesis Test

In conducting a hypothesis test, we assume that the null hypothesis is true until proven otherwise. Thus, there are two possible outcomes to any hypothesis test.

Two Possible Outcomes of a Hypothesis Test

- Rejecting the null hypothesis, in which case we have evidence that supports the alternative hypothesis.
- Not rejecting the null hypothesis, in which case we lack sufficient evidence to support the alternative hypothesis.

We cannot prove conclusively that a null hypothesis is *true*, which is why "accepting the null hypothesis" is not listed as a possible outcome. At best, we can only find statistical evidence that the null hypothesis is *false*, in which case we can safely reject it.

Time out to think _____

Our legal principles maintain that criminal suspects are innocent until proven guilty. At the end of a criminal trial, juries can render only two possible verdicts: guilty or not guilty. Putting these ideas into the language of hypothesis testing, state the null hypothesis for a criminal trial. How do the two possible verdicts correspond to the two possible outcomes of a hypothesis test?

➤ **EXAMPLE 7** Gender Choice Outcomes

Describe the two possible outcomes of a hypothesis test concerning the Gender Choice product.

Solution Recall that the null hypothesis states that the proportion of girl babies is the expected 50%, while the alternative hypothesis states that it is greater than 50%. Thus, there are two possible outcomes:

1. We can reject the null hypothesis and accept the alternative hypothesis. In this case, we conclude that the percentage of girl babies is greater than 50% and that the product actually works.

2. We can decide not to reject the null hypothesis. This means we have no grounds for doubting the null hypothesis, but it does *not* prove that the null hypothesis is true. For Gender Choice, it means that, while we lack evidence that the product works, we have not proven that it does *not* work.

Now try Exercises 33–38, part b. ◄

The Null Hypothesis: To Reject or Not to Reject

Suppose you study a sample of 100 babies born to women using the Gender Choice product and find that 64 of the babies are girls. How do you decide whether this constitutes evidence for rejecting the null hypothesis? The answer comes down to probabilities and statistical significance.

According to the null hypothesis, the true proportion of baby girls among the population of Gender Choice users is 50%. However, even if the null hypothesis is true, it is still possible that a single sample of 100 babies could contain 64 girls *by chance alone*. Using statistical methods that we will not cover here, it is possible to calculate that the probability of obtaining a sample with at least 64 girls by chance is about 3 in 1000. Now, recall that statistical significance at the 0.01 level requires that the probability of an event occurring by chance be less than 1 in 100. Because 3 in 1000 (0.003) is less than 1 in 100 (0.01), observing 64 baby girls is statistically significant at the 0.01 level. Because of this high level of statistical significance, we have good reason to reject the null hypothesis. This, in turn, means that we have strong support for the alternative hypothesis that Gender Choice works.

Hypothesis Test Decisions

We decide the outcome of a hypothesis test by comparing the actual sample result (mean or proportion) to the result expected if the null hypothesis is true.

- If the chance of a sample result at least as extreme as the observed result is less than 1 in 100 (or 0.01), the test is significant at the 0.01 level. The test offers strong evidence for rejecting the null hypothesis (and accepting the alternative hypothesis).
- If the chance of a sample result at least as extreme as the observed result is less than 1 in 20 (or 0.05), the test is significant at the 0.05 level. The test offers moderate evidence for rejecting the null hypothesis.
- If the chance of a sample result at least as extreme as the observed result is greater than 1 in 20, the test is not significant. The test does not provide sufficient grounds for rejecting the null hypothesis.

►EXAMPLE 8 Birth Weight Significance

A county health official believes that the mean birth weight of male babies at a local hospital is greater than the national average of 3.39 kilograms. A random sample of 145 male babies born at that hospital has a mean birth weight of 3.61 kilograms.

by the way

In statistics, the actual probability of obtaining a particular sample result (under the assumption that the null hypothesis is true) is called a *p-value*. The *p*-value for the sample in Example 8 is the given probability of 0.032.

Assuming that the mean birth weight of all male babies born at the hospital is the national average of 3.39 kilograms, a calculation shows that the probability of selecting a sample with a mean birth weight of at least 3.61 kilograms is 0.032. Formulate the null and alternative hypotheses. Then discuss whether the sample provides evidence for rejecting or not rejecting the null hypothesis.

Solution The null hypothesis is the claim that the mean birth weight of all male babies born at this hospital is the national average of 3.39 kilograms. The alternative hypothesis is the claim of the health official—that the hospital mean is higher than the national average. Thus, we have

null hypothesis: mean birth weight = 3.39 kilograms

alternative hypothesis: mean birth weight > 3.39 kilograms

Under the assumption that the null hypothesis is true, the chance of observing a sample with a mean of at least 3.61 kilograms is 0.032. This is less than 0.05 but greater than 0.01, so the result is statistically significant at the 0.05 level but not at the 0.01 level. Statistical significance at the 0.05 level provides moderate evidence for rejecting the null hypothesis, in which case the county official's claim would be supported.

Now try Exercises 39–44.

EXERCISES 6D

Review Questions

1. What is statistical inference? Why is it important?

2. Describe the meaning of statistical significance. What does it mean for a result to be statistically significant at the 0.05 level? at the 0.01 level?

3. How does the idea of statistical significance apply to the question of whether results from a sample can be generalized to conclusions about a population? Explain.

4. Explain why you should usually expect that a proportion found in a single sample will be close to, but not exactly, the true proportion in a population.

5. Briefly describe the use of the formula for margin of error. Give an example in which you interpret the margin of error in terms of 95% confidence.

6. What is the purpose of a hypothesis test? How do we formulate the null and alternative hypotheses for a test?

7. What are the two possible conclusions of a hypothesis test? Explain, and also explain why accepting the null hypothesis is *not* a possible outcome.

8. Briefly discuss how the idea of statistical significance helps us decide whether to reject or not reject a null hypothesis.

Does It Make Sense?

Decide whether each of the following statements makes sense (or is clearly true) or does not make sense (or is clearly false). Explain your reasoning.

9. The study found that more people were cured by the new drug than by the old drug, but the result was not statistically significant.

10. There can no longer be any doubt that our product, the Magic Diet Pill, really works. We financed a study in which the pill led to more weight loss than a placebo, with results significant at the 0.05 level.

11. Both agencies conducted their surveys carefully, both asked the same question and found about the same proportion of "yes" responses, and both interviewed the same number of people. However, the margin of error was smaller for Agency A's survey than for Agency B's survey.

12. If you want to reduce the margin of error in your pre-election survey, you should use a larger sample.

13. We began the trial with the null hypothesis stating that the toxic waste dump had no effect on disease rates

among the residents nearby. Thus, the alternative hypothesis stated that the toxic waste dump did have an effect on disease rates among the residents nearby.

14. We began the trial with the null hypothesis stating that the toxic waste dump had no effect on the residents nearby, and our trial proved this hypothesis to be true.

Basic Skills & Concepts

Subjective Significance. For each event in Exercises 15–20, state whether you think the difference between what occurred and what you would expect by chance is statistically significant. Explain.

15. In 100 tosses of a coin, you get 10 tails.

16. In 500 tosses of a coin, you get 245 heads.

17. An 85% free-throw shooter (in basketball) hits 24 out of 30 free throws.

18. An 85% free-throw shooter (in basketball) hits 30 free throws in a row.

19. A detective in Detroit finds that 115 of the 145 guns used in crimes citywide during the past week were sold by the same gun shop, which is one of several dozen gun shops in the city.

20. A transportation department study finds that, of 134 tires experiencing blowouts that led to the vehicle operator's death, 112 were made at the same factory.

21. **Human Body Temperature.** A study by University of Maryland researchers measured the body temperatures of 106 individuals. The mean for the sample was 98.20°F. The accepted value for human body temperature is 98.60°F. If we assume that the mean body temperature is actually 98.6°F, the probability of getting a sample with a mean of 98.20°F or less turns out to be less than 1 in 1 million. Is this result significant at the 0.05 level? at the 0.01 level? Would it be reasonable to conclude that the accepted value for human body temperature is wrong? Explain.

22. **Seat Belts and Children.** In a study of children injured in automobile crashes (*American Journal of Public Health*, Vol. 82, No. 3), those wearing seat belts had a mean stay of 0.83 day in an intensive care unit. Those not wearing seat belts had a mean stay of 1.39 days. The probability of this difference in means occurring by chance turns out to be less than 1 in 10,000. Is this result significant at the 0.05 level? at the 0.01 level? Would it be reasonable to conclude that seat belts reduce the severity of injuries? Explain.

23. **SAT Preparation.** A study of 75 students who took an SAT preparation course (*American Education Research Journal*, Vol. 19, No. 3) concluded that the mean improvement on the SAT was 0.6 point. If we assume that the preparation course has no effect, the probability of getting a mean improvement of 0.6 point by chance is 0.08, or 8 in 100. Discuss whether this preparation course results in statistically significant improvement.

24. **Weight by Age.** A National Health Survey determined that the mean weight of a sample of 804 men aged 25 to 34 was 176 pounds, while the mean weight of a sample of 1657 men aged 65 to 74 was 164 pounds. The difference is significant at the 0.01 level. Interpret this result.

Margin of Error. Find the margins of error and 95% confidence intervals for each survey result described in Exercises 25–32. Briefly interpret each result in terms of a 95% confidence interval.

25. According to a September 2000 Gallup poll of 1012 people, about one-third (32%) of Americans keep a dog for protection.

26. A Harris poll of 1012 adults resulted in the conclusion that 61% of Americans believe that doctors have "great prestige" (the highest percentage of any profession).

27. Based on telephone surveys of 513 people who have flown within the past year, the Gallup Organization concluded that 69% of travelers are satisfied with airline service.

28. A survey of 1012 adults by Opinion Research Corporation International of Princeton, New Jersey, concluded that

76% of all drivers engage in at least one distracting activity (for example, drinking coffee or talking on cell phones) while driving.

29. A Yankelovich survey of 1012 Americans age 18 or over concluded that 62% with Internet access at work use the Internet for personal reasons at least once a day.

30. A Yankelovich survey of 2818 Americans, conducted in 14 U.S. cities, found that while 73% of respondents knew that health experts recommend drinking eight or more eight-ounce servings of water each day, 51% admitted falling short of that goal.

31. In a Yankelovich survey of 1012 people, 37% of women said they would prefer to spend their weekend leisure time working on a home improvement project, such as gardening or painting.

32. A survey of 2133 adults by American Management Association found that 54% of all employers monitor the Internet connections of their employees.

Formulating Hypotheses. In Exercises 33–38, do the following for each case:

a. State the null and alternative hypotheses for a hypothesis test.

b. Describe the two possible outcomes of the test, using the context of the given situation.

33. A principal claims that the mean verbal SAT score for students in her school is greater than the national average of 500.

34. The Food and Drug Administration claims that the amount of vitamin C in tablets produced by a company is less than the advertised 500 milligrams.

35. A consumer group claims that the amount of preservative added to a brand of potato chips exceeds the amount listed on the packages.

36. The director of public works for a small town claims that the mean water usage among households exceeds 1675 gallons per month, the level required for the water supply to last through the year.

37. An agricultural research company claims that the yield for its new breed of corn is greater than 540 bushels per acre, the yield of the nearest competitor.

38. A chemical company wishes to defend its claim that the levels of pollution downstream from its plant are less than the minimum level specified by the Environmental Protection Agency.

Hypothesis Tests. Each of Exercises 39–44 describes the results of a hypothesis test in terms of the probability of obtaining a particular sample. In each case, use the given context to formulate the null and alternative hypotheses. Then discuss whether the sample provides evidence for rejecting or not rejecting the null hypothesis.

39. The owner of a car rental company claims that the mean annual mileage for the population of all cars in his fleet is more than 11,725 miles (which is the mean annual mileage for all cars in the United States). A random sample of $n = 225$ cars from his fleet has a mean annual mileage of 12,000 miles. Assuming that the mean annual mileage for all cars in his fleet is 11,725 miles, the probability of selecting a random sample with a mean annual mileage of 12,000 miles or more is 0.01.

40. A Senate candidate claims that a majority of voters support her. A poll of 400 voters finds that the proportion of voters who support the candidate is 0.51 (51%). Assuming that the proportion of people in the population who support her is $p = 0.5$, the probability of selecting a sample in which the proportion is 0.51 or more is 0.345.

41. A hospital administrator finds that the mean hospital stay for a sample of 81 women after childbirth is 2.3 days. She claims that the mean stay at her hospital is greater than the national average of 2.1 days. Assuming that the average at her hospital is the same as the national average, the probability of observing a sample with a mean of 2.3 days or more is 0.17.

42. According to the Energy Information Administration, the mean gas mileage of all automobiles in 1998 was 21.4 miles per gallon. An environmental activist claims that sport utility vehicles (SUVs) get less than the mean gas mileage. A random sample of 40 SUVs has a mean gas mileage of 19.8 miles per gallon. Assuming that the true mean is 21.4 miles per gallon, the probability of observing a sample with a mean of 19.8 miles per gallon or less is 0.0019.

43. A marketing company claims that the mean household income of HDTV owners across the population is greater than $40,000. A random sample of 1700 households with HDTVs shows that the mean household income is $41,182. Assuming that the true mean is $40,000, the probability of selecting a sample with a mean income of $41,182 or more is 0.007.

44. A Roper poll used a sample of 100 randomly selected car owners. Within the sample, the mean time for ownership of a single car was 7.01 years. Assuming that the actual mean ownership time is 7.5 years, the probability of selecting a sample with a mean time of ownership of 7.01 or less is 0.10.

Further Applications

45. **Nielsen Ratings.** Nielsen ratings are based on televisions in 5000 households; Nielsen estimates that 13,000 people live in these households. Suppose Nielsen reports that *NFL Monday Night Football* had 35% of the TV audience. Interpret this result with a 95% confidence interval, based on a sample size of 13,000 people.

46. **Hazing Athletes.** A study by researchers at Alfred University concluded that 80% of all student-athletes in this country have been subjected to some form of hazing. The study was based on responses from 1400 athletes. Interpret this result with a 95% confidence interval.

47. **Better Margin of Error.** Suppose you want to decrease the margin of error by a factor of 2—for example, from 4% to 2%. By what factor must you increase the sample size? Explain.

48. **Better Margin of Error.** Suppose you want to decrease the margin of error by a factor of 10—for example, from 1% to 0.1%. By what factor must you increase the sample size? Explain.

49. **Gallup Description.** The Gallup Organization includes the following explanation after its summary of opinion polls. Are the figures in the explanation (sample size and margin of error) consistent? Comment on the provision that is made about "error and bias."

The results reported here are based on telephone interviews with a randomly selected national sample of 1,019 adults, 18 years and older. For results based on the whole sample, one can say with 95 percent confidence that the maximum error attributable to sampling and other random effects is plus or minus 3 percentage points. In addition to sampling error, question wording and practical difficulties in conducting surveys can introduce error or bias into the findings of public opinion polls.

Web Projects

Find useful links for Web Projects on the text Web site: www.aw.com/bennett-briggs

50. **Vital Significance.** Visit a Web site that has vital statistics, which generally concern birth and death rates. Choose a question that interests you, such as whether there are significant differences in the numbers of births at different times of year or whether there are differences in the infant mortality rates between selected states. Collect the relevant data and determine subjectively whether you think the observed differences are significant. Discuss any implications of your findings.

51. **Network Polls.** All the major television networks conduct regular polls on a variety of issues. Visit the Web sites of the major networks and gather the results of a particular poll. Describe the details of the actual polling procedure. Interpret the results in terms of a 95% confidence interval. Based on what you now know about the margin of error and confidence intervals, do you think the conclusions described on the Web site are accurate? Explain.

52. **Genetically Modified Foods.** Use the Web to learn about current studies of whether genetically modified foods are safe for human consumption and for the environment. Choose a study that interests you. Describe how it is an example of a hypothesis test, and discuss its results and conclusions.

In the News

53. **Statistical Significance.** Find a recent news report in which the idea of statistical significance is used. Write a one-page summary of the study and the result that is considered to be statistically significant. Be sure to discuss whether you believe the result and why.

54. **Significant Experiment?** Find a recent news story about a statistical study that used an experiment to determine whether some new treatment was effective. Based on the available information, briefly discuss what you can conclude about the statistical significance of the results. Given this significance (or lack of), do you think the new treatment is useful? Explain.

55. **Margin of Error.** Find three recent surveys that quote a result as a proportion (or percentage) along with a margin of error. Based on the sample size in each case, calculate the margin of error with the formula given in this unit. Does your calculation agree with the stated margin of error? Explain.

56. **Confidence Intervals.** Find three recent surveys that quote a result as a proportion (or percentage) along with a margin of error. Interpret each result in terms of a 95% confidence interval.

57. **Hypothesis Testing.** Find a news report describing a statistical study that involved a hypothesis test. (News reports rarely state explicitly that a hypothesis test was used, but you should be able to decide from the context whether a hypothesis test was involved.) State the null and alternative hypotheses for the study, and describe how a conclusion was reached.

CHAPTER 6 SUMMARY

UNIT	KEY TERMS	KEY IDEAS AND SKILLS
6A	distribution mean median mode outlier symmetric distribution skewed (left or right) distribution variation	**Measures of center in a distribution:** $$\text{Mean} = \frac{\text{sum of all values}}{\text{total number of values}}$$ Median is the middle value. Mode is the most common value. **Understand** effects of outliers. **Be aware of** potential confusion with the term *average*. **Characterize** the shape of a distribution by number of peaks symmetric, left-skewed, or right-skewed variation
6B	range quartiles five-number summary boxplot standard deviation	**Understand and calculate** measures of variation: range = highest value − lowest value five-number summary (low value, lower quartile, median, upper quartile, high value) $$\text{standard deviation} = \sqrt{\frac{\text{sum of (deviations from the mean)}^2}{\text{total number of data values} - 1}}$$ **Interpret** standard deviation with the range rule of thumb.
6C	normal distribution 68-95-99.7 rule standard score percentile	**Describe** a normal distribution and conditions under which it can be expected. **Understand and apply** the 68-95-99.7 rule for normal distributions. **Use** a standard score table for normal distributions.
6D	statistical significance sampling distribution confidence interval margin of error hypothesis test null hypothesis alternative hypothesis	**Understand** the concept of statistical significance, including significance at the 0.05 level significance at the 0.01 level **Understand** how a margin of error arises from a sampling distribution. **Understand** and conduct a basic hypothesis test and make decisions based on the test.

You can take it as understood
That your luck changes only if it's good.

—Ogden Nash, *Roulette Us Be Gay*

CHAPTER 7

Probability: Living with the Odds

Probability is involved in virtually every decision we make. Sometimes the role of probability is clear, as in deciding whether to buy a lottery ticket or whether to plan a picnic based on the probability of rain. In other cases, probability guides decisions on a deeper level. For example, you might choose a particular college because you believe that it is most likely to meet your personal needs. In this chapter, we will look at just how practical and powerful probability can be in our everyday lives.

UNIT 7A

FUNDAMENTALS OF PROBABILITY

Probability plays such a huge role in our lives that it is essential to understand how it works. In this unit, we'll discuss fundamental concepts required for probability calculations.

Let's begin by considering a toss of two coins. Figure 7.1 shows that there are four different ways the coins could fall. We say that each of these four ways is a different **outcome** of the coin toss. The outcomes are the most basic possible results of the coin toss. But suppose we are interested only in the number of heads. Because the two middle outcomes in Figure 7.1 each have 1 head, we say that these two outcomes represent the same **event**. That is, an event describes one or more possible outcomes that all have the same property of interest—in this case, the same number of heads. Figure 7.1 shows that there are four possible outcomes for the two-coin toss, but only three possible events: 0 heads, 1 head, and 2 heads.

Outcome				
Event	0 heads	1 head	1 head	2 heads

FIGURE 7.1 The four possible outcomes for a toss of two coins. The two middle outcomes both represent the same event of 1 head.

DEFINITIONS

Outcomes are the most basic possible results of observations or experiments.

An **event** consists of one or more outcomes that share a property of interest.

Mathematically, we express probabilities with numbers between 0 and 1. For example, the probability of a coin landing on heads is one-half, or 0.5. If an event is impossible, such as the event of meeting a married bachelor, we assign it a probability of 0. At the other extreme, an event that is certain to occur has a probability of 1. The probability of the Sun rising tomorrow is essentially 1. We use the notation $P(\text{event})$ to mean the probability of an event. We often denote events by letters or symbols. For example, using H to represent the event of heads on a coin toss, we write $P(\text{H}) = 0.5$.

Expressing Probability

The probability of an event, expressed as $P(\text{event})$, is always between 0 and 1 (inclusive). A probability of 0 means the event is impossible and a probability of 1 means the event is certain.

Time out to think —————————————————————————————

Consider families of three children. List the different outcomes (birth orders) that result in the *event* of two boys and one girl. How might you express this probability with the notation introduced above?

With these basic definitions and notation, we are ready to discuss the three basic techniques for finding probabilities. They go by the names *theoretical method*, *empirical method*, and *subjective method*. We'll explore each in turn.

THEORETICAL PROBABILITIES

When we say that the probability of heads on a coin toss is 1/2, we are *assuming* that the coin is fair, so it is equally likely to land heads or tails. In essence, the probability is based on a *theory* of how the coin behaves, so we say that the probability of 1/2 comes from the **theoretical method.**

As another example, consider rolling a single die. Because there are six equally likely outcomes (Figure 7.2), the theoretical probability for each outcome is 1/6. For example, the theoretical probability of rolling a 3 is 1/6. As long as all outcomes are equally likely, we can use the following procedure to calculate theoretical probabilities.

TECHNICAL NOTE
Except when stated otherwise, we assume that dice are fair, six-sided cubes with face values of 1 through 6.

Theoretical Method for Equally Likely Outcomes

Step 1: Count the total number of possible outcomes.

Step 2: Among all the possible outcomes, count the number of ways the event of interest, *A*, can occur.

Step 3: Determine the probability, $P(A)$, from

$$P(A) = \frac{\text{number of ways } A \text{ can occur}}{\text{total number of outcomes}}$$

by the way

Theoretical methods are also called *a priori* methods. The words *a priori* are Latin for "before the fact" or "before experience."

FIGURE 7.2 The six possible outcomes for a roll of one die.

➤ **EXAMPLE 1 Guessing Birthdays**

You select a person at random from a large conference group. What is the probability that the person has a birthday in July? Assume 365 days in a year.

Solution If we assume that births are equally likely to occur on any day of the year, then we can use the three-step theoretical method.

Step 1: Each possible birthday represents an outcome, so there are 365 possible outcomes.
Step 2: July has 31 days, so 31 of the 365 total outcomes represent the event of a July birthday.
Step 3: The probability that a randomly selected person has a July birthday is

$$P(\text{July birthday}) = \frac{31}{365} \approx 0.085$$

which is slightly more than 1 in 12. *Now try Exercises 21–22.* ◄

Time out to think _____

Continuing Example 1, find the probability of selecting a person with a July 4 birthday.

➤ **EXAMPLE 2 Two Boys and a Girl**

What is the probability that a family with three children has two boys and one girl? Assume boys and girls are equally likely.

Solution We apply the theoretical method for equally likely outcomes.

Step 1: There are two possible outcomes for each birth: boy (B) or girl (G). For a family with three children, the total number of possible outcomes (birth orders) is $2 \times 2 \times 2 = 8$ (see the Brief Review of the multiplication principle on p. 433). The eight possible outcomes are the birth orders BBB, BBG, BGB, BGG, GBB, GBG, GGB, and GGG.
Step 2: Of these eight possible outcomes, three have two boys and one girl: BBG, BGB, and GBB.
Step 3: The probability that a family with three children has two boys and one girl is

$$P(2 \text{ boys}) = \frac{3}{8} = 0.375$$

In words, the probability that a family with three children has two boys and one girl is 3/8, or 0.375. *Now try Exercises 23–28.* ◄

EMPIRICAL PROBABILITIES

He that leaves nothing to chance will do few things ill, but will do very few things.

—GEORGE SAVILE
HALIFAX

Suppose that, from historical and geological records, we discover that a river has risen above flood level 9 times in the past 1000 years. We might reasonably claim that the probability of such a flood in any given year is about 9 in 1000, or 0.009. This probability is not based on any theoretical ideas. Instead, it is based on observations. When we determine a probability based on observations or experiments, we use the **empirical method.** Because we could also say that the relative frequency of years with floods is 9 in 1000, the empirical method is sometimes called the *relative frequency method.*

THE MULTIPLICATION PRINCIPLE

Let's review an important method for counting outcomes. Suppose we toss two coins and want to count the total number of possible outcomes. Individually, each coin has two possible outcomes: heads (H) and tails (T). Together, the two coins have $2 \times 2 = 4$ possible outcomes: HH, HT, TH, and TT (see Figure 7.1).

We can extend this thinking to additional coins. For example, if we toss three coins, the total number of possible outcomes is $2 \times 2 \times 2 = 8$, as shown in Figure 7.3. We can generalize these examples with the following rule, often called the **multiplication principle.**

Coin 1	Coin 2	Coin 3	Outcome
H	H	H	HHH
		T	HHT
	T	H	HTH
		T	HTT
T	H	H	THH
		T	THT
	T	H	TTH
		T	TTT

FIGURE 7.3 The first column shows the two possible outcomes for the first coin. Each of these two outcomes branches into two possible outcomes for the second coin (second column). Each of these four outcomes branches into two possible outcomes for the third coin (third column). The last column shows the eight possible outcomes.

The Multiplication Principle
Suppose there are M possible outcomes for one process and N possible outcomes for a second process. The total number of possible outcomes for the two processes combined is M × N. This idea extends to any number of processes. For example, if a third process has R possible outcomes, the total number of possible outcomes for the three processes combined is M × N × R.

Example: How many outcomes are possible when you roll two dice?

Solution: A single die has six possible outcomes: the numbers 1 through 6 (see Figure 7.2). A second die also has six possible outcomes. The total number of possible outcomes for two dice is $6 \times 6 = 36$.

Example: A restaurant menu offers two choices for an appetizer, five choices for a main course, and three choices for a dessert. How many different three-course (appetizer, main course, dessert) meals are possible?

Solution: There are two possible outcomes when selecting an appetizer, five possible outcomes when selecting a main course, and three possible outcomes when selecting dessert. The total number of possible outcomes is $2 \times 5 \times 3 = 30$; that is, 30 different three-course meals are possible.

Example: A college offers 12 natural science classes, 15 social science classes, 10 English classes, and 8 fine arts classes. You must take one class from each category to meet your core requirements. How many possible ways are there to meet the core requirements?

Solution: The selection of a course from a category represents a distinct process. Thus, we find the total number of ways to meet the requirements by multiplying the number of choices in each category: $12 \times 15 \times 10 \times 8 = 14{,}400$. There are 14,400 different ways to choose your courses to meet the requirements.

Now try Exercises 13–20.

Besides being useful when there is no theoretical basis for finding a probability, the empirical method is useful for checking the validity of theoretical probabilities. For example, suppose you toss a coin 100 times and get 75 heads. In this case, the empirical probability for heads is 75/100, or 0.75, which is significantly higher than

by the way

The term "100-year flood" is often used to describe a flood of a magnitude that occurs only once every 100 years, on average. Because this is a probability, it is possible (though rare) to have 100-year floods in two or more consecutive years.

the theoretical probability of 0.5 for a fair coin. You therefore have reason to suspect that the theoretical probability does not apply and that the coin is unfair.

►EXAMPLE 3 500-Year Flood

Geological records indicate that a river has crested above flood level just four times in the past 2000 years. What is the empirical probability that the river will crest above flood level next year?

Solution Based on the observation of 4 floods in 2000 years, the empirical probability is

$$\frac{4}{2000} = \frac{1}{500} = 0.002$$

The empirical probability of a flood next year is 1 in 500, or 0.002. Because a flood of this magnitude occurs on average once every 500 years, it could be called a "500-year flood."

Now try Exercises 29–30. ◄

►EXAMPLE 4 Empirical Coin Tossing

If you are interested only in the number of heads in coin tosses, then the possible events when tossing two coins are 0 heads, 1 head, and 2 heads (see Figure 7.1). Suppose you repeat a two-coin toss 100 times and your results are as follows:

- 0 heads occurs 22 times.
- 1 head occurs 51 times.
- 2 heads occurs 27 times.

Compare the empirical probabilities to the theoretical probabilities. Do you have reason to suspect that the coins are unfair?

Solution Let's begin by finding the theoretical probabilities for two tossed coins. Figure 7.1 shows that there are four possible outcomes for the two coins. Only 1 of these 4 outcomes has 0 heads, so the theoretical probability for the event of 0 heads is 1/4, or 0.25. Similarly, only 1 of the 4 outcomes has 2 heads, so the theoretical probability for the event of 2 heads is also 0.25. Finally, 2 of the 4 outcomes have 1 head, so this event has a probability of 2/4, or 0.5.

We find the empirical probabilities by dividing the number of times each event occurred by the total of 100 tosses. The event of 0 heads occurred 22 times, so its empirical probability is 22/100, or 0.22. Similarly, the empirical probabilities for the events of 1 head and 2 heads are 0.51 and 0.27, respectively. Because each empirical probability is quite close to the expected theoretical probability, we have no reason to suspect that the coins are unfair.

Now try Exercises 31–34. ◄

Time out to think

How does the conclusion of Example 4 resemble the idea of a *hypothesis test,* discussed in Unit 6D?

SUBJECTIVE PROBABILITIES

The third method for determining probabilities is to make a **subjective estimate** using experience or intuition. For example, you could make a subjective estimate of the probability that a friend will be married in the next year. Weather forecasts are often

The principal means for ascertaining truth—induction and analogy—are based on probabilities; so that the entire system of human knowledge is connected with the theory of probability.

—PIERRE SIMON, MARQUIS DE LAPLACE (1819)

subjective probabilities based on the experience of the forecaster combined with current weather data. There is not much we can say mathematically about subjective probabilities because their validity depends on personal experience and judgment.

Three Types of Probability

A **theoretical probability** is based on a model in which all outcomes are equally likely. It is determined by dividing the number of ways an event can occur by the total number of possible outcomes.

An **empirical probability** is based on observations or experiments. It is the relative frequency of the event of interest.

A **subjective probability** is an estimate based on experience or intuition.

> **EXAMPLE 5** Which Type of Probability?

Identify the method that resulted in the following statements.

 a. I'm 100% certain that you'll be happy with this car.
 b. Based on data from the Department of Transportation, the chance of dying in an automobile accident is 1 in 7000 (per year).
 c. The chance of rolling a 7 with a twelve-sided die is 1/12.

Solution

 a. This is a subjective probability based on the opinion of the speaker.
 b. This is an empirical probability because it is based on observations recorded in the data set.
 c. This is a theoretical probability because it is based on a model of a fair twelve-sided die that is equally likely to land on any of its twelve sides.

Now try Exercises 35–36. ◄

PROBABILITY OF AN EVENT *NOT* OCCURRING

Suppose we are interested in the probability that a particular event or outcome does *not* occur. For example, suppose we want to know the probability that a 4 does *not* appear on one roll of a die. We can start by finding the probability of getting a 4. There are 6 possible outcomes for the roll of the die (the numbers 1 through 6), so the probability of rolling a 4 is $P(4) = 1/6$. Now, note that it is certain (which means that the probability is 1) that either we roll a 4 or we do not roll a 4. We can write this fact as

$$P(4) + P(\text{not } 4) = 1$$

It then follows that the probability that a 4 does *not* occur is

$$P(\text{not } 4) = 1 - P(4) = 1 - \frac{1}{6} = \frac{5}{6}$$

This answer makes sense because there are five possible events besides 4 (rolls of 1, 2, 3, 5, and 6). We can generalize this observation for any event.

by the way

Another approach to finding probabilities, called the *Monte Carlo method,* uses computer simulations. This technique essentially finds empirical probabilities, but the experiments are simulations in a computer rather than actual experiments in a lab.

HISTORICAL NOTE

As early as 3600 B.C., rounded bones called *astragali* were used much like dice for games of chance in the Middle East. Our familiar cubical dice appeared about 2000 B.C. in Egypt and in China. Playing cards were invented in China in the 10th century and brought to Europe in the 14th century.

Probability of an Event *Not* Occurring

Suppose the probability of an event A is $P(A)$. Then the probability that event A does *not* occur is $1 - P(A)$.

➤ **EXAMPLE 6** Not Two Boys

What is the probability that a randomly chosen family with three children does *not* have two boys and one girl? Assume boys and girls are equally likely.

Solution In Example 2, we found that the probability of two boys and one girl in a family with three children is 3/8, or 0.375. Therefore, the probability that a family with three children does *not* have two boys and one girl is $1 - 3/8 = 5/8$, or 0.625.

Now try Exercises 37–44. ◀

PROBABILITY DISTRIBUTIONS

TABLE 7.1	Tossing Two Coins
EVENT	PROBABILITY
2 heads, 0 tails	0.25
1 head, 1 tail	0.50
0 heads, 2 tails	0.25
Total	**1**

Consider again the possible outcomes and events for tossing two coins (see Figure 7.1). There are four possible outcomes, each with a probability of 1/4. If we are interested only in the total number of heads, two of the four outcomes—the outcomes HT and TH—represent the same event of 1 head. Therefore, the event of 1 head has a probability of 2/4, or 1/2. Table 7.1 summarizes the probabilities of the three possible events of 0 head, 1 head, and 2 heads. This table is a **probability distribution,** which displays the probability for *every possible event*. The sum of all the probabilities must be 1, because it is certain that one of the events will occur when two coins are tossed.

Making a Probability Distribution

A **probability distribution** represents the probabilities of all possible events. To make a table of a probability distribution:

Step 1: List all possible *outcomes*. Use a table or figure if it is helpful.

Step 2: Identify outcomes that represent the same *event*. Find the probability of each event (using the theoretical method).

Step 3: Make a table in which one column lists each event and another column lists each probability. The sum of all the probabilities must be 1.

➤ **EXAMPLE 7** Tossing Three Coins

Make a table of the probability distribution for the number of heads that occur when three coins are tossed simultaneously.

Solution We apply the three-step process.

Step 1: The diagram in Figure 7.3 (in the Brief Review on p. 433) shows the eight possible outcomes when three coins are tossed. Each outcome is equally likely and has a probability of 1/8.

Step 2: Several of the outcomes represent the same event. For example, the outcomes HTT, THT, and TTH all represent the event of 1 head and 2 tails. This event therefore has a probability of 3/8. Similarly, the outcomes HHT, HTH, and THH all represent the event of 2 heads and 1 tail. This event also has a probability of 3/8.
Step 3: There are four possible events: 0 heads, 1 head, 2 heads, and 3 heads. We have their probabilities from steps 1 and 2. Table 7.2 shows the complete probability distribution. *Now try Exercise 45.* ◄

TABLE 7.2 **Tossing Three Coins**

RESULT	PROBABILITY
3 heads, 0 tails	1/8
2 heads, 1 tail	3/8
1 head, 2 tails	3/8
0 heads, 3 tails	1/8
Total	**1**

Time out to think _____
When four coins are tossed, how many different outcomes are there? If you record the number of heads, how many different events are possible?

►**EXAMPLE 8** Two Dice Distribution

Make a probability distribution for the sum of the dice when two dice are rolled. What is the most probable sum?

Solution Because there are six ways for each die to land, there are $6 \times 6 = 36$ outcomes of rolling two dice. We can enumerate all 36 outcomes by making a table in which we list one die along the rows and one die along the columns. Each of the 36 outcomes is equally likely, so each has a probability of 1/36. For each outcome, the table also shows the sum of the numbers on the two dice.

	1	2	3	4	5	6
1	1 + 1 = 2	1 + 2 = 3	1 + 3 = 4	1 + 4 = 5	1 + 5 = 6	1 + 6 = 7
2	2 + 1 = 3	2 + 2 = 4	2 + 3 = 5	2 + 4 = 6	2 + 5 = 7	2 + 6 = 8
3	3 + 1 = 4	3 + 2 = 5	3 + 3 = 6	3 + 4 = 7	3 + 5 = 8	3 + 6 = 9
4	4 + 1 = 5	4 + 2 = 6	4 + 3 = 7	4 + 4 = 8	4 + 5 = 9	4 + 6 = 10
5	5 + 1 = 6	5 + 2 = 7	5 + 3 = 8	5 + 4 = 9	5 + 5 = 10	5 + 6 = 11
6	6 + 1 = 7	6 + 2 = 8	6 + 3 = 9	6 + 4 = 10	6 + 5 = 11	6 + 6 = 12

The possible *events* are the sums from 2 to 12. We find the probability of each event by counting all the outcomes for each sum and then dividing the number of outcomes by 36. For example, the five highlighted outcomes in the table have a sum of 8, so the probability of a sum of 8 is 5/36. Table 7.3 shows the complete probability distribution. The most likely event is a sum of 7, which has a probability of 6/36, or 1/6.

TABLE 7.3 **Probability Distribution for the Sum of Two Dice**

Event (Sum)	2	3	4	5	6	7	8	9	10	11	12	**Total**
Probability	1/36	2/36	3/36	4/36	5/36	6/36	5/36	4/36	3/36	2/36	1/36	**1**

Now try Exercise 46. ◄

STATING THE ODDS

You've probably noticed that there are several ways to express the idea of *probability.* For example, saying that the *chance* of getting heads on a coin toss is 1/2 is equivalent to saying that the *probability* is 1/2. Another common term for probability is **odds,** but strictly speaking the two terms have different meanings. Odds are often stated as the *ratio* of the probability that a particular event will occur to the probability that it will not occur.

DEFINITION

The **odds for** an event A are

$$\text{odds for event } A = \frac{P(A)}{P(\text{not } A)}$$

The **odds against** an event A are

$$\text{odds against event } A = \frac{P(\text{not } A)}{P(A)}$$

Note: In gambling, the term **odds on** generally means the *odds against.*

Although odds are defined as fractions, we usually state them by reducing the fraction to simplest terms and reading it as a ratio. For example, the probability of rolling a 4 on a single die is 1/6 and the probability of *not* rolling a 4 is 5/6. Thus, the *odds for* rolling a 4 are

$$\frac{P(4)}{P(\text{not } 4)} = \frac{1/6}{5/6}$$
$$= \frac{1}{5}$$

We usually state these *odds for* rolling a 4 as *1 to 5.* Taking the reciprocal, we can say the *odds against* rolling a 4 are *5 to 1.*

In gambling, the "odds against" (or "odds on") usually expresses how much you can gain with a win for each dollar you bet. For example, suppose the odds on a particular horse at a horse race are 3 to 1. Then for each $1 you bet on this horse, you will gain $3 if the horse wins. If you place a $2 bet and the horse wins, you will gain $3 \times \$2 = \6. (You will also receive back the money you bet, so when you collect on your $2 bet you will receive the $6 gain plus your original $2, for a total of $8.)

➤**EXAMPLE 9** Two-Coin Odds

What are the odds for getting two heads in tossing two coins? What are the odds against it?

Solution As we found earlier, the probability of getting two heads in tossing two coins is 1/4. That is, $P(2\text{ heads}) = 1/4$. The probability of not getting two heads in tossing two coins is $P(\text{not }2\text{ heads}) = 1 - 1/4 = 3/4$. Thus, the odds for getting two heads are

$$\frac{P(2\text{ heads})}{P(\text{not }2\text{ heads})} = \frac{1/4}{3/4} = \frac{1}{3}$$

The odds for getting two heads in two coin tosses are 1 to 3. We take the reciprocal of the *odds for* to find the *odds against*. Thus, the odds against getting two heads in tossing two coins are 3 to 1.

Now try Exercises 47–50. ◄

➤**EXAMPLE 10** Horse Race Payoff

At a horse race, the odds on Blue Moon are given as 7 to 2. If you bet $10 and Blue Moon wins, how much will you gain?

Solution The 7 to 2 odds mean that, for each $2 you bet on Blue Moon, you gain $7 if you win. A $10 bet is equivalent to five $2 bets, so you gain $5 \times \$7 = \35. You also get your $10 back, so you will receive $45 when you collect on your winning ticket. *Now try Exercises 51–52.* ◄

by the way

What's a blue moon? Many people now say that a blue moon is the second full moon in a calendar month. But the editors of *Sky and Telescope* recently discovered that this definition came from a mistake in a 1946 issue of their magazine. No one knows the original definition of blue moon, but the phrase "once in a blue moon" has come to mean an extremely low probability of occurrence.

EXERCISES 7A

Review Questions

1. Distinguish between an outcome and an event in probability. Give an example in which the same event can occur through two or more outcomes.

2. Explain the notation P(event). Why must the probability always be between 0 and 1 (inclusive)?

3. Briefly describe the differences among theoretical, empirical, and subjective techniques for finding probabilities. Give an example of each.

4. How is the probability of an event *not* occurring related to the probability that it *does* occur? Why?

5. What is a probability distribution? Explain how to make a table of a probability distribution.

6. Explain the common usage of the term *odds* and its usage in gambling.

Does It Make Sense?

Decide whether each of the following statements makes sense (or is clearly true) or does not make sense (or is clearly false). Explain your reasoning.

7. When I toss four coins, there are four different outcomes that all represent the event of one head and three tails.

8. The probability that my sister will get into the college of her choice is 3.7.

9. I estimate that the probability of my getting married in the next 3 years is 0.7.

10. Because either there is life on Mars or there is not, the probability of life on Mars is 0.5.

11. The probability that Jonas will win the race is 0.6 and the probability that he will not win is 0.5.

12. Based on data showing that we've had snow on Christmas in 27 of the past 100 years, the probability of snow this year is 0.27.

Basic Skills & Concepts

Review of the Multiplication Principle. Exercises 13–20 use skills covered in the Brief Review on p. 433.

13. How many different choices of car do you have if a particular model comes in eight colors and three styles (sedan, station wagon, or hatchback)?

14. A local ski sale features eight types of skis, six types of bindings, and seven types of boots. How many different ski/binding/boot packages are available?

15. Your car radio has six preset buttons, each of which can be programmed in three modes (AM, FM1, and FM2). How many different stations can you preset?

16. Three districts elect representatives to the County Commission. There are two candidates in District 1, three candidates in District 2, and four candidates in District 3. How many different Commissions are possible?

17. The local paint store offers wallpaper in eight colors, each of which comes in four different patterns. How many different styles of wallpaper are available?

18. You are required to take five courses, one each in humanities, sociology, science, math, and music. You have a choice of four humanities courses, three sociology courses, five science courses, two math courses, and three music courses. How many different sets of five courses are possible?

19. You have a choice of six TV sets that can be matched with any of five VCR players. How many different TV/VCR systems can you assemble?

20. The car model you are considering comes with or without air conditioning, with or without a sun roof, with or without a CD-ROM player, and in eight stock colors. How many different versions of the car are available?

Theoretical Probabilities. In Exercises 21–28, use the theoretical method to determine the probability of the given outcomes and events. State any assumptions that you need to make.

21. Meeting someone with a phone number that ends in 0

22. Meeting someone born between midnight and 1:00 a.m.

23. Randomly selecting a two-child family with two girl children

24. Randomly selecting a three-child family with exactly two girl children

25. Rolling two dice and getting a sum of 5

26. Selecting a blue M&M from a bag containing 10 red M&Ms, 15 blue M&Ms, and 20 yellow M&Ms

27. Selecting a pair of blue socks from a drawer that holds 4 pairs of black socks, 6 pairs of blue socks, and 12 pairs of white socks

28. Guessing a correct answer on a multiple-choice question with five possible answers

Empirical Probabilities. Use the empirical method to estimate the probabilities in Exercises 29–34.

29. After recording the forecasts of your local weatherman for 30 days, you conclude that he gave a correct forecast 12 times. What is the probability that his next forecast will be correct?

30. What is the probability of a 100-year flood this year?

31. Halfway through the season, a basketball player has hit 86% of her free throws. What is the probability that her next free throw will be successful?

32. You count 66 tails when you toss a coin 100 times. If you don't know whether the coin is fair, what is the probability that the next toss will be a head?

33. A doctor diagnosed pneumonia in 100 patients and 34 of them actually had pneumonia. What is the probability that the next patient diagnosed with pneumonia will actually have pneumonia?

34. What is the probability of a 200-year flood this year?

Subjective Probability. For Exercises 35–36, state your estimate of the subjective probability, and explain how you arrived at your estimate.

35. The probability that you'll graduate from college within the next four years

36. The probability that you'll get an A on the final exam in this course

Event *Not* Occurring. Determine the probability of the events in Exercises 37–44. State any assumptions that you use.

37. What is the probability of not rolling a 4 with a fair die?

38. What is the probability of not tossing two heads with two fair coins?

39. What is the probability that a 76% free-throw shooter will miss his next free throw?

40. What is the probability of meeting a person not born in January, February, or March?

41. What is the probability of not rolling a sum of 7 with two fair dice?

42. What is the probability of a sunny day after a forecast of a 20% chance of rain?

43. What is the probability of not rolling an odd number with a fair die?

44. What is the probability of not rolling a double-6 with two fair dice?

Probability Distributions. In Exercises 45–46, make a probability distribution for the given set of events.

45. The number of heads when four fair coins are tossed

46. The sums that appear when two fair four-sided dice (tetrahedons) are tossed

The Odds. Use the definition given in the text to find both the *odds for* and the *odds against* the events in Exercises 47–50.

47. Rolling a fair die and getting a 1 or a 2

48. Flipping two fair coins and getting two tails

49. Rolling a fair die and getting a 5 or a 6

50. Flipping two fair coins and getting a head and a tail

Gambling Odds. Use the definition of odds in betting for Exercises 51–52.

51. The odds on (against) your bet are 3 to 4. If you bet $20 and win, how much will you gain?

52. The odds on (against) your bet are 5 to 4. If you bet $20 and win, how much will you gain?

Further Applications

53. **Gender Politics.** The following table gives the gender and political party of the 100 delegates at a political convention. Suppose you encounter a delegate at random. What is the probability that the delegate will *not* be a man? What is the probability that the delegate will *not* be a Republican?

	Women	Men
Republicans	21	28
Democrats	25	16
Independents	6	4

54. **Senior Citizen.** In 2000, there were 34.7 million people over 65 years of age out of a U.S. population of 281 million. By 2050, it is estimated that there will be 78.9 million people over 65 years of age out of a U.S. population of 394 million. Would your chances of meeting a person over 65 at random be greater in 2000 or 2050? Explain.

55. **Fair Coins?** You toss three coins 1000 times; the following table shows the results. Compute the empirical probability for each event. Compare the empirical probabilities to the theoretical probabilities for three tossed coins. Do you have reason to believe the coins are unfair? Explain.

Result	Number of occurrences
0 heads	260
1 head	495
2 heads	245
3 heads	0
Total	1000

56. **Marriage Age.** The following table gives percentages of all married women and men who were married for the first time in each age category. What is the probability that a randomly encountered married woman was married, for the first time, between the ages of 35 and 44? Make a probability distribution for ages of first marriage for men and women.

Age at first marriage	Women(%)	Men(%)
Under 20	16.6	6.6
20–24	40.8	36.0
25–29	27.2	34.3
30–34	10.1	14.8
35–44	4.5	7.1
45–64	0.7	1.1
65 and over	0.1	0.1

57. **Deceptive Odds.** Suppose event *A* has a 0.99 probability of occurring and event *B* has a 0.96 probability of occurring—both high probabilities. Compute the odds for event *A* and the odds for event *B*. Comment on the relative difference between the odds for the two events compared to the relative difference between the probabilities. How are the odds deceptive in this case?

58. **Project: Thumb Tack Probabilities.** Find a standard thumb tack and practice tossing it onto a flat surface. Notice that there are two different outcomes: The tack can land with its point facing down or with its point facing up.

 a. Toss the tack 50 times and record the outcomes.

 b. Give the empirical probabilities of the two outcomes based on these results.

 c. If possible, ask several other people to repeat the process. How well do your probabilities agree?

59. **Project: Three-Coin Experiment.** Toss three coins (at once) 50 times and record the outcomes in terms of the number of heads. Based on your observations, give the empirical probability of each result. Do the empirical results agree with the theoretical probabilities? Explain.

60. **Project: Randomizing a Survey.** Suppose you want to conduct a survey involving a sensitive question that not all participants may choose to answer honestly (for example, a question involving cheating on taxes or drug use). Here is a way to conduct the survey and protect the identity of respondents. We will assume that the sensitive question requires a "yes" or "no" answer. First ask all respondents to toss a fair coin. Then give the following instructions:

- If you toss a head, then answer the decoy question: Were you born on an even day of the month? (yes/no).

- If you toss a tail, then answer the real question of the survey (yes/no).

Ask all participants to answer "yes" or "no" to the question they were assigned. Then count the total numbers of "yes" and "no" responses.

a. Choose a particular question that may not produce totally honest responses. Conduct a survey in your class using this technique.

b. Given only the total numbers of "yes" and "no" responses to both questions, explain how you can estimate the numbers of people who answered "yes" and "no" to the real question.

c. Will the results computed in part b be exact? Explain.

d. Suppose the decoy question was replaced by these instructions: If you toss a head, then answer "yes." Can you still determine the number of people who answered "yes" and "no" to the real question?

Web Projects

Find useful links for Web Projects on the text Web site: www.aw.com/bennett-briggs

61. **Blood Groups.** The four major blood groups are designated A, B, AB, and O. Within each group there are two Rh types: positive and negative. Using library resources or the World Wide Web, find data on the relative frequency of blood groups, including the Rh types. Make a table showing the probability of meeting someone in each of the eight blood groups.

62. **Age and Gender.** The proportions of men and women in the population change with age. Using current data from a Web site, construct a table showing the probability of meeting a male or a female in each of these age categories: 0–5, 6–10, 11–20, 21–30, 31–40, 41–50, 51–60, 61–70, 71–80, and over 80.

63. **Accidents.** Use the Web to obtain data to estimate the empirical probabilities of accidents (such as being killed by lightning, by a shark bite, or by a falling airplane part). Are the empirical probabilities consistent with your intuition?

In the News

64. **Probability in the News.** Find a news article or research report that makes use of a probability. Interpret the probability and discuss whether it is theoretical, empirical, or subjective.

65. **Probability in Your Life.** Describe a recent instance from your own life when you used probability to make a decision. What type of probability did you use? How did it help you make the decision?

66. **Gambling Odds.** Find an advertisement for a gambling establishment or lottery in which the term *odds* is used, and interpret its meaning.

COMBINING PROBABILITIES

HISTORICAL NOTE
Chevalier was a title of low-ranking nobility in France. The Chevalier de Mère was a man by the name of Antoine Gombaud.

In 1654, a French con man known as the Chevalier de Mère bet other gamblers even money that he could roll at least one 6 in four rolls of a standard six-sided die. The Chevalier made quite a profit on this bet, despite having calculated his chance of winning incorrectly!

Because the probability of rolling a 6 on one roll of a die is 1/6, the Chevalier falsely reasoned that his chances would be four times as great on four rolls, or $4/6 = 0.67$. In fact, the chance of rolling a 6 within four tries is only 0.52 (as we will show shortly). Nevertheless, because a probability of 0.52 means winning more often than losing, the Chevalier profited over the long run.

As gamblers caught on to this "con game," they were no longer willing to play. The Chevalier therefore proposed a new game in which he rolled *two* dice and bet that he could throw a *double-6* within 24 rolls. He knew that, on any single roll of the two dice, the probability of a double-6 is 1/36 (see Unit 7A, Example 8). He then used his faulty reasoning to guess that, if he rolled the pair of dice 24 times, his odds of rolling at least one double-6 would be $24/36 = 0.67$.

To his dismay, the Chevalier began losing money. To understand his losses, he turned to the mathematician Blaise Pascal, who, in turn, began a correspondence about probability with the mathematician Pierre de Fermat. Pascal and Fermat quickly recognized the error in the Chevalier's reasoning and developed methods for properly combining probabilities. In this unit, we will study some of these methods.

Time out to think

The Chevalier multiplied the single-roll probability of 1/6 by 4 rolls to get his guess that he had a 4/6 chance of rolling a 6 within 4 rolls. Based on this reasoning, what would he have guessed for the probability of rolling a 6 within 6 rolls? How does this show the error in his reasoning?

AND PROBABILITIES

Suppose you toss two dice and want to know the probability that *both* will come up 4. One way to find the probability is to consider the tossing of the two dice as a *single* toss of two dice. Then we can find the probability with the theoretical method from Unit 7A. Because a double-4 is one of 36 possible outcomes for two dice, its probability is 1/36.

Alternatively, we can consider the two dice individually. For each die, the probability of a 4 is 1/6. Rolling a double-4 means rolling a 4 on the first die *and* a 4 on the second die. We find the probability of a double-4 by multiplying the two individual probabilities:

$$P(\text{double } 4) = P(4) \times P(4) = \frac{1}{6} \times \frac{1}{6} = \frac{1}{36}$$

By either method, the probability of rolling a double-4 is 1/36. In general, we call the probability of event *A and* event *B* occurring an ***and* probability** (or *joint probability*).

HISTORICAL NOTE
Blaise Pascal was a precocious Parisian who made fundamental contributions to mathematics and physics. He was deeply religious and his work *Pensées* is an important treatise of western philosophy. Pascal died at the age of 39. Pierre de Fermat was a lawyer, but his passion was mathematics. He spent his life corresponding with mathematicians and preoccupied with difficult problems. One of his most famous problems, known as Fermat's last theorem, was solved only recently.

The advantage of the multiplication technique is that it can easily be extended to other situations involving two or more events. For example, we can use it to find the probability of getting ten heads in a row on a coin toss or of having a baby *and* getting a pay raise in the same year. However, we must distinguish between events that are *independent* and events that are *dependent* upon each other. Let's investigate each case.

Independent Events

The repeated roll of a single die produces **independent events** because the outcome of one roll does not affect other rolls. Similarly, coin tosses are independent events because one coin toss does not affect others. For independent events, we calculate the *and* probability by multiplication.

And Probability: Independent Events

Two events are **independent** if the outcome of one does not affect the probability of the other event. Consider two independent events A and B with individual probabilities $P(A)$ and $P(B)$. The probability that A and B occur together is

$$P(A \text{ and } B) = P(A) \times P(B)$$

This principle can be extended to any number of independent events. For example, the *and* probability of three independent events A, B, and C is

$$P(A \text{ and } B \text{ and } C) = P(A) \times P(B) \times P(C)$$

> **EXAMPLE 1 Consecutive Floods**

Find the probability that a 100-year flood (a flood with a 0.01 probability of striking in a given year) will strike a city in two consecutive years. Assume that a flood in one year does not affect the likelihood of a flood in the next year.

Solution We are told to assume that floods in different years are independent events. Therefore, we find the probability of floods in two consecutive years by multiplying the probability for each individual year:

$$P(2 \text{ consecutive flood years}) = \underbrace{P(\text{flood})}_{\text{year 1}} \times \underbrace{P(\text{flood})}_{\text{year 2}} = 0.01 \times 0.01 = 0.0001$$

The probability of a 100-year flood in two consecutive years is 1 in 10,000. While events with such low probability can certainly occur, remember that this calculation *assumes* that the chance of a flood is 0.01 in any year. If a 100-year flood were to occur in two consecutive years, we might wonder whether the 0.01 probability of a single flood was correct. *Now try Exercise 11.* ◄

> **EXAMPLE 2 Three Coins**

Suppose you toss three coins. What is the probability of getting three tails?

Solution Because coin tosses are independent, we multiply the probability of tails on each individual coin:

$$P(3 \text{ tails}) = \underbrace{P(\text{tails})}_{\text{coin 1}} \times \underbrace{P(\text{tails})}_{\text{coin 2}} \times \underbrace{P(\text{tails})}_{\text{coin 3}} = \frac{1}{2} \times \frac{1}{2} \times \frac{1}{2} = \frac{1}{8}$$

The probability of three tails on three tossed coins is 1/8. *Now try Exercise 12.* ◄

Dependent Events

Suppose you draw candy randomly from a box that initially contains five chocolates and five caramels. Clearly, the probability of getting a chocolate on your first try is 5/10, or 1/2. Now, suppose you get a chocolate on the first try and eat it. What is the probability of getting another chocolate on your second draw?

Because you've already eaten one chocolate, the box now contains only nine pieces of candy, of which four are chocolate. Thus, the probability of getting a chocolate on the second try is 4 out of 9, or 4/9. This is *not* the same as the 1/2 probability on the first draw. Because the outcome of the first event affects the probability of the second event, we say that the events are **dependent events.**

Calculating the probability for dependent events still involves multiplying the individual probabilities, but we must take into account how prior events affect subsequent events. In the case of the candy box, we find the probability of getting two chocolates in a row by multiplying the 1/2 probability on the first draw by the 4/9 probability on the second draw:

$$P(2 \text{ chocolates}) = \underbrace{P(\text{chocolate})}_{\text{first draw}} \times \underbrace{P(\text{chocolate})}_{\text{second draw}} = \frac{1}{2} \times \frac{4}{9} = \frac{2}{9}$$

The probability of drawing two chocolates in a row is 2/9. We can generalize this example to any set of dependent events.

And Probability: Dependent Events

Two events are **dependent** if the outcome of one event affects the probability of the other event. The probability that dependent events *A* and *B* occur together is

$$P(A \text{ and } B) = P(A) \times P(B \text{ given } A)$$

where $P(B$ given $A)$ means "the probability of event *B* given the occurrence of event *A*." This principle can be extended to any number of dependent events. For example, the *and* probability of three dependent events *A*, *B*, and *C* is

$$P(A \text{ and } B \text{ and } C) = P(A) \times P(B \text{ given } A) \times P(C \text{ given } A \text{ and} B)$$

TECHNICAL NOTE
$P(B$ given $A)$ is called a *conditional probability.* Some books denote it as $P(B|A)$.

Time out to think _____

Suppose that instead of eating the first candy you draw, you put it back in the box. Without doing any calculations, compare the probability of drawing two chocolates in a row when you replace the candy in the box to the probability of drawing two chocolates in a row when you eat the first candy. Which probability is larger and why?

▶**EXAMPLE 3** BINGO

The game of BINGO involves drawing labeled buttons from a bin at random, without replacement. There are 75 buttons, 15 for each of the letters B, I, N, G, and O. What is the probability of drawing two B buttons in the first two selections?

Solution BINGO involves dependent events because removing a button changes the contents of the bin. The probability of drawing a B on the first draw is 15/75. If this occurs, the bin has 74 buttons remaining, of which 14 are the letter B. Therefore, the probability of drawing a B button on the second draw is 14/74. The probability of drawing two B buttons in the first two selections is

$$P(\text{B and B}) = \underbrace{P(\text{B})}_{\text{first draw}} \times \underbrace{P(\text{B})}_{\text{second draw}} = \frac{15}{75} \times \frac{14}{74} \approx 0.0378$$

The probability of drawing the letter B on the first two selections is 0.0378, or just under 4 in 100. *Now try Exercise 13.* ◀

▶**EXAMPLE 4** Jury Selection

A three-person jury must be selected at random from a pool of 12 people that has 6 men and 6 women. What is the probability of selecting an all-male jury?

Solution Because there are 6 men in the pool of 12 people, the probability that the first juror is male is 6/12. If the first juror is male, the remaining pool has 5 men among 11 people. Therefore, the probability that the second juror is also male is 5/11. If both the first two jurors are male, the pool is left with 4 men among 10 people, making the probability of a third male juror 4/10. The combined probability is

$P(3 \text{ men})$

$$= \underbrace{P(\text{man})}_{\text{juror 1}} \times \underbrace{P(\text{man given male juror 1})}_{\text{juror 2}} \times \underbrace{P(\text{man given male jurors 1 and 2})}_{\text{juror 3}}$$

$$= \frac{6}{12} \times \frac{5}{11} \times \frac{4}{10} = \frac{120}{1320} \approx 0.091$$

The probability of selecting an all-male jury is roughly 0.09, or 9%. *Now try Exercises 14–18.* ◀

by the way

In the late 1960s, the famed baby doctor Benjamin Spock was convicted by an all-male jury of encouraging draft resistance during the Vietnam War. His defense argued that a jury with women would have been more sympathetic.

EITHER/OR PROBABILITIES

Suppose we want to know the probability that *either* of two events occurs, rather than the probability that both events occur. In that case, we are looking for an ***either/or* probability,** such as the probability that either a #2 bus or a #9 bus will be

the next to arrive at your stop. As with *and* probabilities, there are two cases to consider; we call them *overlapping* and *non-overlapping* events.

Non-Overlapping Events

A coin can land either heads *or* tails, but it can't land on both heads *and* tails at the same time. When two events cannot occur at the same time, they are said to be **non-overlapping** (or *mutually exclusive*). If we make a Venn diagram (see Unit 1C) in which the circles represent different events, non-overlapping events are those for which there is no overlap between the circles. For example, the circles for heads and tails in a coin toss do not overlap because a coin cannot land both heads and tails at the same time (Figure 7.4). We can calculate *either/or* probabilities for non-overlapping events with the following rule.

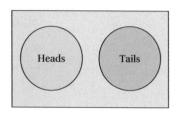

FIGURE 7.4 Venn diagram for non-overlapping events.

Either/Or Probability: Non-Overlapping Events

Two events are **non-overlapping** if they cannot occur together. If A and B are non-overlapping events, the probability that either A or B occurs is

$$P(A \text{ or } B) = P(A) + P(B)$$

This principle can be extended to any number of non-overlapping events. For example, the probability that either event A, event B, or event C occurs is

$$P(A \text{ or } B \text{ or } C) = P(A) + P(B) + P(C)$$

►EXAMPLE 5 Either/Or Dice

Suppose you roll a single die. What is the probability of rolling either a 2 or a 3?

Solution The outcomes 2 and 3 are non-overlapping because a single die can land only one way. Each probability is 1/6 (because there are 6 ways for the die to land), so the combined probability is

$$P(2 \text{ or } 3) = P(2) + P(3) = \frac{1}{6} + \frac{1}{6} = \frac{2}{6}$$

The probability of rolling either a 2 or a 3 is 2/6, or 1/3. *Now try Exercises 19–22.* ◄

Overlapping Events

Suppose you have a standard deck of 52 cards and want to know the probability of drawing either a queen or a club. Because there are 4 queens in the deck, the probability of drawing a queen is 4/52. Similarly, because there are 13 clubs in the deck, the probability of drawing a club is 13/52. The sum of the two probabilities is

$$\frac{4}{52} + \frac{13}{52} = \frac{17}{52}$$

However, this is *not* the probability of drawing a queen *or* a club. The Venn diagram in Figure 7.5 shows why. One circle represents the 4 queens in the deck and the other represents the 13 clubs in the deck. The circles overlap because one

queen *and* club

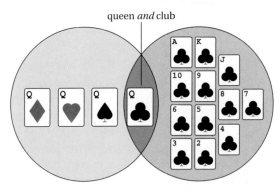

FIGURE 7.5 Venn diagram for overlapping events.
One circle represents queens in a deck of cards and the
other represents clubs. The overlap region contains the
queen of clubs, which belongs to both circles.

card—the queen of clubs—is both a queen *and* a club. That is, the events of draw-
ing a queen and drawing a club are **overlapping events** because both can occur
together. If we simply add the individual probabilities, as above, the queen of clubs
gets counted twice (once as a queen and once as a club). Thus, we can correct the
calculation by subtracting the double-counted probability of drawing the queen of
clubs, which is 1/52 (because the queen of clubs is just 1 of the 52 cards in the deck).
This gives us

$$P(\text{queen or club}) = P(\text{queen}) + P(\text{club}) - P(\text{queen and club})$$

$$= \frac{4}{52} + \frac{13}{52} - \frac{1}{52} = \frac{16}{52}$$

The probability of drawing a queen or a club is 16/52. You can also see this result
from the Venn diagram, which shows 16 cards (out of 52 total) that are either a queen
or a club (or both).

Either/Or Probability: Overlapping Events

Two events are **overlapping** if they *can* occur together. If *A* and *B* are overlap-
ping events, the probability that either *A* or *B* occurs is

$$P(A \text{ or } B) = P(A) + P(B) - P(A \text{ and } B)$$

➤ **EXAMPLE 6** Democrats and Women

You select one person at random from a room with eight people: two Democratic
men, two Republican men, two Democratic women, and two Republican women.
What is the probability that you will select *either* a woman or a Democrat?

Solution The probability of selecting a Democrat is 1/2, because half of the people in the room are Democrats. Similarly, half of the people are women, so the probability of selecting a woman is also 1/2. We cannot simply add these probabilities, because two of the eight people in the room are both Democrats and women. The probability of choosing one of these two people is 2/8, or 1/4. Thus, the overall probability of selecting either a woman or a Democrat is

$$P(\text{woman or Democrat}) = P(\text{woman}) + P(\text{Democrat}) - P(\text{woman and Democrat})$$
$$= \frac{1}{2} + \frac{1}{2} - \frac{1}{4} = \frac{3}{4}$$

The probability that you will select either a woman or a Democrat is 3/4, or 0.75.

Now try Exercises 23–26. ◄

THE *AT LEAST ONCE* RULE

Suppose you toss a coin four times. What is the probability of getting *at least one* head? There are four different ways to get at least one head: by getting 1 head, 2 heads, 3 heads, or 4 heads. Thus, one way to solve this problem is to add the individual probabilities:

$$P(\text{at least one H in 4 tosses}) = P(1\,\text{H}) + P(2\,\text{H}) + P(3\,\text{H}) + P(4\,\text{H})$$

To complete this calculation, we'd need to make a probability distribution that showed the probability for each of the four ways of getting at least one head.

Fortunately, there is an easier way. Because you either *do* or *do not* get at least one head in the four coin tosses, the sum of these two probabilities is 1:

$$P(\text{at least one H in 4 tosses}) + P(\text{no heads in 4 tosses}) = 1$$

(Getting "no heads" means *not* getting at least one head.) Subtracting the probability of no heads from both sides, we find

$$P(\text{at least one H in 4 tosses}) = 1 - P(\text{no heads in 4 tosses})$$

It's easy to find the probability of no heads. The probability of not getting heads on one coin toss is $P(\text{not H}) = 1/2$, so the probability of no heads in four tosses is

$$P(\text{no H in 4 tosses}) = \underbrace{P(\text{not H})}_{\text{toss 1}} \times \underbrace{P(\text{not H})}_{\text{toss 2}} \times \underbrace{P(\text{not H})}_{\text{toss 3}} \times \underbrace{P(\text{not H})}_{\text{toss 4}} = [P(\text{not H})]^4$$

The last step follows because $P(\text{not H})$ is the same for all four tosses. By substituting $P(\text{not H}) = 1/2$, we find that the probability of at least one head in 4 coin tosses is

$$P(\text{at least one H in 4 tosses}) = 1 - P(\text{no H in 4 tosses})$$
$$= 1 - [P(\text{not H})]^4$$
$$= 1 - \left[\frac{1}{2}\right]^4 = 1 - \frac{1}{16} = \frac{15}{16}$$

We can generalize this example to any question of the form "What is the probability of *at least one* particular event in a series of independent trials?"

The *At Least Once* Rule (for Independent Events)

Suppose the probability of an event *A* occurring in one trial is *P(A)*. If all trials are independent, the probability that event *A* occurs *at least once* in *n* trials is

$$P(\text{at least one event } A \text{ in } n \text{ trials}) = 1 - P(\text{no events } A \text{ in } n \text{ trials})$$
$$= 1 - [P(\text{not } A \text{ in one trial})]^n$$

▶EXAMPLE 7 100-Year Flood

What is the probability that a region will experience at least one 100-year flood (a flood that has a 0.01 chance of occurring in any given year) during the next 100 years? Assume that 100-year floods in consecutive years are independent events.

Solution Because there's a 0.01 probability of a flood in any one year, there is a 0.99 chance that a flood will *not* occur in any one year. The *at least once* rule gives us the probability of at least one flood in 100 years:

$$P(\text{at least one flood in 100 years}) = 1 - [P(\text{no flood in one year})]^{100}$$
$$= 1 - [0.99]^{100} \approx 0.634$$

There is a 0.634, or almost 2 out of 3, chance that a 100-year flood will occur at least once in the next 100 years. *Now try Exercises 27–30.* ◀

Time out to think _____

Suppose a region has not experienced a 100-year flood in 100 years. Should you be surprised? Would this fact give you reason to suspect that 100-year floods don't really occur in this region?

▶EXAMPLE 8 Lottery Chances

You purchase 10 lottery tickets, for which the probability of winning some prize on a single ticket is 1 in 10. What is the probability that you will have at least one winning ticket among the 10 tickets?

Lottery: a tax on people who are bad at math.

—MESSAGE CIRCULATED ON THE INTERNET

Solution Because the probability of winning with any one ticket is 0.1, the probability of not winning with one ticket is 0.9. Therefore, the probability of winning at least once with ten tickets is

$$P(\text{at least one winner in 10 tickets}) = 1 - [P(\text{not winning on one ticket})]^{10}$$
$$= 1 - [0.9]^{10} = 0.651$$

The probability of at least one winning ticket among 10 is 0.651. This also means that the probability of all ten tickets being losers is $1 - 0.651 = 0.349$. In other words, you stand a better than 1 in 3 chance of losing on all ten tickets that you buy.

Now try Exercises 31–34. ◀

RETURN TO THE CHEVALIER

We are now ready to return to the story of the Chevalier de Mère. Recall that, in his first game, he bet that he could get at least one 6 within four rolls of a die. We can use the *at least once* rule to calculate his probability of winning. The probability of a 6 on any single roll is 1/6, so the probability of *not 6* is 5/6. Thus, the probability of getting at least one 6 in four rolls is

$$P(\text{at least one 6 in 4 rolls}) = 1 - [P(\text{not 6 on one roll})]^4$$

$$= 1 - \left[\frac{5}{6}\right]^4 \approx 0.518$$

The Chevalier's probability of winning in the first game was 0.518. Although this is considerably lower than the 4/6 (0.667) probability he had guessed, it is still better than even odds. He was likely to come out ahead if he played the game many times.

In his second game, he bet that he could roll a double-6 within 24 tries. The probability of a double-6 on any single roll is 1/36, so the probability of not getting a double-6 on a single roll is 35/36. Therefore, the probability of getting at least one double-6 in 24 rolls is

$$P(\text{at least one double-6 in 24 rolls}) = 1 - [P(\text{not double-6 on one roll})]^{24}$$

$$= 1 - \left[\frac{35}{36}\right]^{24} \approx 0.491$$

The Chevalier's probability of winning in his second game was 0.491, which is slightly less than even odds. Therefore, he was likely to lose more games than he won.

Time out to think

Comment on how the difference between a 52% and a 49% chance of winning can be the difference between rags and riches in gambling. How can casinos exploit this idea?

EXERCISES 7B

Review Questions

1. How did the gambling habits of the Chevalier de Mère help launch the mathematical study of probability?

2. Give an example in which we would be interested in an *and* probability. How do we determine whether the events are independent or dependent? Give examples of each case, and explain how we calculate the probabilities.

3. Give an example in which we would be interested in an *either/or* probability. How do we determine whether the events are overlapping or non-overlapping? Give examples of each case, and explain how we calculate the probabilities.

4. What is the *at least once* rule? Explain how the *at least once* rule can be used to find the correct probabilities in the games of the Chevalier de Mère.

Does It Make Sense?

Decide whether each of the following statements makes sense (or is clearly true) or does not make sense (or is clearly false). Explain your reasoning.

5. The probability of getting heads *and* tails when you toss a coin is 0, but the probability of getting heads *or* tails is 1.

6. If you toss a coin and get heads three times in a row, you're due to get tails on the next toss.

7. The probability of drawing an ace or a spade from a deck of cards is the same as the probability of drawing the ace of spades.

8. I can't believe you chose the lottery numbers 1-2-3-4-5-6. Getting six numbers in a row is much less likely than getting other random numbers.

9. My chance of getting a 5 on the roll of one die is 1/6, so my chance of getting at least one 5 when I roll three dice is 3/6.

10. To find the probability that at least one of my 25 lottery tickets is a winner, I calculated the probability that none of my tickets is a winner and subtracted it from 1.

Basic Skills & Concepts

And Probabilities. Determine whether the events given in Exercises 11–18 are independent or dependent. Then find the probability of the event.

11. Purchasing three winning lottery tickets in a row when each ticket has a 1 in 10 chance of being a winner

12. Rolling four successive 6s with four rolls of a fair die

13. Randomly selecting 3 Americans from a pool of 12 British people and 9 Americans

14. Selecting all women for a six-person jury from a pool of 12 men and 12 women

15. Discovering that your five best friends all have telephone numbers ending in 1

16. Selecting two defective computer chips from a large batch in which the defect rate is 1.5%

17. The next five births at a hospital all being girls

18. Tossing four heads, then four tails on the first eight tosses of a fair coin

Either/Or Probabilities. Determine whether the events in Exercises 19–26 are overlapping or non-overlapping. Then find the probability of the event.

19. Rolling either a 1 or a 2 on a single die

20. Rolling either a 3, a 4, or a 5 on a single die

21. Getting a sum of either 7 or 8 on a roll of two dice

22. Getting a sum of either 2, 3, or 4 on a roll of two dice

23. Selecting either a woman or a Democrat when you select a person randomly from a group composed of 25 Democratic men, 25 Republican men, 25 Democratic women, and 25 Republican women

24. Selecting either a woman or an American when you select a person randomly from a group composed of 20 English women, 25 English men, 25 American women, and 20 American men

25. Drawing either a king or a heart from a standard deck of 52 cards

26. Drawing either an ace or a king from a standard deck of 52 cards

At Least Once Problems. Use the *at least once* rule to find the probabilities of the events described in Exercises 27–34.

27. Getting at least one 6 in five rolls of a single die

28. Getting at least one 1 in four rolls of a single die

29. Getting at least one head when tossing three coins

30. Getting at least one tail when tossing ten coins

31. Getting rain at least once in five days if the probability of rain on each single day is 0.2

32. Getting rain at least once in ten days if the probability of rain on each single day is 0.1

33. Drawing at least one king when you draw cards from a standard deck 20 times; assume you replace the card each time you draw, so there are always 52 cards to draw from.

34. Drawing at least one ace when you draw cards from a standard deck 52 times; assume you replace the card each time you draw, so there are always 52 cards to draw from.

Further Applications

35. **Probability and Court.** The data in the following table show the outcomes of guilty and not-guilty pleas in 1028 criminal court cases.

	Guilty plea	Not-guilty plea
Sent to prison	392	58
Not sent to prison	564	14

SOURCE: Brereton and Casper, Law and Society Review, *Vol. 16, No. 7.*

a. What is the probability that a randomly selected defendant either pled guilty or was sent to prison?

b. What is the probability that a randomly selected defendant either pled not guilty or was not sent to prison?

36. Testing a Drug. An allergy drug was tested by giving 120 people the drug and 100 people a placebo. A control group consisted of 80 people who were given no treatment. The number of people in each group who showed improvement is shown in the table below.

	Allergy drug	Placebo	Control	Total
Improvement	65	42	31	138
No improvement	55	58	49	162
Total	120	100	80	300

a. What is the probability that a randomly selected person in the study was given either the drug or a placebo?

b. What is the probability that a randomly selected person either improved or did not improve?

c. What is the probability that a randomly selected person either was given the drug or improved?

d. What is the probability that a randomly selected person was given the drug and improved?

37. Polling Calls. A telephone pollster has a list of names and telephone numbers for 45 voters, 20 of whom are listed as registered Democrats and 25 of whom are listed as registered Republicans. Calls are made in random order. Suppose you want to find the probability that the first two calls are to Republicans.

a. Are these independent or dependent events? Explain.

b. If you treat them as dependent events, what is the probability that the first two calls are to Republicans?

c. If you treat them as independent events, what is the probability that the first two calls are to Republicans?

d. Compare the results of parts b and c.

38. Dominant and Recessive Genes. Many traits are controlled by a dominant gene, **A**, and a recessive gene, **a**. A child gets two genes, one from each parent. Suppose a child's parents each have the gene combination **Aa**; that is, each parent is equally likely to pass either the **A** or the **a** gene to the child. Make a table showing the probability distribution for the child's possible gene combinations, which are **AA**, **Aa**, and **aa**. If the combinations **AA** and **Aa** both result in the same dominant trait (say, brown hair) and **aa** results in the recessive trait (say, blond hair), what is the probability that a child will have the dominant trait? What is the probability that the child will have the recessive trait?

39. Better Bet for the Chevalier. Suppose that the Chevalier de Mère had bet that he could roll a double-6 within 25 rolls, rather than 24. In that case, what would have been his probability of winning? Had he made this bet, would he still have lost over time? Explain.

40. Lottery Odds. The probability of a $2 winner in a particular state lottery is 1 in 10, the probability of a $5 winner is 1 in 50, and the probability of a $10 winner is 1 in 500.

a. What is the probability of getting either a $2, $5, or $10 winner? Compare to the probability of getting only a $2 winner.

b. If you buy 50 lottery tickets, what is the probability that you will get at least one $5 winner?

c. If you buy 500 lottery tickets, what is the probability that you will get at least one $10 winner?

41. Miami Hurricanes. Studies of the Florida Everglades show that, historically, the Miami region is hit by a hurricane about every 40 years.

a. Based on the historical record, what is the empirical probability that Miami will be hit by a hurricane next year?

b. What is the probability that Miami will be hit by hurricanes in two consecutive years?

c. What is the probability that Miami will be hit by at least one hurricane in the next 10 years?

42. HIV Among College Students. Suppose that 2% of the students at a particular college are infected with HIV.

a. If a student has ten sexual partners over a period of time, what is the probability that at least one of these partners is infected with HIV?

b. If a student has 20 different sexual partners over a period of time, what is the probability that at least one partner is infected with HIV?

Web Projects

Find useful links for Web Projects on the text Web site: www.aw.com/bennett-briggs

43. Lottery Chances. Find a lottery Web site and study the probabilities of winning. Based on your findings, what is the probability of winning at least once if you play ten times? Explain any assumptions you make.

44. **HIV Probabilities.** Research the empirical probabilities of HIV among various groups in the population (for example, groups based on ethnicity, nationality, or sexual orientation). Based on your findings, discuss the relative dangers of sexual promiscuity within various populations.

In the News

45. **Combined Probability in the News.** Find a recent news story in which two or more probabilities were combined in some way. Describe how and why they were combined.

46. **Combined Probability in Your Life.** Cite a recent situation or decision in your life that involved an *and* probability or an *either/or* probability. How did the probability affect the situation or decision?

UNIT 7C

THE LAW OF LARGE NUMBERS

If you toss a coin once, you cannot predict exactly how it will land; you can only state that the probability of a head is 0.5. If you toss the coin 100 times, you still cannot predict precisely how many heads will occur. However, you can reasonably expect to get heads *close to* 50% of the time. If you toss the coin 1000 times, you can expect the proportion of heads to be even closer to 50%. In general, the more times you toss the coin, the closer the percentage of heads will be to exactly 50%. While individual events may be unpredictable, large collections of events should show some pattern or regularity. This principle is called the **law of large numbers** (or the *law of averages*).

> **The Law of Large Numbers**
>
> The **law of large numbers** applies to a process for which the probability of an event A is $P(A)$ and the results of repeated trials are independent. It states:
> *If the process is repeated through many trials, the proportion of the trials in which event A occurs will be close to the probability P(A). The larger the number of trials, the closer the proportion should be to P(A).*

We can illustrate the law of large numbers with a die rolling experiment. The probability of a 1 on a single roll is $P(1) = 1/6 = 0.167$. To avoid the tedium of rolling the die many times, we can let a computer *simulate* random rolls of the die. Figure 7.6 shows a computer simulation of rolling a single die 5000 times. The horizontal axis gives the number of rolls, and the height of the curve gives the proportion of 1s. Although the curve bounces around when the number of rolls is small, the proportion of 1s approaches the probability of 0.167, in agreement with the law of large numbers.

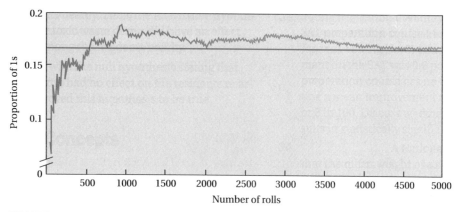

FIGURE 7.6 Results of computer simulation of rolling a die. As the number of rolls grows large, the proportion of 1s gets close to the probability of getting a 1 on a single roll, which is 0.167.

➤**EXAMPLE 1** Roulette

A roulette wheel has 38 numbers: 18 are black, 18 are red, and the numbers 0 and 00 are green.

 a. What is the probability of getting a red number on any spin?
 b. If patrons in a casino spin the wheel 100,000 times, how many times should you expect a red number?

Solution

 a. There are 18 red numbers among 38 total, so the probability of red on a single spin is 18/38, or 0.474.
 b. The law of large numbers tells us that as the game is played more and more times, the proportion of times that the wheel comes up red should get closer to 0.474. Thus, in 100,000 tries, the wheel should come up red close to 47.4% of the time, or about 47,400 times.
 Now try Exercises 13–14. ◄

EXPECTED VALUE

Suppose the InsureAll Company sells a special type of insurance that pays you $100,000 in the event that you quit your job because of serious illness. Based on data from past claims, the empirical probability that a policyholder will make a claim for loss of job is 1 in 500. Should the insurance company expect to earn a profit if it sells the policies for $250 each?

 If InsureAll sells only a few policies, we cannot predict the profit or loss. For example, selling 100 policies for $250 each generates revenue of 100 × $250 = $25,000. If none of the 100 policyholders files a claim, InsureAll makes a tidy profit. But if even one policyholder makes a $100,000 claim, InsureAll faces a huge loss.

 However, if InsureAll sells a large number of policies, the law of large numbers tells us that the proportion of policies for which claims must be paid should be very close to the 1 in 500 probability for a single policy. For example, if InsureAll

sells 1 million policies, it should expect that the number of policyholders making the $100,000 claim will be close to

$$\underbrace{1,000,000}_{\text{number of policies}} \times \underbrace{\frac{1}{500}}_{\substack{\text{probability of} \\ \$100,000 \text{ claim}}} = 2000$$

Paying these 2000 claims will cost

$$2000 \times \$100,000 = \$200 \text{ million}$$

This cost of $200 million represents an *average* of $200 for each of the 1 million policies. Thus, if the policies sell for $250 each, InsureAll should expect to earn an average of $250 − $200 = $50 per policy. This amounts to a profit of $50 million on sales of 1 million policies.

We can find this same answer with a more formal procedure. The InsureAll example involves two distinct events, each with a particular *probability* and *value* for the company:

1. In the event that a person buys a policy, the value to the company is the $250 price of the policy. The probability of this event is 1 because everyone who buys a policy pays the $250.
2. In the event that a person files a claim, the value to the company is −$100,000. It is negative because the company loses $100,000 in this case. The probability of this event is 1/500.

If we multiply the value of each event by its probability and then add the results, we find the average or **expected value** of each insurance policy:

$$\text{expected value} = \underbrace{\$250}_{\substack{\text{value of policy sale}}} \times \underbrace{1}_{\substack{\text{probability of} \\ \text{earning } \$250 \text{ on sale}}} + \underbrace{(-\$100,000)}_{\substack{\text{value of} \\ \text{claim}}} \times \underbrace{\frac{1}{500}}_{\substack{\text{probability of} \\ \text{paying claim}}}$$

$$= \$250 - \$200 = \$50$$

This expected profit of $50 per policy is the same answer we found earlier. Keep in mind that this value is expected only if InsureAll sells enough policies for the law of large numbers to come into play.

Time out to think

Should InsureAll expect to see a profit of $50 on each individual policy? Should it expect a profit of $50,000 on 1000 policies? Explain.

Expected Value

Consider two events, each with its own value and probability. The **expected value** is

$$\text{expected value} = \left(\begin{array}{c}\text{event 1} \\ \text{value}\end{array}\right) \times \left(\begin{array}{c}\text{event 1} \\ \text{probability}\end{array}\right) + \left(\begin{array}{c}\text{event 2} \\ \text{value}\end{array}\right) \times \left(\begin{array}{c}\text{event 2} \\ \text{probability}\end{array}\right)$$

This formula can be extended to any number of events by including more terms in the sum.

►**EXAMPLE 2** Lottery Expectations

Suppose that $1 lottery tickets have the following probabilities: 1 in 5 to win a free ticket (worth $1), 1 in 100 to win $5, 1 in 100,000 to win $1000, and 1 in 10 million to win $1 million. What is the expected value of a lottery ticket? Discuss the implications. (Note: Winners do *not* get back the $1 they spend on the ticket.)

Solution The easiest way to proceed is to make a table of all the relevant events with their values and probabilities. We are calculating the expected value of a lottery ticket to *you,* so the ticket price has a negative value (it costs you money) while the values of winnings are positive.

Event	Value	Probability	Value × probability
Ticket purchase	−$1	1	$-\$1 \times 1 = -\1
Win free ticket	$1	$\frac{1}{5}$	$\$1 \times \frac{1}{5} = \0.20
Win $5	$5	$\frac{1}{100}$	$\$5 \times \frac{1}{100} = \0.05
Win $1000	$1000	$\frac{1}{100,000}$	$\$1000 \times \frac{1}{100,000} = \0.01
Win $1 million	$1,000,000	$\frac{1}{10,000,000}$	$\$1,000,000 \times \frac{1}{10,000,000} = \0.10
			Sum: −$0.64

The expected value is the sum of all the products *value* × *probability*, which the table shows to be −$0.64. Thus, averaged over many tickets, you should expect to lose 64¢ for each lottery ticket that you buy. If you buy, say, 1000 tickets, you should expect to *lose* about $1000 \times \$0.64 = \640. *Now try Exercises 15–18.* ◄

Time out to think

Many states use lotteries to finance worthy causes such as parks, recreation, and education. Lotteries also tend to keep state taxes at lower levels. On the other hand, research shows that lotteries are played primarily by people with low incomes. Do you think lotteries are good social policy? Do you think lotteries are good economic policy?

DILBERT reprinted by permission of United Feature Syndicate, Inc.

►**EXAMPLE 3** Art Auction

You are at an art auction, trying to decide whether to bid $50,000 for a particular painting. You believe that you have a 1/2 probability of reselling the painting to a

client in New York for $70,000 and a 1/4 probability of reselling the painting to a client in San Francisco for $80,000. Otherwise, you will have to keep the painting—an option that has benefits, but not monetary ones. What is the expected value of a $50,000 bid?

Solution There are three relevant events to consider:

- With probability 1, you will pay $50,000 to buy the painting. This option represents a monetary loss to you, so it has a negative value.
- With probability 1/2, you will sell the painting for $70,000. This option is a gain, so it has a positive value.
- With probability 1/4, you will sell the painting for $80,000. This option is also a gain, so it has a positive value.

Thus, the expected value of a $50,000 bid is

$$\text{expected value} = \left(\begin{array}{c}\text{value of}\\\text{event 1}\end{array}\right) \times \left(\begin{array}{c}\text{prob of}\\\text{event 1}\end{array}\right) + \left(\begin{array}{c}\text{value of}\\\text{event 2}\end{array}\right) \times \left(\begin{array}{c}\text{prob of}\\\text{event 2}\end{array}\right) + \left(\begin{array}{c}\text{value of}\\\text{event 3}\end{array}\right) \times \left(\begin{array}{c}\text{prob of}\\\text{event 3}\end{array}\right)$$

$$= (-\$50,000 \times 1) + \left(\$70,000 \times \frac{1}{2}\right) + \left(\$80,000 \times \frac{1}{4}\right)$$

$$= -\$50,000 + \$35,000 + \$20,000 = \$5000$$

With a $50,000 bid, you could expect to gain $5000 by the purchase. It's important to remember that expected value refers to expected gain or loss over *many* such purchases. If this is your only purchase, it might be better to reconsider or to assign a high value to admiring the painting in your own house! *Now try Exercises 19–22.* ◄

THE GAMBLER'S FALLACY

Consider a simple game involving a coin toss: You win $1 if the coin lands heads and you lose $1 if it lands tails. Now, suppose you toss the coin 100 times and get 45 heads and 55 tails, putting you $10 in the hole. Are you "due" for a streak of better luck?

You probably recognize that the answer is *no*. Each coin toss is independent of previous tosses, so your past bad luck has no bearing on your future chances. However, many gamblers—especially compulsive gamblers—guess just the opposite. They believe that when their luck has been bad, it's due for a change. This mistaken belief is often called the **gambler's fallacy** (or the *gambler's ruin*).

> **DEFINITION**
>
> The **gambler's fallacy** is the mistaken belief that a streak of bad luck makes a person "due" for a streak of good luck.

Time out to think

A farmer knows that at this time of year in his part of the country, the probability of rain on a given day is 0.5. It hasn't rained in 10 days and he needs to decide whether to start irrigating. Is he justified in postponing irrigation because he is due for a rainy day? Explain.

One reason people succumb to the gambler's fallacy is a misunderstanding of the law of large numbers. In our coin toss game, the law of large numbers tells us that the proportion of heads tends to be closer to 50% for larger numbers of games. But this does *not* mean that you are likely to recover early losses. The following example shows why.

►**EXAMPLE 4** Continued Losses

You are playing the coin toss game, in which you win $1 for heads and lose $1 for tails. After 100 tosses, you are $10 in the hole because you have 45 heads and 55 tails. You continue playing until you've tossed the coin 1000 times, at which point you've gotten 480 heads and 520 tails. Does the result agree with what we expect from the law of large numbers? Have you gained back any of your losses? Explain.

Solution The proportion of heads in your first 100 tosses was 45%. After 1000 tosses, the proportion of heads has increased to 480 out of 1000, or 48%. This agrees with the law of large numbers, because the proportion grew closer to 50%. However, you've now won $480 for the 480 heads while losing $520 for the 520 tails, for a net loss of $40. Thus, your losses have actually *increased* from $10 to $40, despite the fact that the proportion of heads grew closer to 50%. *Now try Exercises 23–25.* ◄

Streaks

Another common misunderstanding that contributes to the gambler's fallacy involves expectations about streaks. Suppose you toss a coin six times and see the outcomes HHHHHH (all heads). Then you toss it six more times and see the outcomes HTTHTH. Most people would say that the latter outcome is "natural" while the streak of all heads is surprising. However, both outcomes are equally likely. In fact, all possible sets of six outcomes are equally likely: The probability of any *single* particular outcome is 1/2, so all sets of six particular outcomes have the same probability of $(1/2)^6 = 1/64$.

►**EXAMPLE 5** Hot Hand at the Craps Table

The popular casino game of craps involves rolling dice. Suppose you are playing craps and suddenly find yourself with a "hot hand": You roll a winner on ten consecutive bets. Is your hand really "hot"? Should you increase your bet because you are on

HISTORICAL NOTE
One of the most famous compulsive gamblers in history was the Russian writer Fyodor Dostoyevsky. He played at roulette throughout Europe in the 19th century. He once reported that he "won 10,000 francs at first, . . . but I got carried away and dropped half my winnings." His experience with gambling formed the basis of his work *The Gambler*.

Luck is a mighty queer thing. All you know about it for certain is that it's bound to change.

—BRET HARTE, *THE OUTCASTS OF POKER FLAT*

a hot streak? Assume that you are making bets with a 0.486 probability of winning on a single play (the best odds available in craps).

Solution First, let's consider the chance of winning ten times in a row. With a probability of 0.486 on a single play, the chance of ten straight wins is

$$(0.486)^{10} \approx 0.000735$$

Thus, your streak of ten straight wins has a probability of only about 7 in 10,000, which might make you think that your hand really is "hot." However, look around the casino. If it is a large casino, several hundred people may be playing at the craps tables, and there may be *tens of thousands* of individual rolls each night. Thus, it is almost inevitable that someone will have a "hot streak" of ten straight wins. Indeed, even longer hot streaks are likely during any given night. Your apparent "hot hand" is a mere coincidence and your probability of winning on your next bet is still only 0.486. Increasing your bet based on a mere coincidence would be foolish.

Now try Exercises 26–28. ◄

The House Edge

A casino makes money because games are set up so that the expected earnings of patrons are negative (that is, losses). Because the casino earns whatever patrons lose, the casino's earnings are positive. The amount that the casino, or *house*, can expect to earn per dollar bet is called the **house edge.** That is, the house edge is the expected value *to the casino* of a particular bet.

The house edge varies from game to game. It tends to be greatest in games where big winnings are possible, such as slot machines. It tends to be least in games where strategy can improve your odds, such as blackjack. But the house edge is always present, and gamblers who believe they can beat the house regularly are sadly succumbing to the gambler's fallacy.

► **EXAMPLE 6** The House Edge in Roulette

The game of roulette is usually set up so that betting on red is a 1 to 1 bet. That is, you win the same amount of money as you bet if red comes up. Betting on a single number is a 35 to 1 bet. That is, you win 35 times as much as you bet if your number comes up. What is the house edge in each of these two cases? If patrons wager $1 million on such bets, how much should the casino expect to earn?

Solution For each case, the house edge is the casino's expected earnings on the bet. The probability of red is 18/38 (see Example 1), so the probability of not red is 20/38. Let's call the *value* of red for the casino −$1, indicating that the casino pays the amount of the patron's bet when red comes up. The value of not red for the casino is $1, because it wins the patron's bet if red does not come up. The house edge when patrons bet on red is

$$\left(\underbrace{\$1}_{\substack{\text{value to casino} \\ \text{of not red}}} \times \underbrace{\frac{20}{38}}_{\substack{\text{probability} \\ \text{of not red}}} \right) + \left(\underbrace{-\$1}_{\substack{\text{value to casino} \\ \text{of red}}} \times \underbrace{\frac{18}{38}}_{\substack{\text{probability} \\ \text{of red}}} \right) \approx \$0.053$$

That is, the casino can expect to gain $0.053 per dollar gambled. Thus, if patrons bet $1 million on red, the casino can expect to earn about $1 million × 0.053 = $53,000.

The probability of winning a bet on a single number is 1/38 because there are 38 numbers on the roulette wheel. The 35 to 1 payoff offered by the casino on this bet means the casino pays the patron 35 times the amount of the bet. Thus, its value to the casino is −$35. The probability that the bet does not win is 37/38. The casino gains the patron's bet in this case, so its value is $1. Thus, the house edge when patrons bet on a single number is

$$\underbrace{\left(\$1 \times \frac{37}{38}\right)}_{\substack{\text{value to casino of not} \\ \text{hitting the number}}\;\;\;\substack{\text{probability of not} \\ \text{hitting the number}}} + \underbrace{\left(-\$35 \times \frac{1}{38}\right)}_{\substack{\text{value to casino of} \\ \text{hitting the number}}\;\;\;\substack{\text{probability of} \\ \text{hitting the number}}} \approx \$0.053$$

This is the same house edge as for the red bet. Again, if patrons bet $1 million on single numbers, the house can expect to win $53,000. *Now try Exercises 29–30.* ◄

Time out to think

Given the house edge, is it possible for a casino to lose money on its gambling results alone? Explain.

■ EXERCISES 7C

Review Questions

1. Explain the meaning of the law of large numbers. Does this law say anything about what will happen in a single observation or experiment? Why or why not?

2. In 10 tosses of a fair coin, should you be surprised to see 6 heads? In 1000 tosses of a fair coin, should you be surprised to see 600 heads? Explain in terms of the law of large numbers.

3. What is an expected value, and how is it computed? Should we always expect to get the expected value? Why or why not?

4. What is the gambler's fallacy? Give an example.

5. Explain why the probability is the same for any particular set of ten coin toss outcomes. How does this idea affect our thinking about streaks?

6. What is the house edge in gambling? Explain how it virtually guarantees that a casino will win more money than it loses.

Does It Make Sense?

Decide whether each of the following statements makes sense (or is clearly true) or does not make sense (or is clearly false). Explain your reasoning.

7. The expected value to me of each raffle ticket I purchased is −$0.85.

8. The expected value of each insurance policy our company sells is $150, so if we sell 10 more policies our profits will increase by $1500.

9. If you toss a coin four times, it's much more likely to land in the order HTHT than HHHH. (H stands for heads and T for tails.)

10. I haven't won in my last 25 pulls on the slot machine, so I'm due to win on the next pull.

11. I haven't won in my last 25 pulls on the slot machine, so I must be having a bad day and I'm sure to lose if I play again.

12. I've lost $750 so far today on roulette. I'm going to play a little longer so that I can reduce my losses to $500.

Basic Skills & Concepts

13. **Understanding the Law of Large Numbers.** Suppose you toss a fair coin 10,000 times. Should you expect to get exactly 5000 heads? Why or why not? What does the law of large numbers tell you about the results you are likely to get?

14. **Speedy Driver.** Suppose a person who has a habit of driving fast has never had an accident or traffic citation.

What does it mean to say that "the law of averages [large numbers] will catch up with him"? Is it true? Explain.

Expected Value in Games. In Exercises 15–18, find the expected value (to you) of the described game. Would you expect to win or lose money in 1 game? in 100 games? Explain.

15. You are given 5 to 1 odds against tossing three heads in three tosses of a fair coin, meaning you win $5 if you succeed and you lose $1 if you fail.

16. You are given 10 to 1 odds against tossing three heads in three tosses of a fair coin, meaning you win $10 if you succeed and you lose $1 if you fail.

17. You are given 5 to 1 odds against rolling two even numbers with the roll of two fair dice, meaning you win $5 if you succeed and you lose $1 if you fail.

18. You are given 10 to 1 odds against rolling a double number (for example, two 1s or two 2s) with the roll of two fair dice, meaning you win $10 if you succeed and you lose $1 if you fail.

Insurance Claims. In Exercises 19–20, find the expected value (to the company) per policy sold. If the company sells 100,000 policies, what is the expected profit or loss? Explain.

19. An insurance policy sells for $1000. Based on past data, an average of 1 in 100 policyholders will file a $20,000 claim, an average of 1 in 200 policyholders will file a $50,000 claim, and an average of 1 in 500 policyholders will file a $100,000 claim.

20. An insurance policy sells for $500. Based on past data, an average of 1 in 50 policyholders will file a $5000 claim, an average of 1 in 100 policyholders will file a $10,000 claim, and an average of 1 in 200 policyholders will file a $30,000 claim.

21. **Expected Wait.** Suppose you arrive at a bus stop randomly, so all arrival times are equally likely. The bus arrives regularly every 30 minutes without delay (say, on the hour and on the half hour). What is the expected value of your waiting time? Explain.

22. **Expected Wait.** A bus arrives at a bus stop at noon, 12:20, and 1:00. You arrive at the bus stop at random times between noon and 1:00 every day, so all arrival times are equally likely.

 a. What is the probability that you will arrive at the bus stop between noon and 12:20? What is your mean waiting time in that case? (Exercise 21 may be helpful.)

 b. What is the probability that you will arrive at the bus stop between 12:20 and 1:00? What is your mean waiting time in that case?

 c. Overall, what is your expected waiting time for the bus?

 d. Would your expected waiting time be longer or shorter if the bus arrived at equally spaced intervals (say, noon, 12:30, 1:00)? Explain.

23. **Gambler's Fallacy and Coins.** Suppose you play a coin toss game in which you win $1 if a head appears and lose $1 if a tail appears. In the first 100 coin tosses, heads comes up 47 times and tails comes up 53 times.

 a. What percentage of times has heads come up in the first 100 tosses? What is your net gain or loss at this point?

 b. Suppose you toss the coin 200 more times (a total of 300 tosses), and at that point heads has come up 48% of the time. Is this increase in the percentage of heads consistent with the law of large numbers? Explain. What is your net gain or loss at this point?

 c. How many heads would you need in the next 100 tosses in order to break even after 400 tosses? Is this likely to occur?

 d. Suppose that, still behind after 400 tosses, you decide to keep playing because you are "due" for a winning streak. Explain how this belief would illustrate the gambler's fallacy.

24. **Gambler's Fallacy and Dice.** Suppose you roll a die with the following rules: You win $1 if the die comes up with an even number and you lose $1 if it comes up odd.

 a. Suppose you get 45 even numbers in your first 100 rolls. How much money have you won or lost?

 b. On the second 100 rolls, your luck improves and you roll 47 even numbers. How much money have you won or lost over 200 rolls?

 c. Your luck continues to improve, and you roll 148 even numbers in your next 300 rolls. How much money have you won or lost over your total of 500 rolls?

 d. How many even numbers would you have to roll in the next 100 rolls to break even? Is this likely? Explain.

 e. What were the percentages of even numbers after 100, 200, and 500 rolls? Explain how this illustrates the law of large numbers, even while your losses increased.

25. **Can You Catch Up?** Suppose you toss a fair coin 100 times, getting 42 heads and 58 tails, which is 16 more tails than heads.

a. Explain why, on your next toss, the *difference* between the numbers of heads and tails is as likely to grow to 17 as it is to shrink to 15.

b. Extend your explanation from part a to explain why, if you toss the coin 1000 more times, the final difference between the numbers of heads and tails is as likely to be larger than 16 as it is to be smaller than 16.

c. Suppose you are betting on heads with each coin toss. After the first 100 tosses, you are well on the losing side. Explain why, if you continue to bet, you will most likely remain on the losing side. How is this answer related to the gambler's fallacy?

26. **Lottery Draw.** Consider a lottery game in which six balls are drawn randomly from a set of balls numbered 1 through 42. One week, the winning combination consists of balls numbered 21, 4, 17, 33, 40, and 9. The next week, the winning balls are numbered 1, 2, 3, 4, 5, and 6. Is the second winning set more or less likely than or just as likely as the first? Explain.

27. **Coin Streak.** You toss a coin 1000 times and record the results. You then look through your results and find that at one point you got heads ten times in a row. What is the probability of getting heads ten times in ten tosses? Should you be surprised to find this "streak" in your set of 1000 tosses? Explain.

28. **Hot Batter.** You play softball and have a batting average of .333, meaning you average 1 hit in every 3 times at bat. During your 25-game season, you have one game in which you get a hit all three times you bat. Do you think you were "hot" that day, or was it just luck? Explain your reasoning.

29. **House Edge in Blackjack.** In a large casino, the house wins on its blackjack tables with a probability of 50.7%. All bets at blackjack are 1 to 1: If you win, you gain the amount you bet; if you lose, you lose the amount you bet.

a. What is the expected value to you of a single game? What is the house edge?

b. If you played 100 games of blackjack in an evening, betting $1 on each hand, how much should you expect to win or lose? Explain.

c. If you played 100 games of blackjack in an evening, betting $5 on each hand, how much should you expect to win or lose? Explain.

d. If patrons bet $1,000,000 on blackjack in one evening, how much should the casino expect to earn? Explain.

30. **Profitable Casino.** Averaged over all games and all bets being played, a particular casino has a house edge of

0.07, or 7¢ per dollar gambled. If a total of $100 million is wagered in the casino over the course of a year, what is the casino's total profit? Explain.

Further Applications

Powerball. Exercises 31–32 state a jackpot value for the Multi-State Powerball lottery, for which recent prizes and probabilities of winning (on a single $1 ticket) are shown in the following table. (The table has two entries each for $100 and $7 because these prizes can be won in two different ways.) For the given jackpot, find the expected value of the winnings for a single lottery ticket. Then state how much you can expect to win or lose if you spend $365 per year on the lottery. Briefly discuss your results.

Prize	Probability
Jackpot	1 in 80,089,128
$100,000	1 in 1,953,393
$5000	1 in 364,042
$100	1 in 8879
$100	1 in 8466
$7	1 in 207
$7	1 in 605
$4	1 in 188
$3	1 in 74

31. The jackpot is $30 million.

32. The jackpot is $100 million.

Big Game. Exercises 33–34 state a jackpot value for the Multi-State Big Game lottery, for which recent prizes and probabilities of winning (on a single $1 ticket) are shown in the following table. (The table has two entries for $5 because this prize can be won in two different ways.) For the given jackpot, find the expected value of the winnings for a single lottery ticket. Then state how much you can expect to win or lose if you spend $365 per year on the lottery. Briefly discuss your results.

Prize	Probability
Jackpot	1 in 76,275,360
$150,000	1 in 2,179,296
$5000	1 in 339,002
$150	1 in 9686
$100	1 in 7705
$5	1 in 220
$5	1 in 538
$2	1 in 102
$1	1 in 62

33. The jackpot is $3 million.

34. The jackpot is $25 million.

35. **Extra Points in Football.** Football teams have the option of trying to score either 1 or 2 extra points after a touchdown. They can get 1 point by kicking the ball through the goal posts or 2 points by running or passing the ball across the goal line. In the NFL in 1999, 1-point kicks were successful 94% of the time, while 2-point attempts were successful only 37% of the time. In either case, failure means zero points. Calculate the expected values of the 1-point and 2-point attempts (separately). Based on these expected values, which option makes more sense in most cases? Explain.

36. **Average American Age.** The following table gives the distribution of Americans by age categories. Use the midpoint of each age category (for example, 6.5 for the 0–13 category) to find the expected value of American ages; use 75 for the 65 and over category. Explain why the expected value is also the mean age of the population.

Age category	Percentage
0–13	20.0%
14–24	15.3%
25–34	13.6%
35–44	16.3%
45–64	22.2%
65 and over	12.6%

37. **Household Size.** It is estimated that 57% of Americans live in households with 1 or 2 people, 32% live in households with 3 or 4 people, and 11% live in households with 5 or more people. Using 1.5 for the expected number of people in households with 1 or 2 people, 3.5 for households with 3 or 4 people, and 6 for households of 5 or more people, find the expected number of people in an American household. How is this related to the mean household size?

38. **Mail Sweepstakes.** You receive a notice in the mail saying that you will become eligible to win $1 million in a sweepstakes simply by filling out and returning the enclosed card. As it turns out, you are one of 10 million people who return the card, which makes your odds of winning 1 in 10 million.

 a. Assume you spend 37¢ to mail the card back to the sweepstakes. Calculate your expected earnings (or losses) in sending back the card.

 b. Suppose you decide to purchase three magazine subscriptions, which are offered to you as an option when you return the card. Later, you learn that you could have purchased the same subscriptions through a student discount service for $7 less than you paid. What are your expected earnings now?

39. **Psychology of Expected Values.** In 1953, French economist Maurice Allais studied how people assess risk. Here are two survey questions that he used:

 Decision 1

 Option A: 100% chance of gaining $1,000,000

 Option B: 10% chance of gaining $2,500,000; 89% chance of gaining $1,000,000; and 1% chance of gaining nothing

 Decision 2

 Option A: 11% chance of gaining $1,000,000 and 89% chance of gaining nothing

 Option B: 10% chance of gaining $2,500,000 and 90% chance of gaining nothing

 Allais discovered that for decision 1, most people chose option A, while for decision 2, most people chose option B.

 a. For each decision, find the expected value of each option.

 b. Are the responses given in the surveys consistent with the expected values?

 c. Give a possible explanation for the responses in Allais's surveys.

Web Projects

Find useful links for Web Projects on the text Web site: www.aw.com/bennett-briggs

40. **Lotteries.** Go to the Web site for all U.S. lotteries and study the summary of state and multi-state lottery odds and prizes. Pick five lotteries and determine the expected value for winnings in each case. Discuss your results.

41. **The Morality of Gambling.** Republican Senator Richard Lugar attacked state-supported lotteries and the trend toward increasing legalization of gambling, saying "the spread of gambling is a measure of the moral erosion taking place in our country. . . . It says that if you play enough, you can hit the jackpot and be freed from the discipline of self-support through a job or the long commitment to ongoing education." Do a Web search on "gambling and morality" to find other opinions on this issue. Write a short summary of your findings and your own opinion, which you should defend clearly.

In the News

42. Law of Large Numbers. Describe a news story in which the law of large numbers is mentioned. Is the term used in an accurate way? Explain.

43. Personal Law of Large Numbers. Describe a situation in which you personally have made use of the law of large numbers, either correctly or incorrectly. Why did you use the law of large numbers in this situation? Was it helpful?

44. The Gambler's Fallacy. Describe a situation in which you or someone you know has fallen victim to the gambler's fallacy. How should the situation have been dealt with?

UNIT 7D

ASSESSING RISK

A smooth-talking but honest salesman comes to you, offering a new product:

> *I can't reveal the details yet, but you will love this product! It will improve your life in more ways than you can count. Its only downside is that it will eventually kill everyone who uses it. Will you buy one?*

Not likely! After all, could any product be so great that you would die for it? A few weeks later, the salesman shows up again:

> *No one was buying, so we've made some improvements. Your chance of being killed by the product is now only 1 in 10. Ready to buy?*

Despite the improvement, most people still would send the salesman home and wait for his inevitable return:

> *Okay, this time we've really perfected it. We've made it so safe that it would take 20 years for it to kill as many people as live in San Francisco. And it can be yours for a mere $20,000.*

You may be surprised to realize that, if you are like most Americans, you'll jump at this offer. The product is, after all, the automobile. It does indeed improve our lives in many ways, and $20,000 is a fairly typical price. And, given that roughly 40,000 Americans are killed in auto accidents each year, it kills the equivalent of the population of San Francisco (roughly 800,000) in about 20 years.

As this example shows, we continually make tradeoffs between benefits and risks. In this unit, we'll see how ideas of probability can help us quantify risk, thereby allowing us to make informed decisions about such tradeoffs.

RISK AND TRAVEL

Are you safer in a small car or a sport utility vehicle? Are cars today safer than those 30 years ago? If you need to travel across country, are you safer flying or driving? To answer these and many similar questions, we must quantify the risk involved in travel. We can then make decisions appropriate for our own personal circumstances.

> *The cost of living is going up and the chance of living is going down.*
>
> —FLIP WILSON, COMEDIAN

by the way

Roughly 115 people die in automobile accidents per day, or 1 every 13 minutes. Automobile accidents are the leading cause of death in all age groups between 6 and 27. Vehicle occupants account for 85% of automobile fatalities, while pedestrians and cyclists account for 15%.

Travel risk is often expressed in terms of an **accident rate** or **death rate.** For example, suppose an annual accident rate is 750 accidents per 100,000 people. This means that, within a group of 100,000 people, on average 750 will have an accident over the period of a year. The statement also gives an empirical probability: It says that the probability of a person being involved in an accident is 750 in 100,000, or 0.0075.

However, caution is needed in working with accident and death rates. For example, travel risks are sometimes stated *per 100,000 people,* as above, but other times they are stated *per trip* or *per mile.* If we use death rates *per trip* to compare the risks of flying and driving, we neglect the fact that airplane trips are typically much longer than automobile trips. Similarly, if we use accident rates *per person,* we neglect the fact that most automobile accidents involve only minor injuries.

➤EXAMPLE 1 Is Driving Getting Safer?

Figure 7.7 shows the number of automobile fatalities and the total number of miles driven (among all Americans) for each year over a period of more than three decades. In terms of death rate per mile driven, how has the risk of driving changed?

(a)

(b)

FIGURE 7.7 (a) Annual automobile fatalities. (b) Total miles driven annually. Both sets of data are for the United States only. *SOURCE: National Transportation Safety Board.*

Solution Figure 7.7a shows that the number of fatalities annually decreased from about 51,000 in 1966 to about 42,000 in 2000. Meanwhile, Figure 7.7b shows that the number of miles driven increased from about 900 billion (9×10^{11}) to about 2750 billion (2.75×10^{12}). Thus, the death rates per mile for the beginning and end of the period were

1966: $\dfrac{51{,}000 \text{ deaths}}{9 \times 10^{11} \text{ miles}} \approx 5.7 \times 10^{-8}$ death per mile

2000: $\dfrac{42{,}000 \text{ deaths}}{2.75 \times 10^{12} \text{ miles}} \approx 1.5 \times 10^{-8}$ death per mile

Note that, because $10^8 = 100$ million, 5.7×10^{-8} death per mile is equivalent to 5.7 deaths per 100 million miles. Thus, over 34 years, the death rate per 100 million miles dropped from 5.7 to 1.5. By this measure, driving became much safer over the period. Most researchers believe the improvements resulted from better automobile design

and from safety features, such as shoulder belts and air bags, that are much more common today. *Now try Exercises 9–12.* ◄

►**EXAMPLE 2** Which Is Safer: Flying or Driving?

Over the past 20 years in the United States, the average (mean) number of deaths in commercial airplane accidents has been roughly 100 per year. (The actual number varies significantly from year to year.) Currently, airplane passengers in the United States travel a total of about 7 billion miles per year. Use these numbers to calculate the death rate per mile of air travel. Compare the risk of flying to the risk of driving.

Solution Assuming 100 deaths and 7 billion miles in an average year, the risk of air travel is

$$\frac{100 \text{ deaths}}{7 \times 10^9 \text{ miles}} \approx 1.4 \times 10^{-8} \text{ death per mile}$$

This risk is equivalent to 1.4 deaths per 100 million miles, which is roughly the same as the risk of 1.5 deaths per 100 million miles for driving. Note that, because the average air trip covers a considerably longer distance than the average driving trip, the risk *per trip* is much higher for air travel, although the risk *per mile* is about the same.

Now try Exercises 13–14. ◄

Time out to think

What other measures might be used to compare the risks of driving and flying? Is there a measure of risk for which flying would appear safer than driving? Overall, do you think it is fair to say that flying is safer than driving?

VITAL STATISTICS

Data concerning births and deaths of citizens, often called *vital statistics,* are very important to understanding risk–benefit tradeoffs. For example, insurance companies use vital statistics to assess risks and set rates. Health professionals study vital statistics to assess medical progress and decide where research resources should be concentrated. Demographers use birth and death rates to predict future population trends.

One important set of vital statistics, shown in Table 7.4, concerns causes of death. These data are extremely general. A more complete table would categorize data by age, sex, and race. Vital statistics are often expressed in terms of deaths per

Only those who risk going too far can possibly find out how far one can go.

—T. S. Eliot

TABLE 7.4	Leading Causes of Death in the United States (in a single recent year)		
CAUSE	DEATHS	CAUSE	DEATHS
Heart disease	725,192	Diabetes	68,399
Cancer	549,838	Pneumonia/Influenza	63,730
Stroke	167,366	Alzheimer's disease	44,536
Pulmonary disease	124,181	Kidney disease	35,525
Accidents	97,860	Septicemia (blood poisoning)	30,680

Source: Centers for Disease Control.

person or per 100,000 people, which makes it easier to compare the rates for different years and for different states or countries.

> **EXAMPLE 3** Interpreting Vital Statistics

Assuming a U.S. population of 281 million (from the 2000 census), find and compare risks per person and per 100,000 people for pneumonia and cancer.

Solution We find the risk per person by dividing the number of deaths by the total population of 281 million:

Pneumonia: $\dfrac{63{,}730 \text{ deaths}}{281{,}000{,}000 \text{ people}} \approx 0.00023$ death per person

Cancer: $\dfrac{549{,}838 \text{ deaths}}{281{,}000{,}000 \text{ people}} \approx 0.0020$ death per person

We can interpret these numbers as empirical probabilities: The probability of death by pneumonia is about 2.3 in 10,000, while the probability of death by cancer is about 20 in 10,000. To put them in terms of deaths per 100,000 people, we simply multiply the per person rates by 100,000. We get a pneumonia death rate of 23 deaths per 100,000 people and a cancer death rate of 200 deaths per 100,000 people. Thus, the probability of death by cancer is almost nine times as great as the probability of death by pneumonia. *Now try Exercises 15–20.* ◄

Time out to think _____

Table 7.4 suggests that the probability of death by stroke is about 70% higher than the probability of death by accident, but these data include all age groups. How do you think the risks of stroke and accident would differ between young people and old people? Explain.

LIFE EXPECTANCY

One of the most commonly cited vital statistics is *life expectancy,* which is often used to compare overall health at different times or in different countries. However, the idea of life expectancy is surprisingly subtle, so let's investigate it. Figure 7.8a shows the overall U.S. death rate, in deaths per 1000 people, for different age groups. We see an elevated risk of death near birth, after which the death rate drops to very low levels. At about 15 years of age, the death rate begins a gradual rise.

Figure 7.8b shows the **life expectancy** of Americans of different ages, defined as the number of years a person of a given age can expect to live on average. As we would expect, life expectancy is higher for younger people because, on average, they have longer left to live. At birth, the life expectancy of Americans today is about 77 years (74.1 years for men and 79.5 years for women).

DEFINITION

Life expectancy is the number of years a person with a given age today can expect to live on average.

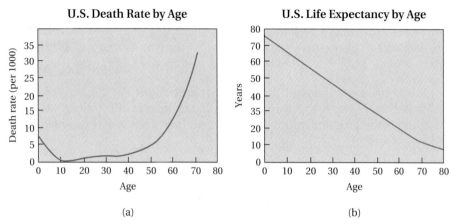

FIGURE 7.8 (a) The overall U.S. death rate (deaths per 1000 people) for different ages. (b) Life expectancy for different ages.
SOURCE: U.S. National Center for Health Statistics.

by the way

Life expectancies vary widely around the world, as well as by sex and race. Canada and Australia have the highest at 78 years (both sexes combined). Among the lowest are Sierra Leone (34 years) and Uganda (43 years).

The subtlety in interpreting life expectancy comes from changes in medical science and public health. Life expectancies are calculated by studying *current* death rates. For example, when we say that the life expectancy of infants born today is 77 years, we mean that the average baby will live to age 77 *if there are no future changes* in medical science or public health. Thus, while life expectancy provides a useful measure of current overall health, it should not be considered a *prediction* of future life spans.

In fact, because of advances in both medical science and public health, life expectancies increased significantly during the past century. As shown in Figure 7.9, U.S. life expectancy at birth rose to its current value from less than 50 years in 1900. If this trend continues, today's infants are likely to live much longer than 77 years on average. On the other hand, declining public health can lead to falling life expectancies. For example, war-torn nations often experience precipitous drops in life expectancy as public health deteriorates.

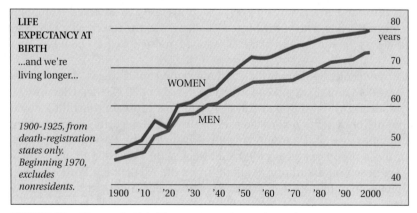

FIGURE 7.9 Changes in U.S. life expectancy during the 20th century.
SOURCE: New York Times and National Center for Health Science Statistics.

The reports of my death are greatly exaggerated.

—MARK TWAIN, FROM LONDON, IN CABLE TO THE ASSOCIATED PRESS

Time out to think _____

Using Figure 7.9, compare the life expectancies of men and women. Briefly discuss these differences. Do they have any implications for social policy? for insurance rates? Explain.

Incidentally, the change in life expectancy shown in Figure 7.9 is truly remarkable: In less than 100 years, life expectancy increased by about 60%. There may be no better single measure of the changes that science and technology have brought to society.

▶EXAMPLE 4 Life Expectancies

Using Figure 7.8b, find the life expectancies of a 20-year-old person and a 60-year-old person. Are the numbers consistent? Explain.

Solution The graph shows that the life expectancy at age 20 is about 57 years and at age 60 is about 21 years. This means that an average 20-year-old can expect to live about 57 more years, to age 77. An average 60-year-old can expect to live about 21 more years, to age 81.

It might at first seem strange that 60-year-olds have a longer average life span than 20-year-olds (81 years versus 77 years). But remember that life expectancies are based on *current* data. If there were no changes in medicine or public health, a 60-year-old would have a greater probability of reaching age 81 than a 20-year-old simply because he or she has already made it to age 60. However, if medicine and public health continue to improve, today's 20-year-olds may live to older ages than today's 60-year-olds. *Now try Exercises 21–26.* ◀

CASE STUDY Life Expectancy and Social Security

As discussed in Unit 4E, Social Security benefits are paid from a combination of taxes collected from today's wage earners (representing most of the "FICA" taxes taken from your paycheck) and the so-called Social Security Trust Fund. Because of the changing age makeup of the U.S. population, the number of retirees collecting Social Security benefits is expected to become much larger in the future, while the number of wage earners paying Social Security taxes is expected to remain relatively constant. Thus, one of the biggest challenges to the future of Social Security is finding a way to make sure there is enough money to pay benefits for future retirees.

Current projections show that, without significant changes, the Social Security program will be unable to pay full benefits after about 2040. Social Security officials have proposed several different ways to solve this problem, including changes in the amounts of benefits paid, increases in the Social Security tax rate, changes in the retirement age, and partially or fully privatizing the Social Security program. Each of these proposals faces political obstacles. But, in addition, all of these proposals are based on assumptions about future life expectancy. Without corresponding changes in retirement age, longer lifespans mean more years of Social Security benefits.

More specifically, recent proposals by Social Security officials have assumed that life expectancy at birth would rise to 79.3 by 2030 and to 81.5 by 2070 (for both sexes).

But these numbers may be far too pessimistic. For example, the Social Security projections assume that American women will not reach a life expectancy of 82 years until 2033—but women in France have *already* achieved this life expectancy. Will it really take the United States some 30 years to catch up to life expectancies in France? One expert panel suggested that more realistic estimates of future life expectancy would be 81 in 2030 and 85.2 in 2070, meaning nearly four more years of Social Security benefits for the average person in 2070. And, if medical science achieves any major breakthroughs in combatting the effects of aging, the Social Security crisis could become far worse.

Time out to think

According to some biologists, there is a good chance that 21st-century advances in medical science will allow most people to live to ages of 100 or more. How would that affect programs like Social Security? What other effects would you expect it to have on society? Overall, do you think large increases in life expectancy will be good or bad for society? Defend your opinion.

■ EXERCISES 7D

Review Questions

1. Briefly explain why quantifying risk is important to decision making.

2. Give an example of a rate used to measure risk in travel. Why are rates more useful than total numbers of deaths or accidents in measuring risk? How are such rates similar to probabilities?

3. What are vital statistics? How are they usually described? Give a few examples.

4. Explain the meaning of the term *life expectancy*. How does life expectancy change with age? How is it affected by changes in the overall health of a population?

Does It Make Sense?

Decide whether each of the following statements makes sense (or is clearly true) or does not make sense (or is clearly false). Explain your reasoning.

5. No one can succeed in selling a product that will kill thousands of people per year.

6. Fewer people are killed in motorcycle accidents than in automobile accidents, so motorcycles must be safer than cars.

7. Your life expectancy is the major factor in determining how long you live.

8. A 60-year-old has a shorter life expectancy than a 20-year-old.

Basic Skills & Concepts

Automobile Safety. Use the following table in Exercises 9–12.

Year	U.S. population (millions)	Traffic fatalities	Licensed drivers (millions)	Vehicle-miles (trillions)
1977	220	47,878	138	1.5
1987	242	46,390	162	1.9
2000	281	41,821	185	2.7

9. Express the 2000 fatality rate in deaths per 100 million vehicle-miles traveled.

10. Express the 2000 fatality rate in deaths per 100,000 population.

11. Express the 2000 fatality rate in deaths per 100,000 licensed drivers.

12. Use three different fatality rates (deaths per 100,000 people, deaths per 100,000 drivers, and deaths per 100 million vehicle-miles) to compare the risk of automobile travel in 1977 and 2000. Discuss your conclusions in detail.

13. **Airline Safety.** In 1999, there were 11 fatalities and 52 accidents in commercial aviation among major (not com-

muter) airlines over 16.5 million flight hours, 6.8 billion miles flown, and 8.2 million departures. Compute the accident and fatality rates in units of accidents or deaths per 100,000 flight hours, per 1 billion miles flown, and per 100,000 departures.

14. **General Aviation Safety.** In 1999, there were 622 deaths and 1908 accidents in general aviation (commercial, private, military) over 27.1 million flight hours. Compute the accident and fatality rates in units of accidents or deaths per 100,000 flight hours. Discuss and compare the safety of commercial aviation (see Exercise 13) and general aviation.

Cause of Death. For Exercises 15–20, consider Table 7.4 and assume a U.S. population of 281 million.

15. What is the empirical probability of death by diabetes during a single year? How much greater is the risk of death by diabetes than the risk of death by septicemia?

16. How much greater is the risk of death by accident than the risk of death by kidney disease?

17. What is the death rate due to kidney disease in deaths per 100,000 of the population?

18. What is the death rate due to heart disease in deaths per 100,000 of the population?

19. If you lived in a typical city of 500,000, how many people would you expect to die of a stroke in a year?

20. If you lived in a typical city of 500,000, how many people would you expect to die of pulmonary disease in a year?

Mortality Rates. Use the graphs in Figure 7.8 to answer Exercises 21–26.

21. Estimate the death rate for 50- to 55-year-olds. Assuming that there were about 15.8 million 50- to 55-year-olds, how many people in this bracket could be expected to die in a year?

22. Approximately what is the death rate for 25- to 35-year-olds? If the population in this age category is 44 million, how many 25- to 35-year-olds can be expected to die in a year?

23. To what age could the average 50-year-old expect to live?

24. At what average age will today's 70-year-olds die? Assume no changes to life expectancies over the next 20 years.

25. Suppose that a life insurance company insures 1 million 50-year-old people. The cost of the premium is $200 per year, and the death benefit is $50,000. What is the expected profit or loss for the insurance company?

26. Suppose that a life insurance company insures 5000 40-year-old people. (Assume a death rate of 3 per 1000 people.) The cost of the premium is $200 per year, and the death benefit is $50,000. How much can the company expect to gain (or lose) in a year?

Further Applications

27. **High/Low U.S. Birth Rates.** The highest and lowest birth *rates* in the United States in 2000 were in Utah and Maine, respectively. Utah reported 47,368 births with a population of about 1.8 million people. Maine reported 13,603 births with a population of about 1.3 million people. Use these data to answer the following questions.

 a. How many people were born per day in Utah? in Maine?

 b. What was the birth rate in Utah in births per 1000 people? in Maine?

28. **High/Low U.S. Death Numbers.** In 2000, there were 227,381 deaths in California, the highest number in the United States. The lowest number of deaths in any state was for Alaska, with 2911 deaths. The populations of California and Alaska were approximately 30 million and 550,000, respectively.

 a. Compute the 2000 death *rates* for California and Alaska in deaths per 1000.

 b. Based on the fact that California and Alaska had the highest and lowest death numbers, respectively, does it follow that California and Alaska had the highest and lowest death *rates,* respectively? Why or why not?

29. **U.S. Birth and Death Rates.** In 2000, the U.S. population was about 281 million. The overall birth rate was 14.8 births per 1000, and the overall death rate was 8.6 deaths per 1000.

 a. Approximately how many births were there in the United States?

 b. About how many deaths were there in the United States?

 c. Based on births and deaths alone (i.e., not counting immigration), about how much did the U.S. population rise during 2000?

d. According to the U.S. Census Bureau, during 2000 the U.S. population actually increased by 3.0 million. Based on this fact and your results from part c, estimate how many people immigrated to the United States. What proportion of the overall population growth was due to immigration?

30. **Sex Ratios.** Figure 7.10 shows the ratio of men to women in various age categories since 1920. Discuss the patterns and trends that you see in this figure, and offer explanations in terms of mortality and life expectancy.

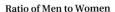
Ratio of Men to Women

FIGURE 7.10

Web Projects

Find useful links for Web Projects on the text Web site: www.aw.com/bennett-briggs

31. **Cancer vs. Heart Disease.** Heart disease and cancer are the two leading causes of death in this country. Based on mortality data for the last ten years available from the Centers for Disease Control, make a case either for or against the claim that cancer will soon overtake heart disease as the leading cause of death.

32. **Crime Reduction.** According to the FBI, the 1990s showed an overall decrease in crime in this country. Criminologists cannot agree on which of many factors are most important in accounting for this trend. Find annual data on both homicides and homicide rates. Discuss the trends you observe, and compare the homicide rate to various disease death rates.

33. **Travel Safety.** The Bureau of Transportation Statistics collects safety data for many forms of travel (for example, train and bus). Visit the Web site and do a comparative study of the safety of automobiles and another form of transportation. Write a short report summarizing your conclusions.

34. **Life Expectancy.** Go to the Web and find as much information as possible about how life expectancies are determined. Use several of the life expectancy calculators that are available on the Web. Do they seem to give accurate or realistic results? Write a short summary of your findings.

In the News

35. **Travel Safety.** Find a recent news report discussing some aspect of travel safety (such as risk of accidents in automobiles or airplanes, the efficacy of child car seats, or the effects of driving while talking on a cellular phone). Summarize any statistics given about risks, and give your overall opinion regarding the safety issue under discussion.

36. **Vital Statistics.** Find a recent news report that gives new data about any aspect of vital statistics. Summarize the report and the statistics, and discuss any personal or social implications of the new data.

37. **Life Expectancies.** Find a recent news report that discusses life expectancies in any context. Summarize the report, and discuss the social implications of the given life expectancy data.

UNIT 7E

COUNTING AND PROBABILITY

As we have seen, computing theoretical probabilities requires counting outcomes. In previous units, we considered problems in which the counting was straightforward or the multiplication principle could be used (see the Brief Review on p. 433). Here, we'll discuss counting tools that will enable us to solve a much wider range of probability problems and to explore the nature of coincidences.

ARRANGEMENTS WITH REPETITION

Suppose that license plates in your state display seven numerals (Figure 7.11). How many different license plates are possible? There are 10 possibilities (the numerals 0 through 9) for the first position and the same 10 possibilities for the second position, so there are $10 \times 10 = 100$ possibilities for the first two positions combined. Similarly, because there are also 10 possibilities for the third position, there are $10 \times 10 \times 10 = 1000$ possibilities for the first three positions. For all seven positions, each with 10 possibilities, the total number of possibilities is

$$10 \times 10 \times 10 \times 10 \times 10 \times 10 \times 10 = 10^7$$

FIGURE 7.11 For each of the seven positions on the license plate, there are ten choices, resulting in 10^7 possible license plates.

That is, this system allows for 10 million different license plates.

The license plate problem involves making selections from a *single* group of choices—in this case, the numerals 0–9. However, the same numerals may be selected over and over again. This type of counting problem, in which we repeatedly select from the same group of choices, is called **arrangements with repetition.**

Arrangements with Repetition

If we make r selections from a group of n choices, a total of $\underbrace{n \times n \times \cdots \times n}_{r \text{ times}} = n^r$ different arrangements are possible.

▶ **EXAMPLE 1** Arrangements with Repetition

 a. How many seven-symbol license plates are possible if both numerals *and* letters can be used in any order?
 b. How many six-character passwords can be made by combining lowercase letters, uppercase letters, numerals, and the characters @, $, and &?

Solution

 a. With 10 numerals and 26 letters in the alphabet, there are 36 choices for each of the seven symbols. Thus, we are selecting $r = 7$ symbols with $n = 36$ choices for each symbol, so the number of seven-symbol license plates is

$$n^r = 36^7 = 78{,}364{,}164{,}096$$

There are more than 78 billion possible seven-symbol license plates.

b. Each password character can be selected from the set of 26 lowercase letters, 26 uppercase letters, 10 numerals, and the three symbols @, $, and &. Thus, there are 65 characters to choose from. For six-character passwords, we select $r = 6$ symbols with $n = 65$ choices for each symbol. The total number of possible passwords is

$$n^r = 65^6 = 75,418,890,625$$

More than 75 billion passwords are possible with these choices.

Now try Exercise 23. ◄

Time out to think _____

Compare the numbers of possibilities in Example 1 to the human population. Will a state using seven-symbol license plates ever run out of possibilities? Do you think a six-symbol password is secure against hackers?

PERMUTATIONS AND COMBINATIONS

Suppose you coach a team of four swimmers. How many different ways can you put together a four-person relay team?

You can choose any of the four swimmers for the first leg. Once you've chosen the first swimmer, you have three swimmers left to choose for the second leg. Thus, the number of possible choices for the first two legs combined is $4 \times 3 = 12$. You then have two swimmers to choose for the third leg. After the first three legs are filled, you have only one choice for the last leg. The total number of possible arrangements for the relay is

$$4 \times 3 \times 2 \times 1 = 24$$

Figure 7.12 shows all 24 possible relay orders with a tree diagram in which the four swimmers are labeled A, B, C, and D. The first row shows the four possibilities for the first leg. The second row shows the three possible second-leg choices for each of the

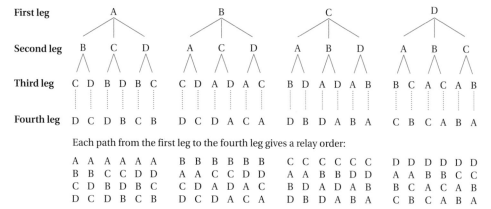

FIGURE 7.12 Tree diagram for the relay possibilities with four swimmers A, B, C, and D. There are four choices for the first leg. Each possible choice for the first leg leads to three remaining choices for the second leg. After the swimmers have been chosen for the first two legs, two choices remain for the third leg. Once the first three legs have been chosen, only one swimmer remains for the fourth leg.

by the way

The word *permute* (root of *permutation*) comes from a Latin term meaning "to change throughout."

first-leg choices. The third row shows each of the prior possibilities branching into two possibilities for the third leg, and the fourth row shows the single remaining choice for the last leg in each case. Counting across the bottom row, we see the total of 24 possibilities. Each of the 24 different relay orders is called a **permutation** of the four letters A, B, C, and D.

Permutations

Mathematically, we are dealing with **permutations** whenever all selections come from a single group of items, no item may be selected more than once, and the *order of arrangement matters* (for example, ABCD is considered different from DCBA). The total number of permutations possible with a group of n items is $n!$, where $n! = n \times (n - 1) \times \cdots \times 2 \times 1$ is read "n factorial."

►EXAMPLE 2 Class Schedules

A middle school principal needs to schedule six different classes—algebra, English, history, Spanish, science, and gym—in six different time periods. How many different class schedules are possible?

Solution This is a case of permutations because all selections come from the same group of classes, each class is scheduled only once, and the order of the classes matters (because a schedule that begins with Spanish is different from one that begins with gym). The principal may choose any of the six classes for the first period, leaving five choices for the second period. After the first two periods are filled, there are four choices left for the third period. Similarly, there are three choices for the fourth period, two choices for the fifth period, and only one choice for the sixth period. Altogether, the number of possible schedules is

$$6! = 6 \times 5 \times 4 \times 3 \times 2 \times 1 = 720$$

The principal can schedule the six classes in 720 different ways.

Now try Exercise 24. ◄

Using Your Calculator

Factorials

Many calculators have a special key for factorials, usually labeled $n!$. For example, to calculate $6!$, press

$6 \; \boxed{n!} \; =$

FIGURE 7.13 With ten swimmers from which to choose a four-person relay team, you have ten choices for the first position, nine for the second position, eight for the third position, and seven for the fourth position. This results in
$10 \times 9 \times 8 \times 7 = 5040$
possible relay teams.

The Permutations Formula

Suppose you coach a team of ten swimmers, from which you must put together a four-person relay team. How many possibilities do you have in this case?

This time, you have ten swimmers from which to choose for the first leg of the relay. Once you make the first choice, you have nine swimmers from which to choose for the second leg, then eight choices for the third leg, and seven choices for the fourth leg (Figure 7.13). The total number of relay possibilities is

$$10 \times 9 \times 8 \times 7 = 5040$$

Each of the 5040 relays represents a different *permutation* because each relay swimmer is selected from the same group of ten swimmers, no swimmer can swim more than once, and the order of the swimmers in the relay is important. However, in this case the relays make use of only four of the total of ten swimmers. Therefore, the number of possible relays made from "ten swimmers selected four at a time" is 5040.

FACTORIALS

Products of the form $4 \times 3 \times 2 \times 1$ come up so frequently in counting problems that they have a special name. Whenever a positive integer n is multiplied by all the preceding positive integers, the result is called **n factorial** and is denoted $n!$ (the exclamation mark is read as "factorial"). For example:

$$1! = 1$$
$$2! = 2 \times 1 = 2$$
$$3! = 3 \times 2 \times 1 = 6$$
$$4! = 4 \times 3 \times 2 \times 1 = 24$$
$$5! = 5 \times 4 \times 3 \times 2 \times 1 = 120$$

In general,

$$n! = n \times (n - 1) \times (n - 2) \times \cdots \times 2 \times 1$$

Note that $n!$ grows rapidly with n. For example, $20! \approx 2.4 \times 10^{18}$, $40! \approx 8.2 \times 10^{47}$, and $60! \approx 8.3 \times 10^{81}$. In fact, factorials quickly become so large that many calculators cannot handle them above $n = 69$ (if the calculator limit is 10^{100}). Note also that, by definition, $0! = 1$.

Example: Calculate each of the following *without* using the factorial key on your calculator.

a. $\dfrac{6!}{4!}$ b. $\dfrac{25!}{22!}$ c. $\dfrac{200!}{199!}$

Solution:

a. We can write out the entire calculation, but it is easier if we simplify it by canceling like terms in the numerator and denominator:

$$\frac{6!}{4!} = \frac{6 \times 5 \times \cancel{4} \times \cancel{3} \times \cancel{2} \times \cancel{1}}{\cancel{4} \times \cancel{3} \times \cancel{2} \times \cancel{1}} = \frac{6 \times 5 \times \cancel{4!}}{\cancel{4!}}$$
$$= 6 \times 5 = 30$$

b. This calculation is easier if we recognize that $25! = 25 \times 24 \times 23 \times 22!$, giving us

$$\frac{25!}{22!} = \frac{25 \times 24 \times 23 \times \cancel{22!}}{\cancel{22!}} = 25 \times 24 \times 23 = 13{,}800$$

c. If you tried 200! on your calculator, you would probably get an error because it is too large for most calculators to handle. But this calculation is easy if we recognize that $200! = 200 \times 199!$, so

$$\frac{200!}{199!} = \frac{200 \times \cancel{199!}}{\cancel{199!}} = 200$$

Now try Exercises 11–22.

We can find a general formula for permutations by writing the product $10 \times 9 \times 8 \times 7$ in a slightly different way:

$$10 \times 9 \times 8 \times 7 = \frac{10 \times 9 \times 8 \times 7 \times \cancel{6!}}{\cancel{6!}} = \frac{10!}{(10 - 4)!}$$

Note that, in the final formula on the right,

- the number 10 in the numerator is the number of swimmers from which the coach may choose the four relay participants, and
- the number $10 - 4 = 6$ in the denominator is the number of swimmers who do *not* swim in the relay.

Permutations have their own special notation: We read $_{10}P_4$ as "the number of permutations of ten items selected four at a time." Using this compact notation, we have

$$_{10}P_4 = \frac{10!}{(10 - 4)!} = 5040$$

Generalizing, we have a formula for calculating the number of permutations.

The Permutations Formula

If we make r selections from a group of n items, the number of permutations is

$$_nP_r = \frac{n!}{(n-r)!} = \underbrace{n \times (n-1) \times (n-2) \times \cdots \times (n-r+1)}_{r \text{ factors here}}$$

where $_nP_r$ is read as "the number of permutations of n items taken r at a time."

►EXAMPLE 3 Leadership Election

Using Your Calculator

Permutations

Many calculators have a special key for permutations, usually labeled $_nP_r$. For example, to find the number of permutations of $n = 10$ items taken $r = 4$ at a time, press

10 $\boxed{_nP_r}$ 4

You should get the result 5040.

A city has 12 candidates running for three leadership positions. The top vote-getter will become the mayor, the second vote-getter will become the deputy mayor, and the third vote-getter will become the treasurer. How many outcomes are possible for the three leadership positions?

Solution The voters choose $r = 3$ leaders from a group of $n = 12$ candidates. The order in which leaders are chosen matters because each of the three leadership positions is different. Thus, we are looking for the number of permutations of 12 candidates taken 3 at a time:

$$_{12}P_3 = \frac{12!}{(12-3)!} = \frac{12!}{9!} = \frac{12 \times 11 \times 10 \times 9!}{9!} = 12 \times 11 \times 10 = 1320$$

There are 1320 possible outcomes for the three leadership positions.

Now try Exercise 25. ◄

►EXAMPLE 4 Batting Orders

How many ways can the manager of a baseball team form a (9-player) batting order from a roster of 15 players?

Solution The manager forms a batting order by selecting $r = 9$ players from the roster of $n = 15$ players. Thus, the number of possible batting orders is the number of permutations of 15 players taken 9 at a time:

$$_{15}P_9 = \frac{15!}{(15-9)!} = \frac{15!}{6!}$$

$$= 15 \times 14 \times 13 \times 12 \times 11 \times 10 \times 9 \times 8 \times 7 = 1{,}816{,}214{,}400$$

Nearly 2 *billion* batting orders are possible for a baseball team with a roster of 15 players!

Now try Exercise 26. ◄

Time out to think

Major league baseball teams carry a roster of 25 players. Without actually calculating, describe how the calculation of the number of possible batting orders would be different for a major league team than for the team in Example 4.

Combinations

Suppose that the five members of a city council decide they need a three-person committee to study the impact of a new shopping center. The council members are Ursula, Vern, Wendy, Yolanda, and Zeke (U, V, W, Y, and Z, for short). How many committees could be formed from the five council members?

First, note that this problem is *not* a permutation problem because we are interested only in the makeup of the committee, *not* the order in which the names are listed. For example, if we were dealing with permutations, we would consider ZWU to be different from WZU. In this case, however, both arrangements are equivalent because they both represent the same committee (with members Zeke, Wendy, and Ursula). Problems of this type are called **combinations** problems.

One way to count the possible committees is to list them all. Going in alphabetical order, we begin by listing all three-person committees that include Ursula:

UVW UVY UVZ UWY UWZ UYZ

Next, we list the committees that include Vern but not Ursula:

VWY VWZ VYZ

This leaves only one committee that includes neither Ursula nor Vern:

WYZ

We've found a total of ten three-person committees that can be made from the five council members.

This method of listing all the possible committees works, but it can become quite tedious when the numbers are large. Fortunately, we can get the same answer with an alternative strategy. First, we find the number of *permutations* of the $n = 5$ council members selected $r = 3$ at a time:

$$_5P_3 = \frac{5!}{(5-3)!} = \frac{5!}{2!} = 5 \times 4 \times 3 = 60$$

That is, there are 60 possible permutations when we choose three people from a group of five. However, because order matters for permutations but does not matter for committees, the number of permutations is an *overcount* of the actual number of different committees. More specifically, any three-person committee can be listed in $3! = 3 \times 2 \times 1 = 6$ different orders. For example, the committee consisting of Zeke, Yolanda, and Wendy has six different permutations:

ZYW ZWY YZW YWZ WZY WYZ

Because each three-person committee is counted $3! = 6$ times by the permutations formula, this formula gives us six times the actual number of committees. Thus, we must divide the number of permutations by $3!$ to find the number of committees:

$$\frac{_5P_3}{3!} = \frac{60}{3!} = \frac{60}{3 \times 2 \times 1} = 10$$

This is the same result we obtained by listing the three-person committees. The permutations part of the equation is $_nP_r$, where $n = 5$ and $r = 3$. We then divided this term by $r! = 3!$ to correct for the overcounting. Thus, the general formula for combinations is the permutations formula divided by $r!$.

A committee is a group of important individuals who singly can do nothing, but who together can agree that nothing can be done.

—FRED ALLEN,
COMEDIAN

Using Your Calculator

Combinations

Many calculators have a special key for combinations, usually labeled $_nC_r$. For example, to find the number of combinations of $n = 10$ items taken $r = 4$ at a time, press

10 $\boxed{_nC_r}$ 4

You should get the result 210.

Combinations

Combinations occur whenever all selections come from a single group of items, no item may be selected more than once, and the order of arrangement does *not* matter (for example, ABCD is considered to be the same as DCBA). If we make r selections from a group of n items, the number of possible combinations is

$$_nC_r = \frac{_nP_r}{r!} = \frac{n!}{(n - r)! \times r!}$$

where $_nC_r$ is read as "the number of combinations of n items taken r at a time."

▶ **EXAMPLE 5** Ice Cream Combinations

Suppose that you select 3 different flavors of ice cream in a shop that carries 12 flavors. How many flavor combinations are possible?

Solution We are looking for the number of combinations of $n = 12$ flavors selected $r = 3$ at a time. From the combinations formula, the number of flavor combinations is

$$_{12}C_3 = \frac{12!}{(12 - 3)! \times 3!} = \frac{12!}{9! \times 3!} = \frac{12 \times 11 \times 10 \times \cancel{9!}}{\cancel{9!} \times 3!} = \frac{12 \times 11 \times 10}{3 \times 2 \times 1} = \frac{1320}{6} = 220$$

There are 220 different three-flavor combinations possible from the 12 flavors.

Now try Exercise 27. ◀

▶ **EXAMPLE 6** Poker Hands

How many different five-card poker hands can be dealt from a standard deck of 52 cards? What is the probability of one particular hand, such as a *royal flush* of hearts (a hand consisting of the ace, king, queen, jack, and 10 of hearts)?

Solution This is a combinations problem, because the order of the cards in a hand does *not* matter. The number of combinations of $n = 52$ cards drawn $r = 5$ at a time is

$$_{52}C_5 = \frac{52!}{(52 - 5)! \times 5!} = \frac{52!}{47! \times 5!} = \frac{52 \times 51 \times 50 \times 49 \times 48 \times \cancel{47!}}{\cancel{47!} \times 5 \times 4 \times 3 \times 2 \times 1} = 2,598,960$$

More than 2.5 million hands are possible in five-card poker. Any particular hand, such as the royal flush of hearts, has a probability of about 1 in 2.5 million.

Now try Exercise 28. ◀

Time out to think _____

Consider the following unimpressive poker hand: 5 of clubs, 7 of hearts, 3 of diamonds, 2 of spades, 9 of hearts. How does the probability of this hand, which has no value in poker, compare to the probability of a royal flush of spades, which is the highest-value hand in poker? Explain.

►**EXAMPLE 7** (Not) Winning the Lottery

Suppose you play a lottery in which the winner is chosen by drawing 6 balls at random from a drum containing 52 numbered balls (numbered 1 through 52). What is the probability that your 6 numbers will match the 6 winning numbers?

Solution The order in which the balls are drawn does not matter, so finding the total number of possible outcomes is a combinations problem. The number of combinations for $n = 52$ balls selected $r = 6$ at a time is

$$_{52}C_6 = \frac{52!}{(52 - 6)! \times 6!} = \frac{52!}{46! \times 6!} = 20{,}358{,}520$$

There are more than 20 million possible combinations of 6 numbers, so the chance that your 6 numbers will match the 6 winning numbers is less than 1 in 20 million.

Now try Exercises 29–38. ◄

PROBABILITY AND COINCIDENCE

Probability has important applications to the idea of coincidences. For example, only a few of the many millions of people who play lotteries win multimillion-dollar prizes each year. Because any particular person's probability of winning is so low, winners often attribute their fortune to "good luck" or to some special "system" they used in choosing lottery numbers. However, *someone* was bound to win, and the particular winners actually won by mere coincidence. Said differently, it is virtually certain that *someone* will win a big lottery prize during the next year. Yet the probability that *you* will win the big prize is exceedingly small.

> **Coincidences Are Bound to Happen**
>
> Although a *particular* outcome may be highly unlikely, *some* similar outcome may be extremely likely or even certain to occur. In general, this means that coincidences will inevitably be experienced by someone, even if these coincidences have a low probability for any particular person.

Coincidences come in many different forms. You may be surprised to find that two people at your dinner party have the same birthday or that a friend's mother met your mother on a trip to China. You might flip a coin and get heads ten times in a row or have a dream that seems to predict an event that later occurs. It is tempting to chalk these cases up to something more mysterious than mere coincidence. But the laws of probability dictate that many coincidences are bound to happen, even though the particular form of a coincidence is unpredictable. The following example about birthday coincidences demonstrates one of the more surprising aspects of this idea.

►**EXAMPLE 8** Birthday Coincidence

Suppose there are 25 students in your class.

 a. What is the probability that at least one person in the class has the same birthday as you? (Assume 365 days in a year.)

HISTORICAL NOTE

New Hampshire began the first legalized lottery in the United States in 1964. Since then, most states have adopted lotteries as a source of revenue. Many states also participate in multi-state lottery games (such as the Powerball game or the Big Game). Lotteries have existed in various forms and places for centuries, including the Roman Empire (1st century) and Italy in the Middle Ages.

Calvin and Hobbes by Bill Watterson

b. What is the probability that *some* pair of students in the class of 25 shares the *same* birthday? Compare this result to the result from part a.

Solution

a. This is an *at least once* question (see Unit 7B). We begin by recognizing that, with 365 days in a year, the probability that any particular student has your birthday is 1/365. Therefore, the probability that a particular student does not have your birthday is 364/365. From the *at least once* rule, the probability that at least one of the other 24 other students has your birthday is

$$P(\text{at least one with your birthday}) = 1 - \left[P(\text{not your birthday}) \right]^{24}$$

$$= 1 - \left[\frac{364}{365} \right]^{24}$$

$$= 1 - 0.936 = 0.064$$

The probability that at least one other student shares your birthday is only 0.064, which is about 1 in 16.

b. This question differs from the question in part a because the pair sharing the same birthday does not have to include you; it can be *any* pair of two students. But this is still an *at least once* question, because we are looking for the probability of a shared birthday among at least one pair of students. Thus, the answer will take the following form:

$$P(\text{at least one pair of shared birthdays}) = 1 - P(\text{no shared birthdays})$$

The easiest way to find the probability of no shared birthdays is by calculating the probability that all 25 students have *different* birthdays. We begin by considering just two students from the class. The first student has a birthday on 1 of the 365 days of the year, so the probability that the second student has a different birthday is 364/365. Now consider a third student. If the first two students have different birthdays, then 2 of the 365 days are already "taken." Thus, the probability that the third student's birthday falls on a third different day is 363/365, since 363 days are not yet "taken" for birthdays. Putting the probabilities together, we find that the chance that the first three students all have different birthdays is

$$\underbrace{\frac{364}{365}}_{\substack{\text{probability that first two students} \\ \text{have different birthdays}}} \times \underbrace{\frac{363}{365}}_{\substack{\text{probability that third student} \\ \text{also has different birthday}}}$$

Similarly, if the first three students all have different birthdays, the probability that a fourth student has a different birthday from any of the first three is 362/365. And so on. Finally, if the first 24 students all have different birthdays, then 24 of the 365 days are "taken," leaving $365 - 24 = 341$ possible different birthdays for the 25th student. Thus, the probability that all 25 students have different birthdays is

$$\frac{364}{365} \times \frac{363}{365} \times \cdots \times \frac{341}{365} = \frac{364 \times 363 \times \cdots \times 341}{365^{24}}$$

Although we can write the product in the numerator more compactly as 364!/340!, the factorials are too large for most calculators to handle. Therefore, the better way to proceed is by computing the numerator and denominator separately. The result is

$$\frac{364 \times 363 \times \cdots \times 341}{365^{24}} \approx \frac{1.348 \times 10^{61}}{3.126 \times 10^{61}} \approx 0.431$$

This is the probability that there are no shared birthdays among the 25 students in the class. Thus, the probability that there is at least one pair of shared birthdays is

$$P(\text{at least one pair of shared birthdays}) = 1 - P(\text{no shared birthdays})$$
$$= 1 - 0.431 = 0.569$$

There's a 0.569 probability—which is about a 57% chance—that at least two of the 25 students have the same birthday! Most people are very surprised at how much higher this probability is than the 0.064 probability that at least one of the students shares *your* birthday, but it illustrates the fact that *some* coincidence is far more likely than a *particular* coincidence.

Now try Exercises 39–40. ◄

EXERCISES 7E

Review Questions

1. What are arrangements with repetition? Give an example of a situation in which the n^r formula gives the number of possible arrangements.

2. What do we mean by *permutations*? Explain the meaning of each of the terms in the permutations formula. Give an example of its use.

3. What do we mean by *combinations*? Explain the meaning of each of the terms in the combinations formula. Give an example of its use.

4. Explain what we mean when we say that *some* outcome is much more likely than a particular outcome. How does this idea affect our perception of coincidences?

Does It Make Sense?

Decide whether each of the following statements makes sense (or is clearly true) or does not make sense (or is clearly false). Explain your reasoning.

5. I used the permutations formula to determine how many possible relay orders we could make with the 10 girls on our swim team.

6. I used the combinations formula to determine how many different five-card poker hands are possible.

7. The number of different possible batting orders for 9 players on a 25-person baseball team is so large that there's no hope of trying them all out.

8. It must be my lucky day, because the five-card poker hand I got had only about a 1 in 2.5 million chance of being dealt to me.

9. The probability that two people in a randomly selected group will have the same last name is much higher than the probability that someone will have the same last name as I do.

10. Someone wins the lottery every week, so I figure that if I keep playing eventually I will be the one who wins.

Basic Skills & Concepts

Review of Factorials. Exercises 11–22 use skills covered in the Brief Review on p. 477. Find the values of the following quantities *without* using the factorial key on your calculator (you may use the multiplication key). Show your work.

11. $7!$

12. $\dfrac{7!}{3!}$

13. $\dfrac{7!}{4!3!}$

14. $\dfrac{12!}{5!}$

15. $\dfrac{12!}{8!(12-8)!}$

16. $\dfrac{21!}{20!}$

17. $6!$

18. $\dfrac{6!}{3!}$

19. $\dfrac{6!}{4!3!}$

20. $\dfrac{11!}{5!}$

21. $\dfrac{11!}{4!(11-4)!}$

22. $\dfrac{16!}{15!}$

Counting Possibilities. Answer the questions in Exercises 23–38 using the appropriate counting technique, which may be either arrangements with repetition, permutations, or combinations. Be sure to explain why this counting technique applies to the problem.

23. How many different four-digit house addresses can be formed with the numerals 0–9?

24. Twelve acts at a comedy club all perform on one evening. How many different ways can their appearances be scheduled?

25. How many different five-letter passwords can be formed from the letters A, B, C, D, E, and F if no repetition of letters is allowed?

26. Ten finalists in a talent show must give final performances. Five contestants will perform on the first night of the show. How many different ways can the schedule for the first night be made?

27. How many different five-person subcommittees can be formed from a group of ten people?

28. How many different four-card hands can be dealt from a deck that has 52 cards?

29. You own ten compact disks, but plan to take only six with you on a road trip. How many different sets of six disks can you take?

30. How many different ten-note "tunes" can be created from the notes C, D, E, F, G, A, and B?

31. Supposedly, if a monkey types randomly on a typewriter long enough, it eventually will type the works of Shakespeare. How many different five-letter "words" could a monkey type from the 26 letters of the alphabet?

32. A lock consists of a dial that can be set to any of the numbers 1–30. How many different three-number combinations are possible?

33. How many different three-letter "words" can be formed with the four letters of the genetic code (A, C, G, T)?

34. You deal five cards from a standard deck of cards face up on a table. How many different sequences of cards (in which order matters) can you deal?

35. How many different four-person subcommittees can be formed from a group of 12 people?

36. How many different five-card hands can be dealt from a deck that has only face cards and aces (16 cards altogether)?

37. Suppose that license plates are made with three letters followed by three numerals (0–9). How many different license plates are possible?

38. The coach of the debate club must choose a team of 8 people to travel to the next meet. How many different teams can be chosen from the 20 members of the club?

Birthday Coincidences. Suppose you are part of a group of people at a dinner party. For the situations given in Exercises 39–40, find the probability that at least one of the other guests has *your* birthday and the probability that *some* pair of guests shares the same birthday. Discuss your results. (Assume 365 days in a year.)

39. You are one of 10 people at the party.

40. You are one of 15 people at the party.

Further Applications

41. Ice Cream Shop. Josh and John's Ice Cream Shop offers 12 different flavors of ice cream and 6 different toppings. Answer the following questions by using the appropriate counting technique (multiplication principle, arrangements with repetitions, permutations, or combinations). Explain why you chose the particular counting technique.

 a. How many different sundaes can you create using one of the ice cream flavors and one of the toppings?

 b. How many different triple cones can you create from the 12 flavors if the same flavor may be used more than once? Assume that you specify which flavor goes on the bottom, middle, and top.

 c. Using the 12 flavors, how many different triple cones can you create with 3 *different* flavors if you specify which flavor goes on the bottom, middle, and top?

 d. Using the 12 flavors, how many different triple cones can you create with 3 *different* flavors if you don't care about the order of the flavors on the cone?

42. Telephone Numbers. A seven-digit phone number in the United States consists of a three-digit *exchange* followed by a four-digit number.

 a. The first digit of the exchange is not allowed to be 0 or 1. How many different seven-digit phone numbers can be formed? Can a city of 2 million people be served by a single area code? Explain.

 b. How many exchanges are needed to serve a city of 80,000 people? Explain.

43. Pizza Hype. Luigi's Pizza Parlor advertises 84 different three-topping pizzas. How many individual toppings does Luigi actually use? Ramona's Pizzeria advertises 45 different two-topping pizzas. How many toppings does Ramona actually use? (Hint: In these problems, you are given the total number of combinations, and you must find the number of toppings that are used.)

44. ZIP Codes. The U.S. Postal Service uses both five-digit and nine-digit ZIP codes.

 a. How many five-digit ZIP codes are available to the U.S. Postal Service?

 b. For a U.S. population of 281 million people, what is the average number of people per ZIP code, if all possible ZIP codes are used? Explain.

 c. How many nine-digit ZIP codes are available to the U.S. Postal Service? Could everyone in the United States have his or her own personal nine-digit ZIP code? Explain.

Counting and Probability. In Exercises 45–50, find the probability of the given event.

45. Choosing five numbers that match five randomly selected balls when the balls are numbered 1 through 32

46. Choosing six numbers that match six randomly selected balls when the balls are numbered 1 through 40

47. Being dealt 5 cards from a standard 52-card deck, and four of the cards are an ace, 2, 3, and 4, all of the same suit.

48. Being dealt 5 cards from a standard 52-card deck, and the cards are a 10, jack, queen, king, and ace, all of the same suit.

49. Guessing the top 3 winners (in order) from a group of 10 finalists in a soccer tournament

50. Guessing the top 3 winners (in any order) from a group of 16 finalists in a spelling bee

51. **Hot Streaks.** Suppose that 2000 people are all playing a game for which the chance of winning is 48%.

 a. What is the probability of one person winning five games in a row?

 b. What is the probability of one person winning ten games in a row?

 c. On average, how many of the 2000 people could be expected to have a "hot streak" of five games?

 d. On average, how many of the 2000 people could be expected to have a "hot streak" of ten games?

52. **Joe DiMaggio's Record.** One of the longest standing records in sports is the 56-game hitting streak (in baseball) of Joe DiMaggio. Assume that a player on a long "hot streak" is batting about .400, which is about the best that anyone ever hits over a period of 50 or more games. (A batting average of .400 means the batter gets a hit 40% of the time. Typically, only a handful of players each year hit that well for a period as long as 56 games.)

 a. What is the probability that a player batting .400 will get at least one hit in four at-bats?

 b. Use the result in part a to calculate the probability of a .400 hitter getting a hit in 56 consecutive games, assuming four at-bats per game.

 c. Suppose that, instead of batting .400, a player has a more ordinary average of .300. In that case, what is the probability of the player getting a hit in 56 consecutive games, assuming four at-bats per game?

 d. Considering the results in parts b and c and the fact that baseball has been played for about 100 years, are you surprised that *someone* set a record of hitting in 56 consecutive games? Explain clearly.

53. **Coin Streaks.** Toss a coin 100 times and record your results (heads or tails) in order. What was your longest streak of consecutive heads or tails? Calculate the probability of this streak by itself. For example, if your longest streak was four heads, what is the probability of four heads in four tosses? Should you be surprised that you got such a streak? Explain.

Web Project

Find useful links for Web Projects on the text Web site: www.aw.com/bennett-briggs

54. **Let's Make a Deal.** In September 1990, controversy about a probability question erupted over an item in the "Ask Marilyn" column in *Parade* magazine (carried by many newspapers across the country). Her column answered a question loosely based on a TV game show called "Let's Make a Deal." Here's the question:

 > *Suppose that you're on a game show and you're given the choice of three doors: Behind one door is a car; behind the other two doors are goats. You pick a door, say No. 1, and the host, who knows what's behind the doors, opens another door, say No. 3, which has a goat. He then says to you, "Do you want to change your pick to door No. 2?" Is it to your advantage to switch your choice?*

 Marilyn answered that the probability of winning was higher if the contestant switched. This answer generated a huge number of letters, including a few from mathematicians, claiming that she was wrong. Marilyn answered with the following logic. When you first pick door No. 1, the chance that you picked the one with the car is 1/3. The probability that you chose a door with a goat is 2/3. When the host opens door No. 3 to reveal a goat, it does not change the 1/3 probability that you picked the right door in the first place. Thus, as only one other door remains, the probability that it contains the car is 2/3. Visit a few of the many Web sites devoted to this problem to gain some understanding of its subtleties. Do you agree with Marilyn's logic? If so, try to explain it in your own words. If not, present an alternative approach.

In the News

55. **Lottery Chances.** Find a newspaper article or advertisement about a lottery in which choosing a winner involves combinations. Calculate the probability of winning. Does your calculation agree with the advertised probability? Explain.

56. **Amazing Coincidence.** Discuss a seemingly amazing coincidence that you've read about or experienced yourself. Based on ideas of probability, decide whether the coincidence was bound to happen to someone. Was it really as amazing as it seemed? Explain.

CHAPTER 7 SUMMARY

UNIT	KEY TERMS	KEY IDEAS AND SKILLS
7A	outcome event theoretical probability empirical probability subjective probability probability distribution odds	**Distinguish among** theoretical, empirical, and subjective probabilities. **Determine** theoretical probabilities: $$P(A) = \frac{\text{number of ways } A \text{ can occur}}{\text{total number of outcomes}}$$ **Make** a probability distribution.
7B	*and* probability independent events dependent events *either/or* probability non-overlapping events overlapping events	***And* probability, independent events:** $P(A \text{ and } B) = P(A) \times P(B)$ ***And* probability, dependent events:** $P(A \text{ and } B) = P(A) \times P(B \text{ given } A)$ ***Either/or* probability, non-overlapping events:** $P(A \text{ or } B) = P(A) + P(B)$ ***Either/or* probability, overlapping events:** $P(A \text{ or } B) = P(A) + P(B) - P(A \text{ and } B)$ ***At least once* rule:** $P(\text{at least one event } A \text{ in } n \text{ trials}) = 1 - P(\text{no events } A \text{ in } n \text{ trials})$ $= 1 - [P(\text{not } A \text{ in one trial})]^n$
7C	law of large numbers expected value gambler's fallacy house edge	**Understand and apply** the law of large numbers. **Calculate and interpret** expected values: $$\text{expected value} = \begin{pmatrix}\text{event 1}\\\text{value}\end{pmatrix} \times \begin{pmatrix}\text{event 1}\\\text{probability}\end{pmatrix} + \begin{pmatrix}\text{event 2}\\\text{value}\end{pmatrix} \times \begin{pmatrix}\text{event 2}\\\text{probability}\end{pmatrix}$$
7D	accident rate death rate life expectancy	**Measure** risk in terms of accident or death rates. **Understand and interpret** vital statistics and life expectancy.
7E	arrangements with repetition permutations combinations	**Arrangements with repetition:** For r selections from a group of n choices, the number of arrangements is n^r. **Permutations:** For r selections from a group of n items, $$_nP_r = \frac{n!}{(n-r)!}$$ **Combinations:** For r selections from a group of n items, $$_nC_r = \frac{_nP_r}{r!} = \frac{n!}{(n-r)! \times r!}$$ **Coincidences are bound to happen**—understand why and the implications for probability.

The greatest shortcoming of the human race is [its] inability to understand the exponential function.

—ALBERT A. BARTLETT,
PROFESSOR OF PHYSICS, UNIVERSITY OF COLORADO

CHAPTER

8

Exponential Astonishment

World population is currently growing by some 80 million people per year—enough to fill an entire new United States in just over three years. A growing population means new challenges for our species, which we can meet only if we understand this growth. In this chapter, we will investigate the mathematical laws of growth— specifically, exponential growth. We will focus on what we call *exponential astonishment:* the intuition-defying reality of exponential growth. Although we will emphasize population growth, we will also study many other important topics, including the decay of waste from nuclear power plants, the depletion of natural resources, and the environmental effects of acid rain.

UNIT **8A** **Growth: Linear versus Exponential:** We will contrast the two basic forms of growth: *linear growth* by the same absolute amount in any time period and *exponential growth* by the same relative amount in any time period. We will also explore the remarkable effects of the repeated doublings that characterize exponential growth.

UNIT **8B** **Doubling Time and Half-Life:** Exponential growth is characterized by a constant *doubling time*, while exponential decay is characterized by a constant *half-life*. We will explore both ideas and some of their many applications.

UNIT **8C** **Real Population Growth:** Real population growth is neither strictly linear nor strictly exponential. We will study both the reasons for and the limits to population growth, which have important implications for the future of the human race.

UNIT **8D** **Logarithmic Scales: Earthquakes, Sounds, and Acids:** We will study the meaning and use of three important *logarithmic scales:* the magnitude scale for earthquakes, the decibel scale for sound, and the pH scale for acidity.

UNIT **8A**

GROWTH: LINEAR VERSUS EXPONENTIAL

magine two communities, Straightown and Powertown, each with an initial population of 10,000 people (Figure 8.1). Straightown grows at a constant rate of 500 people per year, so its population reaches 10,500 after 1 year, 11,000 after 2 years, 11,500 after 3 years, and so on. Powertown grows at a constant rate of 5 *percent* per year. Because 5% of 10,000 is 500, Powertown's population also reaches 10,500 after 1 year. In the second year, however, Powertown's population increases by 5% of 10,500, which is 525, to 11,025. In the third year, Powertown's population increases by 5% of 11,025, or by 551 people. Figure 8.1 contrasts the populations of the two towns over a 45-year period. Note that Powertown's population rises ever more steeply and quickly outpaces Straightown's.

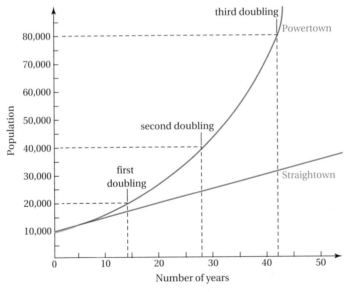

FIGURE 8.1 Straightown grows linearly, while Powertown grows exponentially.

Straightown and Powertown illustrate two fundamentally different types of growth. Straightown grows by the same *absolute* amount —500 people—each year, which is characteristic of **linear growth.** In contrast, Powertown grows by the same *relative* amount—5%—each year, which is characteristic of **exponential growth.**

Two Basic Growth Patterns

Linear growth occurs when a quantity grows by the same *absolute* amount in each unit of time.

Exponential growth occurs when a quantity grows by the same *relative* amount—that is, by the same *percentage*—in each unit of time.

The terms *linear* and *exponential* can also be applied to quantities that decrease with time. For example, *exponential decay* occurs when a quantity decreases by the same relative amount in each unit of time. In the rest of this unit, we will explore the surprising properties of exponential growth.

➤**EXAMPLE 1** Linear or Exponential?

In each of the following situations, state whether the growth (or decay) is linear or exponential, and answer the associated questions.

 a. The number of students at Wilson High School has increased by 50 in each of the past four years. If the student population was 750 four years ago, what is it today?

 b. The price of milk has been rising with inflation at 3% per year. If the price of a gallon of milk was $2 a year ago, what is it now?

 c. Tax law allows you to depreciate the value of your equipment by $200 per year. If you purchased the equipment three years ago for $1000, what is its depreciated value today?

 d. The memory capacity of state-of-the-art computer hard drives is doubling approximately every two years. If a company's top-of-the-line drive holds 300 gigabytes today, what will it hold in six years?

 e. The price of DVD recorders has been falling by about 25% per year. If the price is $200 today, what can you expect it to be in two years?

Solution

 a. The number of students increased by the same absolute amount each year, so this is *linear growth*. Because the student population increased by 50 students per year, in four years it grew by $4 \times 50 = 200$ students, from 750 to 950.

 b. The price rises by the same percent each year, so this is *exponential growth*. If the price was $2 a year ago, it increased by $0.03 \times \$2 = \0.06, making the price $2.06.

 c. The equipment value decreases by the same absolute amount each year, so this is *linear decay*. In three years, the value decreases by $3 \times \$200 = \600, so the value decreases from $1000 to $400.

 d. A doubling is the same as a 100% increase, so the two-year doubling time represents *exponential growth*. With a doubling every two years, the capacity will double three times in six years: from 300 to 600 gigabytes after two years, from 600 to 1200 gigabytes after four years, and from 1200 to 2400 gigabytes after six years.

e. The price decreases by the same percentage each year, so this is *exponential decay.* From $200 today, the price will fall by 25%, or $0.25 \times \$200 = \50, in one year. Therefore, next year's price will be $200 - \$50 = \150. The following year, the price will again fall by 25%, or $0.25 \times \$150 = \37.50, so the price after two years will be $150 - \$37.50 = \112.50. *Now try Exercises 9–16.* ◄

THE IMPACT OF DOUBLINGS

Look again at the graph of Powertown's population in Figure 8.1. After about 14 years, the original population has doubled to 20,000. In the next 14 years, it doubles again to 40,000. It then doubles again, to 80,000, 14 years after that. This type of repeated *doubling,* in which each doubling occurs in the same amount of time, is a hallmark of exponential growth.

The time it takes for each doubling depends on the rate of the exponential growth. In Unit 8B, we'll see how the doubling time depends on the percentage growth rate. Here, we'll explore three parables that show how doublings make exponential growth so very different from linear growth.

Parable 1: From Hero to Headless in 64 Easy Steps

Legend has it that, when chess was invented in ancient times, a king was so enchanted that he said to the inventor, "Name your reward."

"If you please, king, put one grain of wheat on the first square of my chessboard," said the inventor. "Then, place two grains on the second square, four grains on the third square, eight grains on the fourth square, and so on." The king gladly agreed, thinking the man a fool for asking for a few grains of wheat when he could have had gold or jewels. But let's see how it adds up for the 64 squares on a chessboard.

Table 8.1 shows the calculations. Each square gets twice as many grains as the previous square, so the number of grains on any square is a power of 2. The third column shows the total number of grains up to each point, and the last column gives a simple formula for the total number of grains.

TABLE 8.1

SQUARE	GRAINS ON THIS SQUARE	TOTAL GRAINS THUS FAR	FORMULA FOR TOTAL GRAINS
1	$1 = 2^0$	1	$2^1 - 1$
2	$2 = 2^1$	$1 + 2 = 3$	$2^2 - 1$
3	$4 = 2^2$	$3 + 4 = 7$	$2^3 - 1$
4	$8 = 2^3$	$7 + 8 = 15$	$2^4 - 1$
5	$16 = 2^4$	$15 + 16 = 31$	$2^5 - 1$
\vdots	\vdots	\vdots	\vdots
64	2^{63}		$2^{64} - 1$

From the pattern in the last column, we see that the grand total for all 64 squares is $2^{64} - 1$ grains. How much wheat is this? With a calculator, you can confirm that

$2^{64} = 1.8 \times 10^{19}$, or about 18 *billion billion*. Not only would it be difficult to fit so many grains on a chessboard, but this number happens to be larger than the total number of grains of wheat harvested in all human history. The king never finished paying the inventor and, according to legend, instead had him beheaded.

Now try Exercises 17–20.

Parable 2: The Magic Penny

One lucky day, you meet a leprechaun who promises to give you fantastic wealth, but hands you only a penny before disappearing. You head home and place the penny under your pillow. The next morning, to your surprise, you find two pennies under your pillow. The following morning, you find four pennies, and the fourth morning, eight pennies. Apparently, the leprechaun gave you a *magic* penny: While you sleep, each magic penny turns into *two* magic pennies. Table 8.2 shows your growing wealth. Note that "day 0" is the day you met the leprechaun. Generalizing from the table, we find that the amount under your pillow after t days is

$$\$0.01 \times 2^t$$

TABLE 8.2	
DAY	AMOUNT UNDER PILLOW
0	$\$0.01 \times 2^0 = \0.01
1	$\$0.01 \times 2^1 = \0.02
2	$\$0.01 \times 2^2 = \0.04
3	$\$0.01 \times 2^3 = \0.08
4	$\$0.01 \times 2^4 = \0.16

We can use this formula to figure out how long it will be until you have fantastic wealth. After $t = 9$ days, you'll have $\$0.01 \times 2^9 = \5.12, which is barely enough to buy lunch. But by the end of a month, or $t = 30$ days, you'll have $\$0.01 \times 2^{30} = \$10,737,418.24$. That is, you'll be a millionaire within a month, and you'll need a much larger pillow! In fact, if your magic pennies keep doubling, by the end of just 50 days you'll have $\$0.01 \times 2^{50} \approx \11.3 trillion, more than enough to pay off the national debt of the United States. *Now try Exercises 21–24.*

Parable 3: Bacteria in a Bottle

For our third parable, we turn to a story with much more relevance to the future of the human race. Suppose you place a single bacterium in a bottle at 11:00 a.m. It grows and at 11:01 divides into two bacteria. These two bacteria each grow and at 11:02 divide into four bacteria, which grow and at 11:03 divide into eight bacteria, and so on.

Now, suppose the bacteria continue to double every minute, and the bottle is full at 12:00. You may already realize that the number of bacteria at this point must be 2^{60} (because they doubled every minute for 60 minutes), but the important fact is that we have a bacterial disaster on our hands: Because the bacteria have filled the bottle, the entire bacterial colony is doomed. Let's examine this tragedy in greater detail, by asking a few questions about the demise of the colony.

- Question 1: If the tragedy occurred because the bottle was full at 12:00, when was the bottle *half*-full?

Because it took one hour to fill the bottle, many people guess that it was half-full after a half-hour, or at 11:30. However, because the bacteria double in number every minute, they must also have doubled during the last minute. Thus, the bottle went from being half-full to full during the final minute. That is, the bottle was half-full at 11:59, just 1 minute before the disaster.

by the way

The bacteria in a bottle parable was developed by Professor Albert A. Bartlett, University of Colorado, who has given more than 1200 lectures on the lessons of exponential growth around the country over the past 30 years.

- Question 2: Imagine that *you* are a mathematically sophisticated bacterium, and at 11:56 you recognize the impending disaster. You immediately jump on your soapbox and warn that unless your fellow bacteria slow their growth dramatically, the end is just 4 minutes away. Will anyone believe you?

FIGURE 8.2 The population of the bacteria in the bottle.

Facts do not cease to exist because they are ignored.

—ALDOUS HUXLEY

You've done your mathematics correctly; the bottle will indeed be full in just 4 minutes. But how will it appear to others who have not done the calculations? As we've already seen, the bottle will be half-full at 11:59. Continuing to work backward through the doublings each minute, we find that it will be $\frac{1}{4}$ full at 11:58, $\frac{1}{8}$ full at 11:57, and $\frac{1}{16}$ full at 11:56. Thus, if your fellow bacteria look around the bottle at 11:56, they'll see that only $\frac{1}{16}$ of the bottle has been used. In other words, the amount of unused space is *15 times* the amount of used space. You are asking them to believe that, in just the next 4 minutes, they'll fill 15 times as much space as they did in their entire 56-minute history. Unless they do the mathematics for themselves, they are unlikely to take your warnings seriously. Figure 8.2 shows the situation graphically. Note that the bottle remains nearly empty for most of the 60 minutes, but the continued doublings fill it rapidly in the final 4 minutes.

- Question 3: It's 11:59 and, with the bottle half-full, your fellow bacteria are finally taking your warnings seriously. They quickly start a space program, sending little bacterial spaceships out into the lab in search of new bottles. Thankfully, they discover three more bottles (making a total of four, including the one already occupied). Working quickly, they initiate a mass migration by packing bacteria onto spaceships and sending them to the new bottles. They successfully distribute the population evenly among the four bottles, just in time to avert the disaster. Given that they now have four bottles rather than just one, how much time have they gained for their civilization?

Since it took one hour to fill one bottle, you might guess that it would take four hours to fill four bottles. But the bacterial population still *doubles* each minute. If there are enough bacteria to fill one bottle at 12:00, there will be enough to fill two bottles by 12:01 and four bottles by 12:02. The discovery of three new bottles gives them only 2 additional minutes.

- Question 4: Suppose the bacteria continue their space program, constantly looking for more bottles. Is there any hope that further discoveries will allow the colony to continue its exponential growth?

Let's do some calculations. Note that, after n minutes, the bacterial population is 2^n. For example, it is $2^0 = 1$ when the first bacterium starts the colony at 11:00, $2^1 = 2$ at 11:01, $2^2 = 4$ at 11:02, and so on. There are 2^{60} bacteria when the first bottle fills at 12:00, and 2^{62} bacteria when four bottles are full at 12:02. Suppose that, somehow, the bacteria managed to keep doubling every minute until 1:00. By that time, the number of bacteria would be 2^{120}, because it has been 120 minutes since the colony began. Now, we must figure out how much space they'd require for this population.

The smallest bacteria measure approximately 10^{-7} m (0.1 micrometer) across. If we assume that the bacteria are roughly cube-shaped, the volume of a single bacterium is

$$(10^{-7}\,\text{m})^3 = 10^{-21}\,\text{m}^3$$

Therefore, the colony of 2^{120} bacteria would occupy a total volume of

$$2^{120} \times 10^{-21}\, \text{m}^3 \approx 1.3 \times 10^{15}\, \text{m}^3$$

With this volume, the bacteria would cover the entire surface of the Earth in a layer more than 2 meters deep! (See Exercise 27 to calculate this result for yourself.)

In fact, if the doublings continued for just $5\frac{1}{2}$ hours, the volume of bacteria would exceed the volume of the entire universe (see Exercise 28). Needless to say, this cannot happen. The exponential growth of the colony cannot possibly continue for long, no matter what technological advances might be imagined. *Now try Exercises 25–28.*

Time out to think
Some people have suggested that we could find room for an ever-growing human population by colonizing other planets in our solar system. Is this possible?

DOUBLING LESSONS
The three parables reveal at least two key lessons about the repeated doublings that arise with exponential growth. First, if you look back at Table 8.1, you'll notice that the number of grains on each square is nearly equal to the total number of grains on all previous squares combined. For example, the 16 grains on the fifth square are 1 more than the total of 15 grains on the first four squares combined.

Second, all three parables show quantities growing to impossible proportions. We cannot possibly fit all the wheat harvested in world history on a chessboard; $5 trillion worth of pennies is far more than the number of pennies that exist in the world; and a colony of bacteria can't possibly fill the universe.

Key Facts About Exponential Growth

- Exponential growth leads to repeated doublings. With each doubling, the amount of increase is approximately equal to the *sum* of all preceding doublings.
- Exponential growth cannot continue indefinitely. After only a relatively small number of doublings, exponentially growing quantities reach impossible proportions.

■ EXERCISES 8A

Review Questions

1. Describe the basic differences between linear growth and exponential growth.

2. Briefly explain how repeated doublings characterize exponential growth. Describe the impact of doublings, using the chessboard or magic penny parable.

3. Briefly summarize the story of the bacteria in the bottle. Be sure to explain the answers to the four questions asked in the text, and describe why the answers are surprising.

4. Explain the meaning of the two key facts about exponential growth given at the end of this unit. Then create your own example of exponential growth and describe the influence and impact of repeated doubling.

Does It Make Sense?

Decide whether each of the following statements makes sense (or is clearly true) or does not make sense (or is clearly false). Explain your reasoning.

5. Money in a bank account earning compound interest at an annual percentage rate of 3% is an example of exponential growth.

6. Suppose you had a magic bank account in which your balance doubled each day. If you started with just $1, you'd be a millionaire in less than a month.

7. A small town that grows exponentially can become a large city in just a few decades.

8. Human population has been growing exponentially for a few centuries, and we can expect this trend to continue forever in the future.

Basic Skills & Concepts

Linear or Exponential? For each of Exercises 9–16, state whether the growth (or decay) is linear or exponential, and answer the associated question.

9. The population of Danbury is increasing at a rate of 505 people per year. If the population is 15,000 today, what will it be in three years?

10. The population of Winesburg is increasing at a rate of 2% per year. If the population is 100,000 today, what will it be in three years?

11. During the worst periods of hyperinflation in Brazil, the price of food increased at a rate of 30% per month. If your food bill was $100 one month during this period, what was it two months later?

12. The price of a gallon of gasoline is decreasing by 2¢ per week. If the price is $1.60 per gallon today, what will it be in ten weeks?

13. The price of computer memory is decreasing at a rate of 12% per year. If a memory chip costs $80 today, what will it cost in two years?

14. The value of your car is decreasing by 10% per year. If the car is worth $10,000 today, what will it be worth in two years?

15. The value of your house is rising by $10,000 per year. If it is worth $100,000 today, what will it be worth in three years?

16. The value of your house is rising by 10% per year. If it is worth $100,000 today, what will it be worth in three years?

Chessboard Parable. Use the chessboard parable to answer Exercises 17–20. Assume that each grain of wheat weighs 1/7000 pound (see Unit 2B for the definition of a grain as a unit of weight).

17. How many grains of wheat should be placed on square 20 of the chessboard? Find the total number of grains and their total weight (in pounds) at this point.

18. How many grains of wheat should be placed on square 30 of the chessboard? Find the total number of grains and their total weight (in pounds) at this point.

19. What is the total weight of all the wheat when the chessboard is full?

20. According to the U.S. Department of Agriculture, the current world harvest of all wheat, rice, and corn is somewhat less than 2 billion tons per year. Compare this total to the weight of the wheat on the chessboard. (Hint: 1 ton = 2000 pounds.)

Magic Penny Parable. Use the magic penny parable to answer Exercises 21–24.

21. How much money would you have after 15 days?

22. Remember that all your money is in pennies. Suppose that you stacked the pennies after 15 days. How high would the stack rise? (Hint: Find a few pennies and a ruler.)

23. How many days would elapse before you had a total of over $1 billion? (Hint: Proceed by trial and error.)

24. Suppose that you could keep making a single stack of the pennies. After how many days would the stack be long enough to reach the nearest star (beyond the Sun), which is about 4.3 light-years (4.0×10^{13} km) away? (Hint: Proceed by trial and error.)

Bacteria in a Bottle Parable. Use the bacteria parable to answer Exercises 25–26.

25. How many bacteria are in the bottle at 11:50? What fraction of the bottle is full at that time?

26. How many bacteria are in the bottle at 11:10? What fraction of the bottle is full at that time?

27. **Knee-Deep in Bacteria.** The total surface area of the Earth is about 5.1×10^{14} m². Assume that the bacteria

continued their doublings for two hours (as discussed in the text), at which point they were distributed uniformly over the Earth's surface. How deep would the bacterial layer be? Would it be knee-deep, more than knee-deep, or less than knee-deep? (Hint: Remember that a volume divided by an area gives a depth.)

28. **Bacterial Universe.** Suppose the bacteria in the parable continued to double their population every minute. How long would it take until their volume exceeded the total volume of the observable universe, which is about 10^{79} m^3? (Hint: Proceed by trial and error.)

Further Applications

29. **Human Doubling.** Human population in the year 2000 was about 6 billion. Suppose this population increases exponentially with a doubling time of 50 years.

 a. Continue the following table, showing the population at 50-year intervals under this scenario, until you reach the year 3000. Use scientific notation, as shown.

Year	Population
2000	6×10^9
2050	$12 \times 10^9 = 1.2 \times 10^{10}$
2100	$24 \times 10^9 = 2.4 \times 10^{10}$
⋮	⋮

 b. The total surface area of the Earth is about 5.1×10^{14} m^2. Assuming that people could occupy all this area (in reality, most of it is ocean), approximately when would people be so crowded that every person would have only 1 m^2 of space?

 c. Suppose that, when we take into account the area needed to grow food and to find other resources, each

person actually requires about 10^4 m^2 of area to survive. About when would we reach this limit?

 d. Suppose that we learn to colonize other planets and moons in our solar system. The total surface area of the worlds in our solar system that could potentially be colonized (not counting gas planets such as Jupiter) is roughly five times the surface area of the Earth. Under the assumptions of part c, could humanity fit in our solar system in the year 3000? Explain.

Web Projects

Find useful links for Web Projects on the text Web site: www.aw.com/bennett-briggs

30. **Computing Power.** Computing power is often said to increase exponentially. Choose an aspect of computing power (such as processor speed or memory chip capacity) and investigate its growth. Has the growth been exponential? How much longer is the exponential growth likely to continue? Explain.

31. **Web Growth.** Investigate the growth of the Web itself, in terms of both number of users and number of Web pages. Has the growth been linear or exponential? How do you think the growth will change in the future? Explain.

In the News

32. **Linear Growth.** Identify at least two news stories that describe a quantity undergoing linear growth or decay. Describe the growth or decay process.

33. **Exponential Growth.** Identify at least two news stories that describe a quantity undergoing exponential growth or decay. Describe the growth or decay process.

UNIT **8B**

DOUBLING TIME AND HALF-LIFE

E xponential growth leads to repeated doublings and exponential decay leads to repeated *halvings*. However, in most cases of exponential growth or decay, we are given the rate of growth or decay—usually as a percentage—rather than the time required for doubling or halving. In this unit, we'll convert between percentage rates and doubling (or halving) times.

DOUBLING TIME

The time required for each doubling in exponential growth is called the **doubling time.** For example, the doubling time for the magic penny (see Unit 8A) was one day, because your wealth doubled each day. The doubling time for the bacteria in the bottle was one minute.

Given the doubling time, we can easily calculate the value of a quantity at any time. Consider an initial population of 10,000 that grows with a doubling time of 10 years:

- In 10 years, or one doubling time, the population increases by a factor of 2, to a new population of $2 \times 10,000 = 20,000$.
- In 20 years, or two doubling times, the population increases by a factor of $2^2 = 4$, to a new population of $4 \times 10,000 = 40,000$.
- In 30 years, or three doubling times, the population increases by a factor of $2^3 = 8$, to a new population of $8 \times 10,000 = 80,000$.

We can generalize this idea by letting t be the amount of time that has passed and T_{double} be the doubling time. Note that after $t = 30$ years with a doubling time of $T_{\text{double}} = 10$ years, there have been $t/T_{\text{double}} = 30/10 = 3$ doublings. Generalizing, we find that the number of doublings after a time t is t/T_{double}. That is, the size of the population after time t is the initial population times $2^{t/T_{\text{double}}}$.

Calculations with the Doubling Time

After a time t, an exponentially growing quantity with a doubling time of T_{double} increases in size by a factor of $2^{t/T_{\text{double}}}$. The new value of the growing quantity is related to its initial value (at $t = 0$) by

$$\text{new value} = \text{initial value} \times 2^{t/T_{\text{double}}}$$

Time out to think

Consider an initial population of 10,000 that grows with a doubling time of 10 years. Confirm that the above formula gives a population of 80,000 after 30 years, as we found earlier. What does the formula predict for the population after 50 years?

➤ **EXAMPLE 1** Doubling with Compound Interest

Compound interest (Unit 4A) is a form of exponential growth because an interest-bearing account grows by the same percentage each year. Suppose your bank account has a doubling time of 13 years. By what factor does your balance increase in 50 years?

Solution The doubling time is T_{double} = 13 years, so after t = 50 years your balance increases by a factor of

$$2^{t/T_{double}} = 2^{50 \text{ yr}/13 \text{ yr}} = 2^{3.85} = 14.382$$

For example, if you start with a balance of $1000, in 50 years it will grow to $1000 × 14.382 = $14,382. *Now try Exercises 25–32.* ◄

➤ **EXAMPLE 2** World Population Growth

World population doubled from 3 billion in 1960 to 6 billion in 2000. Suppose that world population continues to grow with a doubling time of 40 years. What will the population be in 2030? in 2200?

Solution The doubling time is T_{double} = 40 years. If we let t = 0 represent 2000, the year 2030 is t = 30 years later. If the 2000 population of 6 billion is used as the initial value, the population in 2030 will be

$$\text{new value} = \text{initial value} \times 2^{t/T_{double}}$$
$$= 6 \text{ billion} \times 2^{30 \text{ yr}/40 \text{ yr}}$$
$$= 6 \text{ billion} \times 2^{0.75} = 10.1 \text{ billion}$$

By 2200, which is t = 200 years after 2000, the population will reach

$$\text{new value} = \text{initial value} \times 2^{t/T_{double}}$$
$$= 6 \text{ billion} \times 2^{200 \text{ yr}/40 \text{ yr}}$$
$$= 6 \text{ billion} \times 2^5 = 192 \text{ billion}$$

If world population continues to grow at the same rate it did between 1960 and 2000, it will reach 10 billion by 2030 and 192 billion by 2200.
Now try Exercises 33–34. ◄

Time out to think _____
Do you think that it's really possible for human population to reach 192 billion in 200 years? Why or why not?

THE APPROXIMATE DOUBLING TIME FORMULA

Consider an ecological study of a prairie dog community. The community contains 100 prairie dogs when the study begins, and researchers soon determine that the population is increasing at a rate of 10% per month. That is, each month the population grows to 110% of, or 1.1 times, its previous value (see the "of versus more than" rule in Unit 3A). Table 8.3 tracks the population growth.

DILBERT reprinted by permission of United Feature Syndicate, Inc.

TABLE 8.3	Growth of a Prairie Dog Community		
MONTH	POPULATION	MONTH	POPULATION
0	100	8	$(1.1)^8 \times 100 = 214$
1	$(1.1)^1 \times 100 = 110$	9	$(1.1)^9 \times 100 = 236$
2	$(1.1)^2 \times 100 = 121$	10	$(1.1)^{10} \times 100 = 259$
3	$(1.1)^3 \times 100 = 133$	11	$(1.1)^{11} \times 100 = 285$
4	$(1.1)^4 \times 100 = 146$	12	$(1.1)^{12} \times 100 = 314$
5	$(1.1)^5 \times 100 = 161$	13	$(1.1)^{13} \times 100 = 345$
6	$(1.1)^6 \times 100 = 177$	14	$(1.1)^{14} \times 100 = 380$
7	$(1.1)^7 \times 100 = 195$	15	$(1.1)^{15} \times 100 = 418$

Note that the population nearly doubles (to 195) after 7 months, then nearly doubles again (to 380) after 14 months. This roughly seven-month doubling time is related to the 10% growth rate as follows:

$$\text{doubling time} \approx \frac{70}{\text{percentage growth rate}} = \frac{70}{10/\text{mo}} = 7 \text{ mo}$$

This formula, in which the doubling time is approximately 70 divided by the percentage growth rate, works whenever the growth rate is relatively small (less than about 15%). It is often called the **rule of 70.**

Approximate Doubling Time Formula (Rule of 70)

For a quantity growing exponentially at a rate of $P\%$ per time period, the doubling time is *approximately*

$$T_{\text{double}} \approx \frac{70}{P}$$

This approximation works best for small growth rates and breaks down for growth rates over about 15%.

► **EXAMPLE 3** Population Doubling Time

World population was about 6.0 billion in 2000 and was growing at a rate of about 1.4% per year. What is the approximate doubling time at this growth rate? If this growth rate continues, what will world population be in 2030? Compare to the result in Example 2.

Solution We can use the approximate doubling time formula because the growth rate is much less than 15%. The percentage growth rate of 1.4% per year means we set $P = 1.4/\text{yr}$:

$$T_{\text{double}} \approx \frac{70}{P} = \frac{70}{1.4/\text{yr}} = 50 \text{ yr}$$

The doubling time is approximately 50 years. The year 2030 is $t = 30$ years after 2000, so the population in 2030 will be

$$\text{new value} = \text{initial value} \times 2^{t/T_{\text{double}}}$$
$$= 6.0 \text{ billion} \times 2^{30 \text{ yr}/50 \text{ yr}}$$
$$= 6.0 \text{ billion} \times 2^{0.6} = 9.1 \text{ billion}$$

At a 1.4 % growth rate, world population will be about 9 billion in 2030. This is a billion fewer people than predicted in Example 2, in which the doubling time was assumed to be 40 years rather than 50 years. *Now try Exercises 35–36.* ◄

► **EXAMPLE 4** Solving the Doubling Time Formula

World population doubled in the 40 years from 1960 to 2000. What was the average percentage growth rate during this period? Contrast this growth rate with the 2000 growth rate of 1.4% per year.

Solution We can answer the question by solving the approximate doubling time formula for P. We multiply both sides of the formula by P and divide both sides by T_{double}. You should confirm that this gives the following result:

$$P \approx \frac{70}{T_{\text{double}}}$$

Substituting $T_{\text{double}} = 40$ years, we find

$$P \approx \frac{70}{T_{\text{double}}} = \frac{70}{40 \text{ yr}} = 1.75/\text{yr}$$

The average population growth rate between 1960 and 2000 was about $P\% = 1.75\%$ per year. This is 0.35 percentage point higher than the 2000 growth rate of 1.4% per year. *Now try Exercises 37–40.* ◄

EXPONENTIAL DECAY AND HALF-LIFE

Exponential *decay* occurs whenever a quantity *decreases* by the same percentage in every fixed time period (for example, by 20% every year). In that case, the value of the quantity repeatedly decreases to half its value, with each halving occurring in a time called the **half-life.**

by the way

Plutonium-239 is the chemical element pluto-nium in a form (or *isotope*) with atomic weight 239. Atomic weight is the total number of protons and neutrons in the nucleus. Because all plutonium nuclei have 94 protons, Pu-239 nuclei have $239 - 94 = 145$ neutrons.

You may have heard half-lives described for radioactive materials such as uranium or plutonium. For example, radioactive plutonium-239 (Pu-239) has a half-life of about 24,000 years. To understand the meaning of the half-life, suppose that 100 pounds of Pu-239 is deposited at a nuclear waste site. The plutonium gradually decays into other substances as follows:

- In 24,000 years, or one half-life, the amount of Pu-239 declines to $1/2$ its original value, or to $(1/2) \times 100$ pounds = 50 pounds.
- In 48,000 years, or two half-lives, the amount of Pu-239 declines to $(1/2)^2 = 1/4$ its original value, or to $(1/4) \times 100$ pounds = 25 pounds.
- In 72,000 years, or three half-lives, the amount of Pu-239 declines to $(1/2)^3 = 1/8$ its original value, or to $(1/8) \times 100$ pounds = 12.5 pounds.

We can generalize this idea much as we did earlier for the doubling time. A single halving reduces a quantity by a factor of $1/2$, two halvings reduce it by a factor of $(1/2)^2$, and three halvings reduce it by a factor of $(1/2)^3$. If we let t be the amount of time that has passed and T_{half} be the half-life, then the number of halvings after a time t is t/T_{half}. Thus, the quantity after time t is the original quantity times this factor of $(1/2)^{t/T_{half}}$.

Calculations with the Half-Life

After a time t, an exponentially decaying quantity with a half-life of T_{half} decreases in size by a factor of $(1/2)^{t/T_{half}}$. The new value of the decaying quantity is related to its initial value (at $t = 0$) by

$$\text{new value} = \text{initial value} \times \left(\frac{1}{2}\right)^{t/T_{half}}$$

by the way

Ordinary carbon is carbon-12, which is stable (not radioactive). Carbon-14 is produced in the Earth's atmosphere by high-energy particles coming from the Sun. It mixes with ordinary carbon and there-fore becomes incorporated into living tissue through respiration (breathing).

▶**EXAMPLE 5** Carbon-14 Decay

Radioactive carbon-14 has a half-life of about 5700 years. It collects in organisms only while they are alive. Once they are dead, it only decays. What fraction of the carbon-14 in an animal bone still remains 1000 years after the animal has died?

Solution The half-life is $T_{half} = 5700$ years, so the fraction of the initial amount remaining after $t = 1000$ years is

$$\left(\frac{1}{2}\right)^{t/T_{half}} = \left(\frac{1}{2}\right)^{1000 \text{ yr}/5700 \text{ yr}} = 0.885$$

For example, if the bone originally contained 1 kilogram of carbon-14, the amount remaining after 1000 years is 0.885 kilogram. As you can probably see, we can use this idea to determine the age of bones found at archaeological sites. We'll explore this idea further in Unit 9C. *Now try Exercises 41–44.* ◀

▶**EXAMPLE 6** Plutonium After 100,000 Years

Suppose that 100 pounds of Pu-239 is deposited at a nuclear waste site. How much of it will still be present in 100,000 years?

Solution The half-life of Pu-239 is T_{half} = 24,000 years. Given an initial amount of 100 pounds, the amount remaining after t = 100,000 years is

$$\text{new value} = \text{initial value} \times \left(\frac{1}{2}\right)^{t/T_{half}} = 100 \text{ lb} \times \left(\frac{1}{2}\right)^{100,000 \text{ yr}/24,000 \text{ yr}} = 5.6 \text{ lb}$$

About 5.6 pounds of the original 100 pounds of Pu-239 will still be present in 100,000 years. *Now try Exercises 45–48.* ◄

Time out to think

Plutonium, which is not found naturally on the Earth, is made in nuclear reactors for use both as fuel for nuclear power plants and for nuclear weapons. Based on its half-life, explain why the safe disposal of Pu-239 poses a significant challenge.

THE APPROXIMATE HALF-LIFE FORMULA

The approximate doubling time formula (the rule of 70) found earlier works equally well for exponential decay if we replace the doubling time with the half-life and the percentage growth rate with the percentage decay rate.

> **Approximate Half-Life Formula**
>
> For a quantity decaying exponentially at a rate of $P\%$ per time period, the half-life is *approximately*
>
> $$T_{half} \approx \frac{70}{P}$$
>
> This approximation works best for small decay rates and breaks down for decay rates over about 15%.

> **EXAMPLE 7** Devaluation of Currency

Suppose that inflation causes the value of the Russian ruble to fall at a rate of 12% per year (relative to the dollar). Approximately how long does it take for the ruble to lose half its value?

Solution We can use the approximate half-life formula because the growth rate is less than 15%. The 12% decay rate means we set P = 12/yr:

$$T_{half} \approx \frac{70}{P} = \frac{70}{12/\text{yr}} = 5.8 \text{ yr}$$

The half-life is a little less than six years, meaning that the ruble loses half its value (against the dollar) in six years. *Now try Exercises 49–52.* ◄

EXACT FORMULAS FOR DOUBLING TIME AND HALF-LIFE

The approximate doubling time and half-life formulas are useful because they are easy to remember. However, for more precise work or for cases of larger rates where the approximate formulas break down, we need exact formulas. The formulas are given below. In Unit 9C, we will see how they are derived. These formulas use the **fractional**

HISTORICAL NOTE
The atomic bomb that devastated Nagasaki during World War II generated its destructive power by fission of plutonium-239. (The Hiroshima bomb used uranium-235.)

growth rate, $r = P/100$, rather than the percentage growth rate, $P\%$. For example, if the percentage growth rate is 5% per year, the fractional growth rate is 0.05 per year.

Exact Doubling Time and Half-Life Formulas

For an exponentially growing quantity with a fractional growth rate r, the doubling time is

$$T_{\text{double}} = \frac{\log_{10} 2}{\log_{10}(1 + r)}$$

For an exponentially decaying quantity, we use a *negative* value for r (for example, if the decay rate is 5% per year, we set $r = -0.05$ per year). The half-life is

$$T_{\text{half}} = -\frac{\log_{10} 2}{\log_{10}(1 + r)}$$

Note that *the units of time used for T and r must be the same.* For example, if the fractional growth rate is 0.05 per *month*, then the doubling time will also be measured in months.

►EXAMPLE 8 Large Growth Rate

A population of rats is growing at a rate of 80% per month. Find the exact doubling time for this growth rate and compare it to the doubling time found with the approximate doubling time formula.

Solution The growth rate of 80% per month means $P = 80/\text{mo}$ or $r = 0.8/\text{mo}$. The doubling time is

$$T_{\text{double}} = \frac{\log_{10} 2}{\log_{10}(1 + 0.8)} = \frac{0.301\,030}{\log_{10}(1.8)} = \frac{0.301\,030}{0.255\,273} = 1.18 \text{ mo}$$

The doubling time is almost 1.2 months. Note that this answer makes sense: With the population growing by 80% in a month, we expect it to take a little over a month to grow by 100% (which is a doubling). In contrast, the approximate doubling time formula predicts a doubling time of $70/P = 70/80 = 0.875$ month, which is less than one month. We see that the approximate formula does not work well for large growth rates. *Now try Exercises 53–54.* ◄

►EXAMPLE 9 Ruble Revisited

Suppose the Russian ruble is falling in value against the dollar at 12% per year. Using the exact half-life formula, determine how long it takes the ruble to lose half its value. Compare your answer to the approximate answer found in Example 7.

Solution The percentage decay rate is 12%/yr, so the fractional decay rate is $r = -0.12/\text{yr}$. The half-life is

$$T_{\text{half}} = -\frac{\log_{10} 2}{\log_{10}(1 - 0.12)} = \frac{0.301\,030}{-0.055\,517} = 5.42 \text{ yr}$$

Using Your Calculator

Logarithms

Your calculator should have a button for computing common (base 10) logarithms. It is usually labeled *log*, but be careful because some calculators use this label for natural (base *e*) logarithms. With the right button, you will find $\log_{10} 2 = 0.3010.\ldots$

A brief review . . .

LOGARITHMS

A **logarithm** (or **log,** for short) is a number that represents a power or exponent. In this book, we will focus on only base 10 logs, also called **common logs,** which are defined as follows:

$\log_{10} x$ is *the power to which 10 must be raised to obtain x.*

You may find it easier to remember the meaning with a less technical definition:

$\log_{10} x$ means "10 to what power equals x?"

For example:

$\log_{10} 1000 = 3$ because $10^3 = 1000$
$\log_{10} 10,000,000 = 7$ because $10^7 = 10,000,000$
$\log_{10} 0.1 = -1$ because $10^{-1} = 0.1$
$\log_{10} 30 = 1.4777$ because $10^{1.477} = 30$

Four important rules follow directly from the definition of a logarithm.

1. Taking the logarithm of a power of 10 gives the power. That is,
$$\log_{10} 10^x = x$$

2. Raising 10 to a power that is the logarithm of a number gives back the number. That is,
$$10^{\log_{10} x} = x \quad (x > 0)$$

3. Because powers of 10 are multiplied by adding their exponents, we have the *addition rule* for logarithms:
$$\log_{10} xy = \log_{10} x + \log_{10} y \quad (x > 0 \text{ and } y > 0)$$

4. We can "bring down" an exponent within a logarithm by applying the *power rule* for logarithms:
$$\log_{10} a^x = x \times \log_{10} a \quad (a > 0)$$

Most calculators have a key to compute \log_{10} of any positive number. You should find this key on your calculator and use it to verify that $\log_{10} 1000 = 3$ and $\log_{10} 2 = 0.301030$.

Example: Given that $\log_{10} 2 = 0.301030$, find each of the following:

a. $\log_{10} 8$ **b.** $10^{\log_{10} 2}$ **c.** $\log_{10} 200$

Solution:

a. We notice that $8 = 2^3$. Thus, from Rule 4,
$$\log_{10} 8 = \log_{10} 2^3$$
$$= 3 \times \log_{10} 2 = 3 \times 0.301030 = 0.90309$$

b. From Rule 2,
$$10^{\log_{10} 2} = 2$$

c. We notice that $200 = 2 \times 100 = 2 \times 10^2$. Thus, from Rule 3,
$$\log_{10} 200 = \log_{10}(2 \times 10^2) = \log_{10} 2 + \log_{10} 10^2$$
From Rule 1, we know that $\log_{10} 10^2 = 2$, so
$$\log_{10} 200 = \log_{10}(2 \times 10^2) = 0.301030 + 2 = 2.301030$$

Example: Someone tells you that $\log_{10} 600 = 5.778$. Should you believe it?

Solution: Because 600 is between 100 and 1000, $\log_{10} 600$ must be between $\log_{10} 100$ and $\log_{10} 1000$. From Rule 1, we find that $\log_{10} 100 = \log_{10} 10^2 = 2$ and $\log_{10} 1000 = \log_{10} 10^3 = 3$. Thus, $\log_{10} 600$ must be between 2 and 3, so the claimed answer of 5.778 must be wrong.

Now try Exercises 13–24.

The ruble loses half its value against the dollar in 5.42 years. This exact result is about 0.4 year less than the 5.8 years obtained with the approximate formula. We see that the approximate formula is reasonably accurate for the 12% rate.

Now try Exercises 55–56. ◄

■ EXERCISES 8B

Review Questions

1. What is a doubling time? Suppose a population has a doubling time of 25 years. By what factor will it grow in 25 years? in 50 years? in 100 years?

2. Given a doubling time, explain how you calculate the value of an exponentially growing quantity at any time t.

3. State the approximate doubling time formula and the conditions under which it works well. Give an example.

4. What is a half-life? Suppose a radioactive substance has a half-life of 1000 years. What fraction will be left after 1000 years? after 2000 years? after 4000 years?

5. Given a half-life, explain how you calculate the value of an exponentially decaying quantity at any time t.

6. State the approximate half-life formula and the conditions under which it works well. Give an example.

7. Briefly describe exact doubling time and half-life formulas. Explain all their terms.

8. Give an example in which it is important to use the exact doubling time or half-life formula, rather than the approximate formula. Explain why the approximate formula does not work well in this case.

Does It Make Sense?

Decide whether each of the following statements makes sense (or is clearly true) or does not make sense (or is clearly false). Explain your reasoning.

9. Our town is growing with a doubling time of 25 years, so its population will triple in 50 years.

10. Our town is growing at a rate of 7% per year, so it will double in population about every 10 years.

11. A toxic chemical decays with a half-life of 10 years, so half of it will be gone 10 years from now and all the rest will be gone 20 years from now.

12. The half-life of plutonium-239 is about 24,000 years, so we can expect some of the plutonium produced in recent decades to still be around 100,000 years from now.

Basic Skills & Concepts

Logarithms. Exercises 13–24 use skills covered in the Brief Review on pg. 505. In Exercises 13–22, determine whether each statement is true or false *without* doing any calculations. Explain your reasoning.

13. $10^{0.928}$ is between 10 and 100.

14. $10^{3.334}$ is between 1000 and 10,000.

15. $10^{-5.2}$ is between 100,000 and 1,000,000.

16. $10^{-2.67}$ is between 0.001 and 0.01.

17. $\log_{10} \pi$ is between 0 and 1.

18. $\log_{10} 96$ is between 3 and 4.

19. $\log_{10} 1,600,000$ is between 6 and 7.

20. $\log_{10}(8 \times 10^9)$ is between 12 and 13.

21. $\log_{10} \frac{1}{4}$ is between 0 and 1.

22. $\log_{10} 0.00045$ is between 5 and 6.

23. Given that $\log_{10} 2 = 0.301$, find each of the following *without* a calculator.

 a. $\log_{10} 4$ b. $\log_{10} 20,000$

 c. $\log_{10} 0.5$ d. $\log_{10} 32$

 e. $\log_{10} 0.25$ f. $\log_{10} 0.2$

24. Given that $\log_{10} 5 = 0.699$, find each of the following *without* a calculator.

 a. $\log_{10} 50$ b. $\log_{10} 5000$

 c. $\log_{10} 0.05$ d. $\log_{10} 25$

 e. $\log_{10} 0.20$ f. $\log_{10} 0.04$

Doubling Time. Exercises 25–32 give a doubling time for an exponentially growing quantity. Answer the questions that follow.

25. The doubling time of a population of flies is 3 hours. By what factor does the population increase in 24 hours? in 1 week?

26. The doubling time of a bank account balance is 10 years. By what factor does it grow in 30 years? in 50 years?

27. The doubling time of a city's population is 22 years. How long does it take for the population to quadruple?

28. Prices are rising with a doubling time of 4 weeks. By what factor do prices increase in a year?

29. The initial population of a town is 15,600, and it grows with a doubling time of 20 years. What will the population be in 12 years? in 24 years?

30. The initial population of a town is 15,600, and it grows with a doubling time of 8 years. What will the population be in 12 years? in 24 years?

31. The number of cells in a tumor doubles every 1.5 months. If the tumor begins as a single cell, how many cells will there be after 3 years? after 4 years?

32. The number of cells in a tumor doubles every 6 months. If the tumor begins with a single cell, how many cells will there be after 3 years? after 6 years?

World Population. In 2000, world population was 6 billion. In Exercises 33–34, use the given doubling time to predict the population in 2010, 2060, and 2100.

33. Assume a doubling time of 45 years.

34. Assume a doubling time of 60 years.

35. **Rabbits.** A community of rabbits begins with an initial population of 100 and grows 7% per month. Make a table, similar to Table 8.3, that shows the population for each of the next 15 months. Based on the table, find the doubling time of the population and briefly discuss how well the approximate doubling time formula works for this case.

36. **Mice.** A community of mice begins with an initial population of 1000 and grows 20% per month. Make a table, similar to Table 8.3, that shows the population for each of the next 15 months. Based on the table, find the doubling time of the population and briefly discuss how well the approximate doubling time formula works for this case.

Doubling Time Formula. Use the approximate doubling time formula (rule of 70) in Exercises 37–40. Be sure to discuss whether the formula is valid for the case described.

37. The Consumer Price Index is increasing at a rate of 7% per year. What is its doubling time? By what factor will prices increase in 3 years?

38. A city's population is growing at a rate of 2.0% per year. What is its doubling time? By what factor will the population increase in 75 years?

39. Prices are rising at a rate of 0.7% per month. What is their doubling time? By what factor will prices increase in 1 year? in 8 years?

40. Oil consumption is increasing at a rate of 1.9% per year. What is its doubling time? By what factor will oil consumption increase in a decade?

Half-Life. Exercises 41–48 give a half-life for an exponentially decaying quantity. Answer the questions that follow.

41. The half-life of a radioactive substance is 35 years. If you start with some amount of this substance, what fraction will remain in 70 years? in 140 years?

42. The half-life of a radioactive substance is 250 years. If you start with some amount of this substance, what fraction will remain in 70 years? in 1500 years?

43. The half-life of a drug in the bloodstream is 12 hours. By what factor does the concentration of the drug decrease in 24 hours? in 36 hours?

44. The half-life of a drug in the bloodstream is 4 hours. By what factor does the concentration of the drug decrease in 24 hours? in 36 hours?

45. The current population of a threatened animal species is 1 million, but it is declining with a half-life of 10 years. How many animals will be left in 40 years? in 70 years?

46. The current population of a threatened animal species is 1 million, but it is declining with a half-life of 25 years. How many animals will be left in 40 years? in 70 years?

47. Cobalt-56 has a half-life of 77 days. If you start with 1 kilogram of cobalt-56, how much will remain after 100 days? after 1 year?

48. Radium-226 is a metal with a half-life of 1600 years. If you start with 1 kilogram of radium-226, how much will remain after 1000 years? after 10,000 years?

Half-Life Formula. Use the approximate half-life formula in Exercises 49–52. Be sure to discuss whether the formula is valid for the case described.

49. Urban encroachment is causing the area of a forest to decline at a rate of 10% per year. What is the half-life of the forest? What fraction of the forest will remain in 100 years?

50. A clean-up project is reducing the concentration of a pollutant in the water supply, with an 8% decrease per week. What is the half-life of the concentration of the pollutant? What fraction of the original amount of the pollutant will remain when the project ends after 1 year (52 weeks)?

51. Poaching is causing a population of elephants to decline by 7% per year. What is the half-life for the population? If there are 10,000 elephants today, how many will remain in 50 years?

52. The production of a gold mine is declining by 12% per year. If its current annual production is 4000 kilograms, what will its production be in 10 years?

Exact Formulas. In Exercises 53–56, compare the doubling times found with the approximate and exact doubling time formulas. Then use the exact doubling time formula to answer the given question.

53. Inflation is causing prices to rise at a rate of 10% per year. For an item that costs $1000 today, what will the price be in 3 years?

54. Hyperinflation is driving up prices at a rate of 80% per month. For an item that costs $1000 today, what will the price be in 1 year?

55. A nation of 100 million people is growing at a rate of 3% per year. What will its population be in 50 years?

56. A family of 100 termites invades your house and grows at a rate of 20% per week. How many termites will be in your house after 1 year (52 weeks)?

Further Applications

57. **Plutonium on Earth.** Scientists believe the Earth once had naturally existing plutonium-239. Suppose the Earth had 10 trillion tons of Pu-239 when it formed. Given plutonium's half-life of 24,000 years and the Earth's current age of 4.6 billion years, how much would remain today? Use your answer to explain why plutonium is not found naturally on Earth today.

58. **Nuclear Weapons.** Thermonuclear weapons use tritium for their nuclear reactions. Tritium is a radioactive form of hydrogen (containing 1 proton and 2 neutrons) with a half-life of about 12 years. Suppose a nuclear weapon contains 1 kilogram of tritium. How much will remain in 50 years? Use your answer to explain why thermonuclear weapons require regular maintenance.

Web Projects

Find useful links for Web Projects on the text Web site: www.aw.com/bennett-briggs

59. **National Growth Rates.** Visit the Web site of the Population Reference Bureau, where you will find growth rates, doubling times, and population projections tabulated for many different countries. Select several countries from several continents and record relevant growth data. Comment on whether the doubling times and growth rates are consistent. Discuss how these data are used to make population projections.

60. **World Population Growth.** The Web site for the U.S. Census Bureau contains a wealth of data on world population. Visit this site and gather data on world population growth over the last 50 years. Estimate the population growth rate over each decade. Compute the associated doubling times. Write a two-paragraph statement on the trends that you observe.

In the News

61. **Doubling Time.** Find a news story that gives an exponential growth rate. Find the approximate doubling time from the growth rate and discuss the implications of the growth.

62. **Radioactive Half-Life.** Find a news story that discusses some type of radioactive material. If it is not given, look up the half-life of the material. Discuss the implications for disposal of the material.

UNIT 8C

REAL POPULATION GROWTH

erhaps the most important application of exponential growth concerns human population. From the time of the earliest humans more than 2 million years ago until about 10,000 years ago, human population probably never exceeded 10 million. The advent of agriculture brought about more rapid population growth. Human population reached 250 million by A.D. 1 and continued growing slowly to about 500 million by 1650.

The burst of growth that continues today began with the industrial revolution. Our rapidly developing ability to grow food and exploit natural resources allowed us to build more homes for more people. Meanwhile, improvements in medicine and health science lowered death rates dramatically. As a result, human population began to grow exponentially. Figure 8.3 shows estimated human population over the past 12,000 years.

World Population, 10,000 B.C. to Present

FIGURE 8.3 World population.

To put current world population growth in perspective, consider the following facts:

- Every three years, the world adds as many people as live in the entire United States.
- Each month, world population increases by the equivalent of the population of Switzerland.
- In 1992, 300,000 people died of starvation in Somalia; worldwide, those numbers were replaced by new births in just 29 *hours*.
- During the time it takes to read this chapter, world population will increase by about 10,000 people.

Projections of future population growth have large uncertainties. Nevertheless, current trends suggest that world population will reach 8 billion people by 2020—more

than *four times* the population just a century earlier and nearly double the population at the time most of today's college students were born. Even in the United States, where population is growing more slowly than the world average, the population is expected to increase by *100 million* people within the next 50 years.

➤ EXAMPLE 1 Varying Growth Rate

The average annual growth rate for world population since 1650 has been about 0.7%. However, the annual rate has varied significantly. It peaked at about 2.1% during the 1960s and is currently about 1.4%. Find the approximate doubling time for each of these growth rates. Use each to predict world population in 2050, based on a 2000 population of 6.0 billion.

Solution Using the approximate doubling time formula (Unit 8B), we find the doubling times for the three rates:

For 0.7%: $T_{\text{double}} \approx \dfrac{70}{P} = \dfrac{70}{0.7/\text{yr}} = 100 \text{ yr}$

For 2.1%: $T_{\text{double}} \approx \dfrac{70}{P} = \dfrac{70}{2.1/\text{yr}} = 33 \text{ yr}$

For 1.4%: $T_{\text{double}} \approx \dfrac{70}{P} = \dfrac{70}{1.4/\text{yr}} = 50 \text{ yr}$

To predict world population in 2050, we use the formula

$$\text{new value} = \text{initial value} \times 2^{t/T_{\text{double}}}$$

We set the initial value to 6 billion and note that 2050 is $t = 50$ years after 2000:

For 0.7%: 2050 population = 6.0 billion $\times\ 2^{50\,\text{yr}/100\,\text{yr}}$ = 8.5 billion

For 2.1%: 2050 population = 6.0 billion $\times\ 2^{50\,\text{yr}/33\,\text{yr}}$ = 17.1 billion

For 1.4%: 2050 population = 6.0 billion $\times\ 2^{50\,\text{yr}/50\,\text{yr}}$ = 12.0 billion

Notice the large differences for different growth rates. At the growth rate of the 1960s, world population would exceed 17 billion by 2050—double the population if the growth rate equals the average since 1650. *Now try Exercises 13–16.* ◀

WHAT DETERMINES THE GROWTH RATE?

The world population growth rate is simply the difference between the birth rate and the death rate. For example, suppose that on average there are 8.5 births per 100 people and 6.5 deaths per 100 people per year. Then the population growth rate is

$$\frac{8.5}{100} - \frac{6.5}{100} = \frac{2}{100} = 0.02 = 2\%$$

Overall Growth Rate

The world population growth rate is the difference between the birth rate and the death rate:

$$\text{growth rate} = \text{birth rate} - \text{death rate}$$

Interestingly, birth rates dropped rapidly throughout the world during the past 50 years—the same period that has seen the largest population growth in history. Indeed, worldwide birth rates have never been lower than they are today. Today's rapid population growth comes from the fact that death rates have fallen even more dramatically.

by the way

By 2000, the world birth rate had fallen to about 2.2 per 100 people and the death rate had fallen to about 0.9 per 100 people.

▶**EXAMPLE 2** Birth and Death Rates

In 1950, the world birth rate was 3.7 births per 100 people and the world death rate was 2.0 deaths per 100 people. By 1975, the birth rate had fallen to 2.8 births per 100 people and the death rate to 1.1 deaths per 100 people. Contrast the growth rates in 1950 and 1975.

Solution In 1950, the overall growth rate was

$$\frac{3.7}{100} - \frac{2.0}{100} = \frac{1.7}{100} = 1.7\%$$

In 1975, the overall growth rate was

$$\frac{2.8}{100} - \frac{1.1}{100} = \frac{1.7}{100} = 1.7\%$$

Despite a dramatic fall in birth rates during the 25-year period, the growth rate remained unchanged because death rates fell equally dramatically.

Now try Exercises 17–20. ◀

Time out to think _____

Suppose that medical science finds a way to extend human lifespans significantly. How would this affect the population growth rate?

CARRYING CAPACITY AND REAL GROWTH MODELS

As we saw in Unit 8A, exponential growth cannot continue indefinitely. Indeed, human population cannot continue to grow much longer at its current rate, because we'd be elbow to elbow over the entire Earth in just a few centuries. Theoretical models of population growth therefore assume that human population is ultimately limited by the **carrying capacity** of the Earth—the number of people that the Earth can support.

DEFINITION

For any particular species in a given environment, the **carrying capacity** is the maximum sustainable population. That is, it is the largest population the environment can support for extended periods of time.

Two important models for populations approaching the carrying capacity are (1) a gradual leveling off, known as *logistic growth,* and (2) a rapid increase followed by a rapid decrease, known as *overshoot and collapse.* Let's investigate each model.

Logistic Growth

A **logistic growth** model assumes that population growth gradually slows as the population approaches the carrying capacity. For example, if the carrying capacity is 12 billion people, a logistic model assumes that population growth will slow as this number is approached. The growth rate falls to zero as the carrying capacity is approached, allowing the population to remain steady at that level thereafter.

Logistic Growth

When the population is small relative to the carrying capacity, **logistic growth** is exponential with a fractional growth rate close to the base growth rate r. As the population approaches the carrying capacity, the logistic growth rate approaches zero. The fractional growth rate at any particular time depends on the population at that time as follows:

$$\text{growth rate} = r \times \left(1 - \frac{\text{population}}{\text{carrying capacity}} \right)$$

Note that, when the population is small compared to the carrying capacity, the logistic growth rate is very close to the base growth rate r. The growth rate then gradually decreases to zero as the population approaches the carrying capacity. Figure 8.4 contrasts a logistic growth model with an exponential growth model.

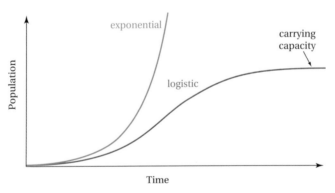

FIGURE 8.4 This graph contrasts exponential growth with logistic growth. Note that logistic growth looks like exponential growth at early times, but then gradually levels off.

➤ **EXAMPLE 3** Are We Growing Logistically?

Assume that the Earth's carrying capacity is 12 billion people. Given that the population growth rate peaked in the 1960s at about 2.1%, when the population was about 3 billion, is it reasonable to assume that human population has been following

a logistic growth pattern since the 1960s? Is it reasonable to assume that population has been growing logistically throughout the past century? Explain.

Solution We need to compare the actual current growth rate of about 1.4% (see Example 1) to the growth rate predicted by a logistic model. First, we must use the 1960s data to find the base growth rate r in the logistic model. You should confirm that solving the logistic growth rate formula for r gives

$$r = \frac{\text{growth rate}}{\left(1 - \dfrac{\text{population}}{\text{carrying capacity}}\right)}$$

Substituting the 1960s growth rate of 2.1% = 0.021 and population of 3 billion, along with a carrying capacity of 12 billion, we find

$$r = \frac{0.021}{\left(1 - \dfrac{3 \text{ billion}}{12 \text{ billion}}\right)} = \frac{0.021}{(1 - 0.25)} = 0.028$$

Now, we use this value of r to predict the growth rate for the current population of about 6 billion:

$$\text{growth rate} = 0.028 \times \left(1 - \frac{6 \text{ billion}}{12 \text{ billion}}\right) = 0.014$$

This logistic model predicts a current growth rate of 1.4%, which matches the actual current growth rate. Thus, based on the given data from the 1960s, it is reasonable to say that human population has been growing logistically since the 1960s. However, human population has not been following logistic growth over longer periods. Logistic growth requires a continually decreasing growth rate, which is inconsistent with the growth rate peaking in the 1960s. In conclusion, it is still too early to know whether the growth rate will continue to decline as population approaches 12 billion or whether we are simply experiencing a temporary change in the exponential growth rate. *Now try Exercises 21–22.* ◀

Overshoot and Collapse

A logistic model assumes that the growth rate automatically adjusts as the population approaches the carrying capacity. However, because of the astonishing rate of exponential growth, real populations often increase beyond the carrying capacity in a relatively short period of time. This phenomenon is called **overshoot.**

When a population overshoots the carrying capacity of its environment, a decrease in the population is inevitable. If the overshoot is substantial, the decrease can be rapid and severe—a phenomenon known as **collapse.** Figure 8.5 contrasts a logistic growth model with overshoot and collapse.

Time out to think _____

The concept of carrying capacity can be applied to any localized environment. Consider the decline of past civilizations such as the ancient Greeks, Romans, Mayans, and Anasazi. Does an overshoot and collapse model describe the fall of any of these or other civilizations? Explain.

by the way

Overshoot and collapse characterizes many predator–prey populations. The population of a predator increases rapidly, causing the prey population to collapse. Once the prey population collapses, the predator population must also collapse because of lack of food. Once the predator population collapses, the prey population can begin to recover—as long as it has not gone extinct.

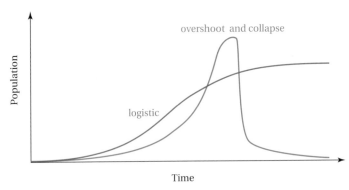

FIGURE 8.5 This graph contrasts logistic growth with overshoot and collapse.

What Is the Carrying Capacity?

Given that human population cannot grow exponentially forever, logistic growth is clearly preferable to any kind of overshoot and collapse. Logistic growth means a sustainable future population, while overshoot and collapse might mean the end of our civilization.

Practical matters

CHOOSING OUR FATE

If human population continued growing at its current rate, humanity would be elbow to elbow on Earth in only a few centuries. Thus, like the growth of bacteria in a bottle (Unit 8A), the exponential growth of human population *will* soon stop. The only questions are *when* and *how*.

First consider the question of *when*. The highest estimates put the Earth's carrying capacity around 40 billion, which we would reach within 150 years at the current growth rate. Thus, regardless of other assumptions, population growth must stop or slow considerably within the next 150 years—which is quite soon on the scale of human history.

As for *how*, there are only two basic ways to slow the growth of a population:

- a decrease in the birth rate or
- an increase in the death rate.

Most people are already choosing the first option, as birth rates now are at historic lows. Indeed, population actually is *decreasing* in a few European nations. Nevertheless, worldwide birth rates still are much higher than death rates, so exponential growth continues.

If a decrease in birth rates doesn't halt the exponential growth, an increase in the death rate will. If population significantly overshoots the carrying capacity by the time this process begins, the increase in the death rate will be dramatic—probably on a scale never before seen. This forecast is not a threat, a warning, or a prophecy of doom. It is simply a law of nature: *Exponential growth always stops*.

As human beings, we can choose to slow our population growth through intelligent and careful decisions. Or, we can choose to do nothing, leaving ourselves to the mercy of natural forces over which we have no more control than we do over hurricanes, tornadoes, earthquakes, or the explosions of distant stars. Either way, it's a choice that each and every one of us must make, and upon which our entire future depends.

The most fundamental question about population growth therefore concerns the carrying capacity. If the carrying capacity is well above the current population, then we have plenty of time to settle into logistic growth and long-term population stability. On the other hand, if the current population is near the carrying capacity, we may need to act very quickly to prevent overshoot and collapse. Unfortunately, any estimate of carrying capacity is subject to great uncertainty, for at least four important reasons:

- The carrying capacity depends on consumption of resources such as energy. However, different countries consume at very different rates. For example, the carrying capacity is much lower if we assume that the growing population will consume energy at the U.S. average rate rather than the Japanese average rate (which is about half the U.S. rate).
- The carrying capacity depends on assumptions about the environmental impact of the average person. A larger average impact on the environment means a lower carrying capacity.
- The carrying capacity can change with both human technology and the environment. For example, estimates of carrying capacity typically consider the availability of fresh water. However, if we can develop new sources of energy (such as fusion), nearly unlimited amounts of fresh water may be obtained through the desalinization of seawater. Conversely, global warming might alter the environment and reduce our ability to grow food, thereby lowering the carrying capacity.
- Even if we could account for the many individual factors in the carrying capacity (such as food production, energy, and pollution), the Earth is such a complex system that precisely predicting the carrying capacity may well be impossible. For example, no one can predict whether or how much the loss of rain forest species affects the carrying capacity.

The history of attempts to guess the carrying capacity of the Earth is full of missed predictions. Among the most famous was that made by English economist Thomas Malthus (1766–1834). In a 1798 paper entitled *An Essay on the Principle of Population as It Affects the Future Improvement of Society,* Malthus argued that food production would not be able to keep up with the rapidly growing populations of Europe and America. He concluded that mass starvation would soon hit these continents. His prediction did not come true, primarily because advances in technology *did* allow food production to keep pace with population growth.

The power of population is indefinitely greater than the power in the Earth to produce subsistence for man.

—THOMAS MALTHUS

Time out to think

Some people argue that while Malthus's immediate predictions didn't come true, his overall point is still valid and human population ultimately faces a disaster due to growing population. What do you think? Defend your opinion.

CASE STUDY The Population of Egypt

Over long periods of time, real population growth patterns tend to be quite complex. Sometimes the growth may look exponential, while at other times it may appear logistic or like overshoot and collapse. It may even appear to be some combination of these possibilities all at once.

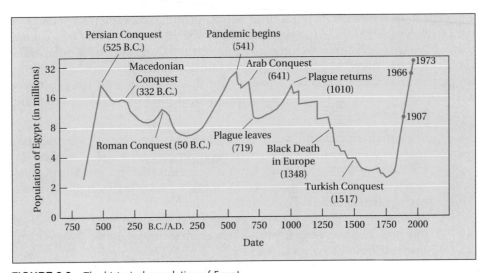

FIGURE 8.6 The historical population of Egypt.

SOURCE: *T. H. Hollingsworth,* Historical Demography *(Ithaca, NY: Cornell University Press, 1969).*

One of the few cases for which long-term population data are available is that of Egypt. Figure 8.6 shows these data, along with a few historical events that affected the population. (The graph uses an exponential vertical scale, with each tick mark double the previous one.) Note the complexity of the pattern. Even with the best mathematical models, it is hard to imagine that scientists in ancient Egypt could have predicted the future population of the region over a period of even a hundred years, let alone several thousand years.

This example offers an important lesson about mathematical models. They are useful for gaining insight into the processes being modeled. However, mathematical models can be used to *predict* future changes only when the processes are relatively simple. For example, it is easy to use mathematical modeling to predict the course of a spaceship because the law of gravity is relatively simple. But the growth of human population is such a complex phenomenon that we have little hope of ever being able to predict it reliably.

◾ EXERCISES 8C

Review Questions

1. Based on Figure 8.3, contrast the changes in human population for the 10,000 years preceding A.D. 1 and the 2000 years since. What has happened over the past few centuries?

2. Briefly describe how the overall growth rate is related to birth and death rates.

3. How do today's birth and death rates compare to those in the past? Why is human population growing?

4. What do we mean by carrying capacity? Why is it so difficult to determine the carrying capacity of the Earth?

5. What is logistic growth? Why would it be good if human population growth followed a logistic growth pattern in the future?

6. What is overshoot and collapse? Under what conditions does it occur? Why would it be a bad thing for the human race?

Does It Make Sense?

Decide whether each of the following statements makes sense (or is clearly true) or does not make sense (or is clearly false). Explain your reasoning.

7. Within the next decade, world population will grow by more than twice the current population of the United States.

8. If birth rates fall more than death rates, the growth rate of world population will fall.

9. The carrying capacity of our planet depends only on our planet's size.

10. Thanks to rapid increases in computing technology, we should be able to pin down the carrying capacity of the Earth to a precise number within just a few years.

11. In the wild, we always expect the population of any animal species to follow a logistic growth pattern.

12. Past history gives us strong reason to believe that human population is following a logistic growth pattern.

Basic Skills & Concepts

Varying Growth Rates. In Exercises 13–16, use the given growth rate to find the approximate doubling time and to predict world population in 2050 (based on a 2000 population of 6.0 billion).

13. Use the average annual growth rate between 1850 and 1950, which was about 0.9%.

14. Use the average annual growth rate between 1950 and 2000, which was about 1.8%.

15. Use the average annual growth rate between 1970 and 2000, which was about 1.6%.

16. Use the current annual growth rate of the United States, which is about 0.7%.

Birth and Death Rates. The following table gives the birth and death rates for four countries in three different years.

Country	Birth rate (per 1000)			Death rate (per 1000)		
	1975	1985	1995	1975	1985	1995
Czech Rep.	19.6	14.5	9.3	11.5	11.8	11.4
Israel	28.2	23.8	21.0	7.1	6.6	6.3
Sweden	12.8	11.8	11.7	10.8	11.3	11.0
U.S.	14.0	15.7	15.1	8.9	8.7	8.8

For each country given in Exercises 17–20, do the following:

a. Find the country's net growth rate due to births and deaths (i.e., neglect immigration) in 1975, 1985, and 1995.

b. Describe in words the general trends in the country's birth and death rates between 1975 and 1995. Based on these trends, predict how the country's population will change over the next 20 years. Do you think your prediction is reliable? Explain.

17. Czech Republic

18. Israel

19. Sweden

20. United States

21. **Logistic Growth.** Consider a population that begins growing exponentially at a base rate of 3.0% per year and then follows a logistic growth pattern. If the carrying capacity is 50 million, find the actual growth rate when the population is 10 million, 30 million, and 45 million.

22. **Logistic Growth.** Consider a population that begins growing exponentially at a base rate of 5.0% per year and then follows a logistic growth pattern. If the carrying capacity is 100 million, find the actual growth rate when the population is 10 million, 50 million, and 90 million.

Further Applications

U.S. Population. In Exercises 23–26, use the given growth rate to estimate U.S. population in 2050 and 2100. Start from the 2000 population of about 280 million. Use the approximate doubling time formula.

23. Use the current U.S. annual growth rate of 0.7%.

24. Use a growth rate of 0.5%.

25. Use a growth rate of 1.0%.

26. Use a growth rate of 0.4%.

27. **Population Growth in Your Lifetime.** Starting from the 6 billion population in 2000, assume that world population maintains its current annual growth rate of 1.4%.

What will be the world population when you are 50 years old? 80 years old? 100 years old?

28. Slower Growth. Repeat Exercise 27, but for a growth rate of 1% rather than 1.4%.

Carrying Capacity. For the carrying capacities in Exercises 29–32, use the 1960s peak annual growth rate of 2.1% and population of 3 billion to predict the current growth rate with a logistic model. Assume a current world population of 6 billion. How do the predicted growth rates compare to the actual growth rate of about 1.4% per year? Is the given carrying capacity too high or too low? Explain.

29. Assume the carrying capacity is 8 billion.

30. Assume the carrying capacity is 10 billion.

31. Assume the carrying capacity is 15 billion.

32. Assume the carrying capacity is 20 billion.

33. Growth Control Mediation. A city with a 1990 population of 100,000 has a growth control policy that limits the increase in residents to 2% per year. Naturally, this policy causes a great deal of dispute. On one side, some people argue that *growth* costs the city its small-town charm and clean environment. On the other side, some people argue that *growth control* costs the city jobs and drives up housing prices. Finding their work limited by the policy, developers suggest a compromise of raising the allowed growth rate to 5% per year. Contrast the populations of this city in 2000, 2010, and 2050 for 2% annual growth and 5% annual growth. Use the approximate doubling formula. If you were asked to mediate the dispute between growth control advocates and opponents, explain the strategy you would use.

Web Projects

Find useful links for Web Projects on the text Web site: www.aw.com/bennett-briggs

34. Population Predictions. Find population predictions from an organization that studies population, such as the United Nations or the U.S. Census Bureau. Read about how the predictions are made. Write a short summary of the methods used to predict future population. Be sure to discuss the uncertainties in the predictions.

35. Carrying Capacity. Find several different opinions concerning the Earth's human population carrying capacity. Based on your research, draw some conclusions about whether overpopulation presents an immediate threat. Write a short essay detailing the results of your research and clearly explaining your conclusions.

36. U.S. Population Growth. Research population growth in the United States to determine the relative proportions of the growth resulting from birth rates and from immigration. Then research both the problems and the benefits of the growing U.S. population. Form your own opinions about whether the United States has a population problem. Write an essay covering the results of your research and stating and defending your opinions.

37. Thomas Malthus. Find more information about Thomas Malthus and his famous predictions about population. Write a short paper on either his personal biography or his work.

38. Extinction. The extinction of a species could be a case of overshoot and collapse. Choose an endangered species and research why it is in decline. Learn how human activity may be changing the carrying capacity for the species. Write a short summary of your findings.

In the News

39. Population Growth. Find a recent news story that concerns population growth. Does the story consider the long-term effects of the growth? If so, do you agree with the claims? If not, discuss a few possible effects of future growth.

40. Immigration. Within the United States, immigration is an important part of the overall population growth. Find a recent story that discusses either the pros or the cons of immigration. Discuss the story and its conclusions.

LOGARITHMIC SCALES: EARTHQUAKES, SOUNDS, AND ACIDS

ou've probably heard earthquake strength described in *magnitudes,* the loudness of sounds described in *decibels,* or the acidity of household cleansers described by *pH.* In all three cases, the measurement scale involves exponential growth, because successive numbers on the scale increase by the same relative amount. For example, a liquid with pH 5 is ten times more acidic than one with pH 6. In this unit, we will explore these three important scales. They are commonly called **logarithmic scales,** which makes sense if you remember that logarithms are powers (see the Brief Review on p. 505).

THE MAGNITUDE SCALE FOR EARTHQUAKES

Earthquakes are a fact of life for much of the world's population (Figure 8.7). In the United States, California and Alaska are most prone to earthquakes, although earthquakes can strike almost anywhere. Most earthquakes are so minor that they can hardly be felt, but severe earthquakes can kill tens of thousands of people. Table 8.4 lists the frequencies of earthquakes of various strengths according to standard categories defined by geologists.

Scientists measure earthquake strength with the earthquake **magnitude scale.** The magnitude is related to the energy released by the earthquake. But

> ### HISTORICAL NOTE
> The original magnitude scale was created by Charles Richter in 1935. This *Richter scale* measured the up and down motion of the ground during a quake. Magnitude 0 was defined as the smallest detectable quake, and each increase of 1 magnitude corresponded to a factor of 10 increase in ground motion. Most earthquakes have nearly the same magnitude on the Richter scale and on the modern magnitude scale, but only the latter quantifies the actual energy released by the earthquake.

FIGURE 8.7 The distribution of earthquakes around the world. Each dot represents an earthquake.

SOURCE: From U.S. Geological Survey.

TABLE 8.4	Earthquake Categories and Their Frequency	
CATEGORY	MAGNITUDE	APPROXIMATE NUMBER PER YEAR (WORLDWIDE AVERAGE SINCE 1900)
Great	8 and up	1
Major	7–8	18
Strong	6–7	120
Moderate	5–6	800
Light	4–5	6000
Minor	3–4	50,000
Very minor	less than 3	magnitude 2–3: 1000 per day
		magnitude 1–2: 8000 per day

each magnitude represents about 32 times as much energy as the previous magnitude. For example, a magnitude 8 earthquake releases 32 times as much energy as a magnitude 7 earthquake. More technically, the magnitude scale is defined as follows.

The Earthquake Magnitude Scale

The magnitude scale for earthquakes is defined so that each magnitude represents about 32 times as much energy as the prior magnitude. More technically, the magnitude, M, is related to the released energy, E, by the following equivalent formulas:

$$\log_{10} E = 4.4 + 1.5M \qquad \text{or} \qquad E = (2.5 \times 10^4) \times 10^{1.5M}$$

The energy is measured in joules (see Unit 2B); magnitudes have no units.

Earthquakes of the same magnitude may cause vastly different amounts of damage depending on *how* their energy is released. Every earthquake releases some of its energy into the Earth's interior, where it is fairly harmless, and some along the Earth's surface, where it shakes the ground up and down. Thus, a moderate earthquake that releases most of its energy along the surface can do more damage than a strong earthquake that releases most of its energy into the interior.

Deaths from earthquakes generally arise indirectly. Ground shaking can cause buildings to collapse, can break gas lines that ignite in fire, and can trigger landslides or tsunamis ("tidal waves"). As a result, the greatest earthquake disasters tend to occur in regions where people cannot afford the high cost of earthquake-resistant construction.

Time out to think

Surface waves from earthquakes make the ground roll up and down like ripples moving outward on a pond. Given this motion, suggest a few ways that buildings can be designed to withstand earthquakes. Do you think it is possible to make a building that could withstand *any* earthquake? Why or why not?

▶ **EXAMPLE 1** The Meaning of One Magnitude

Using the formula for earthquake magnitudes, calculate precisely how much more energy is released for each 1 magnitude on the earthquake scale.

Solution We must look closely at the formula that gives the energy, *E:*

$$E = (2.5 \times 10^4) \times 10^{1.5M}$$

The first term, 2.5×10^4, is a constant number that is the same no matter what value we use for M. The magnitude appears only in the second term, $10^{1.5M}$. Each time we raise the magnitude by 1, such as from 5 to 6 or from 7 to 8, the total energy E increases by a factor of $10^{1.5}$. Thus, each successive magnitude represents $10^{1.5} = 31.623$ times as much energy as the prior magnitude. This is approximately 32 times as much energy, as stated earlier. *Now try Exercises 9–10.* ◀

▶ **EXAMPLE 2** Comparing Disasters

The 1989 San Francisco earthquake, in which 90 people were killed, had magnitude 7.1. Calculate the energy released, in joules. Compare the energy of this earthquake to the energy released in the 1998 Afghanistan earthquake, which had magnitude 6.9 and killed an estimated 5000 people.

Solution The energy released by the San Francisco earthquake was

$$E = (2.5 \times 10^4) \times 10^{1.5M} = (2.5 \times 10^4) \times 10^{1.5 \times 7.1} = 1.1 \times 10^{15} \text{ joules}$$

The San Francisco earthquake was $7.1 - 6.9 = 0.2$ magnitude greater than the Afghanistan earthquake. Thus, it released more energy by a factor of $10^{1.5 \times 0.2} = 10^{0.3} \approx 2$. That is, the San Francisco earthquake released twice as much energy as the Afghanistan earthquake. Nevertheless, the Afghanistan earthquake killed many more people, because more buildings collapsed.

Now try Exercises 11–14. ◀

MEASURING SOUNDS IN DECIBELS

The **decibel scale** is used to compare the loudness of sounds. It is defined so that a sound of 0 decibels, abbreviated 0 dB, represents the softest sound audible to the human ear. Table 8.5 lists the approximate loudness of some common sounds. The loudness of any other sound in decibels can be found with the following formula.

TECHNICAL NOTE

In absolute terms, the intensity of the softest audible sound is about 10^{-12} watt per square meter.

TABLE 8.5	Typical Sounds in Decibels	
DECIBELS	TIMES LOUDER THAN SOFTEST AUDIBLE SOUND	EXAMPLE
140	10^{14}	jet at 30 meters
120	10^{12}	strong risk of damage to human ear
100	10^{10}	siren at 30 meters
90	10^9	threshold of pain for human ear
80	10^8	busy street traffic
60	10^6	ordinary conversation
40	10^4	background noise in average home
20	10^2	whisper
10	10^1	rustle of leaves
0	1	threshold of human hearing
-10	0.1	inaudible sound

The Decibel Scale for Sound

The loudness of a sound in decibels is defined by the following equivalent formulas:

$$\text{loudness in dB} = 10 \log_{10}\left(\frac{\text{intensity of the sound}}{\text{intensity of softest audible sound}}\right)$$

or

$$\frac{\text{intensity of the sound}}{\text{intensity of softest audible sound}} = 10^{\frac{\text{loudness in dB}}{10}}$$

▶ **EXAMPLE 3** Computing Decibels

Suppose a sound is 100 times as intense as the softest audible sound. What is its loudness, in decibels?

Solution We are looking for the loudness in decibels, so we use the first form of the decibel scale formula:

$$\text{loudness in dB} = 10 \log_{10}\left(\frac{\text{intensity of the sound}}{\text{intensity of softest audible sound}}\right)$$

The ratio in parentheses is 100, because we are given that the sound is 100 times as intense as the softest audible sound. Thus, we find

$$\text{loudness in dB} = 10 \log_{10} 100 = 10 \times 2 = 20 \text{ dB}$$

A sound that is 100 times as intense as the softest possible sound has a loudness of 20 dB, which is equivalent to a whisper (see Table 8.5). *Now try Exercises 15–18.* ◄

►EXAMPLE 4 Sound Comparison

How much more intense is a 57-dB sound than a 23-dB sound?

Solution We can compare the loudness of any two sounds by working with the second form of the decibel scale formula. You should confirm for yourself that by dividing the intensity of Sound 1 by the intensity of Sound 2, we find

$$\frac{\text{intensity of Sound 1}}{\text{intensity of Sound 2}} = 10^{\frac{(\text{loudness of Sound 1 in dB}) - (\text{loudness of Sound 2 in dB})}{10}}$$

Substituting 57 dB for Sound 1 and 23 dB for Sound 2, we have

$$\frac{\text{intensity of Sound 1}}{\text{intensity of Sound 2}} = 10^{\frac{(\text{loudness of Sound 1 in dB}) - (\text{loudness of Sound 2 in dB})}{10}} = 10^{\frac{57-23}{10}} = 10^{3.4} = 2512$$

A sound of 57 dB is about 2500 times as intense as a sound of 23 dB.

Now try Exercises 19–20. ◄

The Inverse Square Law for Sound

You've probably noticed that sounds get weaker with distance. If you sit right in front of the speakers at an outdoor concert the sound may be almost deafening, while people a mile away may not hear the music at all. It's easy to understand why.

Figure 8.8 shows the idea. The sound from the speaker gets spread over a larger and larger *area* as it moves farther from the speaker. This area increases with the *square* of the distance from the speaker. For example, at 2 meters the sound is spread over an area $2^2 = 4$ times as large as the area at 1 meter, at 3 meters the sound is spread over an area $3^2 = 9$ times as large as the area at 1 meter, and so on. Thus the intensity of the sound weakens with the square of the distance.

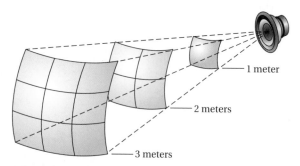

FIGURE 8.8 This figure shows why the intensity of sound decreases with the *square* of the distance from the source. Notice that the area at 2 meters is $2^2 = 4$ times as large as the area at 1 meter, and the area at 3 meters is $3^2 = 9$ times as large.

by the way

What we perceive as sound actually is tiny pressure changes in the air. Sound travels through the air as a wave of such pressure changes, called a sound wave. When a sound wave strikes the eardrum, its energy causes the eardrum to move in response. The brain analyzes the motions of the eardrum, perceiving sound.

ear drum

Because the intensity of sound decreases with the square of the distance, we say that sound follows an **inverse square law.** In other words, the intensity of a sound at a distance d from its source is proportional to $1/d^2$. Many other quantities also follow inverse square laws with distance, including the brightness of a light and the strength of gravity.

The Inverse Square Law for Sound

The intensity of sound decreases with the square of the distance from the source, meaning that the intensity is proportional to $1/d^2$. We therefore say that sound follows an **inverse square law** with distance.

▶**EXAMPLE 5** Sound Advice

How far should you be from a jet to avoid a strong risk of damage to your ear?

Solution Table 8.5 shows that the sound from a jet at a distance of 30 meters is 140 dB, and 120 dB is the level of sound that poses a strong risk of ear damage. Comparing these two sounds, we find that at 30 meters the jet sound is more intense than the sound needed to cause damage by a factor of

$$\frac{\text{intensity of 140 dB sound}}{\text{intensity of 120 dB sound}} = \frac{10^{\frac{140}{10}}}{10^{\frac{120}{10}}} = \frac{10^{14}}{10^{12}} = 10^2 = 100$$

Thus, to be safe, you should be far enough from the jet so that its sound intensity is 100 times less than it is at 30 meters. Because sound intensity follows an inverse square law with distance, moving 10 times farther away makes the intensity $10^2 = 100$ times weaker. Thus, you should be 10×30 m $= 300$ m from the jet to be safe.

Now try Exercises 21–24. ◀

Now try Exercises 21–24.

THE PH SCALE FOR ACIDITY

If you check the labels of many household products, including cleansers, drain openers, and shampoo, you will see that they state a quantity called the **pH.** The pH is used by chemists to classify substances as **neutral, acidic,** or **basic** (also called *alkaline*). By definition,

- Pure water is *neutral* and has a pH of 7.
- *Acids* have a pH lower than 7.
- *Bases* have a pH higher than 7.

Table 8.6 gives a few typical pH values.

Time out to think _____

Check around your house or apartment for labels that state a pH. Are the substances acids or bases?

Chemically, acidity is related to the concentration of positively charged hydrogen ions, which are hydrogen atoms without their electron. The ions themselves are

by the way

The word *acid* comes from the Latin *acidus*, meaning "sour." Some fruits taste sour because they are acidic. Bases are substances that can neutralize the action of acids. Common antacid tablets (for upset stomachs) are bases and work by neutralizing stomach acid.

TABLE 8.6		Typical pH Values	
SOLUTION	pH	SOLUTION	pH
pure water	7	drinking water	6.5–28
stomach acid	2–3	baking soda	8.4
vinegar	3	household ammonia	10
lemon juice	2	drain opener	10–12

by the way

Neutral hydrogen atoms consist of a single, positively charged proton and a single, negatively charged electron. H^+ ions are positive because they are missing the negatively charged electron and hence are simply protons.

denoted H^+ for hydrogen with a positive charge. The concentration of hydrogen ions is denoted $[H^+]$ and is usually measured in units of *moles per liter*. A **mole** is *Avogadro's number* of particles, which is about 6×10^{23} particles.

The pH Scale

The pH scale is defined by the following equivalent formulas:

$$pH = -\log_{10}[H^+] \quad \text{or} \quad [H^+] = 10^{-pH}$$

where $[H^+]$ is the hydrogen ion concentration in moles per liter. Pure water is neutral and has a pH of 7. Acids have a pH lower than 7 and bases have a pH higher than 7.

►**EXAMPLE 6** Finding pH

What is the pH of a solution with a hydrogen ion concentration of 10^{-12} mole per liter? Is it an acid or a base?

Solution Using the first version of the pH formula with a hydrogen ion concentration of $[H^+] = 10^{-12}$ mole per liter, we find

$$pH = -\log_{10}[H^+] = -\log_{10} 10^{-12} = -(-12) = 12$$

(Recall that $\log_{10} 10^x = x$.) A solution with a hydrogen ion concentration of 10^{-12} mole per liter has pH 12. Because this pH is well above 7, the solution is a strong base.

Now try Exercises 25–30. ◄

ACID RAIN

Normal raindrops are mildly acidic, with a pH slightly under 6. However, the burning of fossil fuels releases sulfur or nitrogen that can form sulfuric or nitric acids in the air, and these acids can make raindrops far more acidic than normal, creating the problem known as **acid rain.** Acid rain in the northeastern United States and *acid fog* in the Los Angeles area have been observed with pH as low as 2—the same acidity as pure lemon juice!

Acid rain can kill trees and other plants, doing serious damage to forests. Many forests in the northeastern United States and southeastern Canada have been damaged

by acid rain. Acid rain can also "kill" lakes by making the water so acidic that nothing can survive. Thousands of lakes in the northeastern United States and southeastern Canada have suffered this fate. Surprisingly, you can often recognize a dead lake by its exceptionally clear water—clear because it lacks the living organisms that usually make the water murky.

▶ EXAMPLE 7 Acid Rain versus Normal Rain

In terms of hydrogen ion concentration, how much more acidic is acid rain with a pH of 2 than ordinary rain with a pH of 6?

Solution For acid rain with pH 2, the hydrogen ion concentration is

$$[H^+] = 10^{-pH} = 10^{-2} \text{ mole per liter}$$

For ordinary rain with pH 6, the hydrogen ion concentration is

$$[H^+] = 10^{-pH} = 10^{-6} \text{ mole per liter}$$

Thus, the hydrogen ion concentration in the acid rain is greater than that in ordinary rain by a factor of

$$\frac{10^{-2}}{10^{-6}} = 10^{-2-(-6)} = 10^4$$

That is, the acid rain is 10,000 times as acidic as ordinary rain. *Now try Exercises 31–32.* ◀

■ EXERCISES 8D

Review Questions

1. What is the magnitude scale for earthquakes? What increase in energy is represented by an increase of 1 magnitude on the earthquake scale?

2. What is the decibel scale? Describe how it is defined.

3. What is pH? What pH values define an acid, a base, and a neutral substance?

4. What is acid rain? Why is it a serious environmental problem?

Does It Make Sense?

Decide whether each of the following statements make sense (or is clearly true) or does not make sense (or is clearly false). Explain your reasoning.

5. An earthquake of magnitude 8 will do twice as much damage as an earthquake of magnitude 4.

6. A 120-dB sound is 20% louder than 100-dB sound.

7. If I double the amount of water in the cup, I'll also double the pH of the water in the cup.

8. The lake water was crystal clear, so it could not possibly have been affected by acid rain.

Basic Skills & Concepts

Earthquake Magnitudes. Use the earthquake magnitude scale in Exercises 9–14.

9. How much energy, in joules, is released by an earthquake of magnitude 5?

10. How many times as much energy is released by an earthquake of magnitude 8 as by one of magnitude 6?

11. How much energy, in joules, was released by the 1985 earthquake in Mexico City (magnitude 8.1)?

12. How much energy, in joules, was released by the 1975 earthquake in Tangshan, China (magnitude 7.9), which killed an estimated half-million people?

13. Compare the energy of a magnitude 7 earthquake to that released by a 1-megaton nuclear bomb (5×10^{15} joules).

14. What magnitude earthquake would release energy equivalent to that of a 1-megaton nuclear bomb (5×10^{15} joules)? Which would be more destructive, the bomb or the earthquake? Why?

The Decibel Scale. Use the decibel scale in Exercises 15–20.

15. How many times as loud as the softest audible sound is the sound of busy street traffic?

16. How many times as loud as the softest audible sound is the sound of a siren at 30 meters?

17. What is the loudness, in decibels, of a sound 45 million times as loud as the softest audible sound?

18. What is the loudness, in decibels, of a sound 18 trillion times as loud as the softest audible sound?

19. How much louder (more intense) is a 35-dB sound than a 10-dB sound?

20. Suppose that a sound is 100 times as loud as (more intense than) a whisper. What is its loudness in decibels?

Inverse Square Law. Exercises 21–24 use the inverse square law for sound.

21. How does the intensity of sound from a concert speaker at a distance of 1 meter compare to the intensity at a distance of 5 meters?

22. How does the intensity of sound from a concert speaker at a distance of 1 meter compare to the intensity at a distance of 100 meters?

23. How does the intensity of sound from a concert speaker at a distance of 20 meters compare to the intensity at a distance of 40 meters?

24. How does the intensity of sound from a concert speaker at a distance of 20 meters compare to the intensity at a distance of 100 meters?

The pH Scale. Use the pH scale in Exercises 25–32.

25. If the pH of a solution increases by 2 (e.g., from 4 to 6), how much does the hydrogen ion concentration change? Does the change make the solution more acidic or more basic?

26. If the pH of a solution decreases by 1.5 (e.g., from 5.5 to 4), how much does the hydrogen ion concentration change? Does the change make the solution more acidic or more basic?

27. What is the hydrogen ion concentration of a solution with pH 8.5?

28. What is the hydrogen ion concentration of a solution with pH 3.5?

29. What is the pH of a solution with a hydrogen ion concentration of 0.1 mole per liter? Is this solution an acid or a base?

30. What is the pH of a solution with a hydrogen ion concentration of 10^{-9} mole per liter? Is this solution an acid or a base?

31. How much more acidic is acid rain with a pH of 3 than ordinary rain with a pH of 6?

32. How much more acidic is acid rain with a pH of 2.5 than ordinary rain with a pH of 6?

Further Applications

Logarithmic Thinking. Briefly describe, in words, the effects you would expect in the situations given in Exercises 33–38.

33. An earthquake of magnitude 2.8 strikes the Los Angeles area.

34. You have your ear against a new speaker when it emits a sound with an intensity of 160 dB.

35. A young child (too young to know better) finds and drinks from an open bottle of drain opener with pH 12.

36. An earthquake of magnitude 8.5 strikes the Tokyo area.

37. Your friend is calling to you from across the street in New York City, with a shout that registers 90 dB. Traffic is heavy, and several emergency vehicles are passing by with sirens.

38. A forest situated a few hundred miles from a coal-burning industrial area is subjected regularly to acid rain, with pH 4, for many years.

39. **Sound and Distance**
 a. The decibel level for busy street traffic in Table 8.5 is based on the assumption that you stand very close to

the noise source—say, 1 meter from the street. If your house is 100 meters from a busy street, how loud will the street noise be, in decibels?

b. At a distance of 10 meters from the speakers at a concert, the sound level is 135 dB. How far away should you sit to reduce the level to 120 dB?

c. Imagine that you are a spy in a restaurant. The conversation you want to hear is taking place in a booth across the room, about 8 meters away. The people are speaking softly, so they hear each other's voice at about 20 dB (they are sitting about 1 meter apart). How loud is the sound of their voices when it reaches your table? If you have a miniature amplifier in your ear and want to hear their voices at 60 dB, by what factor must you amplify their voices?

40. **Variation in Sound with Distance.** Suppose that a siren is placed 0.1 meter from your ear.

a. How many times as loud as a siren at 30 meters will the sound you hear be?

b. How loud will the siren next to your ear sound, in decibels?

c. How likely is it that this siren will cause damage to your eardrum? Explain.

41. **Toxic Dumping in Acidified Lakes.** Consider a situation in which acid rain has heavily polluted a lake to a level of pH 4. An unscrupulous chemical company dumps some acid into the lake illegally. Assume that the lake contains 100 million gallons of water and that the company dumps 100,000 gallons of acid with pH 2.

a. What is the hydrogen ion concentration, $[H^+]$, of the lake polluted by acid rain alone?

b. Suppose that the unpolluted lake, without acid rain, would have pH 7. If the lake were then polluted by company acid alone (no acid rain), what hydrogen ion concentration, $[H^+]$, would it have?

c. What is the hydrogen ion concentration, $[H^+]$, after the company dumps the acid into the acid rain–polluted lake (pH 4)? What is the new pH of the lake?

d. If the U.S. Environmental Protection Agency can test for changes in pH of only 0.1 or greater, could the company's pollution be detected?

Web Projects

Find useful links for Web Projects on the text Web site: www.aw.com/bennett-briggs

42. **Earthquakes.** Go to the U.S. Geological Survey Web site and learn more about earthquakes and how they cause damage. Write a short report on some aspect of earthquakes.

43. **Earthquake Disasters.** Find the death tolls for some of the worst earthquake disasters in history. How strongly do the death tolls correlate with the earthquake magnitudes? Discuss the factors that determine the devastation caused by an earthquake.

44. **Acid Rain.** Investigate the problem of acid rain in a region where it has been a particular problem, such as the northeastern United States, southeastern Canada, the Black Forest in Germany, eastern Europe, or China. Write a report on your findings. The report should include a description of the acidity of the rain, the source of the acidity, the damage being caused by the acid rain, and the status of efforts to alleviate this damage.

In the News

45. **Earthquakes.** Find a recent news story dealing with some aspect of earthquakes, such as their destructive power, attempts to predict them, or ways of building to withstand them. How does the magnitude of an earthquake affect the issue? Explain.

46. **Acid Rain.** Find a recent news story dealing with acid rain. Discuss the source of the rain and its environmental impacts.

CHAPTER 8 SUMMARY

UNIT	KEY TERMS	KEY IDEAS AND SKILLS
8A	linear growth exponential growth doublings	In **linear growth,** a quantity grows by the same *absolute* amount in each unit of time. In **exponential growth,** a quantity grows by the same *relative* amount in each unit of time. **Understand** the impact of doublings and why exponential growth cannot continue indefinitely.
8B	doubling time half-life rule of 70	**Doubling time calculations:** After time t, for an exponentially growing quantity with doubling time T_{double}, new value $=$ initial value $\times 2^{t/T_{double}}$ The **approximate doubling time formula** for a quantity growing exponentially at a rate of $P\%$ per time period: $$T_{double} \approx \frac{70}{P}$$ The **exact doubling time formula** for a quantity growing exponentially at a fractional rate r: $$T_{double} = \frac{\log_{10} 2}{\log_{10}(1 + r)}$$ **Half-life calculations:** After time t, for an exponentially decaying quantity with half-life T_{half}, new value $=$ initial value $\times \left(\frac{1}{2}\right)^{t/T_{half}}$ The **approximate half-life formula** for a quantity decaying exponentially at a rate of $P\%$ per time period: $$T_{half} \approx \frac{70}{P}$$ The **exact half-life formula** for a quantity decaying exponentially at a fractional rate (r must be negative): $$T_{half} = -\frac{\log_{10} 2}{\log_{10}(1 + r)}$$
8C	overall growth rate birth rate death rate carrying capacity logistic growth overshoot and collapse	**Logistic growth:** growth rate $= r \times \left(1 - \dfrac{population}{carrying\ capacity}\right)$ **Contrast** exponential growth, logistic growth, and overshoot and collapse. **Understand** factors affecting the carrying capacity.

(Continues on the next page)

| 8D | logarithmic scale
earthquake magnitude
decibel
inverse square law
pH
 acid
 base
 neutral
acid rain | **Earthquake magnitude scale:**
$\log_{10} E = 4.4 + 1.5M$ or $E = (2.5 \times 10^4) \times 10^{1.5M}$
Decibel scale:
loudness in dB $= 10 \log_{10}\left(\dfrac{\text{intensity of the sound}}{\text{intensity of softest audible sound}}\right)$
or
$\dfrac{\text{intensity of the sound}}{\text{intensity of the softest audible sound}} = 10^{\frac{\text{loudness in dB}}{10}}$
Inverse square law for sound: The intensity of a sound *decreases* with the *square* of the distance from the source.
pH scale:
$pH = -\log_{10}[H^+]$ or $[H^+] = 10^{-pH}$ |

Nothing endures but change.

—Heraclitus, c. 500 B.C.

CHAPTER 9

Modeling Our World

It may be impossible to predict the future, but we needn't go forward blindly. Analyzing relationships between variables can help us make educated guesses about the future and perhaps even give us some control over our fates. Mathematics is a tool that allows us to model relationships with enough precision to gain insight into how the world works and how it may change in the future. Just as we used equations to help understand financial options in Chapter 4, we can use mathematical ideas to model many other situations. In this chapter, we discuss the basic principles involved in using mathematics to model our world.

UNIT **9A** **Functions: The Building Blocks of Mathematical Models:** The building blocks of mathematical models are functions, which express mathematical relationships between two or more quantities. We will describe functions in words and pictures, and we will discuss how they are used in mathematical models.

UNIT **9B** **Linear Modeling:** We will write and use equations for linear functions, which describe straight-line graphs. Although these are the simplest of functions, they have many practical applications.

UNIT **9C** **Exponential Modeling:** Writing functions to describe exponential growth and decay allows us to build on the ideas introduced in Chapter 8. We will explore many applications of exponential functions, including models of population growth and techniques for radiometric dating.

FUNCTIONS: THE BUILDING BLOCKS OF MATHEMATICAL MODELS

A real office complex may not look exactly like the scale model used to design it, but the model helps the architects create the design. A road map doesn't look at all like real highways, but it serves as a model of a road system and can be extremely useful when you travel. The purpose of a **mathematical model** is similar to that of an architectural model or a road map: It represents something real and helps us understand it.

Mathematical models directly affect our lives. For example, mathematical models are used to ensure that new bridges won't collapse, to predict how economic changes will affect employment, and to learn why some years have more hurricanes than others. Even if you never need to create a mathematical model yourself, understanding the principles behind mathematical modeling is crucial to understanding current issues.

Mathematical models are based on relationships between quantities that can change, such as wind speed and stress on a bridge or worker productivity and unemployment. These relationships are described by mathematical tools called **functions.** In essence, functions are the building blocks of mathematical models. Some mathematical models consist of only a single function, which we can represent with a simple equation or graph. Other models, such as those used to study Earth's climate, may involve thousands of functions and require supercomputers for their analysis. But the basic idea of a function is the same in all cases. In this unit, we will explore the concept of a function.

LANGUAGE AND NOTATION OF FUNCTIONS

We have used functions several times in this book already, without calling them by name. For example, in Chapter 4, we saw exactly how the *balance* in a savings plan is related to the *interest rate* and the *monthly deposits*. In Chapter 8, we saw how *population* is related to *growth rate* and *time*. These relationships are functions because they tell us specifically how one quantity varies with respect to another quantity. We are now ready to explore the language and notation of functions.

Suppose we want to model the variation in temperature over the course of a day, based on the data in Table 9.1. The first step is to recognize that two quantities are involved in this model: *time* and *temperature*. Our goal is to express the relationship between time and temperature in the form of a function.

The quantities related by a function are called **variables** because they change, or *vary*. In this case, *temperature* is the **dependent variable** because it *depends on* the time of day. *Time* is the **independent variable** because time passes *independently* of the temperature. We say that the temperature varies *with respect to* time. We often write related variables in the form of an ordered pair, with the independent variable first and the dependent variable second:

(*time, temperature*)

She had not understood mathematics until he had explained to her that it was the symbolic language of relationships. "And relationships," he had told her, "contained the essential meaning of life."

—PEARL BUCK,
THE GODDESS ABIDES, PT. I

TABLE 9.1	Temperature Data for One Day		
TIME	TEMPERATURE	TIME	TEMPERATURE
6:00 a.m.	50°F	1:00 p.m.	73°F
7:00 a.m.	52°F	2:00 p.m.	73°F
8:00 a.m.	55°F	3:00 p.m.	70°F
9:00 a.m.	58°F	4:00 p.m.	68°F
10:00 a.m.	61°F	5:00 p.m.	65°F
11:00 a.m.	65°F	6:00 p.m.	61°F
12:00 noon	70°F		

We use special notation to represent functions. For example, we might use x to represent the independent variable and y to represent the dependent variable. Then we write $y = f(x)$, read "y is a function of x." This notation tells us that y is related to x by the function f. In the case of our (*time, temperature*) example, we let t represent time and T represent temperature. We write $T = f(t)$ to indicate that temperature is a function of time.

It may be helpful to think of a function as a box with two slots, one for input and one for output. A value of the independent variable can be put into the box through the input slot. The function inside the box "operates" on the input and produces the corresponding value of the dependent variable, which appears as output from the box.

Functions

A function describes how a **dependent variable** changes *with respect to* one or more **independent variables.** When there are only two variables, we often summarize them as an ordered pair with the independent variable first:

(*independent variable, dependent variable*)

We say that the dependent variable is *a function of* the independent variable. If x is the independent variable and y is the dependent variable, we write the function as

$$y = f(x)$$

HISTORICAL NOTE
Mathematicians worked with functions for centuries before developing a standard notation. In about 1670, the German philosopher Gottfried Leibniz used the notation $f(x)$ to denote a function. But it wasn't until about 1734, when the Swiss mathematician Leonhard Euler adopted the same notation, that $f(x)$ became widely accepted for expressing functions.

Many functions involve changes with respect to time. For example, the weight of a child and the Consumer Price Index both vary with respect to time. But not all functions involve time. For example, mortgage payments are a function of the interest rate, because the payment amount depends on the interest rate. Similarly, the gas mileage of a car is a function of the car's speed, because the mileage is different at different speeds.

Time out to think _____

From your everyday experiences, identify several pairs of variables that appear to be related and might be described by a function. Include at least one pair that does not involve time.

►EXAMPLE 1 Writing Functions

For each situation, express the given function in words. Write the two variables as an ordered pair and write the function with the notation $y = f(x)$.

 a. You are riding in a hot-air balloon. As the balloon rises, the surrounding atmospheric pressure decreases (causing your ears to pop).
 b. You're on a barge headed south down the Mississippi River. You notice that the width of the river changes as you travel southward with the current.

Solution

 a. The pressure *depends on* your altitude, so we say that the pressure changes *with respect to* altitude. *Pressure* is the dependent variable and *altitude* is the independent variable, so the ordered pair of variables is (*altitude, pressure*). If we let A stand for altitude and P stand for pressure, then we write the function as

$$P = f(A)$$

 b. The river width *depends on* your distance from the river's source, so we say that the river width changes *with respect to* the distance from the source. *River width* is the dependent variable and *distance from the source* is the independent variable, so the ordered pair of variables is (*distance from source, river width*). Letting d represent distance and w represent river width, we have the function

$$w = f(d)$$ *Now try Exercises 11–18.*

Time out to think _____

Does dependence imply causality? That is, do changes in the independent variable *cause* changes in the dependent variable? Give a few examples to make your case.

REPRESENTING FUNCTIONS

There are three basic ways to represent a function.

 1. We can represent a function with a **data table,** such as Table 9.1. A table provides detailed information, but can become unwieldy with large quantities of data.

2. We can draw a *picture,* or **graph,** of a function. A graph is easy to interpret and consolidates a great deal of information.

3. We can write a compact mathematical representation of a function in the form of an **equation.**

In the remainder of this unit, we explore the use of graphs to represent functions. We'll discuss equations in Unit 9B.

THE COORDINATE PLANE

The most common way to draw a graph is to use a **coordinate plane,** which is made by drawing two perpendicular number lines. Each number line is called an **axis** (plural: **axes**). Normally, numbers increase to the right on the horizontal axis and upward on the vertical axis. The intersection point of the two axes, where both number lines show the number zero, is called the **origin.** If we are working with general functions, the horizontal axis is called the **x-axis** and the vertical axis is called the **y-axis.**

Points in the coordinate plane are described by two **coordinates** (called an *ordered pair*). The **x-coordinate** gives the point's horizontal position relative to the ori-

gin. Points to the right of the origin have positive x-coordinates and points to the left of the origin have negative x-coordinates. The **y-coordinate** gives the point's vertical position relative to the origin. Points above the origin have positive y-coordinates and points below the origin have negative y-coordinates. We express the location of a particular point by writing its x- and y-coordinates in parentheses in the form (x, y). When working with functions, we always use the x-axis for the independent variable and the y-axis for the dependent variable.

Figure 9.1a shows a coordinate plane with several points identified by their coordinates. Note that the origin is the point $(0, 0)$. Figure 9.1b shows that the axes divide the coordinate plane into four **quadrants,** numbered counterclockwise starting from the upper right.

(a)

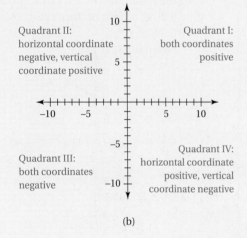

(b)

FIGURE 9.1

Now try Exercises 9–10.

Domain and Range

Before we draw a graph of any function, we must determine the values that we should show on each axis. Mathematically, each axis on a graph extends to infinity in both directions. However, most functions are meaningful only over a small region of the coordinate plane. In the case of the (*time, temperature*) function, based on the data in Table 9.1, negative values of time do not make sense. In fact, the only times of interest in this function are those during which data were being collected—from 6 a.m. to 6 p.m. The times that make sense and are of interest make up the **domain** of the function.

Similarly, the only temperatures of interest for this function are those that occurred between 6:00 a.m. and 6:00 p.m. The lowest temperature recorded in this period was 50°F and the highest was 73°F. We therefore say that temperatures between 50°F and 73°F make up the **range** of the function. More generally, we can make the following definitions, which are illustrated in Figure 9.2.

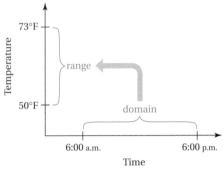

FIGURE 9.2 The values in the domain determine the values in the range.

Domain and Range

The **domain** of a function is the set of values that both make sense and are of interest for the independent variable.

The **range** of a function consists of the values of the dependent variable that correspond to the values in the domain.

Now that we've identified the domain and range, we can draw the graph. We use the horizontal axis for the time, t, and label it *hours after 6:00 a.m.* Thus, $t = 0$ corresponds to the first measurement at 6:00 a.m., and $t = 12$ corresponds to the last measurement at 6:00 p.m. We use the vertical axis for the temperature, T. Figure 9.3a shows the result, with each point plotted individually. To show details more clearly, Figure 9.3b zooms in on the region covered by the domain and the range.

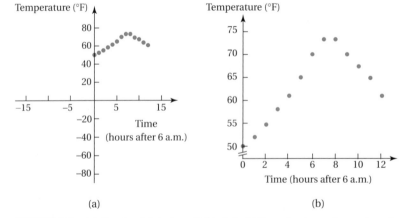

FIGURE 9.3 (a) Graph of data from Table 9.1. (b) Zooming in on the region of interest.

Completing the Model

So far, we have plotted only the thirteen data points from Table 9.1. However, every instant of time has a temperature, and the temperature changes continuously throughout the day. If we want the graph to give us a realistic *model*, we should fill in the gaps between the data points. Because we don't expect any sudden spikes or dips in temperature during the day, it is reasonable to connect the data points with a smooth curve, as shown in Figure 9.4.

The graph is now a model that we can use to predict the temperature at *any* time of day. For example, the model predicts that the temperature was about 67°F at 11:30 a.m. (5.5 hours after 6 a.m.). Keep in mind that this prediction may not be exact. We cannot even check it, because Table 9.1 does not provide data for 11:30 a.m. Nevertheless, the prediction seems reasonable, given our assumptions in drawing the smooth curve. This example illustrates an important lesson that applies to all mathematical models: *A model's predictions can be only as good as the data and the assumptions from which the model is built.*

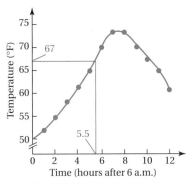

FIGURE 9.4 The graph from Figure 9.3b, but now with a smooth curve connecting the data points.

SUMMARY Creating and Using Graphs of Functions

Step 1: Identify the independent and dependent variables of the function.

Step 2: Identify the domain (values of the independent variable) and the range (values of the dependent variable) of the function. Use this information to choose the scale and labels on the axes. Zoom in on the region of interest to make the graph easier to read.

Step 3: Make a graph using the given data. If appropriate, fill in the gaps between data points.

Step 4: Before accepting any predictions of the model, be sure to evaluate the data and assumptions from which the model was built.

➤ **EXAMPLE 2** Pressure-Altitude Function

Imagine measuring the atmospheric pressure as you rise upward in a hot-air balloon. Table 9.2 shows typical values you might find for the pressure at different altitudes, with the pressure given in units of *inches of mercury.* (This pressure unit is used

TABLE 9.2	Typical Values of Pressure at Different Altitudes	
ALTITUDE (FT)		PRESSURE (IN. MERCURY)
0		30
5,000		25
10,000		22
20,000		16
30,000		10

with barometers, which measure pressure by the height of a column of mercury.) Use these data to graph a function showing how atmospheric pressure depends on altitude. Use the graph to predict the atmospheric pressure at an altitude of 15,000 feet, and discuss the validity of your prediction.

Solution In Example 1, we identified *altitude, A,* as the independent variable and *pressure, P,* as the dependent variable. The *domain* is the set of relevant values of the independent variable, *A*. For the altitude values in Table 9.2, the domain extends from 0 feet (sea level) to 30,000 feet. The *range* is the set of values of the dependent variable, *P*, that correspond to the domain. The range therefore extends from 10 to 30 inches of mercury.

We plot the five data points, as shown in Figure 9.5. Between any two data points, we can reasonably assume that pressure decreases smoothly with increasing altitude. Furthermore, because the Earth's atmosphere doesn't end abruptly, the pressure must decrease more gradually at higher altitudes. Thus, we complete the model by adding a smooth curve to the graph.

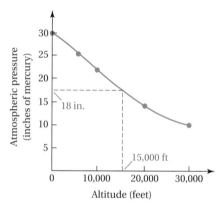

FIGURE 9.5 Pressure-altitude function.

Using this graph, we predict that the atmospheric pressure at 15,000 feet is about 18 inches of mercury. Because we've sketched the function only roughly, we should expect this prediction to be approximately, but not exactly, correct.

Now try Exercise 19. ◄

► **EXAMPLE 3 Hours of Daylight**

The number of hours of daylight varies with the seasons. Use the following data for 40°N latitude (the latitude of San Francisco, Denver, and Washington, D.C.) to model the change in the number of daylight hours with time.

- The number of hours of daylight is greatest on the *summer solstice* (about June 21), when it is about 14 hours for latitude 40°N.
- The number of hours of daylight is smallest on the *winter solstice* (about December 21), when it is about 10 hours for latitude 40°N.
- On the spring and fall *equinoxes* (about March 21 and September 21, respectively), there are about 12 hours of daylight.

According to the model, at what times of year does the number of daylight hours change most gradually? most quickly? Discuss the validity of the model.

Solution We expect the number of hours of daylight to be a function of the time of year. *Time* is the independent variable, because time marches on regardless of other events; we will denote it by *t*. *Hours of daylight* is the dependent variable, because it depends on the time of year; we will denote it by *h*.

The times of interest are all days of the year. Thus, the *domain* is the 365 days of a single year. The *range* extends from 10 hours to 14 hours of daylight. We know from experience that the number of hours of daylight changes smoothly with the seasons, so we can connect the four given data points for each year with a smooth curve. Because the same pattern repeats from one year to the next, we can extend the graph for additional years (Figure 9.6). This type of function, in which a particular pattern repeats over and over, is called a **periodic function.** Because this function is based on simple seasonal patterns, we can expect it to be quite accurate.

by the way

The seasons arise because of the tilt of the Earth's axis. The northern and southern hemispheres alternately get more and less direct sunlight as the Earth orbits the Sun, so the seasons are opposite in the two hemispheres.

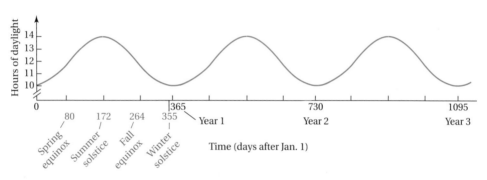

FIGURE 9.6 Hours of daylight over a three-year period, for latitude 40°N.

Note that the curve "tops out" at each summer solstice and "bottoms out" at each winter solstice. Because the curve is relatively flat around the two solstices, the number of hours of daylight varies slowly near the solstices. That is, we have a couple of months with long daylight hours around the summer solstice and a couple of months with short daylight hours around the winter solstice. In contrast, the number of hours of daylight increases rapidly around the time of each spring equinox and decreases rapidly around the time of each fall equinox. In other words, the length of day changes rapidly near the equinoxes. You can observe these facts easily if you pay attention to the changing number of daylight hours over the course of a year.

<u>*Now try Exercise 20.*</u> ◀

That flower of modern mathematical thought, the notion of a function.

—Thomas McCormack

Time out to think

What is the current date? Based on the function shown in Figure 9.6, should the number of hours of daylight be changing rapidly or slowly at this time of year? Try to notice the change from one day to the next, and confirm that it matches the prediction of our model.

EXERCISES 9A

Review Questions

1. What is a mathematical model? Explain this statement: *A model's predictions can be only as good as the data and the assumptions from which the model is built.*

2. What is a function? How do you decide which variable is the independent variable and which is the dependent variable?

3. What are the three basic ways to represent a function?

4. Define *domain* and *range,* and explain how to determine them for a particular function.

Does It Make Sense?

Decide whether each of the following statements makes sense (or is clearly true) or does not make sense (or is clearly false). Explain your reasoning.

5. Scientists at the National Center for Atmospheric Research use mathematical models to learn about the Earth's climate.

6. The demand for concert tickets is a function of their price.

7. I graphed a function showing how my heart rate depends on my running speed. The domain was heart rates from 60 to 180 beats per minute.

8. My mathematical model fits the data perfectly, so I can be confident it will work equally well in any new situations we encounter.

Basic Skills & Concepts

Coordinate Plane Review. Exercises 9–10 use the skills covered in the Brief Review on p. 000.

9. Draw a set of axes in the coordinate plane. Plot and label the following points: $(0, 1)$, $(-2, 0)$, $(1, 5)$, $(-3, 4)$, $(5, -2)$, $(-6, -3)$.

10. Draw a set of axes in the coordinate plane. Plot and label the following points: $(0, -1)$, $(2, -1)$, $(6, 5)$, $(3, -4)$, $(-5, -2)$, $(-6, 2)$.

Related Quantities. For Exercises 11–18, write a short statement that expresses a possible relationship between the given variables.

Example: (age, shoe size).

Solution: As a child ages, shoe size increases. Once the child is full-grown, shoe size remains constant.

11. (*weight of a bag of apples, price of the bag of apples*)

12. (*time, price of movie admission*), where time represents years from 1960 to 2000

13. (*price of a product, demand for a product*)

14. (*distance from Earth, strength of gravity*)

15. (*slope of hill, speed of skateboard*)

16. (*rate of pedaling, speed of bicycle*)

17. (*tax rate, revenue collected*)

18. (*number of cars on road, air quality*)

Further Applications

19. **Pressure Function.** Study Figure 9.5.

 a. Use the graph to estimate the pressure at altitudes of 8000 feet, 17,000 feet, and 25,000 feet.

 b. Use the graph to estimate the altitudes at which the pressure is 27, 20, and 12 inches of mercury.

 c. Estimating beyond the boundaries of the graph, at what altitude do you think the atmospheric pressure reaches 5 inches of mercury? Is there an altitude at which the pressure is exactly zero? Explain your reasoning.

20. **Daylight Function.** Study Figure 9.6, which applies to 40°N latitude.

 a. Use the graph to estimate the number of hours of daylight on April 1 (the 91st day of the year) and October 31 (the 304th day of the year).

 b. Use the graph to estimate the dates on which there are 13 hours of daylight.

 c. Use the graph to estimate the dates on which there are 10.5 hours of daylight.

 d. The graph in Figure 9.6 is valid at 40°N latitude. How do you think the graph would be different at 20°N latitude, 60°N latitude, and 40°S latitude? Why?

Functions from Graphs. Each graph in Exercises 21–22 represents a function. In each case, do the following.

 a. Identify the independent and dependent variables, and describe the domain and range.

 b. Describe the function in words.

21.

22.

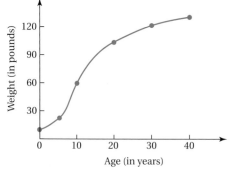

Functions from Data Tables. Each of the data tables in Exercises 23–26 represents a function. In each case, do the following.

a. Identify the independent and dependent variables, and describe the domain and range.

b. Make a clear graph of the function. Explain how you decide on the shape of the curve used to fill in the gaps between the data points.

c. Describe the function in words.

23.

Date	Average high temperature
Jan. 1	42°F
Feb. 1	38°F
Mar. 1	48°F
Apr. 1	58°F
May 1	69°F
June 1	76°F
July 1	85°F
Aug. 1	83°F
Sep. 1	80°F
Oct. 1	69°F
Nov. 1	55°F
Dec. 1	48°F
Dec. 31	44°F

24.

Altitude (ft)	Boiling point of water (°F)
0	212.0
1000	210.2
2000	208.4
3000	206.6
4000	204.8
5000	203.0
6000	201.0
7000	199.3
8000	195.5
9000	193.6

25. Notice that the intervals between data points are not the same in all cases.

Year	Tobacco produced (billions of lb)	Year	Tobacco produced (billions of lb)
1975	2.2	1986	1.2
1980	1.8	1987	1.2
1982	2.0	1988	1.4
1984	1.7	1989	1.4
1985	1.5	1990	1.6

26.

Year	Projected U.S. population (millions)
2005	288
2010	300
2015	312
2020	325
2025	338
2030	351

Rough Sketches of Functions. For each function in Exercises 27–38, use your intuition or additional research, if necessary, to do the following.

a. Describe an appropriate domain and range for the function.

b. Make a rough sketch of a graph of the function. Explain the assumptions that go into your sketch.

c. Briefly discuss the validity of your graph as a model of the true function.

27. (*altitude, temperature*) when climbing a mountain

28. (*day of year, high temperature*) over a two-year period for the town in which you are living

29. (*blood alcohol content, reflex time*) for a single person

30. (*number of pages in a book, time to read the book*) for a single person

31. (*time of day, traffic flow*) at a busy intersection over a full day

32. (*price of gasoline, number of tourists in Yellowstone*)

33. (*time, world record in the 100-meter dash*) between 1970 and 2004

34. (*minutes after lighting, length of candle*)

35. (*time, population of China*), where time is measured in years after 1900

36. (*time of day, elevation of tide*) at a particular seaside port over two days

37. (*angle of cannon, horizontal distance traveled by cannonball*)

38. (*weight of car, average gas mileage*)

Web Projects

Find useful links for Web Projects on the text Web site: www.aw.com/bennett-briggs

39. **Daylight Hours.** Investigate Web sites that give the length of day (hours of daylight) on various days throughout the year for different latitudes (a table of sunrise and sunset times would also work). Make graphs similar to Figure 9.6 showing the variation of hours of daylight over a year for several different latitudes.

40. **Variable Tables.** Find data on the Web for two variables that are clearly related in some way. Make a table (between 10 and 20 entries) of data values. Graph the data and describe in words the function that relates the variables. Hint: Some possible variable pairs are (*time, population of a city*), (*team batting average, average team salary*) for major league baseball teams, and (*blood alcohol content, reaction time*) for a study of effects of alcohol.

In the News

41. **Everyday Models.** Describe three different models (mathematical or otherwise) that you use or encounter frequently in everyday life. What is the underlying "reality" that those models represent? What simplifications are made in constructing those models?

42. **Functions and Variables.** Identify three different variables in recent news stories. For each variable, specify another related variable and then write a paragraph that describes a function relating the two variables. At least one of your three functions should *not* use time as the independent variable.

LINEAR MODELING

In Unit 9A, we represented functions with tables and graphs. We now turn our attention to a more common and versatile way of representing functions: with equations. Although equations are more abstract than pictures, they are easier to manipulate mathematically. Thus, they give us greater power when creating and analyzing mathematical models. We can understand the basic principles involved in mathematical modeling by focusing on the simplest functions: those with straight-line graphs, called *linear functions*.

LINEAR FUNCTIONS

Imagine that we measure the depth of rain accumulating in a rain gauge as a steady rain falls (Figure 9.7a). The rain stops after 6 hours, and we want to describe how the rain depth varied with time during the storm. In this situation, *time* is the independent

FIGURE 9.7 (a) A rain gauge. (b) Graph of a function showing how rain depth varies with time during a storm.

variable and *rain depth* is the dependent variable. Suppose that, based on our measurements with the rain gauge, we find the rain depth function shown in Figure 9.7b. Because the graph is a straight line, we are dealing with a **linear function.**

Rate of Change

The graph shows that, during the storm, the rain depth increased by 1 inch each hour. We say that the **rate of change** of the rain depth with respect to time was 1 inch per hour, or 1 in./hr. This rate of change was constant throughout the storm: No matter which hour we choose to study, the rain depth increased by 1 inch. Thus, we see a key fact about linear functions: *A linear function has a constant rate of change and a straight-line graph.*

Figure 9.8 shows graphs for three other steady rainstorms. The constant rate of change is 0.5 in./hr in Figure 9.8a, 1.5 in./hr in Figure 9.8b, and 2 in./hr in Figure 9.8c.

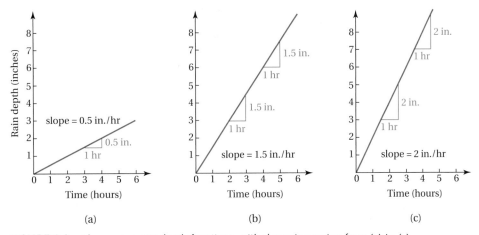

FIGURE 9.8 Three more rain depth functions, with slopes increasing from (a) to (c).

Comparing the three graphs in Figure 9.8 leads to another crucial observation: *The greater the rate of change, the steeper the graph.*

The small triangles on the graphs show the **slope** of each line, defined as the amount that the graph *rises* vertically for a given distance that it *runs* horizontally. That is, *the slope is the rise over the run.* More importantly, Figure 9.8 also shows that the slope is equal to the rate of change.

Linear Functions

A **linear function** has a constant rate of change and a straight-line graph. For all linear functions,

- The rate of change is equal to the slope of the graph.
- The greater the rate of change, the steeper the slope.
- We can calculate the rate of change by finding the slope between any two points on the graph (Figure 9.9):

$$\text{rate of change} = \text{slope} = \frac{\text{change in } \textit{dependent variable}}{\text{change in } \textit{independent variable}}$$

FIGURE 9.9 To find the slope of a straight line, we can look at any two points and divide the change in the dependent variable by the change in the independent variable.

►EXAMPLE 1 Drawing a Linear Model

You hike a 3-mile trail, starting at an elevation of 8000 feet. Along the way, the trail gains elevation at a constant rate of 650 feet per mile. What is the domain for the elevation function? Draw a graph of the linear function that gives your elevation as you hike along the trail. Does this model seem realistic?

Solution Because your elevation depends on the distance you've walked, *distance* is the independent variable and *elevation* is the dependent variable. The domain is 0 to 3 miles, which represents the length of the trail. We are given one data point: (0 mi,

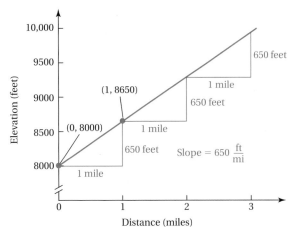

FIGURE 9.10 Linear function for Example 1.

8000 ft) represents the 8000-foot elevation at the start of the trail. We are also given that the rate of change of elevation with respect to distance is 650 feet per mile. Thus, a second point on the graph is (1 mi, 8650 ft). We draw the graph by connecting these two points with a straight line and extending the line over the domain from 0 to 3 miles (Figure 9.10). As we expect, the rate of change is the slope of the graph.

This model assumes that elevation increases at a constant rate along the entire 3-mile trail. While an elevation change of 650 feet per mile seems reasonable as an average, the actual rate of change probably varies from point to point along the trail. Thus, the model's predictions are likely to be reasonable *estimates,* rather than exact values, of your elevation at different points along the trail. *Now try Exercises 11–12.* ◄

▶ **EXAMPLE 2** A Price-Demand Function

A small store sells fresh pineapples. Based on data for pineapple prices between $2 and $7, the storeowners created a model in which a linear function is used to describe how the demand (number of pineapples sold per day) varies with the price (Figure 9.11). For example, the point ($2, 80 pineapples) means that, at a

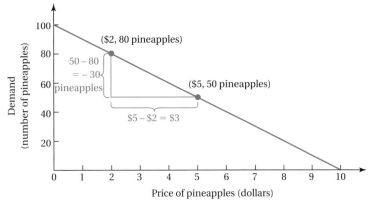

FIGURE 9.11 Linear function for Example 2.

price of $2 per pineapple, 80 pineapples can be sold on an average day. What is the rate of change for this function? Discuss the validity of this model.

Solution The rate of change of the demand function is the slope of its graph. We identify *price* as the independent variable and *demand* as the dependent variable. We can calculate the slope using any two points on the graph. Let's choose Point 1 as ($2, 80 pineapples) and Point 2 as ($5, 50 pineapples). The change in *price* between the two points is $5 − $2 = $3. The change in *demand* between the two points is

$$50 \text{ pineapples} - 80 \text{ pineapples} = -30 \text{ pineapples}$$

The change in demand is negative because demand *decreases* from Point 1 to Point 2. The rate of change is

$$\text{rate of change} = \frac{\text{change in demand}}{\text{change in price}} = \frac{-30 \text{ pineapples}}{\$3} = \frac{-10 \text{ pineapples}}{\$1}$$

The rate of change of the demand function is −10 pineapples per dollar. Thus, for every dollar that the price increases, the number of pineapples sold decreases by 10.

This model seems reasonable within the domain for which the storeowners gathered data: between prices of $2 and $7. Outside this domain, the model's predictions probably are not valid. For example, the model predicts that the store could sell one pineapple per day at a price of $9.90, but could *never* sell a pineapple at a price of $10. On the other extreme, the model predicts that the store could "sell" only 100 pineapples if they were free! As with many models, this price-demand model is useful only in a limited domain. *Now try Exercises 13–16.* ◄

The Change in the Dependent Variable

Consider again the rain depth function in Figure 9.7. Suppose we want to know how much the rain depth changes in a 4-hour period. Because the rate of change for this function is 1 in./hr, the total change after 4 hours is

$$\text{change in rain depth} = \underbrace{1 \frac{\text{in.}}{\text{hr}}}_{\text{rate of change}} \times \underbrace{4 \text{ hr}}_{\text{elapsed time}} = 4 \text{ in.}$$

Notice how the units work out. Note also that the *elapsed time* is the change in the independent variable and the change in rain depth is the change in the dependent variable. We can generalize this idea to other functions.

The Rate of Change Rule

The rate of change rule allows us to calculate the change in the dependent variable from the change in the independent variable:

$$\text{Change in dependent variable} = \left(\begin{array}{c} \text{rate of} \\ \text{change} \end{array} \right) \times \left(\begin{array}{c} \text{change in} \\ \text{independent variable} \end{array} \right)$$

➤ **EXAMPLE 3** Change in Demand

Using the linear demand function in Figure 9.11, predict the change in demand for pineapples if the price increases by $3.

Solution The independent variable is the *price* of the pineapples, and the dependent variable is the *demand* for pineapples. In Example 2, we found that the rate of change of demand with respect to price is –10 pineapples per dollar. Thus, the change in demand for a price increase of $3 is

$$\text{change in } demand = \text{rate of change} \times \text{change in } price$$

$$= -10 \, \frac{\text{pineapples}}{\cancel{\$}} \times \cancel{\$}3$$

$$= -30 \text{ pineapples}$$

This model predicts that a $3 price increase will lead to 30 *fewer* pineapples being sold per day. *Now try Exercises 17–22.* ◄

GENERAL EQUATION FOR A LINEAR FUNCTION

Suppose your job is to oversee an automated assembly line that manufactures computer chips. You arrive at work one day to find a stock of 25 chips that were produced during the night. If chips are produced at a constant rate of 4 chips per hour, how large is the stock of chips at any particular time during your shift?

Answering this question requires finding a function that describes how the number of chips depends on the time of day. We identify *time,* which we'll denote by *t,* as the independent variable. *Number of chips,* which we'll denote by *N,* is the dependent variable. At the start of your shift, $t = 0$ and your initial stock is $N = 25$ chips. Because the stock grows by 4 chips every hour, the *rate of change* of this function is 4 chips per hour. We construct a graph by starting at the initial point (0 hr, 25 chips) and drawing a straight line with a slope of 4 chips per hour (Figure 9.12).

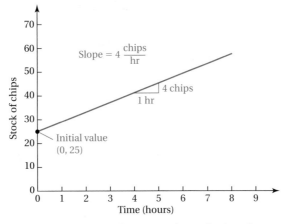

FIGURE 9.12 Linear function with initial value of 25 chips and slope of 4 chips/hr.

She knew only that if she did or said thus-and-so, men would unerringly respond with the complimentary thus-and-so. It was like a mathematical formula and no more difficult, for mathematics was the one subject that had come easy to Scarlett in her schooldays.

—MARGARET MITCHELL, GONE WITH THE WIND

We can now write an equation for the function. First, let's describe the stock of chips at any particular time with a word equation:

number of chips = initial number of chips + change in number of chips

For the change in the number of chips, we have

$$\underbrace{\text{change in } \textit{number of chips}}_{\text{change in dependent variable}} = \underbrace{4 \, \frac{\text{chips}}{\text{hr}}}_{\text{rate of change}} \times \underbrace{\textit{elapsed time}}_{\text{change in independent variable}}$$

Using this result and 25 for the initial number of chips, we get the following equation for the number of chips:

$$\textit{number of chips} = 25 \text{ chips} + \left(4 \, \frac{\text{chips}}{\text{hr}} \times \textit{elapsed time} \right)$$

Simplifying the notation, we replace *elapsed time* by *t* and *number of chips* by *N* to produce the following compact equation:

$$N = 25 + 4t$$

Note that, because we no longer show the units explicitly, we must remember 25 represents a number of chips and 4 represents a rate of change in units of chips/hr. We can use this equation to find the number of chips at any time. For example, after $t = 3.5$ hours, the number of chips is

$$N = 25 + (4 \times 3.5) = 39$$

Time out to think

Use the equation for chip production to find the number of chips produced after 4 hours. Does the result agree with the answer you read from the graph in Figure 9.12?

To generalize from this example to any linear function, note that

- The number of chips, *N*, is the **dependent variable.**
- The time, *t*, is the **independent variable.**
- The initial stock of 25 chips represents the **initial value** of the dependent variable when $t = 0$.
- The term 4 chips/hr is the **rate of change** of *N* with respect to *t*, so 4 chips/hr \times *t* is the *change* in *N*.

General Equation for a Linear Function

dependent variable = initial value + (rate of change \times *independent variable*)

The Algebraic Equation of a Line

If you have taken a course in algebra, you may be familiar with the equation for a linear function in a slightly different form. In algebra, *x* is commonly used for the independent variable and *y* for the dependent variable. For a straight line, the slope is

usually denoted by *m* and the initial value, or **y-intercept,** is denoted by *b*. With these symbols, the equation for a linear function becomes

$$y = mx + b$$

which has the same form as the general equation of a linear function given above. For example, the equation $y = 4x - 4$ represents a straight line with a slope of 4 and a *y*-intercept of -4. As shown in Figure 9.13a, the *y*-intercept tells us where the line crosses the *y*-axis.

Figure 9.13b shows the effects of keeping the same *y*-intercept but changing the slope. A positive slope ($m > 0$) means the line rises to the right. A negative slope ($m < 0$) means the line falls to the right. A zero slope ($m = 0$) means a horizontal line.

Figure 9.13c shows the effects of changing the *y*-intercept for a set of lines that have the same slope. All the lines rise at the same rate, but cross the *y*-axis at different points.

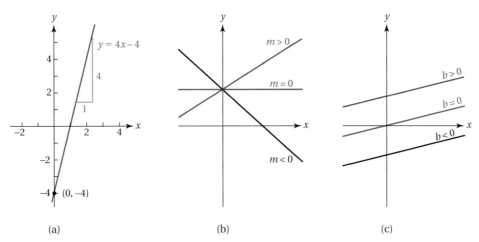

FIGURE 9.13 (a) Graph of $y = 4x - 4$. (b) Lines with same *y*-intercept but different slope. (c) Lines with the same slope but different *y*-intercepts.

▶EXAMPLE 4 Rain Depth Equation

Using the function shown in Figure 9.7, write an equation that describes the rain depth at any time after the storm began. Use the equation to find the rain depth 3 hours after the storm began.

Solution For the rain depth function in Figure 9.7, the rate of change is 1 in./hr and the initial value of the rain depth when the storm begins is 0 inches. Thus, the general equation for this function is

$$\underbrace{rain\ depth}_{\text{dependent variable}} = \underbrace{0\ \text{in.}}_{\text{initial value}} + \underbrace{1\ \frac{\text{in.}}{\text{hr}}}_{\text{rate of change}} \times \underbrace{time}_{\text{independent variable}}$$

The purpose of models is not to fit the data but to sharpen the questions.

—SAMUEL KARLIN, MATHEMATICIAN

We can write this equation more compactly by letting r represent *rain depth* (in inches) and t represent *time* (in hours):

$$r = 0 + (1 \times t), \quad \text{or} \quad r = t$$

Substituting $t = 3$ hours in this equation, we find that the rain depth 3 hours after the storm began is $r = 3$ inches. *Now try Exercises 23–24.* ◄

➤ EXAMPLE 5 Alcohol Metabolism

Alcohol is metabolized by the body (through enzymes in the liver) in such a way that the blood alcohol content (see Unit 2B) decreases linearly. A study by the National Institute on Alcohol Abuse and Alcoholism showed that, for a group of fasting males who consumed four drinks rapidly, the blood alcohol content rose to a maximum of 0.08 g/100 mL about an hour after the drinks were consumed. Three hours later, the blood alcohol content had decreased to 0.04 g/100 mL. Find a linear model that describes the elimination of alcohol after the peak blood alcohol content is reached. According to the model, what is the blood alcohol content five hours after the peak is reached?

Solution We seek a linear function that relates the independent variable *time* to the dependent variable *blood alcohol content* (BAC, for short). To avoid confusion, we'll write the BAC values without the units. If we let $t = 0$ represent the time at which the BAC reached a peak, the study gives us two points: (0 hr, 0.08) and (3 hr, 0.04). We can use these two points to find the rate of change, or slope, of the function:

$$\text{slope} = \frac{\text{change in BAC}}{\text{change in } time} = \frac{(0.04 - 0.08)}{(3 - 0) \text{ hr}} = \frac{-0.04}{3 \text{ hr}} = \frac{-0.0133}{\text{hr}}$$

That is, for each hour, the BAC *decreases* by 0.0133. The initial value for the BAC is 0.08. Thus, the general linear equation for this function becomes

$$\underbrace{\text{BAC}}_{\text{dependent variable}} = \underbrace{0.08}_{\text{initial value}} + \left(\underbrace{-0.0133 \frac{1}{\text{hr}}}_{\text{rate of change}} \times \underbrace{time}_{\text{independent variable}} \right)$$

We can write the equation more compactly as

$$\text{BAC} = 0.08 - (0.0133 \times t)$$

Figure 9.14 shows the graph of this function. The slope is -0.0133 and the vertical intercept is 0.08. To determine the BAC five hours after the peak BAC is reached, we set $t = 5$ and evaluate BAC:

$$\text{BAC} = 0.08 - (0.0133 \times 5) = 0.0135$$

Five hours after the peak BAC is reached (six hours after the drinks were consumed), the BAC is still significant. In fact, it's roughly 15% of the legal limit in most states. *Now try Exercises 25–26.* ◄

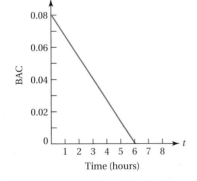

FIGURE 9.14 Linear function for Example 5.

➤ EXAMPLE 6 Price from Demand

Write an equation for the linear demand function in Figure 9.11. Then determine the price that should result in a demand of 80 pineapples per day.

Solution Earlier, we found that the rate of change, or slope, for this function is -10 pineapples per dollar. The initial value, or y-intercept, is 100 pineapples, because that is the demand predicted for a price of \$0. Let's use p for the independent variable *price* and d for the dependent variable *demand*. With the slope of -10 and y-intercept of 100, the equation for this function is

$$d = 100 - 10p$$

Because we are asked to find the price for a given demand, we must solve this equation for the price, p. We do so by first subtracting 100 from both sides, then dividing both sides by -10, and finally interchanging the left and right sides. You should confirm that this gives the following result:

$$p = \frac{d - 100}{-10}$$

Substituting a demand of $d = 80$ pineapples, we find

$$p = \frac{80 - 100}{-10} = \frac{-20}{-10} = 2$$

Based on this model, the price should be set at \$2 if the storeowners want to sell 80 pineapples each day. Note that we must find the units (dollars) for the answer by looking back to the original data. *Now try Exercises 27–28.* ◄

Thinking about . . .

ALGEBRA'S BAGHDAD CONNECTION

After the fall of ancient Rome in the 5th century, European civilization entered the period known as the Dark Ages. However, it was not a dark time in the Middle East, where a new center of intellectual achievement arose in the city of Baghdad (in modern-day Iraq). Jews, Christians, and Muslims in Baghdad worked together in scholarly pursuits during this period.

One of the greatest scholars was a Muslim named Muhammad ibn Musa al-Khworizimi (A.D. 780–850). Al-Khworizimi wrote several books on astronomy and mathematics, including one entitled *Hisab al-jabr walmuqabala*, which translates roughly as "the science of equations." This book preserved and extended the work of the Greek mathematician Diophantus (A.D. 210–290) and thereby laid the foundations of algebra. In fact, the word *algebra* comes directly from the Arabic words *al-jabr* in the book's title. The introduction to the book states its purpose to teach

> *what is easiest and most useful in arithmetic, such as men constantly require in cases of inheritance,*

> *legacies, partition, lawsuits, and trade, and in all their dealings with one another, or where the measuring of lands, the digging of canals, geometrical computations, and other objects of various sorts and kinds are concerned.*

Notice the practical intent of this first algebra book, which stands in stark contrast to the abstract nature of many algebra books today.

In another of his works, al-Khworizimi described the numeral system developed by Hindu mathematicians, thereby popularizing the decimal system and the use of the numeral zero. Although he did not claim credit for the Hindu work, later writers often attributed it to him. That is why modern numerals are known as *Hindu-Arabic* rather than solely Hindu. Some later authors even attributed the numerals to al-Khworizimi personally. In a sloppy writing of his name, the use of Hindu numerals became known as *algorismi*, which later became the English words *algorism* (arithmetic) and *algorithm*. Historians consider al-Khworizimi one of the most important mathematicians of all time, and his work lies at the foundation of modern mathematics.

Linear Functions from Two Data Points

Suppose we have two data points and want to find a linear function that fits them. We can find the equation for this linear function by using the two data points to determine the rate of change (slope) and the initial value for the function. The following three steps summarize the process, which is described for variables (x, y); of course, you can use any symbols you wish.

Creating a Linear Function from Two Data Points

Step 1: Let x be the independent variable and y be the dependent variable. Find the change in each variable between the two given points, and use these changes to calculate the slope, or rate of change:

$$\text{slope} = \frac{\text{change in } y}{\text{change in } x}$$

Step 2: Substitute this slope and the *numerical values* of x and y from either data point into the equation $y = mx + b$. You can then solve for the y-intercept, b, because it will be the only unknown in the equation.

Step 3: Now use the slope and y-intercept to write the equation of the linear function in the form $y = mx + b$.

> **EXAMPLE 7** Crude Oil Use Since 1850

Until about 1850, humans used so little crude oil that we can call the amount zero—at least in comparison to the amount used since that time. By 1960, humans had used a total (cumulative) of 600 billion cubic meters of oil. Create a linear model that describes world oil use since 1850. Discuss the validity of the model.

Solution We seek a function that describes how *total oil used* varies with *time*. We identify *time* as the independent variable, which we'll denote by t and measure in years A.D. *Total oil used* is the dependent variable, so we'll call it y and measure it in billions of cubic meters.

We are given two data points. The first is (1850, 0), indicating that no oil had been used by 1850. The second is (1960, 600), which represents the 600 billion cubic meters used by 1960. We now use the three-step procedure to find an equation for a linear function fitting the two points.

Step 1: Note that we've used t rather than x for the independent variable, because it helps us remember that we are dealing with time. Thus, we let t play the role of x in finding the slope:

$$m = \text{slope} = \frac{\text{change in } y}{\text{change in } t} = \frac{600 - 0}{1960 - 1850} = \frac{600}{110} = 5.45 \text{ (billion m}^3 \text{ per year)}$$

Step 2: We are looking for an equation of the form $y = mt + b$. We already have the slope, $m = 5.45$. If we use the numerical values from the first data point, $t = 1850$ and $y = 0$, this equation becomes

$$0 = 5.45 \times 1850 + b$$

by the way

Annual world oil consumption in 2003 was more than triple annual oil consumption in 1960.

We solve for b by subtracting the product 5.45×1850 from both sides and then interchanging the left and right sides. We find

$$b = -5.45 \times 1850 = -10{,}083 \text{ (billion cubic meters)}$$

Step 3: We now have both the slope, $m = 5.45$, and the y-intercept, $b = -10{,}083$. Thus, the equation for the linear function is

$$y = mt + b = 5.45t - 10{,}083$$

Figure 9.15 shows the graph of this function.

Now that we have the function, we can address the question of its validity as a model for oil consumption. First, note that we draw the graph only for years after 1850. Years prior to 1850, in which we assumed oil consumption to be zero, are not in the domain of the function. The more significant issue is whether the rise in oil consumption has been linear. The rate of oil consumption tends to increase with population. In fact, it has increased more rapidly than population because the average person uses more oil now than in the past. We know that population has risen exponentially since the mid–19th century (see Unit 8C), so a better model for oil consumption would be exponential instead of linear. The blue curve in Figure 9.15 shows an exponential fit to the two data points (we will discuss the creation of exponential models in the next unit). *Now try Exercises 29–34.* ◄

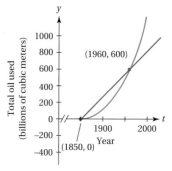

FIGURE 9.15 The red line shows a linear model for the two data points. The blue curve fits the data with an exponential function.

EXERCISES 9B

Review Questions

1. What does it mean to say that a function is linear?

2. Define *rate of change*, and describe how a rate of change is stated in words (that is, using "with respect to").

3. How is the rate of change of a linear function related to the slope of its graph?

4. How do you find the change in the dependent variable, given a change in the independent variable? Give an example.

5. Describe the general equation for a linear function. How is it related to the standard algebraic form $y = mx + b$?

6. Describe the process of creating an equation for a linear function from two data points. How are such models useful?

Does It Make Sense?

Decide whether each of the following statements makes sense (or is clearly true) or does not make sense (or is clearly false). Explain your reasoning.

7. When I graphed the linear function, it turned out to be a wavy curve.

8. I graphed two linear functions, and the one with the greater rate of change had the greater slope.

9. My freeway speed is the rate of change in my distance with respect to time.

10. It's possible to make a linear model from any two data points, but there's no guarantee that the model will fit other data points.

Basic Skills & Concepts

Linear Functions. For each graph shown in Exercises 11–16, do the following.

a. In words, describe the function shown on the graph.

b. Find the slope of the graph and express it as a rate of change (be sure to include units).

c. Briefly discuss the conditions under which a linear function is a realistic model for the given situation.

11.

12.

13.

14.

15.

16.

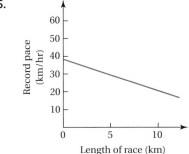

Rate of Change Rule. Each situation in Exercises 17–22 involves a rate of change that you may assume to be constant. Write a statement that describes how one variable varies with respect to the other, give the rate of change numerically (with units), and use the rate of change rule to answer any questions.

Example: Every week your fingernails grow 5 millimeters. How much will your fingernails grow in 2.5 weeks?

Solution: The *length of your fingernails* varies with respect to *time,* with a rate of change of 5 mm/wk. In 2.5 weeks, your fingernails will grow 5 mm/wk × 2.5 wk = 12.5 millimeters.

17. The water depth in a reservoir decreases at a rate of 0.25 inch per hour because of evaporation. How much does the water depth change in 6.5 hours? in 12.5 hours?

18. You drive along the highway at a constant speed of 50 miles per hour. How far do you travel in 2.5 hours? in 5.7 hours?

19. A tree increases its diameter by 0.2 inch per year by adding annual rings. How much does the diameter of the tree increase in 4.5 years? in 20.5 years?

20. A gas station owner finds that for every penny increase in the price of gasoline, she sells 150 fewer gallons of gas per week. How much more or less gas will she sell if she raises

the price by 8 cents per gallon? if she decreases the price by 2 cents per gallon?

21. Snow accumulates during a storm at a constant rate of 4 inches per hour. How much snow accumulates in the first 5.5 hours? in the first 7.8 hours?

22. According to one formula, your maximum heart rate (in beats per minute) is 220 minus your age (in years). How much does your maximum heart rate change from age 35 to age 50? What is your maximum heart rate at age 65?

Linear Equations. Exercises 23–28 describe situations that can be modeled by linear functions. In each case, write an equation for the linear function and use it to answer the given question. Be sure you clearly identify the independent and dependent variables. Then briefly discuss whether a linear model is reasonable for the situation described.

23. The price of a particular model car is $12,000 today and rises with time at a constant rate of $1200 per year. How much will a new car cost in 2.5 years?

24. The world record time in the 100-meter butterfly (swimming) was 53.0 seconds in 1988. Assume that the record falls at a constant rate of 0.05 second per year. What does the model predict for the record in 2010?

25. A snowplow has a maximum speed of 30 miles per hour on a dry highway. Its maximum speed decreases by 0.5 mile per hour for every inch of snow on the highway. According to this model, at what snow depth will the plow be unable to move?

26. The cost of renting a car is a flat $40, plus an additional 10 cents per mile that you drive. How far can you drive for $90?

27. You can rent time on computers at the local copy center for a $5 setup charge and an additional $3 for every 5 minutes. How much time can you rent for $15?

28. In 1980, the population of Boom Town began increasing at a rate of 200 people per year. The 1980 population was 2000 people. What is your projection for the population in the year 2010?

Equations from Two Data Points. In Exercises 29–34, create the required linear function and use it to answer all questions.

29. Suppose that you were 20 inches long at birth and 4 feet tall on your tenth birthday. Based on these two data points, create a linear equation for the function that describes how height varies with age. Use your equation to predict your height at ages 2, 6, 20, and 50. Comment on the validity of this linear model.

30. You can purchase a motorcycle for $4500 or lease it for a down payment of $100 and $120 per month. Find an equation that describes how the cost of the lease depends on time. How long can you lease the motorcycle before you've paid more than its purchase price?

31. A YMCA fundraiser offers raffle tickets for $5 each. The prize for the raffle is a $350 television set, which must be purchased with proceeds from the ticket sales. Find an equation that gives the profit/loss for the raffle as it varies with the number of tickets sold. How many tickets must be sold for the raffle sales to equal the cost of the prize?

32. The Psychology Club plans to pay a visitor $75 to speak at a fundraiser. Tickets will be sold for $2 apiece. Find an equation that gives the profit/loss for the event as it varies with the number of tickets sold. How many people must attend the event for the club to break even?

33. A $1000 washing machine in a laundromat is depreciated for tax purposes at a rate of $50 per year. Find an equation for the depreciated value of the washing machine as it varies with time. When does the depreciated value reach $0?

34. In the town of Paradise Valley, a 3% local sales tax and a 2% state sales tax are charged on all retail sales. Let p be the before-tax amount of a purchase in dollars. Let T be the after-tax amount of the purchase. Find a linear equation that describes how T varies with p. What is the total price of an item that costs $7.50 before taxes?

Further Applications

Algebraic Linear Equations. For each function in Exercises 35–42, find the slope of the graph and the y-intercept. Then sketch the graph for values of x between -10 and 10.

35. $y = 2x + 6$

36. $y = -3x + 3$

37. $y = -5x - 5$

38. $y = 4x + 1$

39. $y = 3x - 6$

40. $y = -2x + 5$

41. $y = -x + 4$

42. $y = 2x + 4$

Linear Graphs. Exercises 43–48 describe situations that can be modeled by linear functions. In each case, draw a graph of the function and use the graph to answer the given question. Be sure you clearly identify the independent and dependent variables. Then briefly discuss whether a linear model is reasonable for the situation described.

43. Your distance traveled varies with time as you drive along a highway at 40 miles per hour for 4 hours. How far do you travel in 2.6 hours?

44. The diameter of a tree increases by 0.2 inch with each passing year. When you started observing the tree, its diameter was 4 inches. Estimate the time at which the tree started growing.

45. The cost of publishing a poster is $2000 for setting up the printing equipment, plus $3 per poster printed. What is the total cost to produce 2000 posters?

46. The amount of sugar in a fermenting batch of beer decreases with time at a rate of 0.1 gram per day, starting from an initial amount of 5 grams. When is the sugar gone?

47. The cost of a particular private school begins with a one-time initiation fee of $2000, plus annual tuition of $10,000. How much will it cost to attend this school for six years?

48. The maximum speed of a semitrailer truck up a steep hill varies with the weight of its cargo. With no cargo, it can maintain a maximum speed of 50 miles per hour. With 20 tons of cargo, its maximum speed drops to 40 miles per hour. At what load does a linear model predict a maximum speed of 0 miles per hour?

49. **Project: Wildlife Management.** A common technique for estimating populations of birds or fish is to tag and release individual animals in two different outings. This procedure is called *catch and release*. If the wildlife remain in the sampling area and are randomly caught, a fraction of the animals tagged during the first outing are likely to be caught again during the second outing. Based on the number tagged and the fraction caught twice, the total number of animals in the area can be estimated.

 a. Consider a case in which 200 fish are tagged and released during the first outing. During a second outing in the same area, 200 fish are again caught and released, of which one-half are already tagged. Estimate N, the *total* number of fish in the entire sampling area. Explain your reasoning.

 b. Consider a case in which 200 fish are tagged and released during the first outing. During a second outing in the same area, 200 fish are again caught and released, of which one-fourth are already tagged. Estimate N, the *total* number of fish in the entire sampling area. Explain your reasoning.

 c. Generalize your results from parts a and b by letting p be the fraction of tagged fish that are caught during the

second outing. Find a formula for the function $N = f(p)$ that relates the total number of fish, N, to the fraction tagged during the second outing, p.

 d. Graph the function obtained in part c. What is the domain? Explain.

 e. Suppose that 15% of the fish in the second sample are tagged. Use the formula from part c to estimate the total number of fish in the sampling area. Confirm your result on your graph.

 f. Locate a real study in which catch and release methods were used. Report on the specific details of the study and how closely it followed the theory outlined in this problem.

Web Projects

Find useful links for Web Projects on the text Web site: www.aw.com/bennett-briggs

50. **Alcohol Metabolism.** Most drugs are eliminated from the blood through an exponential decay process with a constant half-life (see Unit 9C). Alcohol is an exception in that it is metabolized through a linear decay process. Investigate alcohol metabolism on the Web. Find data showing how the blood alcohol content (BAC) decreases over time, and use the data to develop a linear model. Discuss the validity of the model. What assumptions (for example, gender, weight, number of drinks) were used in creating the model?

51. **Property Depreciation.** Go to the IRS Web site and examine the rules for depreciation of some type of property, such as a rental property or a piece of business equipment. Make a linear model that describes the depreciation function.

In the News

52. **Linear Models.** Describe at least two situations from the news or your own life in which predictions must be made and a linear model seems appropriate. Briefly discuss why the linear model works well.

53. **Nonlinear Models.** Describe at least one situation from the news or your own life in which predictions must be made but a linear function is *not* a good model. Briefly discuss the shape (on a graph) that you would expect the function to take.

UNIT 9C

EXPONENTIAL MODELING

I n Unit 9B, we investigated the use of linear functions as mathematical models. The hallmark of linear functions is that they describe quantities that have a constant *absolute* growth rate. As discussed in Chapter 8, another common and important growth pattern is exponential growth, in which the *relative* growth rate is constant. In this unit, we will investigate the exponential function and some of its many applications in mathematical models.

THE EXPONENTIAL FUNCTION

Our first task is to find a general form for an exponential function. Consider a town that begins with a population of 10,000 and grows at a rate of 20% per year. As we saw in Unit 8A, this population grows exponentially because it increases the same relative amount, 20%, each year (see Figure 8.1). In the terminology of Chapter 8, the *percentage growth rate* is $P\% = 20\%$ per year or, equivalently, the *fractional growth rate* is $r = 0.2$ per year.

The initial population is 10,000. With a 20% growth rate, the population at the end of the first year is 20% *more than* the initial value of the population, or $120\% = 1.2$ *times* the initial population (see the "*of* versus *more than*" rule in Unit 3A):

$$\text{population after 1 year} = 10,000 \times 1.2 = 12,000$$

During the second year, the population again grows by 20%, so we can find the population at the end of the second year by multiplying again by 1.2:

$$\begin{aligned} \text{population after 2 years} &= \text{population after 1 year} \times 1.2 \\ &= (10,000 \times 1.2) \times 1.2 \\ &= 10,000 \times 1.2^2 = 14,400 \end{aligned}$$

With another 20% increase in the third year, we find the population at the end of the third year by multiplying once more by 1.2:

$$\begin{aligned} \text{population after 3 years} &= \text{population after 2 years} \times 1.2 \\ &= (10,000 \times 1.2^2) \times 1.2 \\ &= 10,000 \times 1.2^3 = 17,280 \end{aligned}$$

This growth is illustrated in Figure 9.16. The pattern may now be clear. If we let t represent the time in years, we find that

$$\text{population after } t \text{ years} = \text{initial population} \times 1.2^t$$

For example, after $t = 25$ years, the population is $10,000 \times 1.2^{25} = 953,962$.

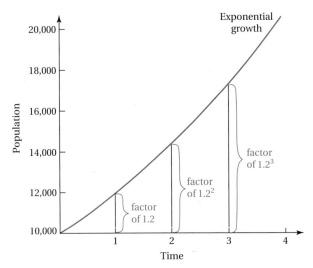

FIGURE 9.16 Exponential growth for three years with a growth rate of 20% per year. Each year, the population increases by a factor of 1.2.

We can extend this idea to *any* exponentially growing quantity Q, to write a general exponential function.

Exponential Functions

An **exponential function** grows (or decays) by the same relative amount per unit time. For any quantity Q growing exponentially with a fractional growth rate r,

$$Q = Q_0 \times (1 + r)^t$$

where

Q = value of the exponentially growing quantity at time t

Q_0 = initial value of the quantity (at $t = 0$)

r = fractional growth rate for the quantity

t = time

Negative values of r correspond to exponential decay. Note that *the units of time used for t and r must be the same.* For example, if the fractional growth rate is 0.05 per *month*, then t must also be measured in months.

We can use the same function for exponential decay if we make r negative. For example, if a quantity decreases by 10% per month, we identify the fractional decay rate as $r = -0.1$ per month.

You may notice that the exponential function is identical to the *compound interest formula* (Unit 4A) if we identify Q as the accumulated balance (called A in the compound interest formula), Q_0 as the starting principal, r as the interest rate, and t as the number of times interest is paid. In other words, compound interest is a form of exponential growth.

ALGEBRA WITH LOGARITHMS

The basic properties of logarithms (see the Brief Review, p. 505) lead to two very useful algebraic techniques. These techniques apply to logarithms with any base, but we will focus on common (base 10) logarithms.

First, we can solve for a variable that appears in an exponent by taking the logarithm of both sides of the equation (as long as both sides are positive) and applying the rule that $\log_{10} a^x = x \log_{10} a$.

Example: Solve for x in the equation $2^x = 50$.

Solution: Taking the logarithm of both sides, we have

$$\log_{10} 2^x = \log_{10} 50$$

By the rule that $\log_{10} a^x = x \log_{10} a$, this equation becomes

$$x \log_{10} 2 = \log_{10} 50$$

Now we divide both sides by $\log_{10} 2$ to get the solution:

$$x = \frac{\log_{10} 50}{\log_{10} 2} = \frac{1.69897}{0.30103} = 5.644$$

Second, we can solve for a variable that appears within a logarithm by making both sides of the equation into powers of 10 and applying the rule that $10^{\log_{10} x} = x$ (as long as $x > 0$).

Example: Solve for x in the equation $2 \log_{10} x = 15$.

Solution: First, we isolate the term containing x by dividing both sides by 2:

$$\log_{10} x = \frac{15}{2} = 7.5$$

Because x is within a logarithmic expression, we make both sides of the equation into powers of 10:

$$10^{\log_{10} x} = 10^{7.5}$$

Applying the rule $10^{\log_{10} x} = x$ (as long as $x > 0$), we find

$$x = 10^{7.5} = 3.1623 \times 10^7$$

Now try Exercises 11–26.

▶ **EXAMPLE 1** U.S. Population Growth

The 2000 census found a U.S. population of about 281 million. Write an equation for the U.S. population that assumes exponential growth at 0.7% per year. Use the equation to predict the U.S. population in 2100.

Solution The quantity Q is the U.S. population. The initial value is the 2000 population, $Q_0 = 281$ million. The percentage growth rate is $P\% = 0.7\%$ per year, so the fractional growth rate is $r = P/100 = 0.007$ per year. Thus, the equation takes the form

$$Q = Q_0 \times (1 + r)^t = 281 \text{ million} \times (1 + 0.007)^t$$
$$= 281 \text{ million} \times (1.007)^t$$

Note that, because the units of r are *per year*, t must be measured in years. When we substitute $t = 100$ years, this exponential function predicts a 2100 population of

$$Q = 281 \text{ million} \times (1.007)^{100} = 564 \text{ million}$$

At its present growth rate of 0.7% per year, the U.S. population will swell to over 560 million by 2100. *Now try Exercises 27–34, part a.* ◀

by the way

Between 1950 and 2000, China's population growth *rate* decreased by more than a factor of 2. However, the actual population increased from about 550 million to 1.2 billion.

► **EXAMPLE 2** Declining Population

China's one-child policy (see Unit 2C, Example 7) was originally implemented with the goal of reducing China's population to 700 million by 2050. China's 2000 population was about 1.2 billion. Suppose China's population declines at a rate of 0.5% per year. Write an equation for the exponential decay of the population. Will this rate of decline be sufficient to meet the original goal?

Solution The quantity Q is China's population. The population is *declining* at a rate of $P\% = 0.5\%$ per year, so the fractional growth rate r should be negative; it is $r = -P/100 = -0.005$ per year. The initial value is the 2000 population, $Q_0 = 1.2$ billion. Thus, the equation takes the form

$$Q = Q_0 \times (1 + r)^t = 1.2 \text{ billion} \times (1 - 0.005)^t = 1.2 \text{ billion} \times (0.995)^t$$

Again, because the units of r are *per year*, t must be measured in years. When we substitute $t = 50$ years, this exponential function predicts a 2050 population of

$$Q = 1.2 \text{ billion} \times (0.995)^{50} = 0.934 \text{ billion}$$

A decline of 0.5% per year will reduce China's population to 0.934 billion, or 930 million, by 2050. In that case, the goal of 700 million will not be met.

Now try Exercises 27–34, part b. ◄

Graphing Exponential Functions

The easiest way to graph an exponential function is to use points corresponding to several doubling times (or half-lives, in the case of decay). We start at the point $(0, Q_0)$ that represents the initial value at $t = 0$. For an exponentially growing quantity, we know that the value of Q is $2Q_0$ (double its initial value) after one doubling time (T_{double}), $4Q_0$ after two doubling times ($2T_{\text{double}}$), $8Q_0$ after three doubling times ($3T_{\text{double}}$), and so on. We simply fit a steeply rising curve between these points, as shown in Figure 9.17a.

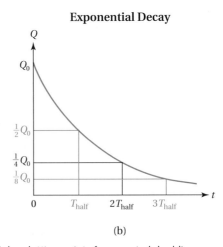

FIGURE 9.17 (a) We can graph exponential growth by plotting points for repeated doublings. (b) We can graph exponential decay by plotting points for repeated halvings.

\mathcal{T}hinking about . . .

DOUBLING TIME AND HALF-LIFE FORMULAS

In Unit 8B, we used both approximate and exact formulas for the doubling time and half-life. Now that we have an equation for the exponential function, we can see where these formulas came from.

For exponential growth, the doubling time is the time required for a quantity to double in size. That is, after a time $t = T_{\text{double}}$, the quantity has grown to twice its initial value, or $Q = 2Q_0$. Substituting these values into the exponential equation gives

$$2Q_0 = Q_0 \times (1 + r)^{T_{\text{double}}}$$

Dividing both sides by Q_0 and interchanging the two sides yields

$$(1 + r)^{T_{\text{double}}} = 2$$

To solve this equation for the doubling time, we must "bring down" the exponent by taking the logarithm of both sides (see the Brief Review on p. 559):

$$\log_{10}[(1 + r)^{T_{\text{double}}}] = \log_{10} 2$$

By the rule that $\log_{10} a^x = x \log_{10} a$, this equation becomes

$$T_{\text{double}} \log_{10}(1 + r) = \log_{10} 2$$

Dividing both sides by $\log_{10}(1 + r)$, we find the exact formula for the doubling time, as given in Unit 8B:

$$T_{\text{double}} = \frac{\log_{10} 2}{\log_{10}(1 + r)}$$

We find the half-life for exponential decay similarly. For decay, the quantity decreases to half its original value, or $Q = (\frac{1}{2})Q_0$, in a time $t = T_{\text{half}}$. Thus, there are two changes in the algebra: We have T_{half} instead of T_{double}, and $\log_{10}(\frac{1}{2})$ instead of $\log_{10} 2$. The result is

$$T_{\text{half}} = \frac{\log_{10}(\frac{1}{2})}{\log_{10}(1 + r)}$$

We can simplify this result by recognizing that $(\frac{1}{2}) = 2^{-1}$. Thus, using the rule that $\log_{10} a^x = x \log_{10} a$, we find that $\log_{10} 2^{-1} = -\log_{10} 2$. With this substitution, the formula takes the form given in Unit 8B:

$$T_{\text{half}} = -\frac{\log_{10} 2}{\log_{10}(1 + r)}$$

Remember that r is negative for exponential decay, so $1 + r$ is less than 1 in this case. (For example, if $r = -0.05$, then $1 + r = 0.95$.) Because the logarithm of a number between 0 and 1 is negative, the formula gives a positive value for the half-life.

For an exponentially decaying quantity, we know that the value of Q falls to $Q_0/2$ (half its initial value) after one half-life (T_{half}), $Q_0/4$ after two half-lives ($2T_{\text{half}}$), $Q_0/8$ after three half-lives ($3T_{\text{half}}$), and so on. Fitting a falling curve to these points gives the graph shown in Figure 9.17b. Note that this curve gets closer and closer to the horizontal axis, but never reaches it.

►EXAMPLE 3 Sensitivity to Growth Rate

The growth rate of the U.S. population has varied substantially during the past century. It depends on the immigration rate, as well as birth and death rates. Starting from a 2000 population of 281 million, project the population in 2100 using growth rates that are just 0.2 percentage point lower and higher than the 0.7% used in Example 1. Make a graph showing the population through 2100 for each growth rate.

Solution A growth rate 0.2 percentage point lower than 0.7% is 0.5%, or $r = 0.005$. At this growth rate, the 2100 population would be

$$Q = 281 \text{ million} \times (1.005)^{100} = 462 \text{ million}$$

A rate 0.2 percentage point higher than 0.7% is 0.9%, or $r = 0.009$, which gives a 2100 population of

$$Q = 281 \text{ million} \times (1.009)^{100} = 688 \text{ million}$$

Within a range of less than half a percentage point, from 0.5% to 0.9%, the projected population for 2100 varies by more than 200 million people (from 462 million to 688 million). Clearly, population projections are very sensitive to changes in the growth rate.

To make the graphs, note that we already have two points for each growth rate: the initial population in 2000, $Q_0 = 281$ million, and the population calculated for 2100. We find a third point in each case from the doubling time, which is the time when the population reaches $2Q_0 = 562$ million. You should use the doubling time formula to confirm that the doubling times are 77 years for $r = 0.009$, 99 years for $r = 0.007$, and 139 years for $r = 0.005$. Figure 9.18 shows the curves for all three growth rates.

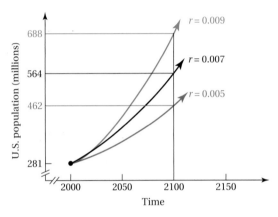

FIGURE 9.18 Future U.S. population growth with three growth rates.

Now try Exercises 27–34, part c. ◀

Alternative Forms of the Exponential Function

Our general equation for the exponential function, $Q = Q_0 \times (1 + r)^t$, contains the growth rate r, but not the doubling time or half-life. Because we are often given the doubling time or half-life, it's useful to rewrite the equation in forms that use T_{double} or T_{half} rather than r.

As before, we let Q represent the quantity at any time t and Q_0 represent the initial value. Then, in terms of the doubling time T_{double}, the exponential function takes the form

$$Q = Q_0 \times 2^{t/T_{double}}$$

This is the same equation we used with the doubling time in Unit 8B, except that now we are using Q for the "new value" and Q_0 for the initial value. As it should, this equation gives $2Q_0$ when $t = T_{double}$, $4Q_0$ when $t = 2T_{double}$, and so on.

For exponential decay, we get a similar function with the half-life:

$$Q = Q_0 \times \left(\frac{1}{2}\right)^{t/T_{\text{half}}}$$

Again, this is the same equation we used with the half-life in Unit 8B, except that now we are using Q for the "new value" and Q_0 for the initial value. You should confirm that this equation gives $Q_0/2$ when $t = T_{\text{half}}$, $Q_0/4$ when $t = 2T_{\text{half}}$, and so on.

Forms of the Exponential Function

- If given the growth or decay rate r, use the exponential function in the form

$$Q = Q_0 \times (1 + r)^t$$

Remember that r is positive for growth and negative for decay.
- If given the doubling time T_{double}, use the exponential function in the form

$$Q = Q_0 \times 2^{t/T_{\text{double}}}$$

- If given the half-life T_{half}, use the exponential function in the form

$$Q = Q_0 \times \left(\frac{1}{2}\right)^{t/T_{\text{half}}}$$

TECHNICAL NOTE
A slightly different form of the exponential function is $Q = Q_0 \times B^t$. If the base B is greater than 1, the function describes exponential growth. If the base B is between 0 and 1, the function describes exponential decay.

SELECTED APPLICATIONS

We have seen several examples in which we used exponential functions to model population growth and decline. We also used exponential equations in Chapter 4 for compound interest calculations. In this section, we'll explore a few more of the many applications in which exponential functions are used to model growth or decay processes.

Inflation

Because prices tend to change with time, price comparisons from one time to another are meaningful only if the prices are adjusted for the effects of inflation (see Unit 3D). We can model the effects of inflation with an exponential function in which r represents the rate of inflation.

▶ **EXAMPLE 4** Monthly and Annual Inflation Rates

The U.S. government reports the rate of inflation (as measured by the Consumer Price Index) both monthly and annually. Suppose that, for a particular month, the *monthly* rate of inflation is reported as 0.8%. What *annual* rate of inflation does this imply? Is the annual rate 12 times the monthly rate? Explain.

Solution We can write an exponential function for the inflation by letting Q_0 represent a price at time $t = 0$ and Q represent a price t months later. Using the monthly inflation rate $r = 0.8\% = 0.008$, we have

$$Q = Q_0 \times (1 + r)^t = Q_0 \times (1 + 0.008)^t = Q_0 \times 1.008^t$$

Note that, because r is given *per month*, t must be measured in months in this equation. Thus, to find the *annual* inflation rate, we substitute $t = 12$ months into the equation. We find that the price, Q, after 12 months is

$$Q = Q_0 \times 1.008^{12} = Q_0 \times 1.1 = 1.1Q_0$$

The price after one year is 1.1 times the initial price, which means that prices rose by 10%. Thus, the *annual* inflation rate is 10%. Note that this is more than 12 times the monthly rate, or $12 \times 0.8\% = 9.6\%$. The reason is that inflation compounds with each passing month, much like the way interest compounds monthly. In this case, the 9.6% found by multiplying the monthly rate by 12 is analogous to the APR quoted for an interest-bearing account (see Unit 4A). The actual annual inflation rate is higher because it is analogous to the annual yield, APY, which also depends on the number of compounding periods. *Now try Exercises 35–38.* ◄

Environment and Resources

Among the most important applications of exponential models are those involving the environment and resource depletion. Global concentrations of many pollutants in the water and atmosphere have increased exponentially, as has consumption of nonrenewable resources such as oil and natural gas. Two basic factors can be responsible for such exponential growth.

1. The *per capita* demand for a resource often increases exponentially. For example, per capita energy consumption in the United States increased exponentially during most of the past century.
2. An exponentially increasing population will mean an exponentially increasing demand for a resource even if per capita demand remains constant.

In most cases, the growth rate is determined by a combination of both factors.

➤ **EXAMPLE 5 Carbon Emissions from China and the United States**

The burning of fossil fuels releases greenhouse gases into the atmosphere. In 1995, the United States emitted about 1.4 billion (1.4×10^9) tons of carbon into the atmosphere, nearly one-fourth of the world total. China was the second largest contributor, emitting about 850 million (8.5×10^8) tons of carbon. However, emissions from China were rising at a rate of about 4% per year, while U.S. emissions were rising at about 1.3% per year. Using these growth rates, project greenhouse gas emissions from the United States and China in 2020. Graph the projected emissions for both countries.

Solution We use 1995 as $t = 0$, so 2020 is $t = 25$ years later. For the United States, we set the 1995 emissions as $Q_0 = 1.4 \times 10^9$ tons of carbon and $r = 0.013$ per year. The exponential function for the U.S. emissions is

$$Q = Q_0 \times (1 + r)^t = (1.4 \times 10^9) \times (1.013)^t$$

Substituting $t = 25$ years, we predict that U.S. carbon emissions in 2020 will be

$$Q = (1.4 \times 10^9) \times (1.013)^{25} = 1.9 \times 10^9 \text{ tons of carbon}$$

For China, we set the 1995 emissions as $Q_0 = 8.5 \times 10^8$ tons of carbon and $r = 0.04$ per year. The exponential function for the Chinese emissions is

$$Q = Q_0 \times (1 + r)^t = (8.5 \times 10^8) \times (1.04)^t$$

Substituting $t = 25$ years, we predict that Chinese carbon emissions in 2020 will be

$$Q = (8.5 \times 10^8) \times (1.04)^{25} = 2.3 \times 10^9 \text{ tons of carbon}$$

Thus, for the given growth scenarios, China will become the largest emitter of greenhouse gases before 2020. We can graph the two functions by finding the doubling time for each country. Using the doubling time formula, we find that the doubling time is 54 years for the United States and 18 years for China. Figure 9.19 shows the graphs for both countries. From the graph, this model predicts that Chinese carbon emissions will surpass U.S. carbon emissions in 2014.

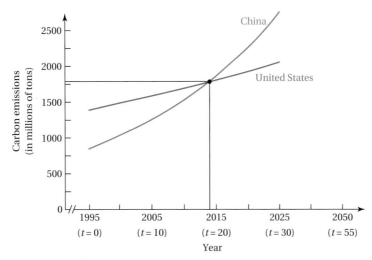

FIGURE 9.19 Exponential growth models for annual carbon emissions from the United States and China.

Now try Exercises 39–40. ◀

Physiological Processes

Many physiological processes are exponential. For example, a cancer tumor grows exponentially, at least in its early stages, and the concentration of many drugs in the bloodstream decays exponentially. (Alcohol is a notable exception, as its concentration decays linearly. See Example 5 in Unit 9B.)

➤EXAMPLE 6 Drug Concentration

Consider an antibiotic that has a half-life in the bloodstream of 12 hours. A 10-milligram injection of the antibiotic is given at 1:00 p.m. How much antibiotic remains in the blood at 9:00 p.m.? Draw a graph that shows the amount of antibiotic remaining as the drug is eliminated by the body.

Solution Because we are given a half-life rather than a growth rate r, we use the form of the exponential function that contains the half-life:

$$Q = Q_0 \times \left(\frac{1}{2}\right)^{t/T_{\text{half}}}$$

In this case, $Q_0 = 10$ milligrams is the initial dose given at $t = 0$ and Q is the amount of antibiotic in the blood t hours later. The half-life is given as $T_{half} = 12$ hours, so the equation is

$$Q = 10 \times \left(\frac{1}{2}\right)^{t/12}$$

We must remember that Q is measured in milligrams and t in hours. At 9:00 p.m., which is $t = 8$ hours after the injection, the amount of antibiotic remaining is

$$Q = 10 \times \left(\frac{1}{2}\right)^{8/12} = 10 \times \left(\frac{1}{2}\right)^{2/3} = 6.3 \text{ mg}$$

Eight hours after the injection, 6.3 milligrams of the antibiotic remain in the bloodstream. Graphing this exponential decay function up to $t = 100$ hours, we see that the amount of antibiotic decreases steadily toward zero (Figure 9.20).

FIGURE 9.20 Exponential decay of an antibiotic with a half-life of 12 hours in the bloodstream.

Now try Exercises 41–42. ◄

Radiometric Dating

As we discussed in Unit 8B, we can take advantage of the exponential nature of radioactive decay to measure the ages of rocks, bones, pottery, or other solid objects that contain radioactive elements. The process is called **radiometric dating.**

The basic idea of radiometric dating is simple: If we know both current and original amounts of the radioactive substance in the object, then we can use an exponential function to calculate the time since the object formed. Of course, we must know the half-life of the substance, but half-lives for nearly all radioactive substances have been carefully measured in laboratories. Thus, the only real difficulty in radiometric dating is determining how much of the radioactive substance was originally in the object. Fortunately, because radioactive materials decay in very specific ways, we can often determine the original amount by studying the overall chemical composition of the object today.

by the way

Most atomic nuclei are *stable,* meaning they do not spontaneously fall apart. By definition, radioactive substances have unstable nuclei that *do* fall apart spontaneously, leaving other nuclei as by-products.

▶ **EXAMPLE 7** The Allende Meteorite

The famous Allende meteorite lit up the skies of Mexico as it fell to Earth on February 8, 1969. Scientists melted and chemically analyzed small pieces of the meteorite and found traces of both radioactive potassium-40 and argon-40. From laboratory studies, we know that potassium-40 becomes argon-40 when it decays. Moreover, because argon-40 is a gas, it could not have been present when the meteorite first solidified (because a meteorite is made from solids, not gases). Thus, the scientists concluded that the argon-40 must have been trapped in the meteorite after it formed, in which case it must have come from decay of potassium-40. From the amounts of the two substances in the meteorite samples, the scientists determined that 8.5% of the potassium-40 originally present in the rock remains today. The half-life of potassium-40 is 1.3 billion (1.3×10^9) years. How old is the rock that makes up the Allende meteorite?

Solution We can model radioactive decay with an exponential function in which Q is the current amount of the radioactive substance and Q_0 is the original amount. Because we are given the half-life, we use the function in the form

$$Q = Q_0 \times \left(\frac{1}{2}\right)^{t/T_{\text{half}}}$$

For radiometric dating, we compare the current amount Q to the original amount Q_0, so the equation is more useful if we divide both sides by Q_0:

$$\frac{Q}{Q_0} = \left(\frac{1}{2}\right)^{t/T_{\text{half}}}$$

In this case, we are given that 8.5% of the original potassium-40 remains, which means that $Q/Q_0 = 8.5\% = 0.085$. If we substitute this value and the half-life of 1.3×10^9 years and interchange the two sides, the previous equation becomes

$$\left(\frac{1}{2}\right)^{t/(1.3 \times 10^9)} = 0.085$$

We can solve for t by taking the log of both sides:

$$\log_{10}\left(\frac{1}{2}\right)^{t/(1.3 \times 10^9)} = \log_{10} 0.085$$

Applying the rule that $\log_{10} a^x = x \log_{10} a$, we find

$$\left(\frac{t}{1.3 \times 10^9}\right) \times \log_{10}\left(\frac{1}{2}\right) = \log_{10} 0.085$$

Finally, we solve for t by multiplying both sides by 1.3×10^9 and dividing both sides by $\log_{10}\left(\frac{1}{2}\right)$:

$$t = 1.3 \times 10^9 \times \frac{\log_{10} 0.085}{\log_{10}\left(\frac{1}{2}\right)} = 4.6 \times 10^9 \text{ yr}$$

We conclude that the Allende meteorite is 4.6 billion years old.

Now try Exercises 43–44. ◀

by the way

Many meteorites date to 4.6 billion years ago, but none have ever been found that are older. Scientists therefore conclude that the solar system itself must have formed 4.6 billion years ago, and hence that the Allende meteorite is a remnant from the formation of the solar system. (The photo shows a meteorite at the American Museum of Natural History in New York City.)

HISTORICAL NOTE
American scientist Willard Libby won the Nobel Prize in Chemistry in 1960 for inventing the method of radiometric dating.

Thinking about . . .

CHANGING RATES OF CHANGE

In this chapter, we have studied two very different families of functions. Linear functions have straight-line graphs and constant rates of change. Exponential functions have graphs that rise or fall steeply, and their rates of change are not constant. This latter fact is illustrated in Figure 9.21a, where we see the graph of an exponential growth function. Superimposed on the graph are several *tangent* lines. These lines touch the graph at just a single point, so they provide a good measure of the steepness of the graph. Note that the tangent lines at $t = 0$, $t = 1$, $t = 2$, and $t = 3$ get progressively steeper (have larger slopes). This means that the rate of change of the function increases as t increases. This is a universal property of exponential growth functions: as the independent variable increases, the rate of change of the function also increases. We say that exponential growth functions "increase at an increasing rate."

An exponential decay function has a different behavior, as shown in Figure 9.21b. In this case, as t increases, the tangent lines become less steep (have smaller slopes). This means that the rate of change decreases as t increases, causing the graph to flatten out. We say that exponential decay functions "decrease at a decreasing rate."

We can now surmise that linear functions are special: They are the *only* functions with constant rates of change. All other functions are more like exponential functions in that they have *variable* rates of change. The rate of change of a function is a very important property, particularly if the function models a real quantity, such as a population or a balance in an investment account. The rate of change tells us not only whether the function is increasing or decreasing, but also how quickly it is increasing or decreasing. This information makes functions even more useful for modeling and prediction.

Given the importance of rates of change, it's not surprising that they are the subject of an entire branch of mathematics called *calculus*. It's fair to say that calculus is the study of rates of change. And because the world around us is constantly changing, it follows that calculus has a lot to say about the world around us.

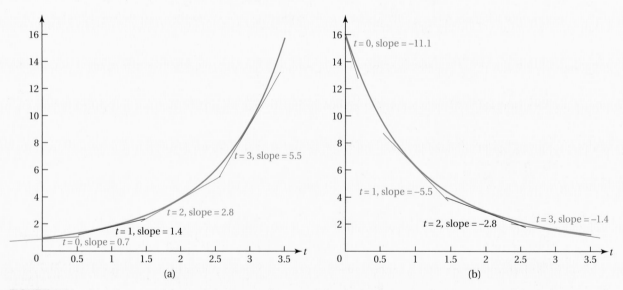

FIGURE 9.21 (a) An exponential growth function with a rate of change that increases. (b) An exponential decay function with a rate of change that decreases (slopes are negative because the function decreases).

EXERCISES 9C

Review Questions

1. Describe the meanings of all the variables in the exponential function. Explain how the function is used for exponential growth and decay.

2. Briefly explain how to find the doubling time and half-life from the exponential equation.

3. Describe how you can graph an exponential function with the help of the doubling time or half-life. What is the general shape of an exponential growth function? What is the general shape of an exponential decay function?

4. Describe the meaning of each of the three forms of the exponential function given in this chapter. Under what circumstance is each form useful?

5. Briefly describe how exponential functions are useful for modeling inflation, environmental and resource issues, physiological processes, and radioactive decay.

6. Briefly describe the process of radiometric dating. What makes it difficult? How can the difficulty be alleviated?

Does It Make Sense?

Decide whether each of the following statements makes sense (or is clearly true) or does not make sense (or is clearly false). Explain your reasoning.

7. After 100 years, a population growing at a rate of 2% per year will have grown by twice as many people as a population growing at a rate of 1% per year.

8. When I used the exponential function to model the decay of the medicine in my bloodstream, the growth rate r was negative.

9. We can use the fact that radioactive materials decay exponentially to determine the ages of ancient bones from archaeological sites.

10. I used the exponential function to figure how much money I'd have in a bank account that earns compound interest.

Basic Skills & Concepts

Review of Logarithms. Exercises 11–26 use the skills covered in the Brief Review on p. 559. In each case, solve the equation for the unknown quantity x.

11. $2^x = 23$

12. $10^x = 43$

13. $3^{2x} = 12$

14. $2 \times 8^x = 24$

15. $3^x = 15$

16. $7^x = 100$

17. $2^{x/3} = 20$

18. $4 \times 10^x = 120$

19. $\log_{10} x = 1.5$

20. $\log_{10} x = -2$

21. $\log_{10} x = -1.5$

22. $\log_{10} 2x = 4$

23. $2 \times \log_{10} x = 8$

24. $\log_{10}(2x) = 2$

25. $\log_{10}(3 + x) = 1$

26. $3 \log_{10}(4x) = 2$

Exponential Growth and Decay Laws. For Exercises 27–34, do the following:

 a. Create an exponential function of the form $Q = Q_0 \times (1 + r)^t$ (where $r > 0$ for growth and $r < 0$ for decay) to model the situation described. Be sure to clearly identify both variables in your function.

 b. Create a table showing the value of the quantity Q for the first 10 units of time (either years, months, weeks, or hours) of growth or decay.

 c. Make a graph of the exponential function.

27. The population of a town with a 1998 population of 85,000 grows at a rate of 2.4% per year.

28. The number of restaurants in a city that had 800 restaurants in 2001 is increasing at a rate of 3% per year.

29. A privately owned forest that had 1 million acres of old growth is being clear cut at a rate of 7% per year.

30. A town with a population of 10,000 is losing residents at a rate of 0.3% per month because of a poor economy.

31. The average price of a home in a town was $125,000 in 1995, but home prices are rising by 7% per year.

32. A certain drug breaks down in the human body at a rate of 15% per hour. The initial amount of the drug in the bloodstream is 8 milligrams.

33. Your starting salary at a new job is $2000 per month, and you get annual raises of 5% per year.

34. You hid 100,000 rubles in a mattress at the end of 1991, when they had a value of $10,000. However, the value of the ruble against the dollar then fell 50% per year.

Further Applications

Annual versus Monthly Inflation. Answer the questions in Exercises 35–36 about monthly and annual inflation rates.

35. If prices increase at a monthly rate of 0.6%, by what percentage do they increase in a year?

36. If the price of gold decreases at a monthly rate of 0.8%, by what percentage does it decrease in a year?

37. **Hyperinflation in Brazil.** During the 1970s and 1980s, Brazil underwent tremendous *hyperinflation*—periods of extraordinarily large inflation in prices. During the worst periods, prices rose as rapidly as 80% per month. At this rate, by what percentage would prices have risen in 1 year? in 1 day?

38. **Hyperinflation in Bosnia.** During the war in Bosnia, inflation sometimes drove prices up at a rate of 100% per month. At this rate, by what percentage would prices rise in 1 year? in 1 day?

39. **Extinction by Poaching.** Suppose that poaching reduces the population of an endangered animal by 10% per year. Further suppose that when the population of this animal falls below 30, its extinction is inevitable (owing to the lack of reproductive options without severe in-breeding). If the current population of the animal is 1000, when will it face extinction? Comment on the validity of the exponential model.

40. **World Oil Production.** World oil production was 518 million tons in 1950. Between 1950 and 1972, production increased at a rate of 7% per year. World oil production in 1998 was approximately 3300 million tons.

 a. What was world oil production in 1972?

 b. Using the result of part a, determine how much oil would have been produced in 1998 if growth in production had continued at a rate of 7% between 1972 and 1998. Compare this result to the actual 1998 figure given above.

 c. Using the result of part a, determine how much oil would have been produced in 1998 if growth in production had proceeded at a rate of 3% between 1972 and 1998. Compare this result to the actual 1998 figure given above.

 d. By trial and error, estimate the annual growth rate in world oil production between 1972 and 1998 with an exponential function.

41. Valium Metabolism. The drug Valium is eliminated from the bloodstream exponentially with a half-life of 36 hours. Suppose that a patient receives an initial dose of 20 milligrams of Valium at midnight.

 a. How much Valium is in the patient's blood at noon the next day?

 b. Estimate when the Valium concentration will reach 10% of its initial level.

42. Aspirin Metabolism. Assume that for the average individual, aspirin has a half-life of 12 hours in the bloodstream. At 12:00 noon, you take a 200-milligram dose of aspirin.

 a. How much aspirin will be in your blood at 6:00 p.m. the same day? at midnight? at 12:00 noon the next day?

 b. Estimate when the amount of aspirin will decay to 5% of its original amount.

Radiometric Dating. Use the radiometric dating formula to answer the questions in Exercises 43–44.

43. Uranium-238 has a half-life of 4.5 billion years.

 a. You find a rock containing a mixture of uranium-238 and lead. You determine that 85% of the original uranium-238 remains; the other 15% decayed into lead. How old is the rock?

 b. Analysis of another rock shows that it contains 55% of its original uranium-238; the other 45% decayed into lead. How old is the rock?

44. The half-life of carbon-14 is about 5700 years.

 a. You find a piece of cloth painted with organic dyes. By analyzing the dye in the cloth, you find that only 77% of the carbon-14 originally in the dye remains. When was the cloth painted?

 b. A well-preserved piece of wood found at an archaeological site has 6.2% of the carbon-14 that it must have had when it was alive. Estimate when the wood was cut.

 c. Is carbon-14 useful for establishing the age of the Earth? Why or why not?

45. Radioactive Waste. A toxic radioactive substance with a density of 2 milligrams per square centimeter is detected in the ventilating ducts of a nuclear processing building that was used 45 years ago. If the half-life of the substance is 20 years, what was the density of the substance when it was deposited 45 years ago?

46. Metropolitan Population Growth. A small city had a population of 85,000 in 2000. Concerned about rapid growth, the residents passed a growth control ordinance limiting population growth to 3% each year. If the population grows at this 3% annual rate, what will the population be in 2010? in 2100? In one or two paragraphs, discuss the implications of your results for the region.

47. Rising Costs. Between 1995 and 2000, the average rate of inflation was about 3% per year (as measured by the Consumer Price Index). If a cart of groceries cost $100 in 1995, what did it cost in 2000?

48. Project: Periodic Drug Doses. It is common to take a drug (such as aspirin or an antibiotic) repeatedly at fixed time intervals. Suppose that an antibiotic has a half-life of 8 hours and a 100-milligram dose is taken every 8 hours.

 a. Write an exponential function that represents the decay of the antibiotic from the moment of the first dose to just *prior* to the next dose (i.e., 8 hours after the first dose). How much antibiotic is in the bloodstream just *prior* to this next dose? How much antibiotic is in the bloodstream just *after* this next dose?

 b. Following a procedure similar to that in part a, calculate the amounts of antibiotic in the bloodstream just prior to and just after the doses at 16 hours, 24 hours, and 32 hours.

 c. Make a graph of the amount of antibiotic in the bloodstream for the first 32 hours after the first dose of the drug. What do you predict will happen to the amount of drug if the doses every 8 hours continue for several days or weeks? Explain.

 d. Consult a pharmacist (or read the fine print on the information sheet enclosed with many medicines) to find the half-lives of some common drugs. Create a model for the metabolism of one drug using the above procedure.

49. Project: Increasing Atmospheric Carbon Dioxide. Between 1860 and 1990, carbon dioxide (CO_2) concentration in the atmosphere rose from roughly 290 parts per million to 350 parts per million. Assume that this growth can be modeled with an exponential function of the form $Q = Q_0 \times (1 + r)^t$.

 a. By experimenting with various values of the fractional growth rate r, find an exponential function that fits the given data for 1860 and 1990.

b. Use this exponential model to predict when the CO_2 concentration will be double its 1990 level.

c. In 1998, the Kyoto treaty aimed to slow the growth in carbon dioxide concentration. Research the current status of this treaty and its effects.

Web Projects

Find useful links for Web Projects on the text Web site: www.aw.com/bennett-briggs

50. Radiometric Dating. Use the Web to conduct additional research on radiometric dating. Choose a specific application of radiometric dating, explain how it is used, and summarize the results.

51. Cost of Living. Gather data on the Consumer Price Index (CPI) between 1990 and the present from the Web. Use the exponential growth law, $Q = Q_0(1 + r)^t$, and the data to find the value of the growth rate r that best fits all of the data. Use the model to predict the CPI in 2010 and 2020. What do these predictions say about the cost of a movie in 2010 and 2020?

52. Carbon Dioxide Emissions. Visit the Web to gather data on U.S. emissions of the greenhouse gas carbon dioxide over the past decade. Use the exponential growth law, $Q = Q_0(1 + r)^t$, together with the data to find the value of the growth rate r that best fits all of the data. Use the model to predict emissions in 2010 and 2020.

53. Resource Consumption. Choose a particular natural resource (such as natural gas or oil) and find data on the Web on the consumption of that resource either nationally or worldwide. Use an exponential growth model together with the data to make the case that consumption of the resource has been increasing exponentially.

54. The Inflation Calculator. Visit the inflation calculator on the Web. Experiment with various amounts of money and various beginning and ending years. Also use the Web to find the prices of several common services or products (such as a gallon of milk, a haircut, a loaf of bread, and admission to the movies). Determine whether the rate of increase in price for these items has been faster or slower than the rate of inflation. Use your experiments to write a three-paragraph explanation of inflation for a friend.

In the News

55. Inflation Rate. Find a news report that states both monthly and annual inflation rates. Using the methods in this unit, check whether the rates agree. Explain.

56. Exponential Process. Find a news account of a process that illustrates either exponential growth or decay. Is the term *exponential* used in the account? How do you know that there is an exponential process at work? Describe how you could use an exponential function to model the process.

57. Radiometric Dating. Find a news report that gives the age of an archaeological site, fossil, or rock based on radiometric dating. Briefly describe how the dating process worked and how accurate we can expect it to be.

CHAPTER 9 SUMMARY

UNIT	KEY TERMS	KEY IDEAS AND SKILLS
9A	mathematical model function dependent variable independent variable domain range periodic function	**Represent** the ordered pair of variables in a function by (*independent variable, dependent variable*). **Understand** the notation $y = f(x)$. **Represent** functions with a data table, graph, or equation. **Create and use** graphs of functions. **Understand** how functions are models.
9B	linear function rate of change slope initial value y-intercept	**Rate of change for a linear function:** $\text{rate of change} = \text{slope} = \dfrac{\text{change in } \textit{dependent variable}}{\text{change in } \textit{independent variable}}$ **Change in the dependent variable:** Change in dependent variable = (rate of change) \times (change in independent variable) **General equation for a linear function:** *dependent variable* = initial value + (rate of change \times *independent variable*) **Algebraic equation of a line:** $y = mx + b$ **Slope from two data points:** $\text{slope} = \dfrac{\text{change in } y}{\text{change in } x}$
9C	exponential function initial value fractional growth rate r radiometric dating	**Equation for an exponential function:** $Q = Q_0 \times (1 + r)^t$ **Equation for an exponential function given the doubling time:** $Q = Q_0 \times 2^{t/T_{\text{double}}}$ **Equation for an exponential function given the half-life:** $Q = Q_0 \times \left(\dfrac{1}{2}\right)^{t/T_{\text{half}}}$

A work of morality, politics, criticism, perhaps even eloquence will be more elegant, other things being equal, if it is shaped by the hand of geometry.

—BERNARD LE BOVIER DE FONTENELLE, 1729

CHAPTER

10

Modeling with Geometry

We live in a three-dimensional world, and much of our understanding of that world is rooted in geometry. Surveying land for a new golf course, analyzing a satellite photograph of the Earth, navigating a ship at sea, and creating a medical image all rely on ideas that originated with the ancient Greeks. This chapter illustrates how both ancient and modern geometry help us understand the world around us.

UNIT **10A** **Fundamentals of Geometry:** Geometry plays a vital role in human culture, in everything from art and architecture to advanced engineering. We will study fundamental ideas of geometry, including formulas for finding the perimeter, area, and volume of common objects.

UNIT **10B** **Problem Solving with Geometry:** Geometry can be used to solve many problems that arise in everyday life. We will explore some examples in this unit.

UNIT **10C** **Fractal Geometry:** In the past few decades, an entirely new type of geometry has become important both in the arts and for understanding nature. We will explore a few of the ideas that underlie this fractal geometry.

UNIT 10A

FUNDAMENTALS OF GEOMETRY

*T*he word *geometry* literally means "earth measure." Many ancient cultures developed geometrical methods to survey flood basins around agricultural fields and to establish patterns of planetary and star motion. However, geometry was always more than just a practical science, as we can see from the use of geometric shapes and patterns in ancient art.

The Greek mathematician Euclid (c. 325–270 B.C.) summarized Greek knowledge of geometry in a 13-volume textbook called *Elements*. The geometry described in Euclid's work, now called **Euclidean geometry,** is the familiar geometry of lines, angles, and planes. In this unit, we'll explore the foundations of Euclidean geometry.

Time out to think

Euclid worked at a university called the Museum, so named because it honored the Muses—the patron goddesses of the sciences and the fine arts. Thus, the Greeks thought that the sciences and fine arts went hand in hand. Do the sciences and the arts still seem so clearly linked today? Why or why not?

POINTS, LINES, AND PLANES

Geometric objects, such as points, lines, and planes, represent *idealizations* that do not exist in the real world (Figure 10.1). A geometric **point** is imagined to have zero size. No real object has zero size, but many real phenomena approximate geometric points. Stars, for example, appear as points of light in the night sky.

A geometric **line** is formed by connecting two points along the shortest possible path. It has infinite length and no thickness. Because no physical object is infinite in length, we usually work with **line segments,** or pieces of a line. A long taut wire is a good approximation to a line segment.

A geometric **plane** is a perfectly flat surface that has infinite length and width, but no thickness. A smooth tabletop is a good approximation to a segment of a plane.

FIGURE 10.1

Representations of a point, a line, and a plane.

Time out to think

Describe at least three additional everyday realizations of points, line segments, and segments of planes. How does each real object compare to its geometric idealization?

Dimension

The **dimension** of an object can be thought of as the number of independent directions in which you could move if you were on the object. If you were a prisoner confined to a point, you would have no place to go, so a point has zero dimensions. A line is one-dimensional because, if you walk on a line, you can move in only one direction. (Forward and backward count as the same direction—one positive and the other negative.) In a plane, you can move in two independent directions, such as north/south and east/west (or any combination of those two directions). Thus, a plane is two-

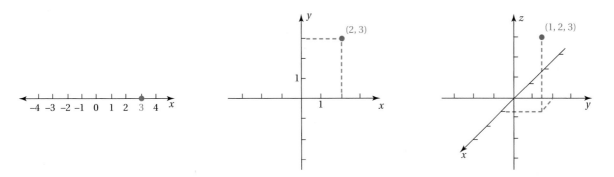

FIGURE 10.2 One coordinate locates a point on a line. Two coordinates locate a point in a plane. Three coordinates locate a point in space.

dimensional. In a three-dimensional **space,** such as the world around us, you can move in three independent directions: north/south, east/west, and up/down.

Another way to think about dimension is by the number of **coordinates** required to locate a point (Figure 10.2). A line is one-dimensional because it requires only one coordinate, such as x, to locate a point. A plane is two-dimensional because it requires two coordinates, such as x and y, to locate a point. Three coordinates, such as x, y, and z, are needed to locate a point in three-dimensional space.

Angles

The intersection of two lines or line segments forms an **angle.** The point of intersection is called the **vertex.** Figure 10.3a shows an arbitrary angle with its vertex at point A, so we call it *angle A*, denoted as $\angle A$. The most common way to measure angles is in *degrees* (°), derived from the ancient base-60 numeral system of the Babylonians. By definition, a full circle encompasses an angle of 360°, so an angle of 1° represents $\frac{1}{360}$ of a circle (Figure 10.3b). To measure an angle, we imagine its vertex as the center of a circle. Figure 10.3c shows that $\angle A$ subtends $\frac{1}{12}$ of a circle. Thus, $\angle A$ measures $\frac{1}{12} \times 360°$, or 30°. Some angles have special names, as shown in Figure 10.4.

- A **right angle** measures 90°.
- A **straight angle** is formed by a straight line and measures 180°.
- An **acute angle** is any angle whose measure is less than 90°.
- An **obtuse angle** is any angle whose measure is between 90° and 180°.

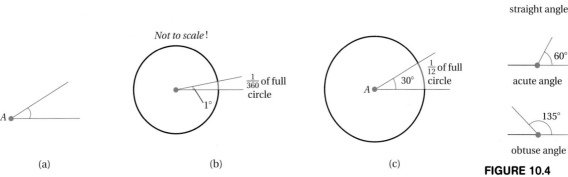

(a) (b) (c)

FIGURE 10.3 Measuring angles.

FIGURE 10.4

HISTORICAL NOTE
Formal geometry usually is traced back to the Greek philosopher Thales (624–546 B.C.). He is regarded as the first person to introduce ideas of abstraction into geometry, envisioning lines of zero thickness and perfect straightness.

Time out to think _____

Draw another acute angle and another obtuse angle. How is the term *acute* in an acute illness related to its meaning in an acute angle? If we say that someone is being obtuse, does the term *obtuse* bear any relation to its meaning in an obtuse angle? Explain.

Two lines or line segments in a plane that meet in a right (90°) angle are said to be **perpendicular** (Figure 10.5a). Two lines or line segments in a plane that are everywhere the same distance apart are said to be **parallel** (Figure 10.5b). Parallel lines in a plane never meet.

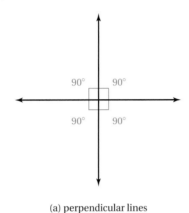

Perpendicular distance between lines is everywhere the same.

(a) perpendicular lines (b) parallel lines

FIGURE 10.5

▶ **EXAMPLE 1 Angles**

Find the degree measure of the angles that subtend

 a. a semicircle (half a circle)
 b. a quarter circle
 c. an eighth of a circle
 d. a hundredth of a circle

TECHNICAL NOTE
Angles can also be measured in *radians*. The angle that subtends a full circle is defined to be 2π radians. Thus,

$$1° = \frac{2\pi}{360} \text{ radians}$$

and

$$1 \text{ radian} = \frac{360°}{2\pi}$$
$$\approx 57.3°$$

Solution

 a. The angle that subtends a semicircle measures $\frac{1}{2} \times 360° = 180°$.
 b. The angle that subtends a quarter circle measures $\frac{1}{4} \times 360° = 90°$.
 c. The angle that subtends an eighth of a circle measures $\frac{1}{8} \times 360° = 45°$.
 d. The angle that subtends a hundredth of a circle measures $\frac{1}{100} \times 360° = 3.6°$.

Now try Exercises 17–30. ◀

PLANE GEOMETRY

Plane geometry is the geometry of two-dimensional objects. Here, we examine problems involving circles and polygons, which are the most common two-dimensional objects.

All points on a circle have the same distance, or **radius,** from the circle's **center** (Figure 10.6). The **diameter** of a circle, which is twice the radius, is the distance across the circle passing through its center.

Any closed shape in the plane made from straight line segments is a **polygon** (Figure 10.7). The root *poly* comes from a Greek word for "many," and a *polygon* is a many-sided figure. A **regular polygon** is a polygon in which all the sides have the same length and all interior angles are equal. Table 10.1 shows several common regular polygons and their names.

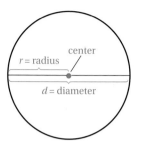

FIGURE 10.6
Definitions for a circle.

FIGURE 10.7 Examples of polygons.

TABLE 10.1	A Few Regular Polygons				
SIDES	NAME	PICTURE	SIDES	NAME	PICTURE
3	Equilateral triangle	△	6	Regular hexagon	⬡
4	Square	□	8	Regular octagon	⯃
5	Regular pentagon	⬠	10	Regular decagon	◯

Time out to think _____
Give several other English words that use the Greek root *poly.* What does *poly* mean in each case?

Among the most important of all polygons is the **triangle,** which comes in many different forms. All three sides of an **equilateral triangle** have equal length (Figure 10.8a), making it a regular polygon. An **isosceles triangle** (Figure 10.8b) has exactly two sides of equal length. A right triangle (Figure 10.8c) contains one right (90°) angle. We will return to right triangles and their many uses in Unit 10B. By sketching some triangles—particularly triangles with one very large angle and two very small angles (Figure 10.8d)—you can probably convince yourself that all triangles have the following property: The sum of the three angle measures is always 180°.

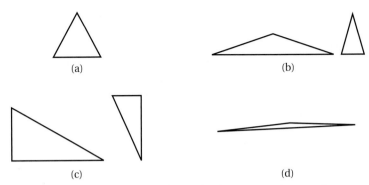

FIGURE 10.8 (a) An equilateral triangle; (b) two isosceles triangles; (c) two right triangles; (d) a triangle that is almost a straight line.

Perimeter

The perimeter of a plane object is simply the length of its boundary (Table 10.2). We can find the perimeter of a polygon by adding the lengths of the individual edges. The perimeter of a circle, called the **circumference,** is related to its diameter or radius by the universal constant π (pronounced "pie"), which has a value of approximately 3.14:

$$\text{circumference of circle} = \pi \times \text{diameter} = \pi \times d$$
$$= 2 \times \pi \times \text{radius} = 2 \times \pi \times r$$

TABLE 10.2 Perimeter and Area Formulas for Familiar Two-Dimensional Objects

OBJECT	PICTURE	PERIMETER	AREA
Circle		$2\pi r = \pi d$	πr^2
Square		$4l$	l^2
Rectangle		$2l + 2w$	lw
Parallelogram		$2l + 2w$	lh
Triangle		$a + b + c$	$\frac{1}{2}bh$

Now try Exercises 31–46.

𝒯hinking about

ARCHIMEDES AND PI

Ancient people recognized that the circumference of any circle is proportional to its radius. The first known attempt to find an exact formula for the circumference was made by the Greek scientist Archimedes (287–212 B.C.). His strategy began with two squares: one *inscribed* in a circle and the other *circumscribed* around the circle (Figure 10.9). He reasoned that the circumference of the circle must be greater than the perimeter of the inscribed square and less than the perimeter of the circumscribed square. Next, he doubled the number of sides of his figures, so that each square became an octagon. He repeated the process, producing inscribed and circumscribed 16-sided figures, 32-sided figures, and so on.

The following table shows the perimeters (in inches) for the inscribed and circumscribed polygons around a circle with a diameter of 1 inch. Notice that the perimeters converge toward each other and

toward π. That is, if we could apply Archimedes' strategy to figures with an infinite number of sides, it would give us an exact value for π.

Number of Sides of Polygons	Perimeter of Inscribed Polygon	Perimeter of Circumscribed Polygon
4	2.8284	4.0000
8	3.0615	3.3137
16	3.1214	3.1826
32	3.1365	3.1517
64	3.1403	3.1441

Archimedes' approximation to π was about 3.14. Because π is irrational, it cannot be written exactly. Today, supercomputers can compute π to billions of digits; the first few digits are $3.141592653589793\ldots$. As of 2002, the most precise approximation to π has 1.2 trillion decimal digits.

FIGURE 10.9 Circles with inscribed (blue) and circumscribed (red) polygons. Left to right, the polygons are squares, octagons, and 16-gons, with perimeters making successively better approximations to the circumference of a circle.

> **EXAMPLE 2** Interior Design

A window consists of a 4-foot by 6-foot rectangle capped by a semicircle (Figure 10.10). How much trim is needed to go around the window?

Solution The trim must line the 4-foot base of the window, two 6-foot sides, and the semicircular cap. The three straight edges have a total length of $4\,\text{ft} + 6\,\text{ft} + 6\,\text{ft} = 16\,\text{ft}$. The perimeter of the semicircular cap is half of the circumference of a full circle with a diameter of 4 feet, or

$$\frac{1}{2} \times \pi \times 4\,\text{ft} \approx \frac{1}{2} \times 3.14 \times 4\,\text{ft} \approx 6.3\,\text{ft}$$

FIGURE 10.10

Thus, the total length of trim needed for the window is about

$$16\,\text{ft} + 6.3\,\text{ft} = 22.3\,\text{ft}$$ *Now try Exercises 47–48.* ◀

Areas

We can find the areas of many geometrical objects with fairly simple formulas (see Table 10.2). For example, the area of any circle is related to its radius by the following formula:

$$\text{area of circle} = \pi \times \text{radius}^2 = \pi \times r^2$$

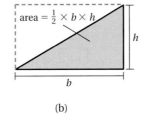

(a) (b)

FIGURE 10.11

Another familiar formula gives the area of a rectangle (Figure 10.11a):

$$\text{area of rectangle} = \text{base} \times \text{height} = b \times h$$

Cutting a rectangle along its diagonal produces two identical right triangles, as shown in Figure 10.11b. As the figure makes clear, the area of each triangle is

$$\text{area of triangle} = \frac{1}{2} \times \text{base} \times \text{height} = \frac{1}{2} \times b \times h$$

In fact, this area formula holds for all triangles.

Similarly, we can find the area formula for a **parallelogram**—a four-sided polygon in which the opposite sides are parallel. Figure 10.12 shows how a parallelogram is transformed into a rectangle with the same area. We can then see that the area formula for a parallelogram is the same as that for a rectangle:

$$\text{area of parallelogram} = \text{base} \times \text{height} = b \times h$$

FIGURE 10.12 We can transform a parallelogram into a rectangle by shifting one triangular segment from one side to the other.

Many other areas can be computed from these basic formulas. For example, we can find the area of any **quadrilateral** (a four-sided polygon) by dividing it into two triangles and adding the areas of the two triangles (Figure 10.13).

FIGURE 10.13 The area of any quadrilateral can be computed as the sum of the areas of two triangles.

➤ **EXAMPLE 3** Building Stairs

You have built a stairway in a new house and want to cover the space beneath the stairs with plywood. Figure 10.14 shows the region to be covered. What is the area of this region?

Solution The region to be covered is triangular, with a base of 12 feet and height of 9 feet. Thus, the area of this triangle is

$$\text{area} = \frac{1}{2} \times b \times h = \frac{1}{2} \times 12 \text{ ft} \times 9 \text{ ft} = 54 \text{ ft}^2$$

The area of the region to be covered is 54 square feet.

Now try Exercises 49–50. ◄

FIGURE 10.14

➤ **EXAMPLE 4** City Park

A one-block city park is bounded by two sets of parallel streets (Figure 10.15). The streets along the block are each 55 yards long and the perpendicular distance between the streets is 39 yards. How much sod should be purchased to cover the entire park in grass?

Solution The city park is a parallelogram with a base of 55 yards and height of 39 yards. Thus, the area of the park is

$$\text{area} = b \times h = 55 \text{ yd} \times 39 \text{ yd} = 2145 \text{ yd}^2$$

The city will need to purchase 2145 square yards of sod.

FIGURE 10.15

Now try Exercises 51–52. ◄

THREE-DIMENSIONAL GEOMETRY

Two of the most important properties of a three-dimensional object, such as a box or a sphere, are its *volume* and its *surface area*. Table 10.3 gives the names of several familiar three-dimensional objects, along with their volume and surface area formulas.

Some of the formulas in Table 10.3 are easily understood. For example, the volume formula for a box or a cube is just the familiar *length* × *width* × *height* formula. The surface area formula for a box or a cube is the result of adding the areas of the six (rectangular) faces of the object.

The volume formula for a cylinder corresponds to multiplying the area of the circular base (πr^2) by the height, h. The surface area formula for a cylinder has two parts. The area of each circular end of the cylinder is πr^2. The area of the curved surface of the cylinder is found by using a clever trick. As shown in Figure 10.16, when the cylinder is cut and unfolded, the curved surface becomes a rectangle. The length of the rectangle is the circumference of the original cylinder ($2\pi r$), and the height of the rectangle is the height, h, of the cylinder. So the area of the curved surface of the cylinder is $2\pi rh$.

by the way

A solid with plane (flat) faces is called a *polyhedron,* from the Greek for "many faces."

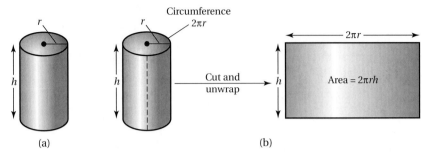

FIGURE 10.16 (a) A circular cylinder has a height h and a circular base with a radius r. (b) To find the area of the curved surface, imagine cutting the cylinder lengthwise and unfolding it to form a rectangle. The area of the rectangle equals the area of the curved surface.

	Surface Area and Volume Formulas for Familiar		
TABLE 10.3	**Three-Dimensional Objects**		

OBJECT	PICTURE	SURFACE AREA	VOLUME
Sphere		$4\pi r^2$	$\frac{4}{3}\pi r^3$
Cube		$6l^2$	l^3
Rectangular prism (box)		$2(lw + lh + wh)$	lwh
Right circular cylinder		$2\pi r^2 + 2\pi rh$	$\pi r^2 h$

Now try Exercises 53–57.

> **EXAMPLE 5** Water Reservoir

A water reservoir has a rectangular base that measures 30 meters by 40 meters, and vertical walls 15 meters high. At the beginning of the summer, the reservoir was filled to capacity. At the end of the summer, the water depth was 4 meters. How much water was used?

[The] knowledge at which geometry aims is of the eternal, and not of the perishing and transient.

—PLATO

Solution The reservoir has the shape of a rectangular prism, so the volume of water in the reservoir is its length times its width times the depth of the water. When the reservoir was filled at the beginning of the summer, the volume of water was

$$30\,\text{m} \times 40\,\text{m} \times 15\,\text{m} = 18{,}000\,\text{m}^3$$

Thinking about . . .

PLATO, GEOMETRY, AND ATLANTIS

Plato (427–347 B.C.) was one of the many ancient Greeks who emphasized the study of geometry and sought to find geometric patterns in nature. He believed that the heavens must exhibit perfect geometric form, and therefore argued that the Sun, Moon, planets, and stars must move in perfect circles—an idea that was not proven false until the early 1600s.

Another of Plato's ideas about geometry and the universe involved the five *perfect solids*. Each of these perfect solids has the special property that all of its faces are the same regular polygon (Figure 10.17). Plato believed that four of the perfect solids represented the four elements that the Greeks thought made up the universe: earth, water, fire, and air. The dodecahedron, he believed, represented the universe as a whole.

Plato presented his ideas in several of his dialogues. In the dialogue *Timaeus,* in which he discussed the role of the perfect solids in the universe, he also invented a moralistic tale about a fictitious land called Atlantis. Interestingly, while Plato's ideas about the universe were abandoned long ago, millions of people today believe that Atlantis really existed. Plato's fiction, in the end, has more adherents than the ideas in which he firmly believed. Commenting on this irony, the popular author Isaac Asimov wrote:

> *If there is a Valhalla for philosophers, Plato must be sitting there in endless chagrin, thinking of how many foolish thousands, in all the centuries since his time . . . who have never read his dialogues or absorbed a sentence of his serious teachings . . . believed with all their hearts in the reality of Atlantis.*

| tetrahedron | cube | octahedron | dodecahedron | icosahedron |
| (4 triangular faces) | (6 square faces) | (8 triangular faces) | (12 pentagonal faces) | (20 triangular faces) |

FIGURE 10.17 The five perfect solids.

At the end of the summer, the amount of water remaining was

$$30 \text{ m} \times 40 \text{ m} \times 4 \text{ m} = 4800 \text{ m}^3$$

Therefore, the amount of water used was $18{,}000 \text{ m}^3 - 4800 \text{ m}^3 = 13{,}200 \text{ m}^3$.

Now try Exercises 58–59. ◀

▶ **EXAMPLE 6** Comparing Volumes

Which holds more soup (Figure 10.18): a can with a diameter of 3 inches and a height of 4 inches or a can with a diameter of 4 inches and a height of 3 inches?

Solution Recall that radius $= \frac{1}{2}$ diameter. Soup cans have the shape of right circular cylinders, so the volumes of the two cans are

Can 1: $V = \pi \times r^2 \times h = \pi \times (1.5 \text{ in.})^2 \times 4 \text{ in.} = 28.27 \text{ in.}^3$

Can 2: $V = \pi \times r^2 \times h = \pi \times (2 \text{ in.})^2 \times 3 \text{ in.} = 37.70 \text{ in.}^3$

FIGURE 10.18

Euclid alone has looked on Beauty bare . . .

—EDNA ST. VINCENT MILLAY, SONNET

The second can, with the larger radius but smaller height, has the larger capacity.

Now try Exercises 60–61. ◄

SCALING LAWS

As we saw in Unit 3B, scaling is a process by which a real object is modeled by a similar object whose dimensions are proportionally larger or smaller. For example, an architect might make a scale model of an auditorium in which all lengths were smaller by a factor of 100. Or a biologist might make a scale model of a cell in which all lengths were larger by a factor of 10,000. In these two examples, the *scale factors* are 100 and 10,000, respectively.

6 ft 5 ft

4 ft $\frac{6}{10}$ ft $\frac{5}{10}$ ft

$\frac{4}{10}$ ft

Actual car Model car

FIGURE 10.19

Suppose that we make an engineering model of a car with a scale factor of 10 (Figure 10.19). The actual car will be 10 times as long, 10 times as wide, and 10 times as tall as the model car. How will the surface area and volume of the actual car scale with the surface area and volume of the model?

Consider the area of the car roof, which is its length times its width. Because the length and width of the actual car roof are each 10 times their size in the model, the area of the actual car roof is

$$\begin{aligned}
\text{actual roof area} &= \text{actual roof length} \times \text{actual roof width} \\
&= (10 \times \text{model roof length}) \times (10 \times \text{model roof width}) \\
&= 10^2 \times \text{model roof length} \times \text{model roof width} \\
&= 10^2 \times \text{model roof area}
\end{aligned}$$

We see that the actual roof area is greater than the model roof area by the *square* of the scale factor ($10^2 = 100$, in this case).

We can do a similar calculation for the volume by considering, for example, the box-shaped (rectangular prism) passenger compartment.

$$\begin{aligned}
\text{actual volume} &= \text{actual length} \times \text{actual width} \times \text{actual height} \\
&= (10 \times \text{model length}) \times (10 \times \text{model width}) \times (10 \times \text{model height}) \\
&= 10^3 \times \text{model length} \times \text{model width} \times \text{model height} \\
&= 10^3 \times \text{model volume}
\end{aligned}$$

Thus, the actual car volume is greater than the model car volume by the *cube* of the scale factor ($10^3 = 1000$, in this case).

Scaling Laws

- *Lengths* always scale with the scale factor.
- *Areas* always scale with the *square* of the scale factor.
- *Volumes* always scale with the *cube* of the scale factor.

▶ **EXAMPLE 7** Doubling Your Size

Suppose that, magically, your size suddenly doubled—that is, your height, width, and depth doubled. For example, if you were 5 feet tall before, you now are 10 feet tall.

 a. By what factor has your waist size increased?
 b. How much more material will be required for your clothes?
 c. By what factor has your weight changed?

Solution

 a. Waist size is like a perimeter. Therefore, your waist size simply doubles, just like your other *linear* dimensions of height, width, and depth. If you had a 30-inch waist before, it is now a 60-inch waist.

 b. Clothing covers surface *area* and therefore scales with the *square* of the scale factor. The scale factor by which you have grown is 2 (doubling), so your surface area grows by a factor of $2^2 = 4$. If your shirt used 2 square yards of material before, it now uses $4 \times 2 = 8$ square yards of material.

 c. Your weight depends on your *volume*, which scales with the *cube* of the scale factor. Your new volume and new weight are therefore $2^3 = 8$ times their old values. If your old weight was 100 pounds, your new weight is $8 \times 100 = 800$ pounds. *Now try Exercises 62–74.* ◀

THE SURFACE AREA TO VOLUME RATIO

Another important concept in scaling is the *relative* scaling of areas and volumes. We define the **surface area to volume ratio** for any object as its surface area divided by its volume:

$$\text{surface area to volume ratio} = \frac{\text{surface area}}{\text{volume}}$$

Because surface area scales with the square of the scale factor and volume scales with the cube of the scale factor, the surface area to volume ratio must scale with the reciprocal of the scale factor:

$$\text{scaling of surface area to volume ratio} = \frac{(\text{scale factor})^2}{(\text{scale factor})^3} = \frac{1}{\text{scale factor}}$$

Thus, when an object is "scaled up," its surface area to volume ratio *decreases*. When an object is "scaled down," its surface area to volume ratio *increases*.

Surface Area to Volume Ratio

- Larger objects have *smaller* surface area to volume ratios than similarly proportioned small objects.
- Smaller objects have *larger* surface area to volume ratios than similarly proportioned large objects.

by the way

Scaling laws explain the appearance of many animals. Pressure, which determines how much weight an animal's body can support, is defined as the weight divided by the area bearing the weight. Because weight scales as the cube and area scales as the square (of length), making an animal larger increases the pressure on its joints. Nature therefore gives larger animals proportionally larger joints to bear the pressure. That is why, for example, elephants have relatively thicker legs than deer.

➤ **EXAMPLE 8** Chilled Drink

Suppose you have a few ice cubes and you want to cool your drink quickly. Should you crush the ice before you put it into your drink? Why or why not?

Solution A drink is cooled by contact between the liquid and the ice surface. Thus, the greater the surface area of the ice, the more rapidly the drink will cool. Because smaller objects have *larger* surface area to volume ratios, the crushed ice will have more total surface area than the same volume of ice cubes. The crushed ice therefore will cool the drink much more quickly. *Now try Exercises 75–80.* ◀

EXERCISES 10A

Review Questions

1. What do we mean by Euclidean geometry?

2. Give a geometric definition for each of the following: point, line, line segment, plane, plane segment, and space. Give an example of an everyday object that can represent each of these geometrical objects.

3. What do we mean by dimension? How is dimension related to the number of coordinates needed to locate a point?

4. Define an angle in geometric terms. What is the vertex? What do we mean when we say an angle subtends some portion of a circle? Distinguish among right angles, straight angles, acute angles, and obtuse angles.

5. What is plane geometry? What does it mean for lines to be perpendicular or parallel in a plane?

6. What is a polygon? How do we measure the perimeter of a polygon? Describe how we calculate the areas of a few simple polygons.

7. What are the formulas for the circumference and area of a circle?

8. Describe how we calculate the volumes and surface areas of a few simple three-dimensional objects.

9. What are the scaling laws for area and volume? Explain.

10. What is a surface area to volume ratio? How does this ratio change if we make an object bigger? smaller?

Does It Make Sense?

Decide whether each of the following statements makes sense (or is clearly true) or does not make sense (or is clearly false). Explain your reasoning.

11. The two highways look like straight lines and they intersect twice.

12. The city park is triangular in shape and has two right angles.

13. My bedroom is a rectangular prism that measures 12 feet by 10 feet by 8 feet.

14. Kara walked around the circular pond to a point on the opposite side, and Jamie swam at the same speed directly across the pond to the same point. Jamie must have arrived before Kara.

15. This basketball is shaped like a right circular cylinder.

16. By building a fence across my rectangular backyard, I can create two triangular yards.

Basic Skills & Concepts

Angles and Circles. In Exercises 17–22, find the degree measure of the angle that subtends the given part of a circle.

17. $\frac{1}{3}$ circle 18. $\frac{1}{12}$ circle

19. $\frac{1}{20}$ circle 20. $\frac{1}{30}$ circle

21. $\frac{1}{90}$ circle 22. $\frac{3}{4}$ circle

Fractions of Circles. In Exercises 23–30, find the fraction of a circle subtended by the given angle.

23. 1° 24. 4°

25. 15° 26. 30°

27. 60° 28. 90°

29. 180° 30. 235°

Circle Practice. In Exercises 31–36, find the circumference and area for each circle described.

31. A circle with a radius of 6 meters

32. A circle with a radius of 4 kilometers

33. A circle with a diameter of 25 centimeters

34. A circle with a radius of 16 meters

35. A circle with a diameter of 9 millimeters

36. A circle with a diameter of 0.5 kilometer

Perimeters and Areas. In Exercises 37–42, use Table 10.2 to find the perimeter and area of the given figure.

37. A square with sides of length 12 miles

38. A rectangle with a length of 5 inches and a width of 23 inches

39. A parallelogram with sides of length 10 feet and 20 feet, and a distance between the 20-foot sides of 5 feet

40. A square with sides of length 0.5 centimeter

41. A rectangle with a length of 2 meters and a width of 8 meters

42. A parallelogram with sides of length 1.5 feet and 6.5 feet, and a distance between the 6.5-foot sides of 0.2 foot

Triangle Geometry. In Exercises 43–46, find the perimeter and area of the triangle shown.

43.

44.

45.

46.

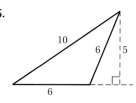

47. Window Space. A picture window has a length of 4 feet and a height of 3 feet, with a semicircular cap on each end (see Figure 10.20). How much metal trim is needed for the perimeter of the entire window, and how much glass is needed for the opening of the window?

FIGURE 10.20

48. A Running Track. A running track has straight legs of length 100 yards that are 50 yards apart (see Figure 10.21).

What is the length of the track, including the circular legs, and what is the area of the infield of the track?

FIGURE 10.21

49. Building Stairs. Refer to Figure 10.14, showing the region to be covered with plywood under a set of stairs. Suppose that the stairs rise at a steeper angle and are 14 feet tall. What is the area of the region to be covered in that case?

50. No Calculation Required. The end views of two different barns are shown in Figure 10.22. Without calculating, decide which end has the greater area. Explain how you know.

FIGURE 10.22

51. Parking Lot. A parking lot is bounded on four sides by streets, as shown in Figure 10.23. How much asphalt (in square yards) is needed to pave the parking lot?

FIGURE 10.23

52. City Park. Figure 10.24 shows a city park in the shape of a parallelogram with a rectangular playground in its center. If all but the playground is covered with grass, what area is covered with grass?

FIGURE 10.24

Three-Dimensional Objects. Use the formulas in Table 10.3 to solve Exercises 53–57.

53. A competition swimming pool is 50 meters long, 25 meters wide, and 2 meters deep. How much water does the pool hold?

54. A large convention center floor measures 40 meters by 60 meters, with a ceiling 10 meters high. How much air does it hold, in cubic meters? in liters?

55. A heat duct in the college library has a circular cross section with a radius of 10 inches and a length of 25 feet. What is the volume of the duct and how much paint (in square feet) is needed to paint the duct?

56. A lacrosse ball has a diameter of 4.5 inches. What are its volume and surface area?

57. Three tennis balls fit perfectly when stacked in a cylindrical can. Without calculating, state which is greater: the circumference of the can or the height of the can. Explain your reasoning.

58. Horse Trough. A water trough for horses is an open rectangular prism 10 feet long, 3 feet wide, and 2 feet deep. How much water does it hold when it is full? How much water does it hold after 60% of the water has evaporated?

59. Water Reservoir. The water reservoir for a city is shaped like a rectangular prism 300 meters long, 100 meters wide, and 15 meters deep. At the end of the day, the reservoir is 70% full. How much water must be added overnight to fill the reservoir?

60. Oil Drums. Which holds more: an oil drum with a radius of 2 feet that is 3 feet high or an oil drum with a radius of 1.5 feet that is 4 feet high?

61. Tree Volumes. Is there more wood in a 50-foot-high tree trunk with a radius of 1.5 feet or in an 80-foot-high tree trunk with a radius of 0.75 foot? Assume that the trees can be regarded as right circular cylinders.

Architectural Model. In Exercises 62–64, suppose you build an architectural model of a new concert hall using a scale factor of 25.

62. How will the height of the actual concert hall compare to the height of the scale model?

63. How will the surface area of the actual concert hall compare to the surface area of the scale model?

64. How will the volume of the actual concert hall compare to the volume of the scale model?

Architectural Model. In Exercises 65–67, suppose you build an architectural model of a new office complex using a scale factor of 50.

65. How will the height of the actual office complex compare to the height of the scale model?

66. How will the amount of paint needed for the exterior of the actual office complex compare to the amount of paint needed for the scale model?

67. Suppose you wanted to fill both the scale model office complex and the actual office complex with marbles. How many times the number of marbles required for the model would be required for the actual building?

Tripling Your Size. In Exercises 68–71, suppose you magically tripled in size—that is, your height, width, and depth tripled.

68. By what factor has your arm length increased?

69. By what factor has your waist size increased?

70. By what factor has the amount of material required for your clothes increased?

71. By what factor has your weight increased?

Comparing People. In Exercises 72–74, consider a person named Sam, who is 10% taller than you but proportioned in exactly the same way. (That is, Sam looks like a larger version of you.)

72. How tall are you? How tall is Sam?

73. What size is your waist? What size is Sam's waist?

74. How much do you weigh? How much does Sam weigh?

Squirrels or People? Use the following facts to work on Exercises 75–76. Squirrels and humans are both mammals that maintain a warm body temperature through metabolism that takes place in the body *volume*. Mammals must constantly generate internal heat to replace the heat that they lose through the *surface area* of their skin.

75. In general terms, how does the surface area to volume ratio of a squirrel compare to that of a human being?

76. Which animal must maintain a higher rate of metabolism to replace the heat lost through the skin: squirrels or humans? Based on your answer, which animal would you expect to eat more food in proportion to its body weight each day? Explain.

Earth and Moon. Use the following facts to work on Exercises 77–78. Both the Moon and the Earth are thought to have formed with similar internal temperatures about 4.6 billion years ago. Both worlds gradually lose this internal heat to space as the heat passes out through their surfaces.

77. The diameter of the Earth is about four times the diameter of the Moon. How does the surface area to volume ratio of the Earth compare to that of the Moon?

78. Based on your answer to Exercise 77, which would you expect to have a hotter interior today: the Earth or the Moon? Why? Use your answer to explain why the Earth remains volcanically active today, while the Moon has no active volcanoes.

79. Comparing Balls. Consider a softball with a radius of approximately 2 inches and a bowling ball with a radius of approximately 6 inches. Compute the surface area and volume for both balls. Then find the surface area to volume ratio for both balls. Which ball has the larger ratio?

80. Comparing Planets. The Earth has a radius of approximately 6400 kilometers, and Mars has a radius of approximately 3400 kilometers (assuming that the planets are spherical). Compute the surface area and volume for both planets. Which planet has the larger surface area to volume ratio?

Further Applications

81. Dimension. Examine a closed book.

 a. How many dimensions are needed to describe the book? Explain.

 b. How many dimensions describe the surface (cover) of the book? Explain.

 c. How many dimensions describe an edge of the book? Explain.

 d. Describe some aspect of the book that represents zero dimensions.

82. Perpendicular and Parallel. Suppose you mark a single point on a line that lies in a fixed plane. Can you draw any other lines in that plane that pass through the point and are *perpendicular* to the original line? Can you draw any other lines that pass through the point and are *parallel* to the original line? Explain.

83. Perpendicular and Parallel. Suppose you draw two parallel lines in a plane. If a third line is perpendicular to one of the two parallel lines, is it necessarily perpendicular to the other as well? Explain.

84. Backyard. Figure 10.25 shows the layout of a backyard that is to be seeded with grass except for the patio and flower garden. What area is to be seeded with grass?

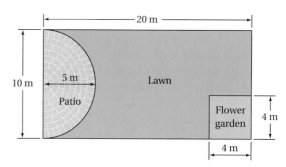

FIGURE 10.25

85. Human Lung. The human lung has approximately 300 million nearly spherical air sacs (alveoli), each with a diameter of about $\frac{1}{3}$ millimeter. The key feature of the air sacs is their surface area, because on their surfaces gas is exchanged between the bloodstream and the air.

 a. What is the total surface area of the air sacs? What is the total volume of the air sacs?

 b. Suppose a single sphere were made that had the same volume as the total volume of the air sacs. What would be the radius and surface area of such a sphere? How would this surface area compare to that of the air sacs?

c. If a single sphere had the same surface area as the total surface area of the air sacs, what would be its radius? Based on your results, comment on the design of the human lung.

86. **Automobile Engine Capacity.** The size of a car engine is often stated as the total volume of its cylinders.

a. American car manufacturers often state engine sizes in cubic inches. Suppose that a six-cylinder car has cylinders with a radius of 2.22 inches and a height of 3.25 inches. What is the engine size?

b. Foreign car manufacturers often state engine sizes in liters. Compare the engine size in part a to that of a foreign car with a 2.2-liter engine.

c. Look up the number of cylinders and the engine size for your car (if you don't own a car, choose a car that you'd like to own). Estimate the dimensions (radius and height) of the cylinders in your car. Explain your work.

87. **The Chunnel.** The world's longest tunnel is the English Channel Tunnel, or "Chunnel," connecting Dover, England, and Calais, France. The Chunnel consists of three separate, adjacent tunnels. Its length is approximately 50 kilometers. Each of the three tunnels is shaped like a half-cylinder with a radius of 4 meters (the height of the tunnel) and a length of 50 kilometers (the length of the tunnel). How much earth (volume) was removed to build the Chunnel?

Web Projects

Find useful links for Web Projects on the text Web site: www.aw.com/bennett-briggs

88. **The Geometry of Ancient Cultures.** Although the work of the ancient Greeks became the basis of modern geometry, many other ancient cultures made extensive use of geometry. Research the use of geometry in an ancient culture of your choice. Some possible areas of focus: (1) study the use of geometry in ancient Chinese art and architecture; (2) investigate the geometry and purpose of Stonehenge; (3) compare and contrast the geometry of the Egyptian pyramids and those of Central America; (4) study the geometry and possible astronomical orientations of Anasazi buildings and communities; or (5) research the use of geometry in the ancient African empire of Aksum (in modern-day Ethiopia).

89. **Surveying and GIS.** Surveying is one of the oldest and most practical applications of geometry. A modern technology for surveying is called Geographical Information Systems, or GIS. Use the Web to learn about GIS. In two pages or less, describe how GIS works, along with a few of its applications.

90. **Platonic Solids.** Why are there five and only five perfect, or Platonic, solids? What special significance did these objects have to the Greeks? Write a short paper on this or any other aspect of Platonic solids.

In the News

91. **Geometry in the News.** Find a recent news report describing a project that requires geometric ideas or techniques. Explain how geometry is used in the project.

92. **Geometric Idealizations.** Describe at least three instances in which the geometric ideals of points, lines, and planes are used in our society.

93. **Circles and Polygons.** Describe at least three instances in which circles or polygons play an important role in your daily life.

94. **Three-Dimensional Objects.** Describe at least three instances in which three-dimensional objects play an important role in your daily life.

PROBLEM SOLVING WITH GEOMETRY

In Unit 10A, we surveyed the geometrical properties of two- and three-dimensional objects. In this unit, we'll investigate a few additional geometrical ideas that can be used to solve many practical problems.

USES OF ANGLES

Many problems can be solved simply by working with angles in an appropriate way. Such problems often require that we measure angles precisely to a fraction of a degree.

We can write fractions of a degree with decimals (for example, 45.23°), but it is more common to subdivide each degree into 60 *minutes of arc* and each minute into 60 *seconds of arc* (see Figure 10.26). We use the symbols ′ and ″ for minutes and seconds of arc, respectively. For example, we read 30°33′31″ as "30 degrees, 33 minutes, and 31 seconds."

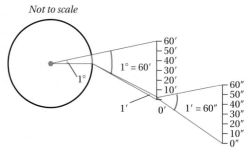

Not to scale

FIGURE 10.26 Each degree is subdivided into 60 minutes of arc, and each minute is subdivided into 60 seconds of arc. This figure is not drawn to scale!

▶ **EXAMPLE 1** Fractional Degrees

a. Convert 3.6° into degrees, minutes, and seconds of arc.
b. Convert 30°33′31″ into decimal form.

Solution

a. Because 3.6° = 3° + 0.6°, we first convert 0.6° into minutes of arc:

$$0.6° \times \frac{60′}{1°} = 36′$$

Thus, 3.6° = 3° + 36′ = 3°36′.

b. Remember that 1 minute of arc is $\frac{1}{60}$°, and 1 second of arc is $\frac{1}{60}$ of a minute of arc, or $\frac{1}{60 \times 60}$°. Thus,

$$30°33′31″ = 30° + 33′ + 31″ = 30° + \left(\frac{33}{60}\right)° + \left(\frac{31}{60 \times 60}\right)°$$

$$= 30° + 0.55° + 0.00861° = 30.55861°$$

Now try Exercises 15–28. ◀

Latitude and Longitude

One common use of angle measures is to locate positions on the Earth (Figure 10.27). **Latitude** measures positions north or south of the equator, which is defined to have latitude 0°. Locations in the Northern Hemisphere have a latitude denoted N (for north) and locations in the Southern Hemisphere have a latitude denoted S (for south). For example, the North Pole and the South Pole have latitude 90°N and 90°S,

respectively. Note that lines of latitude (also called *parallels* of latitude) are actually circles running parallel to the equator.

Longitude measures east-west position, so lines of longitude (also called *meridians* of longitude) are semicircles extending from the North Pole to the South Pole. The line of longitude passing through Greenwich, England, is defined to be longitude 0° and is called the **prime meridian.** Longitudes are usually given with an angle less than (or equal to) 180°, so a location less than halfway around the world to the east of the prime meridian has a longitude designation of E (for east) and a location to the west has a longitude designation of W (for west).

Stating a latitude and a longitude pinpoints a location on Earth. For example, Figure 10.27 shows latitudes and longitudes for Miami, Rome, and Buenos Aires.

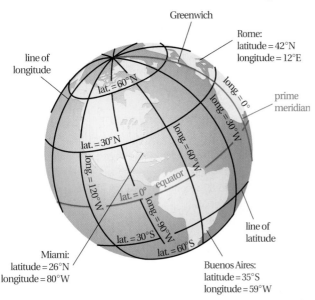

FIGURE 10.27 We can locate any place on the Earth's surface by its latitude and longitude.

➤ **EXAMPLE 2** Latitude and Longitude

Answer each of the following questions, and explain your answers clearly.

a. Suppose you could drill from Miami straight through the center of the Earth and continue in a straight line out the other side of the Earth. At what latitude and longitude would you emerge?

b. Perhaps you've heard that if you dug a straight hole from the United States through the center of the Earth, you'd come out in China. Is this true?

c. Suppose you travel 1° of latitude north or south. How far have you traveled? (Hint: The circumference of the Earth is about 25,000 miles.)

d. Suppose you travel 1° of longitude east or west. How far have you traveled?

Solution

a. The point on Earth directly opposite Miami must have the opposite latitude (26°S rather than 26°N) and must be 180° away in longitude. Let's consider going 180° eastward from Miami's longitude of 80°W. The first 80° eastward would take us to the prime meridian, so we'd have to continue to longitude 100°E to be a full 180° away from Miami. Thus, the position of the point opposite Miami is latitude 26°S and longitude 100°E.

b. It is *not* true. The United States is in the Northern Hemisphere, so the points on the globe opposite the United States must be in the Southern Hemisphere. China is also in the Northern Hemisphere, so it cannot be opposite the United States.

c. If you study Figure 10.27 (or a globe), you'll see that lines of latitude are all parallel to each other. Thus, every degree of latitude represents the same

north or south distance. There are 360° all the way around the Earth's 25,000-mile circumference, so each degree of latitude represents

$$\frac{25{,}000 \text{ mi}}{360°} = 69.4 \text{ mi per degree}$$

Thus, traveling 1° of latitude north or south means traveling a distance of approximately 70 miles.

d. If you study Figure 10.27 (or a globe), you'll see that lines of longitude are *not* parallel to each other. Instead, they are much closer together near the poles than near the equator. Thus, we cannot answer the question without also knowing the latitude. More specifically, 1° of longitude represents about 70 miles along the equator (by the same calculation used in part c), but represents less as you head north or south. *Now try Exercises 29–36.* ◄

Angular Size and Distance

If you hold a quarter right in front of your eye, it can block your entire field of view. But as you move it farther away, it appears to get smaller and it blocks less of your view. The true size of the quarter does not change, of course. Instead, what changes is its **angular size**—the angle that it covers as seen from your eye.

Figure 10.28a shows that, in general, the angular size of an object depends on its true, or physical, size and its distance. To find out precisely how angular size and physical size are related, we can use a little trick. Figure 10.28b shows that as long as an object's angular size is relatively small (less than a few degrees), the object's physical size is approximately the same as the arc length of a piece of circle drawn around it (with your eye at the center). Figure 10.28c shows that, because there are 360° in a full circle and the circle has circumference $2\pi r$, each 1° of angle must subtend a piece of arc that is $2\pi r/360$ in length. Note that the arc length would be twice as long if the angle were 2° instead of 1°, and half as long if the angle were $\frac{1}{2}$° rather than 1°. Generalizing these ideas, we get the following formula for the arc length subtended by any angle A:

$$\text{arc length} = \frac{2\pi r}{360°} \times A,$$

where angle A is measured in degrees, or

$$A = \text{arc length} \times \frac{360°}{2\pi r}$$

Now, if you compare Figures 10.28b and 10.28c, you'll see that the arc length corresponds to the physical size of the object, the angle A corresponds to the angular size of the object, and the radius r corresponds to the object's distance from your eye. Thus, we can write the formula as shown in the box on the next page.

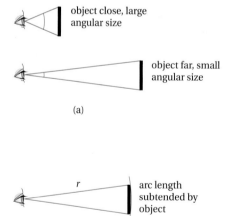

(a) object close, large angular size

object far, small angular size

(b) r — arc length subtended by object — piece of circle with radius r

If this angle is 1°, or 1/360 of a full circle, then

r ... $1°$... this arc length must be 1/360 of the circle's circumference, or $2\pi r/360$

(c)

FIGURE 10.28 Understanding the relationship between angular size and distance.

Angular Size, Physical Size, and Distance

The farther away an object is located from you, the smaller it will appear in angular size. As long as an object's angular size is less than a few degrees, the following formula relates its angular size, physical size, and distance:

$$\text{angular size} = \text{physical size} \times \frac{360°}{2\pi \times \text{distance}}$$

▶ **EXAMPLE 3** Angular Size and Distance

 a. A quarter is about 1 inch in diameter. Approximately how big will it look in angular size if you hold it 1 yard (36 inches) from your eye?

 b. The angular diameter of the Moon as seen from Earth is approximately 0.5°, and the Moon is approximately 380,000 kilometers from Earth. What is the real diameter of the Moon?

Solution

 a. We use the formula relating angular size, physical size, and distance, with 1 inch as the quarter's physical size and 36 inches as its distance:

$$\text{angular size} = (1 \text{ in.}) \times \frac{360°}{2\pi \times (36 \text{ in.})} \approx 1.6°$$

The angular size of a quarter at a distance of 1 yard is about 1.6°.

 b. In this case, we are asked to find the Moon's physical size from its angular size and distance, so we must solve for the formula for physical size. You should confirm that the formula becomes

$$\text{physical size} = \text{angular size} \times \frac{2\pi \times \text{distance}}{360°}$$

Now we substitute 380,000 kilometers for the distance and 0.5° for the angular diameter:

$$\text{physical size} = 0.5° \times \frac{2\pi \times 380,000 \text{ km}}{360°} \approx 3300 \text{ km}$$

The Moon's real diameter is about 3300 kilometers.

Now try Exercises 37–40. ◀

Pitch, Grade, and Slope

In many applications, it's common to express steepness using terms other than the degree measure of the angle. For example, consider a road that rises uniformly 2 feet in the vertical direction for every 20 feet in the horizontal direction (Figure 10.29). The road makes an angle with the horizontal that we could measure in degrees. However, it is more common to say that the road has a **pitch** of 2 in 20, or 1 in 10. It is also common to describe the rise of

2 ft

20 ft

slope = $\frac{1}{10}$
pitch = 2 in 20 = 1 in 10
grade = 10%

FIGURE 10.29 A road that rises 2 feet in the vertical direction for every 20 feet in the horizontal direction has a slope of 1/10, a pitch of 2 in 20, and a grade of 10%.

the road in terms of its **slope,** as discussed in Chapter 9. In this case, the slope, or *rise over run*, is $2/20 = 1/10$. If we express the slope as a percentage, it's called a **grade.** Thus, the grade of the road in this case is $1/10 = 10\%$.

There is no royal road to geometry.

—EUCLID, TO KING PTOLEMY I

▶ EXAMPLE 4 How Steep?

a. Suppose a road has a 100% grade. What is its slope? What is its pitch? What angle does it make with the horizontal?

b. Which is steeper: a road with an 8% grade or a road with a pitch of 1 in 9?

c. Which is steeper: a roof with a pitch of 2 in 12 or a roof with a pitch of 3 in 15?

Solution

a. A 100% grade means a slope of 1 and a pitch of 1 in 1. If you look at a line with a slope of 1 on a graph, you'll see that it is a diagonal line that makes an angle of 45° with the horizontal.

b. A road with a pitch of 1 in 9 has a slope of $1/9 = 0.11$, or 11%. Therefore, this road is steeper than a road with an 8% grade.

c. The 2 in 12 roof has a slope of $2/12 = 0.167$, while the 3 in 15 roof has a slope of $3/15 = 0.20$. The second roof is steeper. *Now try Exercises 41–50.* ◀

FURTHER USES OF TRIANGLES

In Unit 10A, we used the area formula for a triangle to solve practical problems. Here, we investigate two other ways in which we can make use of triangles in modeling real situations.

Using the Pythagorean Theorem

The Pythagorean theorem is surely one of the most famous theorems in mathematics. We already proved the theorem in Unit 1D and put it to use in Unit 2C (see Examples 5 and 6). Because it is so important—and because it is rooted in geometry—we will now explore a few other uses of this theorem.

The Pythagorean Theorem

The Pythagorean theorem applies only to right triangles (those with one 90° angle). For a right triangle with side lengths *a*, *b*, and *c*, in which *c* is the longest side (or *hypotenuse*), the Pythagorean theorem states that

$$a^2 + b^2 = c^2$$

▶ EXAMPLE 5 Distance Measurements

Consider the map in Figure 10.30, showing several city streets in a rectangular grid. The individual city blocks are $\frac{1}{8}$ of a mile in the east-west direction and $\frac{1}{16}$ of a mile in the north-south direction.

FIGURE 10.30 A map of several city blocks, showing the locations of the library and the subway.

a. How far is the library from the subway along the path shown?
b. How far is the library from the subway "as the crow flies" (that is, along a straight diagonal path)?

Solution

a. If you follow the path shown, you'll see that it goes 6 blocks east and 8 blocks north. Thus, the total distance along this path is

$$(6 \times \tfrac{1}{8}\,\text{mi}) + (8 \times \tfrac{1}{16}\,\text{mi}) = \tfrac{3}{4}\,\text{mi} + \tfrac{1}{2}\,\text{mi} = 1\tfrac{1}{4}\,\text{mi}$$

b. As shown in the figure, the distance "as the crow flies" is the hypotenuse of a right triangle. The horizontal and vertical sides of this triangle have lengths $\tfrac{3}{4}$ mile and $\tfrac{1}{2}$ mile, respectively. From the Pythagorean theorem, the length of the hypotenuse, or side c, is

$$c^2 = a^2 + b^2 = (0.75\,\text{mi})^2 + (0.5\,\text{mi})^2 = 0.8125\,\text{mi}^2$$

Taking the square root of both sides, we see that the straight-line distance between the subway and the library is

$$c = \sqrt{0.8125\,\text{mi}^2} = 0.90\,\text{mi}$$

Not surprisingly, the direct path is shorter. *Now try Exercises 51–56.* ◄

Time out to think _____

Using the map in Figure 10.30, see whether there are other routes that have the same length (1.25 miles) as the route shown in bold. If so, how many? Are there shorter routes between the subway and the library that follow the streets?

➤ **EXAMPLE 6** Lot Size

Find the area, in acres, of the mountain lot shown in Figure 10.31.

Solution We can find the area of the triangular lot using the formula $A = \tfrac{1}{2} \times \text{base} \times \text{height}$ (see Unit 10A). We can consider the 250-foot frontage along the stream as the base of the triangle. The height of the triangle is the unlabeled side in Figure 10.31, which we need to figure out. We can do so with the Pythagorean theorem, since we are given that the hypotenuse is the 1200-foot property line. We start by writing the Pythagorean theorem in the form

$$\text{base}^2 + \text{height}^2 = \text{hypotenuse}^2$$

You should confirm that solving this equation for the height gives

$$\text{height} = \sqrt{\text{hypotenuse}^2 - \text{base}^2}$$

Substituting the given values, we find that

$$\text{height} = \sqrt{(1200\,\text{ft})^2 - (250\,\text{ft})^2} \approx 1174\,\text{ft}$$

FIGURE 10.31 A mountain lot shaped like a right triangle, with frontage on a stream.

Now we can use the formula for the area of a triangle:

$$\text{Area} = \tfrac{1}{2} \times \text{base} \times \text{height} = \tfrac{1}{2} \times 250\,\text{ft} \times 1174\,\text{ft} = 146{,}750\,\text{ft}^2$$

Finally, we convert this area to acres, remembering that 1 acre = 43,560 ft^2 (see Unit 2A).

$$146{,}750\,\text{ft}^2 \times \frac{1\,\text{acre}}{43{,}560\,\text{ft}^2} \approx 3.4\,\text{acre}$$

The lot has an area of about 3.4 acres. *Now try Exercises 57–60.* ◄

Similar Triangles

Two triangles are **similar** if they have the same shape, but not necessarily the same size. Having the same shape means that one triangle is a scaled-up or scaled-down version of the other triangle. Similar triangles provide a powerful tool for geometrical modeling.

Figure 10.32 shows two similar triangles. The angles are labeled with capital letters (*A, B, C* and *A′, B′, C′*) and the sides with lowercase letters (*a, b, c* and *a′, b′, c′*). By convention, angle *A* is opposite side *a*, angle *B′* is opposite side *b′*, and so forth. Notice that the labels were chosen so that it is easy to see the similarity. For example, angle *A* and angle *A′* look the same. If you measure the angles and sides in the two similar triangles, you'll find the following key properties.

FIGURE 10.32 These two triangles are similar because they have the same shape. That is, their angles are the same and the ratios of their side lengths are the same.

Similar Triangles

Two triangles are similar if they have the same shape (but not necessarily the same size), meaning that one is a scaled-up or scaled-down version of the other. For two similar triangles,

- corresponding pairs of angles in each triangle are equal. That is, angle *A* = angle *A′*, angle *B* = angle *B′*, and angle *C* = angle *C′*.
- the ratios of the side lengths in the two triangles are all equal:

$$\frac{a}{a'} = \frac{b}{b'} = \frac{c}{c'}$$

▶**EXAMPLE 7** Practice with Similar Triangles

Figure 10.33 shows two similar triangles. Find the lengths of the sides labeled *a* and *c′*.

Solution We start with the property

$$\frac{a}{a'} = \frac{b}{b'} = \frac{c}{c'}$$

FIGURE 10.33 Two similar triangles with two sides of unknown length.

Inserting the known side lengths, we find

$$\frac{a}{4} = \frac{9}{6} = \frac{12}{c'}$$

We can now find the side length a by working with the two terms on the left:

$$\frac{a}{4} = \frac{9}{6} \xrightarrow{\text{multiply both sides by 4}} a = \frac{9 \times 4}{6} = 6$$

Thus, the side length a is 6. To find the side length c', we work with the equation

$$\frac{9}{6} = \frac{12}{c'} \xrightarrow{\text{solve for } c'} c' = \frac{12 \times 6}{9} = 8$$

The side length c' is 8. (You should verify the algebra above.)

Now try Exercises 61–68. ◄

▶ **EXAMPLE 8** Solar Access

Some cities have policies that prevent property owners from constructing new houses and additions that cast shadows on neighboring houses. The intent of these policies is to allow everyone access to the Sun for the use of solar energy devices. Consider the following solar access policy:

> *On the shortest day of the year, a house cannot cast a noontime shadow that reaches farther than the shadow that would be cast by a 12-foot fence on the property line.*

Suppose your house is set back 30 feet from the north property line and that, on the shortest day of the year, a 12-foot fence on that property line would cast a noontime shadow 20 feet in length. If you build an addition to your house, how high can the north side of the remodeled house be under this policy?

Solution The key to solving this problem is drawing a good picture, so that we can see how similar triangles can be used to solve it. Figure 10.34 shows the geometry.

FIGURE 10.34 The figure shows a 12-foot fence on the property line and the 20-foot shadow it casts on the shortest day of the year. Set back 30 feet from the property line, the house is drawn at the maximum height at which its shadow extends to the same point as the fence shadow.

Notice these features of the figure:

- The smaller triangle (at left) shows the 12-foot fence and the 20-foot length of its shadow along the ground.
- The 30-foot setback goes between the fence and the north side of the house. The maximum height of the north side, labeled h, is what we are trying to find.
- Because we are looking for the maximum height of the north side of the house, we want the shadow from this side of the house to reach the same point as the fence shadow. Thus, we extend the shadow line (pointing to the Sun) from the fence to the top of the house. Along the ground, this shadow length is 50 feet (30 feet for the setback plus 20 feet for the fence shadow).
- The end result is a figure with two similar triangles. The smaller one is formed by the 12-foot fence and its 20-foot shadow, and the larger one is formed by the house with unknown height h and its 50-foot shadow.

We can now apply the ratio property for the side lengths of similar triangles. In this case, the ratio of the house height to the fence height must be the same as the ratio of the house shadow length to the fence shadow length:

$$\frac{\text{house height}}{\text{fence height}} = \frac{\text{house shadow length}}{\text{fence shadow length}}$$

Substituting the known values, we find

$$\frac{h}{12\text{ ft}} = \frac{50\text{ ft}}{20\text{ ft}} \xrightarrow{\text{multiply both sides by 12 ft}} h = \frac{50\text{ ft}}{20\text{ ft}} \times 12\text{ ft} = 30\text{ ft}$$

The maximum height of the north side of the house after the remodeling is 30 feet.

Now try Exercises 69–72. ◄

Equations are just the boring part of mathematics. I attempt to see things in terms of geometry.

—STEPHEN HAWKING,
ENGLISH PHYSICIST

OPTIMIZATION PROBLEMS

Optimization problems seek a "best possible" solution. For example, they may seek to find the largest volume or lowest cost that can be obtained under particular conditions. These problems have many uses. We'll look at optimization problems involving *networks* in Chapter 13. To get the general idea, let's consider two geometrical optimization problems here.

➤ **EXAMPLE 9** *Optimizing Area*

You have 132 meters of fence that you plan to use to enclose a corral on a ranch. What shape should you choose if you want the corral to have the greatest possible area? What is the area of this "optimized" corral?

Solution A good way to start this problem is to experiment with a loop of string, making it into various shapes. Notice that a long narrow shape doesn't enclose as much area as a simpler shape like a circle or a rectangle (Figure 10.35). After some experimentation, you can probably convince yourself that the largest area will be found with either a

FIGURE 10.35 All figures shown have the same perimeter, but different areas. The circle has the greatest area. In fact, the circle has the greatest area of all two-dimensional figures with a fixed perimeter.

square or a circle. We can decide between these two possibilities by calculating the areas.

If you use your 132 meters of fence to make a square, each side will have length $132/4 = 33$ meters. The enclosed area of the square will be

$$A = (33 \text{ m})^2 = 1089 \text{ m}^2$$

If you make a circle, the 132 meters of fence will be the circumference of the circle. Because circumference $= 2\pi r$, the radius of the circle will be $r = 132/2\pi$ meters (or 21.008 meters). The area of the circle will therefore be

$$A = \pi r^2 = \pi \left(\frac{132}{2\pi} \text{ m} \right)^2 = 1387 \text{ m}^2$$

We conclude that a circle is the shape with the greatest possible area for a fixed length of perimeter. In this case, your best possible choice is to use the 132 meters of fence to make a circular corral with an area of 1387 square meters.

Now try Exercises 73–76. ◄

►EXAMPLE 10 Optimal Container Design

You are designing a wooden crate (rectangular prism) that must have a volume of 2 cubic meters. The cost of the wood is $12 per square meter. What dimensions give the most economical design? With that optimal design, how much will the material for each crate cost?

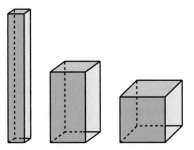

FIGURE 10.36 All three boxes shown have the same volume, but the cube has the *smallest* total surface area.

Solution The wood makes the surfaces of the crate, so to minimize cost we are looking for the minimum *surface area* for a crate with a volume of 2 cubic meters. As in the previous example, you might begin by experimentation, making either little paper boxes or good drawings of boxes (Figure 10.36). Notice that, for a particular amount of paper, a tall skinny box contains much less volume than a cube. Indeed, with experimentation or calculation of volumes you should convince yourself that for a given total surface area, a cube-shaped box has the greatest volume. Turning this around, a cube-shaped box has the minimum surface area for a given volume.

Thus, the crates should be cubes. The volume of a cube is $V = (\text{side length})^3$, so for a volume of 2 cubic meters the side length is

$$\text{side length} = \sqrt[3]{2 \text{ m}^3} = 1.26 \text{ m}$$

A cube has six sides, each with a surface area equal to the square of the side length. Thus, the total surface area of the cube will be

$$\text{surface area} = 6 \times (1.26 \text{ m})^2 = 9.53 \text{ m}^2$$

Since the wood costs $12 per square meter, the total cost for the wood in each crate will be

$$\text{total cost} = 9.53 \text{ m}^2 \times \frac{\$12}{\text{m}^2} \approx \$114$$

Now try Exercises 77–80. ◄

■ EXERCISES 10B

Review Questions

1. How do we describe fractions of a degree of angle?

2. Explain how a position on the Earth can be specified using latitude and longitude.

3. How is angular size related to physical size?

4. Give four different ways to describe the angle that a surface (such as a road) makes with the horizontal.

5. Give at least two examples of ways in which the Pythagorean theorem can be useful in solving a practical problem.

6. Make a sketch of two similar triangles, and explain the properties that make the triangles similar.

7. Give an example of a practical problem that can be solved with similar triangles.

8. What is an optimization problem? Give an example.

Does It Make Sense?

Decide whether each of the following statements makes sense (or is clearly true) or does not make sense (or is clearly false). Explain your reasoning.

9. In December, it is winter at 70°W and 44°S.

10. When I looked at the Moon in the sky, it seemed to be a mile wide.

11. I should have no trouble riding my bike up a road with a grade of 5%.

12. The angles in triangle A have a sum of 180°, as do the angles in triangle B. Therefore, the two triangles are similar.

13. The sides of triangle A are half as long as the corresponding sides of triangle B. Therefore, the two triangles are similar.

14. There are many different boxes that have the same volume, but only one box has the smallest possible surface area.

Basic Skills & Concepts

Angle Conversions I. In Exercises 15–20, convert the given degree measure into degrees, minutes, and seconds of arc.

15. 26.5°

16. 41.333°

17. 15.0161°

18. 0.002°

19. 126.9972°

20. 23.45°

Angle Conversions II. In Exercises 21–26, convert the given angle measure into degrees and decimal fractions of a degree. For example, 30°30′ = 30.5°.

21. 10°15′

22. 10°15′15″

23. 36°45′8″

24. 90°35′27″

25. 1°1′1″

26. 0°22′58″

27. **Minutes in a Circle.** Calculate the number of minutes of arc in a full circle.

28. **Seconds in a Circle.** Calculate the number of seconds of arc in a full circle.

Latitude and Longitude. Consult an atlas or globe as needed to answer the questions in Exercises 29–36.

29. Find the latitude and longitude of Sydney, Australia.

30. Find the latitude and longitude of Ankara, Turkey.

31. Find the latitude and longitude of the location on Earth precisely opposite Honolulu, Hawaii (latitude 21°N, longitude 158°W).

32. Find the latitude and longitude of the location on Earth precisely opposite Omaha, Nebraska (latitude 41°N, longitude 96°W).

33. Which is farther from the North Pole: Buenos Aires, Argentina, or Capetown, South Africa? Explain.

34. Which is farther from the South Pole: Guatemala City, Guatemala, or Lusaka, Zambia? Explain.

35. Buffalo, New York, is at nearly the same longitude as Miami, Florida, but Buffalo's latitude is 43°N while Miami's latitude is 26°N. About how far away is Buffalo from Miami? Explain.

36. Washington, DC, is at about latitude 38°N and longitude 77°W. Lima, Peru, is at about latitude 12°S and longitude 77°W. About how far apart are the two cities? Explain.

Angular Size. Use the formula relating angular size, physical size, and distance in Exercises 37–40.

37. What is the angular size of a quarter viewed from a distance of 10 yards?

38. What is the angular size of a quarter viewed from a distance of 50 yards?

39. The Sun has an angular diameter of about 0.5° and a distance of about 150 million kilometers. What is its true diameter?

40. You are looking at a tree on a hillside that is 1 mile away. You measure the tree's angular height to be about $\frac{1}{3}°$. How tall is the tree?

Slope, Pitch, Grade. In Exercises 41–44, determine which surface is steeper.

41. A roof with a pitch of 1 in 3 or a roof with a slope of $\frac{3}{10}$

42. A road with a 12% grade or a road with a pitch of 1 in 8

43. A railroad track with a 4% grade or a railroad track with a slope of $\frac{1}{20}$

44. A sidewalk with a pitch of 1 in 6 or a sidewalk with a 15% grade

45. **Slope of a Roof.** What is the slope of a 5 in 12 roof? If the roof rises in 10 horizontal feet, how high does it rise vertically?

46. **Grade of a Road.** How much does a road with a 3% grade rise for each horizontal foot? If you drive along this road for 8 miles, how much elevation will you gain?

47. **Pitch of a Roof.** What is the angle (relative to the horizontal) of a 6 in 6 roof? Is it possible to have a 7 in 6 roof? Explain.

48. **Grade of a Path.** What is the approximate grade (expressed as a percentage) of a path that rises 1000 feet every mile?

49. **Grade of a Road.** What is the grade of a road that rises 10 feet for every 120 horizontal feet?

50. **Grade of a Trail.** How much does a trail with a 15% grade rise for each 100 horizontal yards?

Map Distances. Exercises 51–56 refer to the map in Figure 10.37. Assume that the length of each east-west block is $\frac{1}{8}$ mile and the length of each north-south block is $\frac{1}{5}$ mile.

FIGURE 10.37

For the locations given in each exercise, do the following:

 a. Find the shortest possible walking distance (following the streets) between the two locations.

 b. Find the straight-line distance ("as the crow flies") between the two locations.

51. The bus stop and the library

52. The bus stop and the grocery store

53. The bus stop and the theater

54. The theater and the library

55. The grocery store and the library

56. The grocery store and the theater

Acreage Problems. For Exercises 57–60, refer to Figure 10.31, but use the lengths given in the exercise. Find the area in acres of the property under the given assumptions.

57. The stream frontage is 125 feet in length and the property line is 600 feet in length.

58. The stream frontage is 200 feet in length and the property line is 2000 feet in length.

59. The stream frontage is 60 feet in length and the property line is 400 feet in length.

60. The stream frontage is 1200 feet in length and the property line is 3500 feet in length.

Determining Similarity. In Exercises 61–64, determine which pairs of triangles are similar, and explain how you know.

61.

62.

63.

64.

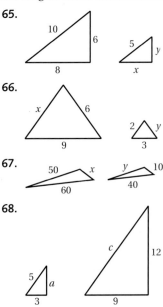

Analyzing Similar Triangles. In Exercises 65–68, determine the lengths of the unknown sides.

65.

66.

67.

68.

Solar Access. In Exercises 69–72, assume that the policy given in Example 8 is in force, and find the maximum allowed height for each house.

69. A 12-foot fence on the property line casts a 20-foot shadow on the shortest day of the year, and the north side of the house is set back 40 feet from the property line.

70. A 12-foot fence on the property line casts a 30-foot shadow on the shortest day of the year, and the north side of the house is set back 40 feet from the property line.

71. A 12-foot fence on the property line casts a 20-foot shadow on the shortest day of the year, and the north side of the house is set back 20 feet from the property line.

72. A 12-foot fence on the property line casts a 30-foot shadow on the shortest day of the year, and the north side of the house is set back 20 feet from the property line.

Optimizing Area. In Exercises 73–76, determine the area of a circular enclosure and a square enclosure made with the given amount of fence. Compare the areas and comment.

73. 120 meters of fence

74. 900 feet of fence

75. 450 meters of fence

76. 2500 yards of fence

77. Designing Cans. Suppose you work for a company that manufactures cylindrical cans. Which will cost more to manufacture: a can with a radius of 3 inches and a height of 4 inches or a can with a radius of 4 inches and a height of 3 inches? Assume that the cost of material for the tops and bottoms is $1.00 per square inch and the cost of material for the curved surfaces is $0.50 per square inch.

78. Designing Plastic Buckets. A company manufactures plastic buckets that are shaped like cylinders without a lid. Which will cost more to manufacture: a bucket with a radius of 0.5 foot and a height of 1.5 feet or a bucket with a radius of 0.75 foot and a height of 1.2 feet? Assume that the plastic material costs $0.50 per square foot, but the bottom of each bucket must have double thickness.

79. Designing Cardboard Boxes. Suppose you are designing a cardboard box that must have a volume of 6 cubic feet. The cost of the cardboard is $0.25 per square foot. What is the most economical design for the box (the one that minimizes the cost), and how much will the material in each box cost?

80. Designing Steel Safes. A large steel safe with a volume of 8 cubic feet is to be designed in the shape of a rectangular prism. The cost of the steel is $8.50 per square foot. What is the most economical design for the safe, and how much will the material in each such safe cost?

Further Applications

81. Reading with the Hubble Space Telescope. The Hubble Space Telescope orbits Earth at an altitude of about 350 kilometers. If it were looking at words (rather than stars), you could use it to read letters with an angular size of 0.1 second of arc. Suppose the Hubble Space Telescope were looking down on Earth. How big would a letter on a page have to be for you to be able to read it with the Hubble Space Telescope? Could you read the letters in a newspaper from orbit? Explain.

82. **Angular Size of a Star.** Imagine a star that is the same size as the Sun (diameter about 1.4 million km) but is located 10 light-years away. What is the star's angular size as seen from Earth? Given that the most powerful telescopes available today can see details no smaller than about 0.01 second of arc, is it possible to see any details on the surface of this star? Explain. (Hint: 1 light-year $\approx 10^{13}$ km.)

83. **Baseball Geometry.** The distance between bases on a baseball diamond (which is really a square) is 90 feet. How far does the catcher throw the ball from home plate to second base (along the diagonal of the diamond)?

84. **Saving Costs.** A phone line runs east along a field for 0.5 mile and then north along the edge of the same field for 0.75 mile. If the phone line cost $3000 per mile to install, how much could have been saved if the phone line had been installed diagonally across the field? Draw a good picture!

85. **Travel Times.** To get to a cabin, you have a choice of either walking west from a parking lot along the edge of a rectangular reservoir for 1.5 miles and then south along the edge of the reservoir for 0.8 mile or swimming directly from the parking lot. If you can walk twice as fast as you can swim, which is the faster route?

86. **Keep off the Grass.** An old principle of public landscaping says "build sidewalks last." In other words, let the people find their chosen paths and then build the sidewalks on the paths. Figure 10.38 shows a campus quadrangle that measures 40 meters by 30 meters. The locations of the doors of the library, chemistry building, and humanities building are shown. How long are the new sidewalks (gold lines) that connect the three buildings?

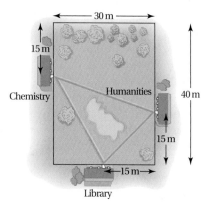

FIGURE 10.38

87. **Water Bed Leak.** Suppose you have, in a second floor bedroom, a water bed that measures 8 ft × 7 ft × 0.75 ft.

One day it leaks and all the water drains into the room below the bedroom.

 a. If the lower room measures 10 ft × 8 ft, how deep is the water in that room (assuming all of the water accumulates in the room)?

 b. What is the weight of the water bed? (Water has a density of 64 pounds per cubic foot.)

88. **Filling a Pool.** A spherical water tank has a radius of 24 feet. Can it hold enough water to fill a swimming pool 50 yards long and 25 yards wide to a constant depth of 6 feet?

89. **Optimal Fencing.** A rancher must design a rectangular corral with an area of 400 square meters. She decides to make a corral that measures 10 meters by 40 meters (which has the correct total area). How much fence is needed for this corral? Has the rancher found the most economical solution? Can you find another design for the corral that requires less fence but still provides 400 square meters of corral?

90. **Optimal Boxes.** You are making boxes and begin with a rectangular piece of cardboard that measures 1.75 meters by 1.25 meters. From each corner of that rectangular piece you cut out a square piece that is 0.25 meter on a side, as shown in Figure 10.39. You fold up the "flaps" to form a box without a lid. What is the surface area of the box? What is its volume? If the corner cuts were squares 0.3 meter on a side, would the resulting box have a larger or smaller volume than the first box? What is your estimate of the size of the square corner cuts that will give a box with the largest volume?

FIGURE 10.39

91. **Optimal Cable.** Telephone cable must be laid from a terminal box on the shore of a large lake to an island. The cable costs $500 per mile to lay underground and $1000 per mile to lay underwater. The locations of the terminal box, the island, and the shore are shown in Figure 10.40.

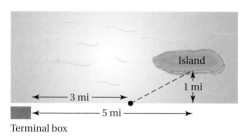

FIGURE 10.40

As an engineer on the project, you decide to lay 3 miles of cable along the shore underground and then lay the remainder of the cable along a straight line underwater to the island. How much will this project cost? Your boss examines your proposal and asks whether laying 4 miles of cable underground before starting the underwater cable would be more economical. How much would your boss's proposal cost? Will you still have a job?

92. **Estimating Heights.** In trying to estimate the height of a nearby building, you make the following observations. If you stand 15 feet away from a light post, you can line up the top of the building with the top of the light post. Furthermore, you can easily determine that the top of the light post is 10 feet above your head and that the building is 50 feet from your sighting position. What is the height of the building? Draw a good picture!

93. **Soda Can Design.** Standard soft drink cans hold 12 ounces, or 355 milliliters, of soda. Thus, their volume is 355 cm³. Assume that soda cans must be right circular cylinders.

 a. The cost of materials for a can depends on its surface area. Through trial and error with cans of different sizes, find the dimensions of a 12-ounce can that has the lowest cost for materials. Explain how you arrive at your answer.

 b. Compare the dimensions of your can from part a to the dimensions of a real soda can from a vending machine or store. Suggest some reasons why real soda cans might not have the dimensions that minimize their use of material.

94. **Melting Ice.** A glacier's surface is approximately rectangular, with a length of about 100 meters and a width of about 20 meters. The ice in the glacier averages about 3 meters in depth. Suppose that the glacier melts into a lake that is roughly circular with a radius of 1 kilometer. Assuming that the area of the lake does not expand significantly, about how much would the water level rise if the entire glacier melted into the lake?

95. **Sand Cones.** Think back to your days in the sandbox and imagine pouring sand through a funnel. The stream of sand forms a cone on the ground that becomes higher and wider as sand is added. The proportions of the sand cone remain the same as the cone grows. With sand, the height of the cone is roughly one-third the radius of the circular base (see Figure 10.41). The formula for the volume of a cone with height h and a base with radius r is $V = \frac{1}{3}\pi r^2 h$.

FIGURE 10.41

 a. If you build a sand cone 2 feet high, how many cubic feet of sand does it hold?

 b. Suppose that you excavate a basement and extract 1000 cubic feet of dirt. If the dirt is stored in a conical pile, how high will the cone be? Assume that a dirt cone has the same proportions as a sand cone.

 c. Estimate the total number of grains of sand in the cone in part a. Explain your assumptions and uncertainties.

96. **The Trucker's Dilemma.** It is a dark and stormy night. Through the beat of the wipers on your truck, you see a rickety country bridge marked "Load Limit 40 Tons." You know your truck, *White Lightning*, like the back of your hand; it weighs 16.3 tons with you and your gear. Trouble is, you are carrying a cylindrical steel water tank that is full of water. The empty weight of the tank is printed on its side: 1750 pounds. But what about the water? Fortunately, the Massey-Fergusson trucker's almanac in your glove box tells you that every cubic inch of water weighs 0.03613 pound. So you dash out into the rain with a tape measure to find the dimensions of the tank: length = 22 ft, diameter = 6 ft 6 in. Back in the truck, dripping and calculating, do you risk crossing? Explain.

97. **The Great Pyramids of Egypt.** Egypt's Old Kingdom began in about 2700 B.C. and lasted for 550 years. During that time, at least six pyramids were built as monuments to the life and afterlife of the pharaohs. These pyramids remain among the largest and most impressive structures constructed by any civilization. The building of the pyramids required a mastery of art, architecture, engineering, and social organization at a level unknown before that time. The collective effort required to complete

the pyramids transformed Egypt into the first nation-state in the world. Of the six pyramids, the best known are those on the Giza plateau outside Cairo, and the largest of those is the Great Pyramid built by Pharaoh Khufu (or Cheops to the Greeks) in about 2550 B.C. With a square base of 756 feet on a side and a height of 481 feet, the pyramid is laced with tunnels, shafts, corridors, and chambers, all leading to and from the deeply concealed king's burial chamber (Figure 10.42). The stones used to build the pyramids were transported, often hundreds of miles, with sand sledges and river barges, by a labor force of 100,000, as estimated by the Greek historian Herodotus. Historical records suggest that the Great Pyramid was completed in approximately 25 years.

FIGURE 10.42

a. To appreciate the size of the Great Pyramid of Khufu, compare its height to the length of a football field (which is 100 yards).

b. The volume of a pyramid is given by the formula $V = \frac{1}{3} \times$ area of base \times height. Use this formula to estimate the volume of the Great Pyramid. State your answer in both cubic feet and cubic yards.

c. The average size of a limestone block in the Great Pyramid is 1.5 cubic yards. How many blocks were used to construct this pyramid?

d. A modern research team, led by Mark Lehner of the University of Chicago, estimated that the use of winding ramps to lift the stones, desert clay, and water would allow placing one stone every 2.5 minutes. If the pyramid workers labored 12 hours a day, 365 days a year, how long would it have taken to build the Great Pyramid? How does this estimate compare with historical records? Why did Lehner's research team conclude that the Great Pyramid could have been completed with only 10,000 laborers, rather than the 100,000 laborers estimated by Herodotus?

e. Constructed in 1889 for the Paris Exposition, the Eiffel Tower is a 980-foot iron lattice structure supported on four arching legs. The legs of the tower stand at the corners of a square with sides of length 120 feet. If the Eiffel Tower were a solid pyramid, how would its volume compare to the volume of the Great Pyramid?

Web Projects

Find useful links for Web Projects on the text Website: www.aw.com/bennett-briggs

98. **Great Circles.** A great circle route is the shortest path between two points on the surface of the Earth. It is the optimal route for an airliner. Pick several pairs of cities somewhere on the Earth. Use the Internet to find the length of the great circle route between the cities in each pair. Estimate the length of other possible routes between the cities to verify that the great circle route is the shortest (this calculation will be easiest if you choose two cities with the same longitude). Explain in words how to find the great circle route between two points on the Earth.

99. **Sphere Packing.** Investigate the sphere packing problem, which has a history dating from the 17th century to the present day. Summarize the problem, early conjectures about its solution, and recent developments. You might also want to explore packing problems with non-spherical objects.

FRACTAL GEOMETRY

*T*he geometry of the previous two units is classical geometry developed largely by the ancient Greeks. While this geometry still serves us well in many applications, it does not work as well with many forms that arise in nature. In recent decades, a new geometry called **fractal geometry** has emerged. It has proved to be so effective at describing nature that it is now used in art and film to create realistic imagery. Figure 10.43 shows an imaginary landscape generated entirely on a computer using fractal geometry.

FIGURE 10.43 A fractal landscape, generated by Anne Burns.

WHAT ARE FRACTALS?

We can investigate fractals by envisioning measurements made with "rulers" of different lengths, such as 10 meters, 1 meter, 1 millimeter, and so on. Each length laid out by a ruler is called an **element** (of length). Suppose we use a ruler of length 10 meters and find that an object is 50 elements long. Then the total length of the object is $50 \times 10 \text{ m} = 500 \text{ m}$. More generally, the length of any object is

$$\text{total length} \approx \text{number of elements} \times \text{length of each element}$$

Measuring the Perimeter of Central Park

Imagine that you are asked to measure the perimeter of Central Park in New York City, which was designed in the shape of a rectangle (Figure 10.44). Suppose you start with a 10-meter ruler. Multiplying the number of elements by their 10-meter length gives you a measurement of the park perimeter.

FIGURE 10.44 The rectangular layout of Central Park.

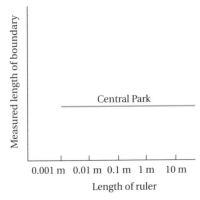

FIGURE 10.45 Ruler length does not affect the measured length of the Central Park perimeter.

FIGURE 10.46 A single long ruler (blue) cannot measure details that can be measured with shorter rulers (red).

Now, suppose that you use a shorter ruler—say, a 1-meter ruler. Your measurement with the 1-meter ruler will not differ much from that with the 10-meter ruler, because you are simply measuring the straight sides of the park. That is, ten 1-meter rulers will fit along the same straight line as one 10-meter ruler. In fact, the length of the ruler will not significantly affect measurements of the Central Park perimeter (Figure 10.45).

Measuring an Island Coastline

Next, imagine that you are asked to measure the perimeter of an island. To avoid problems with tides and waves, imagine that it is winter and the water around the island is frozen. Thus, your task is to measure the length of the coastline defined by the ice-land boundary.

Suppose you begin by using a 100-meter ruler, laying it end to end around the island. The 100-meter ruler (the single blue line segment in Figure 10.46) will adequately measure large-scale features such as bays and estuaries, but will miss features such as promontories and inlets that are less than 100 meters across. Switching to a 10-meter ruler will allow you to follow many features that were missed by the 100-meter ruler. As a result, you'll measure a *longer* perimeter (the short line segments in Figure 10.46).

The 10-meter ruler is still too long to measure all the features along the coastline, so you'll find an even greater perimeter if you switch to a 1-meter ruler. In fact, as you use shorter and shorter rulers to measure the coastline, you'll get longer and longer estimates of the perimeter because shorter rulers measure new levels of detail. Figure 10.47 shows how the measured length of the coastline grows larger when measured with shorter rulers. Whereas the perimeter of Central Park is clearly defined, we cannot agree on the "true" length of the coastline because it depends on the length of the ruler used.

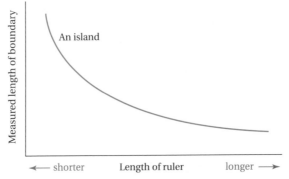

FIGURE 10.47 The measured perimeter of an island gets longer (vertical axis) as the ruler length gets shorter (horizontal axis).

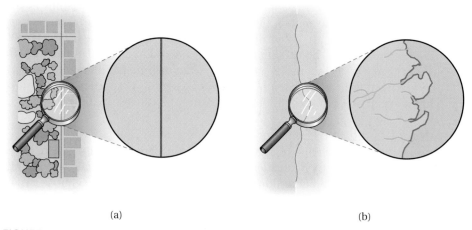

(a) (b)

FIGURE 10.48 (a) A segment of a rectangle looks the same under magnification. (b) A segment of coastline reveals new details under magnification.

Rectangles and Coastlines Under Magnification

Imagine viewing a piece of the rectangular perimeter of Central Park under a magnifying glass. No new details will appear—it is still a straight-line segment (Figure 10.48a). In contrast, if you view a piece of the coastline under a magnifying glass, you will see details that were not visible without magnification (Figure 10.48b).

Objects like the coastline that continually reveal new features at smaller scales are called **fractals.** Many natural objects have fractal structure. For example, coral (Figure 10.49a) and mountain ranges (Figure 10.49b) both reveal more and more features when viewed under greater magnification, and hence are fractals.

(a) (b)

FIGURE 10.49 Coral (a) and mountains (b) show fractal structure.

FRACTAL DIMENSION

The boundary of Central Park is one-dimensional because a single number locates any point. For example, if you tell people to meet 375 meters from the park's northwest corner, going clockwise, they'll know exactly where to go.

1-inch line segment

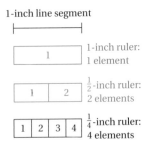

FIGURE 10.50 For a line segment, reducing ruler length by a factor R increases the number of elements by a factor $N = R$.

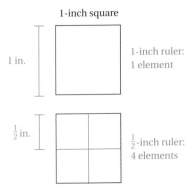

FIGURE 10.51 For a square, reducing ruler length by a factor R increases the number of elements by a factor $N = R^2$.

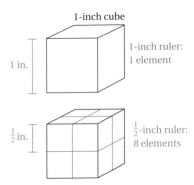

FIGURE 10.52 For a cube, reducing ruler length by a factor R increases the number of elements by a factor $N = R^3$.

In contrast, if you tell people to meet 375 meters along the coastline from a particular point on the island, different people will end up in different places depending on the length of the ruler they use. Thus, the coastline is *not* an ordinary one-dimensional object. But neither is it a two-dimensional object that has area. Instead, we say that the coastline has a **fractal dimension** that lies *between* one and two, indicating that the coastline has some properties that are like length (one dimension) and others that are more like area (two dimensions).

A New Definition of Dimension

If we measure a 1-inch line segment with a 1-inch ruler, we find *one* 1-inch element (Figure 10.50). Using a ruler that is smaller by a factor of 2, or $\frac{1}{2}$ inch in length, we find *two* elements along the 1-inch line segment. If we choose a ruler that is smaller by a factor of 4, or $\frac{1}{4}$ inch in length, we find *four* elements of length along the 1-inch line segment.

Let's use R to represent the *reduction factor* in the length of the ruler and N to represent the factor by which the *number of elements* increases. We can now restate our results for the line segment:

- Reducing the ruler length by a *reduction factor* $R = 2$ (to $\frac{1}{2}$ inch) leads to an increase in the *number of elements* by a factor $N = 2$.
- Reducing the ruler length by a *reduction factor* $R = 10$ (to $\frac{1}{10}$ inch) leads to an increase in the *number of elements* by a factor $N = 10$.

Generalizing for the line segment, we find that decreasing the ruler length by a *reduction factor* R leads to an increase in the *number of elements* by a factor $N = R$.

We can use a similar process to determine the area of a square by counting the number of *area elements* that fit within it (Figure 10.51). Using a ruler that has the same 1-inch length as a side of the square, we can fit a single area element in the square. If we reduce the ruler length by a factor $R = 2$, to $\frac{1}{2}$ inch, we fit $N = 4$ times as many elements in the square. Reducing the ruler length by $R = 4$, to $\frac{1}{4}$ inch, fits $N = 16$ times as many elements in the square. Generalizing for the square, we find that reducing the ruler length by a reduction factor R increases the number of area elements by a factor $N = R^2$.

For a cube, we can count the number of *volume elements* that fit within it (Figure 10.52). A ruler with the same 1-inch length as a side of the cube makes a single volume element that fits in the cube. This time, reducing the ruler length by a factor $R = 2$, to $\frac{1}{2}$ inch, allows us to fit $N = 8$ times as many volume elements in the cube. Reducing the ruler length by $R = 4$, to $\frac{1}{4}$ inch, fits $N = 64$ times as many elements in the cube. Generalizing for the cube, we find that reducing the ruler length by a factor R increases the number of volume elements by a factor $N = R^3$.

Let's summarize our results:

- For a one-dimensional object (such as a line segment), we found $N = R^1$.
- For a two-dimensional object (such as a square), we found $N = R^2$.
- For a three-dimensional object (such as a cube), we found $N = R^3$.

Note that, in each case, the dimension of the object shows up as a power. We can use this idea to define an object's *fractal dimension*.

DEFINITION

The **fractal dimension** of an object is defined as a number D such that

$$N = R^D$$

where N is the factor by which the number of elements increases when we shorten the ruler by a reduction factor R.

►EXAMPLE 1 Finding a Fractal Dimension

In measuring an object, every time you reduce the length of your ruler by a factor of 3, the number of elements increases by a factor of 4. What is the fractal dimension of this object?

Solution Reducing the ruler length by a factor $R = 3$ increases the number of elements by a factor $N = 4$. Thus, we are looking for a fractal dimension D such that

$$4 = 3^D$$

Because D appears as an exponent, we can "bring it down" by taking the logarithm of both sides of the equation (see the Brief Review on p. 559):

$$\log_{10} 4 = \log_{10} 3^D$$

By the rule that $\log_{10} a^x = x \log_{10} a$, the equation becomes

$$\log_{10} 4 = D \log_{10} 3$$

Now we divide both sides by $\log_{10} 3$ and interchange the sides to find

$$D = \frac{\log_{10} 4}{\log_{10} 3} \approx 1.2619$$

The *fractal dimension* of this object is about 1.2619.

Now try Exercises 15–26. ◄

The Snowflake Curve

What kind of object could have a fractal dimension like the one we calculated in Example 1? Let's consider a special object called a **snowflake curve,** generated by a process that begins with a straight-line segment. As shown at the bottom of Figure 10.53, we designate the starting line segment L_0, which has a length of 1. We then generate L_1 with the following three steps:

 1. Divide the line segment L_0 into three equal pieces.
 2. Remove the middle piece.
 3. Replace the middle piece with two segments of the same length arranged as two sides of an equilateral triangle.

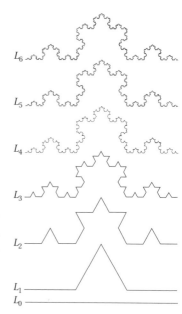

FIGURE 10.53 The generating process for the snowflake curve, from L_0 (bottom) to L_6 (top).

Note that L_1 consists of four line segments and that each has a length of $\frac{1}{3}$ (because L_0 was divided into three equal pieces).

Next, we repeat the three steps on *each* of the four segments of L_1. The result is L_2, which has 16 line segments, each of length $\frac{1}{9}$.

Time out to think

Count the segments shown in Figure 10.53 to confirm that L_2 has 16 line segments. Measure to confirm that each is $\frac{1}{9}$ the length of L_0. Why are the segments $\frac{1}{9}$ the length of L_0? How long are the segments of L_3?

Repeating the three-step process on each segment of the current figure generates L_3, L_4, L_5, and so forth. If we could repeat this process an *infinite* number of times, the snowflake curve, denoted L_∞, would be the ultimate result. Thus, any figure that we actually draw, whether L_6 or $L_{1,000,000}$, is only an approximation to the true snowflake curve.

Now imagine measuring the length of the complete snowflake curve, L_∞. A ruler with a length of 1 would simply lie across the base of the snowflake curve, missing all fine detail and measuring only the straight-line distance between the endpoints. This ruler would yield only one element along the snowflake curve. Reducing the ruler length by a factor $R = 3$, to $\frac{1}{3}$, would make it the same length as each of the four segments of L_1. Thus, it would find four elements along the snowflake curve, or $N = 4$ times as many elements as the first ruler. Reducing the ruler length by another factor $R = 3$, to $\frac{1}{9}$, would make it the length of each of the 16 segments of L_2. Thus, the ruler would find 16 elements along the snowflake curve, or $N = 4$ times as many elements as the previous ruler.

In general, every time we reduce the ruler length by another factor $R = 3$, we find $N = 4$ times as many elements. This is exactly the situation described in Example 1, which means that the snowflake curve has fractal dimension $D \approx 1.2619$.

What does it mean to have a fractal dimension of 1.2619? The fact that the fractal dimension is greater than 1 means that the snowflake curve has more "substance" than an ordinary one-dimensional object. In a sense, the snowflake curve begins to fill the part of the plane in which it lies. The closer the fractal dimension of an object is to 1, the more closely it resembles a collection of line segments. The closer the fractal dimension is to 2, the closer it comes to filling a part of a plane.

►EXAMPLE 2 How Long Is a Snowflake Curve?

How much longer is L_1 than L_0? How much longer is L_2 than L_0? Generalize your results, and discuss the length of a complete snowflake curve.

Solution Recall that L_0 has a length of 1. Figure 10.53 shows that L_1 consists of four line segments, each of length $\frac{1}{3}$. Thus, the length of L_1 is $\frac{4}{3}$ times the length of L_0, or $\frac{4}{3}$:

$$\text{length of } L_1 = \frac{4}{3}$$

L_2 has 16 line segments, each of length $\frac{1}{9}$. Thus, its length is

$$\text{length of } L_2 = \frac{16}{9} = \left(\frac{4}{3}\right)^2$$

Generalizing, we find that

$$\text{length of } L_n = \left(\tfrac{4}{3}\right)^n$$

Because $\left(\tfrac{4}{3}\right)^n$ grows without bound as n gets larger, we conclude that the complete snowflake curve, L_∞, must be infinitely long. *Now try Exercise 27.* ◀

The Snowflake Island

The **snowflake island** is a *region* (island) bounded by three snowflake curves. The process of drawing the snowflake island begins with an equilateral triangle (Figure 10.54). Then we convert *each* of the three sides of the triangle into a snowflake curve, L_∞. We cannot draw the complete snowflake island because it would require an infinite number of steps, but Figure 10.54 shows the results where the sides are L_0, L_1, L_2, and L_6.

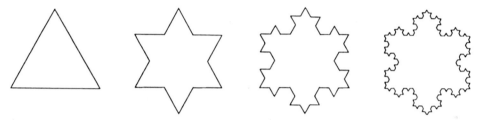

FIGURE 10.54 Left to right, successive approximations to a snowflake island, with sides made from L_0, L_1, L_2, and L_6, respectively.

The snowflake island illustrates some truly extraordinary properties. In Example 2, we found that a single snowflake curve is infinitely long. Thus, the coastline of the snowflake island also must be infinitely long, because it is made of three snowflake curves. However, Figure 10.54 shows that the island's area is clearly contained within the bounds of the page. Thus, we have the intriguing result that a snowflake island is an object with a *finite area* and an *infinitely long boundary.*

Real Coastlines and Borders

Note that each of the four pieces of L_2 in Figure 10.53 looks exactly like L_1, except smaller. Similarly, L_3 consists of four pieces that each look like L_2. In fact, if we magnify *any* piece of the snowflake curve, L_∞, it will look exactly like one of the earlier curves, L_0, L_1, L_2, . . . , used in its generation. Because the snowflake curve looks similar to itself when examined at different scales, we say that it is a **self-similar fractal.** The snowflake curve's self-similarity is due to repeated application of a simple set of rules.

Natural objects, such as real coastlines, also reveal new details under higher magnification. Unlike a self-similar fractal, a natural object isn't likely to look *exactly* the same when magnified. Nevertheless, if we have data about how measured lengths change with different "ruler" sizes, we can still assign a fractal dimension to a natural object.

The first significant data concerning the fractal dimensions of natural objects were collected by L. F. Richardson in about 1960. Richardson's data represented measurements and estimates of the lengths of various coastlines and international borders measured by "rulers" of varying sizes. His data suggest that most coastlines have a fractal dimension of about $D = 1.25$—very close to the fractal dimension of a snowflake curve.

▶ EXAMPLE 3 The Fractal Border of Spain and Portugal

Portugal claims that its international border with Spain is 987 kilometers in length. Spain claims that the border is 1214 kilometers in length. However, the two countries agree on the location of the border. How is this possible?

Solution The border follows a variety of natural objects including rivers and mountain ranges. It is therefore a fractal, with more and more detail revealed on closer and closer examination. Like the snowflake curve, the border will be *longer* if it is measured with a shorter "ruler." Thus, it is quite possible for Spain and Portugal to agree on the border's location but disagree on its length if they measured the border with "rulers" of different length. Spain claims a longer border length, so it must have used a shorter ruler for the measurement. *Now try Exercise 28.* ◀

THE FASCINATING VARIETY OF FRACTALS

The process of repeating a rule over and over to generate a self-similar fractal is called **iteration.** Different sets of rules can produce a fascinating variety of self-similar fractals. Let's look at just a few.

FIGURE 10.55 Several steps (top to bottom) in generating the Cantor set.

Consider a fractal generated from a line segment to which the following rule is applied repeatedly: *Delete the middle third of each line segment of the current figure* (Figure 10.55). With each iteration, the line segments become shorter until eventually the line turns to dust. The limit (after infinitely many iterations) is a fractal called the **Cantor set.** Because this ephemeral structure results from diminishing a one-dimensional line segment, its fractal dimension is less than 1.

Another interesting fractal, called the **Sierpinski triangle,** is produced by starting with a solid black equilateral triangle and iterating with the following rule: *For each black triangle in the current figure, connect the midpoints of the sides and remove the resulting inner triangle* (Figure 10.56). In the complete

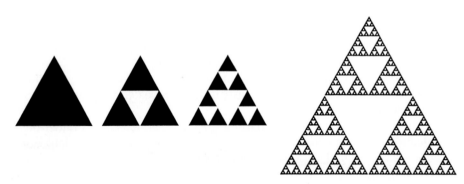

FIGURE 10.56 Several steps (left to right) in generating the Sierpinski triangle.

Sierpinski triangle, which would require *infinitely* many iterations to produce, every remaining black triangle would be infinitesimally small. Therefore, the total area of the black regions in the complete Sierpinski triangle is zero! The fractal dimension of the Sierpinski triangle is between 1 and 2. It is less than 2 because "material" has been removed from the initial two-dimensional triangle.

A closely related object is the **Sierpinski sponge** (Figure 10.57). It is generated by starting with a solid cube and iterating with this rule: *Divide each cube of the current object into 27 identical subcubes and remove the central subcube and the center cube of each face.* The resulting object has a fractal dimension between 2 and 3. It has less than a full three dimensions because material has been removed from the space occupied by the sponge.

FIGURE 10.57 An approximation to a Sierpinski sponge.

In fact, an infinite variety of fractals can be generated by iteration. Some are remarkably beautiful, such as the famous **Mandelbrot set** (Figure 10.58).

All the self-similar fractals we have considered so far are created by repeating the exact same set of rules in each iteration. An alternative approach, called **random iteration,** is to introduce slight random variation in every iteration. The resulting fractals therefore are *not* precisely self-similar, but they are close. Such fractals often appear remarkably realistic. For example, Barnsley's fern (Figure 10.59) is a fractal produced by random iteration, and it looks very much like a real fern.

The fact that fractals so successfully replicate natural forms suggests an intriguing possibility: Perhaps nature produces the many diverse forms that we see around us through simple rules that are applied repeatedly and with a hint of randomness. Because of this observation and because modern computers can test iterative processes, fractal geometry surely will remain an active field of research for decades to come.

FIGURE 10.58 The Mandelbrot set.

FIGURE 10.59 Two views of Barnsley's fern.

■ EXERCISES 10C

Review Questions

1. What is a fractal? Explain why measuring a fractal with a shorter ruler leads to a longer measurement.

2. Why do fractal dimensions fall in between the ordinary dimensions of 0, 1, 2, and 3?

3. Explain the meaning of the factors *R* and *N* used in calculating fractal dimensions.

4. What is the snowflake curve? Explain why we cannot actually draw it, but can only draw partial representations of it.

5. What is the snowflake island? Explain how it can have an infinitely long coastline, yet have a finite area.

6. What do we mean by a self-similar fractal? How is a self-similar fractal, like the snowflake curve, similar to a real coastline? How is it different?

7. Briefly describe what we mean by the process of iteration in generating fractals. Describe the generation of the Cantor set, the Sierpinski triangle, and the Sierpinski sponge. Describe the fractal dimension of each.

8. What is random iteration? Why do objects generated by random iteration make scientists think that fractals are important in understanding nature?

Does It Make Sense?

Decide whether each of the following statements makes sense (or is clearly true) or does not make sense (or is clearly false). Explain your reasoning.

9. I can use a yardstick to find the area of my rectangular patio.

10. I can use a yardstick to measure the length of the mountain skyline accurately.

11. The area of the snowflake island is given by its length times its width.

12. The measured length of fractals increases as the ruler length increases.

13. The edge of this leaf has a fractal dimension of 1.34.

14. This entire leaf, riddled with holes, has a fractal dimension of 1.87.

Basic Skills & Concepts

Ordinary and Fractal Dimensions. Find the dimension of each object in Exercises 15–26, and state whether or not it is a fractal.

15. In measuring the length of the object, when you reduce the length of your ruler by a factor of 3, the number of length elements increases by a factor of 3.

16. In measuring the area of the object, when you reduce the length of your ruler by a factor of 3, the number of area elements increases by a factor of 9.

17. In measuring the volume of the object, when you reduce the length of your ruler by a factor of 3, the number of volume elements increases by a factor of 27.

18. In measuring the length of the object, when you reduce the length of your ruler by a factor of 3, the number of length elements increases by a factor of 4.

19. In measuring the area of the object, when you reduce the length of your ruler by a factor of 3, the number of area elements increases by a factor of 12.

20. In measuring the volume of the object, when you reduce the length of your ruler by a factor of 3, the number of volume elements increases by a factor of 36.

21. In measuring the length of the object, when you reduce the length of your ruler by a factor of 4, the number of length elements increases by a factor of 4.

22. In measuring the area of the object, when you reduce the length of your ruler by a factor of 4, the number of area elements increases by a factor of 16.

23. In measuring the volume of the object, when you reduce the length of your ruler by a factor of 4, the number of volume elements increases by a factor of 64.

24. In measuring the length of the object, when you reduce the length of your ruler by a factor of 4, the number of length elements increases by a factor of 6.

25. In measuring the area of the object, when you reduce the length of your ruler by a factor of 4, the number of area elements increases by a factor of 24.

26. In measuring the volume of the object, when you reduce the length of your ruler by a factor of 4, the number of volume elements increases by a factor of 80.

27. The Quadric Koch Curve and Quadric Koch Island. To draw the *quadric Koch curve* (one of many variations of the snowflake curve), one begins with a horizontal line segment and applies the following rule: Divide each line segment into four equal pieces; replace the second piece with three line segments of equal length that make the shape of a square above the original piece; and replace the third piece with three line segments making a square below the original piece. The quadric Koch curve would result from infinite applications of the rule; the first three stages of the construction are shown in Figure 10.60.

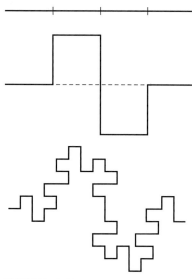

FIGURE 10.60

a. Determine the relation between N and R for the quadric Koch curve.

b. What is the fractal dimension of the quadric Koch curve? Can you draw any conclusions about the length of the quadric Koch curve? Explain.

c. The *quadric Koch island* is constructed by beginning with a square and then replacing each of the four sides of the square with a quadric Koch curve. Explain why the total area of the quadric Koch island is the same as the area of the original square. How long is the coastline of the quadric Koch island?

28. Fractal Dimension from Measurements. An ambitious and patient crew of surveyors has used various rulers to measure the length of the coastline of Dragon Island. The following table gives the measured length of the island, L, and the length of the ruler used, r.

r meters	100	10	1	0.1	0.01	0.001
L meters	315	1256	5000	19,905	79,244	315,479

a. Make a second table with the entries $\log_{10} r$ and $\log_{10} L$.

b. Graph these data ($\log_{10} r$, $\log_{10} L$) on a set of axes. Connect the data points.

c. If the graph of the data is close to a straight line, it is an indication that the coastline is a self-similar fractal. Does the coastline of Dragon Island appear to be a self-similar fractal?

d. What is the approximate slope of the line on your graph? Call the slope s. The fractal dimension of the coastline is $D = 1 - s$. What is the fractal dimension of the coastline of Dragon Island?

Further Applications

29. The Cantor Set. Recall that the Cantor set is formed by starting with a line segment and then successively removing the middle one-third of each segment in the current figure (Figure 10.55). If a ruler the length of the original line segment is used, it detects one element in the Cantor set because it can't "see" details smaller than itself. If the ruler is reduced in size by a factor of $R = 3$, it finds two elements (only solid pieces of line, not holes, are measured). If the ruler is reduced in size by a factor of $R = 9$, how many elements does it find? Based on these results, what is the fractal dimension of the Cantor set? Explain why this number is less than 1.

30. Ordinary Dimensions for Ordinary Objects.

a. Suppose you want to measure the length of the sidewalk in front of your house. Describe a thought process by which you can conclude that $N = R$ for the sidewalk and hence that its fractal dimension is the same as its ordinary dimension of 1.

b. Suppose you want to measure the area of your living room floor, which is square-shaped. Describe a thought process by which you can conclude that $N = R^2$ for the living room and hence that its fractal dimension is the same as its ordinary dimension of 2.

c. Suppose you want to measure the volume of a cubical swimming pool. Describe a thought process by which you can conclude that $N = R^3$ for the pool and hence that its fractal dimension is the same as its ordinary dimension of 3.

31. Fractal Dimensions for Fractal Objects.

a. Suppose you are measuring the length of the stream frontage along a piece of mountain property. You begin with a 15-meter ruler and find just one element along the length of the stream frontage. When you switch to a 1.5-meter ruler, you are able to trace finer details of the stream edge and you find 20 elements along its length. Switching to a 15-centimeter ruler, you find 400 elements along the stream frontage. Based on these measurements, what is the fractal dimension of the stream frontage?

b. Suppose you are measuring the area of a very unusual square leaf with many holes, perhaps from hungry insects, in a fractal pattern (e.g., similar to the Sierpinski triangle, Figure 10.56). You begin with a 10-centimeter ruler and find that it lies over the entire square, making just one element. When you switch to a 5-centimeter ruler, you are better able to cover areas of leaf while skipping areas of holes and you find 3 area elements. You switch to a 2.5-centimeter ruler and find 9 area elements. Based on these measurements, what is the fractal dimension of the leaf? Explain *why* the fractal dimension is less than 2.

c. Suppose you are measuring the volume of a cube cut from a large rock that contains many cavities forming a fractal pattern. Beginning with a 10-meter ruler, you find just one volume element. Smaller rulers allow you to ignore cavities, gauging only the volume of rock material. With a 5-meter ruler, you find 6 volume elements. With a 2.5-meter ruler, you find 36 volume elements. Based on these measurements, what is the fractal dimension of the rock? Explain why a fractal dimension between 2 and 3 is reasonable. (Ignore the practical difficulties caused by the fact that you cannot see through a rock to find all its holes!)

32. Fractal Patterns in Nature. Describe at least five natural objects that exhibit fractal patterns. In each case, explain the structure that makes the pattern a fractal, and estimate its fractal dimension.

33. Natural Fractals Through Branching. One way natural objects reveal fractal patterns is by branching. For example, the intricate structure of the human lung, the web of capillaries in a muscle, the branches or roots of a tree, and the successive division of streams in a river delta all involve branching at different spatial scales. Explain why structures formed by branching resemble self-similar fractals. Further, explain why fractal geometry rather than ordinary geometry leads to a greater understanding of such structures.

Web Projects

Find useful links for Web Projects on the text Web site: www.aw.com/bennett-briggs

34. Fractal Research. Many Web sites are devoted to fractals. Locate at least two sites related to fractals and use them to write a two- to three-page paper on either a specific use of fractals or a technique for generating fractals.

35. Fractal Art. Visit a Web site that features fractal art. Choose a specific piece of art, explain how it was generated, and discuss its visual impact.

CHAPTER 10 SUMMARY

UNIT	KEY TERMS	KEY IDEAS AND SKILLS
10A	Euclidean geometry point, line, plane dimension angle right angle straight angle acute angle obtuse angle area perimeter circle radius diameter circumference polygon triangle square parallelogram surface area volume rectangular prism cube cylinder sphere scaling laws surface area to volume ratio	**Understand** basic concepts of Euclidean geometry. **Know and use** area and perimeter formulas for two-dimensional figures. **Know and use** surface area and volume formulas for three-dimensional figures. **Understand** the use of scaling laws and scale factors: Lengths scale with the scale factor. Areas scale with the square of the scale factor. Volumes scale with the cube of the scale factor. **Understand** the implications of the surface area to volume ratio.
10B	degrees, minutes, seconds latitude, longitude prime meridian, equator angular size pitch slope grade similar triangles optimization	**Understand** various ways to describe angles. **Know** how to locate points and find distances on the Earth with longitude and latitude. **Understand** angular size. **Know** how to use the Pythagorean theorem. **Understand** various ways to measure distances. **Understand and use** properties of similar triangles to solve problems. **Understand** the goal of optimization problems.

(Continues on the next page)

10C	fractal geometry	**Explain** why different rulers can give different length measurements of natural objects.
	fractal dimension	**Fractal dimension:** $N = R^D$, where N is the factor by which the number of elements increases and R is the factor by which the ruler length is reduced.
	snowflake curve	**Understand** the uses of fractal geometry.
	snowflake island	
	self-similar fractal	
	iteration	

[Mathematics] seems to stand for all that is practical, poetry for all that is visionary, but in the kingdom of the imagination you will find them close akin, and they should go together as a precious heritage to every youth.

—FLORENCE MILNER, *SCHOOL REVIEW*, 1898

CHAPTER

Mathematics and the Arts

The connections between art and mathematics go deep into history. The ties are most evident in architecture: The Great Pyramids in Egypt, the Eiffel Tower in France, and modern-day skyscrapers all required mathematics in their design. But mathematics has contributed in equally profound ways to music, painting, and sculpture. In this chapter, we will explore a few of the many connections between mathematics and the arts.

UNIT 11A **Mathematics and Music:** We will explore the connections between mathematics and music, particularly as they apply to the ideas of musical tones and scales.

UNIT 11B **Perspective and Symmetry:** Perspective and symmetry are key ideas in art that are best understood through mathematics. We will explore the role these ideas have played both in classical art and in modern tilings.

UNIT 11C **Proportion and the Golden Ratio:** Proportion is another critical concept in art. We will discuss proportion and its implications by exploring the famous golden ratio.

UNIT 11A

Music is the universal language of mankind.

—HENRY WADSWORTH LONGFELLOW

MATHEMATICS AND MUSIC

he roots of mathematics and music are entwined in antiquity. Pythagoras (c. 500 B.C.) claimed that "all nature consists of harmony arising out of number." He imagined that the planets circled the Earth on invisible heavenly spheres, obeying specific numeric laws and emitting the ethereal sounds known as the "music of the spheres." Thus, he saw a direct connection between geometry and music. Over a thousand years later, the standard curriculum in medieval universities was the *quadrivium* (Latin for "crossroads"), consisting of arithmetic, geometry, music, and astronomy.

SOUND AND MUSIC

Any vibrating object produces sound. The vibrations produce a **wave** (much like a water wave) that propagates through the surrounding air in all directions. When such a wave impinges on the miraculously designed organ called the ear, it is perceived as sound. Of course, some sounds, such as speech and screeching tires, do not qualify as music. Most musical sounds are made by vibrating strings (violins, cellos, guitars, and pianos), vibrating reeds (clarinets, oboes, and some organ pipes), or vibrating columns of air (other organ pipes, horns, and flutes).

One of the most basic qualities of sound is **pitch.** For example, a tuba has a "lower" pitch than a flute, and a violin has a "higher" pitch than a bass guitar. To understand pitch, find a taut string (a guitar string works best, but a stretched rubber band will do). When you pluck the string, it produces a sound with a certain pitch. Next, use your finger to hold the midpoint of the string in place, and pluck either half of the string. Note that a higher-pitched sound is produced, demonstrating an ancient musical principle discovered by the Greeks: *The shorter the string, the higher the pitch.*

It took many centuries to arrive at a precise understanding of pitch. This understanding is attributed to both Galileo (1564–1642) and the French monk Marin Mersenne (1588–1648). Plucking a taut string causes it to vibrate up and down along its length (Figure 11.1a). For example, if the string vibrates up and down 100 times per second, its **frequency** is said to be 100 **cycles per second (cps).** When you pluck only half the string (by holding down its midpoint), the resulting wave is *half* as long as the first wave (Figure 11.1b), and its frequency therefore is twice as much, or 200 cps.

HISTORICAL NOTE

Evidence of music is found in nearly all ancient cultures. Indeed, based on evidence dating back 30,000 years, some archaeologists suspect that music may predate speech.

HISTORICAL NOTE

String and wind instruments were designed at least 5000 years ago in Mesopotamia, and Egyptians played in small ensembles of lyres and flutes by 2700 B.C.

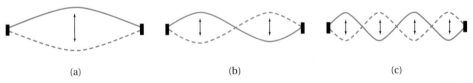

(a) (b) (c)

FIGURE 11.1 A vibrating string at (a) its fundamental frequency, (b) its first harmonic, and (c) its second harmonic.

Thus, pitch is related to the frequency of the vibrating string: *The higher the frequency, the higher the pitch.*

For any particular string, plucking its full length causes it to vibrate at its **fundamental frequency.** Plucking half its length generates a wave with twice the fundamental frequency, called the *first harmonic* (or *overtone*). Halving the length of the string again (Figure 11.1c) doubles the frequency again, so the resulting tone, called the *second harmonic,* has four times the fundamental frequency.

Now we come to another discovery of the ancient Greeks: Doubling the frequency of a tone raises the pitch by an **octave**—perhaps the most appealing and natural combination of notes in music. The piano keyboard (Figure 11.2) is helpful here. An octave is the interval between, say, middle C and the next higher C. For example, middle C on the piano has a frequency of 260 cps, the C above middle C has a frequency of 2×260 cps $= 520$ cps, and the next higher C has a frequency of 2×520 cps $= 1040$ cps. Similarly, the C below middle C has a frequency of about $\frac{1}{2} \times 260$ cps $= 130$ cps.

FIGURE 11.2 Piano keys.

Time out to think _____

The note middle A (above middle C) has a frequency of about 440 cps. What is the frequency of the A an octave higher? What is the frequency of the A an octave below middle A?

SCALES

The musical tones that span an octave comprise a **scale.** The Greeks invented the 7-note (or diatonic) scale that corresponds to the white keys on the piano. In the 17th century, Johann Sebastian Bach adopted a 12-tone scale, which corresponds to both the white and the black keys on a modern piano. With Bach's music, the 12-tone scale spread throughout Europe, becoming a foundation of Western music. Many other scales are possible. For example, 3-tone scales are common in African music, scales with more than 12 tones occur in Asian music, and 19-tone scales are used in contemporary music.

On the 12-tone scale, the frequency separating each tone is called a **half-step,** corresponding to consecutive notes on the piano keyboard. For example, E and F are separated by a half-step, as are F and F# (read "F sharp"). In each half-step, the frequency increases by some *multiplicative* factor; let's call it *f.* Thus, the frequency of C is the frequency of B times the factor *f,* the frequency of C# is the frequency of C times the factor *f,* and so on. The frequencies of the notes across the entire scale are related as follows:

$$\text{C} \underset{f}{\to} \text{C\#} \underset{f}{\to} \text{D} \underset{f}{\to} \text{D\#} \underset{f}{\to} \text{E} \underset{f}{\to} \text{F} \underset{f}{\to} \text{F\#} \underset{f}{\to} \text{G} \underset{f}{\to} \text{G\#} \underset{f}{\to} \text{A} \underset{f}{\to} \text{A\#} \underset{f}{\to} \text{B} \underset{f}{\to} \text{C}$$

Because an octave corresponds to an increase in frequency by a factor of 2, the factor *f* must have the property

$$\underbrace{f \times f \times f \times f \times f \times f \times f \times f \times f \times f \times f \times f}_{\text{12 times}} = f^{12} = 2$$

Thus, f must be the *twelfth root* of two, or $f = \sqrt[12]{2} \approx 1.05946$.

We can now calculate the frequency of every note of a 12-tone scale. Starting from middle C, with its frequency of 260 cps, we multiply by $f = \sqrt[12]{2}$ to find that the frequency of C# is 260 cps $\times f \approx 275$ cps. Multiplying again by f gives the frequency of D as 275 cps $\times f \approx 292$ cps. Continuing in this way generates Table 11.1.

TABLE 11.1		Frequencies of Notes in the Octave Above Middle C	
NOTE	FREQUENCY (CPS)	RATIO TO FREQUENCY OF PRECEDING NOTE	RATIO TO FREQUENCY OF MIDDLE C
C	260	$\sqrt[12]{2} \approx 1.05946$	$1.00000 = 1$
C#	275	$\sqrt[12]{2} \approx 1.05946$	1.05946
D (second)	292	$\sqrt[12]{2} \approx 1.05946$	1.12246
D#	309	$\sqrt[12]{2} \approx 1.05946$	1.18921
E (third)	328	$\sqrt[12]{2} \approx 1.05946$	$1.25992 \approx \frac{5}{4}$
F (fourth)	347	$\sqrt[12]{2} \approx 1.05946$	$1.33484 \approx \frac{4}{3}$
F#	368	$\sqrt[12]{2} \approx 1.05946$	1.41421
G (fifth)	390	$\sqrt[12]{2} \approx 1.05946$	$1.49831 \approx \frac{3}{2}$
G#	413	$\sqrt[12]{2} \approx 1.05946$	1.58740
A (sixth)	437	$\sqrt[12]{2} \approx 1.05946$	$1.68179 \approx \frac{5}{3}$
A#	463	$\sqrt[12]{2} \approx 1.05946$	1.78180
B (seventh)	491	$\sqrt[12]{2} \approx 1.05946$	1.88775
C (octave)	520	$\sqrt[12]{2} \approx 1.05946$	$2.00000 = 2$

by the way

All entries in column 3 are the same because the same factor, $f \approx \sqrt[12]{2}$, separates every pair of notes. The parenthetical terms in column 1 are names used by musicians to describe intervals between the note shown and middle C.

Column 4 of Table 11.1 shows that a few tones have simple ratios of frequency to middle C. For example, the frequency of G is approximately $\frac{3}{2}$ times the frequency of middle C (musicians call this interval a *fifth*), and the frequency of F is approximately $\frac{4}{3}$ times the frequency of middle C (musicians call this interval a *fourth*). According to many musicians, the most pleasing combinations of notes, called **consonant tones,** are those whose frequencies have a simple ratio. Referring to consonant tones, the Chinese philosopher Confucius observed that small numbers are the source of perfection in music.

▶ **EXAMPLE 1 The Dilemma of Temperament**

Because the whole-number ratios in Table 11.1 are not exact, tuners of musical instruments have the problem of *temperament,* which can be demonstrated as follows. Start at middle C with a frequency of 260 cps. Using the whole-number ratios, find the frequency if you raise C by a sixth to A, raise A by a fourth to D, lower D by a fifth to G, and lower G by a fifth to C. Having returned to the same note, have you also returned to the same frequency?

Human speech is like a cracked kettle on which we tap crude rhythms for bears to dance to, while we long to make music that will melt the stars.

—GUSTAVE FLAUBERT

Solution According to Table 11.1, raising a note by a sixth increases its frequency by a factor of $\frac{5}{3}$. Thus, the frequency of A above middle C is $\frac{5}{3} \times 260$ cps = 433.33 cps. Raising this note by a fourth increases its frequency by $\frac{4}{3}$. Thus, D has a frequency of $\frac{4}{3} \times 433.33$ cps = 577.77 cps. Lowering D by a fifth (a factor of $\frac{2}{3}$) to G gives a frequency

of $\frac{2}{3} \times 577.77 = 385.18$ cps. Finally, lowering G by another fifth (a factor of $\frac{2}{3}$) puts us back to middle C, with a frequency of $\frac{2}{3} \times 385.18$ cps $= 256.79$. Note that, by using whole-number ratios, we have not quite returned to the proper frequency of 260 cps for middle C. The problem is that the whole-number ratios are not exact. That is, $\frac{5}{3} \times \frac{4}{3} \times \frac{2}{3} \times \frac{2}{3} = \frac{80}{81}$ is close to, but not *exactly*, 1. *Now try Exercises 12–16.* ◄

MUSICAL SCALES AS EXPONENTIAL GROWTH

The increase in frequencies in a scale is just another example of exponential growth. Each successive frequency is $f \approx 1.05946$ times, or approximately 5.9% more than, the previous frequency. In other words, the frequencies increase at a fixed relative growth rate. Thus, we can use an exponential function (see Unit 9C) to find any frequency on the scale. Suppose we start at a frequency Q_0. Then the frequency Q of the note n half-steps higher is given by

$$Q = Q_0 \times 1.05946^n$$

► **EXAMPLE 2** Exponential Growth on Musical Scales

Use the exponential growth law to find the frequency of the note a fifth above middle C, the note one octave and a fifth above middle C, and the note two octaves and a fifth above middle C.

Solution We let the frequency of middle C be the initial value for the scale; that is, we set $Q_0 = 260$ cps. Table 11.1 shows that the note a fifth above middle C is G, which is seven half-steps above middle C. Thus, we let $n = 7$ in the exponential law and find that the frequency of G is

$$Q = Q_0 \times 1.05946^7 = 390 \text{ cps}$$

The note one octave and a fifth above middle C is $12 + 7 = 19$ half-steps above middle C. Letting $n = 19$, we find that the frequency of this note is

$$Q = Q_0 \times 1.05946^{19} = 779 \text{ cps}$$

The note two octaves and a fifth above middle C is $(2 \times 12) + 7 = 31$ half-steps above middle C. Letting $n = 31$, we find that the frequency of this note is

$$Q = Q_0 \times 1.05946^{31} = 1558 \text{ cps}$$ *Now try Exercises 17–19.* ◄

FROM TONES TO MUSIC

Although the simple frequencies of "pure" tones are the building blocks of music, the sounds of music are far richer and more complex. For example, a plucked violin string does much more than produce a single frequency. The vibration of the string is transferred through the bridge of the violin to its top, and the ribs transfer those vibrations to the back of the instrument. With the top and back of the violin in oscillation, the entire instrument acts as a resonating chamber, which excites and amplifies the higher harmonics of the original tone.

Similar principles generate rich and complex sounds in all instruments. On the left, Figure 11.3 shows a typical sound wave that might be produced by an instrument. It isn't a simple wave like those pictured in Figure 11.1. Instead, it consists of a combination of simple waves that are the harmonics of the fundamental. In fact, the complex wave is the *sum* of the three simple waves shown on the right in

Of course, Beethoven's Eroica Symphony or Michelangelo's ceiling of the Sistine Chapel can be reduced to an adequate mathematical formula. There is, however, one little catch. The only person capable of making such an all-inclusive analysis must be able to feel the emotions that Beethoven and Michelangelo felt, to think in sound and paint as they did.

—J. MURRAY BARBOUR, MUSIC HISTORIAN

FIGURE 11.3 The complex sound wave on the left is the sum of the three simple waves on the right.

Figure 11.3. The fact that a musical sound can be expressed as a sum of simple harmonics is surely the deepest connection between mathematics and music. The French mathematician Jean Baptiste Joseph Fourier first enunciated this principle in about 1810. It was one of the most profound discoveries in mathematics.

Although mathematics helps in understanding music, many mysteries remain. For example, in about 1700, an Italian craftsman known as Stradivarius made what are still considered to be the finest violins and cellos ever produced. Despite years of study by mathematicians and scientists, no one has succeeded in reproducing the unique sounds of a Stradivarius instrument.

THE DIGITAL AGE

When we talk about sound waves and imagine music to consist of waves, we are working with the **analog** picture of music. Until about 25 years ago, all musical recordings (phonograph cylinders, records, and tape recordings) were based on the analog picture of music. Storing music in the analog mode requires storing analogs of sound waves. For example, on records, the grooves in the vinyl surface are etched with the shape of the original musical sound wave. If you have listened to analog recordings, you know that this shape can easily become distorted or damaged.

Most recent musical recordings use a **digital** picture of music. A computer takes the analog version of a piece of music and chops it into very short time intervals (fractions of a second). The time intervals are used much like bins in a histogram. For each time interval, the computer measures the frequencies of the sound waves and records them numerically. This process, called *digitizing,* transforms the music into a list of numbers. The list can then be saved on a storage device such as a compact disc (CD). Inside a playback device such as a CD player, a computer uses a reverse process to convert the numbers back into sound waves so that they can be heard. As long as the time intervals are short enough, the music sounds smooth and authentic, with no evidence of digitizing.

Digital technology provides easy and endless ways to "process" music. Through techniques of **digital signal processing,** the sounds of a musical recording can be modified with computers. For example, extraneous sounds (background noises or "hiss") can be detected and removed. Changing the music by amplifying certain frequencies or attenuating others also is possible. Moreover, once digital music can be modified, it is a short step to creating music. Instruments called **synthesizers** can create and imitate a tremendous variety of sounds digitally without strings, brass tubes, or reeds. In the digital age, the dividing line between mathematics and music all but vanishes.

EXERCISES 11A

Review Questions

1. What is pitch? How is it related to the frequency of a musical note?

2. Define *fundamental frequency, first harmonic,* and *octave.* Why are these concepts important in music?

3. What is a 12-tone scale? How are the frequencies of the notes on a 12-tone scale related to one another?

4. Explain how the notes of the scale are generated by exponential growth.

5. How do the wave forms of real musical sounds differ from the wave forms of simple tones? How are they related?

6. What is the difference between an analog and a digital recording of music? What are the advantages of digital recording?

Does It Make Sense?

Decide whether each of the following statements makes sense (or is clearly true) or does not make sense (or is clearly false). Explain your reasoning.

7. If I pluck this string more often, then it will have a higher pitch.

8. Jack made the length of the string one-fourth of its original length, and the pitch went up two octaves.

9. Exponential growth is found even in musical scales.

10. A piano has 88 keys, so it must have a range of about 7 octaves.

11. The scratch on Jill's U2 phonograph record can be removed by a digital filter.

Basic Skills & Concepts

12. **Octaves.** Starting with a tone having a frequency of 110 cycles per second, find the frequencies of the tones that are one, two, three, and four octaves higher.

13. **Octaves.** Starting with a tone having a frequency of 880 cycles per second, find the frequencies of the tones that are one, two, three, and four octaves lower.

14. **Notes of a Scale.** Find the frequencies of the 12 notes of the scale that starts at the A above middle C; this A has a frequency of 437 cycles per second.

15. **Notes of a Scale.** Find the frequencies of the 12 notes of the scale that starts at the G above middle C; this G has a frequency of 390 cycles per second.

16. **The Dilemma of Temperament.** Start at middle A, with a frequency of 437 cps. Using the whole-number ratios in Table 11.1, find the frequency if you raise A by a fifth to E. What is the frequency if you raise E by a fifth to B? What is the frequency if you lower B by a sixth to D? What is the frequency if you lower D by a fourth to A? Having returned to the same note, have you also returned to the same frequency? Explain.

17. **Exponential Growth and Scales.** Starting at middle C, with a frequency of 260 cps, find the frequency of the following notes:

 a. five half-steps above middle C

 b. a fifth (seven half-steps) above middle C

 c. an octave and a fourth (five half-steps) above middle C

 d. 36 half-steps above middle C

 e. four octaves and three half-steps above middle C

18. **Exponential Growth and Scales.** Starting at middle A, with a frequency of 437 cps, find the frequency of the following notes:

 a. six half-steps above middle A

 b. a third (four half-steps) above middle A

 c. an octave and a fourth (five half-steps) above middle A

 d. 25 half-steps above middle A

 e. two octaves and two half-steps above middle A

19. **Exponential Decay and Scales.** What is the frequency of the note five half-steps *below* middle A (which has a frequency of 437 cps)? eight half-steps *below* middle A?

Further Applications

20. **Circle of Fifths.** The circle of fifths is generated by starting at a particular musical note and stepping upward by intervals of a fifth (seven half-steps). For example, starting

at middle C, a circle of fifths includes the notes $C \rightarrow G \rightarrow D' \rightarrow A' \rightarrow E'' \rightarrow B'' \rightarrow \ldots$, where each $(')$ denotes a higher octave. Eventually the circle comes back to C several octaves higher.

a. Show that the frequency of a tone increases by a factor of $2^{7/12} = 1.498$ if it is raised by a fifth. (Hint: Recall that each half-step corresponds to an increase in frequency by a factor of $f = \sqrt[12]{2}$.)

b. By what factor does the frequency of a tone increase if it is raised by two fifths?

c. Starting with middle C, at a frequency of 260 cycles per second, find the frequencies of the other notes in the circle of fifths.

d. How many fifths are required for the circle of fifths to return to a C? How many octaves are covered by a complete circle of fifths?

e. What is the ratio of the frequencies of the C at the beginning of the circle and the C at the end of the circle?

21. Circle of Fourths. A circle of *fourths* is generated by starting at any note and stepping upward by intervals of a fourth (five half-steps). By what factor is the frequency of a tone increased if it is raised by a fourth? How many fourths are required to complete the entire circle of fourths? How many octaves are covered in a complete circle of fourths?

22. Rhythm and Mathematics. In this unit, we focused on musical sounds, but rhythm and mathematics are also closely related. For example, in "4/4 time," there are four *quarter notes* in a measure. If two quarter notes have the duration of a *half note*, how many half notes are in one measure? If two *eighth notes* have the duration of a quarter note, how many eighth notes are in one measure? If two *sixteenth notes* have the duration of an eighth note, how many sixteenth notes are in one measure?

Web Projects

Find useful links for Web Projects on the text Web site: www.aw.com/bennett-briggs

23. Mathematics and Music. Many Web sites are devoted to connections between music and mathematics. Visit one such site and study a connection in some depth. Write a one- to two-page essay that describes this connection between mathematics and music.

24. Internet Music. Many Web sites make music available digitally. Some of these Web sites have come under attack for violating copyright laws. Find one such Web site. Describe how it works both from a user's standpoint and from a commercial standpoint. Do you think the site's use of music is legal? Is it ethical? Defend your opinions.

25. Mathematics and Composers. Many musical composers, both classical and modern, have used mathematics in their compositions. Research the life of one such composer. Write a one- to two-page essay discussing the role of mathematics in the composer's life and music.

In the News

26. Numbers and Music. Look carefully at cassette or CD liner notes, music catalogs, or other materials related to musical products. Find as many different uses of numbers as possible in these materials, and explain how the numbers are related to the music.

27. Digital Music. The ease with which digital music can be copied has increased the importance of the copyright issue, especially with respect to music available on the Web. Find a recent article on issues of copying digital music. Discuss the article and its conclusions.

UNIT 11B

PERSPECTIVE AND SYMMETRY

e now turn our attention to connections between mathematics and the visual arts, such as painting, sculpture, and architecture. At least three aspects of the visual arts relate directly to mathematics: *perspective, symmetry,* and *proportion.* In this unit, we will see how Renaissance mathematicians and artists discovered perspective and then we will explore the idea of symmetry. We'll save *proportion* for Unit 11C.

PERSPECTIVE

People of all cultures have used geometrical ideas and patterns in their artwork. The ancient Greeks developed strong ties between the arts and mathematics because both endeavors were central to their view of the world. Much of the Greek outlook was lost during the Middle Ages, but the Renaissance brought at least two new developments that made mathematics an essential tool of artists. First, there was a renewed interest in natural scenes, which led to a need to paint with realism. Second, many of the artists of the day also worked as engineers and architects.

The desire to paint landscapes with three-dimensional realism brought Renaissance painters face to face with the matter of perspective. In their attempts to capture depth and volume on a two-dimensional canvas, these artists made a science of painting. The painters Brunelleschi (1377–1446) and Alberti (1404–1472) are generally credited with developing, in about 1430, a system of perspective that involved geometrical thinking. Alberti's principle that a painting "is a section of a projection" lies at the heart of drawing with perspective.

Suppose you want to paint a simple view looking down a hallway with a checkerboard tile floor. Figure 11.4 shows a *side view* of the artist's eye, the canvas,

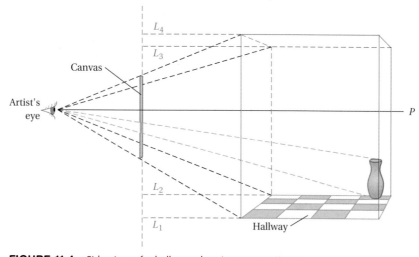

FIGURE 11.4 Side view of a hallway, showing perspectives.

and the hallway. Note the four lines labeled L_1, L_2, L_3, and L_4. The two side walls of the hallway intersect the floor and the ceiling along these four lines. These lines are important because they are parallel to each other in the scene and perpendicular to the canvas (or the plane containing the canvas).

Let's now look at the scene as the artist sees and paints it. The artist looks down the hallway with the point of view shown in Figure 11.5.

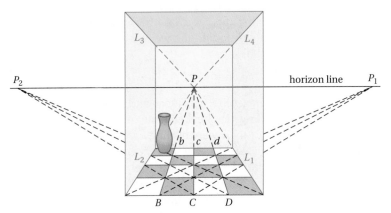

FIGURE 11.5 An artist's view of the hallway from Figure 11.4, showing perspectives. Adapted from M. Kline, *Mathematics in Western Culture.*

The lines L_1, L_2, L_3, and L_4, which are parallel in the actual scene, are not parallel in the painting. In fact, they all meet at a single point, labeled P, which is called the **principal vanishing point.** Thus, we have the first principle of perspective discovered by the Renaissance painters:

> *All lines that are parallel in the real scene and perpendicular to the canvas must intersect at the principal vanishing point of the painting.*

Note that other lines that are parallel to lines L_1, L_2, L_3, and L_4, such as the lines going straight down the hallway along the floor tiles, also meet at the principal vanishing point. For example, the line connecting the points B and b intersects P, as do the line connecting C and c and the line connecting D and d.

What happens to lines that are parallel in the actual scene but not perpendicular to the canvas, such as the diagonal lines along the floor tiles? If you study Figure 11.5 carefully, you'll see that such lines intersect at their own vanishing points, all on the horizontal line passing through the principal vanishing point. This line is called the **horizon line.** For example, the right-slanting diagonals of the floor tiles are parallel in the actual scene. But, in the painting, these lines meet at the vanishing point labeled P_1 on the horizon line. Similarly, the left-slanting diagonals meet at a vanishing point called P_2 on the horizon line. In fact, all sets of lines that are mutually parallel in the real scene (except those parallel to the horizon line) must meet at their own vanishing point on the horizon line. *Now try Exercises 15–20.*

FIGURE 11.6 *The Last Supper,* by Leonardo da Vinci, shown with several lines that are parallel in the real scene and therefore converge at the principal vanishing point behind Christ.

Let no one who is not a mathematician read my works.

—LEONARDO DA VINCI

Time out to think

Imagine looking along a set of long parallel lines that stretches far into the distance, such as a set of train tracks or a set of telephone lines. The lines will appear to your eyes to get closer to each other as you look into the distance. If you were painting a picture of the scene, where would you put the principal vanishing point? Why?

Leonardo da Vinci (1452–1519) contributed greatly to the science of perspective. We can see da Vinci's mastery of perspective in many of his paintings. If you study *The Last Supper* (Figure 11.6), you will notice several parallel lines in the actual scene intersecting at the principal vanishing point of the painting, which is directly behind the central figure of Christ.

The German artist Albrecht Dürer (1471–1528) further developed the science of perspective. Near the end of his life, he wrote a popular book that stressed the use of geometry and encouraged artists to paint according to mathematical principles. Figure 11.7 is one of Dürer's woodcuts, showing an artist using his

FIGURE 11.7 Woodcut by Albrecht Dürer.

principles of perspective. A string from a point on the lute is attached to the wall at the point corresponding to the artist's eye. At the point where the string passes through the frame, a point is placed on the canvas. As the string is moved to different points on the lute, a drawing of the lute is created in perfect perspective on the canvas.

FIGURE 11.8 Sketch by Jan Vredeman de Vries.

The artist Jan Vredeman de Vries (1527–1604) summarized much of the science of perspective in a book he published in 1604. Figure 11.8 shows a sketch from his book that illustrates how thoroughly perspective can be analyzed.

Time out to think _____

Identify some of the vanishing points in the sketch in Figure 11.8. What parallel lines from the real scene converge at each vanishing point?

Perspective drawing is sometimes abused deliberately. The painting *False Perspective* (Figure 11.9) by English artist William Hogarth (1697–1764) reminds us that perspective is essential in art. Note where the fishing line of the near man lands, and how the woman in the window appears to be lighting the pipe of a man on a distant hill. The familiar work of Maurits C. Escher (1898–1972) also confounds us with its use and abuse of perspective. His drawing *Belvedere* (Figure 11.10) illustrates good use of perspective, yet the pillars of the structure are cleverly drawn to show impossible positions.

FIGURE 11.9 *False Perspective*, by William Hogarth.

FIGURE 11.10 *Belvedere*, by M. C. Escher.

FIGURE 11.11 Leonardo da Vinci's sketch showing the symmetry of the human body.

FIGURE 11.12 This Islamic tiling shows many elaborate symmetries. It is located in the Medersa Attarine, a 14th-century school of Muslim theology in Fez, Morocco.

SYMMETRY

The term *symmetry* has many meanings. Sometimes it refers to a kind of balance. For example, *The Last Supper* (see Figure 11.6) is symmetrical because the disciples are grouped in four groups of three, with two groups on either side of the central figure of Christ. A human body is symmetrical because a vertical line drawn through the head and navel divides the body into two (nearly) identical parts (Figure 11.11).

Symmetry can also refer to repetition of patterns. Native American pottery is often decorated with simple borders that use repeating patterns. Similar symmetries are found in African, Muslim, and Moorish art such as that shown in Figure 11.12.

In mathematics, *symmetry* is a property of an object that remains unchanged under certain operations. For example, a circle still looks the same if it is rotated about its center. A square still looks the same if it is flipped across one of its diagonals (Figure 11.13). There are many such mathematical symmetries, often quite subtle. However, three symmetries are easy to identify:

FIGURE 11.13

- **Reflection symmetry:** An object remains unchanged when reflected across a straight line. For example, the letter A has reflection symmetry about a vertical line, while the letter H has reflection symmetry about a vertical and a horizontal line (Figure 11.14).

FIGURE 11.14 Reflection symmetry. `

FIGURE 11.15
Rotation symmetry.

- **Rotation symmetry:** An object remains unchanged when rotated through some angle about a point. For example, the letters O and S have rotation symmetry because they are unchanged when rotated 180° (Figure 11.15).
- **Translation symmetry:** A pattern remains the same when shifted, say to the right or left. The pattern . . . XXX . . . (with the X's continuing in both directions) has translation symmetry because it still looks the same if we shift it to the left or to the right (Figure 11.16). (In mathematics and physics, *translating* an object means moving it in a straight line, without rotating it.)

FIGURE 11.16 Translation symmetry.

▶ **EXAMPLE 1** Finding Symmetries

Identify the types of symmetry in each star in Figure 11.17.

Solution

(a)

(b)

FIGURE 11.17

a. The five-pointed star has five lines about which it can be flipped (reflected) without changing its appearance, so it has five reflection symmetries (Figure 11.18a). Because it has five vertices that all look the same, it can be rotated by $\frac{1}{5}$ of a full circle, or 360°/5 = 72°, and it still looks the same. Similarly, its appearance remains unchanged if it is rotated by 2 × 72° = 144°, 3 × 72° =216°, or 4 × 72° = 288°. Thus, this star has four rotational symmetries.

b. The six-pointed star has six reflection lines about which it can be flipped (reflected) without changing its appearance, so it has six reflection symmetries (Figure 11.18b). Because of its six vertices, it has rotation symmetry when rotated by $\frac{1}{6}$ of a full circle, or 360°/6 = 60°. It also has symmetry if rotated by 2 × 60° = 120°, 3 × 60° = 180°, 4 × 60° = 240°, or 5 × 60° = 300°. Thus, this star has five rotational symmetries.

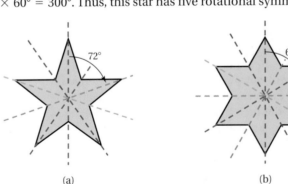

(a) (b)

FIGURE 11.18 Each line represents a reflection symmetry. Rotation symmetries require rotation by multiples of the indicated angles.

Now try Exercises 21–27. ◀

FIGURE 11.19 *The Vision of the Empyrean*, by Gustave Dore.

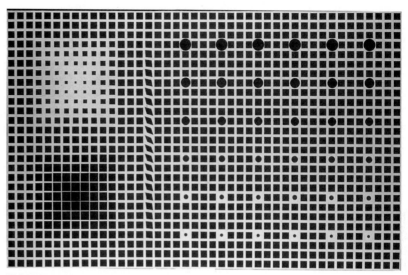

FIGURE 11.20 *Supernovae*, by Victor Vasarely.

SYMMETRY IN ART

Gustave Dore's (1832–1883) engraving *The Vision of the Empyrean* offers a dramatic illustration of rotation symmetry (Figure 11.19). This grand image of the cosmos can be rotated by many different angles and, at least on a large scale, appears much the same.

Sometimes, it is the *departures* from symmetry that make art effective. The 20th-century work *Supernovae* (Figure 11.20), by the Hungarian painter Victor Vasarely (1908–1997), might have started as a symmetric arrangement of circles and squares, but the gradual deviations from that pattern make a powerful visual effect.

Given the strong ties between mathematics and art, you may not be surprised that mathematical algorithms (recipes) can generate art on computers. Figure 11.21 shows an intricate Persian rug design generated on a computer by Anne Burns. The algorithm can be varied to give an endless array of patterns and symmetries.

TILINGS

A form of art called *tilings* (or *tessellations*) involves covering a flat area, such as a floor, with geometrical shapes. Tilings usually have regular or symmetric patterns. Tilings are found in ancient Roman mosaics, stained glass windows, and the elaborate courtyards of Arab mosques—as well as in many modern kitchens and bathrooms.

FIGURE 11.21 A "Persian rug," generated on a computer by Anne Burns.

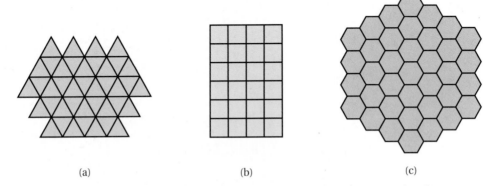

(a) (b) (c)

FIGURE 11.22 The three possible tilings made from regular polygons: (a) triangles, (b) squares, and (c) hexagons.

More precisely, a **tiling** is an arrangement of *polygons* (see Unit 10A) that interlock perfectly with no overlapping. The simplest tilings use just one type of regular polygon. Figure 11.22 shows three such tilings made with equilateral triangles, squares, and regular hexagons, respectively. Note that there are no gaps or overlaps between the polygons in any of the three cases. In each case, the tiling is made by translating (shifting) the same basic polygon in various directions. Thus, these tilings have translation symmetry.

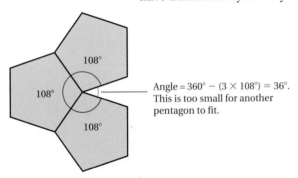

Angle $= 360° - (3 \times 108°) = 36°$. This is too small for another pentagon to fit.

FIGURE 11.23 Regular pentagons cannot make a tiling.

What happens if you try to make a tiling with, say, regular pentagons? If you try, you'll find that it simply does not work. The interior angles of a regular pentagon measure 108°. Thus, as Figure 11.23 shows, the angle that remains when three regular pentagons are placed next to each other is too small to fit another regular pentagon. In fact, a mathematical theorem states that tilings with a single regular polygon are possible only with equilateral triangles, squares, and hexagons (as in Figure 11.22).

Of course, more tilings are possible if we remove the restriction of using only a single type of regular polygon. For example, if we allow different regular polygons, but still require that the arrangement of polygons look the same around each vertex (intersection point), it is possible to make the eight different tiling patterns shown in Figure 11.24.

Time out to think

Verify that each of the tilings in Figure 11.24 uses only regular polygons. How many different regular polygons are used in each of these tilings? Verify that the same arrangement of polygons appears around each vertex. (Look carefully at the polygons; there are no circles in this figure.)

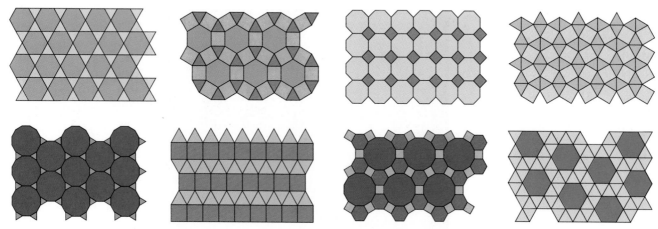

FIGURE 11.24 Eight tilings, each made by combining two or more types of regular polygons.

Tilings that use *irregular* polygons (those with sides of different lengths) are endless in number. As an example, suppose we start with an arbitrary triangle that has no special properties (other than three sides). The easiest way to tile a region with this triangle is by translating it parallel to two of its sides, as shown in Figure 11.25. We shift the original triangle to the right so that the new triangle touches the original triangle at a single point. We also shift the original triangle down so that the new triangle touches the original triangle at a single point. Then we repeat these right/left and up/down translations as many times as we like. The gaps created in this process are themselves triangles that interlock perfectly with the translated triangles to create a tiling.

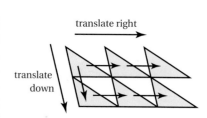

FIGURE 11.25 A tiling made by translating a triangle in two directions.

Figure 11.26 shows another example. This time, we begin by reflecting an arbitrary triangle to produce a wing-shaped object, and then we translate this object up/down and right/left.

All of the tilings discussed so far are called *periodic tilings* because they have a pattern that is repeated throughout the tiling. In recent decades, mathematicians have explored tilings that are *aperiodic,* meaning that they do not have a pattern that repeats throughout the entire tiling. Figure 11.27 shows an aperiodic tiling created by British mathematician Roger Penrose. If you look at the center of the figure, there appears to be a fivefold symmetry (a rotational symmetry that you would find in a pentagon). However, if the figure were extended indefinitely in all directions, the same pattern would never be repeated.

Tilings can be beautiful and practical for such things as floors and ceilings. However, recent research also shows that tilings may be very important in nature. In particular, many molecules and crystals apparently have patterns and symmetries that can be understood with the same mathematics used to study tilings in art.

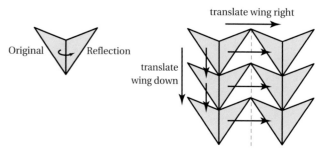

FIGURE 11.26 A tiling made by first reflecting a triangle to make a "wing," then translating the wing in two directions.

FIGURE 11.27 An aperiodic tiling by Roger Penrose.

➤ **EXAMPLE 2** Quadrilateral Tiling

FIGURE 11.28

Create a tiling by translating the quadrilateral shown in Figure 11.28. As you translate the quadrilateral, make sure that the gaps left behind have the same quadrilateral shape.

Solution We can find the solution by trial and error, translating the quadrilateral in different directions until we have correctly shaped gaps. Figure 11.29 shows the solution. Note that the translations are along the directions of the two diagonals of the quadrilateral. The gaps between the translated quadrilaterals are themselves quadrilaterals that interlock perfectly to complete the tiling.

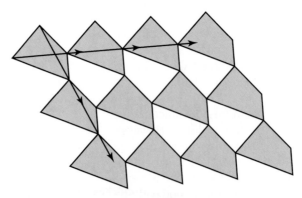

FIGURE 11.29 A quadrilateral tiling.

Now try Exercises 28–33. ◄

EXERCISES 11B

Review Questions

1. Describe the ideas of perspective and symmetry.

2. How is the principal vanishing point in a picture determined?

3. What is the horizon line? How is it important in a painting showing perspective?

4. Briefly describe and distinguish among reflection symmetry, rotation symmetry, and translation symmetry. Draw a simple picture that shows each type of symmetry.

5. What is a tiling? Draw a simple example.

6. Briefly explain why there are only three possible tiling patterns that consist of a single regular polygon. What are the three patterns?

7. Briefly explain why more tilings are possible if we remove the restriction of using regular polygons.

8. What is the difference between periodic and aperiodic tilings?

Does It Make Sense?

Decide whether each of the following statements makes sense (or is clearly true) or does not make sense (or is clearly false). Explain your reasoning.

9. The principal vanishing point of the painting is so far away that it cannot be seen in the painting.

10. Sid does not need to use perspective in her ground-level painting of a flat desert because the desert is two-dimensional.

11. Jane wants near objects to look nearby and far objects to look far away in her painting, so she should use perspective.

12. Kenny likes symmetry, so he prefers the letter R to the letter O.

13. Susan found a sale on octagonal (eight-sided) floor tiles, so she bought them to tile her kitchen.(Assume she uses no other tiles on her floor.)

14. Frank always liked the symmetry of the Washington Monument (in Washington, D.C.).

Basic Skills & Concepts

15. **Vanishing Points.** Consider the simple drawing of a road and a telephone pole in Figure 11.30.

FIGURE 11.30

a. Locate a vanishing point for the drawing. Is it the principal vanishing point?

b. With proper perspective, draw three more telephone poles receding into the distance.

16. **Correct Perspective.** Consider the two boxes shown in Figure 11.31. Which one is drawn with proper perspective relative to a single vanishing point? Explain.

FIGURE 11.31

17. **Drawing with Perspective.** Make the square, circle, and triangle in Figure 11.32 into three-dimensional solid objects: a box, a cylinder, and a triangular prism, respectively.

The given objects should be used as the front faces of the three-dimensional objects and all figures should be drawn with correct perspective relative to the given vanishing point *P*.

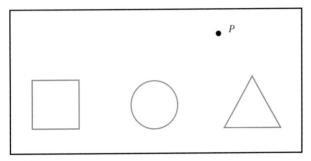

FIGURE 11.32

18. **Drawing MATH with Perspective.** Make the letters M, A, T, and H in Figure 11.33 into three-dimensional solid letters. The given letters should be used as the front faces of three-dimensional letters as deep as the T is wide, and all letters should be drawn with correct perspective relative to the given vanishing point *P*.

FIGURE 11.33

19. **Proportion and Perspective.** The drawing in Figure 11.34 shows two poles drawn with correct perspective relative to a single vanishing point. As you can check, the first pole is 2 centimeters tall in the drawing and the second pole is 2 centimeters away from the first pole in the drawing.

 a. Draw three more identical poles that are equally spaced in the drawing.

 b. What are the heights of these three new poles in the drawing?

FIGURE 11.34

 c. In the actual scene, would these five poles be equally spaced? Explain.

20. **Two Vanishing Points.** Figure 11.35 shows a road receding into the distance. In the direction of the arrow, draw a second road that intersects the first road. Be sure that the vanishing points of the two roads lie on the horizon line.

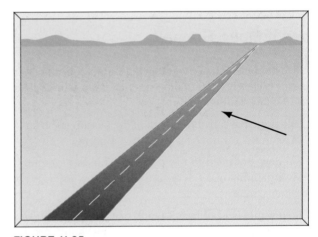

FIGURE 11.35

21. **Symmetry in Letters.** Find all of the capital letters of the alphabet that have

 a. right/left reflection symmetry (such as A)

 b. top/bottom reflection symmetry (such as H)

 c. both right/left and top/bottom reflection symmetry

 d. a rotational symmetry

22. Star Symmetries.

a. How many reflection symmetries does a four-pointed star have (Figure 11.36a)? How many rotational symmetries does a four-pointed star have?

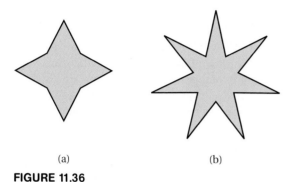

(a) (b)

FIGURE 11.36

b. How many reflection symmetries does a seven-pointed star have (Figure 11.36b)? How many rotational symmetries does a seven-pointed star have?

23. Symmetries of Geometric Figures.

a. Draw an equilateral triangle (all three sides have equal length). How many degrees can the triangle be rotated about its center so that it remains unchanged in appearance? (There are several correct answers.)

b. Draw a square (all four sides have equal length). How many degrees can the square be rotated about its center so that it remains unchanged in appearance? (There are several correct answers.)

c. Draw a regular pentagon (all five sides have equal length). How many degrees can the pentagon be rotated about its center so that it remains unchanged in appearance? (There are several correct answers.)

d. Can you see a pattern in parts a, b, and c? How many degrees can a regular n-gon be rotated about its center so that it remains unchanged in appearance? How many different angles answer this question for an n-gon?

Identifying Symmetries. Identify all of the symmetries in the figures of Exercises 24–27.

24.

25.

26.

Pattern continues in both directions.

27.

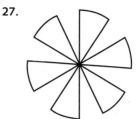

Tilings from Translating Triangles. In Exercises 28–29, make a tiling from each triangle using translations only, as in Figure 11.25.

28. 29.

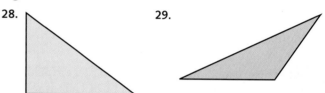

Tilings from Translating and Reflecting Triangles. For Exercises 30–31, make a tiling from the given triangle using translations and reflections, as in Figure 11.26.

30. The triangle in Exercise 28

31. The triangle in Exercise 29

Tilings from Quadrilaterals. In Exercises 32–33, make a tiling from the quadrilateral using translations, as in Figure 11.29.

32. 33.

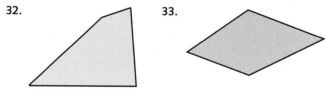

Further Applications

34. Desargues' Theorem. An early theorem of projective geometry, proved by French architect and engineer Girard Desargues (1593–1662), says that if two triangles (*ABC* and *abc* in Figure 11.37) are drawn so that the straight lines joining corresponding vertices (*Aa*, *Bb*, and *Cc*) all meet in a point *P* (corresponding to a vanishing point), then the corresponding sides (*AC* and *ac*, *AB* and *ab*, *BC*

and *bc*), if extended, will meet in three points that all lie on the same line *L*. Draw two triangles of your own in such a way that the conditions of Desargues' theorem are satisfied. Verify that the conclusions of the theorem are true.

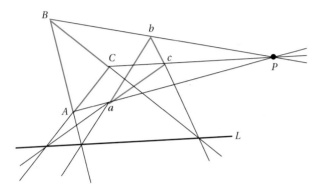

FIGURE 11.37

35. **Why Quadrilateral Tilings Work.** Consider the tiling with quadrilaterals shown in Figure 11.29. Look at any of the points at which four quadrilaterals meet in the tiling. Call such a point *P*. Given that the sum of the inside angles of any quadrilateral is 360°, show that the sum of the angles around the point *P* is also 360°, thus proving that the quadrilaterals interlock perfectly.

36. **Tiling with a Rhombus.** A rhombus is a quadrilateral in which all four sides have the same length and opposite sides are parallel. Show how a tiling can be made from a rhombus using translations only (as in Figure 11.25) and an initial reflection and then translations (as in Figure 11.26).

37. **Project: Symmetry and Proportion in Art.** Find one piece of pre-20th-century art and one piece of 20th-century art that you like. Use as many ideas from this unit as possible (involving perspective and symmetry) to write a two- to three-page analysis and comparison of these two pieces of art.

Web Projects

Find useful links for Web Projects on the text Web site: www.aw.com/bennett-briggs

38. **Art and Mathematics.** Many Web sites are devoted to connections between art and mathematics. Visit one such site and study a connection in some depth. Write a one- to two-page essay that describes this connection between mathematics and art.

39. **Art Museums.** Many art museums have Web sites where you can explore their collections. Visit an art museum on the Web, and study its online collection. Describe a few pieces in which ideas of perspective or symmetry are important.

40. **Escher.** Look on the Web for art by M. C. Escher. Choose one of his works and write a short essay about his use of perspective in the piece.

41. **Penrose Tilings.** Many Web sites are devoted to Penrose tilings (aperiodic tilings). Visit one such site and learn more about the nature and uses of these tilings. Write a short essay describing your findings.

In the News

42. **Perspective in Life.** Describe at least three ways in which the use of visual perspective affects your life.

43. **Symmetry in Life.** Find at least three objects from your daily life that exhibit some type of mathematical symmetry. Describe the symmetry in each case.

UNIT **11C**

PROPORTION AND THE GOLDEN RATIO

n Unit 11B, we studied how mathematics enters into art through the ideas of symmetry and perspective. In this unit, we turn our attention to the third major mathematical idea involved with art: proportion.

The importance of proportion was expressed well by the astronomer Johannes Kepler (1571–1630):

> *Geometry has two great treasures: one is the theorem of Pythagoras; the other, the division of a line into extreme and mean ratio. The first we may compare to gold; the second we may name a precious jewel.*

The senses delight in things duly proportioned.

—St. Thomas Aquinas (1225–1274)

Kepler's statement about *the division of a line into extreme and mean ratio* describes one of the oldest principles of proportion. It dates back to the time of Pythagoras (c. 500 B.C.), when scholars asked the following question: How can a line segment be divided into two pieces that have the most appeal and balance?

Although this was a question of beauty, there seemed to be general agreement on the answer. Suppose a line segment is divided into two pieces, as shown in Figure 11.38. We call the length of the long piece L and the length of the short piece 1.

FIGURE 11.38

The Greeks claimed that the most visually pleasing division of the line had the property that the ratio of the length of the long piece to the length of the short piece is the same as the ratio of the length of the entire line segment to the length of the long piece. That is,

$$\frac{L}{1} = \frac{L + 1}{L}$$

This statement of proportion can be solved (see Exercise 19) to find that L has a special value, denoted by the Greek letter ϕ (phi, pronounced "fie" or "fee"), which is

$$\phi = \frac{1 + \sqrt{5}}{2} = 1.61803\ldots$$

> **by the way**
> The Greek letter ϕ is the first letter in the Greek spelling of Phydias, the name of a Greek sculptor who may have used the golden ratio in his work.

The number ϕ is more commonly called the **golden ratio** (or *golden section*); it is also sometimes called the *divine proportion*. Note that the golden ratio is an irrational number often approximated as 1.6, or $\frac{8}{5}$. Figure 11.39 shows that, for any line segment divided in two pieces according to the golden ratio, the ratio of the long piece to the short piece is

FIGURE 11.39

$$\frac{x}{y} = \frac{1.61803\ldots}{1} = \phi \approx \frac{8}{5}$$

▶EXAMPLE 1 Calculating a Golden Ratio

Suppose the line segment in Figure 11.40 is divided according to the golden ratio. If the length of the longer piece labeled x is 5 centimeters, how long is the entire line segment?

FIGURE 11.40

Solution Because the line segment is divided in the golden ratio, we know that $x/y = \phi$. We can solve for y by multiplying both sides by y and dividing both sides by ϕ:

$$\frac{x}{y} = \phi \quad \rightarrow \quad y = \frac{x}{\phi}$$

Substituting $x = 5$ cm and the approximate value 1.6 for ϕ, we find

$$y = \frac{5\text{ cm}}{\phi} \approx \frac{5\text{ cm}}{1.6} \approx 3.1\text{ cm}$$

The entire segment has a length of $x + y$, so its total length is approximately 5 cm + 3.1 cm = 8.1 cm. *Now try Exercises 11–12.* ◀

THE GOLDEN RATIO IN ART HISTORY

Although the ancient Greeks struggled with the notion of irrational numbers, they embraced the golden ratio. For example, the *pentagram* (Figure 11.41) was the seal of the mystical Pythagorean Brotherhood. This five-pointed star inscribed in a circle produces a pentagon at the center. The golden ratio occurs in at least ten different ways in the pentagram. For example, if the length of each side of the pentagon is 1, then the length of each arm of the star is ϕ.

Time out to think
Using a ruler, find at least one other place in the pentagram of Figure 11.41 where the ratio of the lengths of two line segments is the golden ratio.

FIGURE 11.41 A pentagram.

FIGURE 11.42 A golden rectangle—the side lengths have a ratio $\phi \approx \frac{8}{5}$.

From the golden ratio it is a short step to another famous Greek expression of proportion, the **golden rectangle**—a rectangle whose long side is ϕ times as long as its short side. A golden rectangle can be of any size, but its sides must have a ratio of $\phi \approx \frac{8}{5}$. Figure 11.42 shows a golden rectangle.

The golden rectangle had both practical and mystical importance to the Greeks. It became a cornerstone of their philosophy of *aesthetics*—the study of beauty. There is considerable speculation about the uses of the golden rectangle in art and architecture in ancient times. For example, it is widely claimed that many of the great monuments of antiquity, such as the Pyramids in Egypt, were designed in accordance with the golden rectangle. And, whether by design or by chance, the proportions of the Parthenon (in Athens, Greece) match those of the golden rectangle very closely (Figure 11.43).

The golden rectangle also appears in many more recent works of art and architecture. The book *De Divina Proportione*, illustrated by Leonardo da Vinci in 1509, is filled with references to and uses of ϕ. Indeed, da Vinci's unfinished painting *St. Jerome* (Figure 11.44) places the central figure inside an imaginary golden rectangle. More recently, the French impressionist painter Georges Seurat is said to have used the golden ratio on every canvas. The abstract geometric paintings of the 20th-century Dutch painter Piet Mondrian (Figure 11.45) are filled with golden rectangles.

FIGURE 11.43 The proportions of the Parthenon closely match those of a golden rectangle.

by the way

The Parthenon was completed in about 430 B.C. as a temple to Athena Parthenos, the Warrior Maiden. It stands on the Acropolis (which means "the uppermost city"), about 500 feet above Athens.

FIGURE 11.44 *St. Jerome,* by Leonardo da Vinci, has the central figure positioned as though he were inside a golden rectangle.

FIGURE 11.45 *Place de la Concorde,* by Piet Mondrian, is filled with golden rectangles.

Today, the golden rectangle appears in many everyday items. For example, photographs, note cards, cereal boxes, posters, and windows often have the proportions of the golden rectangle. But the question remains as to whether the golden rectangle is really more pleasing. In the late 19th century, the German psychologist Gustav Fechner (1801–1887) studied the question statistically. He showed several rectangles with various length to width ratios to hundreds of people and recorded their choices for the most and least visually pleasing rectangles. The results, given in Table 11.2, show that almost 75% of the participants chose rectangles that were very close to the golden rectangle.

TABLE 11.2 Fechner's Data

LENGTH TO WIDTH RATIO	MOST PLEASING RECTANGLE (PERCENTAGE RESPONSE)	LEAST PLEASING RECTANGLE (PERCENTAGE RESPONSE)
1.00	3.0	27.8
1.20	0.2	19.7
1.25	2.0	9.4
1.33	2.5	2.5
1.45	7.7	1.2
1.50	20.6	0.4
$\phi \approx 1.62$	**35.0**	**0.0**
1.75	20.0	0.8
2.00	7.5	2.5
2.50	1.5	35.7

Time out to think _____

Do you think that the golden rectangle is visually more pleasing than other rectangles? Explain.

▶ **EXAMPLE 2** Household Golden Ratios

The following household items were found to have the dimensions shown. Which item comes closest to having the proportions of the golden ratio?

- Standard sheet of paper: 8.5 in. \times 11 in.
- 8×10 picture frame: 8 in. \times 10 in.
- 35-mm slide: 35 mm \times 23 mm

Solution The ratio of the sides of a standard sheet of paper is $11/8.5 = 1.29$, which is 20% less than the golden ratio. The ratio of the sides of a standard picture frame is $10/8 = 1.25$, which is 23% less than the golden ratio. The ratio of the sides of a 35-mm slide is $35/23 = 1.52$, which is 6% less than the golden ratio. Of the three objects, the 35-mm slide is closest to being a golden rectangle. *Now try Exercises 13–18.* ◀

THE GOLDEN RATIO IN NATURE

The golden ratio appears to be common in the "artwork" of nature. One striking example is spirals created from golden rectangles. We begin by dividing a golden rectangle to make a square on its left side, as shown in Figure 11.46a. If you measure the sides of the remaining smaller rectangle to the right of the square, you'll see that it is a smaller golden rectangle.

We now repeat this splitting process on the second golden rectangle, this time making the square on the top (instead of the left) of the golden rectangle.

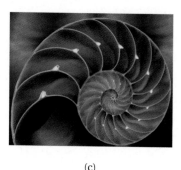

(a) (b) (c)

FIGURE 11.46 (a) A logarithmic spiral begins with a golden rectangle that is divided to make a square on the left. (b) The process is repeated in each successive golden rectangle, and a spiral is created by connecting the corners of all the squares. (c) A chambered nautilus shell looks much like a logarithmic spiral.

This split makes a third, even smaller, golden rectangle. Continuing to split each new golden rectangle in this manner generates the result shown in Figure 11.46b. Now, we connect opposite corners of all the squares with a smooth curve. The result is a continuous curve called a **logarithmic spiral** (or *equiangular spiral*). This spiral very closely matches the spiral shape of the beautiful chambered nautilus shell (Figure 11.46c).

Another intriguing connection between the golden ratio and nature comes from a problem in population biology, first posed by a mathematician known as Fibonacci in 1202. Fibonacci's problem essentially asked the following question about the reproduction of rabbits.

> *Suppose that a pair of baby rabbits takes one month to mature into adults, then produces a new pair of baby rabbits the following month and each subsequent month. Further suppose that each newly born pair of rabbits matures and gives birth to additional pairs with the same reproductive pattern. If no rabbits die, how many pairs of rabbits are in the population at the beginning of each month?*

Figure 11.47 shows the solution to this problem for the first six months. Note that the number of pairs of rabbits at the beginning of each month forms a sequence that begins 1, 1, 2, 3, 5, 8. Fibonacci found that the numbers in this sequence continue to grow with the following pattern:

$$1, 1, 2, 3, 5, 8, 13, 21, 34, 55, \ldots$$

This sequence of numbers is known as the **Fibonacci sequence.**

If we let F_n denote the nth Fibonacci number, then we have $F_1 = 1$, $F_2 = 1$, $F_3 = 2$, $F_4 = 3$, and so forth. The most basic property of the Fibonacci sequence is that the next number in the sequence is the sum of the previous two numbers. For example, note that

$$F_3 = F_2 + F_1 = 1 + 1 = 2 \quad \text{and} \quad F_4 = F_3 + F_2 = 2 + 1 = 3$$

We can express this rule in general as $F_{n+1} = F_n + F_{n-1}$.

by the way

Fibonacci, also known as Leonardo of Pisa, is credited with popularizing the use of Hindu-Arabic numerals in Europe. His book *Liber Abaci* ("Book of the Abacus"), published in 1202, explained their use and the importance of the number zero.

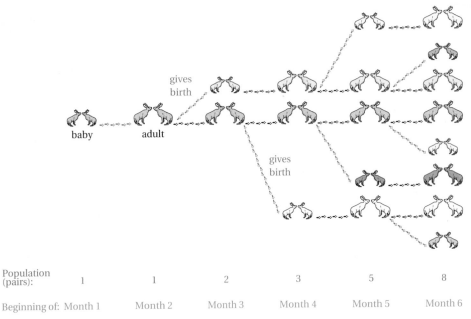

Population (pairs):	1	1	2	3	5	8
Beginning of:	Month 1	Month 2	Month 3	Month 4	Month 5	Month 6

FIGURE 11.47

Time out to think _____

Confirm that the above rule works for Fibonacci numbers F_3 through F_{10}. Use the rule to determine the eleventh Fibonacci number (F_{11}).

The connection between the Fibonacci numbers and the golden ratio becomes clear when we compute the ratios of successive Fibonacci numbers, as shown in Table 11.3. Note that, as we go further out in the sequence, the ratios of successive Fibonacci numbers get closer and closer to the golden ratio $\phi = 1.61803\ldots$.

TABLE 11.3	Ratios of Successive Fibonacci Numbers
$F_3/F_2 =\ \ 2/1 = 2.0$	$F_{11}/F_{10} =\ \ \ \ \ 89/55 = 1.618182$
$F_4/F_3 =\ \ 3/2 = 1.5$	$F_{12}/F_{11} =\ \ \ \ 144/89 = 1.617978$
$F_5/F_4 =\ \ 5/3 = 1.667$	$F_{13}/F_{12} =\ \ 233/144 = 1.618056$
$F_6/F_5 =\ \ 8/5 = 1.600$	$F_{14}/F_{13} =\ \ 377/233 = 1.618026$
$F_7/F_6 =\ \ 13/8 = 1.625$	$F_{15}/F_{14} =\ \ 610/377 = 1.618037$
$F_8/F_7 = 21/13 = 1.6154$	$F_{16}/F_{15} =\ \ 987/610 = 1.618033$
$F_9/F_8 = 34/21 = 1.61905$	$F_{17}/F_{16} = 1597/987 = 1.618034$
$F_{10}/F_9 = 55/34 = 1.61765$	$F_{18}/F_{17} = 2584/1597 = 1.618034$

There are many examples of the Fibonacci sequence in nature. The heads of sunflowers and daisies consist of a clockwise spiral superimposed on a counterclock-

wise spiral (both of which are logarithmic spirals), as shown in Figure 11.48. The number of individual florets in each of these intertwined spirals is a Fibonacci number—for example, 21 and 34 or 34 and 55. Biologists have also observed that the number of petals on many common flowers is a Fibonacci number (for example, irises have 3 petals, primroses have 5 petals, ragworts have 13 petals, and daisies have 34 petals). The arrangement of leaves on the stem of many plants also exhibits the Fibonacci sequence. And spiraling Fibonacci numbers can be identified on pine cones and pineapples.

(a)

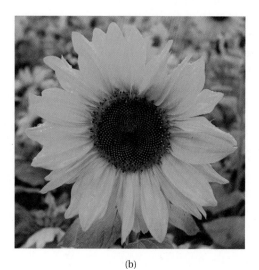
(b)

FIGURE 11.48

EXERCISES 11C

Review Questions

1. Explain the golden ratio in terms of proportions of line segments.

2. How is a golden rectangle formed?

3. What evidence suggests that the golden ratio and golden rectangle hold particular beauty?

4. What is a logarithmic spiral? How is it formed from a golden rectangle?

5. What is the Fibonacci sequence?

6. What is the connection between the Fibonacci sequence and the golden ratio? Give some examples of the Fibonacci sequence in nature.

Does It Make Sense?

Decide whether each of the following statements makes sense (or is clearly true) or does not make sense (or is clearly false). Explain your reasoning.

7. Maria cut her 4-foot walking stick into two 2-foot sticks in keeping with the golden ratio.

8. Dan attributes his love of dominoes to the fact that dominoes are golden rectangles.

9. The circular pattern in the floor is attractive because it exhibits the golden ratio.

10. Each year, Juliet's age is another Fibonacci number.

Basic Skills & Concepts

11. Golden Ratio. Draw a line segment 4 inches long. Now subdivide it according to the golden ratio. Verify your work by computing the ratio of the whole segment length to the long segment length and the ratio of the long segment length to the short segment length.

12. Golden Ratio. A line is subdivided according to the golden ratio, with the smaller piece having a length of 2 meters. What is the length of the entire line?

13. Golden Rectangles. Measure the sides of each rectangle in Figure 11.49, and compute the ratio of the long side to the short side for each rectangle. Which ones are golden rectangles?

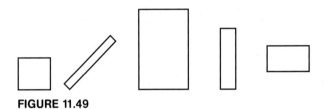

FIGURE 11.49

Dimensions of Golden Rectangles. Exercises 14–17 give the length of one side of a golden rectangle. Find the length of the other side. Notice that the other side could be either longer or shorter than the given side. Use the approximation $\phi \approx 1.62$ for your work.

14. 1.5 inches **15.** 2.5 meters

16. 6.4 kilometers **17.** 0.12 centimeter

18. Everyday Golden Rectangles. Find at least three everyday objects with rectangular shapes (for example, billboards, cereal boxes, windows). In each case, measure the side lengths and calculate the ratio. Are any of these objects golden rectangles? Explain.

Further Applications

19. Finding ϕ. The property that defines the golden ratio is
$$\frac{L}{1} = \frac{L+1}{L}$$

a. Show that, if we multiply both sides by L and rearrange, this equation becomes
$$L^2 - L - 1 = 0$$

Confirm that substituting the value of ϕ for L satisfies this equation.

b. The quadratic formula states that, for any equation of the form $ax^2 + bx + c = 0$, the solutions are given by
$$x = \frac{-b + \sqrt{b^2 - 4ac}}{2a}$$
and
$$x = \frac{-b - \sqrt{b^2 - 4ac}}{2a}$$

Use the quadratic formula to solve for L in the formula for the golden ratio. Show that one of the roots is ϕ.

20. Properties of ϕ.

a. Enter $\phi = (1 + \sqrt{5})/2$ into your calculator. Show that $1/\phi = \phi - 1$.

b. Now compute ϕ^2. How is this number related to ϕ?

21. Logarithmic Spirals. Draw a rectangle that is 10 centimeters on a side. Follow the procedure described in the text for subdividing the rectangle until you can draw a logarithmic spiral. Do all work carefully, and show your measurements with your work.

22. The Lucas Sequence. A sequence called the *Lucas sequence* is closely related to the Fibonacci sequence. The Lucas sequence begins with the numbers $L_1 = 1$ and $L_2 = 3$ and then uses the same relation $L_{n+1} = L_n + L_{n-1}$ to generate L_3, L_4, \ldots.

a. Generate the first ten Lucas numbers.

b. Compute the ratio of successive Lucas numbers L_2/L_1, L_3/L_2, L_4/L_3, and so on. Can you determine if these ratios approach a single number? What number is it?

23. Graphing Fechner's Data. Consider Gustav Fechner's data shown in Table 11.2. Make a histogram that displays the responses for both the most pleasing and the least pleasing rectangle proportions.

24. Project: The Golden Navel. An old theory claims that, on average, the ratio of the height of a person to the height of his/her navel is the golden ratio. Collect "navel ratio data" from as many people as possible. Graph the ratios in a histogram, find the average ratio over your entire sample, and discuss the outcome. Do your data support the theory?

25. Project: Mozart and the Golden Ratio. In a paper called "The Golden Section and the Piano Sonatas of Mozart" (*Mathematics Magazine*, Vol. 68, No. 4, 1995), John Putz gives the lengths (in measures) of the first part (*a*) and the second part (*b*) of the movements of the 19 Mozart piano sonatas. Some of the data are given on the next page.

a = length of first part	b = length of second part
38	62
28	46
56	102
56	88
24	36
77	113
40	69
46	60
15	18
39	63
53	67

a. The ratio of the length of the whole movement to the length of the longer segment is $(a + b)/b$. Add a third column to the table and compute this ratio for the given data.

b. Make a histogram of the ratios that you computed in part a (choosing an appropriate bin size). Comment on how well they are approximated by ϕ.

c. Read the article by John Putz. Do you believe that Mozart composed with the golden ratio in mind?

26. **Project: Debunking the Golden Ratio.** Find the article "Misconceptions about the Golden Ratio" by George Markowsky (*College Mathematics Journal*, Vol. 23, No. 1, 1992). Choose at least one of the misconceptions that Markowsky discusses, summarize it, and then explain whether you find his argument convincing. Discuss your opinion of whether the golden ratio has been consciously used by artists and architects in their work.

Web Projects

Find useful links for Web Projects on the text Web site: www.aw.com/bennett-briggs

27. **Golden Controversies.** Many Web sites are devoted to the controversy concerning the role of the golden ratio in art. Find a specific argument on one side of this controversy and summarize it in a one- to two-page essay.

28. **Fibonacci Numbers.** Learn more about Fibonacci numbers and possible occurrences of the Fibonacci sequence in nature. Write a short essay about one aspect of Fibonacci numbers that you find interesting.

In the News

29. **Proportion in Life.** Describe at least three ways in which the use of visual proportion affects your life.

30. **The Golden Ratio.** Find a recent picture of a new building or architectural design. Study the picture and decide whether the golden ratio is involved in any way.

CHAPTER 11 SUMMARY

UNIT	KEY TERMS	KEY IDEAS AND SKILLS
11A	sound wave pitch frequency harmonic octave musical scale digital recording	**Understand** how a plucked string produces sound. **Measure** frequency in cycles per second, and find harmonics of the frequency. **Understand** the musical scale and the ratios of frequencies among musical notes. **Understand** how the frequencies of a scale exhibit exponential growth. **Explain** the difference between analog and digital representations of music.
11B	perspective vanishing point horizon line symmetry reflection rotation translation	**Understand** the use of perspective in painting. **Find** symmetries in paintings and tilings. **Create** tilings with regular or irregular polygons.
11C	proportion golden ratio golden rectangle Fibonacci sequence	**The golden ratio:** $$\phi = \frac{1 + \sqrt{5}}{2} = 1.61803\ldots$$ **The Fibonacci sequence:** 1, 1, 2, 3, 4, 5, 13, 21, 34, 55, . . . **Understand** claimed uses of the golden ratio in both art and nature.

I must study politics and war that my sons may have liberty to study mathematics and philosophy. My sons ought to study mathematics and philosophy, geography, natural history, naval architecture, navigation, commerce and agriculture in order to give their children a right to study painting, poetry, music, architecture, statuary, tapestry, and porcelain.

—JOHN ADAMS, LETTER TO ABIGAIL ADAMS,
MAY 12, 1780

CHAPTER

12

Mathematics and Politics

Mathematics and politics have been part of human culture for millennia, so it's not surprising that the two have become interwoven through history. We have already studied several instances in which mathematics plays an important role in political decision making, including the case of the federal budget (Unit 4E). But the connections between mathematics and politics go much deeper, all the way to the heart of the democratic process. In this chapter, we'll discuss the prominent role of mathematics in voting and apportionment. Mathematics not only helps us count votes but also sheds light on the question of whether some voting and apportionment systems are fairer than others.

UNIT **12A** **Voting: Does the Majority Always Rule?** In elections with more than two candidates, there are several acceptable ways to choose a winner. We will study these methods and see that different methods can lead to different winners.

UNIT **12B** **Theory of Voting:** We will study issues of fairness in voting, leading to the surprising conclusion that no single system is absolutely fair in all cases.

UNIT **12C** **Apportionment: The House of Representatives and Beyond:** As with voting, there are several acceptable ways to apportion seats in the House of Representatives. We will study these methods and, again, find that no single method is always fair.

UNIT 12A

VOTING: DOES THE MAJORITY ALWAYS RULE?

The 2000 presidential election should have shattered any illusions about the simplicity of voting. The election was not decided for more than a month after election day. In the end, the outcome turned on the results in Florida, where George W. Bush's official margin of victory over Al Gore was fewer than 600 votes out of more than 6 million votes cast in Florida. The difficulties of counting large numbers (see Unit 3C) mean that this result undoubtedly has significant uncertainty associated with it—which is why people will probably never stop arguing about who really received more votes in Florida.

Mathematically, the issue of counting votes and associated uncertainties falls under the science of statistics. However, this is only one of the mathematical issues associated with voting. In this unit, we will investigate various types of voting systems. Then, in Unit 12B, we'll explore issues of fairness related to these voting systems.

MAJORITY RULE

The simplest type of voting involves only two choices. With two choices, the most common way of deciding the vote is by **majority rule:** The choice receiving more than 50% of the votes wins. While we take this rule for granted, it has three properties worth noting:

- Every vote has the same *weight*. That is, no person's vote counts more or less than any other person's vote.
- There is *symmetry* between the candidates: If all the votes were reversed, the loser would become the winner.
- If a vote for the loser were changed to a vote for the winner, the outcome of the election would not be changed.

We can illustrate these properties with a simple example. Suppose that Sharon and Robert are running for captain of the debate team. Sharon wins with 10 votes to Robert's 9 votes. The first property, that all votes have the same weight, means we care only about the vote totals—not about the names of the people who cast the individual votes. The second property tells us that if we reverse all the votes, so that Sharon receives 9 votes and Robert receives 10, then Robert becomes the winner. The third property tells us that if a vote for Robert (the loser) is changed to a vote for Sharon (the winner), Sharon still wins: This change raises Sharon's vote total to 11 and reduces Robert's to 8.

by the way

In 1952, Kenneth May proved that majority rule is the *only* voting system for two candidates that satisfies all three properties.

Majority Rule

Majority rule says that the choice receiving more than 50% of the vote is the winner.

Time out to think

Suppose that a vote for the *winner* is changed to a vote for the loser. Can the outcome of the election change in that case? Explain.

U.S. Presidential Elections

United States presidential elections involve a twist on majority rule. The **popular vote** in a presidential election reflects the total number of votes received by each candidate. However, presidents are elected by **electoral votes.** As mandated by the Constitution, each state gets as many electoral votes as the state has members of Congress (senators *plus* representatives). When you cast a ballot for president, you actually are casting votes for your state's *electors*. The electors meet to cast their votes a few weeks later.

In most states, the electoral votes are determined by a winner-take-all system: *All* the electoral votes of that state go to the candidate with the most popular votes in that state. (As of 2003, the only exceptions to the winner-take-all system are Nebraska and Maine.) The overall winner of the presidential election is chosen by majority rule among the electoral votes. If no candidate receives a majority of the electoral votes, the House of Representatives chooses the president. This has happened twice in U.S. history: The House elected Thomas Jefferson following the undecided election of 1800 and John Quincy Adams following the undecided election of 1824.

▶ **EXAMPLE 1** 2000 Presidential Election

Table 12.1 shows the 2000 presidential election official results. Discuss the outcomes of the popular and electoral votes.

TABLE 12.1	2000 Presidential Election	
	POPULAR VOTE	ELECTORAL VOTE
GEORGE W. BUSH	50,456,141	271
AL GORE	50,996,039	266
RALPH NADER	2,882,807	0
OTHERS	1,069,559	0
ABSTENTIONS	——	1*
TOTAL	105,404,546	538

An elector from the District of Columbia pledged to Al Gore abstained in protest over the District's lack of representation in Congress.

Solution The difference in popular vote totals between Gore and Bush was

$$50{,}996{,}039 - 50{,}456{,}141 = 539{,}898$$

That is, Gore won the popular vote over Bush by a margin of more than a half million votes. However, because of the small percentages of the total vote that went to Nader

HISTORICAL NOTE

The 1800 election pitted the Republican party ticket of Thomas Jefferson and Aaron Burr against the Federalist party ticket of John Adams and Charles Pickney. Voters at the time could vote for either the presidential or the vice-presidential candidate. Jefferson and his running mate Burr ended up tied, with 73 electoral votes each, while Adams and Pickney received 65 and 64 electoral votes, respectively. It took 7 days and 36 ballots before the House of Representatives finally decided the election in favor of Jefferson. The 12th Amendment to the U.S. Constitution, passed in 1804, instituted separate balloting for president and vice president.

HISTORICAL NOTE

Prior to 2000, the winner of the popular vote had lost the electoral vote just twice in U.S.history.In 1876,Samuel J. Tilden won the popular vote by some 264,000 votes over Rutherford B.Hayes, but Hayes won the electoral vote 185 to 184. In 1888, Benjamin Harrison won the popular vote by about 90,000 votes over Grover Cleveland, but Cleveland won the electoral vote by a margin of 233 to 168.

and other candidates, neither Gore nor Bush won a majority of the popular vote. Gore's percentage of the popular vote was

$$\frac{50,996,039}{105,404,546} = 0.4838 = 48.38\%$$

Bush's percentage of the popular vote was

$$\frac{50,456,141}{105,404,546} = 0.4787 = 47.87\%$$

In the electoral vote, Bush won 271 out of the total 538 votes available, which gave him a slight majority of

$$\frac{271}{538} = 0.5037 = 50.37\%$$

Because he won a majority of the electoral vote, Bush became the president of the United States. <u>*Now try Exercises 15–19.*</u> ◄

Time out to think _____

Do you think that it would be fairer if presidential elections were decided by popular vote rather than electoral vote? Why or why not?

Variations on Majority Rule

United States senators are generally allowed to speak about a legislative bill for as long as they wish before it is brought to a vote. If a senator opposes a particular bill, but fears that it will gain a majority vote, the senator may choose to speak continuously (at least during the hours that the issue is under consideration)—and thereby prevent the vote from ever taking place. This technique for preventing a vote on a bill is called a **filibuster.** The filibuster can be ended only by a vote of 3/5, or 60%, of the senators. This situation, in which 60% rather than 50% of the votes are required to end a filibuster, illustrates that majority rule is not always followed strictly.

Whenever a candidate or issue must receive *more* than a majority of the vote to win—such as 60% of the vote, 75% of the vote, or a unanimous vote—we say that a **super majority** is required. Criminal trials offer an example: All states require a super majority vote of the jury to reach a verdict, and many states require that the vote be unanimous. A jury that cannot reach this super majority or unanimous agreement is called a *hung jury.* When a jury is hung, the judge generally declares a mistrial, and the case must be tried again or dropped.

The U.S. Constitution requires super majority votes for many specific issues. For example, an international treaty can be ratified only by a 2/3 super majority vote of the Senate. Amending the Constitution first requires a 2/3 super majority vote on the amendment in both the House and the Senate, and then the amendment must also be approved by 3/4 of the states.

Another variation on majority rule occurs with **veto** power. For example, a bill proposed in the U.S. Congress normally becomes law if it receives a majority vote in both the House and Senate. However, if the president vetoes the bill, it can become law only if it then receives a 2/3 super majority vote in both the House and the Senate to override the veto. The courts can also effectively veto a popular vote.

by the way

The U.S. Constitution also specifies that it may be amended if 2/3 of the states call for "a convention for proposing amendments." However, such a call for a convention has never occurred.

For example, even if a proposition in a state election receives a huge majority of the popular vote, it will not become law if the courts declare that the proposition violates the U.S. Constitution.

► EXAMPLE 2 Majority Rule?

Evaluate the outcome in each of the following cases.

a. Of the 100 senators in the U.S. Senate, 59 favor a new bill on campaign finance reform. The other 41 senators are adamantly opposed and start a filibuster. Will the bill pass?

b. A criminal conviction in a particular state requires a vote by 3/4 of the jury members. On a nine-member jury, seven jurors vote to convict. Is the defendant convicted?

c. A proposed amendment to the U.S. Constitution has passed both the House and the Senate with more than the required 2/3 super majority. Each state holds a vote on the amendment. The amendment garners a majority vote in all but 14 of the 50 states. Is the Constitution amended?

d. A bill limiting the powers of the president has the support of 73 out of 100 senators and 270 out of 435 members of the House of Representatives. But the president promises to veto the bill if it is passed. Will it become law?

Solution

a. The filibuster can be ended only by a vote of 3/5 of the Senate, or 60 out of the 100 senators. Thus, the 59 senators in favor of the bill cannot stop the filibuster. The bill will not become law.

b. Seven out of nine jurors represents a super majority of $7/9 = 77.8\%$. This percentage is more than the required 3/4 (75%), so the defendant is convicted.

c. The amendment must be approved by 3/4, or 75%, of the states. But the 14 states that do not support the amendment represent $14/50 = 28\%$ of the 50 states. Thus, only $100\% - 28\% = 72\%$ of the states voted to ratify the amendment, which is not enough for it to become part of the Constitution.

d. The bill has the support of $73/100 = 73\%$ of the Senate and $270/435 = 62\%$ of the House. But overriding a presidential veto requires a super majority vote of $2/3 = 66.7\%$ in both the House and the Senate. The 62% support in the House is not enough to override the veto, so the bill will not become law. (Note: The law requires a vote of 2/3 of those present in Congress, so this example assumes that all members are present.) *Now try Exercises 20–21.* ◄

VOTING WITH THREE OR MORE CHOICES

Consider the following results in a three-candidate election for governor:

Candidate	Percentage of vote
Smith	32%
Jones	33%
Wilson	35%

No candidate has a majority, so who should become governor?

by the way

Senate rules do not limit what a senator can say. During some filibusters, senators have read novels aloud in order to keep speaking. In 1964, the Civil Rights Act was passed by Congress only after 75 days of filibuster by southern senators.

One possibility is to award the governorship to the person who received the most votes, called a **plurality** of the vote. Wilson has the greatest percentage and becomes governor if we decide this election by plurality. An alternative way to decide the election is to hold a **runoff** between the top two vote getters. In that case, another election must be held between Jones and Wilson.

The plurality method for deciding the election is simpler (and cheaper) than the runoff method. However, it can lead to a situation in which the majority of the electorate opposes the new governor. For example, suppose that all of Smith's supporters prefer Jones to Wilson. In that case, the combined 65% of the electorate that voted for Smith and Jones would choose Jones over Wilson. By the plurality method, Wilson would become governor despite being opposed by 65% of the voters. By the runoff method, Jones would easily defeat Wilson to become governor.

▶ **EXAMPLE 3** 1992 Presidential Election

Table 12.2 shows the results for the three major candidates in the 1992 U.S. presidential election. Analyze the outcome. Is it possible that Bush could have won if Perot had not run?

TABLE 12.2	1992 Presidential Election Results			
	POPULAR VOTE	POPULAR PERCENTAGE	ELECTORAL VOTE	ELECTORAL PERCENTAGE
CLINTON	44,909,889	43.3%	370	68.8%
BUSH	39,104,545	37.7%	168	31.2%
PEROT	19,742,267	19.0%	0	0%
TOTAL	103,756,701	100%	538	100%

Solution Clinton won a *plurality* of the popular vote, but there was no majority winner of the popular vote. However, Clinton won a large majority of the electoral vote. If Perot had not run, the 19% of the popular votes that he received would presumably have been divided in some way between Clinton and Bush. Note that Bush needed an additional 12.3% of the vote to reach a 50% total (37.7% + 12.3% = 50%). Thus, if $12.3/19 \approx 65\%$ of the Perot voters preferred Bush to Clinton, Bush could have won the popular vote in Perot's absence. Of course, winning the popular vote does not guarantee winning the electoral vote. *Now try Exercise 22.* ◀

Time out to think

Do you think that a runoff is fairer than choosing the winner by plurality in a three-way race? Why or why not?

Preference Schedules

So far, we have examined situations in which each voter is allowed to vote for only a single candidate. However, there are many cases in which voters are asked to rank candidates in order of preference. This type of election is used for sports polls, sports

awards (such as the Heisman Trophy), consumer surveys, beauty pageants, and the Academy Awards for motion pictures.

Imagine a club of 55 people that holds an election among five candidates for president. For simplicity, let's call the candidates A, B, C, D, and E. Each ballot asks the voter to rank these candidates in order of preference (Figure 12.1).

Table 12.3 shows a special type of table, called a **preference schedule,** that summarizes the election results. Each column shows a different ranking among the candidates, and the number at the bottom of the column is the number of people who chose that ranking. For example, the first column shows that 18 voters ranked the candidates in the order A, D, E, C, B, and the second column shows that 12 voters ranked the candidates in the order B, E, D, C, A.

Study Table 12.3 carefully. Which candidate should be declared the winner? As we'll soon see, a winner cannot be declared without at least some controversy.

FIGURE 12.1 Sample ballot for the club election.

TABLE 12.3	Preference Schedule for the Club Election					
FIRST	A	B	C	D	E	E
SECOND	D	E	B	C	B	C
THIRD	E	D	E	E	D	D
FOURTH	C	C	D	B	C	B
FIFTH	B	A	A	A	A	A
	18	12	10	9	4	2

TECHNICAL NOTE With five candidates, there are 5! = 120 possible orders in which they can be ranked. This example uses only 6 of these 120 possible orders.

Choosing a Winner by Plurality

Let's begin by counting *first-place* votes in Table 12.3:

- A received 18 first-place votes (column 1).
- B received 12 first-place votes (column 2).
- C received 10 first-place votes (column 3).
- D received 9 first-place votes (column 4).
- E received 4 + 2 = 6 first-place votes (columns 5 and 6).

The supporters of Candidate A can argue that A received a **plurality** *(the most first-place votes)* and should be declared the winner.

Choosing a Winner by a Single Runoff

"Not so fast!" yell the supporters of Candidate B. They suggest a runoff between A and B because they were the top two candidates in first-place votes. Table 12.3 shows that, although 18 people voted for A in first place, the other 37 people all ranked A in last place. These 37 people will vote for B over A in a runoff. Thus, B wins the runoff easily, by a vote of 37 to 18. Supporters of Candidate B can now proclaim victory for their candidate.

Choosing a Winner by Sequential Runoffs

Now the supporters of Candidate C chime in. They claim that a single runoff is unfair because it ignores rankings below the top two. They suggest a series of runoffs called a **sequential runoff:** The candidate with the fewest first-place votes is eliminated at each stage, and runoffs continue until someone claims a majority. By studying the results in Table 12.3, we can determine the outcomes of these runoffs without actually holding separate runoff elections.

We've already found that Candidate E received the fewest first-place votes, so we eliminate E for the first runoff. With E gone, other candidates move up in each ranking. For example, the first column of Table 12.3 shows that 18 people chose the rank order A, D, E, C, B. Without E, this ranking becomes A, D, C, B. Similarly, the ranking without E in the second column of Table 12.3 becomes B, D, C, A.

Table 12.4 shows all the rankings after eliminating E, which represent the results of the first runoff. Note that Candidate B gained 4 first-place votes that previously had gone to E (second-to-last column) and Candidate C gained 2 first-place votes that previously had gone to E (last column). Because no one has yet claimed a majority of the first-place votes, we need a second runoff.

TABLE 12.4	Rankings after Eliminating E for the First Runoff					
FIRST	A	B	C	D	B	C
SECOND	D	D	B	C	D	D
THIRD	C	C	D	B	C	B
FOURTH	B	A	A	A	A	A
	18	12	10	9	4	2

Candidate D has the fewest first-place votes (9) in Table 12.4, so D is eliminated for the second runoff. Table 12.5 shows the results. Note that the 9 people who previously had D in first place (column 4 of Table 12.4) now have C in first place (column 4 of Table 12.5).

TABLE 12.5	Rankings after Eliminating D for the Second Runoff					
FIRST	A	B	C	C	B	C
SECOND	C	C	B	B	C	B
THIRD	B	A	A	A	A	A
	18	12	10	9	4	2

Candidate C is now the leader, with $10 + 9 + 2 = 21$ first-place votes (columns 3, 4, and 6). But this total is still short of a majority, so we need one more runoff. Because B now has the fewest first-place votes ($12 + 4 = 16$), we eliminate B. Table 12.6 shows the new results. Note that C gets the first-place votes of all 16 peo-

TABLE 12.6	Rankings after Eliminating B for the Third Runoff					
FIRST	A	C	C	C	C	C
SECOND	C	A	A	A	A	A
	18	12	10	9	4	2

ple who previously ranked B first. Candidate C now has 37 first-place votes (all but the 18 votes in column 1), which is a majority. Thus, Candidate C is the winner by the sequential runoff method.

Choosing a Winner by a Point System

Now it is the turn of Candidate D's supporters to argue that the winner should be selected by a **point system.** Because there are five candidates, first-place votes are worth 5 points, second-place votes are worth 4 points, and so on, down to 1 point for fifth-place votes. We can calculate the total number of points won by each candidate by multiplying the number of votes in each column of Table 12.3 by the assigned point value and adding these products together.

A gets $(18 \times 5) + (12 \times 1) + (10 \times 1) + (9 \times 1) + (4 \times 1) + (2 \times 1) = 127$ points

B gets $(18 \times 1) + (12 \times 5) + (10 \times 4) + (9 \times 2) + (4 \times 4) + (2 \times 2) = 156$ points

C gets $(18 \times 2) + (12 \times 2) + (10 \times 5) + (9 \times 4) + (4 \times 2) + (2 \times 4) = 162$ points

D gets $(18 \times 4) + (12 \times 3) + (10 \times 2) + (9 \times 5) + (4 \times 3) + (2 \times 3) = 191$ points

E gets $(18 \times 3) + (12 \times 4) + (10 \times 3) + (9 \times 3) + (4 \times 5) + (2 \times 5) = 189$ points

Candidate D is now the winner, by virtue of having the largest number of points.

A system that assigns points for every ranking, as in the above example, is called a **Borda count.** A standard Borda count assigns 1 point for last place and an additional point for each higher place. Many common ranking systems use modified Borda counts. For example, voters in college football polls usually rank their top 25 teams (leaving many other teams unranked). A first-place ranking is worth 25 points, second place is worth 24 points, and so on, to 1 point for twenty-fifth place. At a meet between two college swim teams, the winner may be decided by a point system that awards 4 points for first place, 2 points for second place, and 1 point for third place.

Time out to think

Look for a situation in which a point system is used to determine a winner, such as a football poll or a sports competition. How are points assigned to the different places? Would the outcome be the same if the winner had instead been chosen by a plurality?

► EXAMPLE 4 Majority Loser

The seven sportswriters of Seldom County rank the county's three women's volleyball teams according to the preference schedule in Table 12.7. Determine the winner by plurality and by a Borda count. Discuss the results.

TABLE 12.7	Volleyball Team Rankings		
FIRST	A	C	B
SECOND	B	B	C
THIRD	C	A	A
	4	1	2

Solution By plurality, Team A is the winner, having received 4 out of the 7 first-place votes (which is not just a plurality, but also a majority). For the Borda count, we assign 3 points for a first-place vote, 2 points for a second-place vote, and 1 point for a third-place vote, and find the total number of points.

Team A gets $(4 \times 3) + (1 \times 1) + (2 \times 1) = 15$ points

Team B gets $(4 \times 2) + (1 \times 2) + (2 \times 3) = 16$ points

Team C gets $(4 \times 1) + (1 \times 3) + (2 \times 2) = 11$ points

So Team B is the winner by the Borda count method. Although the majority of the sportswriters chose Team A as the best team, Team B gets ranked first by the Borda count. The fact that the Borda count winner can be different from the majority winner is a well-known shortcoming of the method. *Now try Exercises 23–27.* ◄

Choosing a Winner by Pairwise Comparisons

So far, we have discussed methods by which each of Candidates A, B, C, and D can be declared the winner. Now, Candidate E's supporters point out an important fact about the election rankings. Suppose, they say, that the vote had been only between E and A, without the other candidates. Column 1 of Table 12.3 has A ranked higher than E, so these 18 voters would choose A over E. However, E is ranked higher than A in all the other columns, so E would get the remaining 37 votes (out of the 55 total). That is, E beats A by 37 to 18 in a one-on-one contest.

Now suppose that the vote had been only between E and B. Candidate E is ranked higher than B in columns 1, 4, 5, and 6 (of Table 12.3), so E gets $18 + 9 + 4 + 2 = 33$ votes. The remaining 22 votes go to B. Thus, E beats B by 33 to 22. A similar analysis of the race between only E and C shows that E is the winner by 36 to 19, and E comes out ahead in a one-on-one contest with D by 28 to 27.

Because Candidate E beats every other candidate in one-on-one contests, E's supporters now declare victory in the election. This method of deciding an election by analyzing the outcomes of one-on-one contests is called the **method of pairwise comparisons.** More generally, the winner by this method is the candidate who wins the most one-on-one contests.

The method of pairwise comparisons is also called the *Condorcet method* because it was invented by French mathematician and political leader Marie Jean Antoine Nicholas de Caritat, Marquis de Condorcet (1743–1794). Condorcet, as he is usually called, invented this method because he thought it would lead to fairer elections. However, this method suffers a major drawback, of which Condorcet was aware: It does not always produce a clear winner.

HISTORICAL NOTE

Condorcet did pioneering work in probability and calculus before becoming a leader of the French Revolution. He argued forcefully for equal rights for women, for universal free education, and against capital punishment. In 1794, as extremists took control of the revolution, he was arrested because of his aristocratic background. He died in prison the next day, in a death labeled a suicide by his captors.

➤ **EXAMPLE 5** Condorcet Paradox

Table 12.8 shows a three-candidate preference schedule. Can you find a winner through pairwise comparisons? Explain.

TABLE 12.8			
FIRST	A	C	B
SECOND	B	A	C
THIRD	C	B	A
	14	12	10

Solution With three candidates, three pairwise comparisons are possible: A versus B, B versus C, and A versus C. Columns 1 and 2 show A ranked ahead of B, but B is ahead of A in column 3. Thus, A beats B in a one-on-one contest by 26 to 10. In the B versus C contest, B is ranked

ahead of C in columns 1 and 3, so B beats C by 24 to 12. Similarly, A is ranked ahead of C only in column 1, so C beats A by 22 to 14. Summarizing, we have the following results for the pairwise comparisons:

- A beats B
- B beats C
- C beats A

Because A beats B and B beats C, the first two results suggest that A should also beat C. However, the third result shows that A actually loses to C. Thus, the results are paradoxical, and there is no way to establish a clear winner.

Now try Exercises 28–29. ◄

Time out to think _____

Does Example 5 have a plurality winner? Does it have a winner by a Borda count? Explain.

SUMMARY: CHOOSING A WINNER IS NOT SO EASY

Analyzing the preference schedule in Table 12.3 by five different methods, we found that all five candidates could reasonably claim to be the winner! Other election results may not be so ambiguous, but the point should be clear: Different people can reasonably disagree about who should be declared the winner in an election with more than two candidates. In fact, as we will see in the next unit, a mathematical theorem *proves* that there is *no absolutely fair* way of deciding elections among more than two candidates.

Voting Methods with Three or More Candidates

Plurality method: The candidate with the most first-place votes wins.

Single (top-two) runoff method: The two candidates with the most first-place votes have a runoff. The winner of this runoff is the winner of the election.

Sequential runoff method: A series of runoffs is held, eliminating the candidate with the *fewest* first-place votes at each stage. Runoffs continue until one candidate has a *majority* of the first-place votes and is declared the winner.

Point system (Borda count): Points are awarded according to the rank of each candidate on each ballot (first, second, third, . . .). The candidate with the most points wins.

Method of pairwise comparisons (Condorcet method): The candidate who wins the most pairwise (one-on-one) contests is the winner of the election.

EXERCISES 12A

Review Questions

1. What is majority rule? When can it definitively decide an election?

2. Contrast the popular vote and electoral vote in a presidential election.

3. What is a filibuster? What percentage of the vote is required to end one?

4. What is a super majority? Give several examples in which a super majority is required to decide a vote.

5. What is a veto? How does a veto affect the idea of majority rule?

6. Describe how a three-way election can be decided either by plurality or by runoff. Will both methods necessarily give the same results? Explain.

7. What is a preference schedule? Give an example of how to make one.

8. Using the preference schedule in Table 12.3, explain how the vote can be decided in five different ways.

Does It Make Sense?

Decide whether each of the following statements makes sense (or is clearly true) or does not make sense (or is clearly false). Explain your reasoning.

9. In an election with only two candidates, both candidates received more than 50% of the votes.

10. Susan received only 43% of the vote, but she won a plurality.

11. Herman won a plurality of the vote, but Hanna won the election in a sequential runoff.

12. Fred and Fran tied by the point system (Borda count), but Fran won by the method of pairwise comparisons (Condorcet method).

13. Candidate Reagan won the popular vote for the U.S. presidency and also won the electoral vote.

14. The defendant was found not guilty, even though six of the nine jurors voted for conviction.

Basic Skills & Concepts

15. **1876 Presidential Election.** In the 1876 presidential election, a total of 8,430,783 votes were counted. Samuel

J. Tilden received 4,300,590 votes to Rutherford B. Hayes's 4,036,298 votes. However, Hayes received 185 electoral votes to Tilden's 184. Contrast the outcomes of the popular and electoral votes.

16. **1888 Presidential Election.** In the 1888 presidential election, a total of 11,379,131 votes were counted. Benjamin Harrison received 5,543,892 votes to Grover Cleveland's 5,434,488 votes. However, Cleveland received 233 electoral votes to Harrison's 168. Contrast the outcomes of the popular and electoral votes.

Close Presidential Elections. The following table shows the results for the top two candidates in three other close presidential elections. Use these data in Exercises 17–19.

Year	Major candidates	Electoral vote	Popular vote
1880	James A. Garfield	214	4,449,053
	Winfield S. Hancock	155	4,442,035
1916	Woodrow Wilson	277	9,129,606
	Charles E. Hughes	254	8,538,221
1976	Jimmy Carter	297	40,830,763
	Gerald R. Ford	240	39,147,973

17. Contrast the outcomes of the popular and electoral votes for the 1880 election.

18. Contrast the outcomes of the popular and electoral votes for the 1916 election.

19. Contrast the outcomes of the popular and electoral votes for the 1976 election.

20. **Super Majorities.**

 a. Of the 100 senators in the U.S. Senate, 57 favor a new bill on health care reform. The opposing senators start a filibuster. Is the bill likely to pass?

 b. A criminal conviction in a particular state requires a vote by 2/3 of the jury members. On a 12-member jury, 7 jurors vote to convict. Will the defendant be convicted?

 c. A proposed amendment to the U.S. Constitution has passed both the House and the Senate with more than the required 2/3 super majority. Each state holds a vote on the amendment, and it receives a majority vote in all but 15 of the 50 states. Is the Constitution amended?

 d. A tax increase bill has the support of 70 out of 100 senators and 260 out of 435 members of the House of Rep-

resentatives. The president promises to veto the bill if it is passed. Is it likely to become law?

21. **Super Majorities.**

 a. According to the bylaws of a corporation, a 2/3 vote of the shareholders is needed to approve a merger. Of the 15,890 shareholders voting on a certain merger, 10,580 approve of the merger. Will the merger happen?

 b. A criminal conviction in a particular state requires a vote by 3/4 of the jury members. On a 15-member jury, 11 jurors vote to convict. Will the defendant be convicted?

 c. A proposed amendment to the U.S. Constitution has passed both the House and the Senate with more than the required 2/3 super majority. Each state holds a vote on the amendment, and it receives a majority vote in 32 of the 50 states. Is the Constitution amended?

 d. A tax increase bill has the support of 68 out of 100 senators and 292 out of 435 members of the House of Representatives. The president promises to veto the bill if it is passed. Is it likely to become law?

22. **1912 Presidential Election.** The 1912 U.S. presidential election featured three major candidates, with the vote split as follows. For this exercise, assume that all votes were cast for one of these three candidates. (Actually, other candidates split 7.6% of the total vote.)

Candidate	Popular votes	Electoral votes
Woodrow Wilson	6,286,214	435
Theodore Roosevelt	4,126,020	88
William Taft	3,483,922	8

 a. Calculate each candidate's percentage of the popular vote. Who won a plurality? Did any candidate win a majority?

 b. Calculate each candidate's percentage of the electoral vote. Who won a plurality? Did any candidate win a majority?

 c. Suppose that Taft had dropped out of the election. Is it possible that Roosevelt would have won the popular vote? Could Roosevelt have become president in that case? Explain.

 d. Suppose that Roosevelt had dropped out of the election. Is it possible that Taft would have won the popular vote? Could Taft have become president in that case? Explain.

Preference Schedules. For the preference schedules in Exercises 23–27, answer the following questions.

 a. How many votes were cast?

 b. Find the plurality winner. Did the plurality winner also receive a majority? Explain.

 c. Find the winner by a runoff of the top two candidates.

 d. Find the winner by a sequential runoff.

 e. Find the winner by a Borda count.

 f. Find the winner, if any, by the method of pairwise comparisons.

 g. Summarize the results of the various methods of determining a winner. Based on these results, is there a clear winner? If so, why? If not, which candidate should be selected as the winner, and why?

23.

First	B	D	C	A	D	C
Second	D	A	D	D	A	A
Third	C	C	A	C	B	B
Fourth	A	B	B	B	C	D
	20	15	10	8	7	6

24.

First	B	D	D	C	E
Second	A	B	B	A	A
Third	C	A	E	B	D
Fourth	D	C	C	D	B
Fifth	E	E	A	E	C
	9	7	6	4	3

25.

First	A	A	B	B	C	C
Second	B	C	A	C	A	B
Third	C	B	C	A	B	A
	30	5	20	5	10	30

26.

First	A	B	D
Second	B	A	C
Third	C	D	B
Fourth	D	C	A
	10	10	10

27.

First	E	B	D
Second	D	C	A
Third	A	E	B
Fourth	B	A	C
Fifth	C	D	E
	40	30	20

28. Condorcet Winner. If a candidate wins *all* head-to-head (two-way) races with other candidates, that candidate is called the *Condorcet winner*. A Condorcet winner automatically wins by the method of pairwise comparisons (Condorcet method). Consider the following preference schedule for a four-candidate election. Is there a Condorcet winner? Explain.

First	B	B	A	A
Second	A	A	C	D
Third	C	D	D	C
Fourth	D	C	B	B
	30	30	30	20

29. Condorcet Paradox. Consider the following preference schedule. Can you find a winner through pairwise comparisons? Explain.

First	C	B	A
Second	A	C	B
Third	B	A	C
	8	9	10

Further Applications

30. Three-Candidate Elections. Consider an election in which the votes were cast as follows.

Candidate	Percentage of vote
Able	32%
Best	40%
Crown	28%

a. Who won a plurality? Does any candidate have a majority? Explain.

b. What percentage of Crown's votes would Able need to win a runoff election?

31. Three-Candidate Elections. Consider an election in which the votes were cast as follows.

Candidate	Percentage of vote
Davis	21%
Earnest	38%
Fillipo	41%

a. Who won a plurality? Does any candidate have a majority? Explain.

b. What percentage of Davis's votes would Earnest need to win a runoff election?

32. Three-Candidate Elections. Consider an election in which the votes were cast as follows.

Candidate	Number of votes
Grand	130
Height	150
Inviglio	185

a. Who won a plurality? Does any candidate have a majority? Explain.

b. How many of Grand's votes would Height need to win a runoff election?

33. Three-Candidate Elections. Consider an election in which the votes were cast as follows.

Candidate	Number of votes
Joker	225
King	360
Lord	285

a. Who won a plurality? Does any candidate have a majority? Explain.

b. How many of Joker's votes would Lord need to win a runoff election?

34. Ballots to Schedule. Ten voters are asked to rank three brands of soda: A, B, and C. The ten voters turn in the following ballots, showing their preferences in order:

$$\text{ABC} \quad \text{CBA} \quad \text{CAB} \quad \text{ACB} \quad \text{ABC}$$

$$\text{CAB} \quad \text{ABC} \quad \text{ABC} \quad \text{CAB} \quad \text{ACB}$$

Make a preference schedule for these ballots.

35. Ballots to Schedule. Twelve voters are asked to rank four brands of candy bar: A, B, C, and D. The twelve voters turn in the following ballots, showing their preferences in order:

$$\text{ABCD} \quad \text{CBAD} \quad \text{CADB} \quad \text{ACDB} \quad \text{DABC} \quad \text{ABCD}$$

$$\text{CADB} \quad \text{ABCD} \quad \text{ABCD} \quad \text{CADB} \quad \text{DABC} \quad \text{ACDB}$$

Make a preference schedule for these ballots.

36. Pairwise Comparisons Question. Suppose 15 voters rank four candidates. How many pairs of candidates must be examined to carry out the method of pairwise comparisons? How many pairs of candidates must be examined to carry out the method of pairwise comparisons with five candidates?

37. Borda Question. In a preference schedule with five candidates and 25 voters, what is the total number of points awarded to the candidates using the usual Borda count weights?

38. Borda Question. Suppose 25 voters rank four candidates: A, B, C, and D. With the usual Borda count weights, A receives 60 points, B receives 50 points, and C receives 10 points. How many points does D have? Who wins the election? Explain.

Web Projects

Find useful links for Web Projects on the text Web site: www.aw.com/bennett-briggs

39. Election 2000. Some people claim that the results of the 2000 presidential election were affected by voting irregularities or vote counting problems. Pick one such controversy and examine the evidence on both sides. Write a short report discussing the controversy and defending your personal opinion as to whether the problem affected the election outcome. Also discuss how similar problems might be prevented in the future.

40. Academy Awards. The election process for the Academy Awards (for films) involves several stages and several different voting methods. Use the Web site for the Academy Awards to investigate the full election procedure for the Academy Awards. Describe the procedure and comment on its fairness.

41. Sports Polls. Most sports have regular polls while the sport is in season. Choose a particular sport and investigate how teams are ranked. Describe the methods used to rank teams, give some typical results, and discuss the fairness of the method.

42. Elections Around the World. Many countries have methods much different than the American system for choosing a president or prime minister. Choose a particular country and use the Web to learn about national elections. Describe in detail the voting procedure used and discuss its strengths and shortcomings.

In the News

43. Election Aftermath. Find an article dealing with some aspect of the aftermath of the 2000 presidential election. Summarize the article and discuss how the closeness of the election still affects our lives.

44. Foreign Elections. Find a news article that describes the outcome of a national election in a foreign country. Summarize the outcome of the election. Is the method for choosing a winner mentioned?

45. Non-Political Election. Find an example in the news of an election that is not for a political office. Summarize the method used to choose a winner and describe the outcome of the election.

UNIT 12B

THEORY OF VOTING

In Unit 12A, we saw that different methods of counting votes can lead to different results. In fact, mathematicians, economists, and political scientists have discovered many other surprises about voting. In this unit, we investigate just a few of these surprises.

WHICH METHOD IS FAIREST?

When there are only two candidates in an election, the clear winner is the candidate who gets a majority of the votes. However, as we saw in Unit 12A, choosing a winner can be much more difficult in elections with three or more candidates.

We have studied five different methods for deciding an election (see the Summary on p. 665). Sometimes, all five methods give the same winner. Other times, different methods give different winners. In extreme cases, such as the election results in Table 12.3, the five methods produce five different winners! Thus, the key question becomes, Which method is the fairest?

Criteria of Fairness

Judgments of fairness are necessarily subjective. Nevertheless, mathematicians and political scientists have come up with four basic **fairness criteria** that must be met by a fair voting system.

<table>
<tr><td>

by the way

Voting theorists refer to these four criteria as, respectively, the *majority criterion*, the *Condorcet criterion*, the *monotonicity criterion*, and the *independence of irrelevant alternatives criterion*.

</td><td>

Fairness Criteria

Criterion 1: If a candidate receives a majority of the first-place votes, that candidate should be the winner.

Criterion 2: If a candidate is favored over every other candidate in pairwise races, that candidate should be declared the winner.

Criterion 3: Suppose that Candidate X is declared the winner of an election, and then a second election is held. If some voters rank X higher in the second election than in the first election (without changing the order of other candidates), then X should also win the second election.

Criterion 4: Suppose that Candidate X is declared the winner of an election, and then a second election is held. If voters do not change their preferences but one (or more) of the losing candidates drops out, then X should also win the second election.

</td></tr>
</table>

Because the election in Table 12.3 (Unit 12A) gave five different winners by five different voting methods, you may already be guessing that none of these methods can always satisfy all four fairness criteria. However, to be sure that these criteria are clear, let's examine a few more examples in which we test each criterion.

➤ EXAMPLE 1 An Unfair Plurality

TABLE 12.9

FIRST	A	B	C
SECOND	B	C	B
THIRD	C	A	A
	5	4	2

Consider the preference schedule in Table 12.9. Suppose the winner is chosen by plurality. Does this method satisfy the four fairness criteria?

Solution We see that Candidate A has the most first-place votes and is the plurality winner. Let's apply the four fairness criteria to this election.

Criterion 1: Because no candidate received a majority of first-place votes in this election, the first criterion does not apply.

Criterion 2: To check this criterion, we need to examine pairwise races between the candidates. There are three pairs to consider: A versus B, A versus C, and B versus C. Here are the results:

- A is ranked ahead of B in column 1 (5 votes), but behind B in columns 2 and 3 (4 + 2 = 6 votes). Thus, B beats A by 6 to 5.
- A is ranked ahead of C in column 1 (5 votes), but behind C in columns 2 and 3 (4 + 2 = 6 votes). Thus, C beats A by 6 to 5.
- B is ranked ahead of C in columns 1 and 2 (5 + 4 = 9 votes), but behind C in column 3 (2 votes). Thus B beats C by 9 to 2.

By Criterion 2, B should be the winner because B beats both A and C in pairwise contests. But B is not the plurality winner, so the plurality method is unfair in this case.

Criterion 3: We check this criterion by imagining a second election in which some voters move A higher in their rankings, but do not change the order in which they rank B and C. Column 1 cannot change because A already is ranked first. Moving A higher in column 2 or 3 may give A even more first-place votes, but cannot reduce A's number of first-place votes. Thus, A would still win a plurality, and Criterion 3 is satisfied.

Criterion 4: This time we imagine a second election in which one of the losers drops out. Suppose that C were to drop out. Then B would move up to first place in column 3, picking up the 2 first-place votes in this column. Because B already has 4 first-place votes in column 2, B would then have 6 first-place votes and would win the election by 6 to 5 over A. Thus, the plurality method is unfair in this case because A would lose to B if C were to drop out.

In summary, the plurality method violates two of the four fairness criteria in this election. The basic problem is clear if we look back at the criteria that are violated: Most voters prefer Candidate B over Candidate A, but A is the winner by plurality.

Now try Exercises 9–14. ◄

►**EXAMPLE 2** An Unfair Runoff

Consider the preference schedule in Table 12.10. Who wins by a single runoff? Which of the four fairness criteria are violated?

Solution The first-place totals are 13 for A (column 2), 11 for B (column 4), and 9 + 5 = 14 for C (columns 1 and 3). Thus, the runoff is between A and C. The 11 voters who chose B (column 4) ranked A second, so A picks up these first-place votes and wins the runoff over C by 24 to 14. Now let's evaluate the fairness criteria.

TABLE 12.10				
FIRST	C	A	C	B
SECOND	B	C	A	A
THIRD	A	B	B	C
	9	13	5	11

Criterion 1 does not apply because no candidate received a majority in the original election. Criterion 2 does not apply either, because no candidate wins all the pairwise comparisons. (You should confirm that, in pairwise comparisons, B beats A, A beats C, and C beats B.)

Criterion 3 says that A should still be the winner if A picks up additional first-place votes. But suppose that the five voters in column 3 move A up to first place. Then A's first-place total increases to 18 while C's first-place total decreases to 9 (B's first-place total remains at 11). Because C now has the fewest first-place votes, the runoff is between A and B. Note that B ranks ahead of A in columns 1 and 4, so B gets the

9 + 11 = 20 votes from these columns in the runoff; this is a majority of the 38 total votes, making B the winner. Thus, Fairness Criterion 3 is violated because a gain in first-place votes ends up costing A the election.

Criterion 4 says that A should still be the winner if one (or more) of the losing candidates drops out. Candidate B was eliminated in the original runoff election, so the results are not affected if B drops out. However, if C drops out, B picks up the 9 first-place votes from column 1 and A picks up the 5 first-place votes from column 3. This change increases A's first-place total to 13 + 5 = 18 and B's first-place total to 9 + 11 = 20, making B the winner instead of A. Thus, Fairness Criterion 4 is also violated. *Now try Exercises 15–21.* ◄

►**EXAMPLE 3** A Fair Election

Consider the preference schedule in Table 12.11. Does the plurality method satisfy the four fairness criteria?

TABLE 12.11

FIRST	A	B	C
SECOND	B	C	B
THIRD	C	A	A
	10	4	2

Solution The winner by the plurality method is A, with 10 votes. Because this plurality is also a majority of the 16 votes cast, the first criterion is satisfied. Examining pairwise matchups shows that A beats B by 10 to 6, and A beats C by 10 to 6. Thus, A is the pairwise winner and the second criterion is satisfied. The third criterion also is satisfied because, if we imagine a second election in which A picks up additional votes, it will only increase A's majority. Finally, the fourth criterion asks what happens in a second election in which one (or more) of the losers drops out. Eliminating either B or C cannot reduce A's majority, so A still wins. In summary, the plurality method satisfies all four fairness criteria in this particular election. *Now try Exercises 22–33.* ◄

Time out to think

Example 1 shows an election decided by plurality that *does not* meet all four fairness criteria. Example 3 shows an election decided by plurality that *does* meet the criteria. Note that the plurality is also a *majority* in Example 3. Is it generally true that an election decided by a majority will be fair? Explain.

Arrow's Impossibility Theorem

Suppose we continue to test the fairness criteria on many different elections decided by any of the five methods we have discussed. Sometimes, as in Example 3 above, we will find that all four criteria are satisfied and we can declare the election to be fair. Other times, as in Examples 1 and 2, the results will violate one or more of the fairness criteria.

Despite the fact that some elections are clearly fair and others are not, we can find some general rules. For example, elections decided by plurality always satisfy Fairness Criteria 1 and 3, but sometimes violate Criteria 2 and 4. A similar analysis of the other voting methods is summarized in Table 12.12.

Table 12.12 carries the disconcerting message that none of the five voting systems always gives a fair outcome. During the two centuries following the American and French revolutions, political theorists attempted to devise a better voting system—one that would always give fair results. Unfortunately, their quest

by the way

Kenneth Arrow received the 1972 Nobel Prize in Economics for his mathematical analysis of voting systems that led him to discover the impossibility theorem.

TABLE 12.12	Fairness Criteria and Voting Systems				
	PLURALITY	TOP-TWO RUNOFF	SEQUENTIAL RUNOFF	BORDA COUNT	PAIRWISE COMPARISON
CRITERION 1	Y	Y	Y	N	Y
CRITERION 2	N	N	N	N	Y
CRITERION 3	Y	N	N	Y	Y
CRITERION 4	N	N	N	N	N

NOTE: Y = criterion always holds, N = criterion may be violated.

No one pretends that democracy is perfect or all-wise. Indeed it has been said that democracy is the worst form of government except for all those that have been tried from time to time.

—WINSTON CHURCHILL

was futile, because a perfect voting system will never be found. In 1952, economist Kenneth Arrow proved mathematically that it is *impossible* to find a voting system that always satisfies all four fairness criteria. This result is one of the landmark applications of mathematics to social theory and is now known as **Arrow's impossibility theorem.**

> **Arrow's Impossibility Theorem**
>
> No voting system can satisfy all four fairness criteria in all cases.

Time out to think

Arrow's impossibility theorem tells us that no voting system is perfect. However, some systems may still be superior to others. Among the voting systems we have discussed, which would *you* choose for electing a president? Why?

APPROVAL VOTING

Democratic voting systems have traditionally been based on the principle of *one person, one vote.* However, in light of the fact that no voting system can be perfect, some political theorists have proposed alternative methods of voting. Arrow's impossibility theorem assures us that none of these methods can be perfect either, but a new method *might* give fair results more often than traditional methods.

One such alternative voting system, called **approval voting,** asks voters to specify whether they approve or disapprove of each candidate. Voters may approve as many candidates as they like, and the candidate with the most approval votes wins.

As an example, consider a race for governor among Candidates A, B, and C. With approval voting, the ballot would look something like Figure 12.2.

Suppose that voter opinions about the candidates are as follows (perhaps because Candidates A and B are closely aligned politically, while Candidate C is on the opposite end of the political spectrum):

FIGURE 12.2 A sample ballot for approval voting.

- 32% want A as their first choice, but would also approve of B.
- 32% want B as their first choice, but would also approve of A.
- 1% want A as their first choice, and approve of neither B nor C.
- 35% want C as their first choice, and approve of neither A nor B.

On the approval ballot, we will find that

A is approved by 32% + 32% + 1% = 65%.

B is approved by 32% + 32% = 64%.

C is approved by 35%.

By approval voting, A becomes the new governor and C is clearly in last place. However, if this election had been decided by plurality, C would have been elected by virtue of the most first-place votes, with 35%.

Time out to think _____

Do you think that approval voting is a better way to decide this election than plurality? Do you think it is better in general? Defend your opinion.

▶ **EXAMPLE 4** A Drawback to Approval Voting

The opinions of voters in a particular three-way election for governor are as follows:

- 26% want A as their first choice, but would also approve of B.
- 25% want A as their first choice, and approve of neither B nor C.
- 15% want B as their first choice, and approve of neither A nor C.
- 18% want C as their first choice, but would also approve of B.
- 16% want C as their first choice, and approve of neither A nor B.

Contrast the results if the election is decided by approval voting and by plurality.

Solution By approval voting, we find the following results:

A is approved by 26% + 25% = 51%.

B is approved by 26% + 15% + 18% = 59%.

C is approved by 18% + 16% = 34%.

Thus, B is the winner by approval voting. However, if we count only first choices, we find the following outcome by the plurality method:

A is the first choice for 26% + 25% = 51%.

B is the first choice for 15%.

C is the first choice for 18% + 16% = 34%.

A *majority* of voters want A as their first choice for governor. However, because a larger majority finds B "acceptable," B wins in the approval voting.

Now try Exercises 34–35.

Time out to think _____

Example 4 shows that while approval voting ensures that the winning candidate is acceptable to the largest number of people, another candidate might be the first choice of the majority. Do you think this is a serious drawback to approval voting? Why or why not?

VOTING POWER

So far, we have taken for granted that every voter has the same amount of power to influence an election as every other voter. However, this is not necessarily the case. For example, shareholders in a corporation generally are given votes in proportion to the number of shares they own. A shareholder with 10 shares gets 10 votes, and a shareholder with 1000 shares gets 1000 votes. Thus, not every voter at a shareholder meeting is equally empowered to influence the outcome of an election.

A similar situation arises in politics when voters form groups, or **coalitions,** that agree to vote the same way on a particular issue. Such coalition-building can drastically affect the power of individual voters to influence the outcome of a vote. Several techniques have been devised for measuring the effective power wielded by different voters when not all voters have the same power. Although we will not discuss the details of these techniques here, we can give a brief taste of the type of situation that can arise. Suppose, for example, that the 100 members of the U.S. Senate happened to be divided as follows: 49 Democrats, 49 Republicans, and 2 Independents.

Moreover, suppose that all 49 Democrats favor a particular bill, while all 49 Republicans are opposed. The two Independents may not care which way the vote goes, but if they vote together their two votes will decide the outcome of the election. Both the Democrats and the Republicans are likely to work hard to woo the votes of the Independents, perhaps by agreeing to support other bills that the Independents favor. Thus, the power of the two Independents to influence decision making is much greater than their 2 votes out of 100 would seem to imply.

► **EXAMPLE 5** Missing the Big Vote

A small corporation has four shareholders. The 10,000 shares in this corporation are divided among the shareholders as follows:

> Shareholder A owns 2650 shares (26.5% of the company).
> Shareholder B owns 2550 shares (25.5% of the company).
> Shareholder C owns 2500 shares (25% of the company).
> Shareholder D owns 2300 shares (23% of the company).

The corporation has scheduled a key vote about buying out another company. Each shareholder's vote is counted in proportion to the number of shares the person owns. Suppose that Shareholder D misses the vote. Does it matter?

Solution At first, it would seem disastrous for a major shareholder to miss the chance to influence the future of the company. However, note that any two of the three largest shareholders can form a majority by voting together:

> A and B: 26.5% + 25.5% = 52%
> A and C: 26.5% + 25% = 51.5%
> B and C: 25.5% + 25% = 50.5%

In contrast, Shareholder D cannot be part of a majority unless two other shareholders vote the same way. But in that case, the other two shareholders already determine the outcome by themselves. Thus, Shareholder D has *no power* to affect the election outcome.

Now try Exercise 36. ◄

by the way

Another way to measure voting power was invented in the 1950s by Georgetown University law professor John Banzhaf. Now called the *Banzhaf power index,* it analyzes voting power in terms of the number of ways an individual or group can cast a critical vote that changes an election outcome. By this measure, California's voting power is quite high despite its large number of voters per elector. The reason is the winner-take-all system, in which California's 54 electoral votes (55 starting in 2004) have a much better chance of swaying the election than the fewer electoral votes of small states.

> EXAMPLE 6 Electoral Power

In the 2000 presidential election, California had 54 electoral votes and Wyoming had 3 electoral votes. Their state populations were approximately 33,900,000 and 494,000, respectively. In terms of electoral votes per resident, contrast the voting power of California and Wyoming.

Solution In California, where 33,900,000 people had 54 electoral votes, the number of people represented by each electoral vote was

$$\frac{33{,}900{,}000 \text{ people}}{54 \text{ electoral votes}} \approx 628{,}000 \text{ people per electoral vote}$$

In Wyoming, where 494,000 people had 3 electoral votes, the number of people represented by each electoral vote was

$$\frac{494{,}000 \text{ people}}{3 \text{ electoral votes}} \approx 165{,}000 \text{ people per electoral vote}$$

Thus, each elector in California represented

$$\frac{628{,}000}{165{,}000} \approx 3.8$$

times as many people as an elector in Wyoming. By this analysis, Wyoming voters had nearly four times as much power in the electoral vote as did California voters.

Now try Exercises 37–41. ◄

EXERCISES 12B

Review Questions

1. Briefly summarize each of the four fairness criteria. For each one, give an example that describes why an election would be unfair if the criterion were violated.

2. What is Arrow's impossibility theorem? Summarize its meaning and its importance.

3. What is approval voting? How is it different from the traditional idea of one person, one vote?

4. Give an example in which different voters may wield different amounts of power in an election.

Does It Make Sense?

Decide whether each of the following statements makes sense (or is clearly true) or does not make sense (or is clearly false). Explain your reasoning.

5. Karen won a one-on-one election against each of the other candidates. She feels that she should win even though she lost in a runoff election.

6. Kai won a majority of the votes and yet lost the election by the point system (Borda count). He feels that he should be declared the winner of the election.

7. Wendy demanded a second election because she lost to Walt in a close plurality vote. In the second election, some of Wendy's supporters voted for Walt, but Wendy managed to win. (Assume other votes stayed the same.)

8. Approval voting is the best method for elections because it satisfies all four fairness criteria.

Basic Skills & Concepts

9. **Plurality and Criterion 1.** Explain in words why the plurality method always satisfies Fairness Criterion 1.

10. Plurality and Criterion 2. Consider the preference schedule shown below. Which candidate is the plurality winner? Does this choice satisfy Fairness Criterion 2? Explain.

First	A	B	C
Second	B	C	B
Third	C	A	A
	3	2	2

11. Plurality and Criterion 2. Devise a preference schedule with three candidates (A, B, and C) and 11 voters in which C is the plurality winner and yet A beats C and B beats C in one-on-one races. Explain your work.

12. Plurality and Criterion 3. Explain in words why the plurality method always satisfies Fairness Criterion 3.

13. Plurality and Criterion 4. Suppose the plurality method is used on the following preference schedule. Is Fairness Criterion 4 satisfied? Explain.

First	A	B	C
Second	B	C	B
Third	C	A	A
	6	2	5

14. Plurality and Criterion 4. Devise a preference schedule with three candidates and nine votes such that the plurality method violates Fairness Criterion 4. Explain your work.

15. Runoff Methods and Criterion 1. Explain in words why both the single runoff and the sequential runoff methods always satisfy Fairness Criterion 1.

16. Sequential Runoff and Criterion 2. Suppose the sequential runoff method is used on the following preference schedule. Is Fairness Criterion 2 satisfied? Explain.

First	A	B	C
Second	B	C	B
Third	C	A	A
	10	7	2

17. Sequential Runoff and Criterion 2. Suppose the sequential runoff method is used on the following preference schedule. Is Fairness Criterion 2 satisfied? Explain.

First	A	B	C
Second	B	C	B
Third	C	A	A
	10	7	8

18. Sequential Runoff and Criterion 2. Devise a preference schedule with three candidates and three rankings such that the sequential runoff method violates Fairness Criterion 2. Explain your work.

19. Sequential Runoff and Criterion 3. Suppose the sequential runoff method is used on the following preference schedule. Is Fairness Criterion 3 satisfied? Explain.

First	A	B	A	C
Second	B	C	C	A
Third	C	A	B	B
	7	8	4	10

20. Sequential Runoff and Criterion 4. Suppose the sequential runoff method is used on the following preference schedule. Is Fairness Criterion 4 satisfied? Explain.

First	A	B	C
Second	C	C	B
Third	B	A	A
	8	6	3

21. Sequential Runoff and Criterion 4. Devise a preference schedule with three candidates such that the sequential runoff method violates Fairness Criterion 4. Explain your work.

22. Point System and Criterion 1. Consider the preference schedule below for three candidates. Which candidate wins by the point system (Borda count)? Is Fairness Criterion 1 satisfied? Explain.

First	A	B
Second	B	C
Third	C	A
	3	2

23. Point System and Criterion 1. Devise your own preference schedule with three candidates, three rankings, and seven voters in which the point system (Borda count) violates Fairness Criterion 1. Explain your work.

24. Point System and Criterion 2. Suppose the point system (Borda count) is used on the following preference schedule. Is Fairness Criterion 2 violated? Explain.

First	A	B	C
Second	B	C	B
Third	C	A	A
	5	2	2

25. Point System and Criterion 2. Devise a preference schedule with four candidates such that the point system violates Fairness Criterion 2.

26. Point System and Criterion 3. Explain in words why the point system (Borda count) always satisfies Fairness Criterion 3.

27. Point System and Criterion 4. Suppose the point system (Borda count) method is used on the following preference schedule. Is Fairness Criterion 4 satisfied? Explain.

First	C	B	A
Second	A	C	B
Third	B	A	C
	5	4	3

28. Point System and Criterion 4. Devise a preference schedule with three candidates such that the point system (Borda count) method violates Fairness Criterion 4. Explain your work.

29. Pairwise Comparisons and Criterion 1. Explain in words why the method of pairwise comparisons always satisfies Fairness Criterion 1.

30. Pairwise Comparisons and Criterion 2. Explain in words why the method of pairwise comparisons always satisfies Fairness Criterion 2.

31. Pairwise Comparisons and Criterion 3. Explain in words why the method of pairwise comparisons always satisfies Fairness Criterion 3.

32. Pairwise Comparisons and Criterion 4. Suppose the method of pairwise comparisons is used on the following preference schedule. Is Fairness Criterion 4 satisfied? Explain.

First	A	A	E	C	D
Second	E	C	B	B	B
Third	C	D	A	A	A
Fourth	D	E	C	D	E
Fifth	B	B	D	E	C
	1	1	1	1	1

33. Pairwise Comparisons and Criterion 4. Devise a preference schedule with five candidates such that the method of pairwise comparisons violates Fairness Criterion 4.

34. Approval Voting. Suppose that Candidates A and B have moderate political positions, while Candidate C is relatively conservative. Voter opinions about the candidates are as follows:

- 30% want A as their first choice, but would also approve of B.
- 29% want B as their first choice, but would also approve of A.
- 1% want A as their first choice, and approve of neither B nor C.
- 40% want C as their first choice, and approve of neither A nor B.

a. If all voters could vote only for their first choice, which candidate would win by plurality?

b. Which candidate wins by an approval vote?

35. Approval Voting. Suppose that Candidates A and B have moderate political positions, while Candidate C is quite liberal. Voter opinions about the candidates are as follows:

- 28% want A as their first choice, but would also approve of B.
- 29% want B as their first choice, but would also approve of A.
- 1% want B as their first choice, and approve of neither A nor C.
- 42% want C as their first choice, and approve of neither A nor B.

a. If all voters could vote only for their first choice, which candidate would win by plurality?

b. Which candidate wins by an approval vote?

36. Power Voting. Imagine that a small company has four shareholders who hold 26%, 26%, 25%, and 23% of the company's stock. Assume that votes are assigned in proportion to share holding (e.g., if there are a total of 100 votes, the four people get 26, 26, 25, and 23 votes, respectively). Also assume that decisions are made by strict majority vote. Explain why, although each individual holds roughly one-fourth of the company's stock, the individual with 23% holds *no* effective power in voting.

Electoral Power. Use the table below to answer the questions in Exercises 37–41.

State	Population	Electoral votes
Alaska	627,000	3
Illinois	12,419,000	21
New York	18,976,000	31
Rhode Island	1,048,000	4

37. Which state has more voting power per person: Alaska or Illinois?

38. Which state has more voting power per person: Alaska or New York?

39. Which state has more voting power per person: Rhode Island or Illinois?

40. Which state has more voting power per person: Rhode Island or New York?

41. Rank the four states in order of increasing voting power.

Further Applications

Fairness Criteria. In Exercises 42–46, consider the following preference schedule for four candidates.

First	A	B	C	D
Second	D	A	B	C
Third	B	D	A	A
Fourth	C	C	D	B
	16	10	8	7

42. Suppose the winner is decided by plurality. Analyze whether this choice satisfies the four fairness criteria.

43. Suppose the winner is decided by a single runoff. Analyze whether this choice satisfies the four fairness criteria.

44. Suppose the winner is decided by sequential runoffs. Analyze whether this choice satisfies the four fairness criteria.

45. Suppose the winner is decided by a Borda count (point system). Analyze whether this choice satisfies the four fairness criteria.

46. Suppose the winner is decided by pairwise comparisons. Analyze whether this choice satisfies the four fairness criteria.

Fairness Criteria. In Exercises 47–51, consider the following preference schedule for five candidates (Table 12.3 of Unit 12A).

First	A	B	C	D	E	E
Second	D	E	B	C	B	C
Third	E	D	E	E	D	D
Fourth	C	C	D	B	C	B
Fifth	B	A	A	A	A	A
	18	12	10	9	4	2

47. Suppose the winner is decided by plurality. Analyze whether this choice satisfies the four fairness criteria.

48. Suppose the winner is decided by a single runoff. Analyze whether this choice satisfies the four fairness criteria.

49. Suppose the winner is decided by sequential runoffs. Analyze whether this choice satisfies the four fairness criteria.

50. Suppose the winner is decided by a Borda count (point system). Analyze whether this choice satisfies the four fairness criteria.

51. Suppose the winner is decided by pairwise comparisons. Analyze whether this choice satisfies the four fairness criteria.

52. **Swing Votes.** Suppose that the Senate has the following breakdown in party representation: 49 Democrats, 49 Republicans, and two Independents. Further suppose that all Democrats and Republicans vote along opposing party lines. Assuming that a majority is required to pass a bill, explain why the two Independents, despite holding only 2% of the Senate seats, effectively hold power equal to that of either of the large parties.

Web Projects

Find useful links for Web Projects on the text Web site: **www.aw.com/bennett-briggs**

53. **Other Fairness Criteria.** The fairness criteria discussed in this unit are not exhaustive. Investigate other criteria (such as the favorite-betrayal criterion, the strong adverse results criterion, and the weak defensive strategy criterion). Discuss the merits of these criteria and their shortcomings.

54. **Approval Voting.** Many political systems use approval voting and it is strongly advocated by many political organizations. Investigate the details of approval voting, cite countries in which it is used, and discuss its strengths and weaknesses.

55. **Power Voting and Coalitions.** Use the Web to investigate the political coalitions at the national level in a particular country (for example, Israel). Describe the coalitions, their sizes, and their effective voting power.

In the News

56. **Elections Gone Bad.** Find a news article about an election in which the results were disputed because of corruption or because the election system failed. Discuss the incident in light of the fairness criteria presented in this unit.

57. **Voting Power.** Find a news report about any election in which not all participants had the same voting power. What factors affected voting power?

UNIT **12C**

All legislative Powers herein granted shall be vested in a Congress of the United States, which shall consist of a Senate and House of Representatives.

—ARTICLE 1, SECTION #1, OF THE UNITED STATES CONSTITUTION

APPORTIONMENT: THE HOUSE OF REPRESENTATIVES AND BEYOND

Each state in the United States has two senators and one or more representatives in Congress. Why does the number of representatives vary from state to state, and who decides how many of the 435 seats in the House of Representatives each state gets? These questions involve **apportionment,** because the seats in the House of Representatives must be apportioned (divided) among the states in a fair way.

In this unit, we'll discuss the mathematics that underlies apportionment. As with methods for voting, we'll see that no single method of apportionment is unquestionably fairer than others. Nevertheless, the mathematical ideas are very useful and can be applied to many problems beyond those of government.

DEFINITION

For the United States House of Representatives, **apportionment** is a process used to divide the available seats among the states. More generally, apportionment is a process used to divide any set of people or objects among various individuals or groups.

THE CONSTITUTIONAL CONTEXT

Much of the mathematics of apportionment arose from historical attempts to meet the letter and spirit of the United States Constitution. According to the Constitution, the legislative (law-making) branch of the United States government consists of two distinct bodies: the **Senate** and the **House of Representatives** (called "the House" for short).

The Constitution (Article 1, Section 3) specifies a simple method of apportionment for senators: Each state gets two senators. According to the original Constitution, senators were chosen by state legislatures. However, the 17th amendment to the Constitution, adopted in 1913, changed the method of selecting senators to direct election by the people. Senators are elected for six-year terms.

The requirements for the House of Representatives are much more complicated. The Constitution mandates that seats in the House be apportioned to the states according to their populations, subject to a minimum of one seat for every state. It allows Congress to set the total

number of representatives, so long as the total number does not exceed one for each 30,000 people.

The Constitution directs Congress to reapportion seats in the House every ten years, following a census. For example, the current apportionment is based on results from the 2000 census. However, the Constitution does not specify a particular procedure for apportionment. As we will soon see, it is not easy to choose an apportionment procedure.

►EXAMPLE 1 United States House of Representatives

The House currently has 435 representatives, and the 2000 census reported a U.S. population of 281 million. On average, how many people are represented by each representative? Suppose that the total number of representatives were the constitutional limit of one for every 30,000 people. How many representatives would there be in that case?

Solution Dividing the population of 281 million by 435 representatives, we find that the average number of people served by each representative is

$$\frac{\text{population}}{\text{number of representatives}} = \frac{281{,}000{,}000}{435} \approx 646{,}000$$

On average, each representative serves approximately 646,000 people. If there were one representative for every 30,000 people, the total number of representatives would be

$$\frac{\text{population}}{30{,}000} = \frac{281{,}000{,}000}{30{,}000} \approx 9370$$

That is, the constitutional limit allows more than 9000 representatives, in contrast to the actual number of 435. *Now try Exercises 13–14.* ◄

THE APPORTIONMENT PROBLEM

Example 1 shows that each House member represents 646,000 people, on average. Apportionment would be easy if every state's population were a simple multiple of 646,000. For example, a state with a population of 1,292,000, which is 2 × 646,000, would get 2 representatives. Similarly, a state with a population of 6,460,000, which is 10 × 646,000, would get 10 representatives.

Of course, states do not have such convenient populations. For example, Rhode Island has a population of about 1,050,000, which is about 1.6 × 646,000. We might say that Rhode Island is "entitled" to 1.6 representatives, but representatives are people and cannot be divided fractionally. That is, Rhode Island can have one representative or two representatives, but not 1.6 representatives.

The people of Rhode Island would prefer to have two representatives, because that strengthens their voice in the House. However, people of other states might prefer that Rhode Island have only one representative, thereby leaving one more representative for another state.

In essence, the apportionment problem deals with finding a systematic way of deciding whether Rhode Island gets one or two representatives. The system must be as fair to all states as possible, while also ensuring that precisely 435 House seats are awarded in total.

The Standard Divisor and Quota

Look again at how we determined that Rhode Island was "entitled" to 1.6 representatives. First, we divided the total U.S. population by the total number of representatives to find that each representative serves an average of 646,000 people. We then divided Rhode Island's population by 646,000 to find the 1.6 representatives.

In the terminology of apportionment, the 646,000 average is called the **standard divisor** for this problem. The 1.6 representatives is called the **standard quota** of representatives for Rhode Island. It is the number (quota) that Rhode Island would get if it were possible to have fractional representatives. Note that there is only *one* standard divisor. In contrast, each state has its own standard quota.

Standard Divisor and Quota

The **standard divisor** is the average number of people per seat (in the House of Representatives) for the entire population of the United States:

$$\text{standard divisor} = \frac{\text{total U.S. population}}{\text{number of seats}}$$

The **standard quota** for a state is the number of seats it would be entitled to *if* fractional seats were allowed:

$$\text{standard quota} = \frac{\text{state population}}{\text{standard divisor}}$$

The standard divisor and standard quota also apply to apportionment problems besides that of the House of Representatives. Simply replace the number of seats by the number of items to be apportioned and the state and total populations by the relevant populations in the problem.

➤ **EXAMPLE 2** Finding Standard Quotas

The 2000 census found a population of 902,000 for Montana and 494,000 for Wyoming. Find the standard quota for each state. The 2000 apportionment left each of these states with the constitutional minimum of one representative. Compare their standard quotas to their actual representations.

Solution The standard divisor is the same for all states. As we found earlier, it is 646,000 people per representative. For Montana, the standard quota is

$$\text{standard quota} = \frac{\text{state population}}{\text{standard divisor}} = \frac{902,000}{646,000} \approx 1.4$$

For Wyoming, the standard quota is

$$\text{standard quota} = \frac{\text{state population}}{\text{standard divisor}} = \frac{494,000}{646,000} \approx 0.76$$

Montana's standard quota is 1.4, but it has only one representative. Montana's people can rightfully feel underrepresented in Congress. Wyoming also has the constitutional minimum of one representative, even though its standard quota is only 0.76.

by the way

Besides the 100 senators and 435 representatives, the U.S. Congress also includes a resident commissioner from Puerto Rico and delegates from American Samoa, the District of Columbia, Guam, and the Virgin Islands. These five individuals may take part in the floor discussions and vote in committees, but have no vote in the full House.

Thus, Wyoming's people are relatively overrepresented in Congress, at least according to population. *Now try Exercises 15–18.* ◄

► **EXAMPLE 3** School Teacher Apportionment

A small school district is reapportioning its 14 elementary teachers among its three elementary schools, which have the following enrollments: Washington Elementary, 197 students; Lincoln Elementary, 106 students; and Roosevelt Elementary, 145 students. Find the standard quota of teachers for each school.

Solution This problem is just like the House apportionment problem, except we are apportioning teachers rather than representatives. The standard divisor is the average number of students per teacher. We find it by dividing the total student population by the number of teachers:

$$\text{standard divisor} = \frac{\text{total number of students}}{\text{number of teachers}} = \frac{197 + 106 + 145}{14} = 32$$

We find the standard quota for each school by dividing the school's enrollment by the standard divisor. Table 12.13 shows the calculations and results. Note that the total (sum) of the standard quotas is the number of teachers to be apportioned.

TABLE 12.13	Finding Standard Quotas for a School Teacher Apportionment			
SCHOOL	WASHINGTON	LINCOLN	ROOSEVELT	TOTAL
ENROLLMENT	197	106	145	448
STANDARD QUOTA	$\frac{197}{32} = 6.15625$	$\frac{106}{32} = 3.3125$	$\frac{145}{32} = 4.53125$	14

enrollments divided by standard divisor of 32

Now try Exercises 19–20. ◄

The Challenge of Apportionment

The 435 representatives and 50 states make apportionment in the United States House of Representatives fairly complex. For the purposes of understanding apportionment, it's easier to work with a smaller set of states. Suppose there were only four states, A, B, C, and D, with populations as shown in Table 12.14. Note that the total population is 10,000.

TABLE 12.14	Four-State Standard Quotas for 100 Seats				
STATE	A	B	C	D	TOTAL
POPULATION	936	2726	2603	3735	10,000
STANDARD QUOTA	9.36	27.26	26.03	37.35	100

state populations divided by standard divisor of 100

NOTE: *The total population is 10,000 and the standard divisor is 100.*

Suppose these four states decide to create a legislature with 100 seats. Then the standard divisor is

$$\text{standard divisor} = \frac{\text{total population}}{\text{number of seats}} = \frac{10,000}{100} = 100$$

The second row of Table 12.14 shows the standard quotas calculated with this standard divisor. As it must, the total of the standard quotas equals the 100 available seats.

Time out to think _____

Confirm the standard quotas in Table 12.14 by dividing each state's population by the standard divisor.

We cannot use the fractional standard quotas for the actual apportionment. Instead, we must find a way to make integers from the standard quotas. The most obvious solution would be to round all the standard quotas according to standard rounding rules. In that case, all four standard quotas would round down, because all four have fractional parts less than 0.5. However, rounding in this way leads to a total of $9 + 27 + 26 + 37 = 99$ seats, one short of the 100 seats that are supposed to be apportioned. Failure to attain 100 seats with standard rounding rules means we must find a different way to make integers from the standard quotas. And herein lies the challenge of apportionment: There are many reasonable ways to make integers from the standard quotas, but they do not all lead to the same apportionment results.

HAMILTON'S METHOD

The United States conducted its first census in 1790, presenting Congress with its first task of apportionment almost immediately after ratification of the Constitution. Alexander Hamilton, then Secretary of the Treasury, championed a simple apportionment procedure that worked as follows.

Hamilton's Method of Apportionment

After finding the standard quota for each state:

- First, give each state the number of seats found by rounding its standard quota *down*. This number is the state's **minimum quota** (or *lower quota*).
- If there are any extra seats after each state has been given its minimum quota, look at each state's **fractional remainder**—the fraction that remains in the standard quota after subtracting the minimum quota. Give the extra seats one each to the states with the highest fractional remainders, until the seats are all gone.

Let's apply Hamilton's method to the four states in Table 12.14. The easiest way is by extending the table, as shown in Table 12.15. The first two rows (population and

standard quota) are unchanged. The third row shows the minimum quotas found by rounding down the standard quotas. Note that the total of the minimum quotas is 99, leaving 1 extra seat out of the 100 to be allotted. The fourth row shows the fractional remainders after rounding down. The extra seat goes to the state with the largest fractional remainder, which is State A. The last row shows the final apportionment. Note that State A gets one seat more than its minimum quota, while the other states get precisely their minimum quotas.

TABLE 12.15	Applying Hamilton's Method to the Four-State Data from Table 12.14				
STATE	A	B	C	D	TOTAL
POPULATION	936	2726	2603	3735	10,000
STANDARD QUOTA	9.36	27.26	26.03	37.35	100
MINIMUM QUOTA	9	27	26	37	99
FRACTIONAL REMAINDER	0.36 (largest)	0.26	0.03	0.35	1
FINAL APPORTIONMENT	10	27	26	37	100

state populations divided by standard divisor of 100 →
standard quotas rounded down →
remainders after rounding →
the extra seat goes to the state with the largest fractional remainder (State A) →

▶ EXAMPLE 4 Applying Hamilton's Method

Apply Hamilton's method to determine the apportionment of teachers among the schools in Example 3.

Solution Table 12.16 repeats the enrollments and standard quotas found in Example 3. Hamilton's method tells us first to round down all the standard quotas; we get the minimum quotas in the third row. The total of the minimum quotas is 13, one less than the 14 teachers to be apportioned. The school with the largest fractional remainder (fourth row), which is Roosevelt, gets the extra teacher. Thus, in the final apportionment, Washington and Lincoln get their minimum quotas of 6 and 3 teachers, respectively; Roosevelt gets 5 teachers, rather than its minimum quota of 4.

TABLE 12.16	Applying Hamilton's Method to the School Teacher Apportionment from Table 12.13			
SCHOOL	WASHINGTON	LINCOLN	ROOSEVELT	TOTAL
ENROLLMENT	197	106	145	448
STANDARD QUOTA	$\frac{197}{32} = 6.15625$	$\frac{106}{32} = 3.3125$	$\frac{145}{32} = 4.53125$	14
MINIMUM QUOTA	6	3	4	13
FRACTIONAL REMAINDER	0.15625	0.3125	0.53125 (largest)	1
FINAL APPORTIONMENT	6	3	5	14

enrollments divided by standard divisor of 32 →
standard quotas rounded down →
remainders after rounding →
the extra teacher goes to the school with the largest fractional remainder (Roosevelt) →

Now try Exercises 21–22. ◀

The First Presidential Veto

Hamilton's method is simple and appears reasonably fair. Perhaps as a result, both the Senate and the House voted in 1791 to adopt Hamilton's method for apportionment. However, recall that for a bill to become law, it must not only be passed by both the Senate and the House, but also be signed by the president. If the president vetoes a bill, super majority votes of 2/3 in the House and Senate are needed to override the veto (see Unit 12A). In 1792, the bill authorizing Hamilton's method received the first presidential veto in the history of the United States. In his veto message of April 5, 1792, President George Washington wrote:

I have maturely considered the act passed by the two Houses entitled "An act for an apportionment of representatives among the several States according to the first enumeration," and I return it to your House, wherein it originated, with the following objections:

First. The Constitution has prescribed that representatives shall be apportioned among the several States according to their respective numbers, and there is no one proportion or divisor which, applied to the respective numbers of the States, will yield the number and allotment of representatives proposed by the bill.

Second. The Constitution has also provided that the number of representatives shall not exceed 1 for every 30,000, which restriction is by the context and by fair and obvious construction to be applied to the separate and respective numbers of the States; and the bill has allotted to eight of the States more than 1 for every 30,000.

Congress was unable to override the veto, and instead passed a bill authorizing a different apportionment method proposed by Thomas Jefferson, then Secretary of State. We'll discuss Jefferson's method shortly.

Time out to think _____

The Constitution actually says, "The number of representatives shall not exceed one for every thirty thousand, but each state shall have at least one representative . . . " Read Washington's veto statement carefully. Why did he have to add the words "which restriction is by the context and by fair and obvious construction . . . " in order to justify his veto?

Fairness Issues with Hamilton's Method

Although Hamilton's method did not become law at the time it was proposed, it was reintroduced and adopted as the apportionment procedure in 1850. It remained in use until 1900. During this time, several problems with Hamilton's method emerged; the most famous is called the **Alabama paradox.**

After the 1880 census, the Chief Clerk of the Census Office, C. W. Seaton, used Hamilton's method to compute apportionments for various possible House sizes. (Unlike today, in those days the House size often changed with each apportionment.) He discovered that Alabama would get 8 seats in a House of 299 representatives, but only 7 seats in a House of 300 representatives. This curious result seems unfair, because Alabama loses a seat when the total number of seats increases.

The Alabama Paradox

In a fair apportionment system, adding extra seats must not result in fewer seats for any state. The **Alabama paradox** occurs when the total number of available seats increases, yet one state (or more) loses seats as a result. It can occur with Hamilton's method.

Alabama did not actually lose a seat in the 1882 apportionment, because Congress chose a larger house size (325) for which the Alabama paradox did not present itself. But knowing that the paradox was possible reduced support for Hamilton's method, and it was abandoned by Congress in 1900.

Hamilton's method has not been used for House apportionment since 1900, but it has been studied, leading to the discovery of two other paradoxes. The **population paradox** was discovered around 1900, when it was found that Hamilton's method would award a seat to Maine at Virginia's expense, even though Virginia was growing much faster than Maine.

The Population Paradox

When apportionment changes because of population growth, we would expect faster-growing states to gain seats at the expense of slow-growing states. When the opposite occurs—a slow-growing state gains a seat at the expense of a faster-growing state—we have the **population paradox.**

The **new states paradox** was discovered in 1907, when Oklahoma became the 46th state in the United States. Because it was not yet time for reapportionment, Congress decided to add *new* seats for Oklahoma. Based on its population, Oklahoma was clearly entitled to 5 seats, so Congress increased the number of seats in the House from 386 to 391. Surprisingly, calculations with Hamilton's method showed that the addition of Oklahoma's 5 seats would have caused New York to lose a seat while Maine gained one.

The New States Paradox

When additional seats are added to accommodate a new state, we do not expect this addition to change the apportionment for existing states. If it does, we have the **new states paradox.**

> **EXAMPLE 5** Four-State Alabama Paradox

Using Hamilton's method, recompute the apportionment in Table 12.15 if there are 101 seats instead of 100. Does the Alabama paradox occur? Explain.

HISTORICAL NOTE

In the 1876 election, Democrat Samuel J. Tilden won the popular vote but lost the presidency to Republican Rutherford B. Hayes by one electoral vote. The electoral vote was severely disputed, with some states appointing rival slates of electors. Tilden eventually stepped aside in the "Compromise of 1877," which led to Hayes's inauguration on March 4, 1877. Interestingly, if electors had been apportioned by Hamilton's method, as the law required, Tilden would have won in the first place. Instead, electors had been apportioned according to a compromise from a few years earlier, even though the compromise violated the legal requirement.

Solution The total population is 10,000, so with 101 seats the standard divisor is

$$\text{standard divisor} = \frac{\text{total population}}{\text{number of seats}} = \frac{10,000}{101} \approx 99.0099$$

We find the standard quota for each state by dividing its population by this standard divisor. You should confirm that the standard quotas come out as shown in Table 12.17. Note that they are slightly different from the standard quotas shown in Table 12.15 for 100 seats. The third row of Table 12.17 shows the minimum quotas, which are unchanged from Table 12.15. However, the fractional remainders (fourth row) have changed, which changes the final apportionment.

TABLE 12.17 Recomputed Apportionment for Table 12.15 with 101 Seats

STATE	A	B	C	D	TOTAL
POPULATION	936	2726	2603	3735	10,000
STANDARD QUOTA	9.4536	27.5326	26.2903	37.7235	101
MINIMUM QUOTA	9	27	26	37	99
FRACTIONAL REMAINDER	0.4536	0.5326 (2nd largest)	0.2903	0.7235 (largest)	2
FINAL APPORTIONMENT	9	28	26	38	101

state populations divided by standard divisor of 99.0099 → STANDARD QUOTA

standard quotas rounded down → MINIMUM QUOTA

remainders after rounding → FRACTIONAL REMAINDER

the extra seats go to the two states with the largest fractional remainders (D and B) → FINAL APPORTIONMENT

This time, there are two extra seats after the minimum quotas have been filled. By Hamilton's method, these two seats go to States D and B, because they have the two largest fractional remainders. Meanwhile, State A ends up with its minimum quota of 9 seats, even though it had 10 seats in Table 12.15. In other words, increasing the total number of seats from 100 to 101 has cost State A a seat—an illustration of the Alabama paradox. *Now try Exercises 23–26.* ◄

JEFFERSON'S METHOD

Thomas Jefferson's rival apportionment method became law after President Washington vetoed Hamilton's method. Jefferson's method essentially tries to avoid decisions about fractional remainders by seeking minimum quotas that use up all available seats. As we've seen, the minimum quotas used in Hamilton's method—which are simply the standard quotas rounded down—often leave extra seats. Jefferson realized that he could achieve his goal by changing the standard quotas to new values, called **modified quotas,** so that the new set of minimum quotas would use up all the seats.

Finding modified quotas means dividing the state populations by a **modified divisor** that is different (lower) than the standard divisor. The trick in Jefferson's method is choosing a modified divisor that leads to the desired result, which is a set of minimum quotas that uses all the seats. Choosing a modified divisor generally requires trial and error: If your first choice doesn't work, try again until you find one that does.

Jefferson's Method of Apportionment

Begin by finding the standard divisor, standard quotas, and minimum quotas (as in Hamilton's method). If there are no extra seats after each state has been given its minimum quota, then the apportionment is complete. If there *are* extra seats, *start the computations over* as follows.

- Choose a **modified divisor** that is *lower* than the standard divisor. Use it to compute **modified quotas** by dividing the state populations by the modified divisor. Round these modified quotas down to find a new set of minimum quotas.
- If there are just enough seats to fill all the minimum quotas, the apportionment is complete. Otherwise, do one of the following:
 1. If there are still extra seats with the new minimum quotas, start again with a lower modified divisor.
 2. If there are not enough total seats to fill all the minimum quotas, start again with a higher modified divisor (but still smaller than the standard divisor).

Let's apply Jefferson's method to the same four states to which we applied Hamilton's method. The first three rows of Table 12.18 are unchanged from Table 12.15, showing the minimum quotas (third row) that result from the standard divisor of 100. Because the minimum quotas leave an extra seat, Jefferson's method tells us to try a new (lower) divisor. Study Table 12.18 carefully to see how we implement Jefferson's method.

TABLE 12.18 Applying Jefferson's Method to the Four-State Data from Table 12.15

STATE	A	B	C	D	TOTAL
POPULATION	936	2726	2603	3735	10,000
STANDARD QUOTA	9.36	27.26	26.03	37.35	100
MINIMUM QUOTA	9	27	26	37	99
MODIFIED QUOTA (WITH DIVISOR 99)	9.45	27.54	26.29	37.73	101.01
MINIMUM QUOTA (WITH DIVISOR 99)	9	27	26	37	99
MODIFIED QUOTA (WITH DIVISOR 98)	9.55	27.82	26.56	38.11	102.04
MINIMUM QUOTA (WITH DIVISOR 98)	9	27	26	38	100

state populations divided by standard divisor of 100
standard quotas rounded down
state populations divided by *modified divisor* of 99
modified quotas rounded down
state populations divided by *modified divisor* of 98
modified quotas rounded down; all 100 seats are used, so this is final apportionment

- The standard divisor is 100, so we try a lower *modified divisor* of 99. We divide the state populations by this divisor to find the modified quotas in the

fourth row. The fifth row shows the new set of minimum quotas, found by rounding down the modified quotas.

- Because the new set of modified quotas still leaves an extra seat, we try an even lower modified divisor of 98. The last two rows show the modified quotas and minimum quotas found with this divisor. This time they add up to 100, so the apportionment is complete.

Note that State A gets 9 seats by Jefferson's method, as opposed to 10 seats by Hamilton's method (see Table 12.15). Meanwhile, State D gets 38 seats by Jefferson's method, one more than it gets by Hamilton's method.

Is Jefferson's Method Fair?

We've already seen that Hamilton's method is susceptible to several paradoxes that make it appear unfair in some cases. Is Jefferson's method better?

Mathematicians have studied these methods carefully. Jefferson's method avoids all three of the paradoxes found with Hamilton's method (the Alabama, population, and new states paradoxes). However, Jefferson's method sometimes leads to a different problem. Let's investigate.

Remember that standard quotas represent the number of seats that would be apportioned *if* fractional apportionment were allowed. For example, we earlier found a standard quota of 1.6 for Rhode Island, suggesting that it is "entitled" to 1.6 seats. We might therefore expect that any "fair" apportionment method would either round this standard quota up to give Rhode Island two seats or round it down to give Rhode Island one seat. If instead Rhode Island ended up with zero seats or three or more seats, we would probably conclude that the apportionment was unfair. This idea, which holds that the actual apportionment for any state should be its standard quota rounded either up or down, is often called the **quota criterion.** An apportionment that fails the quota criterion is generally considered unfair. As the following example shows, Jefferson's method sometimes fails the quota criterion.

by the way

Hamilton's method always satisfies the quota criterion because it starts with the standard quota rounded down, then at most adds one—which gives the standard quota rounded up. Hamilton's method and other apportionment methods that always satisfy the quota criterion are generically called *quota methods.* In contrast, methods that change the divisor, like Jefferson's method, are called *divisor methods* and sometimes violate the quota criterion.

The Quota Criterion

For a fair apportionment, the number of seats assigned to each state should be its standard quota rounded either up or down to the nearest integer.

▶ **EXAMPLE 6** Jefferson's Method and the Quota Criterion

Consider a four-state legislature with 100 seats, in which the states have the following populations: State A, 680; State B, 1626; State C, 1095; and State D, 6599. Use Jefferson's method to apportion the 100 seats. Is the quota criterion satisfied?

Solution We begin by finding the standard divisor, which is

$$\text{standard divisor} = \frac{\text{total population}}{\text{number of seats}} = \frac{10,000}{100} = 100$$

Table 12.19 shows the computations by Jefferson's method. The second row gives the standard quotas and the third row gives the resulting minimum quotas. There are

three extra seats, so Jefferson's method tells us to try a lower modified divisor. Row 4 shows the modified quotas with a divisor of 98 and the last row shows the new minimum quotas. Because this new set of minimum quotas uses all 100 seats, it represents the final apportionment.

Note that State D has a standard quota of 65.99. By the quota criterion, State D should get either 65 or 66 seats. However, State D ends up with 67 seats, so this apportionment violates the quota criterion.

TABLE 12.19	A Case Where Jefferson's Method Violates the Quota Criterion				
STATE	A	B	C	D	TOTAL
POPULATION	680	1626	1095	6599	10,000
STANDARD QUOTA	6.80	16.26	10.95	65.99	100
MINIMUM QUOTA	6	16	10	65	97
MODIFIED QUOTA (WITH DIVISOR 98)	6.94	16.59	11.17	67.34	102.04
MINIMUM QUOTA (WITH DIVISOR 98)	6	16	11	67	100

state populations divided by standard divisor of 100 → STANDARD QUOTA

standard quotas rounded down → MINIMUM QUOTA

state populations divided by modified divisor of 98 → MODIFIED QUOTA

modified quotas rounded down → MINIMUM QUOTA

Now try Exercises 27–30. ◄

What Became of Jefferson's Method

After being signed into law by President Washington in 1792, Jefferson's method was used through the 1830s. However, violations of the quota criterion occurred in the apportionments following both the 1820 and the 1830 census. Moreover, Congress learned that the violations tend to give extra seats to larger states at the expense of smaller states. As a result, Congress abandoned Jefferson's method for the apportionment following the 1840 census.

OTHER APPORTIONMENT METHODS

As we've discussed, Jefferson's method was abandoned by 1840 and Hamilton's method, adopted in 1850, was abandoned in 1900. Many other apportionment methods have been proposed, but only two others have been used for the United States House of Representatives.

Webster's Method

In 1832, the famous senator and orator Daniel Webster, of Massachusetts, proposed a variation on Jefferson's method. Webster's method was adopted for the apportionment following the 1840 census. It was then abandoned in favor of Hamilton's method in 1850, but resurrected in 1900. Webster's method remained in use until 1940.

Webster's method is very similar to Jefferson's method with one exception: Instead of looking for a set of modified quotas that can all be rounded *down* to give the correct total number of seats, Webster's method seeks modified quotas that give

the correct total number of seats by using standard rounding rules. That is, we round *up* when the fractional part of a modified quota is 0.5 or more, and down when the fractional part is less than 0.5. Note that, whereas the modified divisor in Jefferson's method is always less than the standard divisor, the modified divisor in Webster's method may be either greater or less than the standard divisor.

►EXAMPLE 7 Applying Webster's Method

Consider a four-state legislature with 100 seats, in which the states have the following populations: State A, 948; State B, 749; State C, 649; and State D, 7654. Use Webster's method to apportion the 100 seats.

Solution The total population is 10,000 and the number of seats is 100, so the standard divisor is 100 (as in Example 6). Table 12.20 shows the computations by Webster's method. The second row gives the standard quotas and the third row gives the resulting minimum quotas; there are two extra seats. Webster's method tells us to try a modified divisor and then round the modified quotas to the nearest integer. Row 3 shows the modified quotas with a divisor of 99.85. Finding a divisor that works generally requires some trial and error (not shown here). The last row shows the rounded quotas, which represent the final apportionment because they use all 100 seats.

TABLE 12.20	A Four-State Apportionment with Webster's Method				
STATE	A	B	C	D	TOTAL
POPULATION	948	749	649	7654	10,000
STANDARD QUOTA	9.48	7.49	6.49	76.54	100
MINIMUM QUOTA	9	7	6	76	98
MODIFIED QUOTA (WITH DIVISOR 99.85)	9.4942	7.5013	6.4997	76.6550	100.1502
ROUNDED QUOTA	9	8	6	77	100

state populations divided by standard divisor of 100 →

standard quotas rounded down →

state populations divided by *modified divisor* of 99.85 →

modified quotas rounded to nearest integer →

Now try Exercises 31–32. ◄

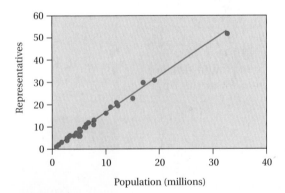

FIGURE 12.3 A scatter plot of number of representatives versus population for the 50 states (2000 data). A perfect straight-line correlation would mean all states were equally represented.

Time out to think

Is the quota criterion satisfied in Example 7? Explain.

The Hill-Huntington Method

All apportionment methods have the goal of giving all states fair representation in the House (Figure 12.3). Toward this end, in 1941 Congress replaced Webster's method with an alternative called the Hill-Huntington method, which remains in use today.

The Hill-Huntington method gets its name from Joseph Hill, who served as Chief Statistician of the Census Bureau, and Harvard mathematician Edward Huntington.

After Hill proposed the method in 1911, mathematicians debated whether it was fairer than Webster's method, while Congress debated both the fairness and the political aspects of its use. For the 1920 census, Congress continued to use Webster's method. For the 1930 census, Webster's method and the Hill-Huntington method led to the same apportionments, so controversy was temporarily averted. But in the apportionment following the 1940 census, Webster's method gave one extra seat to Michigan while the Hill-Huntington method gave the seat to Arkansas. Michigan tended to vote Republican while Arkansas tended to vote Democratic. Because Democrats had a majority in Congress, they made a political decision: They voted to use the Hill-Huntington method, thereby giving the extra seat to Democratic-leaning Arkansas. President Roosevelt (also a Democrat) signed the bill into law.

The Hill-Huntington method is almost identical to Webster's method, except it uses a different rule to decide whether modified quotas should be rounded up or down. In Webster's method, rounding follows the usual rule in which we round up if the fractional part is 0.5 or more, and round down if it is less than 0.5. In the Hill-Huntington method, rounding is based instead on the **geometric mean** of the integers on either side of the modified quota.

DEFINITIONS

The **geometric mean** of any two numbers x and y is $\sqrt{x \times y}$. The more familiar mean, $(x + y)/2$, is called the **arithmetic mean.** In general, we assume a "mean" is an arithmetic mean unless told otherwise. (Thus, the mean we have used elsewhere in this book has always been the arithmetic mean.)

If the modified quota is less than the geometric mean of the two nearest integers, it gets rounded down. If it is more than the geometric mean, it gets rounded up. As an example, suppose a state has a modified quota of 2.47, which we could potentially round to either 2 or 3. Under Webster's method, we would round it down to 2, because the fractional part of 2.47 is less than 0.5. Under the Hill-Huntington method, we first find the geometric mean of 2 and 3, which is

$$\sqrt{2 \times 3} = \sqrt{6} \approx 2.45$$

Because the modified quota of 2.47 is *greater* than this geometric mean, it gets rounded up to 3 in the Hill-Huntington method.

Note that the geometric mean of two consecutive integers is always less than their arithmetic mean. For example, the geometric mean of 2 and 3 is 2.45, which is less than their arithmetic mean of 2.5 by 0.05. However, the two means tend to be closer for larger consecutive integers. For example, the geometric mean of 10 and 11 is $\sqrt{10 \times 11} = \sqrt{110} \approx 10.488$, which is only 0.012 less than their arithmetic mean of 10.5. In practical terms, use of the geometric mean in the Hill-Huntington method therefore increases the chance that extra seats will go to smaller states rather than larger states.

➤ **EXAMPLE 8** Applying the Hill-Huntington Method

Apply the Hill-Huntington method to the case described in Example 7. Is the resulting apportionment the same as by Webster's method?

Solution Table 12.21 shows the calculations. The first three rows are the same as in Table 12.20, because they are computed the same way by all methods. The fourth row shows modified quotas with a modified divisor of 100.06. Again, finding a modified divisor that works generally requires some trial and error, which is not shown here. To determine the rounded quotas, we first find the geometric means (row 5). The Hill-Huntington method then tells us to round down if the modified quota is less than the geometric mean and to round up otherwise. The last row shows the rounded quotas; because they use all 100 seats, the apportionment is complete. Note that States C and D end up with different numbers of seats in this apportionment than in Webster's apportionment in Example 7.

TABLE 12.21	Applying the Hill-Huntington Method to the Four-State Data from Table 12.20				
STATE	A	B	C	D	TOTAL
POPULATION	948	749	649	7654	10,000
STANDARD QUOTA	9.48	7.49	6.49	76.54	100
MINIMUM QUOTA	9	7	6	76	98
MODIFIED QUOTA (WITH DIVISOR 100.06)	9.47	7.49	6.49	76.49	99.94
GEOMETRIC MEAN	$\sqrt{9 \times 10}$ ≈ 9.487	$\sqrt{7 \times 8}$ ≈ 7.483	$\sqrt{6 \times 7}$ ≈ 6.481	$\sqrt{76 \times 77}$ ≈ 76.498	
ROUNDED QUOTA	9	8	7	76	100

Annotations (left margin):
- state populations divided by standard divisor of 100 → STANDARD QUOTA
- standard quotas rounded down → MINIMUM QUOTA
- state populations divided by *modified divisor* of 100.06 → MODIFIED QUOTA
- geometric mean of integers on either side of modified quota → GEOMETRIC MEAN
- modified quotas rounded down if less than geometric mean, up otherwise → ROUNDED QUOTA

Now try Exercises 33–34. ◀

Time out to think _____

Consider the results in Examples 7 and 8. Which apportionment method would State C prefer? Which would State D prefer? Explain.

IS THERE A BEST METHOD FOR APPORTIONMENT?

The history of apportionment is a quest for fairness. Hamilton's method proved to be unfair because of the three paradoxes (Alabama, population, and new states). Jefferson's method proved to be unfair because it can violate the quota criterion. How do Webster's method and the Hill-Huntington method compare?

Mathematicians have studied the methods by running simulations of many possible apportionments. Results show that, like Jefferson's method, Webster's method and the Hill-Huntington method can also violate the quota criterion. However, the violations occur much less frequently with Webster's or Hill-Huntington than with Jefferson's method, making them arguably fairer.

Practical matters

REDISTRICTING

Apportionment determines the number of representatives for each state. Politically, however, this number is only the beginning of the practical issues associated with electing representatives. In states with more than one seat in the House, the seats are divided among congressional districts within the state. For example, a state with six seats has six districts, with the people in each district electing one representative. Just as apportionment is recomputed every ten years following a census, so are districts redrawn with new boundaries at these times. Although few people follow the details of this redistricting, it can have significant political impact.

The Constitution leaves it to the states to decide how to draw the boundaries of their districts. In most states, redistricting is up to the state legislature, which can use the process for political advantage. To see how, consider a state that has six House seats and in which half the people are Democrats and half are Republicans. If you ignored redistricting, you might guess that this state would have three Democratic representatives and three Republican representatives.

However, suppose the state elects its six representatives in six districts (District 1 through District 6) with 1 million people each. The even division of voter preferences among the two parties means that there are 3 million Democrats and 3 million Republicans in the state. Now, to take an extreme case, suppose the legislature draws the boundaries of District 1 so that all of its 1 million people are Democrats. Clearly, District 1 will elect a Democratic representative. But the remaining five districts will have 3 million Republicans and only 2 million Democrats. If they are

drawn so that their populations reflect this general distribution, every one of these five districts will have Republicans outnumbering Democrats by a 3 to 2 margin. The end result: Despite having equal numbers of Democratic and Republican voters, the state ends up with five Republican representatives and only one Democratic representative. Of course, the same idea could occur in reverse, giving Democrats the advantage rather than Republicans.

The practice of forming voting districts to give an unfair advantage to a political party is so common that it has its own name—*gerrymandering*. (The term originated in 1812 when Massachusetts, under Governor Elbridge *Gerry*, created a district with roughly the shape of a sala*mander*.) Since the 1960s, the Supreme Court has found gerrymandering illegal, ruling that it violates the Constitution's intention to provide equal representation to all citizens. Nevertheless, the issue of fairness is sufficiently subjective to leave the party that controls redistricting in any state with broad influence over the makeup of the state's House delegation. Similar redistricting considerations apply to state and local representation. Today, political parties use sophisticated mathematical models to help draw districts in a way most favorable to their interests. (But if they go too far, opposing parties may claim illegal gerrymandering and challenge the district boundaries in court.)

Redistricting is more complex mathematically than apportionment, but it offers the same basic lesson: Democracy is much more difficult in practice than it is in principle, which is why it is so important to be politically and mathematically aware if you hope to influence the future of your town, state, country, or world.

For example, if Jefferson's method had remained in use, nearly every apportionment since 1850 would have violated the quota criterion. In contrast, no violations have occurred with Webster's method or the Hill-Huntington method. Simulations show that, by chance alone, Webster's and Hill-Huntington are expected to violate the quota criterion only once in several hundred apportionments. Interestingly, Webster's appears slightly less prone to violations of the quota criterion than Hill-Huntington, suggesting that it is slightly fairer.

Is there any apportionment system that is unquestionably fair in all circumstances? Such an apportionment system would have to satisfy the quota criterion in all cases, while also being immune to the three paradoxes that affect Hamilton's method. Unfortunately, a theorem proved by mathematicians M. L. Balinsky and H. P. Young states that such a system is impossible.

In essence, the **Balinsky and Young theorem** for apportionment is analogous to Arrow's impossibility theorem for voting (see Unit 12B). It tells us that, in the end, we cannot choose between apportionment procedures on the basis of fairness alone. As a result, apportionment will always involve political decisions, and we can expect it to be a subject of continued debate.

EXERCISES 12C

Review Questions

1. What is apportionment? What does it mean for the U.S. House of Representatives?

2. Briefly describe the nature of the apportionment problem. How does this problem arise from the requirements of the Constitution?

3. Explain how Hamilton's method apportions seats. Briefly describe the history of Hamilton's method.

4. What is the Alabama paradox? What other paradoxes affect Hamilton's method? Why do these paradoxes make the method seem unfair in the cases where they arise?

5. Explain how Jefferson's method apportions seats. Briefly describe the history of Jefferson's method.

6. What is the quota criterion? Why are violations of this criterion considered unfair?

7. Briefly describe how Webster's method and the Hill-Huntington method differ from Jefferson's method.

8. Explain why Webster's method and the Hill-Huntington method are considered fairer than Jefferson's method, but still not fair in all cases. What is the significance of the Balinsky and Young theorem?

Does It Make Sense?

Decide whether each of the following statements makes sense (or is clearly true) or does not make sense (or is clearly false). Explain your reasoning.

9. Mike is the president of a large company with 12 divisions. He plans to use an apportionment method to decide how many staff support persons to assign to each division.

10. Charlene is the head judge in a figure skating competition. She plans to use an apportionment method to decide how the judges' points should be allocated to the skaters.

11. The Hill-Huntington method is superior to other apportionment methods because it requires more advanced mathematics.

12. Ten new teachers were hired in the Meadowlark School District this year. Horizon School has the same proportion of the district's students as it did last year, but it now has three fewer teachers. The apportionment method used to allocate teachers was not fair.

Basic Skills & Concepts

13. **Representation in Congress.** If the population of the United Stated increased to 300 million and the number of representatives remained 435, how many Americans, on average, would each representative serve? With this population, if the number of representatives were set at the constitutional limit of one representative for every 30,000 people, how many representatives would there be in Congress?

14. **Representation in Congress.** If the population of the United States increased to 310 million and the number of representatives increased to 445, how many Americans, on average, would each representative serve? With this population, if the number of representatives were set at the constitutional limit of one representative for every 30,000 people, how many representatives would there be in Congress?

State Representation. The following table shows populations and numbers of seats in the House for four states, based on the 2000 census (which took effect with the 2002 Congress). For the states indicated in Exercises 15–18, find the standard quota and compare it to the actual number of seats for the state. Then explain whether the state is relatively under- or overrepresented in the House. Use the census 2000 value for the total U.S. population, which was 281 million.

State	Population	Seats in House
Connecticut	3,406,000	5
Georgia	8,186,000	13
Florida	15,982,000	25
Ohio	11,353,000	18

15. Connecticut

16. Georgia

17. Florida

18. Ohio

19. **Standard Quotas in Business.** A large company has four divisions with 250, 320, 380, and 400 employees, respectively. A total of 35 computer technicians must be allocated to the four divisions according to their size. Find the standard quota for each division.

20. **Standard Quotas in Education.** Capital University has five colleges with 560, 1230, 1490, 1760, and 2340 students, respectively. A total of 18 academic advisors must be allocated to the five colleges according to their size. Find the standard quota for each college.

21. **Hamilton's Method.** Use Hamilton's method to determine the allocation of computer technicians in Exercise 19.

22. **Hamilton's Method.** Use Hamilton's method to determine the allocation of academic advisors in Exercise 20.

Alabama Paradox. In Exercises 23–26, assume that 100 delegates must be apportioned to states with the given populations. Determine the number of representatives for each state using Hamilton's method. Then assume that the number of delegates is increased to 101. Determine the new number of representatives for each state using Hamilton's method. State whether the change in total delegates results in the Alabama paradox.

23. A: 950; B: 670; C: 246

24. A: 2540; B: 1140; C: 6330

25. A: 770; B: 155; C: 70; D: 673

26. A: 562; B: 88; C: 108; D: 242

Jefferson's Method. In Exercises 27–30, apply Jefferson's method to states with the given populations. Assume that 100 delegates are to be apportioned. In each case, state whether the quota criterion is satisfied.

27. A: 98; B: 689; C: 212 (modified divisor 9.83)

28. A: 1280; B: 631; C: 2320 (modified divisor 42.00)

29. A: 69; B: 680; C: 155; D: 75 (modified divisor 9.60)

30. A: 1220; B: 5030; C: 2460; D: 690 (modified divisor 92.00)

31. **Webster's Method.** Use Webster's method to determine the allocation of computer technicians in Exercise 19.

32. **Webster's Method.** Use Webster's method to determine the allocation of academic advisors in Exercise 20.

33. **Hill-Huntington Method.** Use the Hill-Huntington method to determine the allocation of computer technicians in Exercise 19.

34. **Hill-Huntington Method.** Use the Hill-Huntington method to determine the allocation of academic advisors in Exercise 20.

Further Applications

The New States Paradox. In Exercises 35–36, the populations of three states are given. Determine how 100 delegates should be apportioned among these states using Hamilton's method. Then suppose that a new state with a population of 500 is added to the system, along with 5 new seats. Determine the apportionment for the four states (assuming the populations of the first three states remain the same). Does the addition of the new state decrease the representation of any of the original states?

35. A: 1140; B: 6320; C: 250

36. A: 5310; B: 1330; C: 3308

Comparing Methods. Exercises 37–40 list populations for three or four states. In each case, assume 100 delegates are to be apportioned and do the following:

a. Find the apportionment by Hamilton's method.

b. Find the apportionment by Jefferson's method.

c. Find the apportionment by Webster's method.

d. Find the apportionment by the Hill-Huntington method.

e. Compare the results of the various methods. Which methods give the same results? Do any of the results

violate the quota criterion? Overall, which method do you think is best for this apportionment? Why?

37. A: 535; B: 344; C: 120

38. A: 144; B: 443; C: 389

39. A: 836; B: 2703; C: 2626; D: 3835

40. A: 1234; B: 3498; C: 2267; D: 5558

Non-House Apportionments. Exercises 41–44 describe apportionment problems for situations besides the House of Representatives. In each case, do the following:

 a. Find the apportionment by Hamilton's method.

 b. Find the apportionment by Jefferson's method.

 c. Find the apportionment by Webster's method.

 d. Find the apportionment by the Hill-Huntington method.

 e. Compare the results of the various methods.

41. A high school is creating a student committee to allocate use of classrooms after hours. The committee is to consist of 10 student members, chosen from three interest groups: social groups, which have 48 members; political groups, which have 97 members; and athletic groups, which have 245 members.

42. A city plans to purchase 16 new emergency vehicles. They are to be apportioned among 485 members of the police department, 213 members of the fire department, and 306 members of the paramedic squad.

43. A chain of hardware stores is reapportioning its 25 managers among stores in four locations according to their monthly gross sales. The sales at the four stores are as follows: Boulder, $2.5 million; Denver, $7.6 million; Broomfield, $3.9 million; Ft. Collins, $5.5 million.

44. The city parks commission plans to build nine new parks, to be apportioned among three neighborhoods by population. The neighborhoods are Greenwood, population 4300; Willowbrook, population 3040; and Cherryville, population 2950.

Web Projects

Find useful links for Web Projects on the text Web site: www.aw.com/bennett-briggs

45. 2000 Census Apportionment. How did your home state fare in the apportionment based on the 2000 census? How have changes in population affected your state's standard quota since the 2000 apportionment? Overall, do you think the apportionment was fair to your state?

46. House Districts. Once a state knows its number of seats in the U.S. House, it is still up to the state to draw boundaries for the districts of various representatives. This process can be very contentious. Find the district boundaries for your state. Do the boundaries make sense geographically, or do they appear to be politically motivated? Explain.

47. Redistricting Roulette. The Center for Voting and Democracy offers a game that you can play to study the process of redistricting. Try the game; then write a short report on what you learned.

In the News

48. Apportionment News. Find a news story that concerns the most recent apportionment for the U.S. House of Representatives. Summarize the story and the role of apportionment methods in any controversies.

49. Local Apportionment. Find a recent news story concerning an apportionment problem that applies locally or at the state level. Discuss the apportionment procedure used. Does the apportionment seem fair?

50. Your State's Representatives. How many seats does your state have in the House of Representatives? How does this number compare to your state's standard quota? Based on your findings, is your state over- or underrepresented in the House? Explain.

CHAPTER 12 SUMMARY

UNIT	KEY TERMS	KEY IDEAS AND SKILLS
12A	majority rule popular vote electoral vote filibuster super majority veto preference schedule plurality	**Understand** U.S. presidential elections. **Understand** variations on majority rule. **Apply** five methods for deciding an election with three or more candidates: plurality method single (top-two) runoff sequential runoff point system (Borda count) pairwise comparisons (Condorcet method)
12B	fairness criteria Arrow's impossibility theorem approval voting	**Apply** the four fairness criteria. **Know** that no voting method can satisfy all fairness criteria in all cases (Arrow's impossibility theorem). **Understand** approval voting as an alternative voting system. **Understand** variations in voting power when not all voters have equal weight.
12C	apportionment standard divisor standard quota minimum quota modified divisor modified quota Alabama paradox population paradox new states paradox quota criterion	**Know** the history of apportionment mathematics. **Apply** four apportionment methods: Hamilton's method Jefferson's method Webster's method Hill-Huntington method **Know** the potential flaws of each method. **Understand** the significance of the Balinsky and Young theorem.

One accurate measurement is worth a thousand expert opinions.

—GRACE HOPPER, MATHEMATICIAN AND FIRST WOMAN TO REACH RANK OF REAR ADMIRAL IN THE U.S. NAVY

CHAPTER **13**

Mathematics and Business

The most obvious connections between mathematics and business come through issues of finance and budgets, some of which we have studied already. But mathematics is important to business in many more ways. For example, computers have transformed business in the past two decades, and mathematics is used in the design of both hardware and software. More recently, the Internet and other widespread networks have globalized business and created a revolution in the flow of information. In this chapter, we will study the mathematics of networks and examine some of the ways these ideas are changing the world of business.

UNIT **13A** **Network Analysis:** We will begin the study of networks with a famous problem, the bridges of Königsberg. We will then explore fundamental properties of networks.

UNIT **13B** **The Traveling Salesman Problem:** We will examine the search for efficient network strategies by studying the traveling salesman problem, a classic and largely unsolved problem of network analysis.

UNIT **13C** **Scheduling Problems:** Many practical problems in business and management involve scheduling. We will focus on the critical path method, which can be used to design schedules for everything from house construction to moon landings.

UNIT 13A

NETWORK ANALYSIS

A **network** is a collection of points or objects that are interconnected in some way. The telephone system is a network in which each individual telephone can be connected to any other telephone through the phone lines. The Internet is a network that connects individual computers. These networks are extremely complex, as they must allow any two telephones or computers to connect on demand—that is, whenever a user wants to make a connection. Nevertheless, the principles involved in **network analysis** are relatively simple and easy to visualize. We will devote most of this chapter to understanding properties and uses of networks.

THE BRIDGES OF KÖNIGSBERG

The origins of network analysis go back to the early 18th century, with a problem that concerned the Prussian town of Königsberg (now Kaliningrad, Russia) on the Pregel River. Königsberg had seven bridges straddling the Pregel River (Figure 13.1a). A popular pastime was to try to find a path that started and ended at the same point and crossed each bridge *exactly once*. Most citizens of Königsberg believed that such a tour was impossible, but no one was sure. The problem came to the attention of the great Swiss mathematician Leonhard Euler (pronounced "oiler").

Euler recognized that he could approach the problem by using the simplified model of the town shown in Figure 13.1b.

- Each of the seven bridges is represented by a line or curve, called an **edge,** labeled with a lowercase letter (*a*, *b*, *c*, . . .).

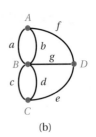

FIGURE 13.1 (a) A map of the Pregel River flowing through Königsberg. Capital letters label distinct land masses and lowercase letters label bridges. (b) A network representation of the bridges of Königsberg; vertices represent the land masses (capital letters) and edges represent the bridges (lowercase letters).

- Each of the four "pieces" of land to which bridges connect—the two islands, the north shore, and the south shore—is represented by a dot, or **vertex,** labeled with a capital letter (A, B, C, \dots).

This *network* looks different from the actual map of the town, but it captures the essential aspects of the problem. Three bridges land at different places on the north side of the river, but in the network these three bridges all connect to vertex A, which represents the entire north side of the river. And because the problem concerns only the *connections* among the four pieces of land, the lengths and shapes of the edges needn't be precise or to scale. The problem of finding a path through the town that starts and ends at the same point and crosses all the bridges exactly once is equivalent to finding a path through the network that starts and ends at the same vertex and traverses each edge exactly once.

Time out to think _____

Suppose you are on the south shore of the river in Königsberg and wish to reach the north shore via the eastern island. Can you make this trip by crossing only two bridges? Describe your path both on the real map in Figure 13.1a and on the network in Figure 13.1b.

In other applications, network vertices might represent power stations, railway terminals, jobs on an assembly line, or computers, while edges represent connections between them. Networks don't have to represent physical connections. For example, a network could have vertices representing businesses and edges representing economic ties between them.

➤ **EXAMPLE 1** Intranet

Figure 13.2 shows the layout of computers, servers, and cables in a small office *intranet* (a network within an office). Draw a network to represent this intranet.

FIGURE 13.2 Connections between computers in an office intranet.

Solution Each of the computers and servers becomes a vertex in the network, labeled with a capital letter. Each of the cables becomes an edge, labeled with a lowercase letter. Figure 13.3 shows one way of drawing the resulting network.

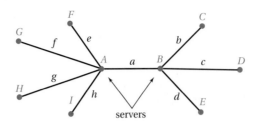

FIGURE 13.3 Network representing the connections in Figure 13.2. Vertices represent computers (capital letters) and edges represent cables connecting computers (lowercase letters).

Now try Exercise 17. ◄

➤ **EXAMPLE 2** Geographic Network

Figure 13.4 shows a map of southwestern Europe. Draw a network in which vertices represent countries and edges represent the relationship "share a common border." What could such a network be used for?

Solution The network shown in Figure 13.5 makes it easier to see which countries share borders. For example, Luxembourg's borders with Belgium, Germany, and France are clear. One possible use of this network would be to plan trade routes that minimize border crossings.

FIGURE 13.4 Map of southwestern Europe.

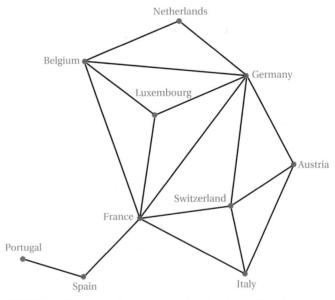

FIGURE 13.5 Network representing borders in southwestern Europe. Vertices represent countries and edges represent the relationship "share a common border."

Now try Exercises 18–19. ◄

Euler Circuits

An **Euler circuit** is a path through a network that starts and ends at the same point and traverses every edge exactly once. Thus, the bridges of Königsberg problem comes down to the question of whether its network has an Euler circuit.

Before we tackle the Königsberg bridge network, let's look for Euler circuits in some simpler networks. It's easy to find an Euler circuit in Figure 13.6a: Start from any vertex, and go around the edges in either a clockwise or a counterclockwise direction. In doing so, we start and end at the same point and traverse every edge exactly once. We can also find an Euler circuit for Figure 13.6b: For example, start at vertex E and follow the edges in the order $a \to b \to c \to d \to e \to f \to g$. This path takes us back to vertex E, and we traverse every edge exactly once.

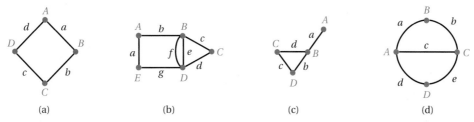

| (a) | (b) | (c) | (d) |

FIGURE 13.6 Networks (a) and (b) have Euler circuits, but networks (c) and (d) do not.

Time out to think _____

Draw an Euler circuit on each of Figures 13.6a and 13.6b by putting arrows on the appropriate edges. Can you find an Euler circuit for Figure 13.6b that traverses the edges in an order besides $a \to b \to c \to d \to e \to f \to g$?

Now consider the network in Figure 13.6c. We cannot find an Euler circuit through this network because, no matter where we start, we have to go *out and back* along edge *a* in order to traverse it at all. Thus, there is no way to traverse all the edges *once* while starting and ending at the same point. A similar problem prevents us from finding an Euler circuit for Figure 13.6d: We can draw an Euler circuit around the lower or upper *half* of the network, but no matter where we start, we cannot get back to the starting point without either missing an edge or traversing an edge twice.

Is there a general rule that tells us when an Euler circuit exists? Note that all the vertices in Figures 13.6a and 13.6b—the networks that have Euler circuits—have either two or four edges connected to them. For example, vertex B in Figure 13.6b is connected to the four edges b, c, e, and f. In contrast, vertex A in Figure 13.6c has only one edge connected to it, and vertices A and C in Figure 13.6d have three edges connected to them. In fact, the key to whether a network has an Euler circuit lies in whether its vertices have even or odd numbers of edges: It has an Euler circuit if all the vertices have an even number of edges.

> **Euler Circuits**
>
> An **Euler circuit** is a path that starts and ends at the same vertex and traverses *every* edge exactly *once*. An Euler circuit exists for a network *if* each vertex has an even number of edges.

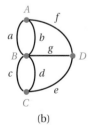
►EXAMPLE 3 The Optimal Mail Route

An ingenious mail carrier wants to improve her efficiency. Her assignment is to deliver mail along the shaded sidewalks of the city blocks shown in Figure 13.7. Can she park her truck in one place and do the job without walking any of the sidewalks more than once? If so, how?

■ = delivery route
□ = not on delivery route

FIGURE 13.7 A mail carrier's route. Capital letters represent street intersections and lowercase letters represent sidewalks on the route.

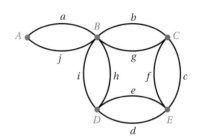

FIGURE 13.8 Network representation for Figure 13.7. Vertices are street intersections and edges are sidewalks on the route.

Solution The mail carrier's question essentially asks whether there is an Euler circuit for her route. We begin by transforming the map into a network. We represent each intersection as a vertex, and each shaded sidewalk along which she must deliver mail as an edge (Figure 13.8). Note that we show a distinct edge for each side of each street since she must walk each side separately to deliver the mail.

Every vertex in this network has an even number of edges. Therefore, an Euler circuit exists and her desired route is possible. One of many possible Euler circuits is the path $a \to b \to c \to d \to e \to f \to g \to h \to i \to j$. That is, by parking her truck at intersection A and delivering mail along the Euler circuit, she can complete her route without retracing her steps. *Now try Exercises 20–21.* ◄

Answering the Bridges of Königsberg Problem

FIGURE 13.1b
(repeated)

Is it possible to cross each of the seven bridges of Königsberg exactly once while starting and ending in the same place? Take another look at the bridge network (Figure 13.1b) shown again to the left. Vertices A, C, and D each have three edges, and vertex B has five edges. Every vertex fails to meet the requirement of having an even number of edges. Therefore, this network has no Euler circuit, so the people of Königsberg were searching in vain for a path that crossed each bridge once and returned to its starting point. This may seem an anticlimactic end to the bridges of Königsberg story, but Euler's work on the problem started the broad field of mathematical research called *network theory* (or *graph theory*).

Finding Euler Circuits

If a network has an Euler circuit, how can we find it? One approach is by trial and error. But there is a more systematic method, or *algorithm,* that we call the **burning bridges rule.**

The Burning Bridges Rule for Finding Euler Circuits

You may begin your circuit from any vertex in the network. However, as you choose edges to follow, never use an edge that is the *only* connection to a part of the network that you have not already visited.

In other words, never use an edge that is the last "bridge" to territory that hasn't been visited.

Time out to think

In finding or traversing an Euler circuit, does it matter at which vertex you start? Explain.

► **EXAMPLE 4** Applying the Burning Bridges Rule

Find an Euler circuit for the network shown in Figure 13.9.

FIGURE 13.9

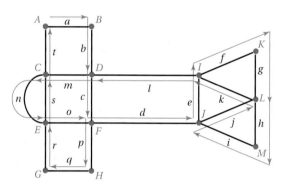

FIGURE 13.10 The path following the arrows, which begins and ends at vertex *A*, is an Euler circuit for the network in Figure 13.9.

Solution First, we should confirm that the network has an Euler circuit. All the vertices have either two or four edges. Because these numbers are even, the network has an Euler circuit. Now we find the Euler circuit by applying the burning bridges rule. Starting arbitrarily at vertex *A*, we begin tracing the edges in the order $a \to b \to c \to d \to e$. At most vertices we have several choices for a path, but so far the choices have been unimportant. Having arrived at vertex *I*, we have a choice of turning left and using edge *l* or going right along edge *f* or edge *k*. Using edge *l* violates the burning bridges rule because it is the only remaining link to edges *f* through *k*— edges that have not yet been covered. Therefore, we must proceed along either *f* or *k*. Once we choose *f* or *k*, we can proceed in several different ways, one of which is shown in Figure 13.10. *Now try Exercises 22–23.* ◄

NETWORK TERMINOLOGY

Before we consider other network applications, a bit of terminology will be useful. We have already discussed Euler circuits. More generally, a **circuit** is any path *within* a network that begins and ends at the same vertex without using any edge more than once. The network in Figure 13.11 contains many distinct circuits, such as the path connecting the vertices $A \rightarrow B \rightarrow C \rightarrow A$ or the path connecting the vertices $A \rightarrow G \rightarrow F \rightarrow D \rightarrow B \rightarrow A$. Note that neither of these circuits is an *Euler circuit* because it does not traverse *every* edge once.

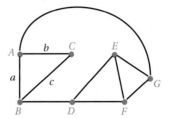

FIGURE 13.11 Among the many circuits in this network are the paths connecting vertices *A–B–C–A* and *D–E–G–F–D*.

Time out to think _____

Find at least four other circuits in Figure 13.11. Does this network have an Euler circuit? Why or why not?

The two networks in Figure 13.12 are said to be **complete** because each vertex is connected to *every* other vertex. A complete network always includes many circuits. For example, a path around the outer edges forms a circuit.

A **tree** is a network in which all the vertices are connected but in which there are no circuits (Figure 13.13). Among their many uses, trees can be used to represent genealogies and family trees.

(a) (b)

FIGURE 13.12 Two complete networks, in which each vertex is connected to every other vertex: (a) has order 4, because it has 4 vertices; (b) has order 5, because it has 5 vertices.

FIGURE 13.13 This network is a tree, because all its vertices are connected but there are no circuits within it.

The **order** of any network is its number of vertices. For example, the two complete networks in Figure 13.12 have order 4 and order 5, respectively; the tree network in Figure 13.13 has order 9. Each vertex in a network may be described further by its **degree:** the number of edges connected to it. For example, all the vertices in the network of Figure 13.12b have degree 4.

Network Terminology

A **circuit** is a path within a network that begins and ends at the same vertex without using any edges more than once.

In a **complete network,** every vertex is connected to every other vertex.

A **tree** is a network in which all of the vertices are connected and no circuits appear.

The **order** of a network is its number of vertices.

The **degree** of a vertex within a network is the number of edges connected to it.

> **EXAMPLE 5** Network Classification

Describe each network shown in Figure 13.14. Identify the order of the network and degree of each of its vertices.

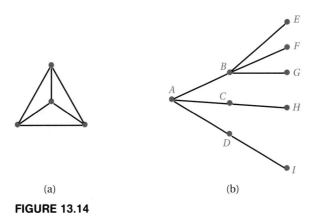

(a) (b)

FIGURE 13.14

Solution

 a. This network is *complete* because each vertex is connected to all of the other vertices. It has *order* 4 because it has 4 vertices. Each vertex has *degree* 3 because it is connected to the other 3 vertices.

 b. This network is a *tree* because all the vertices are connected but it contains no circuits. It has order 9 because it has 9 vertices. Vertex A has degree 3 because it is connected to 3 other vertices. Vertex B has degree 4, vertices C and D have degree 2, and the remaining vertices have degree 1.

<div align="right">Now try Exercises 24–29. ◄</div>

MINIMUM COST SPANNING NETWORKS

One practical application of networks is in minimizing the cost of connecting places or objects. For example, imagine that the local telephone company needs to connect the seven towns of Sunshine County, shown in Figure 13.15a, with new high-speed telephone lines. Figure 13.15b shows a network representation of Sunshine County with distances between towns indicated on the edges. Note that no edges are shown between towns where lakes or mountains would make it difficult to install wires.

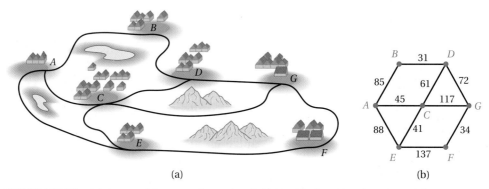

(a) (b)

FIGURE 13.15 (a) A map of the seven towns (capital letters) in Sunshine County and the routes between them along which telephone lines could be strung. (b) Network representation of Sunshine County; vertices represent the towns (*A* through *G*) and edges represent potential routes for telephone lines. The numbers on each edge, generally called costs or weights, are the distances in miles along each route.

Let's assume that the cost of putting up new lines is proportional to the distance. In that case, the numbers on the edges also represent the cost of connecting towns. In general, numbers on edges are called **costs** or **weights,** even when they don't have anything to do with money.

The telephone company's objective is to find a set of edges in the original network that

- either directly or indirectly links every town to every other town

and

- has the minimum possible cost (or length).

A set of edges that meets the first criterion, linking all the towns, is called a **spanning network.** Two of the many possible spanning networks for Sunshine County are highlighted in Figure 13.16. In both networks, it is possible to get from any vertex to any other vertex using only highlighted edges. The total cost of a spanning network is the sum of the individual costs on its edges. The total cost of the spanning network in Figure 13.16b is substantially lower than the total cost of the one in Figure 13.16a. In

fact, if you tried every possible spanning network, you'd find that the one shown in Figure 13.16b has the lowest cost of all of them. It is called the **minimum cost spanning network** for this problem.

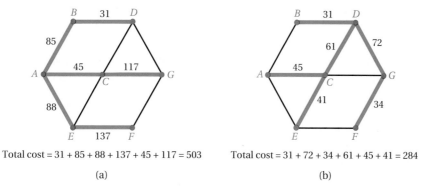

Total cost = 31 + 85 + 88 + 137 + 45 + 117 = 503 Total cost = 31 + 72 + 34 + 61 + 45 + 41 = 284

(a) (b)

FIGURE 13.16 Two spanning networks—networks that link all the vertices—for Sunshine County are shown in blue. The total cost of each spanning network is the sum of the individual costs on its edges. The total cost for spanning network (a) is 503, which is much higher than the total cost of 284 for spanning network (b).

➤ **EXAMPLE 6** Spanning Network Costs

Find another spanning network for Sunshine County and calculate its total cost. Compare this cost to the total cost of the spanning networks shown in Figure 13.16.

Solution The network shown in Figure 13.17 is another spanning network because it connects all the towns. We find its total cost of 416 by adding the costs of its edges. This spanning network has a lower total cost than the one in Figure 13.16a, but the cost is still substantially higher than that of the minimum cost spanning network in Figure 13.16b.

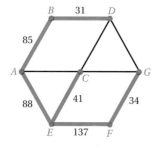

Total cost = 85 + 31 + 88 + 41 + 137 + 34 = 416

FIGURE 13.17 Another spanning network for Sunshine County. *Now try Exercises 30–32.* ◄

Finding Minimum Cost Networks

Fortunately, there are ways to find the minimum cost spanning network without trying every possibility. One simple procedure is called **Kruskal's algorithm.**

Kruskal's Algorithm for Finding Minimum Cost Networks

Step 1: Make a list of the edges from the least expensive to the most expensive.

Step 2: Begin with the least expensive edge. Highlight it to indicate that it is part of the minimum cost spanning network. Continue to select edges in order of increasing cost until every vertex is connected, either directly or indirectly, to every other vertex.

Step 3: If a closed circuit has been created within the spanning network, remove the most expensive edge. The final result is the minimum cost spanning network.

➤ **EXAMPLE 7** Minimum Cost Power Lines

Figure 13.18a shows a network of towns (vertices) and possible power-line connections (edges). The cost of building a power line (in millions of dollars) along each edge also is shown. How should the power lines be built to connect the towns at minimum cost? What is the total cost?

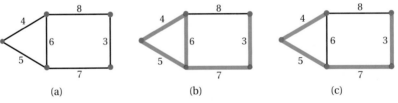

(a) (b) (c)

FIGURE 13.18 (a) A network showing costs (on edges) for possible power-line connections among a set of towns (vertices). (b) The results (bold edges) of Step 2 of Kruskal's algorithm. (c) The final result of Kruskal's algorithm is the minimum cost spanning network.

Solution We follow the three-step process.

Step 1: In order of increasing cost, the edges have costs of 3, 4, 5, 6, 7, and 8.
Step 2: We start by highlighting the edge with cost 3. Then we continue to highlight the edges with costs of 4, 5, 6, and 7, as shown in Figure 13.18b. All the vertices are now connected.
Step 3: The procedure created a closed circuit with the edges with costs 4, 5, and 6. We remove the edge with cost 6 because it is the most expensive edge in the circuit. The final result is shown in Figure 13.18c. This is the minimum cost spanning network. Its total cost is $4 + 5 + 7 + 3 = 19$. Because costs in this problem represent millions of dollars, the cost of building this power-line network will be $19 million.

Now try Exercises 33–34. ◄

EXERCISES 13A

Review Questions

1. What is a network? What is network analysis?

2. Briefly describe the problem of the bridges of Königsberg. What was its eventual resolution found by Euler?

3. Explain how a network can be used to represent the Königsberg bridges problem. What do the edges and vertices represent? Give some examples of what edges and vertices can represent in other networks.

4. What is an Euler circuit? How can we tell whether one exists for a network?

5. Describe the burning bridges rule for finding Euler circuits, and give an example of its application.

6. Define and draw an example of a complete network of order 4 and a tree of order 7.

7. What is a circuit? What makes an Euler circuit special?

8. How do we define the degree of a vertex? What must the degree of the vertices be in order for a network to have an Euler circuit?

9. What do we mean by a spanning network? How do we find the cost of a spanning network? What is a minimum cost spanning network?

10. Describe Kruskal's algorithm and give an example of its use.

Does It Make Sense?

Decide whether each of the following statements makes sense (or is clearly true) or does not make sense (or is clearly false). Explain your reasoning.

11. Garret plans to use a network to analyze trade relations among the countries of the Western Hemisphere.

12. Erin plans to use network analysis to represent all the games scheduled in a season of the local volleyball league.

13. A salesman who wants to visit each town in his district exactly once needs to find an Euler circuit.

14. Jeff plans to use a tree network to represent the games in a single-elimination softball tournament.

15. Aubrey needs to find the most efficient path to walk through a neighborhood to distribute campaign literature. She plans to use a minimum cost spanning tree.

16. Wes created a network whose order is less than the degree of every vertex.

Basic Skills & Concepts

17. **City Streets.** Consider the map of streets and intersections in Figure 13.19.

FIGURE 13.19

a. Let edges represent the streets and let vertices represent the intersections, and draw the network that results.

b. Use the same city map, but assume that each street has a sidewalk on both sides of it. Draw the network that results when edges are used to represent sidewalks instead of streets.

18. **Layout of an Art Gallery.** The floor plan of a small art gallery is shown in Figure 13.20. Draw a network in which vertices represent rooms and edges represent doors.

FIGURE 13.20

19. **Friendships and International Trade.** Among the members of a fourth-grade class, Amy trades baseball cards with Beth, Cate, and Daniel; Beth trades with Daniel and Cate. Draw a network that represents these trading relations. Edges in the network represent the relationship "trades with." Based on this example, explain how networks might also be used to describe international trade relations.

20. **Euler Circuits.** The town of Sleepy Waters is entwined around the Tranquility River with eight bridges and three islands, as shown in Figure 13.21.

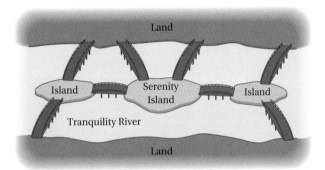

FIGURE 13.21

 a. How many vertices are needed to draw a network representing the town? How many edges are needed?

 b. Draw a network that represents the town.

 c. Is it possible to begin on Serenity Island and walk in a closed loop that crosses each bridge exactly once and returns to the starting point? Explain.

21. **Euler Circuits.** Baytown is located on the coast with three islands nearby. Ferry boats can be taken between the town and the islands along the routes shown in Figure 13.22.

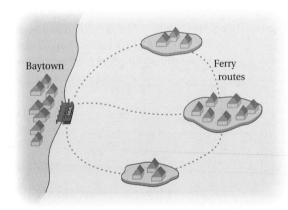

FIGURE 13.22

 a. How many vertices are needed to draw Baytown and the islands? How many edges are needed to draw the ferry routes? Explain.

 b. Draw a network that represents the town, islands, and ferry routes.

 c. Is it possible to begin at Baytown and make a closed loop that uses each ferry boat exactly once and returns to Baytown? Explain.

22. **Reading Meters.** A gas meter reader must visit all the houses on the streets of the map in Figure 13.23.

FIGURE 13.23

 a. Draw a network for this map in which each vertex represents an intersection and each edge represents a street that the meter reader must walk.

 b. Does this network have an Euler circuit? If so, draw it. If not, why not?

 c. Explain why an Euler circuit would be useful to the meter reader.

23. **Checking Parking Meters.** An attendant must make hourly inspections of parking meters along the sidewalks (shaded strips) on the map in Figure 13.24. Convert the

☐ = inspection route
☐ = not on inspection route

FIGURE 13.24

map to a network in which edges represent sidewalks and vertices represent intersections. Can you find a circuit beginning and ending at City Hall that covers each sidewalk exactly once? If so, show it on your network.

Network Terminology. For the networks shown in Exercises 24–29, answer the following questions.

a. What is the order of the network?

b. What is the degree of each vertex of the network?

c. Does the network have a special form (such as a tree or a complete network)? If so, what?

d. Does the network have an Euler circuit? Explain.

e. In those cases in which an Euler circuit does exist, apply the burning bridges rule to find one. In each case, explain the steps of the method and note the vertices at which critical decisions must be made.

24.

25.

26.

27.

28.

29.

30. **Spanning Networks.** Consider the network in Figure 13.25 and the three spanning networks labeled I, II, and III. Find the cost of each of the three spanning networks. Which of the three spanning networks has the minimum cost?

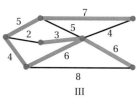

FIGURE 13.25

Kruskal's Algorithm. Use Kruskal's algorithm to find the minimum cost spanning network for the networks shown in Exercises 31–32.

31.
32.
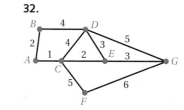

33. **Community Planning.** A mountain community is being planned and provisions must be made for water. The map in Figure 13.26 shows the building sites (the vertices of the network) and the feasible routes for water pipelines (the edges of the network). The weights on the vertices give the distances between sites in kilometers. A single well can supply all the houses and be drilled on any of the housing sites. Assume that the cost of supplying water to all the homes depends on the total length of pipe used. Find the minimum cost spanning network that supplies water to all the houses.

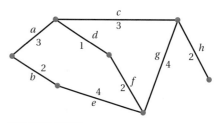

FIGURE 13.26

34. **Local Area Networks.** Consider the computer network shown in Figure 13.27, in which vertices represent workstations and edges represent feasible connections. The

weight on each edge gives the cost of the wire needed for that link. Find the minimum cost spanning network that connects all the computers.

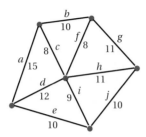

FIGURE 13.27

Further Applications

35. Drawing Networks.

a. Draw a network of order 6 whose vertices all have an odd degree.

b. Draw a tree of order 7 in which at least one vertex has an even degree.

c. Draw a network of order 5 with no vertices having an odd degree.

36. Drawing Networks.

a. Draw a network of order 5 whose vertices all have an even degree.

b. Draw a network of order 6 in which all vertices have degree 2.

c. Draw a network of order 8 in which all vertices have degree 3.

37. Complete Network. Draw a complete network of order 5. How many edges does it have? Show that the number of edges in the network satisfies the following rule: A complete network with n vertices has $n \times (n - 1)/2$ edges.

38. Complete Network. Draw a complete network of order 6. How many edges does it have? Show that the number of edges in the network satisfies the following rule: A complete network with n vertices has $n \times (n - 1)/2$ edges.

39. Neighboring States. Find a map of Tennessee and its immediate neighbors. Letting vertices represent states and edges represent the relationship "is a neighbor of," draw a network showing all connections among Tennessee and its neighbors. What is the order of the network? What is the degree of the vertex corresponding to Tennessee?

40. Neighboring States. Find a map of Missouri and its immediate neighbors. Letting vertices represent states and edges represent the relationship "is a neighbor of," draw a network showing all connections among Missouri and its neighbors. What is the order of the network? What is the degree of the vertex corresponding to Missouri?

41. Family Trees. A family tree is a network in which each vertex is a person or couple and each edge represents a parent-child or couple relationship.

a. Sebastian and Emily are the parents of three children: Paul, Victor, and Alice. Draw a family tree showing only these five people.

b. Victor is married to Melinda, and they have two children: Debra and Katy. Add Melinda and the two children to the family tree from part a.

c. Make a family tree for your own family that includes all your siblings, your parents and their siblings, and your grandparents.

42. Soccer Tournament. In the three weeks that remain in the season, the five teams of the Skyline Soccer League (A, B, C, D, and E) must play the games indicated in the following table. (An X means the two teams play and a blank means they do not play.) Draw a network that shows the teams as vertices and relationships "plays against" as edges. What is the order of the network you have drawn? What is the degree of each vertex of the network? Is the network complete? Explain.

	A	B	C	D	E
A		x	x		x
B	x			x	
C	x			x	x
D		x	x		x
E	x		x	x	

43. Soccer Schedule. As a wise coach who knows some network theory, you have drawn the network in Figure 13.28 to show the games remaining in the final four weeks of the soccer season. The vertices represent teams and the

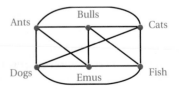

FIGURE 13.28

edges represent the relationship "plays against." Convert the network to a table, similar to the table in Exercise 42, that shows all the games to be played.

44. **Traversable Networks.** For any network, an Euler circuit must traverse all the edges *and* return to the starting vertex. Sometimes a path traverses all the edges but doesn't return to the starting vertex. A network with this property is said to be *traversable.* Here is the rule that determines whether a network is traversable:

If a network has two vertices with an odd number of edges and all other vertices have an even number of edges, it is traversable but does not have an Euler circuit.

a. Explain why a traversable network can have two vertices with an odd degree.

b. Consider the networks shown in Exercises 24–29. For each one, determine whether it is traversable. If it is, show a traversing path.

Web Projects

Find useful links for Web Projects on the text Web site: www.aw.com/bennett-briggs

45. **Personal Tour.** Find a map on the Web of a place you would like to visit, such as a park, zoo, or museum. Represent the map as a network and plan a personal tour that ensures you see all the things you are interested in seeing.

46. **Network Theory.** Network theory (also called graph theory) has an endless supply of problems, many of them still unsolved. There are Web sites that maintain collections of network theory problems. Visit such a site and choose a specific problem. Describe the problem and, if possible, its history, applications, and methods of solution.

In the News

47. **Networks in the News.** Find a news article that refers to a network or involves a topic that could be described in terms of a network. Describe the network, and identify the components that correspond to vertices and edges.

48. **Networks in Your Life.** Find a map that shows possible routes of importance to you, such as campus footpaths you could take between classes or roads you could take on a five-store shopping trip. How could a network be used to represent the possibilities and to help you choose a route? Explain.

UNIT 13B

THE TRAVELING SALESMAN PROBLEM

Imagine that you work as a traveling salesperson. Your job requires you to make presentations about your products to companies located in five cities. How do you find a route that allows you to visit all the cities once and return home with the minimum amount of travel?

This is an example of a famous problem in mathematics called the **traveling salesman problem.** We can represent this problem as a network by letting vertices represent the cities. Edges, labeled with distances, represent possible routes between pairs of cities (Figure 13.29). The solution to the traveling salesman problem is the shortest path that starts and ends in the same place and visits each city exactly once. That is, it is the shortest *circuit* (closed path) through a network that visits all the *vertices* exactly once. In this unit, we will investigate the traveling salesman problem. As you might expect, we will find that it has many important applications that go far beyond traveling salesmen.

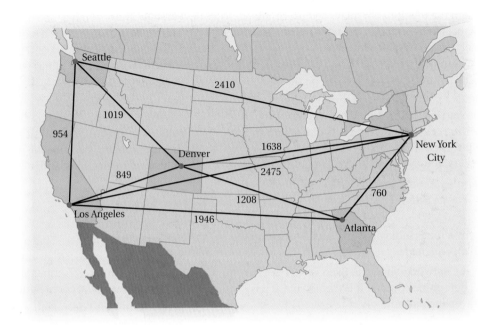

FIGURE 13.29 A map of airline routes among five cities. Each route is labeled with its distance in miles.

HAMILTONIAN CIRCUITS

A path that passes through every *vertex* of a network exactly once and returns to the starting vertex is called a **Hamiltonian circuit,** after the 19th-century Irish mathematician William Rowan Hamilton (1805–1865). The networks in Figures 13.30a and 13.30b have Hamiltonian circuits on the paths indicated with arrows. Note that, while Hamiltonian circuits pass through every *vertex,* they do not necessarily traverse every *edge.* Many networks do not have Hamiltonian circuits. For example, the network in Figure 13.30c does not have a Hamiltonian circuit because it is impossible to find a circuit that passes through all vertices without passing through at least one vertex twice.

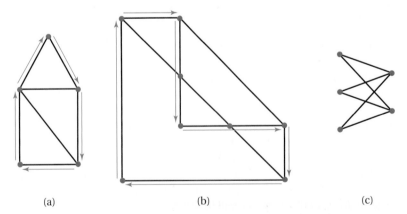

(a) (b) (c)

FIGURE 13.30 A Hamiltonian circuit is a path that passes through every vertex once and returns to the starting vertex. The paths indicated by arrows in (a) and (b) are Hamiltonian circuits. Network (c) has no Hamiltonian circuits.

> **Hamiltonian Circuits**
>
> A **Hamiltonian circuit** is a closed path through a network that passes through every vertex exactly once and returns to the starting vertex.

Time out to think _____

Is it possible to find other Hamiltonian circuits in the networks of Figures 13.30a and 13.30b (other than those created by reversing the circuits shown)?

➤ **EXAMPLE 1** Hamiltonian Circuit for Sunshine County

Find at least two Hamiltonian circuits for the Sunshine County network shown in Figure 13.15b. Also find one circuit that visits all the vertices but is *not* a Hamiltonian circuit. (The costs on the edges do not affect whether a circuit is Hamiltonian.)

Solution Figures 13.31a and 13.31b show two Hamiltonian circuits for the Sunshine County network. The circuits can start at any vertex, but they must end at that same vertex and must visit every other vertex exactly once along the way. The circuit in Figure 13.31c also visits all the vertices. However, it is *not* a Hamiltonian circuit because it visits vertex C more than once.

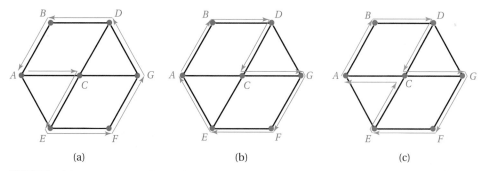

| (a) | (b) | (c) |

FIGURE 13.31 This figure shows the Sunshine County network of Figure 13.15 (p. 710). The paths in (a) and (b) are Hamiltonian circuits; the path in (c) is not because it visits vertex C more than once. *Now try Exercises 9–12.* ◀

Finding Hamiltonian Circuits

In Unit 13A, we found a test for whether a network has an Euler circuit and a method for finding an Euler circuit. Unfortunately, finding Hamiltonian circuits is much more difficult. In fact, there is no known rule for *efficiently* determining whether a Hamiltonian circuit exists in a network, except in a few special cases. For example, a single-loop network always has a Hamiltonian circuit (Figure 13.32a), as does a complete network (Figure 13.32b). A tree never has a Hamiltonian circuit because it contains no circuits at all (Figure 13.32c).

Aside from such special cases, trial and error is the only way to find Hamiltonian circuits. Moreover, some networks have many Hamiltonian circuits. For example, a complete network of order 4 (that is, 4 vertices) has three different Hamiltonian circuits (Figure 13.33).

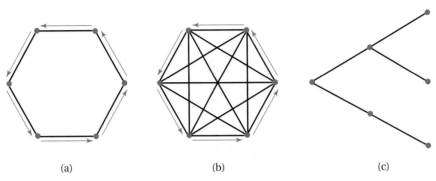

FIGURE 13.32 (a) A single-loop network always has a Hamiltonian circuit that follows the loop. (b) A complete network—in this case, of order 6—always has Hamiltonian circuits. The one shown here simply follows the outer edges. (c) A tree never has a Hamiltonian circuit.

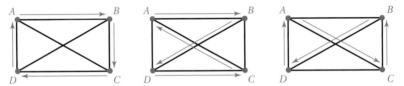

FIGURE 13.33 The three Hamiltonian circuits for a complete network of order 4. By testing complete networks of other orders, we find that, in general, a complete network of order n has $(n-1)!/2$ Hamiltonian circuits.

You might imagine that complete networks of higher order would have more Hamiltonian circuits—and they do. In fact, the following simple formula tells us the number of Hamiltonian circuits in a complete network.

Hamiltonian Circuits in Complete Networks

The number of Hamiltonian circuits in a complete network of order n is

$$\frac{(n-1)!}{2}$$

(Recall that $(n-1)! = (n-1) \times (n-2) \times \cdots \times 2 \times 1$.)

You can verify this formula by finding all the Hamiltonian circuits in complete networks of various orders. The formula is also easy to derive if you think about the number of possible ways of making a trip among all the vertices in a network. If you start at any one vertex in a network of n vertices, you have $n-1$ choices for your first stop, $n-2$ choices for your second stop, and so on. Thus, counting the total number of ways you can make a circuit among the n vertices is a permutations problem (see Unit 7E). Because you begin with $n-1$ choices for your first stop, the total number

of permutations is $(n - 1)!$. Finally, you divide this number by 2 because the counting includes both forward and backward trips on the same route, but direction doesn't matter for Hamiltonian circuits.

Time out to think _____

Use the above formula to confirm that a complete network of order $n = 4$ has three Hamiltonian circuits, as we found in Figure 13.33.

➤ **E X A M P L E 2** Complete Network of Order 5

How many Hamiltonian circuits are there in a complete network of order 5? Draw such a network, and identify all of its Hamiltonian circuits.

Solution Figure 13.34 shows the complete network of order $n = 5$, which has

$$\frac{(5 - 1)!}{2} = \frac{4!}{2} = \frac{4 \times 3 \times 2 \times 1}{2} = 12$$

Hamiltonian circuits. A complete network of order 5 has every vertex connected to every other vertex.

FIGURE 13.34 The twelve Hamiltonian circuits for a complete network of order 5.

Now try Exercises 13–14. ◄

SOLVING TRAVELING SALESMAN PROBLEMS

In the traveling salesman problem, each edge of a circuit is labeled with a length (or distance). The solution to a traveling salesman problem is the shortest path (smallest total of the lengths) that starts and ends in the same place and visits each city once. Thus, the solution is the *shortest* Hamiltonian circuit in a network.

Suppose you're planning a summer vacation during which will you visit five national parks. Let's make the problem more exotic by assuming that you will rent a private airplane at one park and travel between parks by air. If you have to pay for the plane according to the number of miles flown, what is the lowest cost path? Although sales are not involved in this problem, it is mathematically equivalent to the traveling salesman problem: You want to begin and end at the same place, and you seek the shortest path that visits all five parks.

Table 13.1 lists the five national parks and the distances between them. We can represent the data with a network in which vertices represent the national parks and edges represent the air routes between them (Figure 13.35). The edges are labeled with distances in miles, but the network does not need to be drawn to scale.

TABLE 13.1	**Air Miles Between National Parks**				
	BRYCE	CANYONLANDS	CAPITOL REEF	GRAND CANYON	ZION
BRYCE	—	136	69	108	52
CANYONLANDS	136	—	75	202	188
CAPITOL REEF	69	75	—	151	123
GRAND CANYON	108	202	151	—	101
ZION	52	188	123	101	—

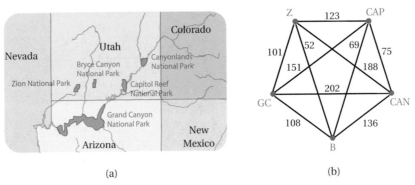

(a) (b)

FIGURE 13.35 (a) Map of five national parks in the western United States. (b) Complete network representing the parks (vertices) and all air routes (edges) between them. The numbers on the edges are distances in miles. (B = Bryce Canyon, GC = Grand Canyon, CAP = Capitol Reef, Z = Zion, and CAN = Canyonlands.)

➤ EXAMPLE 3 Hamiltonian Circuits in the National Park Network

How many Hamiltonian circuits are there in the national park network shown in Figure 13.35? Two possible circuits are shown in Figure 13.36. Describe each in words, and find its total length.

Solution The network in Figure 13.35 is a complete network of order 5, because it is possible to fly from any national park to any of the other four. Hence, it has 12 Hamiltonian circuits (see Example 2). If we start from Bryce, the circuit in Figure 13.36a shows a trip taken in the order Canyonlands, Grand Canyon, Capitol Reef, Zion, and back to Bryce (or the reverse order). Its total length is

$$136 \text{ mi} + 202 \text{ mi} + 151 \text{ mi} + 123 \text{ mi} + 52 \text{ mi} = 664 \text{ mi}$$

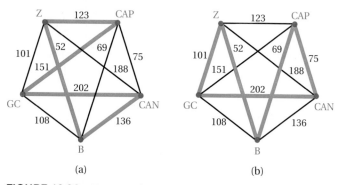

FIGURE 13.36 Two Hamiltonian circuits for the national park network of Figure 13.35. The circuit in (a) has a total cost (length) of 664 miles, while (b) has a total cost (length) of 499 miles.

The circuit in Figure 13.36b goes from Bryce to Capitol Reef, Canyonlands, Grand Canyon, Zion, and back to Bryce (or the reverse order). Its total length is

$$69 \text{ mi} + 75 \text{ mi} + 202 \text{ mi} + 101 \text{ mi} + 52 \text{ mi} = 499 \text{ mi}$$

Now try Exercise 15. ◀

An Explosion of Possibilities

Example 3 above shows that the circuit in Figure 13.36b is substantially shorter than the circuit in Figure 13.36a. But is it the *shortest* circuit for the national park problem? The only way to know for sure is to calculate the length of *all* 12 Hamiltonian circuits for this network. If you do so, you'll find that the circuit in Figure 13.36b is indeed the shortest one. Thus, it is the solution to this particular version of the traveling salesman problem because it is the lowest cost trip among the five national parks.

Unfortunately, finding and checking all the Hamiltonian circuits for a problem is worse than just tedious. The number of Hamiltonian circuits in many real problems is so great that even the fastest imaginable computers could never check all of them.

▶**EXAMPLE 4** Airline Planning for 15 Cities

The CEO of your company wants to take the company jet on a 15-city tour, beginning and ending at the company headquarters in Chicago (one of the 15 cities; see Figure 13.37). How many possible routes must be checked to find the shortest route among the 15 cities? Suppose you have a computer that can find and measure 1 million Hamiltonian circuits per second. How much computer time will it take to find the shortest route?

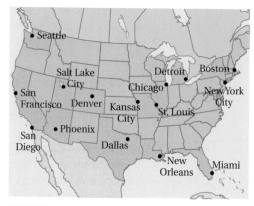

FIGURE 13.37

Solution In principle, it is possible to fly between any pair of cities. Thus, a network representing this problem is complete, because every vertex is connected to every other. For a complete network of order $n = 15$, the number of Hamiltonian circuits is

$$\frac{(n-1)!}{2} = \frac{(15-1)!}{2} = \frac{14!}{2} \approx 4.4 \times 10^{10} = 44 \text{ billion}$$

If your computer can find and measure 1 million circuits each second, checking 44 billion circuits will take

$$\frac{4.4 \times 10^{10} \text{ circuits}}{10^6 \text{ circuits}/\text{s}} = 44{,}000 \text{ s}$$

Converting to hours, we have

$$44{,}000 \text{ s} \times \frac{1 \text{ min}}{60 \text{ s}} \times \frac{1 \text{ hr}}{60 \text{ min}} \approx 12 \text{ hr}$$

That is, it will take about half a day of computer time to check all the Hamiltonian circuits and find the shortest route. *Now try Exercises 16–17.* ◀

▶ **EXAMPLE 5** Airline Planning for 30 Cities

Just as you finish planning the 15-city tour in Example 4, the CEO informs you that she actually needs to visit 30 cities. Is it possible to find the shortest route by checking all the possibilities? Explain.

Solution In this case, the problem involves a complete network of order $n = 30$. Thus, it has

$$\frac{(n-1)!}{2} = \frac{(30-1)!}{2} = \frac{29!}{2} \approx 4.4 \times 10^{30}$$

possible Hamiltonian circuits. Checking 1 million circuits per second, it would take

$$\frac{4.4 \times 10^{30} \text{ circuits}}{10^6 \text{ circuits}/\text{s}} = 4.4 \times 10^{24} \text{ s}$$

or

$$4.4 \times 10^{24} \text{ s} \times \frac{1 \text{ min}}{60 \text{ s}} \times \frac{1 \text{ hr}}{60 \text{ min}} \times \frac{1 \text{ day}}{24 \text{ hr}} \times \frac{1 \text{ yr}}{365 \text{ days}} \approx 1.4 \times 10^{17} \text{ yr}$$

to check every circuit and find the shortest route. Note that 10^{17} years is some 10 million (10^7) times the current age of the universe (about 10 billion, or 10^{10}, years). Clearly, this is not a viable way to plan the CEO's trip. *Now try Exercises 18–19.* ◀

Nearly Optimal Solutions to Traveling Salesman Problems

No known method besides trial and error is guaranteed to find the shortest Hamiltonian circuit in an arbitrary network. Unfortunately, as Example 5 shows, checking all the possible Hamiltonian circuits is impossible in many real problems. However, there are some methods that usually (but not always) find **near-optimal solutions**—that is, solutions with a total length *close* to that of the shortest circuit. One such method that can be used for complete networks is called the **nearest neighbor method**.

The Nearest Neighbor Method

Beginning at any vertex, travel to the nearest vertex that has not yet been visited. Continue this process of visiting "nearest neighbors" until the circuit is complete.

The nearest neighbor method and refinements of it can find near-optimal solutions to gigantic traveling salesman problems, and do it in relatively little time. One of the largest problems solved to date involves a network of the 13,509 American cities with populations of over 500 people. Figure 13.38 shows the near-optimal solution found by one group of scientists. The pathway represents an efficient route that visits each of the 13,509 cities once and returns to the starting point.

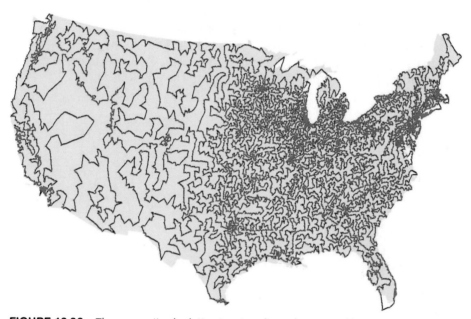

FIGURE 13.38 The near-optimal solution to a traveling salesman problem in which we seek the shortest path among 13,509 cities with populations over 500. Note that such cities are more closely packed in the eastern United States than in the west.

Courtesy of Bill Cook, Georgia Institute of Technology; David Applegate, AT&T Labs–Research; Robert Bixby, Rice University; and Vasek Chvatal, Rutgers University.

> **by the way**
>
> A traveling salesman problem with over 15,000 cities was solved efficiently in 2001. Researchers are now considering a world traveling salesman problem with 1,904,711 cities.

▶ **EXAMPLE 6** Nearest Neighbor Method for the National Park Network

Apply the nearest neighbor method to the national park network (Figure 13.35), beginning at Bryce. Does this solution seem to be "nearly optimal"?

Solution The nearest neighbor to Bryce is Zion, 52 miles away. From Zion, the nearest neighbor not already visited is Grand Canyon, 101 miles away. From Grand Canyon, the nearest unvisited neighbor is Capitol Reef, 151 miles away. Finally, from

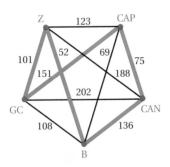

FIGURE 13.39 The solution to the national park network found by applying the nearest neighbor method starting at Bryce. This solution has a total length of 515 miles.

Capitol Reef we have only one place left to visit: Canyonlands, 75 miles away. We then return to Bryce. The circuit is shown in Figure 13.39. Its total length is

$$52 \text{ mi} + 101 \text{ mi} + 151 \text{ mi} + 75 \text{ mi} + 136 \text{ mi} = 515 \text{ mi}$$

This is slightly longer than the 499-mile length of the shortest circuit for this network (Figure 13.26b), but only by 16 miles. This solution is *nearly* optimal, because its cost is only 3% higher than the cost of the optimal solution.

Now try Exercises 20–21. ◄

►CASE STUDY Computer Circuit Board

A computer circuit board has thousands of small holes at precise locations so that computer components can be inserted. The holes are drilled by a special machine, or robot (Figure 13.40a). In principle, the robot could start drilling holes at any location on the board, but it must finish back at its starting point so that it is ready for the next board. In other words, the robot's task is a *traveling salesman problem:* It must start and end in the same place while drilling at the location of every required hole. Finding the shortest possible path is important because the drilling machine is expensive to operate, and a shorter path allows the machine to drill more boards each day.

(a) (b)

FIGURE 13.40 (a) The problem of getting a robot to drill holes efficiently in a computer circuit board is essentially a traveling salesman problem. (b) This path is a near-optimal solution showing how a robot should move among 3038 holes that must be drilled in a computer circuit board.

Courtesy of Bill Cook, Georgia Institute of Technology; David Applegate, AT&T Labs–Research; Robert Bixby, Rice University; and Vasek Chvatal, Rutgers University.

A network representing this problem would have a vertex for each of the thousands of holes that must be drilled. Moreover, the network would be complete because, in principle, the robot can go from any one hole location to any other. Example 5 showed that we could never test all the Hamiltonian circuits in a complete network with just 30 vertices, and a complete network with thousands of vertices is a far more complex problem. Fortunately, near-optimal solutions, such as the one shown in Figure 13.40b for a circuit board with 3038 holes, can be found.

ℙractical matters

OPERATIONS RESEARCH

The traveling salesman problem and other network problems are part of the branch of mathematics called **operations research** (or *management science*). While the subject may not sound familiar to many people, it could be the area of mathematics that has the most immediate impact on our everyday lives, often in hidden ways.

Operations research seeks strategies for carrying out complex tasks in optimal ways. Depending on the problem at hand, *optimal* may mean most profitable, most cost-effective, or most efficient. With this broad scope, operations research finds use in all areas of business, management, planning, and decision making. The increasing globalization of travel and commerce makes operations research ever more important to governments and businesses, as even relatively small improvements in efficiency can have a big impact on costs, revenues, or profits.

Operations research originated in Britain during World War II, as a way to make strategic decisions about the deployment of radar, the use of artillery, and other military tactics. Following World War II, the ideas spread to the United States, where operations research was applied to many non-military problems. Its growth has gone hand in hand with the increasing power of computers, which allow new optimization strategies to be tested and refined.

Today, operations research is having a growing impact on our lives. For example, airlines use its techniques to solve the so-called fleet assignment problem: What is the optimal way to assign aircraft to flight routes to minimize costs and make best use of aircraft? Delta Airlines claims to have saved over $300 million in three years because of these methods. Nearly all routing problems, whether they involve public transportation, commercial shipping, snowplows, or garbage trucks, ultimately involve ideas from operations research. For example, the state of North Carolina used operations research to design a new school bus routing system. The process saved the state $25.2 million in capital costs and $27.9 million in operating costs.

Manufacturers rely on operations research to maximize the output of assembly lines and make the best use of human and material resources—always with an eye on minimizing costs and maximizing profits. As we will see in Unit 13C, any complex planning and scheduling problem, from the construction of highway interchanges to the scheduling of college courses to the launch of a space shuttle, requires the use of operations research if it is to have a chance of success. As you look around, try to see places where operations research is affecting your life.

EXERCISES 13B

Review Questions

1. Give a simple example of the traveling salesman problem in its traditional form. List a few types of problems that are mathematically equivalent to traveling salesman problems but don't involve traveling salesmen.

2. What is a Hamiltonian circuit? How does it differ from an Euler circuit? How many Hamiltonian circuits can be drawn in a complete network?

3. How is the solution to a traveling salesman problem related to Hamiltonian circuits? Why are traveling salesman problems so difficult to solve?

4. What do we mean by a near-optimal solution to a traveling salesman problem? Describe the nearest neighbor method for finding a near-optimal solution. Give an example of its use.

Does It Make Sense?

Decide whether each of the following statements makes sense (or is clearly true) or does not make sense (or is clearly false). Explain your reasoning.

5. Travis needs to solve a traveling salesman problem to plan his visits to his six cousins.

6. Justin wants to visit the houses on every street of his neighborhood without walking along the same street twice. He needs to solve a traveling salesman problem.

7. Tonight, Roger plans to test the Hamiltonian circuits in a complete graph with eight vertices to find the shortest circuit.

8. Barbara isn't a salesperson, so the traveling salesman problem really isn't relevant to her life.

Basic Skills & Concepts

Practice with Hamiltonian Circuits. For Exercises 9–12, consider the networks shown. In each case, follow the edges according to the arrows, and determine whether that path forms a Hamiltonian circuit. Justify your reasoning and explain your answer carefully. If a path does not form a Hamiltonian circuit, can you find a different path that does?

9.

10.

11.

12.

13. **Number of Hamiltonian Circuits.**

 a. How many different Hamiltonian circuits are there in a complete network of order $n = 8$?

 b. How many different Hamiltonian circuits are there in a complete network of order $n = 18$?

 c. Suppose a computer could check Hamiltonian circuits at a rate of one per second. How long would it take to check all the circuits in each of the networks in parts a and b?

14. **Number of Hamiltonian Circuits.**

 a. How many different Hamiltonian circuits are there in a complete network of order $n = 10$?

 b. How many different Hamiltonian circuits are there in a complete network of order $n = 20$?

 c. Suppose a computer could check Hamiltonian circuits at a rate of one per second. How long would it take to check all the circuits in each of the networks in parts a and b?

15. **Traveling the National Parks.** Suppose that you are taking a trip to four national parks: Bryce, Canyonlands, Capitol Reef, and Grand Canyon. Use the data in Table 13.1 to answer the following questions.

 a. Draw a network that shows the four national parks and the distances between them.

 b. What is the total length of a trip that visits the parks in this order: Bryce, Canyonlands, Capitol Reef, Grand Canyon, and back to Bryce?

 c. Using the nearest neighbor method and starting at Grand Canyon, find a circuit that passes through the three other parks and returns to Grand Canyon. Describe your work.

 d. Repeat part c, but start and end at Bryce.

 e. Is either of the circuits in parts c and d the shortest possible circuit? Explain.

Solving Traveling Salesman Problems by Computer. In Exercises 16–19, suppose you want to find the solution to a traveling salesman problem with the indicated number of cities.

 a. How many Hamiltonian circuits will need to be tested?

 b. With a computer that can test 1 million circuits per second, how long will the job take?

 c. Put the answer to part b in perspective by comparing it to some well-known length of time.

16. 10 cities

17. 16 cities

18. 20 cities

19. 25 cities

20. **The Mail Must Go Through.** The following table shows the distances between pairs of towns in a rural county.

	Avila	Barila	Camilla	Dorrito	Eldorado
Avila	—	12	9	5	7
Barila	12	—	10	12	6
Camilla	9	10	—	15	8
Dorrito	5	12	15	—	9
Eldorado	7	6	8	9	—

a. Make a network showing the five towns and the distances between them.

b. As a mail truck driver, you must visit each town once every weekday. Find the shortest route that visits each town exactly once and returns to the starting town.

21. **A Traveling Salesperson Problem.** The network in Figure 13.41 shows the flight times (in hours) between cities that a salesperson must visit. Starting once from each city in the network, apply the nearest neighbor method to find a circuit that visits each city exactly once. Do you get different circuits for different starting cities? Can you determine whether any of your circuits is the shortest possible circuit? Explain.

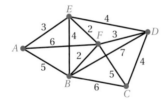

FIGURE 13.41

Further Applications

22. **Five-City Vacation.** A section of a road atlas looks like Figure 13.42, showing distances between various pairs of the cities St. Louis, Chicago, Kansas City, Memphis, and Louisville. For the connections shown, find the shortest circuit that passes through each city exactly once and

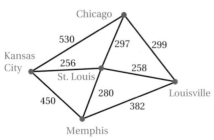

FIGURE 13.42

starts and finishes in the same city. Some connections between cities are missing. Would a shorter circuit be possible if all connections between the cities were available?

23. **Car Shuttles.** Each week Abe, Barbara, Carl, or Dolores hosts a bridge game at his or her house. Everyone is eager to save gasoline, so the host always picks up and drops off the other three players. The location of each player's house and the distances between them are shown in Figure 13.43. What path should Abe use to pick up his guests so as to minimize the distance traveled? Would Barbara, Carl, or Dolores use a different route? Explain.

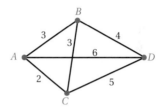

FIGURE 13.43

24. **Taxi Routing.** A taxi driver based at a train station (the blue square) must deliver five riders to five different hotels. The locations of the hotels are shown in Figure 13.44, and the travel times (in minutes) are indicated on the edges. Find the route that minimizes the total travel time for the taxi, beginning and ending at the train station. Assume that no time is spent waiting at the hotels.

FIGURE 13.44

25. **Overnight Delivery.** An overnight delivery company must be able to deliver packages by air between any pair of 35 cities. The cost of flying its planes is $15,000 per hour. Its current schedule requires a total of 890 hours of flight time each day (for the entire fleet of planes). Suppose a more efficient set of routes can be found that reduces the total flight requirement to 860 hours per day. How much money will the company save each day? each year?

26. **Interstate Trucking.** A computer manufacturer uses trucks to deliver computers throughout the United States. Its fleet of trucks currently drives an average of 75,000 miles per day, at a cost of approximately $0.95 per mile. Suppose an operations research project shows that the company can accomplish the same deliveries but reduce its total truck driving by 2%. How much will the company save each year?

Web Projects

Find useful links for Web Projects on the text Web site: www.aw.com/bennett-briggs

27. **Traveling Salesman Problem History.** Visit the Web to read about the history of the traveling salesman problem. Give your own summary of the history, including specific problems that led to progress and the current status of solution methods.

28. **Interactive Traveling Salesman Problem.** The Web has live interactive demonstrations of solutions of the traveling salesman problem. The simulations allow you to

choose networks of various sizes and to use several different methods. Visit such a Web site, experiment, and report on your findings.

29. **Operations Research.** Use the Web to investigate at least three companies that use operations research to improve their efficiency or output. Describe the methods used and some of the results.

In the News

30. **Routing Problems.** Find a news article about a project that clearly involves complex routing of people or vehicles. Could the project be formulated as a traveling salesman problem? Discuss.

31. **Operations Research.** Given its widespread use, it's not difficult to find news reports about projects and problems that involve operations research. Find such a report and identify the particular problem (for example, routing trains or manufacturing car parts). In what ways is efficiency important in the problem?

UNIT **13C**

SCHEDULING PROBLEMS

Another type of problem that relies on networks is planning a project with many stages or tasks. Some of the tasks may be performed simultaneously, whereas others have a definite ordering. Problems like this arise in everything from planning space flights to designing shopping centers. Finding an efficient way to schedule the various tasks can have a huge impact on the cost and time for a large project.

A HOUSE BUILDING PROJECT

We can investigate the principles of scheduling problems by considering a relatively small project: building a house. Figure 13.45 shows a simplified **flowchart** for the various tasks in building a house. The flowchart is a network in which the edges represent tasks and the vertices represent completion points for the tasks. Time flows from left to right through this network, specifying the precedence of the various tasks.

To interpret this network, note the following:

- Each edge is labeled with the number of months needed to complete the task. For example, the label on the *finance* edge indicates that two months are required to obtain financing for the house.

- In some phases of the project, only one task can be undertaken at a time. For example, only the *design* task can be carried out between vertices *A* and *B*.
- During other phases, two or more tasks can be carried out concurrently. For example, between points *C* and *F*, *construction* can take place at the same time as *paint*, *appliances*, and *carpet* are ordered.

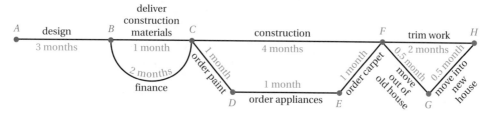

FIGURE 13.45 This network represents the flow of a house building project. Vertex *A* is the start of the project and other vertices represent completion points for various tasks. The tasks are represented by the edges, and each edge is labeled with the time required for the task.

> **EXAMPLE 1** Interpreting the House Building Network

Interpret the *construction, order paint, order appliances,* and *order carpet* edges (tasks) in the house building network shown in Figure 13.45.

Solution The *construction* edge indicates that the project plan allows four months for construction. The *order paint, order appliances,* and *order carpet* edges show that the plan allows one month for each of these tasks. (Note that this example assumes these three tasks are done one after the other. In some real projects, they might be done simultaneously.) The fact that these edges connect the same two vertices (*C* and *F*) as the *construction* edge means that the paint, appliances, and carpet should be ordered *during* the period of construction. It also means that all these tasks must be completed before the project can continue after construction. The fact that only three months are needed for ordering these supplies, while four months are needed for construction, means that a month can be spared in ordering these supplies without delaying the project schedule.

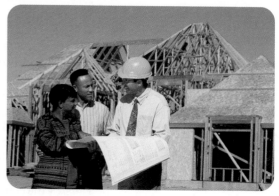

◄

In parts of a network where two or more tasks take place simultaneously, the project cannot continue until *all* the simultaneous tasks are complete. For example, the project in Figure 13.45 cannot go beyond point *F* until *all* the tasks connecting vertices *C* and *F* are complete. The tasks that require the most time between stages (vertices) in the project are called the **limiting tasks** for that period. For example, *construction* is the limiting task between stages *C* and *F* because it takes four months, while *order paint, order appliances,* and *order carpet* require a total of three months.

The time required to complete the project is dictated by the limiting tasks. We can find the completion time by looking at the path through the network that

includes all the limiting tasks. This is called the **critical path** for the network, and the completion time for the project is the length of the critical path.

Limiting Tasks and Critical Path

When two (or more) tasks can occur at the same time between two stages of the project, the task that requires the most time is called the **limiting task.**

The **critical path** through the network is the path that includes all the limiting tasks. The length of the critical path is the completion time for the project.

➤ **EXAMPLE 2** Limiting Tasks and Critical Path

 a. What is the limiting task between stages *B* and *C* in Figure 13.45?
 b. What is the limiting task between stages *F* and *H* in Figure 13.45?
 c. Find the critical path for the house building project in Figure 13.45.

Solution

 a. Between stages *B* and *C*, the tasks *deliver construction materials* and *finance* take place simultaneously. *Finance* is the limiting task because it takes two months, while *deliver construction materials* takes only one month.

 b. Between stages *F* and *H*, *trim work* takes place simultaneously with *moving out of the old house* and *moving into the new house*. *Trim work* is the limiting task because it takes two months, while the moves take one month combined.

 c. We start at point *A*. The design phase takes us to *B* in three months. The limiting task between *B* and *C* is *finance*, so we follow the *finance* edge to *C*. Between *C* and *F*, *construction* is the limiting task, so we follow the *construction* edge to *F*. We complete the critical path by following the *trim work* edge to *H*, since *trim work* is the limiting task between *F* and *H*. The critical path is shown in Figure 13.46. The total time along the critical path, which is the completion time for the project, is $3 + 2 + 4 + 2 = 11$ months.

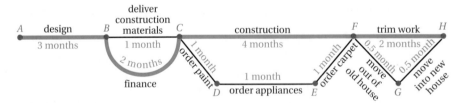

FIGURE 13.46 The critical path for the house building project. The edges in this path require a total time of 11 months, so this is the time needed to complete the project.

Now try Exercises 11–18. ◄

Time out to think _____

Find the total time along the path formed by edges *design, deliver construction materials, construction,* and *trim work.* Explain why, although this path adds up to a shorter time than the critical path, it is *not* the completion time for the project.

IS THE PROJECT ON SCHEDULE?

We can investigate the scheduling requirements in more depth by exploring the conditions under which the project remains on schedule and the conditions under which it falls behind.

Earliest Start and Finish Times

The **earliest start time (EST)** for a task is the earliest time after the beginning of the project at which the task can start (assuming the project is on schedule). For example, the earliest start time for the *finance* task in Figure 13.45 is three months (after the project begins) because it cannot start until the *design* task is complete.

The **earliest finish time (EFT)** for a task is the earliest time after the beginning of the project at which the task may be completed (assuming the project is on schedule). For example, the earliest finish time for the *finance* task is five months: the three months before it can start and the two months required for its completion. The following general rules translate these definitions into the language of the network.

Finding Earliest Start and Finish Times

- The earliest start time (EST) of a task leaving a particular vertex is the *largest* of the earliest finish times of the tasks entering that vertex.
- The earliest finish time (EFT) of a task is the earliest start time of that task *plus* the time required for the task. That is,

$$EFT = EST + \text{time for task}$$

►**EXAMPLE 3** Finding Earliest Start and Finish Times

Find the earliest start and finish times for the *deliver construction materials* and *construction* tasks in Figure 13.45.

Solution The *deliver construction materials* task leaves from vertex *B.* The only task entering vertex *B* is *design,* which has an earliest finish time of three months. Thus, the earliest start time for *deliver construction materials* is three months. Because *deliver construction materials* takes one month to complete, its earliest finish time is

$$EFT = \underbrace{3 \text{ months}}_{EST} + \underbrace{1 \text{ month}}_{\substack{\text{time required} \\ \text{for } task}} = 4 \text{ months}$$

The *construction* task leaves from vertex *C,* which is entered by two tasks: *deliver construction materials* and *finance.* Construction cannot begin until both of those tasks are completed. Because *finance* is the limiting task entering vertex *C,* its earliest

"I CAN REMEMBER WHEN ALL WE NEEDED WAS SOMEONE WHO COULD CARVE AND SOMEONE WHO COULD SEW."

Copyright © 1998, 2004 by Sidney Harris.

finish time of five months is also the earliest *start* time for construction. Construction requires four months, so its earliest finish time is four months after its earliest start time, or

$$5 \text{ months} + 4 \text{ months} = 9 \text{ months}$$

All of these values and more are summarized in Table 13.2, which shows earliest start times (EST) in column 2 and earliest finish times (EFT) in column 4.

Now try Exercises 19–26, part a. ◄

Latest Start and Finish Times

The **latest start time (LST)** for a task is the latest time at which it can start *without delaying* the overall project. Similarly, the **latest finish time (LFT)** of a task is the time by which it must be completed to avoid delays. Again, we can express these ideas with two rules.

Finding Latest Start and Finish Times

- The latest finish time (LFT) of a task *entering* a particular vertex is the *smallest* of the latest start times of the tasks leaving that vertex.
- The latest start time (LST) of a task is the latest finish time of that task *minus* the time required for the task. That is,

 LST = LFT − time for task

The rules for finding earliest and latest start and finish times can be applied to any individual task in a project. However, it is often easiest to find *all* of the earliest start and finish times by starting at the *beginning* of the project and considering each task in *forward* order. Similarly, it is easiest to find the latest start and finish times by starting at the *end* of the project and considering each task in *reverse* order.

►**EXAMPLE 4** Finding Latest Start and Finish Times

Find the latest start and finish times for the *trim work, move into new house, move out of old house,* and *order carpet* tasks.

Solution The minimum time required to complete the entire project is 11 months (see Example 3). Therefore, 11 months is the latest finish time for the *trim work* task, since this task enters the final vertex. Because *trim work* takes 2 months to complete, its latest start time is 2 months prior to its latest finish time, or $11 - 2 = 9$ months.

The task *move into new house* also enters the final vertex, so it also has a latest finish time of 11 months. Because this task requires only 0.5 month to complete, its latest start time is $11 - 0.5 = 10.5$ months.

The task *move out of old house* enters vertex *G.* The only task leaving that vertex is *move into new house.* Thus, the latest start time of 10.5 months for *move into new*

TABLE 13.2	Scheduling Demands for House Building Project					
TASK	EST	LST	EFT	LFT	SLACK TIME	CRITICAL PATH?
design	0	0	3	3	0	Yes
deliver construction materials	3	4	4	5	1	No
finance	3	3	5	5	0	Yes
construction	5	5	9	9	0	Yes
order paint	5	6	6	7	1	No
order appliances	6	7	7	8	1	No
order carpet	7	8	8	9	1	No
trim work	9	9	11	11	0	Yes
move out of old house	9	10	9.5	10.5	1	No
move into new house	9.5	10.5	10	11	1	No

house is the latest finish time for *move out of old house*. Because *move out of old house* requires 0.5 month, its latest start time is 10.5 − 0.5 = 10 months.

The *order carpet* task enters vertex *F*, which has two tasks leaving it: *move out of old house* and *trim work*. Thus, *order carpet* must be completed before both those other tasks have begun. Because *trim work* has the smaller latest start time of 9 months, this is the latest finish time for *order carpet*. We subtract the 1 month required for *order carpet* to find its latest start time of 9 − 1 = 8 months. Table 13.2 shows all of the latest start times (LST) and latest finish times (LFT).

Now try Exercises 19–26, part b. ◄

Slack Time

As shown in Table 13.2, the earliest start time for *deliver construction materials* is three months, and its latest start time is four months. In other words, this task cannot begin earlier than three months into the project, but there is no harm to the overall schedule if it begins as much as four months into the project. We say that *deliver construction materials* has a **slack time** of one month. Note that this one month of slack time also is the difference between the earliest and latest finish times for *deliver construction materials*. In general,

$$\text{slack time} = \text{LST} - \text{EST} = \text{LFT} - \text{EFT}$$

The next to last column in Table 13.2 shows the slack time for each of the tasks in the house building project. Note that tasks on the critical path have no slack time because they must be completed on schedule if the overall project is to remain on schedule.

►EXAMPLE 5 Impact of Slack Time

The task *move out of old house* is supposed to take 0.5 month. Suppose that, unexpectedly, it takes a full month. What is the impact on the overall schedule? Next, suppose that *construction* takes five months to complete, rather than the expected four months. What is the impact on the overall schedule?

Solution The task *move out of old house* has one month of slack time. Therefore, an "overrun" of 0.5 month for this task has no impact on the overall schedule—as long as everything else remains on time. The *construction* task is on the critical path and therefore has no slack time. A one-month "overrun" on this task will delay the entire project by one month. *Now try Exercises 27–32.* ◄

APPLICATIONS OF SCHEDULING PROBLEMS

The house building example illustrates the basic principles of a **scheduling problem.** The method we used to analyze it is called the **critical path method** (CPM). In practice, scheduling problems can be far more complex than our simple example. Imagine the problem of scheduling the various tasks in building a new freeway overpass (where traffic must be rerouted during certain phases of construction), in making sure a big-budget movie is ready for release by the holiday season, or in coordinating a military response to a terrorist attack.

The house building example has one key ingredient that is not always present in real scheduling problems: It contains good estimates of the time required for each task. Knowing these times allows us to find the critical path—which is necessary for CPM analysis. When task times are uncertain, as is the case when a project has never been tried before, a variation on CPM called the *project evaluation and review technique* (PERT) must be used. PERT was developed in the 1950s and 1960s to address the scheduling problems in the newly developing space program, including the Apollo project Moon landings. Today, large computer programs are used to run CPM- or PERT-based analyses of nearly all complex projects. Indeed, CPM and PERT are important components of most major projects undertaken today by businesses and governments.

Although we have only touched on CPM analysis—and haven't done any problems using PERT—remember that both are extensions of the network ideas discussed in this chapter. It's a long way from the bridges of Königsberg to the Moon, but the road is paved with the mathematics of networks.

► **EXAMPLE 6** Major Project Delays

Suppose that a new multi-level freeway overpass is being built in an area with heavy daily traffic. The project has been contracted to a team that has total labor costs of $500,000 per day. In addition, the traffic detours required during the project cost local businesses an estimated $200,000 per day in lost revenues and extra driving time. The project was budgeted to be completed in two months. Suppose instead that unexpected delays make it take four months. What is the total cost of the delays?

Solution A two-month delay is about 60 days. Because the labor rate is $500,000 per day, this results in a budget overrun of

$$\frac{\$500,000}{day} \times 60 \ days = \$30 \text{ million}$$

Moreover, it costs local businesses

$$\frac{\$200,000}{day} \times 60 \ days = \$12 \text{ million}$$

Overall, this two-month delay costs some $42 million. ◄

EXERCISES 13C

Review Questions

1. Describe the meanings of the edges and vertices in the house building project example.

2. What is a limiting task? How are limiting tasks related to the critical path for the project?

3. Explain the meaning and calculation of each of the following: earliest start time, earliest finish time, latest start time, and latest finish time.

4. What is slack time? Why do tasks on the critical path always have zero slack time?

5. Explain what we mean by CPM and PERT. How are they important?

6. Give at least two examples of projects—either ones that you are working on personally or ones that are taking place on your campus or in your town—for which the methods of scheduling problems would be applicable.

Does It Make Sense?

Decide whether each of the following statements makes sense (or is clearly true) or does not make sense (or is clearly false). Explain your reasoning.

7. Jim plans to use a critical path analysis to organize his ten-city summer vacation.

8. MaryAnn is using critical path analysis to plan the remodeling of the upstairs of her house.

9. Completing the electrical wiring is on the critical path for a housing project. A two-week delay in this task will not have any effect on the completion of the project.

10. In the schedule for a housing project, completing the plumbing has two weeks of slack time. A two-week delay in this task will not have any effect on the completion of the project.

Basic Skills & Concepts

Scheduling a Paint Job. For Exercises 11–14, refer to the schedule for a painting job shown in Figure 13.47. The estimated time for the completion of each task (in hours) is shown on the corresponding edge of the network.

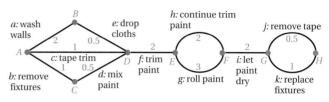

FIGURE 13.47

11. Which tasks are on the critical path between A and E?

12. Which tasks are on the critical path between E and H?

13. Give the tasks on the critical path for the entire project.

14. What is the length of the critical path for the entire project?

Building a Hotel. Exercises 15–18 refer to the scheduling network shown in Figure 13.48 for building a new hotel. Completion times, in months, for the tasks are shown on the edges.

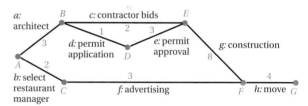

FIGURE 13.48

15. Which tasks are on the critical path between A and E?

16. Which tasks are on the critical path between B and G?

17. Give the tasks on the critical path for the entire project.

18. What is the length of the critical path for the entire project?

Start and Finish Times for a Paint Job. For Exercises 19–22, refer to Figure 13.47.

19. For the tasks d: *mix paint* and f: *trim paint*, find
 a. the EST
 b. the LST

20. For the tasks g: *roll paint* and j: *remove tape*, find
 a. the EST
 b. the LST

21. For the tasks *d: mix paint* and *f: trim paint,* find

 a. the EFT

 b. the LFT

22. For the tasks *g: roll paint* and *j: remove tape,* find

 a. the EFT

 b. the LFT

Start and Finish Times for Building a Hotel. For Exercises 23–26, refer to Figure 13.48.

23. For the tasks *d: permit application* and *f: advertising,* find

 a. the EST

 b. the LST

24. For the tasks *c: contractor bids* and *g: construction,* find

 a. the EST

 b. the LST

25. For the tasks *d: permit application* and *f: advertising,* find

 a. the EFT

 b. the LFT

26. For the tasks *c: contractor bids* and *g: construction,* find

 a. the EFT

 b. the LFT

Slack Time in a Paint Job. For Exercises 27–29, refer to Figure 13.47.

27. What is the effect of a one-hour delay in task *c: tape trim?*

28. What is the effect of a one-hour delay in task *g: roll paint?*

29. Make a table similar to Table 13.2 showing the EST, LST, EFT, LFT, and slack time for each task of the project.

Slack Time in Building a Hotel. For Exercises 30–32, refer to Figure 13.48.

30. What is the effect of a one-month delay in task *c: contractor bids?*

31. What is the effect of a one-month delay in task *d: permit application?*

32. Make a table similar to Table 13.2 showing the EST, LST, EFT, LFT, and slack time for each task of the project.

Further Applications

Doing the Laundry. The following table gives the critical times (in minutes) for the job of washing two loads of laundry (one person). Refer to this table in Exercises 33–37.

Task	EST	LST	EFT	LFT	Slack time
a: sort colors and whites	0	0	15	15	
b: wash colors	15	15	45	45	
c: remove stains from whites	15	25	35	45	
d: dry colors	45	45	90	90	
e: wash whites	45	60	75	90	
f: fold colors	90	120	105	135	
g: dry whites	90	90	135	135	
h: fold whites	135	135	150	150	

33. Fill in the slack time column in the table.

34. Which tasks are on the critical path?

35. How much time is needed for the entire project?

36. Which pairs of tasks take place at the same time? Which tasks must be done alone?

37. Use the table to make a network for the project.

Cooking a Pasta Dinner. The following table gives the critical times (in minutes) for the job of cooking a pasta dinner (alone). Refer to this table in Exercises 38–42.

Task	EST	LST	EFT	LFT	Slack time
a: boil water	0	0	5	5	
b: wash vegetables	0	1	4	5	
c: cook pasta	5	5	17	17	
d: heat up sauce	5	7	15	17	
e: make salad	5	8	14	17	
f: set table	17	17	22	22	

38. Fill in the slack time column in the table.

39. Which tasks are on the critical path?

40. How much time is needed for the entire project?

41. Which pairs of tasks take place at the same time? Which tasks must be done alone?

42. Use the table to make a network for the project.

Web Projects

Find useful links for Web Projects on the text Web site:
www.aw.com/bennett-briggs

43. **Critical Path Method.** Use the Web to learn about a specific scheduling project that has been addressed using the critical path method. Give a detailed description of the project and the results of using the critical path method.

In the News

44. **Scheduling Problems.** Find a news article about a project that involves major scheduling challenges. Identify some of the aspects of scheduling that were critical. Speculate on the consequences of failing to find an effective solution to the scheduling problem.

CHAPTER 13 SUMMARY

UNIT	KEY TERMS	KEY IDEAS AND SKILLS
13A	network edge, vertex order, degree circuit tree complete network spanning network cost (weight) Euler circuit	**Draw** networks to represent real sets of objects or places. **Understand** the bridges of Königsberg problem and why it has no solution. **Apply** the burning bridges rule to find Euler circuits. **Identify** the order of a network and the degree of each vertex. **Calculate** costs of spanning networks. **Use** Kruskal's algorithm to find a minimum cost spanning network.
13B	traveling salesman problem Hamiltonian circuit near-optimal solution operations research nearest neighbor method	**Describe** the traveling salesman problem. **Understand** Hamiltonian circuits: What are they? How many exist in a complete network of order n? **Apply** the nearest neighbor method to find near-optimal solutions to traveling salesman problems.
13C	limiting task critical path earliest start time (EST) earliest finish time (EFT) latest start time (LST) latest finish time (LFT) slack time CPM, PERT	**Understand** how to use a network for a scheduling problem. **Find** a critical path. **Explain** why scheduling problems are so important.

Credits

Prologue

p. 1: EPA Photo/Jamam Nasrallah.

Chapter 1

p. 13: © Chuck Savage/Corbis. *p. 14:* © Corbis. *p. 20:* © Jeff Bennett. *p. 21:* © Bettmann/Corbis. *p. 23:* © PhotoDisc. *p. 38:* Figure 2.1, p. 29 from THE COSMIC PERSPECTIVE by Jeffrey Bennett, Megan Donahue, Nicholas Schneider and Mark Voit. Copyright © 1999 by Addison Wesley Longman. Reprinted by permission of Pearson Education, Inc. *p. 46:* © Digital Vision. *p. 55:* Figure 13.16, p. 379 from THE COSMIC PERSPECTIVE by Jeffrey Bennett, Megan Donahue, Nicholas Schneider and Mark Voit. Copyright © 1999 by Addison Wesley Longman. Reprinted by permission of Pearson Education, Inc. *p. 65:* © PhotoDisc. *p. 66:* © PhotoDisc. *p. 77:* © PhotoDisc. *p. 79:* © PhotoDisc.

Chapter 2

p. 83: © PhotoDisc. *p. 91:* © ThinkStock. *p. 93:* © PhotoDisc. *p. 94:* © PhotoDisc. *p. 95:* © Digital Vision. *p. 98:* © PhotoDisc. *p. 99:* © AP/WideWorld. *p. 103:* © Corbis. *p. 106:* © PhotoDisc. *p. 108:* © PhotoDisc. *p. 116:* © Sara Anderson. *p. 127:* © PhotoDisc.

Chapter 3

p. 133: © PhotoDisc Blue. *p. 149:* © Digital Vision. *p. 150:* © PhotoDisc Blue. *p. 155:* © PhotoDisc. *p. 157:* © PhotoDisc Blue. *p. 158:* © PhotoDisc. *p. 160:* © Jeff Bennett. *p. 161:* NASA. *p. 172:* © Sara Anderson. *p. 175:* NASA. *p. 177:* PhotoDisc. *p. 179:* © Digital Vision. *p. 181:* © Corbis. *p. 189:* © Corbis. *p. 190:* © PhotoDisc Blue. *p. 197:* © PhotoDisc. *p. 200:* © Corbis. *p. 201:* (Fig. 3.5) From *The New York Times,* April 7, 1995. © 1995 The New York Times. Used with permission. All Rights Reserved. *p. 207:* (Fig. 3.6) From *The New York Times,* June 26, 2003. © 2003 The New York Times. Used with permission. All Rights Reserved.

Chapter 4

p. 209: © AFP/Corbis. *p. 210:* © Oxford. *p. 222:* © PhotoDisc. *p. 240:* © Corbis. *p. 243:* © Corbis. *p. 244:* © PhotoDisc. *p. 247:* © PhotoDisc. *p. 249:* © Yeshiva University. *p. 257:* © Digital Vision. *p. 259:* © PhotoDisc. *p. 267:* © ThinkStock. *p. 269:* © Bettmann/Corbis. *p. 273:* © PhotoDisc Blue. *p. 276:* © PhotoDisc. *p. 287:* © PhotoDisc.

Chapter 5

p. 301: © PhotoDisc. *p. 303:* © Digital Vision. *p. 305:* © PhotoDisc. *p. 307:* © Bettmann/Corbis. *p. 314:* © PhotoDisc. *p. 316:* © PhotoDisc. *p. 318:* © PhotoDisc. *p. 320:* © Corbis. *p. 324:* Photo by Michael Okoniewski, Copyright © 1994. *p. 327:* © PhotoDisc. *p. 334:* (Fig. 5.5) From Bennett/Briggs/Triola, STATISTICAL REASONING FOR EVERYDAY LIFE 2E, page 101. © 2003 Pearson Education, Inc., publishing as Pearson Addison Wesley. Reprinted by permission of Pearson Education. *p. 335:* (Figs. 5.7 and 5.8) From Bennett/Briggs/Triola, STATISTICAL REASONING FOR EVERYDAY LIFE 2E, page 103. © 2003 Pearson Education, Inc., publishing as Pearson Addison Wesley. Reprinted by permission of Pearson Education. *p. 338:* (Fig. 5.12) From *The New York Times,* September 30, 1995. © 1995 The New York Times. Used with permission. All Rights Reserved. *p. 345:* © PhotoDisc. *p. 348:* (Fig. 5.20) From *The New York Times,* October 3, 1995. © 1995 The New York Times. Used with permission. All Rights Reserved. *p. 349:* (Fig. 5.21) From *The New York Times,* August 20, 2000. © 2000 The New York Times. Used with permission. All Rights Reserved. *p. 352:* (Fig. 5.25) From *The New York Times,* October 6, 1999. © 1999 The New York Times. Used with permission. All Rights Reserved. *p. 353:* (Fig. 5.26) Copyright October 18, 1999 USA TODAY. Reprinted with permission. *p. 355:* (Fig. 5.28) *Time Almanac* with Information Please. *p. 356:* (Fig. 5.31) From *The New York Times,* April 4, 2000. © 2000 The New York Times. Used with permission. All Rights Reserved. *p. 359:* (Fig. 5.38) From *The New England Journal of Medicine,* 2/25/99. Copyright © 1999 Massachusetts Medical Society. All rights reserved. *p. 366:* © Corbis. *p. 367:* (top) © Bettmann/Corbis

(bottom) © PhotoDisc. *p. 368:* From THE COSMIC PERSPECTIVE by Jeffrey Bennett, Megan Donahue, Nicholas Schneider and Mark Voit. Copyright © 1999 by Addison Wesley Longman. Reprinted by permission of Pearson Education, Inc.

Chapter 6

p. 379: © Digital Vision. *p. 382:* © Duomo/Corbis. *p. 388:* © Reuters NewMedia Inc./Corbis. *p. 392:* © AFP/Corbis. *p. 400:* © Digital Vision. *p. 407:* © PhotoEdit. *p. 415:* © Corbis. *p. 416:* © Digital Vision. *p. 419:* © PhotoDisc. *p. 423:* © Jeff Bennett. *p. 424:* © PhotoDisc.

Chapter 7

p. 429: © PhotoDisc. *p. 432:* © PhotoDisc. *p. 434:* FEMA. *p. 435:* © Corbis. *p. 439:* © PhotoDisc. *p. 444:* © Greg Fiume/ NewSport/Corbis. *p. 458:* © AFP/Corbis. *p. 459:* © PhotoDisc. *p. 465:* © PhotoDisc. *p. 480:* © PhotoDisc. *p. 485:* © Jeff Bennett. *p. 486:* © Bettmann/Corbis.

Chapter 8

p. 489: © AP Photo/Luis Romero. *p. 493:* Courtesy of Professor Albert Allen Bartlett. *p. 502:* © Richard A. Cooke/Corbis. *p. 503:* © Bettmann/Corbis. *p. 507:* © PhotoDisc. *p. 508:* © PhotoDisc. *p. 510:* (top) © PhotoDisc (bottom) © Morton Beebe/Corbis. *p. 511:* © PhotoDisc. *p. 521:* © Roger Ressmeyer/Corbis. *p. 522:* © Bettmann/Corbis. *p. 524:* © PhotoDisc. *p. 526:* © Digital Vision. *p. 527:* © PhotoDisc.

Chapter 9

p. 531: NASA. *p. 534:* © PhotoDisc. *p. 537:* © PhotoDisc. *p. 538:* © Corbis. *p. 552:* © Digital Vision. *p. 560:* © PhotoDisc Red. *p. 567:* © Jonathan Blair/Corbis. *p. 570:* © PhotoDisc.

Chapter 10

p. 575: © Bill Ross/Corbis. *p. 584:* © PhotoDisc. *p. 597:* © PhotoDisc. *p. 600:* © PhotoDisc Blue. *p. 604:* © PhotoDisc. *p. 609:* (Fig. 10.43) Anne Burns, Professor of Mathematics, Long Island University, C. W. Post Campus. *p. 611:* (Fig. 10.49a) © PhotoDisc. (Fig. 10.49b) © William Briggs. *p. 613:* (Fig. 10.53) From THE SCIENCE OF FRACTAL IMAGES, Heinz-Otto Peitgen & Dietmar Saupe, Eds., © 1988, Springer-Verlag, N.Y. Reprinted with permission. *p. 615:* (Fig. 10.54) "4 stages in construction

of a snowflake island" from Benoit B. Mandelbrot, THE FRACTAL GEOMETRY OF NATURE, © 1982 by Benoit B. Mandelbrot. New York. W. H. Freeman & Co., page 43. *p. 616:* (Fig. 10.56) From THE SCIENCE OF FRACTAL IMAGES, Heinz-Otto Peitgen & Dietmar Saupe, Eds., © 1988, Springer-Verlag, N.Y. Reprinted with permission. *p. 617:* (Fig. 10.57) "Menger Sponge" from Benoit B. Mandelbrot, THE FRACTAL GEOMETRY OF NATURE, © 1982 by Benoit B. Mandelbrot. New York. W. H. Freeman & Co., page 45. (Fig. 10.58) © Professor Robert Devaney, Department of Mathematics, Boston University. (Fig. 10.59) Figure 3.10.8(b) from FRACTALS EVERYWHERE, by Michael Barnsley, copyright © 1988 by Academic Press. Reprinted with permission from Elsevier.

Chapter 11

p. 623: © Reuters NewMedia Inc./Corbis. *p. 624:* © Archivo Iconografico, S.A./Corbis. *p. 625:* © PhotoDisc Blue. *p. 633:* (Fig. 11.6) Scala/Art Resource, NY. (Fig. 11.7) © The Granger Collection. *p. 634:* (Fig. 11.8) From "Perspective, Jan Vredeman de Vries." Reprinted by permission of Dover Publications, Inc., NY. (Fig. 11.9) © Corbis/Bettmann. (Fig. 11.10) M. C. Escher's "Belvedere" © 2003 Cordon Art B.V.-Baarn-Holland. All Rights Reserved. *p. 635:* (Fig. 11.11) © The Granger Collection. (Fig. 11.12) © John R. Jones; Papilio/Corbis. *p. 637:* (Fig. 11.19) © The Granger Collection. (Fig. 11.20) Tate Gallery, London/Art Resource, NY. (Fig. 11.21) Anne Burns, Professor of Mathematics, Long Island University, C. W. Post Campus. *p. 640:* (Fig. 11.27) Courtesy of Paul J. Steinhardt, Princeton University. *p. 647:* (Fig. 11.43) © Corbis. (Fig. 11.44) Scala/Art Resource, NY. (Fig. 11.45) © 2004 Mondrian/ Holtzman Trust/Artists Rights Society (ARS), New York. *p. 649:* (Fig. 11.46c) © PhotoDisc. *p. 651:* (Fig. 11.48b) © PhotoDisc.

Chapter 12

p. 655: © Corbis. *p. 656:* © Reuters NewMedia Inc./Corbis. *p. 657:* © PhotoDisc. *p. 659:* © Bettmann/Corbis. *p. 669:* PhotoDisc. *p. 680:* © Reuters NewMedia Inc./Corbis. *p. 681:* Bettmann/Corbis. *p. 688:* PhotoDisc.

Chapter 13

p. 701: © PhotoDisc. *p. 706:* © PhotoDisc. *p. 710:* © PhotoDisc. *p. 724:* © PhotoDisc. *p. 731:* © PhotoDisc. *p. 736:* NASA.

Answers to Selected Exercises

Note: Answers do not include Review Questions, Web Projects, or In the News.

CHAPTER 1

Unit 1A

5. Does not make sense.
7. Makes sense.
9. Does not make sense.
11. a. *Premise:* 80% of the population supports the Governor's proposal. *Conclusion:* The proposal must be a good idea.
 b. The mere fact that a lot of people support the proposal does not make it a good idea.
13. a. *Premise:* There's no proof that global warming will have any bad consequences for our society. *Conclusion:* We have no reason to be concerned about global warming.
 b. This is an appeal to ignorance because it cites lack of knowledge about one thing to assert the truth of the opposite thing. In this case, a lack of proof that bad consequences will occur does not mean that the consequences won't occur.
15. a. *Premise:* He refused to testify by pleading the Fifth Amendment. *Conclusion:* He is guilty.
 b. This fallacy offers limited choice because being guilty is only one of many possible reasons why someone might refuse to testify.
17. a. *Premise:* The witness is a convicted drug dealer. *Conclusion:* The witness cannot be trusted.
 b. This is a personal attack because the prosecutor focuses on an aspect of the witness's character that may not have anything to do with the reliability of his testimony in this case.
19. a. *Premise:* There's too much pollution already. *Conclusion:* We should not build more highways.
 b. The issue in highway building is more likely to be traffic than pollution, so focusing on pollution is a diversion.
21. *Premise:* Following President Reagan's defense buildup, the Soviet Union began the process of democratization that ultimately led to its breakup. *Conclusion:* Reagan is responsible for the changes that led to the demise of the Soviet Union. False cause.
23. *Premise:* The mother either is still using drugs or has used drugs in the past. *Conclusion:* The mother should not have custody. Limited choice.
25. *Premise:* Mom smoked when she was my age. *Conclusion:* I won't heed her advice to stop smoking. Personal attack.

27. *Premise:* No one has ever proven that telepathy doesn't exist. *Conclusion:* I believe in telepathy. Appeal to ignorance.
29. *Premise:* Senator Smith is one of the biggest recipients of campaign contributions from the National Rifle Association. *Conclusion:* Senator Smith's bill cannot help the cause of gun control. Personal attack.
31. *Premises:* The percentage of the population over 18 who smoke has decreased from 40% to about 20%. The percentage of overweight people has increased from 25% to 35%. *Conclusion:* Quitting smoking leads to overeating. False cause or hasty generalization.
33. *Premise:* My little boy loves dolls and my little girl loves trucks. *Conclusion:* There's no truth to the claim that boys are more interested in mechanical toys while girls prefer maternal toys. Appeal to ignorance or hasty generalization.
35. *Premise:* Bush favors repealing the estate tax, which falls most heavily on the rich. *Conclusion:* Bush thinks the rich need to be richer; vote for Gore. Straw man.

Unit 1B

7. Does not make sense.
9. Makes sense.
11. Does not make sense.
13. Proposition.
15. Proposition.
17. Not a proposition.
19. London is not the capital of England. Original is true. Negation is false.
21. Caesar was not a Roman. Original is true. Negation is false.
23. The city council supported the police chief.
25. Congress upheld the veto (opposed the bill).
27. The anti-discrimination policy is supported by the Constitution.

29.

q	r	q and r
T	T	T
T	F	F
F	T	F
F	F	F

31. Quebec is the capital of Canada. Moscow is the capital of Russia. The first proposition is false and the second is true, and the conjunction is false.

33. Ben is married. Ben is a bachelor. The propositions cannot both be true, and the conjunction is false.
35. Grass is green. Skies are blue. Both propositions are true (at least sometimes), and the conjunction is true.

37.

q	r	s	q and r and s
T	T	T	T
T	T	F	F
T	F	T	F
T	F	F	F
F	T	T	F
F	T	F	F
F	F	T	F
F	F	F	F

39. Exclusive.
41. Probably exclusive.
43. Exclusive.

45.

r	s	r or s
T	T	T
T	F	T
F	T	T
F	F	F

47.

p	not p	p and (not p)
T	F	F
F	T	F

49.

p	q	r	p or q or r
T	T	T	T
T	T	F	T
T	F	T	T
T	F	F	T
F	T	T	T
F	T	F	T
F	F	T	T
F	F	F	F

51. Seattle is the capital of the United States. Paris is the capital of France. The first proposition is false and the second is true, and the disjunction is true.
53. Bill Clinton was a U.S. president. Martin Luther King was a U.S. president. The first proposition is true and the second is false, and the disjunction is true.
55. $8 \times 4 = 32$. $7 \times 3 = 22$. The first proposition is true and the second proposition is false, and the disjunction is true.
57. Search on earthquakes and California.

59. Search on Martin Luther King and assassination.

61.

p	r	if p then r
T	T	T
T	F	F
F	T	T
F	F	T

63. *Hypothesis:* San Francisco is in California. *Conclusion:* San Francisco is in the Pacific time zone. Both propositions are true, and the implication is true.
65. *Hypothesis:* Paris is the capital of China. *Conclusion:* London is the capital of England. The hypothesis is false; the conclusion and the implication are true.
67. *Hypothesis:* Sparrows can fly. *Conclusion:* Sparrows are birds. Both propositions are true, and the implication is true.
69. *Hypothesis:* $2 + 2 = 5$. *Conclusion:* Primates are not mammals. Both propositions are false, and the implication is true.
71. If a person is a resident of Miami, then that person is a resident of the United States. Proposition is true.
73. If you are a musician, then you play the saxophone. Proposition is false.
75. If you have prostate cancer, then you are a male. Proposition is true.
77. *Converse:* If Marco lives in the United States, then he lives in Chicago; false. *Inverse:* If Marco does not live in Chicago, then he does not live in the United States; false. *Contrapositive:* If Marco does not live in the United States, then he does not live in Chicago; true.
79. *Converse:* If I get wet, then I went swimming; false. *Inverse:* If I don't go swimming, then I won't get wet; false. *Contrapositive:* If I don't get wet, then I didn't go swimming; true.
81. *Converse:* If there is gas in the tank, then the engine is running; false. *Inverse:* If the engine is not running, then there is no gas in the tank; false. *Contrapositive:* If there is no gas in the tank, then the engine is not running; true.
83. If a young man has not wept, then he is a savage. If an old man will not laugh, then he is a fool.
85. If a person has no vices, then that person has few virtues.
87. If Sue lives in Cleveland, then she lives in Ohio.
89. If Sue lives in Ohio, then she lives in Cleveland.
91. **a.** Yes. **b.** Yes. **c.** Yes. **d.** Yes. **e.** No. **f.** Yes.
93. If a computer exhibits human-level intelligence, then it must have the requisite processing power.

95.

p	q	p or q	not (p or q)	(not p) and (not q)
T	T	T	F	F
T	F	T	F	F
F	T	T	F	F
F	F	F	T	T

The propositions are equivalent.

97.

p	q	p or q	not (p or q)	(not p) or (not q)
T	T	T	F	F
T	F	T	F	T
F	T	T	F	T
F	F	F	T	T

The propositions are not equivalent.

99.

p	q	r	p or q	(p or q) and r	p and r	q and r	(p and r) or (q and r)
T	T	T	T	T	T	T	T
T	T	F	T	F	F	F	F
T	F	T	T	T	T	F	T
T	F	F	T	F	F	F	F
F	T	T	T	T	F	T	T
F	T	F	T	F	F	F	F
F	F	T	F	F	F	F	F
F	F	F	F	F	F	F	F

The propositions are equivalent.

Unit 1C

7. Does not make sense.
9. Does not make sense.
11. Does not make sense.
13. Rational.
15. Rational.
17. Natural.
19. Integers.
21. Rational.
23. Rational.
25. Real.
27. Natural.
29. {January, February, . . . , November, December}.
31. {Washington, Adams, Jefferson}.
33. {California, Colorado, Connecticut}.
35. {100, 101, . . . , 199, 200}.

37.

39.

41.

43.
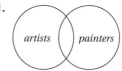

45. *Subject:* bachelors. *Predicate:* men.

47. *Subject:* U.S. presidents. *Predicate:* men.
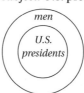

49. No fish are flying animals. *Subject:* fish. *Predicate:* flying animals.

51. All nurses are people who know CPR. *Subject:* nurses. *Predicate:* people who know CPR.

53.

55.

57.

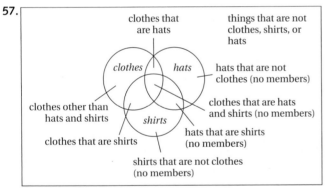

59. a. 28. **b.** 15. **c.** 13. **d.** 13.

61.

63.

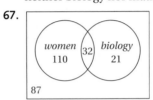

65.

math 9 | 7 | biology 16

11

16 students are taking only biology; 11 students are taking neither biology nor math.

67.

women 110 | 32 | biology 21

87

69.

coffee 4806 | 91 | gall stones 385

14,068

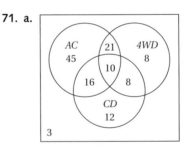

71. a.

AC 45 | 21 | 4WD 8

10

16 | 8

CD 12

3

b. 31 (includes those who chose CD also).
c. 61. **d.** 120. **e.** 45.

73.

75.

77.

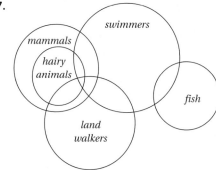

There could not be hairy fish. There could be hairy animals that swim. There could be walking mammals. There could be hairy animals that walk on land.

79.

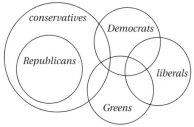

There could be conservative Democrats. There could be liberal Green Party members. There could not be liberal Republicans.

81. a.

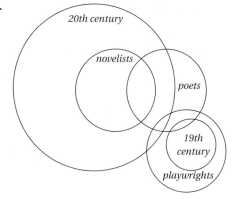

b. No. **c.** Yes. **d.** Yes.

83. a. 16 different sets of options.
 b. Not possible.
 c. 32 different sets of options.
 d. 2^n different sets of options.

Unit 1D

 9. Does not make sense.
11. Makes sense.
13. Does not make sense.
15. Inductive.
17. Inductive.
19. Deductive.
21. Inductive.
23. Premises are true; weak; conclusion is false.
25. Premises could be true; weak; conclusion is false.
27. Premises could be true; moderately strong; conclusion could be true.
29. All islands are tropical lands; valid. The first premise is false, and the argument is not sound.

31. All salty foods are foods that cause high blood pressure (HBP); valid. The premises could be true, in which case the argument is sound.

33. All states in the EST zone are states that are east of the Mississippi River; valid. The premises are true, and the argument is sound.

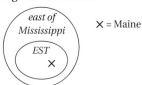

35. All opera singers are people who can whistle a Mozart tune; valid. The premises could be true, in which case the argument is sound.

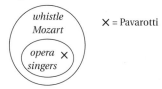

37. Affirming the hypothesis; valid. The premises are true, and the argument is sound.

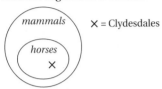
X = Clydesdales

39. Affirming the hypothesis; valid. If the premises are true, then the argument is sound.

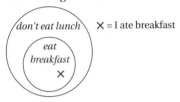
X = I ate breakfast

41. Denying the hypothesis; not valid. The argument is not sound.

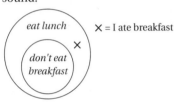
X = I ate breakfast

43. If interest rates decline, then the bond market improves. Affirming the conclusion; not valid. The argument is not sound.

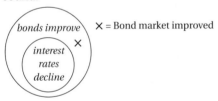
X = Bond market improved

45. The argument is in standard form, and it is valid.
47. If taxpayers have more disposable income, then they will fuel the economy. If taxes are cut, then the economy will be fueled. Valid.
49. True.
51. Not true by counterexample.
53. Example of valid and sound:
 Premise: All living mammals breathe.
 <u>Premise: All monkeys are mammals.</u>
 Conclusion: All living monkeys breathe.
55. Example of valid and not sound:
 Premise: All mammals fly. (false)
 <u>Premise: All monkeys are mammals. (true)</u>
 Conclusion: All monkeys fly. (false)
57. Example of not valid with true premises and conclusion:
 Premise: All mammals breathe. (true)
 <u>Premise: All mammals have hair. (true)</u>
 Conclusion: All hairy animals breathe. (true)

59. Affirming the conclusion (invalid):
 Premise: If I am in Phoenix, then I am in Arizona.
 <u>Premise: I am in Arizona.</u>
 Conclusion: I am in Phoenix.
61. Denying the conclusion (valid):
 Premise: If I am in Phoenix, then I am in Arizona.
 <u>Premise: I am not in Arizona.</u>
 Conclusion: I am not in Phoenix.

Unit 1E

5. Makes sense.
7. Does not make sense.
9. Makes sense.
11. **a.** The state constitution is less strict than the U.S. Constitution.
 b. A *yes* vote would tighten (strengthen) laws against obscenity.
 c. False.
13. (1) We have enough money to afford a house. (2) We will be staying in the same place long enough to make buying a house sensible.
15. (1) Not all countries get along well at present. (2) The United Nations can help nations learn to get along.
17. We are looking for possible unstated motives that may be the unstated "real reason" for opposition to the tax cuts. Among the possibilities: (1) a desire to maintain or increase government spending programs, which are more likely to be cut if revenues fall; (2) a belief that the tax system should redistribute wealth from the wealthy to the poor, which is less likely to occur with tax cuts.
19. **a.** No. **b.** No. **c.** Yes. **d.** No. **e.** Yes.
21. **b.** Jack appears to qualify.
 c. Jill appears to qualify.
 d. No, Ivan must also satisfy either (c) or (d) under provision (1).
 e. Yes, Joan satisfies three of the conditions under provision (1).
23. Plan A: If you go, it costs $1600; if you cancel, it costs $400. Plan B: If you go, it costs $2900; if you cancel, it costs nothing. If you go, Plan B costs $1300 more. If you cancel, Plan A costs $400 more.
25. The deal does not state the shipping/handling charges or the "regular club prices."
27. Were there more than 350 people on each jet or both jets combined? And was there any damage or injuries?
29. Is $83 billion the gross domestic product or 6% of the gross domestic product?
31. **a.** They could be consistent because Alice does not specify the time period for her 253 cases.
 b. They could be consistent if Alice obtained many convictions through plea agreements without going to trial.
33. The package has the better value.

35. If you go, Plan 2 costs $250 more. If you cancel, Plan 1 costs $75 more.
37. If you plan to fly 10 times (or a multiple of 10 times), then Plan A ($3150) is better than Plan B ($3250).
39. The speaker is affirming the conclusion, which is a fallacy. Premise: If athletes use drugs, then their performance will improve. Premise: Athletes' performances have improved. Conclusion: Athletes are using drugs. There could be other explanations.
47. One interpretation: The poet was 20 years old when he wrote the poem and will live to age 70.

CHAPTER 2

Unit 2A

7. Does not make sense.
9. Does not make sense.
11. Does not make sense.
13. **a.** 2/3. **b.** 11/6. **c.** 8/3. **d.** 5/6. **e.** 19/20. **f.** 35/52. **g.** 2/17. **h.** 8/15.
15. **a.** 3/10. **b.** 31/250. **c.** 39/50. **d.** 1/200. **e.** 46/25. **f.** 3009/1000. **g.** 1/10,000. **h.** 1001/10,000.
17. **a.** 1.5. **b.** 0.8. **c.** 0.167. **d.** 0.143. **e.** 0.222. **f.** 0.636. **g.** 0.967. **h.** 4.067.
19. Dollars per pound, $/lb.
21. Dollars per square foot, $/ft^2.
23. Pounds per square inch, lb/in^2.
25. Miles per gallon, mi/gal.
27. Grams per cubic centimeter, g/cm^3.
29. 54 ft.
31. 12,320 yd.
33. 8712 ft^2.
35. 15,840 ft.
37. 40,320 min.
39. **a.** 50 m^2; 150 m^3. **b.** 330 in^2; 3960 in^3. **c.** 25,000,000 ft^3.
41. 1 ft^2 = 144 in^2; 1 ft^2/144 in^2 = 1; 144 in^2/1 ft^2 = 1.
43. 720 in^2.
45. 1 ft^3 = 1728 in^3; 1 ft^3/1728 in^3 = 1; 1728 in^3/1 ft^3 = 1.
47. 139,968 in^3.
49. 1 British pound.
51. $235.75.
53. 11,790 yen.
55. $6.43/gal.
57. 56 mi/hr.
59. $5000/acre.
61. $6.06.
63. $20.
65. $3040.
67. 60 mi/hr.
69. 2737.5 hr.
71. Wrong; $0.85.

73. Wrong; $0.22/lb vs. $0.39/lb.
75. 89.3 gal.
77. **a.** 36.4 hr; 28.6 hr. **b.** $87; $103.
79. 2025 ft^3.
81. 4320 L.
83. **a.** 105 dog years. **b.** 15 to 22 weeks.

Unit 2B

7. Makes sense.
9. Makes sense.
11. Does not make sense.
13. Does not make sense.
15. Makes sense.
17. Does not make sense.
19. 10^{11}.
21. 10.
23. 10^{-6}.
25. 1.1 × 10^6.
27. 10^{21}.
29. 10^3.
31. 10^{-10}.
33. 1.001 × 10^8.
35. For example, 5 ft 10 in. = 70 in.
37. 8 lb.
39. 4 liquid pt; 3.44 dry pt.
41. 322,560 million in^3; 187 million ft^3.
43. 1000 times smaller.
45. 1000 times smaller.
47. 1,000,000,000,000 times smaller.
49. 32.8 ft.
51. 0.80 km.
53. 75.7 L.
55. 0.31 in^3.
57. 111.5 m^2.
59. **a.** 7.2°C. **b.** 68°F. **c.** 5.0°F. **d.** −22.0°F. **e.** 21.1°C.
61. **a.** −223.15°C. **b.** −33.15°C. **c.** 283.15 K.
63. 930 watts. Yes.
65. **a.** 4.5 billion joules. **b.** 1.7 kilowatts. **c.** 375 L, 99 gal.
67. 4 g/cm^3; sink.
69. 80 people/mi^2.
71. New Jersey: 1132 people/mi^2. Wyoming: 0.51 person/mi^2.
73. **a.** 1.0 g/100 mL; lethal concentration. **b.** 0.25 g/100 mL; not safe to drive.
75. **a.** Metric mile is 6.8% shorter than a USCS mile.
77. 31,996 ft; 6.06 mi; 9.75 km.
79. 6 karats.
81. 0.5 g.
83. $245.
85. 750 kilowatt-hr; $60 saved.
87. 720 million kilowatt-hr per month; 1.6 million kg of coal; 720,000 homes.

89. 4.3 million joules per day; 50 watts.

91. 1,752,000 kilowatt-hr in a year; 175 households.

Unit 2C

3. Does not make sense.

5. Makes sense (if you didn't let the double negatives fool you!).

7. (10 child, 0 adult); (8 child, 1 adult); (6 child, 2 adult); (4 child, 3 adult); (2 child, 4 adult); (0 child, 5 adult).

9. 0.03 second.

11. 100 km.

15. 646.2 ft.

17. The bow is much longer than it is tall. A circle would work only if the bow's height were about half its horizontal extent.

21. (1 fertilizer, 8 topsoil); (4 fertilizer, 7 topsoil); (7 fertilizer, 6 topsoil); (10 fertilizer, 5 topsoil); (13 fertilizer, 4 topsoil); (16 fertilizer, 3 topsoil); (19 fertilizer, 2 topsoil); (22 fertilizer, 1 topsoil); (25 fertilizer, 0 topsoil).

23. 21.3 ft.

25. 10.0 m.

27. 75 seconds.

29. 4 apples.

31. The blind fiddler is a woman.

33. Select a fruit from the box labeled *Apples and Oranges*.

35. 7 crossings.

37. The flagpoles are next to each other (no distance between them).

39. 3 socks.

41. Prisoner 1 does not see red hats on both Prisoners 2 and 3; if he did, he would know that he had a white hat (and would be freed). Therefore, the hats on Prisoners 2 and 3 must be WW, WR, or RW, and Prisoner 2 knows this! So if Prisoner 2 saw a red hat on Prisoner 3, Prisoner 2 would know that he had a white hat (and would be freed). Therefore, Prisoner 3 knows that he must have a white hat, and he is freed.

CHAPTER 3

Unit 3A

7. Makes sense.

9. Makes sense.

11. Does not make sense.

13. Makes sense.

15. Makes sense.

17. $1/4 = 0.25 = 25\%$.

19. $0.45 = 45\% = 9/20$.

21. $0.33333\ldots = 33.333\ldots\% = 1/3$.

23. $23\% = 0.23 = 23/100$.

25. $1/8 = 0.125 = 12.5\%$.

27. $1.34 = 134\% = 67/50$.

29. $0.65 = 65\% = 13/20$.

31. $99\% = 0.99 = 99/100$.

33. $5; 1/5 = 0.2; 500\%$.

35. $11; 1/11 = 0.091; 1100\%$.

37. $0.156; 6.4; 15.6\%$.

39. $0.455; 2.2; 45.5\%$.

41. $1.24; 0.809; 124\%$.

43. $0.68; 1.47; 68\%$.

45. 22.1%.

47. 41.0%.

49. 18.7%.

51. *Absolute:* Mary's grew more ($10,000 versus $8000). *Relative:* Both grew the same 40%.

53. Absolute change = 205 million; relative change = 270%.

55. Absolute change = −743 newspapers; percentage change = −33.4%.

57. Absolute change = 0.7 million newspapers; relative change = 65.4%.

59. Absolute change = 18 million arrivals; relative change = 37.5%.

61. Absolute change = 295,000 operations; relative change = 141.1%.

63. 140%.

65. 80%.

67. The retail cost is 2 times the wholesale cost.

69. The retail cost is 1.4 times the wholesale cost.

71. −15.5 percentage points; −22.7%.

73. 21 percentage points; 53.8%.

75. 33%; 40%.

77. $752.12.

79. $18,744.

81. 17.3%.

83. 12.6%.

85. False. Her final weight was 1% less than her original weight.

87. False. Your taxes increased 11.3% over two years.

89. Not possible.

91. Possible.

93. Possible.

95. No, you cannot average averages.

97. True.

99. False, some of the hotels with restaurants may also have pools.

101. About $75 billion.

103. $3.6 billion.

105. 69%.

107. 375,000.

109. $13.09.

Unit 3B

9. Makes sense.

11. Makes sense.

13. Does not make sense.

15. a. 5,000,000; 5 million. **b.** 7,000,000,000; 7 billion.
 c. 0.02; 2 hundredths.
 d. 800,000,000,000; 800 billion.
 e. 0.000 000 1; 1 ten-millionth.
 f. 0.0009; 9 ten-thousandths.
17. a. 6×10^2. **b.** 9×10^{-1}. **c.** 5×10^4.
 d. 3×10^{-3}. **e.** 5×10^{-4}. **f.** 7×10^{10}.
19. a. 2.4×10^{10}. **b.** 6.5×10^2. **c.** 4.5×10^{-3}.
 d. 2.2×10^{88}.
21. a. 10^{35} is 10^9, or 1 billion, times larger than 10^{26}.
 b. 10^{27} is 10^{10}, or 10 billion, times larger than 10^{17}.
 c. 1 billion is 10^3, or 1000, times larger than 1 million.
 d. 7 trillion is 10^9, or 1 billion, times larger than 7 thousand.
 e. 2×10^{-6} is 10^3, or 1000, times larger than 2×10^{-9}.
 f. 6.1×10^{29} is 10^2, or 100, times larger than 6.1×10^{27}.
23. 10^{20} joules.
25. 1.2×10^{11} bytes.
27. 10^{-6} m.
29. a. 2×10^6; exact value is 2×10^6.
 b. $3 \times 10^2 = 300$; exact value is 311.0.
 c. -1.44×10^5; exact value is -1.49×10^5.
 d. 10^{10}; exact value is 10^{10}.
 e. 2100; exact value is 2169.
 f. 2×10^5; exact value is 2.00948×10^5.
31. A 10-story building is between 100 and 120 feet high. A football field is 100 yards, or 300 feet, long—almost three times longer than the building is high.
33. It is not unusual for Americans to put 15,000 to 20,000 miles on a single car in a year. Assume that an average American (excluding traveling business people) takes two 5000-mile airplane trips per year. This amounts to 10,000 miles flown.
35. $10 billion.
37. Approximately 90 gallons per year (assuming 2 pints per day).
39. 1000 gal.
41. 16 candy bars.
43. Uranium releases 35,000 times more energy.
45. About 7×10^{-7} liter.
47. About 1.8 million kilograms.
49. 100,000 to 1.
51. 15,840 to 1.
53. 2,000,000 to 1.
55. 328 to 1.
57.

Planet	Model diameter	Model distance from sun
Mercury	0.5 mm	6 m
Venus	1.2 mm	11 m
Earth	1.3 mm	15 m
Mars	0.7 mm	23 m
Jupiter	14.3 mm	78 m
Saturn	12.0 mm	143 m
Uranus	5.2 mm	287 m
Neptune	4.8 mm	450 m
Pluto	0.2 mm	590 m

59. 7.1 m; 0.071 mm.
61. 6027 marriages per day.
63. 82 suicides per day.
65. Approximately 6.9×10^5 km (assuming 10 bills per mm).
67. About 24 stadiums.
69. About 9000 gallons per month (assuming 100 million households).
71. a. About 1821 kg/cm³.
 b. 7300 kg, or the mass of a tank.
 c. 7×10^{11} kg/cm³, more than the total mass of Mt. Everest.
73. 1.5 mm; 1.8 mm; 1.3 mm; 1.7 mm; take your height in quarters.

Unit 3C

7. Does not make sense.
9. Does not make sense.
11. Makes sense.
13. Makes sense.
15. a. 2. **b.** 15. **c.** 779. **d.** 4. **e.** 14. **f.** 235.
 g. 2000. **h.** 89. **i.** -14.
17. Three significant digits; it is precise to the nearest tenth (0.1) of a km/hr.
19. Six significant digits; it is precise to the nearest thousandth (0.001) of a second.
21. Two significant digits; it is precise to the nearest 0.00001 mm.
23. Four significant digits; it is precise to the nearest 0.000 000 1 m.
25. One significant digit; it is precise to the nearest 100,000.
27. Five significant digits; it is precise to the nearest 0.0001.
29. 13,000,000.
31. 6,370,000.
33. 8.3×10^{14}.
35. Random error due to miscounting; systematic error due to interpretation of what counts as a sport utility vehicle.
37. Random error due to miscalculation of average; systematic error due to underreporting income on tax returns.
39. Random error due to errors in reading scale; no systematic error if the scale is well calibrated.
41. Random error due to miscounting; systematic error due to overfilling for clerk's friends.
43. (1) is a random error; (2) is a systematic error (most likely due to underreporting).
45. All altitude readings will be about 2780 feet too low; this is a systematic error.
47. Absolute error = $0.50; relative error = 2.7%.
49. Absolute error = $-$17; relative error = -26.2%.
51. Absolute error = -0.25 cup; relative error = -4.5%.
53. Absolute error = -0.3 mi; relative error = -2.5%.
55. Tape measure is more accurate; laser is more precise.
57. Digital scale is more precise and more accurate.

59. 140 liters.

61. 1600 kg^2.

63. 38 mi.

65. $230.

67. Random and systematic errors could be present; the claim is not believable with the given precision.

69. Random and systematic errors could be present; the claim is believable with the given precision.

71. Random and systematic errors could be present; the claim is believable with the given precision.

73. Random and systematic errors could be present; the claim is not believable with the given precision.

Unit 3D

5. Does not make sense.

7. Makes sense.

9. Does not make sense.

11. 317.5.

13. $21.25.

15. 1955: 93.3; 1965: 100.0; 1975: 181.7; 1985: 383.3; 1995: 386.2; 2000: 496.8.

17. About $67,000.

19. 7.6%.

21. $1.78.

23. $2.19.

25. $1 in 1973 had the same purchasing power as $4.05 in 2002.

27. Palo Alto: $1,049,000; Sioux City: $135,000; Boston: $523,000.

29. Spokane: $260,000; Denver: $290,000; Juneau: $333,000.

31. Health care spending increased by a factor of 18.75; the CPI increased by a factor of about 4.05. The $80 billion spent in 1973 is equivalent to about $324 billion in 2002 dollars.

33. Cost of college relative change 325%; CPI relative change 118%.

35. Home prices relative change 64%; CPI relative change 32%.

37. $2.78.

39. In 1996, actual dollars are 1996 dollars.

41. Yes, it is consistent ($6.27 either way).

43. 1968.

Unit 3E

5. Makes sense.

7. Does not make sense.

9. Does not make sense.

11. Josh; Josh; Jude.

13. **a.** New Jersey; Nebraska.
 b. The percentage of nonwhites is significantly less in Nebraska than in New Jersey.

15. **a.** Whites: 0.18%; nonwhites: 0.54%; total: 0.19%.
 b. Whites: 0.16%; nonwhites: 0.34%; total: 0.23%.
 c. The rate for both whites and nonwhites was higher in New York than in Richmond, yet the overall rate was

higher in Richmond than in New York. The percentage of nonwhites was significantly less in New York than in Richmond.

17. 16%.

19. **b.** 216 people were accused of lying; of these, 18 were actually lying and 198 were telling the truth; 91.7% were falsely accused.
 c. 1784 people were telling the truth according to the polygraph; of these, 1782 were actually telling the truth and 2 were lying; 99.9% of those found to be telling the truth were actually telling the truth.

21. The housing program will see an increase of 1%, which represents an increase in actual dollars. However, the increase is less than the increase in the CPI and thus effectively amounts to a cut.

23. **a.** Drug B had the better cure rate for men (98.0% vs. 50.0%) and for women (11.2% vs. 5.0%).
 b. Drug A had the better cure rate overall (45.0% vs. 27.0%).
 c. Because the drugs had such different effects on men and women, the individual cure rates should be cited and used for evaluating the drugs.

25. Within each category, the percentage of women hired was greater than the percentage of men hired. Overall, the hiring rate was higher for men (55%) than for women (41.7%). Note that the number of blue-collar jobs was over five times greater than the number of white-collar jobs.

27. Answers will vary.

29. Answers will vary.

31. **a.** Absolute difference = $12,627;
 income $41,000 → savings 0.5% of income;
 income $530,000 → savings 2.4% of income.
 b. Absolute difference = $12,234;
 income $41,000 → savings 2.9% of income;
 income $530,000 → savings 2.5% of income.

CHAPTER 4

Unit 4A

9. Does not make sense.

11. Does not make sense.

13. Makes sense.

15. $x = 10$.

17. $y = 5$.

19. $z = 6$.

21. $x = 5$.

23. $x = 9$.

25. $n = 3$.

27. $x = 4$.

29. $w = -2$.

31. $625.

33. $3200.

35.

	Yancy		Samantha	
Year	Interest	Balance	Interest	Balance
0	——	$500	——	$500
1	$25	$525	$25	$525
2	$25	$550	$26.25	$551.25
3	$25	$575	$27.56	$578.81
4	$25	$600	$28.94	$607.75
5	$25	$625	$30.39	$638.14

Yancy's balance increases by $125, or 25%. Samantha's balance increases by $138.14, or 27.6%.

37. $2687.83.
39. $162,822.98.
41. $42,918.71.
43. $1418.34.
45. $15,205.70.
47. $2848.95.
49. $31,663.68.
51. 2.53%.
53. 3.30%.
55. $1040.81; $1221.40; $2225.54; APY = 4.08%.
57. $10,618.37; $13,498.59; $33,201.17; APY = 6.18%.
59. $5335.80; $6920.15; $18,346.48; APY = 6.72%.
61. $6140.
63. $5497.
65. $48,678.
67. $19,910.
69. After 10 years, Chang has $705.30; after 30 years, he has $1403.40. After 10 years, Kio has $722.52; after 30 years, she has $1508.74. Kio has $17.22, or 2.4%, more than Chang after 10 years. Kio has $105.34, or 7.5%, more than Chang after 30 years.
71. 6.77%; 6.80%; 6.82%.

73.

	Account 1		Account 2	
Year	Interest	Balance	Interest	Balance
0	—	$1000	—	$1000
1	$55	$1055	$57	$1057
2	$58	$1113	$59	$1116
3	$61	$1174	$63	$1179
4	$65	$1239	$67	$1246
5	$68	$1307	$70	$1317
6	$72	$1379	$74	$1391
7	$76	$1455	$79	$1470
8	$80	$1535	$83	$1553
9	$84	$1619	$87	$1640
10	$89	$1708	$93	$1733

Account 1 has increased in value by $708, or 70.8%. Account 2 has increased in value by $733, or 73.3%.

75. After 5 years, Bernard has $1946.64; after 20 years, he has $3505.80. After 5 years, Carla has $1797.60; after 20 years, she has $3805.33.
77. Plan A: $13,597; Plan B: $15,526.
79. 14.3 years.
81. 68.1 years.

Unit 4B

9. Makes sense.
11. Does not make sense.
13. Does not make sense.
15. 64.
17. 256.
19. 81.
21. 1/16.
23. 64.
25. 78,125.
27. $x = \pm 10$.
29. $x = 3$.
31. $x = 8$.
33. $x = 32$.
35. $936.85.
37. $3757.16.
39. $76,301.01; $24,000.
41. $86,144.21; $43,200.
43. $329.96.
45. $256.13.
47. Monthly deposit = $829.59.
49. 87.3%; 13.4%.
51. 105.5%; 7.5%.
53. −44.4%; −17.8%.
55. 68.0%; 6.7%.
57. $1,789,006; $792,418; $22,670; $6804.
59. Maytag; MYG; +$0.31; $24.11.
61. Monsanto.
63. $520.
65. Maytag; it has the highest dividend yield.
67. a. Yes; price = 12 × earnings. **b.** $2.04. **c.** About right.
69. a. No. **b.** None (had a loss). **c.** Expensive.
71. a. Yes; price = 16 × earnings. **b.** $1.69. **c.** Expensive.
73. 2.11%.
75. 5.00%.
77. $40.95.
79. $70.87.
81. $665.50.
83. Yolanda: balance = $15,528.23; deposits = $12,000. Zach: balance = $15,093.47; deposits = $12,000.
85. Juan: balance = $32,775.87; deposits = $24,000. Maria: balance = $33,736.06; deposits = $25,000.
87. Balance of $15,848.11 is not enough.

89. Balance of $29,081.87 is not enough.
91. $2.20 per share or 37.9% per share.
93. Total return $= 439,900\%$;
annual return $= 18.3\%$; much higher than
the average for large companies.

Unit 4C

7. Makes sense.
9. Does not make sense.
11. Makes sense.
13. a. Starting principal $= \$40,000$;
APR $= 7\%$;
number of payments per year $= 12$;
loan term $= 20$ years;
payment amount $= \$310$.
b. Number of payments $= 240$;
amount paid $= \$74,400$.
c. Interest payments $= \$34,400$;
payments toward principal $= \$40,000$.
15. a. $241.26. **b.** $57,902.40.
c. Interest payments $= \$32,902.40$;
payments toward principal $= \$25,000$.
17. a. $1048.82. **b.** $377,575.20.
c. Interest payments $= \$227,575.20$;
payments toward principal $= \$150,000$.
19. a. $1014.27. **b.** $182,568.60.
c. Interest payments $= \$82,568.60$;
payments toward principal $= \$100,000$.
21. a. $166.07. **b.** $5978.52.
c. Interest payments $= \$978.52$;
payments toward principal $= \$5000$.
23. a. $477.83. **b.** $86,009.40.
c. Interest payments $= \$36,009.40$;
payments toward principal $= \$50,000$.
25. Monthly payments $= \$768.91$.

End of month	Interest	Payment toward principal	New principal
1	$708.33	$60.58	$99,939.42
2	$707.90	$61.01	$99,878.41
3	$707.47	$61.44	$99,816.97

27. Monthly payments: $308.77; $241.79; $202.76. You can
afford only option 3.
29. $229.20; $2750.40.
31. $94.19; $3390.84.

33.

Month	Payment	Expenses	Interest	New balance
0				$1200.00
1	$200	$75	$18.00	$1093.00
2	$200	$75	$16.40	$984.40
3	$200	$75	$14.77	$874.17
4	$200	$75	$13.11	$762.28
5	$200	$75	$11.43	$648.71
6	$200	$75	$9.73	$533.44
7	$200	$75	$8.00	$416.44
8	$200	$75	$6.25	$297.69
9	$200	$75	$4.47	$177.16
10	$200	$75	$2.66	$54.82

A partial 11th payment will pay off the loan.

35.

Month	Payment	Expenses	Interest	Balance
0				$300.00
1	$300	$175	$4.50	$179.50
2	$150	$150	$2.69	$182.19
3	$400	$350	$2.73	$134.92
4	$500	$450	$2.02	$86.94
5	0	$100	$1.30	$188.24
6	$100	$100	$2.82	$191.06
7	$200	$150	$2.87	$143.93
8	$100	$80	$2.16	$126.09

37. Option 1: monthly payments $= \$733.76$; total
payments $= \$264,153.60$. Option 2: monthly
payments $= \$927.01$; total payments $= \$166,861.80$.
39. Option 1: monthly payments $= \$810.49$; total
payments $= \$291,776.40$. Option 2: monthly
payments $= \$1061.89$; total payments $= \$191,140.20$.
41. Choice 1: monthly payments $= \$587.01$; closing
costs $= \$1200$. Choice 2: monthly payments $= \$559.37$;
closing costs $= \$2800$.
43. Choice 1: monthly payments $= \$545.74$; closing
costs $= \$2000$. Choice 2 monthly payments $= \$518.88$;
closing costs $= \$3600$.
45. a. $224.93. **b.** $316.69.
c. Total payments for 20-year loan $= \$53,983.20$; total
payments for 10-year loan $= \$38,002.80$.
47. Savings of about $250 per month with the ARM; about
$188 per month more with the ARM.
49. $62,141; $77,676.
51. a. $95.57; $130.17; $158.34. **b.** $67,444.20.
c. $325.43; $78,103.

Unit 4D

11. Does not make sense. **13.** Makes sense.

15. Makes sense. **17.** Makes sense.

19. Gross income = $49,600; AGI = $46,100; taxable income = $35,650.

21. Gross income = $93,650; AGI = $87,450; taxable income = $75,350.

23. Itemize; your itemized deductions total $11,945, which is more than your standard deduction.

25. Gross income = $33,550; AGI = $33,050; taxable income = $25,250 (take standard deduction).

27. Gross income = $35,900; AGI = $35,900; taxable income = $22,000 (take standard deduction).

29. $5660. **31.** $16,734.25. **33.** $17,020.

35. $17,995. **37.** $500 **39.** $0. **41.** $280.

43. Cheaper to own (home cost including savings through mortgage interest deduction = $1406, less than rent of $1600).

45. Maria's true cost = $6700; Steve's true cost = $8500.

47. FICA tax = $2142; total tax = $4447; tax rate = 15.9%.

49. FICA tax = $3427; total tax = $9300; tax rate = 20.2%.

51. FICA tax = $3687; total tax = $10,597; tax rate = 22.0%.

53. Pierre: taxes = $33,296, rate = 27.7%; Katarina: taxes = $13,990, rate = 11.7%.

55. You contribute $400 per month, but tax savings mean your take-home pay goes down by only $340.

57. You contribute $800 per month, but tax savings mean your take-home pay goes down by only $600.

59. Combined tax at single rates = $9475; tax filing jointly = $9195; no marriage penalty (marriage benefit = $280).

61. Combined tax at single rates = $92,094; tax filing jointly = $99,246.50; marriage penalty = $7152.50.

Unit 4E

9. Makes sense.

11. Does not make sense.

13. Does not make sense.

15. a. Surplus. **b.** Surplus. **c.** Yes.

17. About $47,000 per worker.

19. a. $63,000.

b. Outlays = $1,113,000; deficit = $63,000; debt = $836,000.

c. $69,000.

d. Outlays = $869,000; surplus = $231,000; debt = $605,000.

21. Deficit of about $176.5 billion; deficit of about $168 billion.

23. $852 billion. **25.** $56 billion.

27. $221 billion. **29.** $110 billion deficit.

31. Cut government spending, borrow money, raise taxes.

33. About 222,000 years.

35. $7.7 trillion; $11.5 trillion.

37. $2.343 trillion; deficit = $432 billion.

39. About $493 billion per year.

41. 897 years.

CHAPTER 5

Unit 5A

9. Does not make sense.

11. Does not make sense.

13. Makes sense.

15. *Population:* registered voters in California; *sample:* 1026 people selected for interviews; *population parameters:* approval ratings of candidates among all registered voters; *sample statistics:* approval ratings of candidates among those in the sample.

17. *Population:* all new computers; *sample:* five computers selected for measurements; *population parameter:* speed of all new computers on specific tasks; *sample statistic:* speed of the selected computers on specific tasks.

19. *Population:* all adult Americans; *sample:* 1010 selected adults; *population parameters:* rankings of prestigious professions among all adult Americans; *sample statistics:* rankings of prestigious professions by those in the sample.

21. *Step 1:* Population is all students at the school; goal is to determine annual pizza consumption. *Step 2:* Choose a representative sample. *Step 3:* Determine annual pizza consumption for those in the sample. *Step 4:* Infer annual pizza consumption for all students. *Step 5:* Assess results and formulate conclusion.

23. *Step 1:* Population is adult American women; goal is to determine mean height of adult American women. *Step 2:* Choose a representative sample. *Step 3:* Determine mean height for those in the sample. *Step 4:* Infer mean height for all adult American women. *Step 5:* Assess results and formulate conclusion.

25. *Step 1:* Population is all alkaline batteries in digital cameras; goal is to determine mean lifetime of these batteries. *Step 2:* Choose a representative sample. *Step 3:* Determine mean lifetime of alkaline batteries in the sample. *Step 4:* Infer mean lifetime of alkaline batteries in digital cameras. *Step 5:* Assess results and formulate conclusion.

27. Not representative (age could be an important factor); not representative; not representative; representative.

29. Stratified.

31. Stratified.

33. Simple random.

35. Observational, not case-control.

37. Observational, not case-control.

39. Experiment; treatment group consists of plants treated with fertilizer; control group consists of plants treated with no fertilizer; placebo and blinding are not needed.

41. Observational.

43. Observational, case-control.

45. Experiment.

47. 50.5% to 55.5%; yes.

49. 4.0% to 4.4%; 4.1% to 4.5%; no.

51. Good evidence.

53. Good evidence.

55. **a.** Population is American parents; population parameter is percentage of parents who respond yes to the question.
 b. Sample is 748 parents surveyed; sample statistic is percentage of parents who respond yes to the question.
 c. 59.4% to 66.6%.
57. **a.** Population is all people of working age; population parameter is percentage of population that is unemployed.
 b. Sample is people in 60,000 households; sample statistic is percentage of those in sample who are unemployed.
 c. 6.2% to 6.6%.
59. **a.** Population is all Americans; population parameter is percentage of Americans who respond yes to the question.
 b. Sample is 1546 people surveyed; sample statistic is percentage of people in the sample who respond yes to the question.
 c. 48% to 54%.

Unit 5B

5. Does not make sense.
7. Does not make sense.
9. You should doubt the results of this study.
11. You should doubt the results of this study.
13. You should doubt the results of this study.
15. You should doubt the results of this study.
17. You should doubt the results of this study.
19. You should doubt the results of this study.
21. You should doubt this claim.
23. You should doubt this claim.
25. Possible conflict of interest.
27. Possible selection bias.
29. Possible selection bias.
31. Possible conflict of interest.
33. Answers will vary.
35. What are the sample and sample size? How is "great deal of confidence" measured? Which military leaders are considered?
37. What are the sample and sample size? Is self-reporting accurate?
39. What are the sample and sample size?
41. 98% of all movies is different from 98% of top rental movies.
43. It may be true, but the sample is very small and there could be bias in the study.

Unit 5C

7. Does not make sense.
9. Does not make sense.
11. Does not make sense.

13. Does not make sense.

15.

Grade	Freq.	Rel. freq.	Cum. freq.
A	4	0.17	4
B	7	0.29	11
C	8	0.33	19
D	3	0.13	22
F	2	0.08	24
Total	24	1.00	24

17. Qualitative.
19. Qualitative.
21. Qualitative.
23. Qualitative.

25.

Bin	Freq.	Rel. freq.	Cum. freq.
95–99	2	0.08	2
90–94	2	0.08	4
85–89	6	0.24	10
80–84	3	0.12	13
75–79	4	0.16	17
70–74	1	0.04	18
65–69	4	0.16	22
60–64	0	0.00	22
55–59	1	0.04	23
50–54	1	0.04	24
<50	1	0.04	25
Total	25	1.00	25

27.

29.

31.

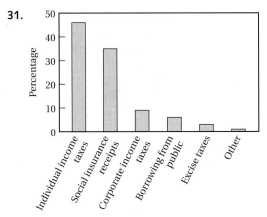

33.

Bin	Frequency
20–29	0
30–39	12
40–49	13
50–59	5
60–69	3
70–79	1
Total	34

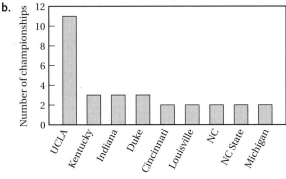

37. a. The teams are qualitative categories.

b.

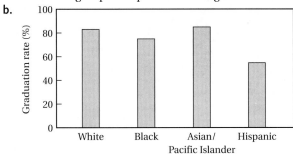

39. a. The ethnic groups are qualitative categories.

b.

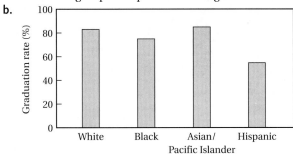

41. a. The land masses are qualitative categories.

b.

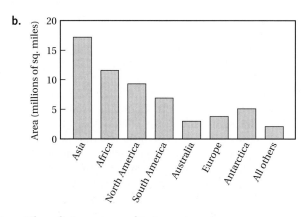

43. a. The religions are qualitative categories.

b.

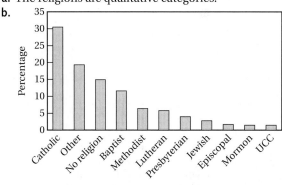

45. a. The years are quantitative categories.

b.

Unit 5D

9. Does not make sense.

11. Makes sense.

13. a. China, India, Russia.
 b. China, India, Russia.

15. a. Cancer increased; tuberculosis decreased; cardiovascular disease increased and then decreased, for an overall decrease; pneumonia decreased.
 b. The death rate for cardiovascular disease reached a maximum of about 500 (deaths per 100,000) in 1950.
 c. The death rate for cancer in 2000 was about 210 (deaths per 100,000).

d. If the lines of the graph were extended, cancer would overtake cardiovascular disease as the leading cause of death.

17. **a.** 1980: about 8%; 1990: about 15%; 1995: about 15%; 2000: about 12%.
 b. 1960: about 52%; 1980: about 23%; 2000: about 17%.
 c. 1960: about 27%; 1980: about 47%; 2000: about 57%.

19. There are significant regional differences; probabilities tend to be much higher in the northern states.

21. **a.** Each bar represents the percentage of the population for a particular age group in a particular year.
 b. There has been a general decrease in the percentage of people in the youngest age category.
 c. The data are truly three-dimensional.

23. Number of events for women; total number of women participants; total number of all participants.

25. The area of the 2000 TV is about 16 times greater than the area of the 1980 TV. The volume of the 2000 TV is about 64 times greater than the volume of the 1980 TV.

27. The zero point for earnings is not on the graph.

29.
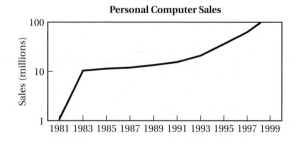

31. Most of the growth occurred after 1950.

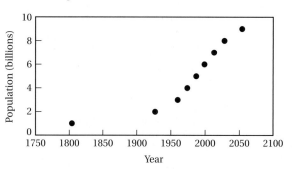

33.
Amount of Alcohol and Auto Fatalities

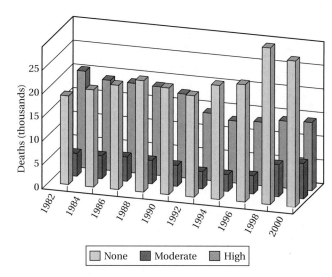

35. It is difficult to include U.S. data on the same graph.

Annual Deaths Due to Firearms (excluding U.S., which has 30 times more deaths than Germany and Canada)

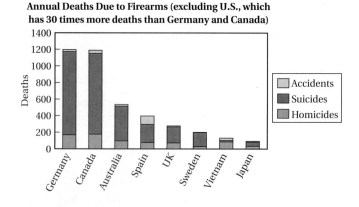

Unit 5E

7. Makes sense.
9. Does not make sense.
11. Makes sense.

13. a. Weight of car, city gas mileage of car.
 b. Strong negative correlation.
 c. Heavier cars get lower city gas mileage.
15. a. Salary level, percent of income given to charity.
 b. Weak negative correlation.
 c. There's a slight indication that lower-salary employees give a larger percentage of their income to charity, but the trend is very weak.
17. Pounds, calories; weak positive correlation.
19. Revolutions per minute, miles per hour; strong positive correlation.
21. % alcohol, seconds; strong positive correlation.
23. Pounds, dollars; weak positive correlation.
25. a.

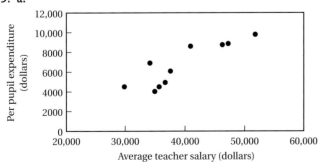

 b. Weak positive correlation.
27. a.

 b. Weak negative correlation.
29. a.

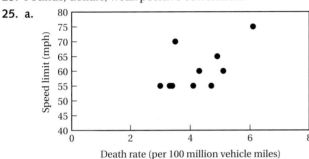

 b. Strong positive correlation.
31. There is a positive correlation between the crime rate and the number of people in prison; common underlying cause.

33. There is a positive correlation between gasoline prices and the number of airline passengers; possible direct cause.
35. The cost of deer hunting permits is positively correlated with deer sightings; direct cause.
37. a. Guideline 1. **b.** Guideline 2. **c.** Guideline 3.
 d. The above observations suggest that perhaps your headaches are caused by bad air in the building in which you work (there is a "disease" called sick building syndrome).
39. People do not decide to become conductors until relatively late in life—say, 30 years of age or over. Having already survived this many years, a person has a longer-than-average life expectancy.

CHAPTER 6

Unit 6A
7. Makes sense.
9. Makes sense.
11. Does not make sense.
13. Mean = 671.37; median = 672.2; mode = 672.2.
15. Mean = 0.188; median = 0.165; mode = 0.16.
17. Mean = 9.5; median = 10; modes = {5, 10, 15}.
19. Mean = 0.81237 pound; median = 0.8161 pound; 0.7901 is an outlier; mean = 0.81608 pound; median = 0.8163 pound.
21. Median.
23. Median.
25. Mean or median.
27. a. One peak. **b.** Left-skewed. **c.** Small variation.
29. a. Two peaks. **b.** Symmetric. **c.** Large variation.
31. a. One peak. **b.** Symmetric. **c.** Moderate variation.
33. a. One peak. **b.** Right-skewed. **c.** Large variation.
35. Distribution has two peaks (bimodal), is not symmetric, and has high variation.
37. Distribution has one peak, is symmetric, and has moderate variation.

Unit 6B
7. Does not make sense.
9. Makes sense.
11. Makes sense.
13. Mean = 7.2; median = 7.2.
15. a. Atlanta: mean = 5.86 minutes, median = 5.5 minutes, range = 4.0 minutes. Boston: mean = 6.77 minutes, median = 6.5 minutes, range = 10.5 minutes.
 b.

	Atlanta	Boston
Low value	4.0	1.5
Lower quartile	5.0	5.0
Median	5.5	6.5
Upper quartile	7.0	9.5
High value	8.0	12.0

(Boxplot not shown.)

 c. Standard deviation: Atlanta, 1.27 minutes;
 Boston, 3.25 minutes.

 d. Range rule of thumb approximation: Atlanta, 1 minute;
 Boston, 2.6 minutes.

17. a. First seven: mean = 58.3 years, median = 57 years,
 range = 4 years. Last seven: mean = 57.4 years,
 median = 56 years, range = 23 years.

 b.

	First seven	Last seven
Low value	57	46
Lower quartile	57	52
Median	57	56
Upper quartile	61	64
High value	61	69

 (Boxplot not shown.)

 c. Standard deviation: first seven, 1.9 years; last seven,
 7.8 years.

 d. Range rule of thumb approximation: first seven,
 1.00 years; last seven, 5.75 years.

19. a.

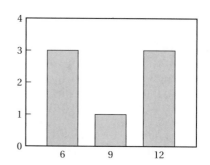

 b.

	Set 1	Set 2	Set 3	Set 4
Low value	9	8	8	6
Lower quartile	9	8	8	6
Median	9	9	9	9
Upper quartile	9	10	10	12
High value	9	10	10	12

 (Boxplot not shown.)

 c. Standard deviations = 0.000, 0.816, 1.000, 3.000.

21. The means are nearly equal, but the variation is signifi-
cantly greater for the second shop than for the first. If you
want a reliable delivery time, choose the first shop.

23. A lower standard deviation means more certainty in the
return and less risk.

25. Batting averages are less varied today. Because the mean
is unchanged, batting averages above 0.350 are (slightly)
less common today.

Unit 6C

 5. Makes sense.

 7. Does not make sense.

 9. Makes sense.

11. (a) and (c) are normal; (c) has the larger standard deviation.

13. Normal.

15. Normal.

17. Not normal.

19. a. 84%. **b.** 97.5%. **c.** 16%. **d.** 2.5%.
 e. 16%. **f.** 2.5%. **g.** 84%. **h.** 68%.

21. 16%.

23. 51; 2.5%.

25. 0.

27. -1.125.

29. a. $z = 1$; 84.13 percentile.
 b. $z = -1.5$; 6.68 percentile.
 c. $z = -2$; 2.28 percentile.

31. a. $z = 1.6$. **b.** $z = -1.5$. **c.** $z = -1.1$.

33. 68%.

35. 16%; traveling may not be safe.

37. a. $z = 0.56$; 71 percentile.
 b. $z = -0.24$; 40 percentile.
 c. $z = -0.44$; 33 percentile.
 d. $z = 0.08$; 53 percentile.

39. Not possible.

41. About 9%.

Unit 6D

9. Makes sense.

11. Does not make sense.

13. Makes sense.

15. Significant.

17. Not significant.

19. Significant.

21. Significant at the 0.05 and 0.01 levels.

23. Not significant at the 0.05 or 0.01 level.

25. Margin of error = 0.03, or 3 percentage points; confidence interval is 29% to 35%.

27. Margin of error = 0.04, or 4 percentage points; confidence interval is 65% to 73%.

29. Margin of error = 0.03, or 3 percentage points; confidence interval is 59% to 65%.

31. Margin of error = 0.03, or 3 percentage points; confidence interval is 34% to 40%.

33. **a.** Null hypothesis: mean SAT score = 500; alternative hypothesis: mean SAT score > 500.
 b. Rejecting the null hypothesis means there is evidence that SAT scores in the school are greater than 500. Failing to reject the null hypothesis means there is insufficient evidence to conclude that SAT scores in the school are greater than 500.

35. **a.** Null hypothesis: amount of preservative = advertised amount; alternative hypothesis: amount of preservative > advertised amount.
 b. Rejecting the null hypothesis means there is evidence that the amount of preservative exceeds the advertised amount. Failing to reject the null hypothesis means there is insufficient evidence to conclude that the amount of preservative exceeds the advertised amount.

37. **a.** Null hypothesis: yield = 540 bushels per acre; alternative hypothesis: yield > 540 bushels per acre.
 b. Rejecting the null hypothesis means there is evidence that the yield exceeds 540 bushels per acre. Failing to reject the null hypothesis means there is insufficient evidence to conclude that the yield exceeds 540 bushels per acre.

39. Null hypothesis: mean annual mileage of cars in the fleet = 11,725 miles; alternative hypothesis: mean annual mileage of cars in the fleet > 11,725 miles. The result is significant at the 0.01 level and provides good evidence for rejecting the null hypothesis.

41. Null hypothesis: mean stay = 2.1 days; alternative hypothesis: mean stay > 2.1 days. The result is not significant at the 0.05 level, and there are no grounds for rejecting the null hypothesis.

43. Null hypothesis: mean income = $40,000; alternative hypothesis: mean income > $40,000. The result is signifi-

cant at the 0.01 level and provides good evidence for rejecting the null hypothesis.

45. Margin of error = 0.009, or 0.9 percentage point; confidence interval is 34.1% to 35.9%.

47. The sample size must be four times as great.

49. The margin of error, 3 percentage points, is consistent with the sample size.

CHAPTER 7

Unit 7A

7. Makes sense.

9. Makes sense.

11. Does not make sense.

13. 24 choices.

15. 18 stations.

17. 32 styles.

19. 30 systems.

21. 0.1.

23. $1/4 = 0.25$.

25. $4/36 = 0.11$.

27. $6/22 = 0.27$.

29. $12/30 = 0.4$.

31. 0.86.

33. $34/100 = 0.34$.

37. $5/6 = 0.83$.

39. 0.24.

41. $5/6 = 0.83$.

43. $3/6 = 0.5$.

45.

Result	Probability
0 H	1/16
1 H	4/16
2 H	6/16
3 H	4/16
4 H	1/16

47. Odds for are 1 to 2; odds against are 2 to 1.

49. Odds for are 1 to 2; odds against are 2 to 1.

51. Gain of $15.

53. $52/100 = 0.52$; $51/100 = 0.51$.

55. The coins appear to be unfair. There should be about 125 occurrences of 0 heads and 3 heads.

57. Odds for $A = 99$ to 1; odds for $B = 24$ to 1.

Unit 7B

5. Makes sense.

7. Does not make sense.

9. Does not make sense.

11. Independent; 0.001.

13. Dependent; 0.0632.

15. Independent; 0.00001.

17. Independent; 0.0313.

19. Non-overlapping; 0.333.
21. Non-overlapping; 0.306.
23. Overlapping; 0.75.
25. Overlapping; 0.308.
27. 0.598.
29. 0.875.
31. 0.672.
33. 0.798.
35. a. 0.986. **b.** 0.619.
37. a. Dependent. **b.** 0.303. **c.** 0.309.
 d. Nearly equal.
39. 0.506.
41. a. $1/40 = 0.025$. **b.** 0.000625. **c.** 0.224.

Unit 7C

 7. Makes sense.
 9. Does not make sense.
11. Does not make sense.
13. You shouldn't expect to get exactly 5000 heads. The proportion of heads should approach 0.5 as the number of tosses increases.
15. Expected value $= -\$0.25$; outcome of one game cannot be predicted; over 100 games, you should expect to lose.
17. Expected value $= \$0.50$; outcome of one game cannot be predicted; over 100 games, you should expect to win.
19. Expected value $= \$350$; expected profit $= \$35,000,000$.
21. 15 minutes.
23. a. 47% heads; lost $6. **b.** Lost $12.
 c. You need 56 heads, which is possible but not likely.
 d. The probability of a head on any toss is always 0.5.
25. a. If you toss a head, the difference becomes 15; if you toss a tail, the difference becomes 17.
 b. On each toss, the difference in heads and tails is equally likely to increase or decrease. After 1000 tosses, the difference is equally likely to be greater than 16 or less than 16.
 c. Once you have fewer heads than tails, the deficit of heads is likely to remain.
27. $P(10 \text{ heads in } 10 \text{ tosses}) = 1/1024$; you should not be surprised.
29. a. $-\$0.014$. **b.** Lose $1.40. **c.** Lose $7.
 d. $14,000.
31. Expected value $= -\$0.43$; expected loss $= \$157$.
33. Expected value $= -\$0.78$; expected loss $= \$285$.
35. 1-point expected value $= 0.94$; 2-point expected value $= 0.74$.
37. 2.6 people.
39. a. Decision 1: Option A expected value $= \$1,000,000$;
 Option B expected value $= \$1,140,000$.
 Decision 2: Option A expected value $= \$110,000$;
 Option B expected value $= \$250,000$.

 b. Consistent for Decision 2; not consistent for Decision 1.

Unit 7D

 5. Does not make sense.
 7. Does not make sense.
 9. 1.5 fatalities per 100 million miles.
11. 22.6 deaths per 100,000 drivers.
13. 0.32 accident per 100,000 flight hours, 7.6 accidents per billion miles flown, 0.63 accident per 100,000 departures, 0.067 death per 100,000 flight hours, 1.6 deaths per billion miles flown, 0.13 death per 100,000 departures.
15. 0.00024; 2.2 times as great as for septicemia.
17. 12.6 deaths per 100,000.
19. Approximately 298 people.
21. About 5 deaths per 1000 people; 79,000 deaths.
23. 80 years of age.
25. $50 million loss.
27. a. Utah: 130 births per day; Maine: 37 births per day.
 b. Utah: 26.3 births per 1000 people; Maine: 10.5 births per 1000 people.
29. a. 4.2 million births. **b.** 2.4 million deaths.
 c. 1.8 million people. **d.** 1.2 million; 40%.

Unit 7E

 5. Makes sense.
 7. Makes sense.
 9. Makes sense.
11. 5040.
13. 35.
15. 495.
17. 720.
19. 5.
21. 330.
23. $10^4 = 10,000$ addresses.
25. 720 passwords.
27. 252 subcommittees.
29. 210 sets.
31. $26^5 = 11,881,376$ words.
33. $4^3 = 64$ words.
35. 495 subcommittees.
37. 17,576,000 plates.
39. 0.0244; 0.117.
41. a. 72 sundaes. **b.** 1728 cones. **c.** 1320 cones.
 d. 220 cones.
43. 9 toppings; 10 toppings.
45. 1/201,376.
47. 1/649,740.
49. 1/720.
51. a. 0.025. **b.** 0.00065. **c.** 50 people.
 d. 1.3 people.

CHAPTER 8

Unit 8A

5. Makes sense.
7. Makes sense.
9. Linear; 16,515 people.
11. Exponential; $169.
13. Exponential; $61.95.
15. Linear; $130,000.
17. 524,288 grains; 1,048,575 grains; about 150 pounds.
19. 1.3×10^{12} tons.
21. $327.68.
23. About 37 days.
25. 2^{50} bacteria; 1/1024 full.
27. 2.6 meters; more than knee-deep.
29. **a.** (Only 100-year intervals are shown here.)

Year	Population
2000	6.0×10^{9}
2100	2.4×10^{10}
2200	9.6×10^{10}
2300	3.8×10^{11}
2400	1.5×10^{12}
2500	6.1×10^{12}
2600	2.5×10^{13}
2700	9.8×10^{13}
2800	3.9×10^{14}
2900	1.6×10^{15}
3000	6.3×10^{15}

b. Between 2800 and 2850. **c.** About 2150.

Unit 8B

9. Does not make sense.
11. Does not make sense.
13. False.
15. False.
17. True.
19. True.
21. False.
23. **a.** 0.602. **b.** 4.301. **c.** -0.301. **d.** 1.505.
 e. -0.602. **f.** -0.699.
25. $2^8 = 256$; 2^{56}.
27. 44 years.
29. 23,645 people; 35,839 people.
31. 16,777,216 cells; 4,294,967,296 cells.
33. 7.0 billion; 15.1 billion; 28.0 billion.

35.

Month	Beginning-of-month population	End-of-month population
1	100	107
2	107	114
3	114	123
4	123	131
5	131	140
6	140	150
7	150	161
8	161	172
9	172	184
10	184	197
11	197	210
12	210	225
13	225	241
14	241	258
15	258	276

Doubling time is just over 10 months.
37. 10 years; 1.23.
39. 100 months; 1.09; 1.95.
41. 0.25; 0.0625.
43. 0.25; 0.125.
45. 62,500 animals; 7813 animals.
47. 0.41 kg; 0.037 kg.
49. 7 years; 0.00005.
51. 10 years; 313 elephants.
53. Approximate: 7 years; exact: 7.27 years; price = $1331.
55. Approximate: 23.33 years; exact: 23.45 years; population = 438 million.
57. Essentially zero.

Unit 8C

7. Makes sense.
9. Makes sense.
11. Does not make sense.
13. Doubling time = 78 years; 2050 population = 9.4 billion.
15. Doubling time = 44 years; 2050 population = 13.2 billion.
17. **a.** 8.1, 2.7, -2.1 per 1000.
 b. Birth rate decreased, death rate constant; decrease in population.
19. **a.** 2.0, 0.5, 0.7 per 1000.
 b. Birth rate decreased slightly, death rate increased slightly; slight population growth.
21. 2.4%; 1.2%; 0.3%.
23. 396 million; 560 million.
25. 459 million; 754 million.
27. Answers will vary: 2050 population = 12.0 billion; 2080 population = 18.2 billion; 2100 population = 24.1 billion.
29. 0.84% per year; carrying capacity too low.
31. 1.58% per year; carrying capacity too high.

33. 2% growth rate: 122,000; 149,000; 328,000.
5% growth rate: 164,000; 269,000; 1,950,000.

Unit 8D

5. Does not make sense.
7. Does not make sense.
9. 7.9×10^{11} joules.
11. 3.5×10^{16} joules.
13. Bomb releases 6.3 times as much energy.
15. 10^8 times as loud.
17. 77 dB.
19. 316 times as loud.
21. 25 times as strong at 1 m.
23. 4 times as strong at 20 m.
25. Decreases by a factor of 100; more basic.
27. 3.2×10^{-9} mole per liter.
29. pH = 1; acid.
31. 1000 times.
33. Minor event; few effects.
35. Serious effects.
37. Shout could not be heard.
39. a. 40 dB. **b.** 57 m. **c.** 1.94 dB; factor of 640,000.
41. a. 10^{-4} mole per liter. **b.** 10^{-5} mole per liter; pH = 5.
 c. 1.1×10^{-4} mole per liter; pH = 3.96.
 d. The change in part b could be detected; the change in part c could not be detected.

CHAPTER 9

Unit 9A

5. Makes sense.
7. Does not make sense.
9.

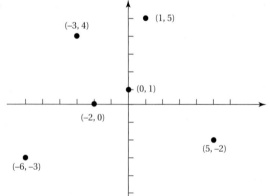

11. As the weight of the bag of apples increases, the price of the bag also increases.
13. As the price of a product increases, the demand (the number of items that can be sold) generally decreases.
15. If you are coasting down a hill on a skateboard, the steeper the hill, the faster you will coast.

17. If the tax rate increases (for example, a 6% sales tax increases to 7%), then the amount of revenue generated by the tax will also increase (assuming overall sales do not decrease).
19. a. 23 in., 15 in., and 12 in.
 b. 4000 ft, 12,000 ft, and 25,000 ft.
 c. It appears that the pressure reaches 5 inches of mercury at an altitude of about 50,000 ft. The pressure approaches zero, but theoretically never reaches zero.
21. a. The independent variable is time, measured in years, and the dependent variable is world population. The domain is the years between 1950 and 2000. The range is all populations between about 2.5 billion and 6 billion.
 b. The function shows a steadily increasing world population between 1950 and 2000 (and beyond).
23. a. The variables are (time, temperature) or (date, temperature). The domain is all days over the course of a year. The range is temperatures between 38° and 85°.
 b.

c. The temperature increases during the first half of the year and then decreases during the second half of the year.
25. a. The variables are (year, tobacco production). The domain is the ten years between 1975 and 1990. The range is between about 1.2 and 2.2 billion pounds.
 b.

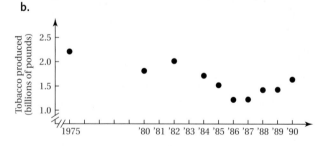

c. In this case, because the data give yearly figures for production, it would not be appropriate to fill in the graph between the data points.
27. The domain of the function (altitude, temperature) is all altitudes of interest—say, 0 ft to 15,000 ft (or 0 m to 4000 m). The range is all temperatures associated with the altitudes in the domain; the interval 30°F to 90°F (or about 0°C to 30°C) would cover all temperatures of interest.

With some reliable data, this graph is a good model of how the temperature varies with altitude.

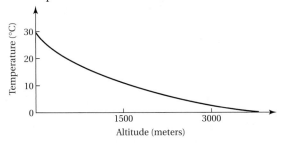

29. The domain of the (blood alcohol content, reflex time) function consists of all reasonable BACs (in g/100 mL); for example, numbers between 0 and 0.25 would be appropriate. The range would consist of the reflex times associated with those BACs. The validity of this graph as a model of alcohol impairment will depend on how accurately reflex times can be measured.

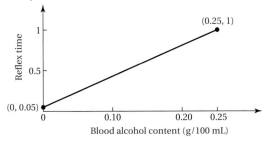

31. The domain of the function (time of day, traffic flow) consists of all times over a full day. It could be all numbers between 0 hours and 24 hours or all times between, say, 6:00 a.m. on Monday and 6:00 a.m. on Tuesday. The range consists of all traffic flows (in units of number of cars per minute) at the times during the day. We would expect light traffic flow at night, medium traffic flow during the midday hours, and heavy traffic flow during the two rush hours. The graph of this function would be a good model only if based on reliable data.

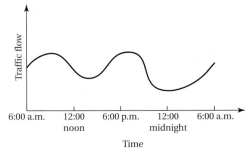

33. The domain of the function (time, world record in the 100-meter dash) consists of all years from 1970 to 2004. The range consists of the world records for the 100-meter

dash at all times between 1970 and 2004. Some research would be needed to find the exact records and the days on which they were lowered. But the graph would show a slow decrease (from about 10.2 seconds to 9.8 seconds for men and from about 11.0 seconds to 10.4 seconds for women) over this period of time.

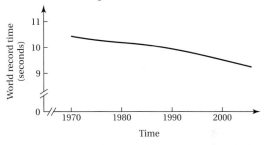

35. The domain of the function (time, population of China) is all years from 1900 to, say, 2004. The range consists of the populations of China during those years—roughly 400 million to 1.3 billion. With accurate yearly data, this graph would be a good model of population growth in China.

37. The domain of the function (angle of cannon, horizontal distance traveled by cannonball) is all cannon angles between 0° and 90°. The range would consist (literally) of all ranges of the cannon (the horizontal distance traveled by the cannonball) for the various angles in the domain. It is well known that a projectile has maximum range when the angle is about 45°, so the graph below shows a peak at about 45°. This is a good qualitative model.

Unit 9B

7. Does not make sense.

9. Makes sense.

11. a. Rain depth increases linearly with time.
 b. $\frac{4}{3}$ inch per hour.

 c. Good model if rainfall rate is constant for four hours.
13. a. On a long trip, distance from home decreases linearly
 with time.
 b. −71.4 miles per hour.
 c. Good model if speed is constant for seven hours.
15. a. Shoe size increases linearly with the height of the
 individual.
 b. 0.1375 size per inch.
 c. The model is a rough approximation at best.
17. The water depth decreases with respect to time at a rate
 of 0.25 inch per hour. The rate of change is −0.25 in./hr.
 In 6.5 hours, the water depth decreases 1.625 inches. In
 12.5 hours, the water depth decreases 3.125 inches.
19. The tree diameter increases by 0.2 inch per year. The rate
 of change is 0.2 inch per year. In 4.5 years, the tree diame-
 ter increases 0.9 inch. In 20.5 years, the tree diameter
 increases 4.1 inches.
21. The snow depth increases by 4 inches per hour. The rate
 of change is 4 inches per hour. In 5.5 hours, the snow
 depth will increase by 22 inches. In 7.8 hours, the snow
 depth will increase by 31.2 inches.
23. The independent variable is time (or t) measured in years.
 We will let $t = 0$ represent today. The dependent variable
 is price (or p). The equation of the price function is
 $p = 12{,}000 + 1200t$. The price of the car in 2.5 years will
 be \$15,000. This function doesn't give a good model of car
 prices.
25. The variables in this problem are (snow depth, maximum
 speed), or (d, s), where snow depth is measured in inches
 and maximum speed is measured in mph. The equation
 for the function is $s = 30 − 0.5d$. The plow has zero maxi-
 mum speed when the snow depth reaches 60 inches, or
 5 feet. The plowing rate most likely is not a constant, so
 this model is an approximation.
27. The variables in this problem are (time, rental cost), or
 (t, r), where t is measured in minutes. The equation for
 the function is $r = 5 + 0.60t$. The number of minutes you
 can rent for the \$10 you have left after you pay the setup
 charge is 16.67. This function gives a very good model of
 rental costs, provided all of the costs are quoted correctly.
29. $h = 20 + 2.8a$; 25.6 in.; 36.8 in; 76.0 in.; 160.0 in. Model is
 accurate for small ages only.
31. $P = −350 + 5n$; 70 tickets.
33. $V = 1000 − 50t$; 20 years.

35. The equation $y = 2x + 6$ describes a straight line with
 y-intercept of $(0, 6)$ and slope 2.

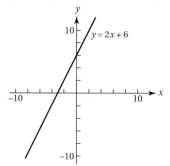

37. The equation $y = −5x − 5$ describes a straight line with
 y-intercept of $(0, −5)$ and slope $−5$.

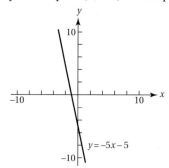

39. The equation $y = 3x − 6$ describes a straight line with
 y-intercept of $(0, −6)$ and slope 3.
41. The equation $y = −x + 4$ describes a straight line with
 y-intercept of $(0, 4)$ and slope $−1$.

Sketches for Exercises 39 and 41:

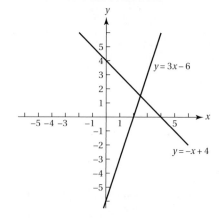

43. The variables are (time, distance traveled).

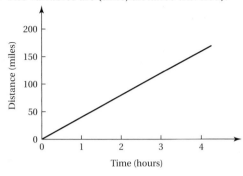

In 2.6 hours, the distance traveled is 104 miles. Provided the speed is really a constant, this linear equation gives a good model of the trip.

45. The variables are (number of posters, cost).

The cost of printing 2000 posters is $8000. This function probably gives a fairly realistic estimate of printing costs.

47. The variables in this problem are (time, cost).

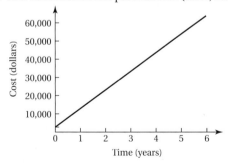

The cost of six years of school is $62,000. Provided costs do not change during the six-year period, this function is an accurate model of the cost.

Unit 9C
7. Does not make sense.
9. Makes sense.

11. 4.52.
13. 1.13.
15. 2.46.
17. 12.97.
19. 31.62.
21. 0.032.
23. 10,000.
25. 7.
27. a. $Q = 85{,}000 \times (1.024)^t$.
b.

Year	Population
0	85,000
1	87,040
2	89,129
3	91,268
4	93,458
5	95,701
6	97,998
7	100,350
8	102,759
9	105,225
10	107,750

c.

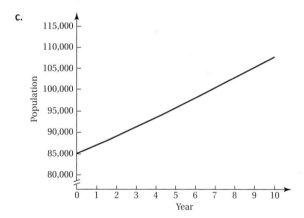

29. a. $Q = 10^6 \times (0.93)^t$.
b.

Year	Millions of acres
0	1
1	0.93
2	0.86
3	0.80
4	0.75
5	0.70
6	0.65
7	0.60
8	0.56
9	0.52
10	0.48

c.

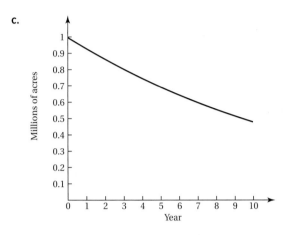

31. a. $Q = \$125,000 \times (1.07)^{t}$.

b.

Year	Average price
0	$125,000
1	$133,750
2	$143,113
3	$153,130
4	$163,850
5	$175,319
6	$187,591
7	$200,723
8	$214,773
9	$229,807
10	$245,894

c.

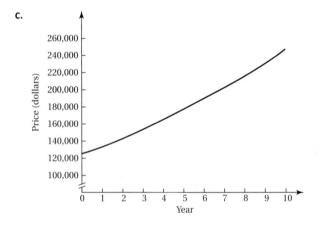

33. a. $Q = 2000 \times (1.05)^{t}$.

b.

Year	Monthly salary
0	$2000.00
1	$2100.00
2	$2205.00
3	$2315.25
4	$2431.01
5	$2552.56
6	$2680.19
7	$2814.20
8	$2954.91
9	$3102.66
10	$3257.79

c.

35. 7.4%.

37. 115,583% per year; 1.98% per day.

39. 33.28 years.

41. a. 15.9 mg. **b.** 120 hours.

43. a. 1.06 billion years old. **b.** 3.88 billion years old.

45. 9.5 mg/cm^2.

47. $115.93.

CHAPTER 10

Unit 10A

11. Does not make sense.

13. Makes sense.

15. Does not make sense.

17. 120°.

19. 18°.

21. 4°.

23. $\frac{1}{360}$.

25. $\frac{1}{24}$.

27. $\frac{1}{6}$.

29. $\frac{1}{2}$.

31. 37.7 m; 113.1 m^2.

33. 78.5 cm; 490.9 cm^2.

35. 28.3 mm; 63.6 mm^2.

37. 48 mi; 144 mi^2.

39. 60 ft; 100 ft^2.
41. 20 m; 16 m^2.
43. 24 units; 24 units2.
45. 33 units; 30 units2.
47. 17.4 ft; 19.1 ft^2.
49. 84 ft^2.
51. 27,600 yd^2.
53. 2500 m^3.
55. 54.5 ft^3; 130.9 ft^2.
57. Circumference is greater.
59. 135,000 m^3.
61. The first tree (353.4 ft^3 vs. 141.4 ft^3).
63. 625 times as great.
65. 50 times as great.
67. 125,000 times as many.
69. 3.
71. 27.
73. Answers will vary. Example: 32 in.; 35.2 in.
75. Squirrels have a higher surface area to volume ratio.
77. The Moon's surface area to volume ratio is four times as great as the Earth's.
79. Softball: 50.3 in.2, 33.5 in^3, ratio = 1.5; bowling ball: 452.4 in^2, 904.8 in^3, ratio = 0.5.
81. a. 3. **b.** 2. **c.** 1. **d.** A corner of a page.
83. The third line would necessarily be perpendicular to both lines.
85. a. 1.05×10^8 mm^2 = 105 m^2; 5.8×10^6 mm^3.
 b. 112 mm; 1.56×10^5 mm^2; air sacs have about 670 times as much surface area.
 c. 2.9 m; for their relatively small volume, the air sacs have a very large surface area.
87. 0.0038 km^3 = 3.8×10^6 m^3.

Unit 10B

9. Does not make sense.
11. Makes sense.
13. Makes sense.
15. 26°30′0″.
17. 15°0′58″.
19. 126°59′50″.
21. 10.2500°.
23. 36.7522°.
25. 1.0169°.
27. 21,600′.
29. 34°S, 151°E.
31. 21°S, 22°E.
33. Buenos Aires.
35. 1200 mi.
37. 0.16°.
39. 1.31×10^6 km.
41. Roof with a 1 in 3 pitch.
43. Railroad track with a 1/20 slope.

45. 5/12 = 0.42; 4.2 ft.
47. 45°; yes, it is possible.
49. 8.3%
51. a. 0.95 mi. **b.** 0.78 mi.
53. a. 0.85 mi. **b.** 0.65 mi.
55. a. 0.98 mi. **b.** 0.71 mi.
57. 0.84 acre.
59. 0.27 acre.
61. Similar.
63. Not similar.
65. $x = 4$, $y = 3$.
67. $x = 15$, $y = 33.3$
69. 36 ft.
71. 24 ft.
73. Circular: 1146 m^2; square: 900 m^2.
75. Circular: 16,114 m^2; square: 12,656 m^2.
77. Can 1 costs $94.25; can 2 costs $138.23.
79. Cube; $4.95.
81. 17 cm; no.
83. 127.3 ft.
85. Walking is faster.
87. a. 0.525 ft deep. **b.** 2688 lb.
89. 100 m; no; a square 20 m × 20 m fence would require 80 m of fence.
91. $3736; $3414.
93. a. Radius = 3.8 cm, height = 7.7 cm.
95. a. 75.4 ft^3. **b.** 4.7 ft high.
 c. Assuming 10,000 grains per cubic inch, there are about 1.3 billion grains.
97. a. About 1.6 times the length of a football field.
 b. 91,636,272 ft^3; 3,393,936 yd^3.
 c. 2,262,000 blocks. **d.** 21.5 years.
 e. About 5% of the volume of the Great Pyramid.

Unit 10C

9. Makes sense.
11. Does not make sense.
13. Makes sense.
15. The dimension is 1 and the object is ordinary (non-fractal).
17. The dimension is 3 and the object is ordinary (non-fractal).
19. The dimension is 2.262 and the object is a fractal.
21. The dimension is 1 and the object is ordinary (non-fractal).
23. The dimension is 3 and the object is ordinary (non-fractal).
25. The dimension is 2.292 and the object is a fractal.
27. a. When the ruler size is reduced by a factor of $R = 4$, there are $N = 8$ times as many elements.
 b. 1.5.
 c. The objects that lead to the quadric island all have the same area, because each time a piece of the boundary juts out (adding area) another piece juts in (removing the same amount of area). The length of the coastline of the quadric island is infinite.

29. When the ruler is reduced by a factor of $R = 9$, there will be 4 elements found; fractal dimension = 0.631.

31. a. 1.301. **b.** 1.585. **c.** 2.585.

33. The branching in many natural objects has the same pattern repeated on many different scales. This is the process by which self-similar fractals are generated. Euclidean geometry is not equipped to describe the repetition of patterns on many scales.

CHAPTER 11

Unit 11A

7. Does not make sense.

9. Makes sense.

11. Does not make sense.

13. 440 cps; 220 cps; 110 cps; 55 cps.

15.

Note	Frequency (cps)
G	390
G#	413
A	438
A#	464
B	491
C	521
C#	552
D	584
D#	619
E	656
F	695
F#	736
G	780

17. a. 347 cps. **b.** 390 cps. **c.** 694 cps. **d.** 2080 cps.
e. 4946 cps.

19. 327 cps; 275 cps.

21. By a factor of 1.335; 12 fourths; 5 octaves.

Unit 11B

9. Does not make sense.

11. Makes sense.

13. Does not make sense.

15. a. A vanishing point of the picture is that point at which the edges of the road appear to meet. According to the definition given in the book, this vanishing point is not the principal vanishing point. The principal vanishing point is the apparent intersection point in the picture of all the lines that are parallel in the real scene and perpendicular to the canvas. The road in this scene is not perpendicular to the canvas.

b.

17.

19. a.

b. 1.28 cm; 1.02 cm; 0.82 cm. **c.** No.

21. a. A, H, I, M, O, T, U, V, W, X, Y. **b.** B, C, D, E, H, I, K, O, X.
c. H, I, O, X. **d.** H, I, N, O, S, X, Z.

23. a. 120°; 240°. **b.** 90°; 180°; 270°.
c. 72°; 144°; 216°; 288°.
d. A regular polygon with n sides can be rotated through $360° \div n$ and multiples of this angle, and its appearance remains the same; $n - 1$ different angles for an n-gon.

25. It has reflection symmetries; it can be reflected across a vertical line through its center or a horizontal line through its center, and its appearance remains the same. It has rotation symmetry; it can be rotated through 180°, and its appearance remains the same.

27. It has rotation symmetry with angles of $360° \div 6 = 60°$ and multiples of 60°. It can also be reflected across six lines through its center and retain its appearance.

29.

31.

33.

35. The angles around a point *P* are precisely the angles that appear inside of a single quadrilateral. Thus, the angles around *P* have a sum of 360°, and the quadrilaterals around *P* fit together perfectly.

Unit 11C
7. Does not make sense.
9. Does not make sense.
11. The longer segment should have a length of 2.47 inches, and the shorter segments should have a length of 1.53 inches.
13. Third and fifth rectangles.
15. 4.05 m or 1.54 m.
17. 0.19 cm or 0.074 cm.

CHAPTER 12

Unit 12A
9. Does not make sense.
11. Makes sense.
13. Makes sense.
15. Although Tilden narrowly won the popular vote, Hayes narrowly won the electoral vote and became president.
17. Garfield narrowly won the popular vote and easily won the electoral vote to become president.
19. Carter narrowly won the popular vote and comfortably won the electoral vote to become president.

21. a. The percentage of shareholders favoring the merger is $10,580/15,890 = 66.58\%$, which is just shy of the required $2/3 = 66.67\%$ vote needed.
 b. A 3/4 vote of a 15-member jury requires at least 12 votes; thus, there will be no conviction.
 c. Only 64% of the states support the amendment, so it fails to pass.
 d. The override will get a $68/100 = 68\%$ vote in the Senate and a $292/435 = 67.1\%$ vote in the House, so the veto can be overturned.
23. a. A total of 66 votes were cast.
 b. D is the plurality winner (but not by a majority).
 c. D is the winner of the top-two runoff.
 d. The winner by sequential runoff is D.
 e. D is the winner by the Borda count.
 f. D wins by the method of pairwise comparisons.
 g. As the winner by all five methods, candidate D is clearly the winner of the election.
25. a. A total of 100 votes were cast.
 b. C is the plurality winner (but not by a majority).
 c. A is the winner of the top-two runoff.
 d. The winner by sequential runoff is A.
 e. B is the winner by the Borda count.
 f. B wins by the method of pairwise comparisons.
 g. There is no clear winner.
27. a. A total of 90 votes were cast.
 b. E is the plurality winner (but not by a majority).
 c. B is the winner of the top-two runoff.
 d. Since only three candidates received first-place votes, the sequential runoff and the top-two runoff methods are the same.
 e. E is the winner by the Borda count.
 f. No winner by the pairwise comparisons method.
 g. Candidates B and E each win two of the five methods, so the outcome is debatable.
29. A wins over B, B wins over C, and C wins over A. This curious situation, in which A is preferred to B, who is preferred to C, who is preferred to A, is called the *Condorcet paradox* or *voting paradox*.
31. a. Fillipo won a plurality but not a majority.
 b. Earnest would need 57.15% of Davis's votes.
33. a. King has a plurality but not a majority.
 b. To overtake King, Lord would need 150 extra votes, which is $150/225 = 66.7\%$ of Joker's votes.
35.

First	A	A	C	C	D
Second	B	C	A	B	A
Third	C	D	D	A	B
Fourth	D	B	B	D	C
	4	2	3	1	2

37. 375 points.

Unit 12B

5. Makes sense.
7. Does not make sense.
9. Assume a candidate receives a majority. Then she or he is the only candidate to receive a plurality and wins by the plurality method. Thus, Criterion 1 is satisfied.
11. The following preference schedule is just one possible example.

First	B	A	C	C
Second	A	B	A	B
Third	C	C	B	A
	2	4	2	3

13. Criterion 4 is violated when C drops out.
15. Assume a candidate receives a majority. In either runoff method, votes are redistributed as candidates are eliminated. But it is impossible for another candidate to accumulate enough votes to overtake a candidate who already has a majority.
17. Criterion 2 is violated.
19. Criterion 3 is violated (if the four voters in the third column of the table move C above A).
21. The following preference schedule is just one possible example.

First	A	B	C
Second	B	A	B
Third	C	C	A
	4	3	5

23. The following preference schedule is just one possible example.

First	A	B	C
Second	B	C	B
Third	C	A	A
	4	2	1

25. The following preference schedule is just one possible example.

First	A	D	C
Second	B	B	B
Third	C	A	D
Fourth	D	C	A
	4	8	3

27. Criterion 4 is violated.
29. Assume candidate A wins a majority of the first-place votes. Then in every head-to-head race with another candidate, A must win (by a majority). Thus, A wins every head-to-head race and is the winner by pairwise comparisons, so Criterion 1 is satisfied.

31. Suppose candidate A wins by the method of pairwise comparisons and in a second election moves up above candidate B in at least one ballot. A's position relative to B remains the same or improves and A's position and B's position relative to the other candidates remain the same, so A must win the second election.
33. The following preference schedule is just one possible example.

First	A	A	B	E	E
Second	B	C	A	B	D
Third	C	D	C	A	B
Fourth	D	E	D	C	A
Fifth	E	B	E	D	C
	1	3	2	1	2

35. a. Candidate C wins by plurality.
 b. Candidate B wins by approval voting.
37. Alaska.
39. Rhode Island.
41. New York, Illinois, Rhode Island, Alaska.
43. Criterion 1 does not apply; Criteria 2 and 3 are satisfied; in this case (but not in general), Criterion 4 is satisfied.
45. Criterion 1 does not apply; Criterion 2 is satisfied; the point system always satisfies Criterion 3; Criterion 4 is satisfied.
47. Criterion 1 does not apply; Criterion 2 is violated; the plurality method always satisfies Criterion 3; Criterion 4 is violated.
49. Criterion 1 does not apply; Criterion 2 is violated; Criterion 3 is satisfied; Criterion 4 is violated.
51. Criterion 1 does not apply; the pairwise comparisons method always satisfies Criteria 2 and 3; Criterion 4 is satisfied.

Unit 12C

9. Makes sense.
11. Does not make sense.
13. 689,655 people per representative; 10,000 representatives.
15. 5.27; underrepresented.
17. 24.74; overrepresented.
19. 6.48, 8.30, 9.85, 10.37.
21. 7, 8, 10, 10.
23. 51, 36, 13; 52, 36, 13; no paradox.
25. 46, 9, 4, 41; 47, 9, 4, 41; no paradox.
27. 9, 70, 21; quota criterion is violated because state B's standard quota is 68.97.
29. 7, 70, 16, 7; quota criterion is satisfied.
31. 7, 8, 10, 10; modified divisor 38.4.
33. 7, 8, 10, 10; modified divisor 38.4.
35. 15, 82, 3; 15, 81, 3, 6; State B's representation decreases.
37. a. 54, 34, 12. b. 54, 34, 12; modified divisor 9.90.
 c. 54, 34, 12; modified divisor 9.99.
 d. 54, 34, 12; modified divisor 9.99.

39. a. 9, 27, 26, 38. **b.** 8, 27, 26, 39; modified divisor 98.3.
 c. 8, 27, 26, 39; modified divisor 99.5.
 d. 8, 27, 26, 39; modified divisor 99.5.
41. a. 1, 3, 6. **b.** 1, 2, 7; modified divisor 35.0.
 c. 1, 3, 6; modified divisor 38.0.
 d. 1, 3, 6; modified divisor 38.0.
43. a. 3, 10, 5, 7. **b.** 3, 10, 5, 7; modified divisor 0.76.
 c. 3, 10, 5, 7; modified divisor 0.77.
 d. 3, 10, 5, 7; modified divisor 0.78.

CHAPTER 13

Unit 13A
11. Makes sense.
13. Does not make sense.
15. Does not make sense.
17. a.

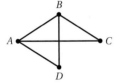

 b. The network would look the same as above except that every edge would be replaced by *two* edges.

19.

21. a. 4 vertices and 5 edges.
 b.

 c. No, the network has no Euler circuit.
23.

 There are many Euler circuits.
25. a. The network has order 8 (a total of 8 vertices).
 b. Vertices *A, D, E, F, G,* and *H* have degree 3; vertices *B* and *C* have degree 4.
 c. The network has no special form.
 d. With vertices of odd degree, there are no Euler circuits.
27. a. The network has order 8 (a total of 8 vertices).
 b. All vertices have degree 4.

 c. The network has no special form (actually it is called *bipartite* since every vertex of the top set of four vertices is connected to every vertex of the bottom set of four vertices).
 d. Because all vertices have even degree, there are Euler circuits.
29. a. The network has order 5.
 b. Vertex *A* has degree 1; vertices *C* and *D* have degree 2; vertex *E* has degree 3; vertex *B* has degree 4.
 c. The network has no special form.
 d. With vertices of odd degree, there are no Euler circuits.
31. Edges *AB, BC, BE, EF,* and *BD*; length 14.
33. Edges *d, b, f, h, a, c*; length 13.
35. Answers may vary. *Examples:*

37. The complete network of order 5 is shown below.

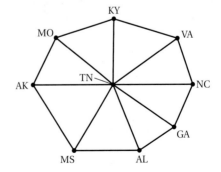

 The network has 10 edges, which agrees with the rule:
$$\frac{n \times (n-1)}{2} = \frac{5 \times (5-1)}{2} = 10$$

39. Order of network = 9, degree of TN = 8.

41. a.

b.

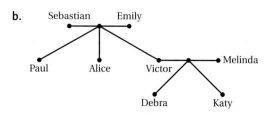

43.

—	A	B	C	D	E	F
A	—	X	X	X	X	
B	X	—	X		X	X
C	X	X	—	X		X
D	X		X	—	X	X
E	X	X		X	—	X
F		X	X	X	X	—

Unit 13B

5. Makes sense.

7. Makes sense if the edges have lengths or weights.

9. The circuit shown is not a Hamiltonian circuit because it doesn't return to the starting point. It can be made into a Hamiltonian circuit by rotating the arrow along the vertical edge so that it points to the leftmost vertex.

11. The circuit shown is not a Hamiltonian circuit because it doesn't return to the starting point. It cannot be made into a Hamiltonian circuit.

13. a. 2520 Hamiltonian circuits.
 b. Approximately 1.78×10^{14} Hamiltonian circuits.
 c. 42 minutes; approximately 5.6 million years.

15. a. The network will look exactly like Figure 13.39 of the text except that the vertex for Zion will be missing, along with all of the edges connected to it. The resulting network will have four vertices and a total of six edges.
 b. 470 miles.
 c. Grand Canyon to Bryce to Capitol Reef to Canyonlands to Grand Canyon; 454 miles.
 d. Bryce to Capitol Reef to Canyonlands to Grand Canyon to Bryce; 454 miles.
 e. There are three different Hamiltonian circuits possible in this network of order four. They have lengths 454, 470, and 558, so we have found the Hamiltonian circuit with the shortest length in parts c and d.

17. a. 6.5×10^{11} circuits. **b.** 650,000 sec. **c.** A week.

19. a. 3.1×10^{23} circuits. **b.** 9.8 billion years.
 c. About 2/3 the age of the universe.

21. The nearest neighbor method finds the circuit *FEABCDF*, which has the shortest length, 23.

23. The nearest neighbor method gives the circuit *ACBDA*, which has length 15. The other two possible circuits, *ACDBA* with length 14 and *ABCDA* with length 17, can be found by trial and error.

25. $450,000 per day; $164,250,000 per year.

Unit 13C

7. Does not make sense.

9. Does not make sense.

11. Tasks *a*, *e*, and *f*.

13. Tasks *a*, *e*, *f*, *g*, *i*, and *k*.

15. Tasks *a*, *d*, and *e*.

17. Tasks *a*, *d*, *e*, *g*, and *h*.

19. a. Task *d*: 1 hr; task *f*: 2.5 hr.
 b. Task *d*: 2 hr; task *f*: 2.5 hr.

21. a. Task *d*: 1.5 hr; task *f*: 4.5 hr.
 b. Task *d*: 2.5 hr; task *f*: 4.5 hr.

23. a. Task *d*: 3 mo; task *f*: 2 mo.
 b. Task *d*: 3 mo; task *f*: 12 mo.

25. a. Task *d*: 4 mo; task *f*: 5 mo.
 b. Task *d*: 4 mo; task *f*: 15 mo.

27. A one-hour delay in this task will not delay the project.

29.

Task	EST	LST	EFT	LFT	Slack
a	0.0	0.0	2.0	2.0	0
b	0.0	1.0	1.0	2.0	1.0
c	0.0	1.5	1.0	2.5	1.5
d	1.0	2.0	1.5	2.5	1.0
e	2.0	2.0	2.5	2.5	0
f	2.5	2.5	4.5	4.5	0
g	4.5	4.5	7.5	7.5	0
h	4.5	5.5	6.5	7.5	1.0
i	7.5	7.5	9.5	9.5	0
j	9.5	10.0	10.0	10.5	0.5
k	9.5	9.5	10.5	10.5	0

31. A one-month delay in this task will delay the project one month.

33. The slack times are *a*, 0; *b*, 0; *c*, 10; *d*, 0; *e*, 15; *f*, 30; *g*, 0; *h*, 0.

35. 150 minutes.

37.

```
       b: wash colors    d: dry colors    f: fold colors
              30               45               15         h: fold
a: sort                                                    whites
  15                                                          15
       c: remove stains  e: wash whites   g: dry whites
              20               30               45
```

39. The critical path is *a*, *c*, and *f*.

41. Tasks (*a*, *b*) and (*c*, *d*, *e*) take place at the same time; task *f* must be done alone.

Index

Index of Applications

Note: CS=Case Study, E=Example,
IE=In-text Example, P=Problem, PR=Project,
PM=Practical Matters.